The Physics of Deformation and Flow

**McGRAW-HILL
INTERNATIONAL
BOOK COMPANY**

New York
St. Louis
San Francisco
Auckland
Bogotá
Guatemala
Hamburg
Johannesburg
Lisbon
London
Madrid
Mexico
Montreal
New Delhi
Panama
Paris
San Juan
São Paulo
Singapore
Sydney
Tokyo
Toronto

The Physics of Deformation and Flow

E. W. BILLINGTON

and

A. TATE

Royal Armament Research and Development Establishment
Fort Halstead, U.K.

This book was set in Times Roman 327

British Library Cataloging in Publication Data

Billington, E W
The physics of deformation and flow.
1. Rheology
2. Deformations (Mechanics)
3. Fluid dynamics
I. Title II. Tate, A
531′.382 QC189.5 80-40133

ISBN 0-07-005285-9

THE PHYSICS OF DEFORMATION AND FLOW

1 2 3 4 MPMP 8 3 2 1

Printed and bound in the United States of America.

CONTENTS

PREFACE

The purpose of this book is to present the modern continuum mechanics approach to the macroscopic response of a material to external loads and, where possible, to relate the constitutive equations involved to the material's microstructural elements. The resulting theory is applied to a wide spectrum of materials with particular attention being paid to their nonlinear behavior. Because of its wide scope, it has proved impossible to deal in depth with the microstructural mechanics approach to each type of material. However, an outline of the presently available theories is given and references to the literature cited so that any particular theory can be pursued in greater depth. The experimental basis for constitutive equations relating to particular types of material is presented in detail. The basis of this approach is the belief that a stage has now been reached in the development of materials science when it would appear appropriate to attempt some integration of macroscopic and microscopic viewpoints, and success in such an attempt would seem most likely to follow from a wide survey of the present position. It is hoped that this book will go some way toward providing such a survey and that it may prove a useful supplement for both students and research scientists or engineers who may be committed to a more limited viewpoint.

The text requires a basic understanding of vectors and tensors and their associated calculus. Although a number of books on tensor analysis are available, workers in the field of deformation and flow are still often unfamiliar with these notations. For this reason, a brief mathematical resume of the results required in the text has been given in Chapter 1. The coverage is intended to be sufficient to make the text accessible to material scientists and engineers. Wherever possible tensor equations have been expressed in direct notation. This has the advantage of recording physical statements in a form unencumbered by mathematical details needed only for calculations. Where necessary, or to clarify some particular aspect, we present equations using both the direct notation and the indicial notation.

Chapters 2 and 3 cover the kinematics and physical principles of continuum mechanics, and are basic to any discussion of the subject. In Chapter 4 there is given a brief resume of the elements of statistical mechanics. This is intended to provide an introduction to those aspects basic to any attempt to relate the microstructural properties of a material to its macroscopic mechanical response. Chapter 5 is concerned with the invariance principles governing the formulation of constitutive equations. In this chapter we follow the approach set out in the authoritative works of Truesdell and Toupin and Truesdell and Noll, which are listed in the references, and which provide an advanced treatment of those aspects of continuum mechanics discussed in this book.

The general deformation and flow properties of various idealized materials are discussed in Chapters 6 to 11. The principal objective of these six chapters has been to present a discussion of the physical properties of matter with particular reference to the interdependence of the continuum and structural viewpoints. Included in these chapters are sections describing the experimental characterization of a materials mechanical response, these sections containing a description of typical published results together with sections which describe the microstructural properties of the materials. The SI system of units has been used throughout the book. Discussion of the particle aspects of the subject has been restricted to certain specific topics. The choice of these topics has been determined by our attempt to indicate how the physics of deformation and flow draws upon other branches of science, such as thermodynamics, rate processes, metal physics, and in particular the theories of crystal dislocation mechanics and the liquid state, in order to obtain a physically satisfying description of the mechanical properties of particular types of material. This has necessitated a rather broad interpretation of the subject, leading in some places to an overlap with the neighboring domains of metallurgy, chemistry, mechanical engineering, and chemical engineering.

The remaining three chapters are concerned with a discussion of various technical applications of the theories considered in the previous six chapters. Choice of material for these three chapters reflects the interest of the authors in the subject.

We are indebted to a large number of colleagues for their kind assistance during the preparation of this book. Although the subject has deep historical roots, we have restricted our references to the present century. Full historical notes and references to the literature can be found in the works of Truesdell and Toupin and Truesdell and Noll, and we wish to acknowledge here our indebtedness to the writings of these authors.

June 1979

E.W.B
A.T.

ACKNOWLEDGMENTS

The authors wish to thank the following for permission to use figures in this book:

The Institute of Physics for Figs. 7-3, 7-4, 7-5, 10-8, 12-1, 12-3, 12-4, 13-4, 13-5, and 13-6; J. G. Oldroyd, D. J. Strawbridge, B. A. Toms, and the Institute of Physics for Figs. 9-1, 9-6, and 9-7; R. D. Heidenreich, W. Shockley, and the Institute of Physics for Fig. 11-16; E. Orowan and the Institute of Physics for Figs. 14-10 and 14-14; the Royal Society for Figs. 6-2, 9-2, 9-3, 9-4, 9-5, 10-9, 13-1, 13-2, and 13-3; R. S. Rivlin, D. W. Saunders, and the Royal Society for Figs. 8-2, 8-3, 8-4, 8-6, 8-7, 8-8, and 8-9; M. J. Doyle, A. Maranci, E. Orowan, S. T. Stork, and the Royal Society for Fig. 14-16; W. G. Johnston, J. J. Gilman, and the American Institute of Physics for Fig. 11-14; D. F. Stein, J. R. Low, and the American Institute of Physics for Fig. 11-15; W. P. Rees, B. E. Hopkins, H. R. Tipler, and the Metals Society for Fig. 14-11; N. J. Petch and the Metals Society for Fig. 14-12; M. Zaid, B. Paul, and the Franklin Institute for Fig. 13-40; E. M. Lenoe, R. A. Heller, and the American Society of Rheology for Figs. 8-10 and 8-11; J. T. Jones, D. Wyllie, and the Institution of Mechanical Engineers for Fig. 9-9; D. L. Holt, S. G. Babcock, S. J. Green, C. J. Maiden, and the American Society for Metals for Figs. 13-8 and 13-9; L. R. G. Treloar and the Oxford University Press for Fig. 8-5; the Oxford University Press for Figs. 13-37 and 13-38; A. Guinier and Freeman (San Francisco) for Fig. 7-6; J. W. Johnson, D. G. Holloway, and the *Philosophical Magazine* for Fig. 14-15; *Acta Metallurgica* for Figs. 13-11 and 13-12; E. H. Andrews and the *Journal of Materials Science* for Figs. 14-23, 14-24, 14-25, 14-26, 14-29, and 14-30; E. H. Andrews, Y. Fukahori, and the *Journal of Materials Science* for Figs. 14-28, 14-31 and 14-32; H. Markovitz, L. J. Elyash, F. J. Padden, T. W. De Witt, and the *Journal of Colloid Science* for Fig. 9-8; C. J. Maiden, S. J. Green, and the *Journal of Applied Mechanics* for Fig. 13-10; R. F. Recht and the *Journal of Applied Mechanics* for Fig. 13-42; J. Harding and Wiley Interscience for Fig. 13-13; P. M. Naghdi and Pergamon Press for Fig. 12-2; the *International Journal of Engineering Science* for Figs. 13-35 and 13-36; S. Clyens, W. Johnson, and the *International Journal of Mechanical Science* for Fig. 13-7; J. Awerbuch, S. R. Bodner, and the *International Journal of Solids and Structures* for Fig. 13-43.

CONVENTIONS AND NOTATION

In general, when using direct notation, capital letters denote tensors and lower-case letters denote vectors; however, this system is not rigidly adhered to, there being several exceptions. These exceptions relate to physical quantities denoted by symbols whose usage is sufficiently general to justify their retention. Hence, in direct notation the distinction between vectors and tensors has been made by use of a different type of letter. With this as a basis, the following conventions regarding the type of letters used for symbols apply throughout the book:

Points: $\boldsymbol{XYZ}$, $\boldsymbol{xyz}$
Scalars: A, a
Vectors: $\boldsymbol{A}$, $\boldsymbol{a}$
Tensors: $\mathbf{A}$, $\mathbf{a}$
Sets, bodies, and regions: $\mathscr{A}$
Groups: $\boldsymbol{g}$
Tensor-valued functions of tensor variables: $\mathfrak{g}$

Only the frequently used symbols are listed below. Various symbols have had to be used to denote more than one quantity. The section where a quantity is defined or first appears is given in brackets.

$\mathbf{A}_n$		nth Rivlin–Ericksen tensor (2-4)
	A	Helmholtz free energy (4-1)
$\mathbf{B}$,	B^{ij}	Left Cauchy–Green tensor (2-2)
	B_r	Reference configuration of a body $\mathscr{B}$ (2-1)
	$\mathscr{B}$	Body (2-1)
$\mathbf{C}$,	$C_{\alpha\beta}$	Right Cauchy–Green tensor (2-2)

$\tilde{\mathbf{C}}$,	$\tilde{C}^{\alpha\beta\gamma\delta}$	Elasticity tensor (6-1-1)
$\mathbf{D}$,	D_{ij}	Stretching tensor (2-4)
	D	Deformation constant (8-5-1)
$\mathbf{E}$,	$E_{\alpha\beta}$	Green–St. Venant strain tensor (2-6)
	E	Young's modulus (6-1-1)
$\mathbf{F}$,	$F^i{}_\alpha$	Deformation gradient (2-2)
$\mathbf{F}_{(t)}$,	$F^i_{(t)\alpha}$	Relative deformation gradient (2-2)
$\boldsymbol{F}$,	F_i	Force (3-2)
$\mathbf{G}'$,	$G'^i{}_j$	Function of the deviatoric stress tensor, $\mathbf{T}'$ (10-3-1)
$\boldsymbol{G}_\alpha$		Base vectors (referential frame) (2-1)
$\boldsymbol{G}$,	G^i	Body couple density (3-5)
G		Tensor-valued functional (5-1)
	G	Shear modulus of rigidity (6-1-1)
	$\mathscr{G}$	Constitutive functional of a material (5-1)
$\boldsymbol{H}$,	H^i	Angular momentum (3-2)
	$H(R_t)$	Rate at which thermal energy is being fed into the system (3-8)
	H	Hamiltonian function (4-2)
	$J_i'\ (i = 2, 3)$	Invariants of the deviatoric stress tensor $\mathbf{T}'$ (10-2-1)
	$J = \sqrt{g}\mathscr{J}$	Jacobian of deformation (2-2)
	$\mathscr{J}$	Jacobian (1-11)
	$K(R_t)$	Kinetic energy of the material occupying the region R_t (3-8)
	K_c	Chemical equilibrium constant (4-1)
$\mathbf{L}$,	L_{ij}	Velocity gradient (2-2)
	L	Lagrangian function (4-2)
$\mathbf{L}_n(t)$		nth acceleration gradient (2-2)
L		Vector-valued functional (5-1)
$\mathbf{M}$,	M^{ij}	Couple stress tensor (3-5)
$\boldsymbol{M}$,	M^i	Bending moment (6-5-4)
	M	Mass (3-1)
$\mathbf{N}$		(9-1-5)
$\boldsymbol{N}$,	N^α	Unit normal vector (2-5)
	P	Pressure (7-1-1)
	$P(R_t)$	Mechanical power (rate that work is being done on the material) (3-8)
	$\mathscr{P}$	Part of a body (2-2)
$\mathbf{Q}$		Arbitrary orthogonal tensor (1-9)
	$R^{(c)}_{\alpha\beta\gamma\delta}$	Riemann–Christoffel tensor (1-8)
$\mathbf{R}$,	$R^i{}_\alpha$	Local rotation tensor (2-3)
$\tilde{\mathbf{R}}$		Infinitesimal rotation tensor (2-7)
$\mathbf{S}$		Displacement gradient (6-2)
$\tilde{\mathbf{S}}$		Infinitesimal displacement gradient (6-2-1)
	S	Entropy (3-9)
	S_t	Material surface in the current configuration (2-5)

$\mathbf{T}$,	T^{ij}	Cauchy's stress tensor (3-4)
$\mathbf{T}^0$,	$(T^0)^{\alpha i}$	Nominal stress tensor (3-7)
$(\mathbf{T}^0)^T$		First Piola–Kirchhoff stress tensor (3-7)
$\tilde{\mathbf{T}}$,	$\tilde{T}^{\alpha\beta}$	Second Piola–Kirchhoff stress tensor (3-7)
$\mathbf{T}_E$		Extra stress tensor (5-4)
	$\mathscr{T}$	Energy required to create unit area of new interface (14-7-1)
	$\mathscr{T}_0$	Surface energy (14-7-2)
$\mathbf{U}$,	$U_{\alpha\beta}$	Right stretch tensor (2-3)
	$U(R_t)$	Internal energy of the material occupying the region R_t (3-8)
	U_n	Energy of a system which is in a state n (4-1)
$\mathbf{V}$,	V^{ij}	Left stretch tensor (2-3)
$\mathbf{W}$,	W_{ij}	Spin tensor (2-1)
	W	Strain energy density (6-1-2)
$\mathbf{X}$,	X^α	Point in the reference configuration (2-1)
	X	Typical particle (2-1)
	Y	Yield stress in simple uniaxial tension (and compression) (10-1)
	Z	Partition function (4-1)
	X, Y, Z	Particles; also cartesian coordinates of $\boldsymbol{X}$
$\boldsymbol{a}$,	a^i	Acceleration (2-1)
$\boldsymbol{b}$,	b^i	Burgers vector (11-1-3)
$\mathbf{c}$,	c^i	Surface couple density (3-5)
$\mathbf{e}$,	e_{ij}	Almansi–Hamel strain tensor (2-6)
$\boldsymbol{f}$,	f_i	Body force per unit mass (3-2)
$\mathfrak{f}$		Response function (5-7)
	$f^{(N)}$	Probability density function (4-4)
	$f(\mathbf{T})$	Yield function (10-2-1)
$\boldsymbol{g}_i$		Base vectors (spatial frame) (2-1)
	$g^{(2)}$	Radial distribution function (4-4)
$\mathfrak{g}$		Response function of an elastic material (5-1)
$\boldsymbol{h}$,	h^i	Total angular momentum density (3-5)
	h	Rate of flow of thermal energy per unit area into the material occupying the region R_t (3-8)
	$h(\hbar = h/2\pi)$	Planck's constant (4-1)
$\boldsymbol{j}$		Flux per unit area (3-10)
	k	Yield stress in pure shear (10-1)
	k	Boltzmann's constant (4-1)
	k	Sum of the degrees of freedom of all N particles (4-2)
	k	Thermal conductivity (7-3-3)
$\boldsymbol{m}$,	m^i	Internal angular momentum (3-5)
	m	Mass of molecule (4-2)
$\boldsymbol{p}$,	p^i	Linear momentum (3-2)

$\mathfrak{p}$		Response function (5-6)
$\boldsymbol{q}$,	q^i	Heat vector (directed opposite to the flux of heat) (3-8)
$\mathfrak{q}$		Response function (5-6)
	q_k	Generalized coordinates of Lagrange (4-2)
$\boldsymbol{r}_i$		Instantaneous position of the ith particle of mass m_i (4-2)
	r	Heat supply or source per unit mass and time (3-8)
	s	Elapsed time (2-1)
$\mathbf{t}$,	t_{ij}	Deviator of the square of the deviatoric stress tensor $\mathbf{T}'$ (10-3-3)
$\boldsymbol{t}$,	t^i	Stress or traction vector (3-2)
	t	Present time (2-1)
$\boldsymbol{u}$,	u^i	Displacement vector (2-7)
$\boldsymbol{v}$,	v^i	Velocity (2-1)
$\boldsymbol{w}$,	w^i	Angular velocity (2-1)
	w_n	Probability that a system is in a state n (4-1)
$\boldsymbol{x}, \boldsymbol{y}, \boldsymbol{z}$,	x^i, y^i, z^i	Points in space (spatial coordinates) (2-1)
	x, y, z	Cartesian coordinates of $\boldsymbol{x}$ (1-7)
	z	Partition function per unit volume (4-1)
$\boldsymbol{\Gamma}$,	Γ^i	Applied torque (3-2)
$\mathbf{\Pi}$,	Π^{ij}	Momentum flux density tensor (7-1-1)
	Φ_v	Viscous dissipation function (7-1-2)
	Ω	Generalized form of the Helmholtz free energy (4-1)
	$\Omega^{(2,2)}$	Reduced collision cross section for the hard-core part of the potential (7-7-5)
	Θ	Polar coordinate in the referential description (8-5-1)
$\boldsymbol{\alpha}^{(p)}$		Dislocation density tensor (11-3-1)
	$\alpha_i\,(i = 0, 1, 2)$	Elastic response coefficients (8-1-2)
	$\beta_i\,(i = 0, 1, 2)$	Elastic response coefficients (8-1-2)
	γ	Shear strain (2-6)
	γ	Entropy production per unit mass (3-9)
	δ_{ij}	Kronecker delta (1-2)
	∂	Boundary operator (1-11)
$\boldsymbol{\epsilon}$,	ϵ_{ij}	Natural strain (10-3-1)
	ε	Internal energy per unit mass (3-8)
	ζ	Bulk viscosity (7-1-2); also, stress ratio (10-4-3)
	η	Entropy per unit mass (3-9)
	η	Shear, or dynamic viscosity (7-1-2)
	θ	Absolute temperature (3-9)
	θ	Polar coordinate in the material description (1-7)
$\boldsymbol{\vartheta}$		Deformation function (2-1)
	κ	Bulk modulus (6-1-1)
	κ	Work-hardening parameter (10-3-1)
	$\varkappa$	Reference configuration (2-1)
	λ	Lamé coefficient of linear elasticity (6-1-1)

	$\lambda_a\ (a = 1, 2, 3)$	Principal stretches (2-3)
	μ	Chemical potential (4-1)
	μ	Mobility of a particle (7-6-2)
	ν	Kinematic viscosity (7-2-2)
$\boldsymbol{\nu}$,	ν^i	Unit normal vector (1-10)
	ν	Poisson's ratio (6-1-1)
	μ, ν	Lode's variables (10-4-3)
$\boldsymbol{\xi}$,	ξ^i	Place at time τ (2-2)
	ρ	Mass density in the spatial, i.e., current configuration (3-1)
	ρ_r	Mass density in the reference configuration (3-1)
	$\sigma_a\ (a = 1, 2, 3)$	Principal stresses (3-6)
	σ	Strain energy function (8-2)
	τ	Past time (2-1)
	$\tau_{(s)}$	Principal shearing stress (3-6)
$\boldsymbol{\phi}$,	ϕ^i	Function identifying a motion (2-1)
	ϕ	Polar coordinate (1-7)
	φ	Scalar potential for wave propagation (6-4-1)
$\boldsymbol{\chi}$,	χ^i	Configuration (2-1)
$\hat{\boldsymbol{\chi}}_{(t)}$,	$\hat{\chi}^i_{(t)}$	Relative deformation function (2-2)
	ψ	Helmholtz free energy per unit mass (3-9)
$\boldsymbol{\omega}$,	ω^i	Vorticity vector (7-2-1)
	ω	Dimensionless ratio of J'_2 and J'_3 (10-2-3)
	$\mathcal{O}$	Orthogonal group (5-3)
	$\mathcal{g}$	Isotropy group (5-3)

CHAPTER

ONE

SCALARS, VECTORS, AND TENSORS

1-1 INTRODUCTION

This chapter is devoted to the basic mathematical formalism required to describe those aspects of continuum mechanics presented in succeeding chapters. All relativistic or quantum effects are neglected as being incompatible with the energetic and spatial scale of present interest. Thus the space which will be of primary concern is the familiar three-dimensional euclidean space of common experience. Physical laws and relations must have the same form irrespective of which particular geometrical representation of space is chosen; tensors are entities which have a mathematical structure designed to possess just this property, and the basic properties of tensors are here explored. A more complete discussion is found in the references given at the end of this chapter.

1-2 NOTATION

To write out the equations of mathematical physics in terms of their coordinate components is very cumbersome. It is therefore important to develop a mathematical shorthand which will both reduce the labor and display more clearly the essential physical features of the equations. Throughout the text scalars and points are denoted by italic letters, for example, A. Vectors (and sometimes points) are denoted by boldface italic letters, whereas boldface letters are used for tensors and linear transformations. Components of vectors and tensors, as is customary, have indicial subscripts to denote covariant components and indicial superscripts to denote contravariant components (see Sec. 1-5). As the space of interest is three dimensional the indices take the values 1, 2, and 3. If the coordinate system is rectangular cartesian, and consequently there is no difference between covariant and contravariant components, then all the indices are written as subscripts.

Covariant derivatives (see Sec. 1-8) are denoted by a comma; for example, $a_{i,j}$ is the covariant derivative of the vector a_i.

Indices take the values 1, 2, and 3 unless stated otherwise. If an index is repeated in a term it will be understood that this term is to be summed with respect to the repeated index over the range 1, 2, and 3. If curvilinear coordinates are being used the summation will involve one subscript and one superscript; otherwise the term will not necessarily retain its tensorial character. No index should appear more than twice in an expression and where necessary indices must be relabeled to ensure this does not occur.

As an example, consider the expression

$$\begin{aligned} A^1{}_1u^1 + A^1{}_2u^2 + A^1{}_3u^3 &= v^1 \\ A^2{}_1u^1 + A^2{}_2u^2 + A^2{}_3u^3 &= v^2 \\ A^3{}_1u^1 + A^3{}_2u^2 + A^3{}_3u^3 &= v^3 \end{aligned} \tag{1-2-1}$$

Using the above conventions, this can be written simply as

$$A^i{}_ju^j = v^i \tag{1-2-2}$$

The index j in the above expression must, by the convention, be summed over the range 1, 2, and 3, but replacing j by any other letter gives the same result; hence an index such as j is referred to as a dummy index. If it is required to eliminate v^i between Eq. (1-2-2) and

$$B^j{}_iv^i = w^j \tag{1-2-3}$$

one must first replace the index j in Eq. (1-2-2) by k, say, in order that three j's do not appear as in $B^j{}_iA^i{}_ju^j$. The correct form is thus

$$B^j{}_iA^i{}_ku^k = w^j \tag{1-2-4}$$

Next it is convenient to introduce the Kronecker delta symbol δ_{ij}, which is defined by the set of equations

$$\delta_{ij} = \begin{cases} 1 & \text{if } i = j \\ 0 & \text{if } i \neq j \end{cases} \tag{1-2-5}$$

Using the summation convention, note that $\delta_{ii} = 3$. Sometimes the Kronecker delta is referred to as a substitution tensor because in any expression in which it occurs i and j can be treated as interchangeable, i.e., the expression $a_i\delta_{ij}$ can be written a_j because the only nonzero term occurs when $i = j$.

Finally, it is useful to define a permutation symbol e_{ijk} by the set of equations

$$\begin{aligned} e_{123} = e_{231} = e_{312} &= +1 \\ e_{132} = e_{321} = e_{213} &= -1 \end{aligned} \tag{1-2-6}$$

All other values of i, j, and k make e_{ijk} zero.

The indicial notation used in Eq. (1-2-2) is a considerable reduction from Eq. (1-2-1), but the presence of the indices still emphasizes the component form. Equations (1-2-1) can be expressed even more succinctly in direct notation:

$$\mathbf{A}\boldsymbol{u} = \boldsymbol{v} \tag{1-2-7}$$

where $\mathbf{A}$ represents the array or matrix of coefficients

$$[A^i{}_j] = \begin{bmatrix} A^1{}_1 & A^1{}_2 & A^1{}_3 \\ A^2{}_1 & A^2{}_2 & A^2{}_3 \\ A^3{}_1 & A^3{}_2 & A^3{}_3 \end{bmatrix}$$

and $\boldsymbol{u}$ and $\boldsymbol{v}$ are column vectors

$$[u^j] = \begin{bmatrix} u^1 \\ u^2 \\ u^3 \end{bmatrix} \qquad [v^i] = \begin{bmatrix} v^1 \\ v^2 \\ v^3 \end{bmatrix}$$

In direct notation, $\delta^i{}_j$, which has the matrix form

$$[\delta^i{}_j] = [\delta_{ij}] = \begin{bmatrix} 1 & 0 & 0 \\ 0 & 1 & 0 \\ 0 & 0 & 1 \end{bmatrix}$$

is denoted by $\mathbf{I}$.

The matrix formed by interchanging rows and columns of a matrix $\mathbf{A}$ is called its transpose and will be denoted by $\mathbf{A}^T$. Thus, if

$$[A_{ij}] = \begin{bmatrix} A_{11} & A_{12} & A_{13} \\ A_{21} & A_{22} & A_{23} \\ A_{31} & A_{32} & A_{33} \end{bmatrix}$$

then

$$[A_{ji}] = \begin{bmatrix} A_{11} & A_{21} & A_{31} \\ A_{12} & A_{22} & A_{32} \\ A_{13} & A_{23} & A_{33} \end{bmatrix} \tag{1-2-8}$$

The sum of the diagonal elements of a matrix is called its trace and will be denoted by

$$\operatorname{tr}\mathbf{A} = A^i{}_i = A^1{}_1 + A^2{}_2 + A^3{}_3 \tag{1-2-9}$$

Using the permutation symbol the determinant of the matrix $\mathbf{A}$ is defined by

$$\det[A^i{}_j] = \begin{vmatrix} A^1{}_1 & A^1{}_2 & A^1{}_3 \\ A^2{}_1 & A^2{}_2 & A^2{}_3 \\ A^3{}_1 & A^3{}_2 & A^3{}_3 \end{vmatrix} = e_{ijk}A^i{}_1A^j{}_2A^k{}_3 \tag{1-2-10}$$

or, equivalently,

$$e_{pqr}\det\mathbf{A} = e_{ijk}A^i{}_pA^j{}_qA^k{}_r \tag{1-2-11}$$

Thus

$$\det \mathbf{A} = \det \mathbf{A}^T \tag{1-2-12}$$

Another form which Eq. (1-2-11) can take is

$$\begin{vmatrix} A^i{}_p & A^i{}_q & A^i{}_r \\ A^j{}_p & A^j{}_q & A^j{}_r \\ A^k{}_p & A^k{}_q & A^k{}_r \end{vmatrix} = e_{ijk}e_{pqr} \det \mathbf{A} \tag{1-2-13}$$

Setting $\mathbf{A} = \mathbf{I}$ gives the important identity

$$e_{ijk}e_{pqr} = \begin{vmatrix} \delta_{ip} & \delta_{iq} & \delta_{ir} \\ \delta_{jp} & \delta_{jq} & \delta_{jr} \\ \delta_{kp} & \delta_{kq} & \delta_{kr} \end{vmatrix} \tag{1-2-14}$$

An important special relation is obtained by setting $i = p$:

$$e_{ijk}e_{iqr} = \delta_{jq}\delta_{kr} - \delta_{jr}\delta_{kq} \tag{1-2-15}$$

1-3 FUNCTIONS AND FUNCTIONALS

A function or mapping F relates each element x of a set, which collectively will be denoted by X, called the domain of the function, with a single element y of a set Y, which is called the range of the function. This relationship is usually written as

$$y = F(x) \tag{1-3-1}$$

The element y is called the image of x under the transformation F.

If, for all x_1, x_2 in X, $x_1 \neq x_2$ implies $F(x_1) \neq F(x_2)$ then F is said to be one–one into, while if every member y of the set Y corresponds to the image of some x the transformation is said to be onto. A transformation which is one–one into and onto is said to be one–one. If F is one–one then there exists an inverse function F^{-1} which is also one–one and we can write $x = F^{-1}(y)$.

A functional is defined as a function whose domain X is a set of elements each of which is itself a function $x(t)$ over the real variable t from a to b. A functional F will be written in the form

$$y = F\left(x \left(\overset{b}{\underset{a}{t}} \right) \right) = \overset{b}{\underset{t=a}{F}} \left(x(t) \right) \tag{1-3-2}$$

An example of a functional is the integral

$$y = \int_a^b x(t)\,dt \tag{1-3-3}$$

1-4 GROUPS

A group G is defined as a system of elements A, B, C, etc., which is closed under a single-valued binary operation which is associative, and relative to which G

contains an element I satisfying the identity law, and with each element another element (called its inverse) satisfying the inverse law:

Associative law:	$A(BC) = (AB)C$	for all A, B, C
Identity law:	$AI = IA = A$	for all A
Inverse law:	$AA^{-1} = A^{-1}A = I$	for each A and some A^{-1}

The inverse of a product is the product of the inverses taken in reverse order. For

$$(AB)(B^{-1}A^{-1}) = A(BB^{-1})A^{-1} = AIA^{-1} = I$$

then

$$(AB)^{-1} = B^{-1}A^{-1} \tag{1-4-1}$$

A group whose operation is commutative, that is, $AB = BA$, is called an abelian group.

A subgroup S of a group G is defined as a subset of G which is itself a group with respect to the binary operation of G. In any group the set consisting of the identity I above is a subgroup. The whole group is also a subgroup of itself. Subgroups of G other than the "improper" subgroups I and G are called proper subgroups.

In any group G, AXA^{-1} is called the conjugate of X under conjugation by A. If A is a fixed element of G then $\bar{X} = AXA^{-1}$ is also in G and as $X = A^{-1}\bar{X}A$ this is a one–one relation mapping G onto itself. This mapping preserves the group relations, for

$$\bar{X}\bar{Y} = (AXA^{-1})(AYA^{-1}) = AXYA^{-1} = \overline{XY} \tag{1-4-2}$$

Thus any subgroup S of G is transformed into another subgroup $\bar{S}$ of G. This is written symbolically as

$$\bar{S} = ASA^{-1} \tag{1-4-3}$$

$\bar{S}$ is said to be conjugate to S. If a subgroup S is identical with all its conjugate subgroups, i.e.,

$$ASA^{-1} = S \qquad \text{for every } A \text{ in } G \tag{1-4-4}$$

it is called an invariant or self-conjugate subgroup, or it is said to be normal in G.

1-5 VECTOR SPACES

Scalars and vectors are two types of quantity frequently met in elementary physics. Scalars are quantities, such as mass and charge, which have magnitude but no associated direction, while vectors have both magnitude and direction. Examples of vectors are directed line segments, velocity, electric current, and the flow of heat. Addition of vectors is by the so-called parallelogram law, while multiplying a vector by a real positive number merely increases its magnitude. These laws can be generalized into the following definition.

A vector space (often called a linear space) $\mathscr{V}$ over a field of scalars $\mathscr{F}$ is a set of elements, called vectors, such that any two vectors $\boldsymbol{u}$ and $\boldsymbol{v}$ of $\mathscr{V}$ determine a unique vector $\boldsymbol{u}+\boldsymbol{v}$ as sum and any vector $\boldsymbol{u}$ from $\mathscr{V}$ and any scalar a from $\mathscr{F}$ determine a scalar product $a\boldsymbol{u}$ in $\mathscr{V}$ with the following properties:

1. $\mathscr{V}$ is an abelian group under addition with an identity element $\boldsymbol{o}$.
2. Distributive laws: $a(\boldsymbol{u}+\boldsymbol{v}) = a\boldsymbol{u}+a\boldsymbol{v}$; $(a+b)\boldsymbol{u} = a\boldsymbol{u}+b\boldsymbol{u}$. .
3. $(ab)\boldsymbol{u} = a(b\boldsymbol{u})$; $1\boldsymbol{u} = \boldsymbol{u}$.

The term "field" implies a set of algebraic axioms of composition, and examples of fields are real numbers, rational numbers, and complex numbers.

A subspace $\mathscr{U}$ of a vector space $\mathscr{V}$ is a subset of $\mathscr{V}$ which is itself a vector space with respect to the operations of addition and scalar multiplication in $\mathscr{V}$.

A set of vectors $\boldsymbol{v}_1, \boldsymbol{v}_2, \ldots, \boldsymbol{v}_m$ are said to be linearly independent (over $\mathscr{F}$) if and only if, for all scalars a_i in $\mathscr{F}$,

$$a_1\boldsymbol{v}_1 + a_2\boldsymbol{v}_2 + \cdots + a_m\boldsymbol{v}_m = \boldsymbol{o} \tag{1-5-1a}$$

implies

$$a_1 = a_2 = \cdots = a_m = 0 \tag{1-5-1b}$$

Vectors which are not linearly independent are called linearly dependent. A basis of a vector space is defined as a linearly independent subset of vectors which generates (spans) the whole space. Any vector $\boldsymbol{v}$ in the space $\mathscr{V}$ can be expressed as a linear combination of the vectors of a given basis in one and only one way. To show this, let the vectors of the basis be $\boldsymbol{g}_1, \boldsymbol{g}_2, \ldots, \boldsymbol{g}_k$ and let

$$\boldsymbol{v} = a_1\boldsymbol{g}_1 + a_2\boldsymbol{g}_2 + \cdots + a_k\boldsymbol{g}_k \tag{1-5-2}$$

Also,

$$\boldsymbol{v} = b_1\boldsymbol{g}_1 + b_2\boldsymbol{g}_2 + \cdots + b_k\boldsymbol{g}_k \tag{1-5-3}$$

Subtracting

$$(a_1 - b_1)\boldsymbol{g}_1 + (a_2 - b_2)\boldsymbol{g}_2 + \cdots + (a_k - b_k)\boldsymbol{g}_k = \boldsymbol{o} \tag{1-5-4}$$

But as $\boldsymbol{g}_1, \boldsymbol{g}_2, \ldots, \boldsymbol{g}_k$ are linearly independent, by Eq. (1-5-1),

$$a_1 = b_1, a_2 = b_2, \ldots, a_k = b_k \tag{1-5-5}$$

The numbers $a_1, a_2, \ldots, a_k$ are called the components of $\boldsymbol{v}$ with respect to the basis $\boldsymbol{g}_1, \boldsymbol{g}_2, \ldots, \boldsymbol{g}_k$.

It can be shown that if a vector space $\mathscr{V}$ has a basis consisting of a finite number of elements then any other basis for $\mathscr{V}$ will have the same number of elements. This number is called the dimension of $\mathscr{V}$.

In the elementary theory of vectors there exists a type of multiplication called scalar multiplication. The scalar product of two vectors $\boldsymbol{u}$ and $\boldsymbol{v}$ is denoted by $\boldsymbol{u} \cdot \boldsymbol{v}$ and is sometimes referred to as an inner product. Inner or scalar products have the following properties:

1. $(\boldsymbol{u}+\boldsymbol{v}) \cdot \boldsymbol{w} = \boldsymbol{u} \cdot \boldsymbol{w}+\boldsymbol{v} \cdot \boldsymbol{w}$
2. $(a\boldsymbol{u}) \cdot \boldsymbol{v} = a(\boldsymbol{u} \cdot \boldsymbol{v})$
3. $\boldsymbol{u} \cdot \boldsymbol{v} = \boldsymbol{v} \cdot \boldsymbol{u}$
4. $\boldsymbol{u} \cdot \boldsymbol{u} > 0$ unless $\boldsymbol{u} = \boldsymbol{o}$

A vector space within which a scalar product having the above properties is defined is called a euclidean vector space. The scalar product of base vectors $\boldsymbol{g}_i$ and $\boldsymbol{g}_j$ are denoted by g_{ij}:

$$\boldsymbol{g}_i \cdot \boldsymbol{g}_j = g_{ij} \tag{1-5-6}$$

Now let the vector $\boldsymbol{u} = u^i\boldsymbol{g}_i$ (summation convention used) and $\boldsymbol{v} = v^j\boldsymbol{g}_j$. Then

$$\boldsymbol{u} \cdot \boldsymbol{v} = g_{ij}u^i v^j \tag{1-5-7}$$

The vectors $\boldsymbol{u}$ and $\boldsymbol{v}$ are said to be orthogonal if

$$\boldsymbol{u} \cdot \boldsymbol{v} = 0 = g_{ij}u^i v^j \tag{1-5-8}$$

The square of the magnitude $|\boldsymbol{v}|$ of the vector $\boldsymbol{v}$ is defined by

$$|\boldsymbol{v}|^2 = \boldsymbol{v} \cdot \boldsymbol{v} = g_{ij}v^i v^j \tag{1-5-9}$$

It can also be shown that two important inequalities are valid in a euclidean vector space:

Schwarz inequality: $$|\boldsymbol{u} \cdot \boldsymbol{v}| \leqslant |\boldsymbol{u}|\,|\boldsymbol{v}| \tag{1-5-10}$$

Triangle inequality: $$|\boldsymbol{u}+\boldsymbol{v}| \leqslant |\boldsymbol{u}|+|\boldsymbol{v}| \tag{1-5-11}$$

The Schwarz inequality allows the angle ϕ between two vectors $\boldsymbol{u}$ and $\boldsymbol{v}$ to be defined by

$$\cos\phi = \frac{\boldsymbol{u} \cdot \boldsymbol{v}}{|\boldsymbol{u}|\,|\boldsymbol{v}|} \tag{1-5-12}$$

A vector $\boldsymbol{u}/|\boldsymbol{u}|$ whose magnitude is unity is said to be normalized. A set of vectors which are all mutually orthogonal and each of which is normalized is said to form an orthonormal set. It can be shown that in a euclidean vector space of finite dimension one can always find a basis which is orthonormal. If such a basis has base vectors $\boldsymbol{i}_1, \boldsymbol{i}_2, \ldots, \boldsymbol{i}_n$ then

$$\boldsymbol{i}_i \cdot \boldsymbol{i}_j = \delta_{ij} \tag{1-5-13}$$

and the kth component of a vector $\boldsymbol{v}$ equals its projection $\boldsymbol{v} \cdot \boldsymbol{i}_k$ on the kth base vector. Thus if $\boldsymbol{v} = v^j\boldsymbol{i}_j$ then

$$\boldsymbol{v} \cdot \boldsymbol{i}_k = v^j\boldsymbol{i}_j \cdot \boldsymbol{i}_k = v^j\delta_{jk} = v^k \tag{1-5-14}$$

A basis $\boldsymbol{g}^j$ which is dual to the basis $\boldsymbol{g}_j$ is defined by the relations

$$\boldsymbol{g}^j \cdot \boldsymbol{g}_k = \delta^j{}_k = \delta_{jk} \tag{1-5-15}$$

and the scalar product of dual bases is defined by

$$g^{ij} = \boldsymbol{g}^i \cdot \boldsymbol{g}^j \tag{1-5-16}$$

In terms of the dual basis

$$v = v_i g^i \tag{1-5-17}$$

The following relations are easily proved:

$$g_j = g_{jk} g^k \qquad g^j = g^{jk} g_k \qquad g^{ij} g_{jk} = g^i{}_k = \delta^i{}_k = \delta_{ik} \tag{1-5-18}$$

$$v_j = g_{jk} v^k \qquad v^j = g^{jk} v_k \tag{1-5-19}$$

The components v^i are called the contravariant components and the components v_i the covariant components of the vector $\boldsymbol{v}$. Comparing Eqs. (1-5-13) and (1-5-15), it can be seen that an orthonormal basis is dual to itself and for such a basis there is no distinction between covariant and contravariant components.

The scalar product of $\boldsymbol{u}$ and $\boldsymbol{v}$ can now be written in the form

$$\boldsymbol{u} \cdot \boldsymbol{v} = (u_j g^j) \cdot (v^k g_k) = u_j v^k (g^j \cdot g_k) = u_j v^k \delta_{jk} = u_j v^j \tag{1-5-20}$$

In a euclidean vector space of three dimensions another form of product called the vector product of two vectors $\boldsymbol{u}$ and $\boldsymbol{v}$ is defined by

1. $\boldsymbol{u} \cdot (\boldsymbol{u} \times \boldsymbol{v}) = \boldsymbol{v} \cdot (\boldsymbol{u} \times \boldsymbol{v}) = 0$

   i.e., the direction of the vector product is orthogonal to each of the vectors in the product.
2. The direction of $\boldsymbol{u} \times \boldsymbol{v}$ is that of a right-hand screw.
3. If the angle between $\boldsymbol{u}$ and $\boldsymbol{v}$ is ϕ,

$$|\boldsymbol{u} \times \boldsymbol{v}| = |\boldsymbol{u}|\,|\boldsymbol{v}| \sin \phi$$

It follows from the definition that

$$\boldsymbol{u} \times \boldsymbol{v} = -\boldsymbol{v} \times \boldsymbol{u}$$

$$(a\boldsymbol{u}) \times \boldsymbol{v} = a(\boldsymbol{u} \times \boldsymbol{v})$$

$$\boldsymbol{u} \times (\boldsymbol{v} + \boldsymbol{w}) = \boldsymbol{u} \times \boldsymbol{v} + \boldsymbol{u} \times \boldsymbol{w}$$

If $\boldsymbol{i}_1$, $\boldsymbol{i}_2$, $\boldsymbol{i}_3$ form an orthonormal right-handed system of vectors then it can be shown by inspection that

$$\boldsymbol{i}_j \times \boldsymbol{i}_k = e_{jkl} \boldsymbol{i}_l \tag{1-5-21}$$

where e_{jkl} is the permutation symbol.

1-6 LINEAR TRANSFORMATIONS

A linear transformation $\mathbf{T}$ of a vector space $\mathscr{V}$ into another vector space $\mathscr{W}$ is defined as a correspondence which maps each vector $\boldsymbol{u}$ of $\mathscr{V}$ on a vector $\mathbf{T}(\boldsymbol{u})$ of $\mathscr{W}$ such that, for all vectors $\boldsymbol{u}$ and $\boldsymbol{v}$ of $\mathscr{V}$ and all scalars a, b of a field $\mathscr{F}$,

$$\mathbf{T}(a\boldsymbol{u} + b\boldsymbol{v}) = a\mathbf{T}(\boldsymbol{u}) + b\mathbf{T}(\boldsymbol{v}) \tag{1-6-1}$$

To emphasize the linearity the brackets are often dropped:

$$w = \mathbf{T}(\boldsymbol{u}) = \mathbf{T}\boldsymbol{u} \tag{1-6-2}$$

If there is a transformation $\mathbf{S}$ of $\boldsymbol{u}$ into $\boldsymbol{v}$ given by $\boldsymbol{v} = \mathbf{S}\boldsymbol{u}$ and there is a transformation $\mathbf{T}$ of $\boldsymbol{v}$ into $\boldsymbol{w}$ given by $\boldsymbol{w} = \mathbf{T}\boldsymbol{v}$, then the transformation taking $\boldsymbol{u}$ into $\boldsymbol{w}$ is defined as the product of the two transformations and is written $\mathbf{TS}$. Thus

$$w = \mathbf{T}\boldsymbol{v} = \mathbf{T}(\mathbf{S}\boldsymbol{u}) = \mathbf{TS}\boldsymbol{u} \tag{1-6-3}$$

The product of two linear transformations is linear for

$$\mathbf{TS}(a\boldsymbol{u}+b\boldsymbol{v}) = \mathbf{T}(a\mathbf{S}\boldsymbol{u}+b\mathbf{S}\boldsymbol{v}) = a\mathbf{TS}\boldsymbol{u}+b\mathbf{TS}\boldsymbol{v} \tag{1-6-4}$$

The identity or unit linear transformation $\mathbf{I}$ satisfies the equation

$$\mathbf{I}\boldsymbol{v} = \boldsymbol{v} \qquad \text{for all } \boldsymbol{v} \text{ in } \mathscr{V} \tag{1-6-5}$$

If there exists a linear transformation $\mathbf{T}^{-1}$ such that $\mathbf{TT}^{-1} = \mathbf{T}^{-1}\mathbf{T} = \mathbf{I}$, then $\mathbf{T}$ is said to be nonsingular or invertible and $\mathbf{T}^{-1}$ is called its inverse. Otherwise $\mathbf{T}$ will be called singular. Considering three linear transformations acting on an arbitrary vector,

$$\mathbf{U}(\mathbf{TS})\boldsymbol{u} = \mathbf{U}(\mathbf{T}(\mathbf{S}\boldsymbol{u})) = (\mathbf{UT})\mathbf{S}\boldsymbol{u} \tag{1-6-6}$$

Thus linear transformations are associative with respect to multiplication and under multiplication form a group. It follows from Eq. (1-4-1) that

$$(\mathbf{TS})^{-1} = \mathbf{S}^{-1}\mathbf{T}^{-1} \tag{1-6-7}$$

If $\mathbf{T}$ is an endomorphism, i.e., it maps $\mathscr{V}$ into itself, then it is possible to define the transpose of a matrix $\mathbf{T}$, which will be written $\mathbf{T}^T$, by the relation

$$\boldsymbol{v}\cdot(\mathbf{T}\boldsymbol{u}) = \boldsymbol{u}\cdot(\mathbf{T}^T\boldsymbol{v}) \tag{1-6-8}$$

for all pairs of vectors $\boldsymbol{u}$ and $\boldsymbol{v}$ in $\mathscr{V}$. From Eq. (1-6-8),

$$(\mathbf{TS})^T = \mathbf{S}^T\mathbf{T}^T \tag{1-6-9}$$

A transformation is said to be symmetric if

$$\mathbf{T} = \mathbf{T}^T \tag{1-6-10}$$

and antisymmetric if

$$\mathbf{T} = -\mathbf{T}^T \tag{1-6-11}$$

Any transformation can always be written as the sum of a symmetric and antisymmetric part for

$$\mathbf{T} = \tfrac{1}{2}(\mathbf{T}+\mathbf{T}^T)+\tfrac{1}{2}(\mathbf{T}-\mathbf{T}^T)$$

A transformation $\mathbf{T}$ is orthogonal if it preserves the absolute value of every vector $\boldsymbol{u}$ so that

$$|\mathbf{T}\boldsymbol{u}| = |\boldsymbol{u}|$$

Thus

$$(\mathbf{T}\boldsymbol{u}) \cdot (\mathbf{T}\boldsymbol{u}) = \boldsymbol{u} \cdot \boldsymbol{u} \tag{1-6-12}$$

However, by Eq. (1-6-8), this can be written as

$$\boldsymbol{u} \cdot \mathbf{T}^T\mathbf{T}\boldsymbol{u} = \boldsymbol{u} \cdot \boldsymbol{u}$$

for every vector $\boldsymbol{u}$ in $\mathscr{V}$. Thus $\mathbf{T}$ is orthogonal if

$$\mathbf{T}^T\mathbf{T} = \mathbf{I} \qquad \text{or} \qquad \mathbf{T}^T = \mathbf{T}^{-1} \tag{1-6-13}$$

Let the vector space $\mathscr{V}$ have a basis $\boldsymbol{g}_1, \boldsymbol{g}_2, \ldots, \boldsymbol{g}_n$ and the vector space $\mathscr{W}$ have a basis $\boldsymbol{e}_1, \boldsymbol{e}_2, \ldots, \boldsymbol{e}_m$. Applying $\mathbf{T}$ to each vector $\boldsymbol{g}_a$ gives

$$\boldsymbol{h}_a = \mathbf{T}\boldsymbol{g}_a \qquad \text{for } a = 1, 2, \ldots, n \tag{1-6-14}$$

but each $\boldsymbol{h}_a$ can be expressed in terms of the basis vectors; thus

$$\boldsymbol{h}_a = T^i{}_a \boldsymbol{e}_i = \mathbf{T}\boldsymbol{g}_a \tag{1-6-15}$$

This set of scalars $T^i{}_a$ can be regarded as a matrix:

$$[T^i{}_a] = \begin{bmatrix} T^1{}_1 & \cdots & T^1{}_n \\ \vdots & & \vdots \\ T^m{}_1 & \cdots & T^m{}_n \end{bmatrix} \tag{1-6-16}$$

The linear transformation $\mathbf{T}$ thus has the above matrix representation relative to the given bases. The coordinate representation of the equation

$$\boldsymbol{v} = \mathbf{T}\boldsymbol{u} \tag{1-6-17}$$

is then given by

$$\begin{aligned} v^i \boldsymbol{e}_i &= \mathbf{T}u^a \boldsymbol{g}_a = T^i{}_a u^a \boldsymbol{e}_i \\ \text{or} \qquad v^i &= T^i{}_a u^a \end{aligned} \tag{1-6-18}$$

The contravariant components of $\mathbf{T}$ are given by

$$\mathbf{T}\boldsymbol{g}^a = T^{ia}\boldsymbol{e}_i \tag{1-6-19}$$

and the covariant components by

$$\mathbf{T}\boldsymbol{g}_a = T_{ia}\boldsymbol{e}^i \tag{1-6-20}$$

From the definition in Eq. (1-6-8), it can be shown that if $\mathbf{T}$ has the representation T_{ia} then $\mathbf{T}^T$ is represented by T_{ai} in accord with Eq. (1-2-8), i.e.,

$$(T^T)_{ai} = T_{ia} \qquad \text{or} \qquad (T^T)_a{}^i = T^i{}_a \tag{1-6-21}$$

Often the linear transformation will be into $\mathscr{V}$. In this case the matrix T_{ij} will be square. The symmetric part of T_{ij} can now be written $T_{ij}{}^s$ where

$$T_{ij}{}^s = \tfrac{1}{2}(T_{ij} + T_{ji}) \tag{1-6-22}$$

and the antisymmetric part $T_{ij}{}^A$ is

$$T_{ij}{}^A = \tfrac{1}{2}(T_{ij} - T_{ji}) \tag{1-6-23}$$

Clearly,

$$T_{ij} = T_{ij}^{s} + T_{ij}^{A} \tag{1-6-24}$$

The component form of an orthogonal matrix **T** as given in Eq. (1-6-13) becomes

$$\begin{aligned} \mathbf{T}^T\mathbf{T} &= \mathbf{I} \\ (T^T)^i{}_j T^j{}_k &= \delta_{ik} \\ \text{or} \qquad T_j^i T^j{}_k &= \delta_{ik} \end{aligned} \tag{1-6-25}$$

Now it can be shown that

$$\det(\mathbf{AB}) = \det \mathbf{A} \det \mathbf{B}$$

Thus, taking the determinant of both sides of Eq. (1-6-25) and using Eq. (1-2-12) shows that, for an orthogonal matrix,

$$\det \mathbf{T} = \pm 1 \tag{1-6-26}$$

If $\det \mathbf{T} = +1$ the orthogonal transformation is called a proper one, while if $\det \mathbf{T} = -1$ it is called an improper orthogonal transformation.

It is sometimes useful to consider the tensor or dyadic product. This is formed from two vectors, one from a vector space $\mathscr{V}$ of dimension n and one from a vector space $\mathscr{W}$ of dimension m, say. It exists in a new vector space of dimension nm denoted by $\mathscr{W} \otimes \mathscr{V}$ and is called the tensor-product space of $\mathscr{W}$ and $\mathscr{V}$. For a set of vectors $\boldsymbol{v}_i$ in $\mathscr{V}$ and a set of vectors $\mathbf{w}_i$ in $\mathscr{W}$, the tensor product, or open product of two vectors, denoted $\boldsymbol{w} \otimes \boldsymbol{v}$, is a tensor called a dyad, defined by the requirement that

$$(\boldsymbol{w} \otimes \boldsymbol{v})\boldsymbol{u} = \boldsymbol{w}(\boldsymbol{v} \cdot \boldsymbol{u}) \tag{1-6-27}$$

for all vectors $\boldsymbol{u}$ in $\mathscr{V}$. The tensor product $\boldsymbol{w} \otimes \boldsymbol{v}$ has the following properties:

$$\begin{aligned} \boldsymbol{w} \otimes (\boldsymbol{v}_1 + \boldsymbol{v}_2) &= \boldsymbol{w} \otimes \boldsymbol{v}_1 + \boldsymbol{w} \otimes \boldsymbol{v}_2 \\ (\boldsymbol{w}_1 + \boldsymbol{w}_2) \otimes \boldsymbol{v} &= \boldsymbol{w}_1 \otimes \boldsymbol{v} + \boldsymbol{w}_2 \otimes \boldsymbol{v} \\ a(\boldsymbol{w} \otimes \boldsymbol{v}) &= a\boldsymbol{w} \otimes \boldsymbol{v} = \boldsymbol{w} \otimes a\boldsymbol{v} \end{aligned} \tag{1-6-28}$$

where a is an arbitrary scalar.

If $\mathscr{W}$ has a basis $\boldsymbol{e}_1, \ldots, \boldsymbol{e}_m$ and $\mathscr{V}$ a basis $\boldsymbol{g}_1, \ldots, \boldsymbol{g}_n$ then the set of elements $\boldsymbol{g}_{ia} = \boldsymbol{e}_i \otimes \boldsymbol{g}_a$ is a basis for $\mathscr{W} \otimes \mathscr{V}$. Thus as $\boldsymbol{v} = v^a \boldsymbol{g}_a$ and $\boldsymbol{w} = w^i \boldsymbol{e}_i$ the component form of the tensor product is

$$\boldsymbol{w} \otimes \boldsymbol{v} = w^i v^a \boldsymbol{e}_i \otimes \boldsymbol{g}_a \tag{1-6-29}$$

A linear transformation **T** can now be represented in the alternative form

$$\mathbf{T} = T^i{}_a \boldsymbol{e}_i \otimes \boldsymbol{g}^a = T^{ia} \boldsymbol{e}_i \otimes \boldsymbol{g}_a \tag{1-6-30}$$

where, by using Eq. (1-6-15), the components can be shown to be given by

$$T^i{}_a = \boldsymbol{e}^i \cdot (\mathbf{T}\boldsymbol{g}_a) \qquad \text{or} \qquad T_{ia} = \boldsymbol{e}_i \cdot (\mathbf{T}\boldsymbol{g}_a) \qquad \text{or} \qquad T^{ia} = \boldsymbol{e}^i \cdot (\mathbf{T}\boldsymbol{g}^a) \tag{1-6-31}$$

It is useful to define a set of scalars g_i^a, g_{ia}, and g^{ia} by the relations

$$g_i^a = \boldsymbol{e}_i \cdot \boldsymbol{g}^a \qquad g_{ia} = \boldsymbol{e}_i \cdot \boldsymbol{g}_a \qquad g^{ia} = \boldsymbol{e}^i \cdot \boldsymbol{g}^a \tag{1-6-32}$$

Consider now a change of basis from $\boldsymbol{e}_1, \ldots, \boldsymbol{e}_m$ to $\boldsymbol{e}'_1, \ldots, \boldsymbol{e}'_m$ in $\mathscr{W}$ and from $\boldsymbol{g}_1, \ldots, \boldsymbol{g}_n$ to $\boldsymbol{g}'_1, \ldots, \boldsymbol{g}'_n$ in $\mathscr{V}$ given by the relations

$$\boldsymbol{e}'_{i'} = A_{i'}{}^{i}\boldsymbol{e}_i \qquad \text{and} \qquad \boldsymbol{g}'_{a'} = B_{a'}{}^{a}\boldsymbol{g}_a \tag{1-6-33}$$

Then, using Eq. (1-6-15),

$$T^{i'}{}_{a'}\boldsymbol{e}'_{i'} = T^{i'}{}_{a'}A_{i'}{}^{i}\boldsymbol{e}_i = \mathbf{T}\boldsymbol{g}'_{a'} = \mathbf{T}B_{a'}{}^{a}\boldsymbol{g}_a = T^{i}{}_{a}B_{a'}{}^{a}\boldsymbol{e}_i \tag{1-6-34}$$

Hence,

$$T^{i'}{}_{a'} = T^{i}{}_{a}B_{a'}{}^{a}(A^{-1})^{i'}{}_{i} \tag{1-6-35}$$

Similarly,

$$g_{i'}{}^{a'} = A_{i'}{}^{j}(B^{-1})_{b}{}^{a'}g_{j}{}^{b} \tag{1-6-36}$$

Noting that

$$A_{i'}{}^{j}(A^{-1})^{i'}{}_{i} = \delta_{ij} \qquad \text{and} \qquad B_{a'}{}^{a}(B^{-1})_{b}{}^{a'} = \delta_{ab}$$

we find

$$T^{i'}{}_{a'}g_{i'}{}^{a'} = T^{i}{}_{a}g_{i}{}^{a} \tag{1-6-37}$$

The above product is therefore invariant to basis changes; it is called the trace of the tensor $\mathbf{T}$ and is written $\operatorname{tr}\mathbf{T}$. Thus

$$\operatorname{tr}\mathbf{T} = T^{i}{}_{a}g_{i}{}^{a} \tag{1-6-38}$$

If $\mathbf{T}$ is endomorphic and maps $\mathscr{V}$ into itself then

$$g_{i}{}^{a} = \boldsymbol{g}_i \cdot \boldsymbol{g}^a = \delta_{i}{}^{a} \tag{1-6-39}$$

and the trace takes the simpler form

$$\operatorname{tr}\mathbf{T} = T^{i}{}_{i} = T^{ij}g_{ji} = T_{ij}g^{ji} \tag{1-6-40}$$

Taking the determinant of the matrix equations (1-6-35) and (1-6-36) and noting that

$$\det A_{i'}{}^{j}(A^{-1})^{i'}{}_{i} = 1 \qquad \text{and} \qquad \det B_{a'}{}^{a}(B^{-1})_{b}{}^{a'} = 1$$

we find

$$\det(T^{i'}{}_{a'}g_{i'}{}^{a'}) = \det(T^{i}{}_{a}g_{i}{}^{a}) \tag{1-6-41}$$

Thus $\det(T^{i}{}_{a}g_{i}{}^{a})$ is also invariant to basis changes; it is called the determinant of $\mathbf{T}$ and is written $\det\mathbf{T}$. Thus

$$\det\mathbf{T} = \det(T^{i}{}_{a}g_{i}{}^{a}) = \det g_{i}{}^{a}\det T^{i}{}_{a} \tag{1-6-42}$$

As

$$g^{i}{}_{a}\boldsymbol{e}_i = \boldsymbol{g}_a \tag{1-6-43}$$

taking the scalar product with $\boldsymbol{e}_j$ gives

$$g^{i}{}_{a}\boldsymbol{e}_i \cdot \boldsymbol{e}_j = \boldsymbol{e}_j \cdot \boldsymbol{g}_a = g^{i}{}_{a}e_{ij} = g_{ja} \tag{1-6-44}$$

in agreement with Eq. (1-6-32). Taking the scalar product of Eq. (1-6-43) with $\boldsymbol{g}_b$

gives

$$g^i{}_a \boldsymbol{e}_i \cdot g^j{}_b \boldsymbol{e}_j = \boldsymbol{g}_a \cdot \boldsymbol{g}_b \tag{1-6-45}$$

Thus

$$g^i{}_a g^j{}_b e_{ij} = g_{ab} \tag{1-6-46}$$

Taking the determinant of both sides of this gives

$$\det g^i{}_a = \left(\frac{\det g_{ab}}{\det e_{ij}}\right)^{1/2} \tag{1-6-47}$$

Again, if we are considering endomorphic transformations of $\mathscr{V}$ into itself,

$$g^i{}_a = \delta^i{}_a \qquad \det g^i{}_a = 1 \qquad \det \mathbf{T} = \det T^i{}_j = \det g_{ij} \det T^{ij} \tag{1-6-48}$$

It should be noted that the summation convention in general only produces another tensor if mixed indices are used, i.e., one covariant and one contravariant index, and thus $\operatorname{tr} \mathbf{T} \neq T_{ii} \neq T^{ii}$, as Eq. (1-6-40) shows. Similarly,

$$\det \mathbf{T} = \det T^i{}_j \neq \det T_{ij} \neq \det T^{ij}$$

1-7 EUCLIDEAN POINT SPACE

With each point P of the familiar space of elementary geometry can be associated a position vector $\boldsymbol{OP}$ giving the position of P relative to an origin O. It can be shown that the set of all position vectors forms a three-dimensional euclidean vector space $\mathscr{E}$. A basis $\boldsymbol{g}_i$ of $\mathscr{E}$ together with an origin O are said to form a reference frame for the euclidean point space. Let the position vector $\boldsymbol{OP}$ be denoted by $\boldsymbol{x}$. If there is a functional relation between the vectors $\boldsymbol{v}$ of a vector space $\mathscr{V}$ and the position $\boldsymbol{x}$ then $\boldsymbol{v}$ is called a vector field and is written $\boldsymbol{v}(\boldsymbol{x})$. If an orthonormal basis $\boldsymbol{i}_k$ is chosen for $\mathscr{E}$ then

$$\boldsymbol{x} = \bar{x}_k \boldsymbol{i}_k \tag{1-7-1}$$

and the $\bar{x}_k$ are called rectangular cartesian coordinates of P. A new set of variables can be introduced which satisfy relations of the form

$$x^i = f_i(\bar{x}_1, \bar{x}_2, \bar{x}_3) \qquad \text{for } i = 1, 2, 3 \tag{1-7-2}$$

where the functions f_i are single valued, possess continuous derivatives, and are independent of each other. Such a set of variables is called a curvilinear coordinate. A necessary and sufficient condition for the independence of the f_i is that the determinant

$$\begin{vmatrix} \dfrac{\partial f_1}{\partial \bar{x}_1} & \dfrac{\partial f_1}{\partial \bar{x}_2} & \dfrac{\partial f_1}{\partial \bar{x}_3} \\ \dfrac{\partial f_2}{\partial \bar{x}_1} & \dfrac{\partial f_2}{\partial \bar{x}_2} & \dfrac{\partial f_2}{\partial \bar{x}_3} \\ \dfrac{\partial f_3}{\partial \bar{x}_1} & \dfrac{\partial f_3}{\partial \bar{x}_2} & \dfrac{\partial f_3}{\partial \bar{x}_3} \end{vmatrix} \neq 0 \tag{1-7-3}$$

This determinant is called the jacobian of the functions f_i. If the above conditions are satisfied, Eqs. (1-7-2) can be solved in terms of a unique set of inverse functions.

A set of base vectors $\boldsymbol{g}_i$ are now chosen such that the vectorial distance between P at $\mathbf{x}$ and a neighboring point Q at $\mathbf{x}+d\mathbf{x}$ is given by

$$d\mathbf{x} = \boldsymbol{g}_i\, dx^i \tag{1-7-4}$$

or

$$\boldsymbol{g}_i = \frac{\partial \mathbf{x}}{\partial x^i} \tag{1-7-5}$$

The magnitude of the distance PQ is often written [using Eq. (1-5-6)] as ds:

$$\begin{aligned} ds^2 = d\mathbf{x}\cdot d\mathbf{x} &= \boldsymbol{g}_i\cdot\boldsymbol{g}_j\, dx^i\, dx^j \\ &= g_{ij}\, dx^i\, dx^j \end{aligned} \tag{1-7-6}$$

where g_{ij} is an important quantity characterizing the geometrical properties of space and is called the fundamental or metric tensor. From Eq. (1-7-4) it is clear that the base vector $\boldsymbol{g}_i$ is tangential to the coordinate line through P for which all coordinates except x^i are kept constant. These vectors are called the covariant base vectors. If the coordinate lines are orthogonal then the base vectors will be orthogonal and, by Eq. (1-5-8),

$$g_{ij} = \boldsymbol{g}_i\cdot\boldsymbol{g}_j = 0 \tag{1-7-7}$$

In general, the base vectors are not unit vectors, for the magnitude of $\boldsymbol{g}_i$ is given by

$$\begin{aligned} |\boldsymbol{g}_i|^2 &= \boldsymbol{g}_i\cdot\boldsymbol{g}_i = g_{ii} \\ |\boldsymbol{g}_i| &= \sqrt{g_{ii}} \end{aligned} \qquad (i \text{ not summed}) \tag{1-7-8}$$

For future reference the metric tensor of some simple coordinate systems are now given.

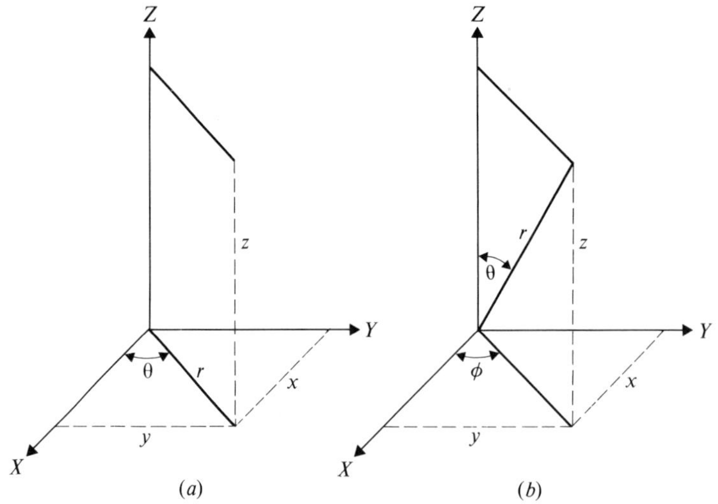

Figure 1-1 Some simple coordinate systems. (*a*) Cylindrical polar coordinates. (*b*) Spherical polar coordinates.

Rectangular cartesian coordinates (see Fig. 1-1*a*)

$$ds^2 = dx^2 + dy^2 + dz^2$$

$$g_{xx} = g_{yy} = g_{zz} = 1$$

All other components are zero.

Cylindrical polar coordinates (see Fig. 1-1*a*)

$$ds^2 = dr^2 + (r\,d\theta)^2 + dz^2$$

$$g_{rr} = g^{rr} = g_{zz} = g^{zz} = 1 \qquad g_{\theta\theta} = r^2 \qquad g^{\theta\theta} = \frac{1}{r^2}$$

All other components are zero.

Spherical polar coordinates (see Fig. 1-1*b*)

$$ds^2 = dr^2 + (r\,d\theta)^2 + (r \sin\theta\, d\phi)^2$$

$$g_{rr} = g^{rr} = 1 \qquad g_{\theta\theta} = r^2 \qquad g^{\theta\theta} = \frac{1}{r^2}$$

$$g_{\phi\phi} = (r \sin\theta)^2 \qquad g^{\phi\phi} = \frac{1}{(r\sin\theta)^2}$$

All other components are zero.

If a new coordinate system x'_i is chosen it will have a new associated set of base vectors $\boldsymbol{g}'_i$ which are related, using Eqs. (1-7-5), as

$$\boldsymbol{g}'_j = \frac{\partial \boldsymbol{x}}{\partial x^{j'}} = \frac{\partial \boldsymbol{x}}{\partial x^i}\frac{\partial x^i}{\partial x^{j'}} = \frac{\partial x^i}{\partial x^{j'}}\boldsymbol{g}_i \tag{1-7-9}$$

In particular, for a rectangular cartesian set of coordinates $\bar{x}_k$ and orthonormal base vectors $\boldsymbol{i}_k$,

$$\boldsymbol{g}_i = \frac{\partial \bar{x}_k}{\partial x^i}\boldsymbol{i}_k \tag{1-7-10}$$

Hence,

$$g_{ij} = \frac{\partial \bar{x}_k}{\partial x^i}\frac{\partial \bar{x}_p}{\partial x^j}\boldsymbol{i}_k \cdot \boldsymbol{i}_p = \frac{\partial x_k}{\partial x^i}\frac{\partial \bar{x}_p}{\partial x^j}\delta_{kp} = \frac{\partial \bar{x}_k}{\partial x^i}\frac{\partial \bar{x}_k}{\partial x^j} \tag{1-7-11}$$

Taking the determinant of both sides of the above equation

$$\det g_{ij} = \det\frac{\partial \bar{x}_k}{\partial x^i}\det\frac{\partial \bar{x}_k}{\partial x^j} \tag{1-7-12}$$

or letting

$$\det g_{ij} = g \qquad \text{and} \qquad \det\frac{\partial \bar{x}_k}{\partial x^i} = \left|\frac{\partial \bar{x}}{\partial x}\right|$$

which is the jacobian of the coordinate transformation, then

$$\left|\frac{\partial \bar{x}}{\partial x}\right|^2 = g \tag{1-7-13}$$

We will only consider coordinate transformations for which the jacobian is positive and in this case

$$\left|\frac{\partial \bar{x}}{\partial x}\right| = \sqrt{g} \tag{1-7-14}$$

It is useful to find the elementary volume ΔV of the parallelepiped whose sides are $dx^1\boldsymbol{g}_1, dx^2\boldsymbol{g}_2, dx^3\boldsymbol{g}_3$. Now it is shown in elementary vector analysis that the volume of the parallelepiped whose sides are parallel to the vectors $\boldsymbol{a}$, $\boldsymbol{b}$, and $\boldsymbol{c}$ is $(\boldsymbol{a}\times\boldsymbol{b})\cdot\boldsymbol{c}$ and thus

$$\Delta V = (\boldsymbol{g}_1 \times \boldsymbol{g}_2)\cdot \boldsymbol{g}_3\, dx^1\, dx^2\, dx^3 \tag{1-7-15}$$

But, using Eq. (1-7-10),

$$(\boldsymbol{g}_1 \times \boldsymbol{g}_2)\cdot \boldsymbol{g}_3 = \frac{\partial \bar{x}_p}{\partial x^1}\frac{\partial \bar{x}_q}{\partial x^2}\frac{\partial \bar{x}_r}{\partial x^3}(\boldsymbol{i}_p \times \boldsymbol{i}_q)\cdot \boldsymbol{i}_r \tag{1-7-16}$$

Now, using Eq. (1-5-21),

$$\boldsymbol{i}_p \times \boldsymbol{i}_q = e_{pqs}\boldsymbol{i}_s \tag{1-7-17}$$

where e_{pqs} is the permutation symbol defined by Eq. (1-2-6). Thus

$$(\boldsymbol{i}_p \times \boldsymbol{i}_q)\cdot \boldsymbol{i}_r = e_{pqs}\boldsymbol{i}_s \cdot \boldsymbol{i}_r = e_{pqs}\delta_{sr} = e_{pqr} \tag{1-7-18}$$

and hence

$$(\boldsymbol{g}_1 \times \boldsymbol{g}_2)\cdot \boldsymbol{g}_3 = e_{pqr}\frac{\partial \bar{x}_p}{\partial x^1}\frac{\partial \bar{x}_q}{\partial x^2}\frac{\partial \bar{x}_r}{\partial x^3} \tag{1-7-19}$$

Writing out the six components of the right-hand side of Eq. (1-7-19), it can be seen that

$$e_{pqr}\frac{\partial \bar{x}_p}{\partial x^1}\frac{\partial \bar{x}_q}{\partial x^2}\frac{\partial \bar{x}_r}{\partial x^3} = \begin{vmatrix} \dfrac{\partial \bar{x}_1}{\partial x^1} & \dfrac{\partial \bar{x}_2}{\partial x^1} & \dfrac{\partial \bar{x}_3}{\partial x^1} \\[2ex] \dfrac{\partial \bar{x}_1}{\partial x^2} & \dfrac{\partial \bar{x}_2}{\partial x^2} & \dfrac{\partial \bar{x}_3}{\partial x^2} \\[2ex] \dfrac{\partial \bar{x}_1}{\partial x^3} & \dfrac{\partial \bar{x}_2}{\partial x^3} & \dfrac{\partial \bar{x}_3}{\partial x^3} \end{vmatrix} \tag{1-7-20}$$

This determinant is just the jacobian of the transformation and can be written $|\partial\bar{x}/\partial x|$. Thus, by Eq. (1-7-14),

$$(\boldsymbol{g}_1 \times \boldsymbol{g}_2)\cdot \boldsymbol{g}_3 = \left|\frac{\partial \bar{x}}{\partial x}\right| = \sqrt{g} \tag{1-7-21}$$

and hence

$$\Delta V = \sqrt{g}\, dx^1\, dx^2\, dx^3 \tag{1-7-22}$$

It can easily be seen that the set of vectors defined by $(\boldsymbol{g}_2 \times \boldsymbol{g}_3)/\sqrt{g}$ satisfy the conditions for the dual basis to $\boldsymbol{g}_1$ and so we have

$$\boldsymbol{g}^1 = \frac{\boldsymbol{g}_2 \times \boldsymbol{g}_3}{\sqrt{g}} \qquad \boldsymbol{g}^2 = \frac{\boldsymbol{g}_3 \times \boldsymbol{g}_1}{\sqrt{g}} \qquad \boldsymbol{g}^3 = \frac{\boldsymbol{g}_1 \times \boldsymbol{g}_2}{\sqrt{g}} \tag{1-7-23}$$

These vectors are called the contravariant base vectors. For an orthogonal coordinate system the covariant and contravariant base vectors have the same direction and satisfy the relation

$$\boldsymbol{g}^i = \frac{\boldsymbol{g}_i}{g_{ii}} \qquad (i \text{ not summed}) \tag{1-7-24}$$

and hence

$$g^{ii} = \frac{1}{g_{ii}} \tag{1-7-25}$$

Clearly, the contravariant base vectors are also not unit vectors. As the coordinates may not have the same dimensions (i.e., in spherical polar coordinates r has the dimension of length while θ and ϕ are dimensionless), the components of the vector may not have the same dimensions, which is unacceptable in physical problems. If the vectors are resolved relative to unit vectors in the direction of the base vectors, then all the components will have the same dimensions; these are called the physical components of the vector $\boldsymbol{a}$ and will be written as $a\langle i \rangle$. From Eq. (1-7-8) the unit vector ε_i in the direction of $\boldsymbol{g}_i$ is given by

$$\varepsilon_i = \frac{\boldsymbol{g}_i}{\sqrt{g_{ii}}} \qquad (i \text{ not summed}) \tag{1-7-26}$$

Thus

$$\boldsymbol{a} = a^{\langle i \rangle} \varepsilon_i = a^i \boldsymbol{g}_i \tag{1-7-27}$$

Comparing Eqs. (1-7-26) and (1-7-27),

$$a^{\langle i \rangle} = \sqrt{g_{ii}}\, a^i = \sqrt{g_{ii}}\, g^{ij} a_j \tag{1-7-28}$$

This technique can be extended to find the physical components of higher-order tensors (see, for example, Sec. 3-6).

In transforming from a coordinate system x^i to a new system $x^{i'}$ the base vectors change, as given by Eq. (1-7-9), and so the resolved components must also change. To find how they change note

$$\boldsymbol{a} = a^{j'} \boldsymbol{g}_{j'} = a^{j'} \frac{\partial x^i}{\partial x^{j'}} \boldsymbol{g}_i = a^i \boldsymbol{g}_i$$

Thus

$$a^i = \frac{\partial x^i}{\partial x^{j'}} a^{j'} \tag{1-7-29}$$

Also from Eq. (1-7-9),

$$g'_{jk} = g'_j \cdot g'_k = \frac{\partial x^i}{\partial x^{j'}} g_i \frac{\partial x^l}{\partial x^{k'}} g_l$$

or

$$g'_{jk} = \frac{\partial x^i}{\partial x^{j'}} \frac{\partial x^l}{\partial x^{k'}} g_{il} \tag{1-7-30}$$

From Eqs. (1-7-29) and (1-7-30) it is easily shown that

$$a_i = \frac{\partial x^{j'}}{\partial x^i} a_{j'} \tag{1-7-31}$$

The transformation relations of Eqs. (1-7-29) and (1-7-31) for contravariant and covariant vector components can be extended and used as the basis for the definition of tensor components of higher order. Thus the entities $a^{i\ldots j}_{k\ldots l}$ form the components of a tensor if, under a coordinate transformation $\boldsymbol{x}$ to $\boldsymbol{x}'$, the new components $a'^{p\ldots q}_{r\ldots s}$ are related to the old components by equations of the form

$$a'^{p\ldots q}_{r\ldots s} = \frac{\partial x'^p}{\partial x^i} \cdots \frac{\partial x'^q}{\partial x^j} \frac{\partial x^k}{\partial x'^r} \cdots \frac{\partial x^l}{\partial x'^s} a^{i\ldots j}_{k\ldots l} \tag{1-7-32}$$

The total number of superscripts and subscripts is called the order or rank of the tensor. If tensors of different rank and type are added or subtracted, the sum or difference is not generally a tensor. Linear sums of tensors of the same type, i.e., the same number of covariant and contravariant components, in which the coefficients are scalars are tensors of the same type.

If such linear sums are set equal to zero they form tensor equations. The importance of tensor analysis is based on the observation that if all the components of a tensor are zero in one frame of reference then by Eq. (1-7-32) they are zero in every frame of reference. Thus if a tensor equation holds in one frame of reference it will hold in all frames of reference.

To find whether a set of functions can form the components of a tensor it is often convenient to use the quotient law. This states that if the inner product of this set of functions with an arbitrary tensor is another tensor then the set of functions do form a tensor. For example, suppose the set of functions $b(ij)$ satisfy the equation

$$a_i b(ij) = c_j \tag{1-7-33}$$

where a_i is an arbitrary covariant vector and c_j is another covariant vector. Then $b(ij)$ is a mixed tensor of the form

$$b(ij) = b^i{}_j \tag{1-7-34}$$

It is useful to have a tensor form of the permutation symbol; this is called a permutation tensor and its covariant components ε_{ijk} are given by

$$\varepsilon_{ijk} = e_{ijk}\sqrt{g} \tag{1-7-35}$$

and the contravariant components by

$$\varepsilon^{ijk} = \frac{e_{ijk}}{\sqrt{g}} \tag{1-7-36}$$

In three dimensions an antisymmetric tensor $\mathbf{T}^A$ has only three independent nonzero components. Thus one can associate it with a vector t^A whose components satisfy

$$(t^A)^k = -\tfrac{1}{2}\varepsilon^{ijk}T_{ij}{}^A \tag{1-7-37}$$

t^A is called the axial vector of $\mathbf{T}^A$. Equivalently,

$$T_{ij}{}^A = -\varepsilon_{ijk}(t^A)^k \tag{1-7-38}$$

The dyadic product of an nth-order tensor $T^{i\ldots j}_{p\ldots r}$ and an mth-order tensor $S^{k\ldots l}_{q\ldots s}$ is defined as the $(m+n)$th-order tensor $\mathbf{T}\otimes\mathbf{S}$ having components $T^{i\ldots j}_{p\ldots r}S^{k\ldots l}_{q\ldots s}$. If a covariant index is given the same letter as a contravariant index, then using the summation convention the resulting entity consists of a sum of terms and can easily be shown to be another tensor whose order is reduced by two from that of the original tensor. In the case of the dyadic product this contraction can clearly be done in many different ways. However, the contraction $T^{i\ldots j}_{p\ldots r}S^{r\ldots l}_{q\ldots s}$, in which the index at the right of the covariant sequence of the left tensor is set equal to the index at the left of the contravariant sequence of the right tensor, is of particular practical importance and is denoted by $\mathbf{TS}$ in direct notation and is of order $(m+n-2)$. A double contraction for second-order tensors leads to a scalar quantity which can be represented by tr $(\mathbf{TS})$.

1-8 COVARIANT DIFFERENTIATION

In a general curvilinear coordinate system the base vectors depend on position and the differential change in magnitude and direction for a differential change in position will also be a vector. Thus

$$\frac{\partial \boldsymbol{g}_i}{\partial x^j} = \Gamma_{ij}{}^p \boldsymbol{g}_p \tag{1-8-1}$$

where the coefficients $\Gamma_{ij}{}^p$ are known as the Christoffel symbols. By Eq. (1-7-5),

$$\frac{\partial \boldsymbol{g}_i}{\partial x^j} = \frac{\partial^2 \mathbf{x}}{\partial x^j\,\partial x^i} = \frac{\partial \boldsymbol{g}_j}{\partial x^i}$$

Thus

$$\Gamma_{ij}{}^p = \Gamma_{ji}{}^p \tag{1-8-2}$$

Now, by Eq. (1-5-6),

$$g_{ij} = \boldsymbol{g}_i \cdot \boldsymbol{g}_j$$

Thus

$$\frac{\partial g_{ij}}{\partial x^k} = \frac{\partial \boldsymbol{g}_i}{\partial x^k}\cdot \boldsymbol{g}_j + \boldsymbol{g}_i \cdot \frac{\partial \boldsymbol{g}_j}{\partial x^k}$$

Using Eq. (1-8-1),

$$\frac{\partial g_{ij}}{\partial x^k} = \Gamma_{ik}{}^p g_{pj} + \Gamma_{jk}{}^p g_{pi} \tag{1-8-3a}$$

Similarly,

$$\frac{\partial g_{jk}}{\partial x^i} = \Gamma_{ji}{}^{p} g_{pk} + \Gamma_{ki}{}^{p} g_{pj} \tag{1-8-3b}$$

and

$$\frac{\partial g_{ki}}{\partial x^j} = \Gamma_{kj}{}^{p} g_{pi} + \Gamma_{ij}{}^{p} g_{pk} \tag{1-8-3c}$$

Subtracting Eq. (1-8-3a) from the other two and noting $\Gamma_{ij}{}^{p}$ is symmetric with respect to the interchange of i and j gives

$$2\Gamma_{ij}{}^{p} g_{pk} = \frac{\partial g_{jk}}{\partial x^i} + \frac{\partial g_{ik}}{\partial x^j} - \frac{\partial g_{ij}}{\partial x^k} \tag{1-8-4}$$

Multiplying by g^{kl},

$$\Gamma_{ij}{}^{l} = \tfrac{1}{2} g^{kl} \left(\frac{\partial g_{jk}}{\partial x^i} + \frac{\partial g_{ik}}{\partial x^j} - \frac{\partial g_{ij}}{\partial x^k} \right) \tag{1-8-5}$$

For future reference the Christoffel symbols of some simple coordinate systems are given.

Rectangular cartesian coordinates All symbols are zero.

Cylindrical polar coordinates The only nonzero symbols are $\Gamma_{\theta\theta}{}^{r} = -r$ and $\Gamma_{r\theta}{}^{\theta} = 1/r$.

Spherical polar coordinates The only nonzero symbols are

$$\Gamma_{\theta\theta}{}^{r} = -r \qquad \Gamma_{\phi\phi}{}^{r} = r\sin^2\theta \qquad \Gamma_{\phi\phi}{}^{\theta} = -\sin\theta\cos\theta$$

$$\Gamma_{r\theta}{}^{\theta} = \frac{1}{r} \qquad \Gamma_{r\phi}{}^{\phi} = \frac{1}{r} \qquad \Gamma_{\theta\phi}{}^{\phi} = \cot\theta$$

The partial derivative of a scalar is a covariant vector. To show this let ψ be a scalar; then, on transforming from a coordinate system x^i to a system x'^i by the usual rules of partial differentiation,

$$\frac{\partial \psi}{\partial x'^i} = \frac{\partial x^j}{\partial x'^i} \frac{\partial \psi}{\partial x^j} \tag{1-8-6}$$

Thus $\partial\psi/\partial x^i$ transforms as a covariant tensor. Now consider a vector $\boldsymbol{a}$ which can be written in the form

$$\boldsymbol{a} = a^i \boldsymbol{g}_i \tag{1-8-7}$$

At a neighboring point let the vector become $\boldsymbol{a} + d\boldsymbol{a}$. Differentiating Eq. (1-8-7),

$$d\boldsymbol{a} = da^i \boldsymbol{g}_i + a^i \, d\boldsymbol{g}_i \tag{1-8-8}$$

By the ordinary rules of calculus,

$$da^i = \frac{\partial a^i}{\partial x^j} dx^j \tag{1-8-9}$$

and, by Eq. (1-8-1),

$$dg_i = \frac{\partial g_i}{\partial x^j} dx^j = \Gamma_{ij}{}^p g_p \, dx^j \tag{1-8-10}$$

Substituting Eqs. (1-8-9) and (1-8-10) into Eq. (1-8-8) and interchanging the roles of i and p in the last term as they are both dummy indices gives

$$d\boldsymbol{a} = \left(\frac{\partial a^i}{\partial x^j} + a^p \Gamma_{pj}{}^i \right) g_i \, dx^j \tag{1-8-11}$$

Under a coordinate transformation $d\boldsymbol{a}$ is invariant, g_i transforms as a covariant vector, and dx^j as a contravariant vector. Thus the term in brackets transforms as a mixed tensor. This term is called the covariant derivative of the contravariant vector a^i and will be denoted by $a^i{}_{,j}$. Thus

$$a^i{}_{,j} = \frac{\partial a^i}{\partial x^j} + a^p \Gamma_{pj}{}^i \tag{1-8-12}$$

Similarly, it can be shown that the covariant derivative of a covariant vector a_i is given by

$$a_{i,j} = \frac{\partial a_i}{\partial x^j} - a_p \Gamma_{ij}{}^p \tag{1-8-13}$$

In general, the covariant derivative of the tensor $a^{i \ldots j}_{k \ldots l}$ is given by

$$a^{i \ldots j}_{k \ldots l,m} = \frac{\partial a^{i \ldots j}_{k \ldots l}}{\partial x^m} + a^{p \ldots j}_{k \ldots l} \Gamma_{pm}{}^i + \cdots + a^{i \ldots p}_{k \ldots l} \Gamma_{pm}{}^j - a^{i \ldots j}_{p \ldots l} \Gamma_{km}{}^p - \cdots - a^{i \ldots j}_{k \ldots p} \Gamma_{lm}{}^p \tag{1-8-14}$$

As all the Christoffel symbols are zero in a rectangular cartesian coordinate system, it is clear from Eq. (1-8-14) that in this particular system the covariant derivatives are just the partial derivatives. As it is always possible to represent a euclidean space by a rectangular cartesian set of coordinates in which the ordinary rules of partial differentiation apply, similar rules must apply to covariant differentiation in curvilinear coordinates, e.g.,

$$(a_{ij} b^k)_{,l} = a_{ij,l} b^k + a_{ij} (b^k{}_{,l}) \tag{1-8-15}$$

In a rectangular cartesian coordinate system $g_{ij} = g^{ij} = g^i{}_j = \delta_{ij}$ and $\varepsilon_{ijk} = \varepsilon^{ijk} - e_{ijk}$ which are constants and therefore their partial differentials are zero. The covariant derivative of each of these tensors is therefore zero in all coordinate systems.

A useful entity in mathematical physics is the differential operator del ∇, which is defined in rectangular cartesian coordinates by the equation

$$\nabla = \boldsymbol{i}_j \frac{\partial}{\partial \bar{x}_j} \tag{1-8-16}$$

In general curvilinear coordinates, this has the form

$$\nabla = g^i \frac{\partial}{\partial x^i} \tag{1-8-17}$$

Let ϕ, $\boldsymbol{a}$, and $\mathbf{T}$ be differentiable scalar, vector, and tensor fields. The gradient, divergence, and curl are given by the relations

$$\begin{aligned} \operatorname{grad} \phi &= \nabla \phi = \phi_{,k} \boldsymbol{g}^k \\ \operatorname{grad} \boldsymbol{a} &= a^k{}_{,m} \boldsymbol{g}_k \otimes \boldsymbol{g}^m \end{aligned} \tag{1-8-18}$$

$$\operatorname{div} \boldsymbol{a} = \nabla \cdot \boldsymbol{a} = \boldsymbol{g}^k \cdot (a^l \boldsymbol{g}_l)_{,k} = a^k{}_{,k} \tag{1-8-19}$$

$$\operatorname{div} \mathbf{T} = T^{km}{}_{,k} \boldsymbol{g}_m \tag{1-8-20}$$

$$\begin{aligned} \operatorname{curl} \boldsymbol{a} &= \nabla \times \boldsymbol{a} = \varepsilon^{ijk} a_{k,j} \boldsymbol{g}_i \\ \operatorname{curl} \mathbf{T} &= \varepsilon^{ijk} T^m_{k,j} \boldsymbol{g}_i \otimes \boldsymbol{g}_m \end{aligned} \tag{1-8-21}$$

The differentiability of $\boldsymbol{a}$ and $\mathbf{T}$ ensures the existence of the necessary derivatives.

It has been shown above how to determine the covariant derivatives of tensors. The resulting entities are also tensors and thus the second covariant derivative can be found by repeating the process, and so on for higher-order derivatives. After some manipulation it can be shown that

$$a_{i,jk} - a_{i,kj} = R^p_{ijk} a_p \tag{1-8-22}$$

where the tensor R^p_{ijk} is called the Riemann–Christoffel tensor and is given by

$$R^p_{ijk} = \left(\frac{\partial \Gamma_{ik}{}^p}{\partial x^j} - \frac{\partial \Gamma_{ij}{}^p}{\partial x^k} + \Gamma_{ik}{}^q \Gamma_{qj}{}^p - \Gamma_{ij}{}^q \Gamma_{qk}{}^p \right) \tag{1-8-23}$$

From Eq. (1-8-22) it can be seen that the necessary and sufficient conditions for covariant differentiation to be commutative is that the Riemann–Christoffel tensor should be identically zero. As a euclidean space can always be represented by a rectangular cartesian coordinate system and as the Christoffel symbols are zero in such a system, the Riemann–Christoffel tensor is identically zero and thus it is zero in all coordinate systems. Thus, in a euclidean space, covariant differentiation is commutative.

The covariant tensor R_{lijk}, which is the associate of the Riemann–Christoffel tensor, is called the curvature tensor:

$$R_{lijk} = g_{lp} R^p_{ijk} \tag{1-8-24}$$

where

$$R_{ijkl} = \frac{1}{2} \left(\frac{\partial^2 g_{il}}{\partial x^j \partial x^k} + \frac{\partial^2 g_{jk}}{\partial x^i \partial x^l} - \frac{\partial^2 g_{ik}}{\partial x^j \partial x^l} - \frac{\partial^2 g_{jl}}{\partial x^i \partial x^k} \right) + g^{pq} (\Gamma_{jkp} \Gamma_{ilq} - \Gamma_{jlp} \Gamma_{ikq}) \tag{1-8-25}$$

and

$$\Gamma_{ijk} = g_{kl} \Gamma_{ij}{}^l \tag{1-8-26}$$

From Eq. (1-8-25) it can be deduced that R_{ijkl} satisfies the relations

$$R_{ijkl} = -R_{jikl} = -R_{ijlk} = R_{klij} \tag{1-8-27}$$

Thus R_{ijkl} has only six independent components which can be taken to be

$$R_{3232} \qquad R_{3231} \qquad R_{3221} \qquad R_{3131} \qquad R_{3121} \qquad R_{2121} \tag{1-8-28}$$

1-9 ISOTROPIC TENSOR FUNCTIONS

Suppose both the bases $\boldsymbol{e}_i$ and $\boldsymbol{g}_a$ of Eq. (1-6-15) are in $\mathscr{V}$. It will be convenient to relabel $\boldsymbol{e}_i, \boldsymbol{g}_i'$ and suppose the transformation relating the two bases is **A**. Then

$$\mathbf{A}\boldsymbol{g}_a = A^i{}_a \boldsymbol{g}_i' \tag{1-9-1}$$

An arbitrary vector $\boldsymbol{v}$ may be described in either of the two bases

$$\boldsymbol{v} = v^a \boldsymbol{g}_a = v'^i \boldsymbol{g}_i' \tag{1-9-2}$$

Thus

$$v'^i = A^i{}_a v^a \tag{1-9-3}$$

But **A** may be applied to $\boldsymbol{v}$ to relate it to a vector $\bar{\boldsymbol{v}}$:

$$\bar{\boldsymbol{v}} = \mathbf{A}\boldsymbol{v} \qquad \text{or} \qquad \bar{v}^a \boldsymbol{g}_a = \mathbf{A} v^b \boldsymbol{g}_b = A^c{}_b v^b \boldsymbol{g}_c \tag{1-9-4}$$

As b and c are dummy indices, replace c by i and b by a:

$$\bar{\boldsymbol{v}} = A^i{}_a v^a \boldsymbol{g}_i = v'^i \boldsymbol{g}_i \tag{1-9-5}$$

Thus the components of $\bar{\boldsymbol{v}}$ in $\boldsymbol{g}_i$ are the same as $\boldsymbol{v}$ in $\boldsymbol{g}_i'$. A linear transformation of a vector can thus be regarded in two ways, either as changing the vector or as a change of basis. Under a change of basis the component representation of a linear transformation also changes. Consider the relation

$$\boldsymbol{u} = \mathbf{T}\boldsymbol{v} \tag{1-9-6}$$

Either

$$u^a \boldsymbol{g}_a = \mathbf{T} v^b \boldsymbol{g}_a = v^b T^a{}_b \boldsymbol{g}_a \qquad \text{or} \qquad u'^i \boldsymbol{g}'_i = \mathbf{T} v'^j \boldsymbol{g}'_j = T'^i{}_j v'^j \boldsymbol{g}'_i \tag{1-9-7}$$

Thus

$$u^a = T^a{}_b v^b \qquad \text{and} \qquad u'^i = T'^i{}_j v'^j$$

but

$$u'^i = A^i{}_a u^a \qquad \text{and} \qquad v'^j = A^j{}_b v^b$$

Hence

$$A^i{}_a T^a{}_b = T'^i{}_j A^j{}_b \tag{1-9-8}$$

or, if $A^j{}_b$ is nonsingular,

$$T'^i{}_j = A^i{}_a T^a{}_b (A^{-1})^b{}_j \tag{1-9-9}$$

Equivalently, if **A** is applied to every vector so that

$$\bar{\boldsymbol{u}} = \mathbf{A}\boldsymbol{u} \qquad \bar{\boldsymbol{v}} = \mathbf{A}\boldsymbol{v} \tag{1-9-10}$$

$$\bar{\boldsymbol{u}} = \bar{\mathbf{T}}\bar{\boldsymbol{v}} \qquad \boldsymbol{u} = \mathbf{T}\boldsymbol{v} \tag{1-9-11}$$

then

$$\bar{\mathbf{T}} = \mathbf{A}\mathbf{T}\mathbf{A}^{-1} \tag{1-9-12}$$

In the $\boldsymbol{g}$ basis, as in Eq. (1-9-5),

$$\bar{\boldsymbol{u}} = u'^i \boldsymbol{g}_i \qquad \text{and} \qquad \bar{\boldsymbol{v}} = v'^j \boldsymbol{g}_j$$

$$u'^i \boldsymbol{g}_i = \bar{\mathbf{T}} v'^j \boldsymbol{g}_j = v'^j \bar{\mathbf{T}}^i{}_j \boldsymbol{g}_i \tag{1-9-13}$$

Using Eq. (1-9-3) and an equivalent expression for u^a yields

$$\bar{\mathbf{T}}^i{}_j = A^i{}_a T^a{}_b (A^{-1})^b{}_j = T'^i{}_j \tag{1-9-14}$$

Again it has been shown that the operation of a linear transformation is equivalent to a change of basis. If the change of basis tensor $\mathbf{Q}$ is orthogonal, then for the transformed vectors

$$\bar{\boldsymbol{u}} = \mathbf{Q}\boldsymbol{u} \qquad \text{and} \qquad \bar{\boldsymbol{v}} = \mathbf{Q}\boldsymbol{v} \tag{1-9-15}$$

the scalar product

$$\bar{\boldsymbol{u}} \cdot \bar{\boldsymbol{v}} = \mathbf{Q}\boldsymbol{u} \cdot \mathbf{Q}\boldsymbol{v} = \boldsymbol{u} \cdot \mathbf{Q}^T \mathbf{Q}\boldsymbol{v} = \boldsymbol{u} \cdot \boldsymbol{v} \tag{1-9-16}$$

is invariant. Thus the magnitude of vectors is unchanged and also the angle between them. Hence an orthogonal transformation merely rotates a vector and an orthonormal basis remains orthonormal.

An isotropic linear transformation $\mathbf{T}$ is one that is unchanged by rotation:

$$\mathbf{T} = \mathbf{Q}\mathbf{T}\mathbf{Q}^T \tag{1-9-17}$$

It can be shown that an isotropic tensor and its matrix are given by the relation

$$\mathbf{T} = a\mathbf{I} \qquad \text{or} \qquad T^i{}_b = a\delta^i{}_b \tag{1-9-18}$$

where a is an arbitrary scalar.

A scalar-valued function of a vector may be written in the form

$$\phi = f(\boldsymbol{v}) = f(v^i, \boldsymbol{g}_i) \tag{1-9-19}$$

An example of such a function would be

$$\phi = a_0 + a_i v^i + a_{ij} v^i v^j + \cdots \tag{1-9-20}$$

On changing to a new basis the value of ϕ is unchanged, for each term is unchanged, but as

$$\phi = a_0 + a_{i'} v^{i'} + a_{i'j'} v^{i'} v^{j'} + \cdots \tag{1-9-21}$$

where

$$a_{i'} = (Q^T)^j_{i'} a_j \qquad a_{i'j'} = (Q^T)^k_{i'} a_{kl} Q^l_{j'} \qquad \text{etc.}$$

the functional form of ϕ in general changes. If the coefficients a_i, etc., are isotropic so that $a_i = a_{i'}$ then ϕ is called an orthogonal scalar invariant.

If $\boldsymbol{v}$ is rotated under an orthogonal transformation $\mathbf{Q}$ to a new vector $\boldsymbol{v}'$, that is, $\boldsymbol{v}' = \mathbf{Q}\boldsymbol{v}$, and if

$$f(\boldsymbol{v}) = f(\mathbf{Q}\boldsymbol{v}) \tag{1-9-22}$$

then $f(\boldsymbol{v})$ is called an isotropic scalar-valued vector function. A scalar-valued function of several vectors $\boldsymbol{v}_1, \boldsymbol{v}_2, \ldots, \boldsymbol{v}_n$ is isotropic if

$$f(\boldsymbol{v}_1, \boldsymbol{v}_2, \ldots, \boldsymbol{v}_n) = f(\mathbf{Q}\boldsymbol{v}_1, \mathbf{Q}\boldsymbol{v}_2, \ldots, \mathbf{Q}\boldsymbol{v}_n) \tag{1-9-23}$$

and f is called a simultaneous orthogonal invariant. A representation theorem due to Cauchy states that a scalar-valued function of vectors is a simultaneous invariant if and only if it can be expressed in the form of a function of the inner products $\boldsymbol{v}_i \cdot \boldsymbol{v}_j$ $(i,j = 1, \ldots, n)$. A scalar-valued function of vectors may also be isotropic relative to a subgroup of the full orthogonal group. Then Eq. (1-9-23) must only hold for those $\mathbf{Q}$ within the subgroup G.

In a similar way a scalar-valued function of a tensor $\mathbf{T}$ may be written

$$\phi = f(\mathbf{T}) = f(T^{ij}, \boldsymbol{g}_i) \tag{1-9-24}$$

and an example of such a function is

$$f(\mathbf{T}) = a_0 + a_{ij}T^{ij} + a_{ijkl}T^{ij}T^{kl} + \cdots \tag{1-9-25}$$

If the functional form of f is unchanged when the basis is changed then f is called an orthogonal invariant. An isotropic scalar-valued tensor function is one whose value is the same for all similar tensors, i.e.,

$$f(\mathbf{T}) = f(\mathbf{QTQ}^T) \tag{1-9-26}$$

Applying Cauchy's theorem to the basis vectors $\boldsymbol{g}_i$ shows that an invariant ϕ must be of the form

$$\phi = f(\mathbf{T}) = f(T^{ij}, g_{ij}) \tag{1-9-27}$$

A scalar-valued tensor function of several tensors $f(\mathbf{T}_1, \ldots, \mathbf{T}_n)$ is isotropic if

$$f(\mathbf{T}_1, \ldots, \mathbf{T}_n) = f(\mathbf{QT}_1\mathbf{Q}^T, \ldots, \mathbf{QT}_n\mathbf{Q}^T) \tag{1-9-28}$$

f being called a simultaneous invariant or simply invariant. The function f is said to be isotropic relative to G where G is a subgroup of the full orthogonal group if Eq. (1-9-28) holds for all $\mathbf{Q}$ in G.

A tensor-valued tensor function can now be represented by

$$\mathbf{\Phi} = \mathbf{F}(\mathbf{T}) \qquad \Phi^{ij} = F^{ij}(T^{ij}, \boldsymbol{g}_i) \tag{1-9-29}$$

where, for example,

$$\Phi^{ij} = a^{ij} + a^{ij}_{kl}T^{kl} + \cdots \tag{1-9-30}$$

is just such a function in component form. The function $\mathbf{F}(\mathbf{T})$ is said to be an isotropic tensor-valued function of a tensor if

$$\mathbf{QF}(\mathbf{T})\mathbf{Q}^T = \mathbf{F}(\mathbf{QTQ}^T) \tag{1-9-31}$$

For several independent tensors $\mathbf{T}_1, \mathbf{T}_2, \ldots, \mathbf{T}_n$, $\mathbf{F}$ is isotropic if

$$\mathbf{QF}(\mathbf{T}_1, \ldots, \mathbf{T}_n)\mathbf{Q}^T = \mathbf{F}(\mathbf{QT}_1\mathbf{Q}^T, \ldots, \mathbf{QT}_n\mathbf{Q}^T) \tag{1-9-32}$$

and $\mathbf{F}$ is isotropic relative to a subgroup G if Eq. (1-9-32) holds for every $\mathbf{Q}$ in G.

The concepts of isotropy and isotropy relative to a subgroup G can be extended to functions whose values and arguments are scalars, vectors, and tensors of any order. For example, a vector-valued function $\boldsymbol{v} = \boldsymbol{v}(\mathbf{T}, \boldsymbol{u})$ is isotropic if

$$\mathbf{Q}\boldsymbol{v}(\mathbf{T}, \boldsymbol{u}) = \boldsymbol{v}(\mathbf{QTQ}^T, \mathbf{Q}\boldsymbol{u}) \tag{1-9-33}$$

and isotropic relative to a subgroup G if Eq. (1-9-33) applies for all $\mathbf{Q}$ in G.

1-10 PRINCIPAL VALUE AND DIRECTIONS, BASIC INVARIANTS, CAYLEY–HAMILTON THEOREM, AND POLAR DECOMPOSITION

Let $\boldsymbol{v}$ be a vector of unit magnitude indicating a direction in space. This direction is called a principal direction of the linear transformation $\mathbf{T}$ if

$$\mathbf{T}\boldsymbol{v} = \lambda \boldsymbol{v} \tag{1-10-1}$$

where λ is a real scalar. In component form this can be written

$$T^i{}_j v^j = \lambda v^i \qquad \text{or} \qquad (T^i{}_j - \lambda \delta^i{}_j) v^j = 0 \tag{1-10-2}$$

In order that Eqs. (1-10-2) have a nontrivial solution,

$$\det (T^i{}_j - \lambda \delta^i{}_j) = 0 \tag{1-10-3}$$

If this determinant is expanded it will give a cubic equation in λ which is known as the characteristic equation and is of the form

$$\mathrm{III}_\mathbf{T} - \mathrm{II}_\mathbf{T}\lambda + \mathrm{I}_\mathbf{T}\lambda^2 - \lambda^3 = 0 \tag{1-10-4}$$

With a little algebraic manipulation it can be shown that

$$\mathrm{I}_\mathbf{T} = T^i{}_i \tag{1-10-5a}$$

$$\mathrm{II}_\mathbf{T} = \tfrac{1}{2}(T^i{}_i T^j{}_j - T^i{}_j T^j{}_i) \tag{1-10-5b}$$

$$\mathrm{III}_\mathbf{T} = \tfrac{1}{6}(T^i{}_i T^j{}_j T^k{}_k - 3T^i{}_j T^j{}_i T^k{}_k + 2T^i{}_j T^j{}_k T^k{}_i) = \det T^i{}_j \tag{1-10-5c}$$

The $\mathrm{I}_\mathbf{T}, \mathrm{II}_\mathbf{T}, \mathrm{III}_\mathbf{T}$ are invariant to basis changes, as is shown by the fact that all indices are dummy indices, and are known as the basic or principal invariants of $\mathbf{T}$. As Eq. (1-10-4) is a cubic equation it has at least one real root and may have three roots which will be denoted by $\lambda_1, \lambda_2, \lambda_3$ and which are called the principal values of $\mathbf{T}$. From Eq. (1-10-4) it follows that

$$\mathrm{I}_\mathbf{T} = \lambda_1 + \lambda_2 + \lambda_3 \tag{1-10-6a}$$

$$\mathrm{II}_\mathbf{T} = \lambda_1\lambda_2 + \lambda_2\lambda_3 + \lambda_3\lambda_1 \tag{1-10-6b}$$

$$\mathrm{III}_\mathbf{T} = \lambda_1\lambda_2\lambda_3 \tag{1-10-6c}$$

It can also be shown by premultiplying both sides of Eq. (1-10-1) by $\mathbf{Q}$ that similar tensors $\mathbf{T}'$ and $\mathbf{T}$, related by

$$\mathbf{T}' = \mathbf{QTQ}^T \tag{1-10-7}$$

where **Q** is an orthogonal tensor, have the same principal values and their principal directions are related by

$$\boldsymbol{v}' = \mathbf{Q}\boldsymbol{v} \tag{1-10-8}$$

If the tensor **T** is an orthogonal tensor **Q**, then the real principal values satisfy the relation

$$\lambda^2 = 1 \tag{1-10-9}$$

As the characteristic equation is a cubic it must have one real root and for an orthogonal tensor this real root must be ± 1. Also, $\det Q^i{}_j = \pm 1$, which from Eqs. (1-10-5*c*) and (1-10-6*c*) is $\lambda_1\lambda_2\lambda_3$. Two cases occur:

1. Proper orthogonal $\det \mathbf{Q} = \lambda_1\lambda_2\lambda_3 = +1$, λ_1, say, is $+1$, and λ_2 and λ_3 are both either $+1$ or -1 or are complex conjugate. If both are positive, **Q** is simply the identity tensor and $\boldsymbol{v}$ is unchanged. In the other cases only the direction $\boldsymbol{v}_1$ remains unchanged and **Q** then represents a rigid rotation about $\boldsymbol{v}_1$.
2. Improper orthogonal $\det \mathbf{Q} = \lambda_1\lambda_2\lambda_3 = -1$, λ_1, say, is -1, and λ_2 and λ_3 are both either -1 or $+1$ or conjugate complex and their product is 1. If all the λ are -1, then **Q** is -1 and is called an inversion tensor since every $\boldsymbol{v}$ is reversed. In the other cases only one direction $\boldsymbol{v}_1$ is such that $\boldsymbol{v}_1$ and $-\boldsymbol{v}_1$ are reversed in sign and **Q** represents an inversion followed by a rigid-body rotation about $\boldsymbol{v}_1$.

Another important case is when the tensor **T** is symmetric and we will then denote it by **S**. Taking the bases as rectangular cartesian, Eq. (1-10-2) takes the form

$$(S_{ij} - \lambda\delta_{ij})v_j = 0 \tag{1-10-10}$$

Again note that there must be at least one real root of Eq. (1-10-4) which can be taken to be λ_1. Let the cartesian coordinates now be chosen so that x_1 lies along the principal direction associated with λ_1. Then, using Eq. (1-10-2),

$$S_{11} = \lambda_1 \qquad S_{12} = S_{21} = S_{31} = S_{13} = 0 \tag{1-10-11}$$

The characteristic equation can thus be written as

$$\begin{vmatrix} (\lambda_1 - \lambda) & 0 & 0 \\ 0 & (S_{22} - \lambda) & S_{23} \\ 0 & S_{23} & (S_{33} - \lambda) \end{vmatrix} = 0$$

which when expanded gives

$$(\lambda_1 - \lambda)[\lambda^2 - (S_{22} + S_{33})\lambda + (S_{22}S_{33} - S_{23}^2)] = 0 \tag{1-10-12}$$

The discriminant of the quadratic expression is

$$(S_{22} + S_{33})^2 - 4(S_{22}S_{33} - S_{23}^2) = (S_{22} - S_{33})^2 + 4S_{23}^2 \tag{1-10-13}$$

which is positive so the remaining roots are also real. Thus all the principal values of a symmetric tensor are real.

If two principal values are distinct then their associated principal directions are orthogonal. To show this let the two distinct principal values be λ_1 and λ_2. Equations (1-10-10) give

$$(S_{ij}-\lambda_1\delta_{ij})v_{1j} = 0 \tag{1-10-14a}$$

$$(S_{ij}-\lambda_2\delta_{ij})v_{2j} = 0 \tag{1-10-14b}$$

Multiply Eq. (1-10-14a) by v_{2i} and Eq. (1-10-14b) by v_{1i} and subtract, interchanging i and j as they are dummy indices, and remembering that S_{ij} is symmetric

$$(\lambda_1-\lambda_2)v_{1j}v_{2j} = 0 \tag{1-10-15}$$

If $\lambda_1 \neq \lambda_2$ then $\mathbf{v}_1 \cdot \mathbf{v}_2 = 0$, that is, $\mathbf{v}_1$ and $\mathbf{v}_2$ are orthogonal. If the three principal values are distinct then the three principal directions form a unique mutually orthogonal set of directions. If two principal values are the same but differ from the third then their associated principal directions must be orthogonal to the third principal direction but are otherwise undefined. If all three principal values are equal then any direction is a principal direction. If the rectangular cartesian coordinates are chosen so that they lie in the principal directions the symmetric tensor takes the particularly simple form

$$[S_{ij}] = \begin{bmatrix} \lambda_1 & 0 & 0 \\ 0 & \lambda_2 & 0 \\ 0 & 0 & \lambda_3 \end{bmatrix} \tag{1-10-16}$$

When written in this form the matrix S_{ij} is said to be diagonalized. Higher powers of $\mathbf{S}$ take the simple form

$$[S_{ij}S_{jk}] = \begin{bmatrix} \lambda_1^2 & 0 & 0 \\ 0 & \lambda_2^2 & 0 \\ 0 & 0 & \lambda_3^2 \end{bmatrix} \tag{1-10-17}$$

and in general

$$[\mathbf{S}^n] = \begin{bmatrix} \lambda_1^n & 0 & 0 \\ 0 & \lambda_2^n & 0 \\ 0 & 0 & \lambda_3^n \end{bmatrix} \tag{1-10-18}$$

Consider the expression $\mathrm{III_S}\mathbf{I}-\mathrm{II_S}\mathbf{S}+\mathrm{I_S}\mathbf{S}^2-\mathbf{S}^3$; the terms in each row which lie on the diagonal are all of the form $\mathrm{III_S}-\mathrm{II_S}\lambda_a+\mathrm{I_S}\lambda_a^2-\lambda_a^3$ where $a = 1, 2,$ or 3 and are therefore zero because the λ_a are roots of Eq. (1-10-4). The expression is therefore identically zero and this leads to the important Cayley–Hamilton theorem

$$\mathbf{S}^3 = \mathrm{III_S}\mathbf{I}-\mathrm{II_S}\mathbf{S}+\mathrm{I_S}\mathbf{S}^2 \tag{1-10-19}$$

This has been demonstrated for symmetric tensors but it can be shown generally that, for any tensor $\mathbf{T}$,

$$\mathbf{T}^3 = \mathrm{III_T}\mathbf{I}-\mathrm{II_T}\mathbf{T}+\mathrm{I_T}\mathbf{T}^2 \tag{1-10-20}$$

By repeated use of the Cayley–Hamilton theorem any polynomial $f(\mathbf{T})$ whose coefficients are scalars can be expressed as a quadratic in $\mathbf{T}$:

$$f(\mathbf{T}) = A_0\mathbf{I}+A_1\mathbf{T}+A_2\mathbf{T}^2 \tag{1-10-21}$$

where A_0, A_1, and A_2 are polynomials of the invariants $\mathrm{I_T}$, $\mathrm{II_T}$, and $\mathrm{III_T}$. If $\mathbf{T}$ possesses three real principal values and a cartesian coordinate system is taken whose axes lie along the principal vectors, $\mathbf{T}$ is diagonalized and this gives the three equations

$$\begin{aligned} f(\lambda_1)-A_2\lambda_1^2-A_1\lambda_1-A_0 &= 0 \\ f(\lambda_2)-A_2\lambda_2^2-A_1\lambda_2-A_0 &= 0 \\ f(\lambda_3)-A_2\lambda_3^2-A_1\lambda_3-A_0 &= 0 \end{aligned} \tag{1-10-22}$$

Solving for A_0, A_1, and A_2 and substituting in Eq. (1-10-21) gives the Lagrange–Sylvester equation

$$\begin{aligned} f(\mathbf{T}) = {} & f(\lambda_1)\frac{(\mathbf{T}-\lambda_2\mathbf{I})(\mathbf{T}-\lambda_3\mathbf{I})}{(\lambda_1-\lambda_2)(\lambda_1-\lambda_3)}+f(\lambda_2)\frac{(\mathbf{T}-\lambda_3\mathbf{I})(\mathbf{T}-\lambda_1\mathbf{I})}{(\lambda_2-\lambda_3)(\lambda_2-\lambda_1)} \\ & +f(\lambda_3)\frac{(\mathbf{T}-\lambda_1\mathbf{I})(\mathbf{T}-\lambda_2\mathbf{I})}{(\lambda_3-\lambda_1)(\lambda_3-\lambda_2)} \end{aligned} \tag{1-10-23}$$

A representation of $\mathbf{T}$ which will be found to be of considerable use in continuum mechanics is the polar decomposition theorem which expresses a general second-rank tensor as a product of a positive symmetric tensor with an orthogonal tensor. Let $\mathbf{C}$ be a positive definite symmetric tensor having principal values λ_a and principal directions $\mathbf{v}_a$. A tensor is a positive definite tensor if

$$\mathbf{C}\mathbf{v}_a\cdot\mathbf{v}_a > 0 \tag{1-10-24}$$

Hence

$$\lambda_a\mathbf{v}_a\cdot\mathbf{v}_a = \lambda_a > 0 \tag{1-10-25}$$

Thus the principal values of a positive definite tensor are positive, and hence from Eqs. (1-10-5*c*) and (1-10-6*c*) the determinant is also positive. Taking a basis along the principal axes, $\mathbf{C}$ is then diagonalized and relative to this same basis the tensor $\mathbf{U}$ whose principal values are $\lambda_a^{1/2}$ satisfies the equation

$$\mathbf{U}^2 = \mathbf{C} \tag{1-10-26}$$

or

$$\mathbf{U} = \mathbf{C}^{1/2} \tag{1-10-27}$$

Let $\mathbf{T}$ be a nonsingular and hence invertible tensor, that is, $\det\mathbf{T} \neq 0$; the polar decomposition theorem then states that $\mathbf{T}$ can be represented in either of the forms

$$\mathbf{T} = \mathbf{QU} = \mathbf{VQ} \tag{1-10-28}$$

where $\mathbf{Q}$ is an orthogonal tensor and $\mathbf{U}$ and $\mathbf{V}$ are positive definite symmetric tensors. Considering $\mathbf{T}$ as a linear transformation relating the vectors $\boldsymbol{v}'$ and $\boldsymbol{v}$, i.e.,

$$\boldsymbol{v}' = \mathbf{T}\boldsymbol{v} \tag{1-10-29}$$

then

$$\mathbf{T}\boldsymbol{v} \cdot \mathbf{T}\boldsymbol{v} = \boldsymbol{v}' \cdot \boldsymbol{v}' > 0 \tag{1-10-30}$$

Thus $\mathbf{T}^T\mathbf{T}$ is a positive definite symmetric tensor which we can denote by $\mathbf{C}$. From Eq. (1-10-26),

$$\mathbf{T}^T\mathbf{T} = \mathbf{U}^2 = \mathbf{C} \tag{1-10-31}$$

Having found $\mathbf{U}$, $\mathbf{Q}$ is formed from the relation

$$\mathbf{Q} = \mathbf{T}\mathbf{U}^{-1} \tag{1-10-32}$$

Now

$$\mathbf{Q}^T\mathbf{Q} = (\mathbf{T}\mathbf{U}^{-1})^T(\mathbf{T}\mathbf{U}^{-1}) = \mathbf{U}^{-1}\mathbf{T}^T\mathbf{T}\mathbf{U}^{-1} = \mathbf{I} \tag{1-10-33}$$

Thus $\mathbf{Q}$ is indeed orthogonal. Suppose two decompositions existed:

$$\mathbf{T} = \mathbf{Q}\mathbf{U} = \mathbf{Q}'\mathbf{U}' \tag{1-10-34}$$

Then

$$\mathbf{U}^2 = \mathbf{T}^T\mathbf{T} = \mathbf{U}'^2 \qquad \text{or} \qquad \mathbf{U} = \mathbf{U}' \qquad \text{and hence} \qquad \mathbf{Q} = \mathbf{Q}' \tag{1-10-35}$$

Thus this decomposition is unique. Similarly it can be shown that $\mathbf{T}$ can be uniquely represented by

$$\mathbf{T} = \mathbf{V}\mathbf{Q}^* \tag{1-10-36}$$

Now

$$\mathbf{T} = \mathbf{Q}^*\mathbf{Q}^{*T}\mathbf{V}\mathbf{Q}^* = \mathbf{Q}^*(\mathbf{Q}^{*T}\mathbf{V}\mathbf{Q}^*) \tag{1-10-37}$$

and, as the decomposition equation (1-10-34) is unique,

$$\mathbf{Q} = \mathbf{Q}^* \qquad \text{and} \qquad \mathbf{U} = \mathbf{Q}^T\mathbf{V}\mathbf{Q} \tag{1-10-38}$$

Letting the principal values and directions of $\mathbf{U}$ be denoted by λ_a and $\boldsymbol{u}_a$, that is, $\mathbf{U}\boldsymbol{u}_a = \lambda_a\boldsymbol{u}_a$, and using Eq. (1-10-38),

$$\mathbf{Q}\mathbf{U}\boldsymbol{u}_a = \lambda_a(\mathbf{Q}\boldsymbol{u}_a) = \mathbf{V}(\mathbf{Q}\boldsymbol{u}_a) \tag{1-10-39}$$

Thus the principal values of $\mathbf{V}$ and $\mathbf{U}$ are the same and the principal directions of $\mathbf{V}$ denoted by $\boldsymbol{v}_a$ are related to those of $\mathbf{U}$ by

$$\boldsymbol{v}_a = \mathbf{Q}\boldsymbol{u}_a \tag{1-10-40}$$

1-11 INTEGRAL THEOREMS

A curve C can be represented parametrically by a single parameter ξ:

$$x^i = x^i(\xi) \qquad \text{for } i = 1, 2, 3 \tag{1-11-1}$$

and C can be regarded as a map of the interval $a \leqslant \xi \leqslant b$ of real numbers. An element of length ds of the curve is defined by

$$ds^2 = d\mathbf{x} \cdot d\mathbf{x} = g_{ij}\, dx^i\, dx^j = \left(g_{ij} \frac{dx^i}{d\xi} \frac{dx^j}{d\xi} \right) d\xi^2 \qquad \text{(1-11-2)}$$

and the arc length

$$s(C) = \int_a^b \left(g_{ij} \frac{dx^i}{d\xi} \frac{dx^j}{d\xi} \right)^{1/2} d\xi \qquad \text{(1-11-3)}$$

Denoting the unit tangent vector to C by $\boldsymbol{\tau}$, the projection of a vector field $\boldsymbol{v}$ along C is given by

$$\boldsymbol{v} \cdot \boldsymbol{\tau} = \boldsymbol{v} \cdot \left(\frac{d\mathbf{x}}{ds} \right) = g_{ij} v^i \frac{dx^j}{ds} \qquad \text{(1-11-4)}$$

The flow of $\boldsymbol{v}$ along C is then given by

$$\int_C \boldsymbol{v} \cdot d\mathbf{x} = \int_C \boldsymbol{v} \cdot \boldsymbol{\tau}\, ds = \int_C g_{ij} v^i\, dx^j = \int_a^b g_{ij} v^i \left(\frac{dx^j}{d\xi} \right) d\xi \qquad \text{(1-11-5)}$$

If C is a closed curve the integral is called the circulation of $\boldsymbol{v}$ and is often written in the form $\oint \boldsymbol{v} \cdot d\mathbf{x}$.

A surface S can be represented parametrically by two parameters ξ_1 and ξ_2:

$$x^i = x^i(\xi_1, \xi_2) \qquad \text{for } i = 1, 2, 3 \qquad \text{(1-11-6)}$$

If ξ_1 is a constant, Eq. (1-11-6) describes a set of curves called surface coordinates lying in the surface S, and similarly if ξ_2 is a constant. It will be assumed that the unit normal vector of the surface is defined everywhere, the condition for this being that not all the determinants Δ_i vanish simultaneously where

$$\Delta_i = e_{ijk} \frac{\partial x^j}{\partial \xi_1} \frac{\partial x^k}{\partial \xi_2} = \frac{\partial(x^j, x^k)}{\partial(\xi_1, \xi_2)} = \begin{vmatrix} \dfrac{\partial x^j}{\partial \xi_1} & \dfrac{\partial x^j}{\partial \xi_2} \\[2ex] \dfrac{\partial x^k}{\partial \xi_1} & \dfrac{\partial x^k}{\partial \xi_2} \end{vmatrix} \qquad \text{(1-11-7)}$$

An increment of length along the ξ_1 surface coordinate will be denoted by $d\boldsymbol{\xi}_1$, and similarly along the ξ_2 surface coordinate by $d\boldsymbol{\xi}_2$:

$$d\boldsymbol{\xi}_1 = \frac{\partial x^i}{\partial \xi_1} \mathbf{g}_i\, d\xi_1 \qquad \text{and} \qquad d\boldsymbol{\xi}_2 = \frac{\partial x^i}{\partial \xi_2} \mathbf{g}_i\, d\xi_2 \qquad \text{(1-11-8)}$$

An element of area $d\mathbf{a}$ is then given by

$$d\mathbf{a} = d\boldsymbol{\xi}_1 \times d\boldsymbol{\xi}_2 = \varepsilon_{ijk} \frac{\partial x^i}{\partial \xi_1} \frac{\partial x^j}{\partial \xi_2} \mathbf{g}^k\, d\xi_1\, d\xi_2 = \sqrt{g}\, \Delta_k \mathbf{g}^k\, d\xi_1\, d\xi_2 \qquad \text{(1-11-9)}$$

where the orientation of $d\mathbf{a}$ is determined by the order of $d\boldsymbol{\xi}_1$ and $d\boldsymbol{\xi}_2$ in the vector product. The area of the surface S is then given by

$$A(S) = \int_D \sqrt{d\mathbf{a} \cdot d\mathbf{a}} = \int_D \sqrt{g g^{kl} \Delta_k \Delta_l}\, d\xi_1\, d\xi_2 \qquad \text{(1-11-10)}$$

where D is the region in the ξ_1, ξ_2 plane whose values specify S. The unit normal $\boldsymbol{n}$ to $d\boldsymbol{a}$ is given by

$$\boldsymbol{n} = \frac{d\boldsymbol{a}}{da} = \Delta_k (g^{lm}\Delta_l\Delta_m)^{-1/2}\boldsymbol{g}^k \tag{1-11-11}$$

The flux of a vector field $\boldsymbol{v}$ across the surface S is defined by

$$\int_S \boldsymbol{v}\cdot\boldsymbol{n}\, da = \int_S \boldsymbol{v}\cdot d\boldsymbol{a} = \int_S v^i\, da_i = \int_D \sqrt{g}v^i\Delta_i\, d\xi_1\, d\xi_2 \tag{1-11-12}$$

A volume V can be represented parametrically by three parameters ξ_1, ξ_2, and ξ_3:

$$x_i = x_i(\xi_1, \xi_2, \xi_3) \tag{1-11-13}$$

If ξ_2 and ξ_3 are held constant, Eq. (1-11-13) describes a curvilinear coordinate within the volume V which will be called the ξ_1 coordinate, and similarly for ξ_2 and ξ_3. Incremental lengths along the coordinates are given by $d\boldsymbol{\xi}_1$, $d\boldsymbol{\xi}_2$, and $d\boldsymbol{\xi}_3$, where

$$d\boldsymbol{\xi}_1 = \frac{\partial x_i}{\partial \xi_1}\boldsymbol{g}_i\, d\xi_1 \qquad d\boldsymbol{\xi}_2 = \frac{\partial x_i}{\partial \xi_2}\boldsymbol{g}_i\, d\xi_2 \qquad d\boldsymbol{\xi}_3 = \frac{\partial x_i}{\partial \xi_3}\boldsymbol{g}_i\, d\xi_3 \tag{1-11-14}$$

An element of volume is now given by

$$\begin{aligned} dv &= (d\boldsymbol{\xi}_1 \times d\boldsymbol{\xi}_2)\cdot d\boldsymbol{\xi}_3 \\ &= e_{ijk}\sqrt{g}\frac{\partial x_i}{\partial \xi_1}\frac{\partial x_j}{\partial \xi_2}\frac{\partial x_k}{\partial \xi_3}\, d\xi_1\, d\xi_2\, d\xi_3 \\ &= \mathscr{J}\sqrt{g}\, d\xi_1\, d\xi_2\, d\xi_3 \end{aligned} \tag{1-11-15}$$

where $\mathscr{J}$ is the jacobian

$$\mathscr{J} = \begin{vmatrix} \dfrac{\partial x_1}{\partial \xi_1} & \dfrac{\partial x_1}{\partial \xi_2} & \dfrac{\partial x_1}{\partial \xi_3} \\ \dfrac{\partial x_2}{\partial \xi_1} & \dfrac{\partial x_2}{\partial \xi_2} & \dfrac{\partial x_2}{\partial \xi_3} \\ \dfrac{\partial x_3}{\partial \xi_1} & \dfrac{\partial x_3}{\partial \xi_2} & \dfrac{\partial x_3}{\partial \xi_3} \end{vmatrix} \tag{1-11-16}$$

Thus, for an arbitrary scalar function ϕ and setting $J = \mathscr{J}\sqrt{g}$,

$$\int_R \phi\, dv = \int_{\bar{D}} \phi J\, d\xi_1\, d\xi_2\, d\xi_3 \tag{1-11-17}$$

where $\bar{D}$ is the region in the ξ_1, ξ_2, ξ_3 space whose values specify R.

We now confine our attention to rectangular cartesian coordinates x_1, x_2, x_3 because, as Bowen and Wang (1976) point out, without euclidean parallelism it is generally impossible to add vectors or tensors at different points of the domain and thus an integral is meaningless. Suppose a function f possesses continuous partial derivatives throughout a regular region R which is bounded by a regular convex surface S as shown in Fig. 1-2. Consider the volume integral

$$\int_R \frac{\partial f}{\partial x_1} dx_1\, dx_2\, dx_3$$

This can be immediately integrated with respect to x_1 to give

$$\int [f(B) - f(A)]\, dx_2\, dx_3$$

Let $\boldsymbol{n}_B$ be the unit vector along the outward normal at B. Then the element $dx_2\, dx_3$ can be written $n_{1B}\, da_B$. As the unit vector $\boldsymbol{n}_A$ is in the opposite direction, the element $dx_2\, dx_3$ can also be written as $-n_{1A}\, da_A$. Thus the integrand can be written as $[f(A)n_{1A}\, da_A + f(B)n_{1B}\, da_B]$, and summing the elements over the whole surface finally gives

$$\int_R \frac{\partial f}{\partial x_1} dv = \int_{\partial R} f n_1\, da \tag{1-11-18}$$

where R is a regular region with the boundary ∂R and $\boldsymbol{n}$ is the outward unit vector normal to ∂R. Similar expressions hold for the x_2 and x_3 components, and hence in general

$$\int_R \frac{\partial f}{\partial x_i} dv = \int_{\partial R} f n_i\, da \tag{1-11-19}$$

Each component of a tensor can be treated in a similar manner to give

$$\int_R \frac{\partial}{\partial x_i} u_{jk\ldots r}\, dv = \int_{\partial R} u_{jk\ldots r} n_i\, da \tag{1-11-20}$$

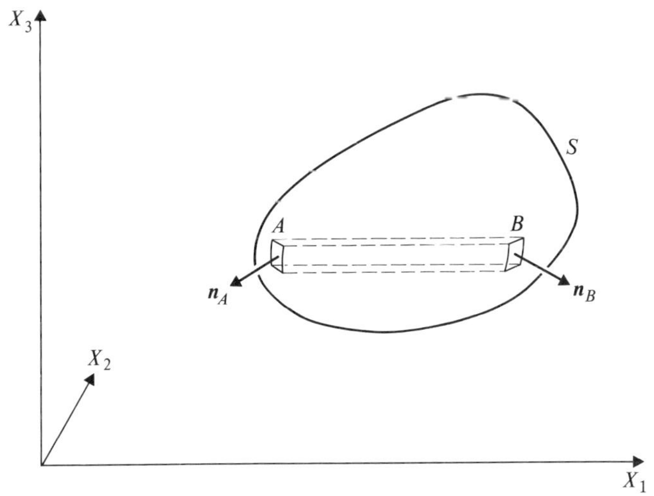

Figure 1-2 Elementary volume within the volume enclosed by the convex surface S.

This is the general form of Green's theorem. Three special cases which are of considerable use follow immediately:

$$\int_{\partial R} f\,d\boldsymbol{a} = \int_R \operatorname{grad} f\,dv \qquad \int_{\partial R} f n_i\,da = \int_R \frac{\partial f}{\partial x_i}\,dv \tag{1-11-21}$$

$$\int_{\partial R} \boldsymbol{u}\cdot\boldsymbol{n}\,da = \int_R \operatorname{div}\boldsymbol{u}\,dv \qquad \int_{\partial R} u_i n_i\,da = \int_R \frac{\partial u_i}{\partial x_i}\,dv \tag{1-11-22}$$

$$\int_{\partial R} \boldsymbol{u}\times\boldsymbol{n}\,da = -\int_R \operatorname{curl}\boldsymbol{u}\,dv \qquad \int_{\partial R} e_{ijk} n_j u_k\,da = \int_R e_{ijk}\frac{\partial}{\partial x_j}u_k\,dv \tag{1-11-23}$$

Setting

$$\boldsymbol{u} = \phi\boldsymbol{\nabla}\psi - \psi\boldsymbol{\nabla}\phi \qquad u_i = \phi\frac{\partial\psi}{\partial x_i} - \psi\frac{\partial\phi}{\partial x_i} \tag{1-11-24}$$

into Eq. (1-11-22) gives another important identity:

$$\int_{\partial R} [\phi\boldsymbol{\nabla}\psi - \psi\boldsymbol{\nabla}\phi]\cdot\boldsymbol{n}\,da = \int_R [\phi\nabla^2\psi - \psi\nabla^2\phi]\,dv \tag{1-11-25}$$

where

$$\nabla^2\psi = \frac{\partial^2\psi}{\partial x_i^2} \tag{1-11-26}$$

A special case of Eq. (1-11-20) which is particularly important in continuum mechanics is the divergence theorem expressed in the form of Eq. (1-11-22). Replacing $\boldsymbol{u}$ in Eq. (1-11-22) by the vector field $(\boldsymbol{u}\cdot\boldsymbol{a})(\mathbf{T}\boldsymbol{b})$ yields

$$\int_{\partial R} [(\boldsymbol{u}\cdot\boldsymbol{a})(\mathbf{T}\boldsymbol{b})]\cdot\boldsymbol{n}\,da = \int_R \operatorname{div}[(\boldsymbol{u}\cdot\boldsymbol{a})(\mathbf{T}\boldsymbol{b})]\,dv \tag{1-11-27}$$

where $\boldsymbol{a}$ and $\boldsymbol{b}$ are arbitrary vectors. From Eqs. (1-6-8) and (1-6-27) there can be obtained the relations

$$[(\boldsymbol{u}\cdot\boldsymbol{a})(\mathbf{T}\boldsymbol{b})]\cdot\boldsymbol{n} = (\boldsymbol{a}\cdot\boldsymbol{u})(\boldsymbol{b}\cdot\mathbf{T}^T\boldsymbol{n}) = \boldsymbol{a}\cdot\{[\boldsymbol{u}\otimes(\mathbf{T}^T\boldsymbol{n})]\boldsymbol{b}\} \tag{1-11-28}$$

$$\operatorname{div}[(\boldsymbol{u}\cdot\boldsymbol{a})(\mathbf{T}\boldsymbol{b})] = \boldsymbol{a}\cdot\{[\boldsymbol{u}\otimes\operatorname{div}\mathbf{T} + (\operatorname{grad}\boldsymbol{u})\mathbf{T}]\boldsymbol{b}\} \tag{1-11-29}$$

Substituting in Eq. (1-11-27) from Eqs. (1-11-28) and (1-11-29) and applying the condition that $\boldsymbol{a}$ and $\boldsymbol{b}$ are arbitrary vectors yields

$$\int_{\partial R} \boldsymbol{u}\otimes(\mathbf{T}^T\boldsymbol{n})\,da = \int_R [\boldsymbol{u}\otimes\operatorname{div}\mathbf{T} + (\operatorname{grad}\boldsymbol{u})\mathbf{T}]\,dv \tag{1-11-30}$$

This form of the divergence theorem finds application in the formulation of the general physical principles underlying continuum mechanics. A special case that finds application is obtained from Eq. (1-11-30) by taking $\boldsymbol{u}$ to be a fixed arbitrary vector to yield

$$\int_{\partial R} \mathbf{T}^T\boldsymbol{n}\,da = \int_R \operatorname{div}\mathbf{T}\,dv \tag{1-11-31}$$

Finally, the relation between surface integrals and curvilinear integrals can be obtained by considering

$$\int_S \operatorname{curl} \boldsymbol{u} \cdot \boldsymbol{n}\, da = \int_S e_{ijk} \frac{\partial u_k}{\partial x_j} da_i \tag{1-11-32}$$

Now, by Eqs. (1-11-7) and (1-11-9),

$$da_i = e_{ipq} \frac{\partial x_p}{\partial \xi_1} \frac{\partial x_q}{\partial \xi_2} d\xi_1\, d\xi_2 \tag{1-11-33}$$

where ξ_1 and ξ_2 are the surface coordinates in the surface S. Substituting Eq. (1-11-33) into Eq. (1-11-32) yields

$$\begin{aligned}\int_S \operatorname{curl} \boldsymbol{u} \cdot \boldsymbol{n}\, da &= \int_D e_{ijk} e_{ipq} \frac{\partial u_k}{\partial x_j} \frac{\partial x_p}{\partial \xi_1} \frac{\partial x_q}{\partial \xi_2} d\xi_1\, d\xi_2 \\ &= \int_D \left(\frac{\partial u_k}{\partial \xi_1} \frac{\partial x_k}{\partial \xi_2} - \frac{\partial u_k}{\partial \xi_1} \frac{\partial x_k}{\partial \xi_2} \right) d\xi_1\, d\xi_2 \end{aligned} \tag{1-11-34}$$

Integrating by parts

$$\int \operatorname{curl} \boldsymbol{u} \cdot \boldsymbol{n}\, da = \int \left[u_k \frac{\partial x_k}{\partial \xi_2} \right]_{\xi_{1A}}^{\xi_{1B}} d\xi_2 - \int \left[u_k \frac{\partial x_k}{\partial \xi_1} \right]_{\xi_{2C}}^{\xi_{2D}} d\xi_1 \tag{1-11-35}$$

where the limits are on the boundaries of the region D and lie on the bounding curve C of S. If the curve C is taken in one sense, i.e., clockwise, and it is remembered that

$$dx_k = \frac{\partial x_k}{\partial \xi_1} d\xi_1 + \frac{\partial x_k}{\partial \xi_2} d\xi_2$$

Eq. (1-11-35) can be written in the form

$$\int_S \operatorname{curl} \boldsymbol{u} \cdot \boldsymbol{n}\, da = \int_C \boldsymbol{u} \cdot d\boldsymbol{s} = \int_C u_k\, dx_k \tag{1-11-36}$$

This relation is known as Stokes' theorem.

GENERAL REFERENCES

Abraham, R. (1966) "Linear and Multilinear Algebra," Benjamin, New York.

Bowen, R. M., and C. C. Wang (1976) "Introduction to Vectors and Tensors," vols. I and II, Plenum Press, New York.

Brillouin, L. (1964) "Tensors in Mechanics and Elasticity," Academic Press, New York.

Buck, R. C. (1965) "Advanced Calculus," McGraw-Hill, New York.

Chadwick, P. (1976) "Continuum Mechanics," Allen and Unwin, London.

Chorlton, F. (1976) "Vector and Tensor Methods," Ellis Horwood, Chichester.

Courant, R. (1965) "Differential and Integral Calculus," vol. II, Interscience, New York.

Ericksen, J. L. (1960) Tensor Fields, in S. Flügge (ed.), "Handbuch der Physik," vol. III/1, pp. 794–858, Springer-Verlag, Berlin.

Jeffreys, H. (1931) "Cartesian Tensors," Cambridge University Press, London.

Jeffreys, H., and B. S. Jeffreys (1950) "Methods of Mathematical Physics," Cambridge University Press, London.
Kellog, O. D. (1929) "Foundations of Potential Theory," Springer, Berlin.
Martin, A. D., and V. J. Mizel (1966) "Introduction to Linear Algebra," McGraw-Hill, New York.
Morse, P. M., and H. Feshbach (1953) "Methods of Theoretical Physics," McGraw-Hill, New York.
Pipes, L. A., and L. R. Harvill (1970) "Applied Mathematics for Engineers and Physicists," McGraw-Hill, New York.
Schouten, J. A. (1951) "Tensor Analysis for Physicists," Clarendon Press, Oxford.
Schouten, J. A. (1954) "Ricci-Calculus," Springer-Verlag, Berlin.
Spain, B. (1953) "Tensor Calculus," Interscience, New York.
Temple, G. (1960) "Cartesian Tensors," Methuen, London.
Tyldesley, J. R. (1975) "An Introduction to Tensor Analysis," Longman, London.

CHAPTER

TWO

KINEMATICS

2-1 THE REFERENTIAL AND SPATIAL DESCRIPTIONS

Although all materials consist basically of atoms and molecules, and on a microscopic scale have a discrete structure, in common experience they can often be regarded as being continuously distributed in space. This is because the distance between the constituent atoms, $\sim 10^{-6}$ to $\sim 10^{-7}$ mm, is much smaller than the dimensions of any macroscopic experimental probe. When the discrete nature of a material can be disregarded it is called a continuum.

The physics of deformation and flow of a continuous material is here developed by way of the concept of a body $\mathscr{B}$. A body $\mathscr{B}$ is a set whose elements can be identified as having a one-to-one correspondence, with the points of a region B of a euclidean point space. The elements of $\mathscr{B}$ are called particles and B is referred to as a configuration of $\mathscr{B}$. A point in B is said to be occupied by a particle of $\mathscr{B}$ if a given particle of $\mathscr{B}$ corresponds with the point.

Kinematics is that aspect of mechanics which describes the motions of bodies without regard to the forces causing these motions. Central to any discussion of kinematics are the primitive concepts of position and time intervals $s = t - \tau$, where t denotes the present time, τ past time ($-\infty < \tau \leqslant t$), and s is the elapsed time which is a positive quantity ($0 \leqslant s < \infty$).

The particles of $\mathscr{B}$ are identified by their time-independent positions relative to a selected origin O. A particular configuration B_r of $\mathscr{B}$, referred to as the reference configuration, is chosen. The position of a typical particle X is labeled by the position $\boldsymbol{X}$ of the point, relative to the origin O, which it occupies in B_r. Thus,

$$\boldsymbol{X} = \kappa(X) \tag{2-1-1}$$

the components of $\boldsymbol{X}$ being called the referential coordinates of X; the terms lagrangian and material are also used to describe these coordinates. The term referential is preferred since this serves to emphasize that the particles of present interest refer only to an element of a body and not a small portion of matter. Although this way of defining particles of a continuum by noting their positions at an initial instant of time is simple to visualize, any set of numbers which could be used to uniquely specify each particle and which vary continuously from one particle to another could be used for a referential coordinate system (see, for example, Lamb, 1932).

Because the vector-valued function κ arises as a direct consequence of introducing the concept of a reference configuration B_r of $\mathscr{B}$, the symbol κ will often be used in the form of a subscript to indicate an implicit dependence upon the reference configuration B_r.

The points of the euclidean point space are identified by their positions relative to a selected origin o which need not coincide with O. If $\boldsymbol{x}$ denotes the position relative to the origin o of the point x occupied by a representative particle X of $\mathscr{B}$ in a given configuration, then it follows that there must exist a function ϑ such that

$$\boldsymbol{x} = \vartheta(X) \tag{2-1-2}$$

the components of $\boldsymbol{x}$ being referred to as the spatial coordinates of X; the terms current and eulerian are also used to describe these coordinates.

In a motion of a body, the configuration changes with the time t. If, with each value of t, there is associated a unique configuration of a body $\mathscr{B}$, the family of configurations is called a motion of $\mathscr{B}$. This definition requires that functions ϕ exist such that

$$\boldsymbol{x} = \phi(X, t) \tag{2-1-3}$$

which identifies a motion as a sequence of correspondences between particles and points. Thus, in a motion of a body $\mathscr{B}$, a representative particle X occupies a succession of points which together form a curve in euclidean point space. This curve is called the path of X and is given parametically by Eq. (2-1-3). Now Eq. (2-1-1) has an inverse which can be substituted into Eq. (2-1-3) to give

$$\boldsymbol{x} = \chi(X, t) = \chi(\kappa^{-1}(\boldsymbol{X}), t) \equiv \chi_\kappa(\boldsymbol{X}, t) \tag{2-1-4}$$

the motion now being given as a function of the position in the reference configuration B_r. Unless it is important to distinguish the configuration, the subscript κ will be omitted and Eq. (2-1-4) can then be expressed in the form

$$\boldsymbol{x} = \chi(\boldsymbol{X}, t) \tag{2-1-5}$$

Physically inadmissable motions are excluded by assuming that Eq. (2-1-5) is continuously differentiable a sufficient number of times with respect jointly to the position and time variables. This smoothness requirement implies that curves, surfaces, and regions in the reference configuration are carried by the motion into material curves, surfaces, and regions in the spatial, i.e., current, configuration, all such subsets of the current configuration being occupied at any instant by the same particles. Although the body $\mathscr{B}$ can occupy the reference configuration, there is no instant at which the body must of necessity occupy the reference configuration which can therefore, if required, be an imagined one.

The various physical properties characteristic of the mechanical response of a body $\mathscr{B}$, regarded as a continuous medium, are represented by scalar, vector, and tensor fields defined either on a reference configuration or on the configurations constituting a motion of $\mathscr{B}$. When defined on a reference configuration, the referential position $\boldsymbol{X}$ and the time t are the independent variables and the fields are said to be given in the referential description. In contrast, when defined on the configurations constituting a motion of $\mathscr{B}$, the independent variables are $\boldsymbol{x}$ and t, and the fields are said to be given in the spatial description. A system of general referential and spatial coordinates in space is set up by adjoining to the respective origins O and o the bases $\boldsymbol{G}_\alpha$ $(\alpha = 1, 2, 3)$ and $\boldsymbol{g}_i$ $(i = 1, 2, 3)$. Referential coordinates in the system $(O, \boldsymbol{G}_\alpha)$ are denoted by X^α $(\alpha = 1, 2, 3)$ and spatial coordinates in the system $(o, \boldsymbol{g}_i)$ by x^i $(i = 1, 2, 3)$. The indicial form of Eq. (2-1-5) and its inverse may now be expressed in the form

$$x^k = \chi^k(X^\alpha, t) \tag{2-1-6a}$$

$$X^\alpha = (\chi^\alpha)^{-1}(x^k, t) \tag{2-1-6b}$$

a spatial description being transferred to the referential description by way of Eq. (2-1-6a) and a referential description being transferred to the spatial description by way of Eq. (2-1-6b). With this notation, a vector field $\boldsymbol{u}$ and a tensor field $\mathbf{A}$ have the form

$$\boldsymbol{u} = u^k \boldsymbol{g}_k = u_k \boldsymbol{g}^k = u^\mu \boldsymbol{G}_\mu = u_\mu \boldsymbol{G}^\mu \tag{2-1-7}$$

$$\begin{aligned} \mathbf{A} &= A^k{}_l \boldsymbol{g}_k \otimes \boldsymbol{g}^l = A_{kl} \boldsymbol{g}^k \otimes \boldsymbol{g}^l = A^{kl} \boldsymbol{g}_k \otimes \boldsymbol{g}_l \\ &= A^\mu{}_\nu \boldsymbol{G}_\mu \otimes \boldsymbol{G}^\nu = A^{\mu\nu} \boldsymbol{G}_\mu \otimes \boldsymbol{G}_\nu = A_{\mu\nu} \boldsymbol{G}^\mu \otimes \boldsymbol{G}^\nu \\ &= A^k{}_\mu \boldsymbol{g}_k \otimes \boldsymbol{G}^\mu = A^{k\mu} \boldsymbol{g}_k \otimes \boldsymbol{G}_\mu = A_{k\mu} \boldsymbol{g}^k \otimes \boldsymbol{G}^\mu \end{aligned} \tag{2-1-8}$$

the various sets of components being given by

$$u^\alpha = \boldsymbol{u} \cdot \boldsymbol{G}^\alpha \qquad u_\alpha = \boldsymbol{u} \cdot \boldsymbol{G}_\alpha \qquad u^i = \boldsymbol{u} \cdot \boldsymbol{g}^i \qquad u_i = \boldsymbol{u} \cdot \boldsymbol{g}_i \tag{2-1-9}$$

$$\begin{matrix} A^i{}_j = \boldsymbol{g}^i \cdot (\mathbf{A}\boldsymbol{g}_j) & A^{ij} = \boldsymbol{g}^i \cdot (\mathbf{A}\boldsymbol{g}^j) & A^{i\alpha} = \boldsymbol{g}^i \cdot (\mathbf{A}\boldsymbol{G}^\alpha) \\ A^\alpha{}_\beta = \boldsymbol{G}^\alpha \cdot (\mathbf{A}\boldsymbol{G}_\beta) & A^{\alpha\beta} = \boldsymbol{G}^\alpha \cdot (\mathbf{A}\boldsymbol{G}^\beta) & A_{\alpha i} = \boldsymbol{G}_\alpha \cdot (\mathbf{A}\boldsymbol{g}_i) \end{matrix} \tag{2-1-10}$$

from which it is evident that vectors may be referred to either of the bases $\boldsymbol{g}_i$, $\boldsymbol{G}_\alpha$ and tensors to either bases or to both.

Let X, Y, and Z be three distinct particles of $\mathscr{B}$ and let $\boldsymbol{X}$, $\boldsymbol{Y}$, $\boldsymbol{Z}$ and $\boldsymbol{x}$, $\boldsymbol{y}$, $\boldsymbol{z}$ be the positions of the points occupied by these particles in the reference and current configurations relative to the origins O and o respectively.

A neighborhood of a given point is the set of all points whose distances from the given point are less than some positive real number. Let B_r be any configuration of a small neighborhood $\mathscr{N}(X)$ of a particle X and let Z denote a typical particle in $\mathcal{N}(X)$, it being noted that X and Z are two distinct particles of $\mathscr{B}$. The vector-valued function κ_X defined by

$$\boldsymbol{Z} = \kappa_X(Z) = \kappa(Z) - \kappa(X) \tag{2-1-11}$$

is called the localization in $\mathcal{N}(X)$ of the configuration B_r. In Eq. (2-1-11), $\boldsymbol{Z}$ is the position vector of the point occupied by the particle Z in the neighborhood $\mathcal{N}(X)$ when the position $\boldsymbol{X}$ of the point occupied by the particle X is taken as the origin. The physical significance of localization in a small neighborhood of a configuration is considered in Sec. 5-2 in relation to the principle of local action. In terms of the spatial description, a neighborhood $\mathcal{N}(\mathscr{P})$ of a point $\mathscr{P}$ may be defined as the subset of points L_i satisfying $\boldsymbol{l}_i = \boldsymbol{p} + u_i\boldsymbol{g}_i$ $(i = 1, 2, 3)$ where $|u_i| < \delta$ and δ is a positive number; $\boldsymbol{l}_i$ and $\boldsymbol{p}$ are the positions relative to the origin o of the points occupied by the particles L_i and $\mathscr{P}$. With regard to a neighborhood considered relative to a subset P of points, an interior point possesses a neighborhood which belongs entirely to P, an exterior point has a neighborhood containing no point of P, and a boundary point has a neighborhood containing at least one point of P and at least one point not belonging to P, it being noted that a boundary point does not of necessity belong to P.

A rigid material is defined as an aggregate of material particles in which the distance between any two particles remains constant with time. Thus, a motion of a body $\mathscr{B}$ is said to be rigid if the distance between the points occupied by every pair of particles is invariable. The rigidity of the motion requires that

$$|\boldsymbol{x} - \boldsymbol{y}| = |\boldsymbol{X} - \boldsymbol{Y}| \qquad |\boldsymbol{y} - \boldsymbol{z}| = |\boldsymbol{Y} - \boldsymbol{Z}| \qquad |\boldsymbol{x} - \boldsymbol{z}| = |\boldsymbol{X} - \boldsymbol{Z}| \tag{2-1-12}$$

from which can be obtained

$$(\boldsymbol{x} - \boldsymbol{z}) \cdot (\boldsymbol{y} - \boldsymbol{z}) = (\boldsymbol{X} - \boldsymbol{Z}) \cdot (\boldsymbol{Y} - \boldsymbol{Z}) \tag{2-1-13}$$

Considering Z to occupy an interior point of the reference configuration, the concept of neighborhood allows the three points with positions $\boldsymbol{Z} + a\boldsymbol{G}_\alpha$ $(\alpha = 1, 2, 3)$ relative to O to be occupied by particles $L_\alpha \in \mathscr{B}$ for a small enough value of the scalar a. Let the positions of L_α in the current configuration be given by $\boldsymbol{z} + a\boldsymbol{g}_\alpha$. Replacing Y by L_α in Eq. (2-1-13) gives

$$(\boldsymbol{x} - \boldsymbol{z}) \cdot \boldsymbol{g}_\alpha = (\boldsymbol{X} - \boldsymbol{Z}) \cdot \boldsymbol{G}_\alpha \tag{2-1-14}$$

and then replacing X, Y by L_α, L_β shows that

$$\boldsymbol{g}_\alpha \cdot \boldsymbol{g}^\beta = \boldsymbol{G}_\alpha \cdot \boldsymbol{G}^\beta \qquad \boldsymbol{G}^\beta \cdot \boldsymbol{G}_\alpha = \delta^\beta{}_\alpha \tag{2-1-15}$$

Equations (1-5-15), (1-5-17), (1-5-19), (2-1-14), and (2-1-15) give

$$(\boldsymbol{x} - \boldsymbol{z}) = [(\boldsymbol{x} - \boldsymbol{z}) \cdot \boldsymbol{g}_\mu]\boldsymbol{g}^\mu = [(\boldsymbol{X} - \boldsymbol{Z}) \cdot \boldsymbol{G}_\mu]\boldsymbol{g}^\mu$$

and use of Eq. (1-6-27) gives

$$(\boldsymbol{x} - \boldsymbol{z}) = (\boldsymbol{g}^\mu \otimes \boldsymbol{G}_\mu)(\boldsymbol{X} - \boldsymbol{Z}) = \mathbf{Q}(\boldsymbol{X} - \boldsymbol{Z}) \tag{2-1-16}$$

where

$$\mathbf{Q} = \boldsymbol{g}^\mu \otimes \boldsymbol{G}_\mu$$

is a proper orthogonal tensor. Setting $\boldsymbol{c} = \boldsymbol{z} - \mathbf{Q}\boldsymbol{Z}$ in Eq. (2-1-16) gives

$$\boldsymbol{x} = \boldsymbol{c}(t) + \mathbf{Q}(t)\boldsymbol{X} \tag{2-1-17}$$

Since $\boldsymbol{z}$ and $\boldsymbol{g}_\alpha$ and hence $\boldsymbol{c}$ and $\mathbf{Q}$ are time dependent, Eq. (2-1-17) is the form which Eq. (2-1-5) takes for a rigid motion.

The velocity $\boldsymbol{v}$ is obtained from Eq. (2-1-17) in the form

$$\boldsymbol{v} = \dot{\boldsymbol{x}} = \dot{\boldsymbol{c}} + \dot{\mathbf{Q}}\boldsymbol{X} \tag{2-1-18}$$

and the acceleration $\boldsymbol{a}$ in the form

$$\boldsymbol{a} = \ddot{\boldsymbol{x}} = \ddot{\boldsymbol{c}} + \ddot{\mathbf{Q}}\boldsymbol{X} \tag{2-1-19}$$

a superimposed dot denoting differentiation with respect to t. In Eqs. (2-1-18) and (2-1-19), both $\boldsymbol{v}$ and $\boldsymbol{a}$ are calculated by forming time derivatives of the spatial position $\boldsymbol{x}$ while holding fixed the referential position $\boldsymbol{X}$. As in fluid mechanics, the field lines of the velocity, given in the spatial description, are here referred to as streamlines. For a general function $\psi(X, t)$ the material time derivative D/Dt, for which $\boldsymbol{X}$ is held constant, is related to the spatial time derivative, for which $\boldsymbol{x}$ is held constant, by

$$\frac{D\psi}{Dt} = \frac{\partial\psi}{\partial t} + (\operatorname{grad}\psi)\boldsymbol{v} = \dot{\psi} \tag{2-1-20}$$

or in component form

$$\frac{D\psi^{i\ldots j}{}_{k\ldots l}}{Dt} = \frac{\partial\psi^{i\ldots j}{}_{k\ldots l}}{\partial t} + \psi^{i\ldots j}{}_{k\ldots l,m}\dot{x}^m \tag{2-1-21}$$

In Eq. (2-1-20) the local rate $\partial\psi/\partial t$ is found by an observer situated at a fixed $\boldsymbol{x}$ and the material rate D/Dt is found by an observer moving with a particle $\boldsymbol{X}$. For a motion in which the local rate vanishes, that is, $\partial\psi/\partial t = 0$ in Eq. (2-1-20), ψ is said to be a steady field, and a uniform field if the convected rate $(\partial\psi/\partial\boldsymbol{x})\boldsymbol{v}$ vanishes.

Using the inverse of Eq. (2-1-17), $\boldsymbol{v}$ and $\boldsymbol{a}$ can be expressed in terms of $\boldsymbol{x}$ and t in the form

$$\boldsymbol{v} = \dot{\boldsymbol{c}} + \mathbf{W}(\boldsymbol{x} - \boldsymbol{c}) \tag{2-1-22}$$

where

$$\mathbf{W} = \dot{\mathbf{Q}}\mathbf{Q}^T \tag{2-1-23}$$

from which can be obtained the relation

$$\ddot{\mathbf{Q}}\mathbf{Q}^T = \dot{\mathbf{W}} - \dot{\mathbf{Q}}\dot{\overline{\mathbf{Q}^T}} = \dot{\mathbf{W}} - (\dot{\mathbf{Q}}\mathbf{Q}^T)(\mathbf{Q}\dot{\overline{\mathbf{Q}^T}}) = \dot{\mathbf{W}} + \mathbf{W}^2$$

enabling $\boldsymbol{a}$ to be expressed in the form

$$\boldsymbol{a} = \ddot{\boldsymbol{c}} + (\dot{\mathbf{W}} + \mathbf{W}^2)(\boldsymbol{x} - \boldsymbol{c}) \tag{2-1-24}$$

The orthogonality relation $\mathbf{Q}\mathbf{Q}^T = \mathbf{I}$ gives $\mathbf{W} = -\mathbf{Q}\dot{\overline{\mathbf{Q}^T}} = -(\dot{\mathbf{Q}}\mathbf{Q}^T)^T$ from which it follows, by reference to Eq. (1-6-11), that $\mathbf{W}$ is skew-symmetric. In terms of the axial vector $\boldsymbol{w}$ of $\mathbf{W}$,

$$v = \dot{c} + w \times (x - c) \tag{2-1-25}$$

$$a = \ddot{c} + \dot{w} \times (x - c) + w \times [w \times (x - c)] \tag{2-1-26}$$

from which it is inferred that $\boldsymbol{w}$ is just the angular velocity of the body $\mathscr{B}$ and $\dot{\boldsymbol{c}}, \ddot{\boldsymbol{c}}$ are the velocity and acceleration respectively of the particle momentarily occupying, at time t, the point with position $\boldsymbol{c}$ relative to o.

2-2 THE DEFORMATION AND VELOCITY GRADIENTS

The term deformation is employed when the function χ and its inverse do not depend upon the time t and when the time-dependent properties of the underlying physical mechanisms do not themselves contribute directly to the observed mechanical response. Expressed in terms of the referential and spatial descriptions, a deformation is a mapping of a reference configuration into a current, i.e., a deformed, configuration. A motion can be regarded as a one-parameter family of deformations, each instant of time corresponding to a particular deformation.

Central to any discussion of modern continuum mechanics is the deformation gradient $\mathbf{F}$ underlying the local analysis of deformation and motion (see, for example, Truesdell and Toupin, 1960; Truesdell and Noll, 1965). This fundamental kinematic tensor is defined by

$$\mathbf{F}(X, t) = (\text{Grad}\,\chi)(X, t) \tag{2-2-1a}$$

$$\mathbf{F} = \text{Grad}\,\mathbf{x} \tag{2-2-1b}$$

which may be expressed by way of the systems $(O, \boldsymbol{G}_\alpha)$ and $(o, \boldsymbol{g}_i)$ of referential and spatial coordinates in the component form

$$\mathbf{F} = F^p{}_\mu \boldsymbol{g}_p \otimes \boldsymbol{G}^\mu \tag{2-2-2a}$$

$$F^i{}_\alpha = x^i{}_{,\alpha} \tag{2-2-2b}$$

In Eq. (2-2-1b), Grad denotes the gradient operator with regard to position in the reference configuration B_r of the body $\mathscr{B}$. Since Eq. (2-1-5) has an inverse, i.e., Eq. (2-1-6b), the smoothness condition requiring that the mapping has continuous derivatives implies that $\mathbf{F}$ has an inverse $\mathbf{F}^{-1}$ defined by

$$\mathbf{F}^{-1}(\mathbf{x}, t) = (\text{grad}\,\chi^{-1})(\mathbf{x}, t) \tag{2-2-3a}$$

$$\mathbf{F}^{-1} = \text{grad}\,\boldsymbol{X} \tag{2-2-3b}$$

which has the component form

$$\mathbf{F}^{-1} = (F^{-1})^\mu{}_p \boldsymbol{G}_\mu \otimes \boldsymbol{g}^p \tag{2-2-4}$$

where

$$(F^{-1})^\alpha{}_i = X^\alpha{}_{,i} \tag{2-2-5}$$

In Eq. (2-2-3*b*), grad denotes the gradient operator with regard to position in the current configuration B_t of the body $\mathscr{B}$. Expressing Eqs. (2-2-1) in the equivalent form

$$d\boldsymbol{x} = \mathbf{F}\, d\boldsymbol{X} \tag{2-2-6}$$

the deformation gradient $\mathbf{F}$ maps the neighborhood of $\mathscr{P}$ in the reference configuration into the configuration χ, given the jacobian

$$J = \det \mathbf{F} \neq 0 \tag{2-2-7}$$

which, for a mapping having continuous derivatives, is the necessary and sufficient condition for invertibility.

If the reference configuration B_r is chosen as the present configuration $\chi(\mathscr{P}, t)$ and the position of $\mathscr{P}$ at an earlier time τ is denoted by ξ, then the motion is represented by

$$\begin{aligned} \xi &= \hat{\chi}(\mathscr{P}, \tau) \\ &= \hat{\chi}(\chi^{-1}(\boldsymbol{x}, t), \tau) = \hat{\chi}(\boldsymbol{x}, t, \tau) \\ &= \hat{\chi}_{(t)}(\boldsymbol{x}, \tau) \end{aligned} \tag{2-2-8}$$

where $\hat{\chi}_{(t)}$ is called the relative deformation function. The relative deformation gradient is defined as

$$\mathbf{F}_{(t)} = \mathbf{F}_{(t)}(\boldsymbol{x}, \tau) = \frac{\partial \hat{\chi}_{(t)}}{\partial \boldsymbol{x}} \tag{2-2-9}$$

Now as the value of the function $\hat{\chi}_{(t)}$ is the place occupied at time τ by the particle which occupies $\boldsymbol{x}$ at time t, we have

$$\mathbf{F}_{(t)}(t) = \mathbf{I} \tag{2-2-10}$$

Consider next the change of reference configuration from B_r to $\bar{B}_r$ where

$$X = \kappa(\mathscr{P}) \qquad \text{and} \qquad \bar{X} = \bar{\kappa}(\mathscr{P}) \tag{2-2-11}$$

Then

$$\bar{X} = \bar{\kappa}(\kappa^{-1}(X)) \tag{2-2-12}$$

The deformation gradients in the two reference frames are related by

$$\mathbf{F} = \frac{\partial \boldsymbol{x}}{\partial \boldsymbol{X}} = \frac{\partial \boldsymbol{x}}{\partial \bar{\boldsymbol{X}}} \frac{\partial \bar{\boldsymbol{X}}}{\partial \boldsymbol{X}} = \bar{\mathbf{F}}\mathbf{G} \tag{2-2-13}$$

where

$$\mathbf{G} = \frac{\partial \bar{\boldsymbol{X}}}{\partial \boldsymbol{X}} \tag{2-2-14}$$

In particular, if $\bar{B}_r$ is taken as the configuration at time t,

$$\mathbf{F}(\tau) = \mathbf{F}_{(t)}(\tau)\mathbf{F}(t) \tag{2-2-15}$$

for

$$\frac{\partial \hat{\chi}}{\partial \boldsymbol{X}} = \frac{\partial \hat{\chi}}{\partial \boldsymbol{x}} \frac{\partial \boldsymbol{x}}{\partial \boldsymbol{X}} \tag{2-2-16}$$

Consider a material curve C_t in the current configuration B_t of a body $\mathscr{B}$ with a material curve C_r in the reference configuration B_r. Let $\boldsymbol{N}$ and $\boldsymbol{v}$ be unit tangent vectors to C_r at X and to C_t at x respectively; then

$$d\boldsymbol{x} = \boldsymbol{v}\,ds \qquad d\boldsymbol{X} = \boldsymbol{N}\,dS \tag{2-2-17}$$

There follow from Eqs. (2-2-17), by way of Eqs. (1-6-8) and (2-2-6), the relations

$$\left(\frac{ds}{dS}\right)^2 = \boldsymbol{N}\cdot(\mathbf{C}\boldsymbol{N}) = [\boldsymbol{v}\cdot(\mathbf{B}^{-1}\boldsymbol{v})]^{-1} \tag{2-2-18}$$

where

$$\mathbf{C} = \mathbf{F}^T\mathbf{F} \qquad C_{\alpha\beta} = F^i{}_\alpha F_{i\beta} \tag{2-2-19}$$

and

$$\mathbf{B} = \mathbf{F}\mathbf{F}^T \qquad B^{ij} = F^i{}_\alpha F^{j\alpha} \tag{2-2-20}$$

are termed deformation tensors.

From Eq. (2-2-15) immediately follows the relation

$$\mathbf{C}(\tau) = \mathbf{F}(\tau)^T\mathbf{F}(\tau) = \mathbf{F}(t)^T\mathbf{F}_{(t)}(\tau)^T\mathbf{F}_{(t)}(\tau)\mathbf{F}(t) = \mathbf{F}(t)^T\mathbf{C}_{(t)}(\tau)\mathbf{F}(t) \tag{2-2-21}$$

and, similarly,

$$\mathbf{B}(\tau) = \mathbf{F}_{(t)}(\tau)\mathbf{B}(t)\mathbf{F}_{(t)}(\tau)^T \tag{2-2-22}$$

while, from Eq. (2-2-10),

$$\mathbf{B}_{(t)}(t) = \mathbf{C}_{(t)}(t) = \mathbf{I} \tag{2-2-23}$$

When configurations which are continually in the process of change are to be examined the rates of change become the most important factor. New kinematical variables that describe the instantaneous time rates of deformation are therefore introduced. Such variables are referred to as rate tensors and, because they are functions of the instantaneous configuration alone and do not depend on the initial configuration, they are not in general the time rates of the deformation tensors introduced above. Differentiating Eq. (2-2-15) with respect to τ and then setting $\tau = t$ gives a rate tensor which will be denoted by $\mathbf{L}$, where

$$\mathbf{L}(t) = \dot{\mathbf{F}}_{(t)}(t) = \dot{\mathbf{F}}(t)\mathbf{F}^{-1}(t) = -\mathbf{F}(t)\,(\overline{\mathbf{F}^{-1}(t)})^{\boldsymbol{\cdot}} \tag{2-2-24}$$

Now, from Eq. (2-2-9),

$$\dot{\mathbf{F}}_{(t)}(\tau) = \frac{\partial}{\partial\tau}\left[\frac{\partial\boldsymbol{\chi}_{(t)}}{\partial\boldsymbol{x}}\right] = \frac{\partial}{\partial\boldsymbol{x}}\dot{\boldsymbol{\chi}}_{(t)}(\boldsymbol{x},\tau) = \frac{\partial\boldsymbol{\xi}}{\partial\boldsymbol{x}} \tag{2-2-25}$$

Setting $\tau = t$ gives $\dot{\boldsymbol{\xi}} = \dot{\boldsymbol{x}} = \boldsymbol{v}$ and thus

$$\mathbf{L}(t) = \dot{\mathbf{F}}_{(t)}(t) = \frac{\partial\boldsymbol{v}}{\partial\boldsymbol{x}} = \operatorname{grad}\boldsymbol{v} \qquad L_{ij} = v_{i,j} \tag{2-2-26}$$

where $\mathbf{L}$ is called the tensor of velocity gradients. A tensor $\mathbf{L}_2$ of acceleration gradients can similarly be defined by

$$\mathbf{L}_2(t) = \ddot{\mathbf{F}}_{(t)}(t) \qquad \text{(2-2-27)}$$

and in general the nth acceleration gradient is given by

$$\mathbf{L}_n(t) = \overset{(n)}{\mathbf{F}}_{(t)}(t) \qquad \text{(2-2-28)}$$

With this notation, there follow from Eqs. (2-2-10), (2-2-26), and (2-2-28) the particular relations

$$\mathbf{L}_0 = \mathbf{I} \qquad \mathbf{L}_1 = \mathbf{L} \qquad \text{(2-2-29)}$$

2-3 STRETCH AND ROTATION

Since $\mathbf{F}$ is invertible, the polar decomposition theorem set out in Sec. 1-10 can be applied, thus giving

$$\mathbf{F} = \mathbf{QU} = \mathbf{VQ} \qquad \text{(2-3-1)}$$

where $\mathbf{Q}$ is orthogonal and $\mathbf{U}$ and $\mathbf{V}$ are positive definite tensors as defined by Eq. (1-10-24). It has been shown that the determinant of a positive definite tensor is positive and as $\det \mathbf{F}$ is positive this implies that $\det \mathbf{Q}$ is positive, that is, $\det \mathbf{Q} = +1$, and $\mathbf{Q}$ is a rotation which will be denoted by $\mathbf{R}$. Thus

$$\mathbf{F} = \mathbf{RU} = \mathbf{VR} \qquad \text{(2-3-2)}$$

or, in indicial notation,

$$F^i{}_\alpha = R^i{}_\beta U^\beta{}_\alpha = V^i{}_j R^j{}_\alpha \qquad \text{(2-3-3)}$$

These equations indicate that the total deformation represented by the tensor $\mathbf{F}$ can be achieved in two ways: either (1) the initial configuration can be deformed to the final shape without rotation by the tensor $\mathbf{U}$ and then rotated to its final position by the rotation tensor $\mathbf{R}$, or (2) the initial configuration can be rotated without change of shape by the rotation tensor $\mathbf{R}$ and then deformed to its final shape by the tensor $\mathbf{V}$.

The principal vectors $\boldsymbol{u}$ of the tensor $\mathbf{U}$ are determined by the equation

$$\mathbf{U}\boldsymbol{u} = \lambda \boldsymbol{u} \qquad \text{(2-3-4)}$$

where the λ's are known as the principal values of $\mathbf{U}$. Equation (2-3-4) can also be written in the form

$$(\mathbf{U} - \lambda \mathbf{I})\boldsymbol{u} = \boldsymbol{o} \qquad \text{(2-3-5)}$$

and the condition that a nontrivial solution exists is that

$$\det(\mathbf{U} - \lambda \mathbf{I}) = 0 \qquad \text{(2-3-6)}$$

On expanding the above determinant a cubic in λ is obtained, called the characteristic equation, which can be written in the form

$$\lambda^3 - \lambda^2 \mathrm{I}_\mathbf{U} + \lambda \mathrm{II}_\mathbf{U} - \mathrm{III}_\mathbf{U} = 0 \qquad \text{(2-3-7)}$$

where I_U, II_U, and III_U are known as the first, second, and third invariants of **U**. There follow from Eqs. (2-3-6) and (2-3-7) the relations

$$\begin{aligned} I_U &= \operatorname{tr}\mathbf{U} = U^\alpha{}_\alpha \\ II_U &= \tfrac{1}{2}(U^\alpha{}_\alpha U^\beta{}_\beta - U^\alpha{}_\beta U^\beta{}_\alpha) \\ III_U &= \det \mathbf{U} \\ &= \tfrac{1}{6}(U^\alpha{}_\alpha U^\beta{}_\beta U^\gamma{}_\gamma - 3U^\alpha{}_\beta U^\beta{}_\alpha U^\gamma{}_\gamma + 2U^\alpha{}_\beta U^\beta{}_\gamma U^\gamma{}_\alpha) \end{aligned} \tag{2-3-8}$$

When **U** is a symmetric tensor it has been shown in Sec. 1-10 that the three roots of the cubic equation (2-3-7) are all real and will be denoted by λ_a ($a = 1, 2, 3$). The three vectors $\boldsymbol{u}_a$ associated with the principal values λ_a are known as the principal vectors and it has been shown that they are all mutually orthogonal. Noting the relations $\mathbf{U} = \mathbf{UI} = \mathbf{U}(\boldsymbol{u}^a \otimes \boldsymbol{u}_a) = (\mathbf{U}\boldsymbol{u}^a) \otimes \boldsymbol{u}_a$, Eq. (2-3-4) gives

$$\mathbf{U} = \sum_{a=1}^{3} \lambda_a (\boldsymbol{u}^a \otimes \boldsymbol{u}_a)$$

which is generally referred to as the spectral representation of the symmetric tensor **U**. It follows immediately from Eq. (2-3-2) that the principal vectors $\boldsymbol{v}_a$ of **V** are related to the principal vectors $\boldsymbol{u}_a$ of **U** by

$$\boldsymbol{v}_a = \mathbf{R}\boldsymbol{u}_a \qquad v^i{}_a = R^i{}_\alpha u^\alpha{}_a \tag{2-3-9}$$

and that the principal values of **U** and **V** are the same. Let $\boldsymbol{N}$ be a unit vector in the initial configuration; then the final length ds of an element of length dS originally in the direction $\boldsymbol{N}$ is obtained by way of Eq. (1-6-8) in the form

$$ds^2 = d\boldsymbol{x} \cdot d\boldsymbol{x} = \mathbf{F}\,d\boldsymbol{X} \cdot \mathbf{F}\,d\boldsymbol{X} = \boldsymbol{N} \cdot (\mathbf{F}^T\mathbf{F}\boldsymbol{N})\,dS^2 = \boldsymbol{N} \cdot (\mathbf{U}^2\boldsymbol{N})\,dS^2 \tag{2-3-10}$$

If $\boldsymbol{N}$ is a principal direction then

$$\left(\frac{ds}{dS}\right)^2 = \lambda_a^2 \tag{2-3-11}$$

Thus the λ_a are the principal stretches, i.e., the extension ratios of elements in the principal directions.

The two tensors **C** and **B**, defined by Eqs. (2-2-19) and (2-2-20), can now be expressed by way of the polar decomposition of Eq. (2-3-2) in the form

$$\mathbf{C} = \mathbf{U}^2 = \mathbf{F}^T\mathbf{F} \tag{2-3-12a}$$

$$\mathbf{B} = \mathbf{V}^2 = \mathbf{F}\mathbf{F}^T \tag{2-3-12b}$$

the tensors **B**, **C**, **U**, and **V** being referred to as deformation tensors; if there is no deformation they are unit tensors. The deformation tensors **C** and **B** are known as the right and left Cauchy–Green tensors respectively. If **F** is replaced by the relative deformation tensor $\mathbf{F}_{(t)}$ the corresponding rotation and deformation tensors are written in the form $\mathbf{R}_{(t)}$, $\mathbf{U}_{(t)}$, $\mathbf{V}_{(t)}$, $\mathbf{C}_{(t)}$, and $\mathbf{B}_{(t)}$ and the same relations between them still apply. For example,

$$\mathbf{F}_{(t)} = \mathbf{R}_{(t)}\mathbf{U}_{(t)} \tag{2-3-13}$$

Now, from Eqs. (2-2-15) and (2-3-12),

$$\mathbf{C}(\tau) = \mathbf{F}(\tau)^T\mathbf{F}(\tau) = \mathbf{F}(t)^T\mathbf{F}_{(t)}(\tau)^T\mathbf{F}_{(t)}(\tau)\mathbf{F}(t) = \mathbf{F}(t)^T\mathbf{C}_{(t)}(\tau)\mathbf{F}(t) \tag{2-3-14}$$

and, similarly,

$$\mathbf{B}(\tau) = \mathbf{F}_{(t)}(\tau)\mathbf{B}(t)\mathbf{F}_{(t)}(\tau)^T \tag{2-3-15}$$

Equation (2-2-10) gives

$$\mathbf{R}_{(t)}(t) = \mathbf{U}_{(t)}(t) = \mathbf{V}_{(t)}(t) = \mathbf{B}_{(t)}(t) = \mathbf{C}_{(t)}(t) = \mathbf{I} \tag{2-3-16}$$

2-4 STRETCHING AND SPIN

If the polar decomposition is differentiated,

$$\frac{\partial}{\partial\tau}\mathbf{F}_{(t)}(\tau) = \frac{\partial}{\partial\tau}(\mathbf{R}_{(t)}(\tau)\mathbf{U}_{(t)}(\tau)) = \dot{\mathbf{R}}_{(t)}(\tau)\mathbf{U}_{(t)}(\tau)+\mathbf{R}_{(t)}(\tau)\dot{\mathbf{U}}_{(t)}(\tau) \tag{2-4-1}$$

Setting $\tau = t$ and using Eq. (2-3-16),

$$\mathbf{L} = \dot{\mathbf{U}}_{(t)}(t)+\dot{\mathbf{R}}_{(t)}(t) \tag{2-4-2}$$

As $\mathbf{U}$ is a symmetric tensor in every configuration $\dot{\mathbf{U}}$ is also a symmetric tensor and differentiating the orthogonality condition for $\mathbf{R}$ shows that $\dot{\mathbf{R}}_{(t)}(t)$ is antisymmetric; $\mathbf{L}$ has thus been decomposed into a symmetric part $\dot{\mathbf{U}}_{(t)}(t)$ written as $\mathbf{D}$ and an antisymmetric part $\dot{\mathbf{R}}_{(t)}(t)$ written as $\mathbf{W}$. Thus

$$\mathbf{L} = \mathbf{D}+\mathbf{W} \tag{2-4-3}$$

where

$$\mathbf{D} = \dot{\mathbf{U}}_{(t)}(t) = \tfrac{1}{2}(\mathbf{L}+\mathbf{L}^T) \tag{2-4-4}$$

$$\mathbf{W} = \dot{\mathbf{R}}_{(t)}(t) = \tfrac{1}{2}(\mathbf{L}-\mathbf{L}^T) \tag{2-4-5}$$

$\mathbf{D}$ is called the stretching tensor or rate of deformation tensor and $\mathbf{W}$ the spin tensor.

To find the significance of $\mathbf{D}$ consider the rate of change of distance between two neighboring material points

$$\frac{D}{Dt}(ds^2) = 2ds\,d\dot{s} = \frac{D}{Dt}dx^i\,dx_i = 2dx^i\,d\dot{x}_i \tag{2-4-6}$$

But

$$d\dot{x}_i = \frac{D}{Dt}dx_i = dv_i = v_{i,j}\,dx^j \tag{2-4-7}$$

Thus

$$2ds\,d\dot{s} = 2v_{i,j}\,dx^i\,dx^j = 2D_{ij}\,dx^i\,dx^j \tag{2-4-8}$$

If $d\boldsymbol{x}$ is instantaneously directed along the unit vector $\boldsymbol{\nu}$,

$$d\boldsymbol{x} = \boldsymbol{\nu}\,ds \qquad dx^i = \nu^i\,ds \tag{2-4-9}$$

and
$$\frac{d\dot{s}}{ds} = D_{ij}v^i v^j \tag{2-4-10}$$

As **D** is a symmetric tensor it will have real principal values D_a corresponding to orthogonal principal directions $\boldsymbol{v}_a$. Applying Eq. (2-4-10) to a principal axis shows that

$$\frac{d\dot{s}_a}{ds_a} = D_a \tag{2-4-11}$$

An elementary box whose sides are normal to the principal axes thus changes its volume at a rate given by

$$\frac{d\dot{v}}{dv} = \frac{d\dot{s}_1}{ds_1} + \frac{d\dot{s}_2}{ds_2} + \frac{d\dot{s}_3}{ds_3} = D_1 + D_2 + D_3 = \operatorname{tr}\mathbf{D} = \operatorname{div}\boldsymbol{v} \tag{2-4-12}$$

The condition for isochoric flow is thus that

$$\operatorname{tr}\mathbf{L} = \operatorname{tr}\mathbf{D} = D^i{}_i = \operatorname{div}\boldsymbol{v} = v^i{}_{,i} = 0 \tag{2-4-13}$$

The rate of change of $d\mathbf{x}$ due to applying **W** alone to $d\mathbf{x}$ produces no change in the distance between material points, for in this case

$$\frac{D}{Dt}ds^2 = d\dot{x}_i\,dx_i = W_{ij}\,dx^i\,dx^j = 0 \tag{2-4-14}$$

and thus the matrix must represent a rigid-body rotation. The axial vector $\boldsymbol{w}$ associated with **W** is called the angular velocity vector w, and $2w$ is usually denoted by $\boldsymbol{\omega}$ and called the vorticity:

$$\boldsymbol{\omega} = \operatorname{curl}\boldsymbol{v} \tag{2-4-15}$$

The nth derivative with respect to τ of the relative right Cauchy–Green tensor $\mathbf{C}_{(t)}(\tau)$ calculated at $\tau = t$ is called the nth Rivlin–Ericksen tensor (Rivlin and Ericksen, 1955):

$$\mathbf{A}_n(t) = \overset{(n)}{\mathbf{C}}_{(t)}(t) = \frac{d^n}{d\tau^n}\mathbf{C}_{(t)}(\tau) \tag{2-4-16}$$

evaluated at $\tau = t$. Differentiation of $\mathbf{C}_{(t)}(\tau)$ gives

$$\overset{(n)}{\mathbf{C}}_{(t)}(\tau) = \sum_{i=0}^{n} \binom{n}{i} \overset{(i)}{\mathbf{F}}_{(t)}(\tau)^T \overset{(n-i)}{\mathbf{F}}_{(t)}(\tau)$$

which, for $\tau = t$, reduces by way of Eqs. (2-2-28) and (2-2-29) to

$$\mathbf{A}_n = \mathbf{L}_n + \mathbf{L}_n{}^T + \sum_{i=1}^{n-1} \binom{n}{i} \mathbf{L}_i{}^T \mathbf{L}_{n-i} \tag{2-4-17}$$

Now $\mathbf{C}_{(t)}(\tau) = \mathbf{U}^2_{(t)}(\tau)$ and, as $\mathbf{D} = \dot{\mathbf{U}}_{(t)}(t)$, $\mathbf{A}_n$ is related to the nth derivative of the stretch tensor. Applying Eq. (2-4-17),

$$\mathbf{A}_1 = \mathbf{L} + \mathbf{L}^T = 2\mathbf{D} \tag{2-4-18}$$

2-5 TRANSPORT PROPERTIES

Various transport formulas have now to be defined in terms of line, surface, and volume integrals. However, before stating the transport formulas it is necessary to derive certain relations for the differentials of the area on a surface and of the volume in a region denoted by da and dv respectively. This is essential, since to establish the transport formulas it is necessary to transfer the spatial description of the differentials da and dv to their respective referential description. This is in order that the range of the integration of the line, surface, and volume integrals shall be a fixed curve, surface, or region in the reference configuration B_r of the body $\mathscr{B}$, the accompanying differentiation with respect to time t being then performed under the integral sign.

Consider a material surface S_t in the current configuration B_t of a body $\mathscr{B}$ with a material surface S_r in the reference configuration B_r. Let the element of area dA at X, defined by the two vectors $d\boldsymbol{X}$ and $\delta\boldsymbol{X}$, be carried into the element of surface da at $x \in S_t$ defined by the two vectors $d\mathbf{x}$ and $\delta\mathbf{x}$. The area dA can be expressed in the form

$$\boldsymbol{N}\,dA = d\boldsymbol{X} \times \delta\boldsymbol{X} = (\boldsymbol{G}_\beta \times \boldsymbol{G}_\gamma)\,dX^\beta \delta X^\gamma \tag{2-5-1}$$

and, similarly,

$$\boldsymbol{v}\,da = d\mathbf{x} \times \delta\mathbf{x} = (\boldsymbol{g}_j \times \boldsymbol{g}_k)\,dx^j \delta x^k \tag{2-5-2}$$

where $\boldsymbol{N}$ and $\boldsymbol{v}$ are unit vectors normal to S_r and S_t at X and x respectively. Equation (2-5-2) can be rearranged, by way of Eqs. (1-5-21) and (1-7-35), into the form

$$v_i\,da = e_{ijk}\sqrt{g}\,dx^j \delta x^k$$

$$\frac{\partial x^i}{\partial X^\alpha} v_i\,da = e_{ijk}\sqrt{g}\left(\frac{\partial x^i}{\partial X^\alpha}\frac{\partial x^j}{\partial X^\beta}\frac{\partial x^k}{\partial X^\gamma}\right) dX^\beta \delta X^\gamma$$

Equations (1-7-20) and (1-6-42) give

$$v_i\,da = e_{\alpha\beta\gamma}\sqrt{g}\left|\frac{\partial x^i}{\partial X^\alpha}\right|\frac{\partial X^\alpha}{\partial x^i}\,dX^\beta \delta X^\gamma$$

$$= \sqrt{\frac{g}{G}}\det\left[\frac{\partial x^i}{\partial X^\alpha}\right]\frac{\partial X^\alpha}{\partial x^i} N_\alpha\,dA$$

which can be expressed, by way of Eqs. (1-6-47) and (2-2-7), in the form

$$\boldsymbol{v}\,da = J(\mathbf{F}^{-1})^T \boldsymbol{N}\,dA \tag{2-5-3}$$

From Eq. (2-5-3) there can be obtained, by way of Eq. (1-6-8), the relations

$$\left(\frac{da}{dA}\right)^2 = J^2 \boldsymbol{N} \cdot (\mathbf{C}^{-1}\boldsymbol{N}) = J^2[\boldsymbol{v} \cdot (\mathbf{B}\boldsymbol{v})]^{-1} \tag{2-5-4}$$

and if no change in area occurs then

$$\boldsymbol{N} \cdot (\mathbf{C}^{-1}\boldsymbol{N}) = J^{-2} \qquad \boldsymbol{v} \cdot (\mathbf{B}\boldsymbol{v}) = J^{2} \tag{2-5-5}$$

which imposes a restriction on $\mathbf{F}$.

Let the element of volume dV at X, defined by the three vectors $d\boldsymbol{X}$, $\delta\boldsymbol{X}$, and $\Delta\boldsymbol{X}$, be carried into the element of volume dv at x defined by the three vectors $d\boldsymbol{x}$, $\delta\boldsymbol{x}$, and $\Delta\boldsymbol{x}$. Here x is now an interior point of a material region R_t in the current configuration B_t of a body $\mathscr{B}$, the particle currently situated at x occupying the point X in the reference configuration B_r. The volume within the parallelepiped determined by the vectors $d\boldsymbol{X}$, $\delta\boldsymbol{X}$, and $\Delta\boldsymbol{X}$ can be expressed in the form

$$\begin{aligned} dV = (d\boldsymbol{X} \times \delta\boldsymbol{X}) \cdot \Delta\boldsymbol{X} &= (\mathbf{G}_\alpha \times \mathbf{G}_\beta) \cdot \mathbf{G}_\gamma \, dX^\alpha \delta X^\beta \Delta X^\gamma \\ &= e_{\alpha\beta\gamma}\sqrt{G}\, dX^\alpha \delta X^\beta \Delta X^\gamma \end{aligned} \tag{2-5-6}$$

Similarly,

$$\begin{aligned} dv &= e_{ijk}\sqrt{g}\, dx^i \delta x^j \Delta x^k \\ &= e_{ijk}\sqrt{g}\frac{\partial x^i}{\partial X^\alpha}\frac{\partial x^j}{\partial X^\beta}\frac{\partial x^k}{\partial X^\gamma} dX^\alpha \delta X^\beta \Delta X^\gamma \\ &= e_{\alpha\beta\gamma}\sqrt{g}\left|\frac{\partial x^i}{\partial X^\alpha}\right| dX^\alpha \delta X^\beta \Delta X^\gamma \\ &= \sqrt{\frac{g}{G}}\det\left[\frac{\partial x^i}{\partial X^\alpha}\right] dV \end{aligned} \tag{2-5-7}$$

and hence, from Eqs. (1-6-42) and (1-6-47),

$$dv = \det \mathbf{F}\, dV \qquad dv = J\, dV \tag{2-5-8}$$

A motion in which $dV = dv$, i.e.,

$$J = 1 \qquad (dV = dv) \tag{2-5-9}$$

is called an isochoric, or volume-preserving, motion.

Using the notation $d\boldsymbol{s}$ for a directed element of arc length, the required set of transport formulas can now be expressed as follows.

Let ϕ be a continuously differentiable scalar field representing some property of a body $\mathscr{B}$. The transport formulas can be expressed, by way of Eq. (2-2-24) and the relations $\dot{\mathbf{F}} = \mathbf{LF}$, $\dot{J} = J \operatorname{tr} \mathbf{L}$, in the form

$$\frac{d}{dt}\int_{C_t} \phi\, d\boldsymbol{s} = \int_{C_t}\left(\frac{D}{Dt}\phi\, d\boldsymbol{s} + \phi \mathbf{L}\, d\boldsymbol{s}\right) \tag{2-5-10}$$

$$\frac{d}{dt}\int_{S_t} \phi \boldsymbol{v}\, da = \int_{S_t}\left[\left(\frac{D}{Dt}\phi + \phi \operatorname{tr}\mathbf{L}\right)\boldsymbol{v} - \phi \mathbf{L}^T \boldsymbol{v}\right] da \tag{2-5-11}$$

$$\frac{d}{dt}\int_{R_t} \phi\, dv = \int_{R_t}\left(\frac{D}{Dt}\phi + \phi \operatorname{tr}\mathbf{L}\right) dv \tag{2-5-12}$$

where C_t, S_t, and R_t denote a material curve, a material surface, and a material region respectively in the current configuration B_t of a body $\mathscr{B}$. Similarly, for a continuously differentiable vector field $\boldsymbol{a}$,

$$\frac{d}{dt}\int_{C_t} \boldsymbol{a}\cdot d\boldsymbol{s} = \int_{C_t}\left(\frac{D}{Dt}\boldsymbol{a}+\mathbf{L}^T\boldsymbol{a}\right)\cdot d\boldsymbol{s} \tag{2-5-13}$$

$$\frac{d}{dt}\int_{S_t} \boldsymbol{a}\cdot \boldsymbol{\nu}\, da = \int_{S_t}\left(\frac{D}{Dt}\boldsymbol{a}+\boldsymbol{a}\operatorname{tr}\mathbf{L}-\mathbf{L}\boldsymbol{a}\right)\cdot \boldsymbol{\nu}\, da \tag{2-5-14}$$

$$\frac{d}{dt}\int_{R_t} \boldsymbol{a}\, dv = \int_{R_t}\left(\frac{D}{Dt}\boldsymbol{a}+\boldsymbol{a}\operatorname{tr}\mathbf{L}\right) dv \tag{2-5-15}$$

2-6 STRAIN TENSORS

The classical strain measures are the Almansi–Hamel strain $\mathbf{e}$ (Almansi, 1911; Hamel, 1912) and the Green–St. Venant strain $\mathbf{E}$ (Green, 1841; St. Venant, 1844).

The Almansi–Hamel strain is defined by

$$2\mathbf{e} = \mathbf{I}-\mathbf{B}^{-1} \qquad 2e_{ij} = g_{ij}-X^{\alpha}{}_{,i}X_{\alpha,j} \tag{2-6-1}$$

where $\mathbf{B}^{-1}$ is the inverse of $\mathbf{B}$ defined by Eq. (2-2-20) and the Green–St. Venant strain is defined by

$$2\mathbf{E} = \mathbf{C}-\mathbf{I} \qquad 2E_{\alpha\beta} = x^{k}{}_{,\alpha}x_{k,\beta}-G_{\alpha\beta} \tag{2-6-2}$$

where $\mathbf{C}$ is defined by Eq. (2-2-19). Consider now the expression

$$2\mathbf{F}^T\mathbf{e}\mathbf{F} = \mathbf{F}^T[\mathbf{I}-(\mathbf{F}^{-1})^T\mathbf{F}^{-1}]\mathbf{F} = \mathbf{C}-\mathbf{I} = 2\mathbf{E}$$

from which the relation between the two classical strain measures is given by

$$\mathbf{E} = \mathbf{F}^T\mathbf{e}\mathbf{F} \tag{2-6-3}$$

The tensors $\mathbf{U}$ and $\mathbf{V}$, when referred to the principal axes as coordinate axes, take the simple form

$$\mathbf{U} = \mathbf{V} = \begin{bmatrix} \lambda_1 & 0 & 0 \\ 0 & \lambda_2 & 0 \\ 0 & 0 & \lambda_3 \end{bmatrix} \tag{2-6-4}$$

and the invariants can be written, using Eq. (2-3-8), in the form

$$\begin{aligned} \mathrm{I}_{\mathbf{U}} &= \lambda_1+\lambda_2+\lambda_3 = \mathrm{I}_{\mathbf{V}} \\ \mathrm{II}_{\mathbf{U}} &= \lambda_1\lambda_2+\lambda_2\lambda_3+\lambda_3\lambda_1 = \mathrm{II}_{\mathbf{V}} \\ \mathrm{III}_{\mathbf{U}} &= \lambda_1\lambda_2\lambda_3 = \mathrm{III}_{\mathbf{V}} \end{aligned} \tag{2-6-5}$$

The definitions given in Eqs. (2-3-12), (2-6-1), and (2-6-2) can be used to establish the following relations between the invariants:

$$\begin{aligned} \mathrm{I}_{\mathbf{B}} &= \mathrm{I}_{\mathbf{C}} = \mathrm{I}_{\mathbf{U}}^2-2\mathrm{II}_{\mathbf{U}} \\ \mathrm{II}_{\mathbf{B}} &= \mathrm{II}_{\mathbf{C}} = \mathrm{II}_{\mathbf{U}}^2-2\mathrm{I}_{\mathbf{U}}\mathrm{III}_{\mathbf{U}} \\ \mathrm{III}_{\mathbf{B}} &= \mathrm{III}_{\mathbf{C}} = \mathrm{III}_{\mathbf{U}}^2 \end{aligned} \tag{2-6-6}$$

$$
\begin{aligned}
2\mathrm{I}_{\mathbf{E}} &= \mathrm{I}_{\mathbf{C}} - 3 \\
4\mathrm{II}_{\mathbf{E}} &= \mathrm{II}_{\mathbf{C}} - 2\mathrm{I}_{\mathbf{C}} + 3 \\
8\mathrm{III}_{\mathbf{E}} &= \mathrm{III}_{\mathbf{C}} - \mathrm{II}_{\mathbf{C}} + \mathrm{I}_{\mathbf{C}} - 1
\end{aligned}
\tag{2-6-7}
$$

$$
\begin{aligned}
2\mathrm{I}_{\mathbf{e}} &= 3 - \frac{\mathrm{II}_{\mathbf{B}}}{\mathrm{III}_{\mathbf{B}}} \\
4\mathrm{II}_{\mathbf{e}} &= 3 - 2\frac{\mathrm{II}_{\mathbf{B}}}{\mathrm{III}_{\mathbf{B}}} + \frac{\mathrm{I}_{\mathbf{B}}}{\mathrm{III}_{\mathbf{B}}} \\
8\mathrm{III}_{\mathbf{e}} &= 1 - \frac{\mathrm{II}_{\mathbf{B}}}{\mathrm{III}_{\mathbf{B}}} + \frac{\mathrm{I}_{\mathbf{B}}}{\mathrm{III}_{\mathbf{B}}} - \frac{1}{\mathrm{III}_{\mathbf{B}}}
\end{aligned}
\tag{2-6-8}
$$

The strain measures defined above are not the only possible measures of strain. Any function of the stretch tensors $\mathbf{U}$ or $\mathbf{V}$ which is bounded and which tends to zero for small deformations can be used as a measure of strain.

If there is no deformation the tensors $\mathbf{e}$ and $\mathbf{E}$ reduce to zero in contrast to the deformation tensors $\mathbf{B}$ and $\mathbf{C}$ which for this limiting condition are unit tensors.

The material derivatives of the strain tensors follow simply from their definitions. Thus, noting that

$$\dot{\mathbf{F}} = \mathbf{LF} \tag{2-6-9}$$

it follows that

$$
\begin{aligned}
\dot{\mathbf{E}} &= \tfrac{1}{2}(\dot{\mathbf{F}}^T\mathbf{F} + \mathbf{F}^T\dot{\mathbf{F}}) = \tfrac{1}{2}[\mathbf{F}^T(\mathbf{F}^T)^{-1}\dot{\mathbf{F}}^T\mathbf{F} + \mathbf{F}^T\dot{\mathbf{F}}\mathbf{F}^{-1}\mathbf{F}] \\
&= \tfrac{1}{2}\mathbf{F}^T(\mathbf{L}^T + \mathbf{L})\mathbf{F} = \mathbf{F}^T\mathbf{DF}
\end{aligned}
\tag{2-6-10}
$$

There follows from Eqs. (2-6-3) and (2-6-10) the relation

$$
\begin{aligned}
\dot{\mathbf{E}} = \mathbf{F}^T\mathbf{DF} &= \dot{\mathbf{F}}^T\mathbf{eF} + \mathbf{F}^T\dot{\mathbf{e}}\mathbf{F} + \mathbf{F}^T\mathbf{e}\dot{\mathbf{F}} \\
&= \mathbf{F}^T(\dot{\mathbf{e}} + \mathbf{L}^T\mathbf{e} + \mathbf{eL})\mathbf{F}
\end{aligned}
\tag{2-6-11}
$$

Hence

$$\dot{\mathbf{e}} + \mathbf{L}^T\mathbf{e} + \mathbf{eL} = \mathbf{D} \tag{2-6-12}$$

There are several simple types of strain that can be used to either approximate or at least give insight into more complex problems.

(a) Homogeneous deformation If the components of the deformation gradient are at most functions of time, the deformation is said to be homogeneous and can be written in the form

$$\boldsymbol{x} = \mathbf{F}(t)\boldsymbol{X} + \boldsymbol{X}_0 \tag{2-6-13}$$

where $\boldsymbol{X}_0$ is a constant vector. It is evident from Eq. (2-6-13) that in this deformation straight lines deform into straight lines, planes into planes, and ellipsoids into ellipsoids. From their definitions it follows that the deformation and strain tensors are all independent of spatial position.

(b) Uniform dilatation If in Eq. (2-6-13) **F** takes the form

$$\mathbf{F} = k\mathbf{I} \tag{2-6-14}$$

then the deformation is referred to as uniform dilatation. Spheres are deformed into spheres and their radii are increased to k of their original length. It has been shown in Sec. 2-5 that the volume change can be expressed in the form

$$dv = \det \mathbf{F}\, dV \tag{2-6-15}$$

which gives, by way of Eq. (2-6-14), the relation

$$\frac{dv}{dV} = k^3 \tag{2-6-16}$$

for the volume change accompanying uniform dilatation.

(c) Simple extension Let the rectangular cartesian coordinates (X, Y, Z) be the reference coordinates of a body $\mathscr{B}$ in the undeformed state and the cartesian coordinates (x, y, z) be the spatial coordinates in the deformed configuration. Take the coordinate axes to be the principal axes of strain. An example of a pure homogeneous deformation is simple extension defined for a stretch ($\mathbf{R} = \mathbf{I}$) in the Z direction by

$$[\mathbf{F}] = \begin{bmatrix} \alpha\lambda_3 & 0 & 0 \\ 0 & \alpha\lambda_3 & 0 \\ 0 & 0 & \lambda_3 \end{bmatrix} \tag{2-6-17}$$

The simple deformations are thus

$$\begin{gathered} x = \alpha\lambda_3 X \qquad y = \alpha\lambda_3 Y \qquad z = \lambda_3 Z \\ \lambda_1 = \lambda_2 = \alpha\lambda_3 \end{gathered} \tag{2-6-18}$$

from which it is evident that simple extension is characterized by the condition that two, but not three, principal stretches are equal. The factor α is related to the transverse contraction ratio v by the definition

$$v = -\frac{\Delta x}{\Delta z} = \frac{1-\alpha\lambda_3}{\lambda_3 - 1} \tag{2-6-19}$$

which can be arranged to give

$$\alpha = \frac{1}{\lambda_3}(1+v) - v \tag{2-6-20}$$

The volume change is given by Eqs. (2-6-15) and (2-6-17) in the form

$$\frac{dv}{dV} = \alpha^2\lambda_3^3 \tag{2-6-21}$$

Equation (2-6-21) shows that the extension is isochoric if and only if $\alpha = \lambda_3^{-3/2}$. Simple extension is considered in relation to nonlinear elasticity in Secs. 8-3-1 and 8-4-1.

(d) Plane strain If the deformation can be expressed in the form

$$x^1 = f(X^1, X^2) \qquad x^2 = g(X^1, X^2) \qquad x^3 = X^3 \tag{2-6-22}$$

it is said to be one of plane strain. It is evident that plane strain is characterized by identical deformations in a family of parallel planes together with no deformation in the direction of their normals. Plane strain is considered in relation to linear elasticity in Sec. 6-3-1.

(e) Simple shear A simple shear of amount γ is given by

$$x = X + \gamma Y \qquad y = Y \qquad z = Z \tag{2-6-23}$$

$$[\mathbf{F}] = \begin{bmatrix} 1 & \gamma & 0 \\ 0 & 1 & 0 \\ 0 & 0 & 1 \end{bmatrix} \tag{2-6-24}$$

For this type of deformation, it is evident from Eqs. (2-6-15) and (2-6-24) that since $\det \mathbf{F}$ is unity there is no change in volume, i.e., simple shear is isochoric. Simple shear is considered in detail in Secs. 8-3-2 and 8-4-2.

2-7 COMPATIBILITY CONDITIONS

The components of a strain tensor, in a simply connected region of space, are not independent but have to satisfy the condition that the metrics in both the initial and final states are euclidean. From Eq. (2-3-10),

$$\begin{aligned} ds^2 = d\mathbf{x} \cdot d\mathbf{x} &= \mathbf{F}\, d\mathbf{X} \cdot \mathbf{F}\, d\mathbf{X} \\ &= F^i{}_\alpha F_{i\beta}\, dX^\alpha\, dX^\beta \\ &= C_{\alpha\beta}\, dX^\alpha\, dX^\beta \end{aligned} \tag{2-7-1}$$

Thus $C_{\alpha\beta}$ can be regarded as the metric of the final configuration referred to referential coordinates. Let the curvature formed from $C_{\alpha\beta}$ and its partial differential with respect to X^α, as given in Eq. (1-8-25), be denoted by $R^{(C)}{}_{\alpha\beta\gamma\delta}$. The condition that the final space be euclidean is given by

$$R^{(C)}{}_{\alpha\beta\gamma\delta} = 0 \tag{2-7-2}$$

As was shown in Sec. 1-8, this gives rise to six independent equations. These equations are known as St. Venant's conditions of compatibility and are supposed to hold everywhere except upon certain singular points, curves, or surfaces.

Another useful way of expressing strain is in terms of the displacement $\boldsymbol{u}$ from the initial to the final configuration. Let a material point be displaced from the position P to the position P' and let the origins of the referential and spatial frames of reference be at O and o respectively, as shown in Fig. 2-1. The vectors $\boldsymbol{X}$, $\boldsymbol{x}$, $\boldsymbol{u}$, and $\boldsymbol{c}$ are as shown in the figure; clearly,

$$\boldsymbol{x} + \boldsymbol{c} = \boldsymbol{X} + \boldsymbol{u} \tag{2-7-3}$$

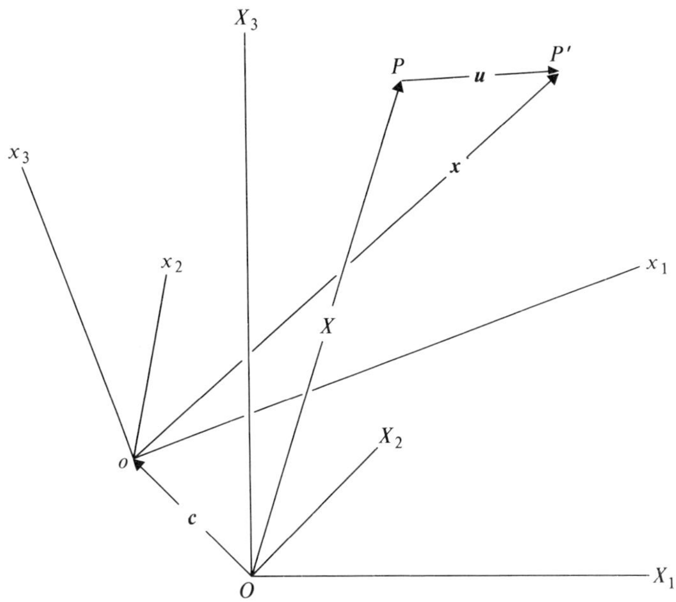

Figure 2-1 Relationship between displacement and referential and spatial reference frames.

At a neighboring point

$$(\mathbf{x}+d\mathbf{x})+\mathbf{c} = (\mathbf{X}+d\mathbf{X})+(\mathbf{u}+d\mathbf{u}) \tag{2-7-4}$$

Thus

$$d\mathbf{x} = d\mathbf{X}+d\mathbf{u} \tag{2-7-5}$$

Now

$$ds^2 = d\mathbf{x}\cdot d\mathbf{x} \qquad \text{and} \qquad dS^2 = d\mathbf{X}\cdot d\mathbf{X}$$

and, from Eq. (2-7-5),

$$ds^2 - dS^2 = d\mathbf{x}\cdot d\mathbf{x} - d\mathbf{X}\cdot d\mathbf{X} = 2e_{ij}\,dx^i\,dx^j - 2d\mathbf{x}\cdot d\mathbf{u} \quad d\mathbf{u}\cdot d\mathbf{u} \tag{2-7-6}$$

But

$$d\mathbf{u} = u^i{}_{,j}\mathbf{g}_i\,dx^j$$

and substituting in Eq. (2-7-6) and changing several dummy indices leads directly to

$$e_{ij} = \tfrac{1}{2}(u_{j,i}+u_{i,j}-u^k{}_{,i}u_{k,j}) \tag{2-7-7}$$

Interchanging the roles of referential and spatial coordinates leads to

$$E_{\alpha\beta} = \tfrac{1}{2}(u_{\beta,\alpha}+u_{\alpha,\beta}+u^\gamma{}_{,\alpha}u_{\gamma,\beta}) \tag{2-7-8}$$

The strain tensors can thus be expressed simply in terms of the displacement gradients $u_{i,j}$ or $u_{\alpha,\beta}$.

If the displacement gradients are infinitesimal, that is, $|u_{i,j}| \ll 1$ so that second-order terms and above may be neglected, then Eq. (2-7-7) reduces to

$$\tilde{e}_{ij} = \tfrac{1}{2}(u_{j,i} + u_{i,j}) \tag{2-7-9}$$

The infinitesimal strain defined by Eq. (2-7-9) is known as Cauchy's strain measure. Referring the x^i and X^α to the same coordinate system gives

$$\tilde{E}_{\alpha\beta} \approx g^i{}_\alpha g^j{}_\beta \tilde{e}_{ij} \tag{2-7-10}$$

and the principal values of both strain tensors are the same; thus, for infinitesimal strain there is no difference between spatial and referential strain measures.

Let the differential elements $d\boldsymbol{X}$ define a sphere of radius R_0 at $\boldsymbol{X}$. For a sphere of constant radius

$$dS^2 = G_{\alpha\beta}\, dX^\alpha\, dX^\beta \equiv R_0^2 = \text{constant} \tag{2-7-11}$$

Now, since

$$ds^2 = d\boldsymbol{x} \cdot d\boldsymbol{x} \qquad dS^2 = d\boldsymbol{X} \cdot d\boldsymbol{X} \tag{2-7-12}$$

it follows from Eq. (2-2-20) that Eq. (2-2-18) can be rearranged to give

$$dS^2 = (B^{-1})_{ij}\, dx^i\, dx^j \tag{2-7-13}$$

Substituting for dS in Eq. (2-7-11) from Eq. (2-7-13) gives

$$R_0^2 = G_{\alpha\beta}\, dX^\alpha\, dX^\beta = (B^{-1})_{ij}\, dx^i\, dx^j \tag{2-7-14}$$

In Eq. (2-7-14), $\mathbf{G}$ is an euclidean metric tensor; hence the term involving $\mathbf{G}$ is positive definite and the quadric of the second form is an ellipsoid. This quadric is called the strain quadric and its principal axes correspond to the principal directions.

The rotation is characterized by the $\tilde{\mathbf{R}}$ tensor where

$$\tilde{R}_{ij} = \tfrac{1}{2}(u_{i,j} - u_{j,i}) \tag{2-7-15}$$

The dual or axial vector is given by

$$\begin{aligned} \omega^i &= \tfrac{1}{2}\varepsilon^{ijk}\tilde{R}_{jk} \\ &= \tfrac{1}{4}\varepsilon^{ijk}(u_{j,k} - u_{k,j}) \\ &= \tfrac{1}{4}\varepsilon^{ijk}u_{j,k} - \tfrac{1}{4}\varepsilon^{ikj}u_{j,k} \qquad \text{(interchanging } j \text{ and } k \text{ in second term as dummy indices)} \\ &= \tfrac{1}{2}\varepsilon^{ijk}u_{j,k} \\ &= \tfrac{1}{2}(\text{curl}\,\boldsymbol{u})^i \end{aligned} \tag{2-7-16}$$

Using Eqs. (2-7-9) and (2-7-15),

$$u_{i,j} = \tilde{e}_{ij} + \tilde{R}_{ij} \tag{2-7-17a}$$

or

$$u_{i,j} = \tilde{e}_{ij} + \varepsilon_{ijk}\omega^k \tag{2-7-17b}$$

To find the compatibility conditions it is noted that as the space is euclidean the curvature tensor is zero and covariant differentiation is commutative. Making use of Eq. (2-7-9),

$$\tilde{e}_{ij,kl} = \tfrac{1}{2}(u_{j,ikl} + u_{i,jkl}) \qquad (2\text{-}7\text{-}18a)$$

$$\tilde{e}_{kl,ij} = \tfrac{1}{2}(u_{l,kij} + u_{k,lij}) \qquad (2\text{-}7\text{-}18b)$$

$$\tilde{e}_{ik,jl} = \tfrac{1}{2}(u_{k,ijl} + u_{i,kjl}) \qquad (2\text{-}7\text{-}18c)$$

$$\tilde{e}_{jl,ik} = \tfrac{1}{2}(u_{l,jik} + u_{j,lik}) \qquad (2\text{-}7\text{-}18d)$$

Adding Eqs. (2-7-18*a*) and (2-7-18*b*) and then subtracting Eqs. (2-7-18*c*) and (2-7-18*d*) shows

$$\tilde{c}_{ijkl} \equiv \tilde{e}_{ij,kl} + \tilde{e}_{kl,ij} - \tilde{e}_{ik,jl} - \tilde{e}_{jl,ik} = 0 \qquad (2\text{-}7\text{-}19)$$

This set of compatibility equations reduces to six independent equations when account is taken of the symmetry of $\tilde{\mathbf{e}}$.

2-8 MULTIPLY-CONNECTED REGIONS AND DISLOCATIONS

At first it appears that the displacement $\boldsymbol{u}$ must be a single-valued continuous function of position if a body is to retain its continuity and be free from cracks, but it will now be shown that this is only true for "simply connected" bodies when the compatibility conditions are satisfied. The connectivity of spaces forms part of the discipline of topology and an exact formulation is very complicated, but providing liberal use is made of our intuitive geometrical picture of space, a simple treatment can be given as presented by Lamb (1932).

A region is "connected" when it is possible to pass from any one point of it to any other by an infinity of paths, each of which lies wholly in the region.

A closed path or loop is called a "circuit" and if it can be continuously deformed to a point without passing out of the region it is said to be a "reducible circuit".

A "simply connected" region is one such that all circuits drawn within it are reducible.

A "barrier" is a surface drawn across the region and limited by the line or lines in which it meets the boundary. Hence a barrier is necessarily a connected surface.

An "n-ply-connected" region is one in which $(n-1)$ barriers are needed to reduce it to a simply connected region while n barriers can destroy the continuity of the region.

To find the conditions for the displacement $\boldsymbol{u}$ to be a single-valued function of position consider a curve L joining two points A and B in a multiply-connected region V and let the coordinate system be chosen to be rectangular cartesian. Although the displacement and its associated rotation may be many valued the strain is assumed to be single valued and continuous, except possibly on certain singular surfaces, lines, or points. The reason for this assumption is that the strain is usually directly related to the stress, e.g., in small displacements they are linearly related, and the stress must be single valued and continuous if it is not to give rise to infinite accelerations.

Integrating Eq. (2-7-17*a*) along the line L and noting that in rectangular cartesian coordinates there is no distinction between covariant and contravariant components, covariant differentiation is equivalent to partial differentiation:

$$u_i(B)-u_i(A) = \int_A^B u_{i,j}\,dx_j = \int_A^B (\tilde{e}_{ij}+\tilde{R}_{ij})\,dx_j \tag{2-8-1}$$

Now

$$\int_A^B \tilde{R}_{ij}\,dx_j = \int_A^B \tilde{R}_{ij}\,d[x_j-x_j(B)] = [\tilde{R}_{ij}(x_j-x_j(B))]_A^B - \int_A^B [x_j-x_j(B)]\,d\tilde{R}_{ij}$$

$$= \tilde{R}_{ij}(A)[x_j(A)-x_j(B)] + \int_A^B [x_j-x_j(B)]\,d\tilde{R}_{ij} \tag{2-8-2}$$

and

$$d\tilde{R}_{ij} = \tilde{R}_{ij,k}\,dx_k \tag{2-8-3}$$

But, from Eq. (2-7-15),

$$\begin{aligned}\tilde{R}_{ij,k} &= \tfrac{1}{2}(u_{i,jk}-u_{j,ik})\\ &= \tfrac{1}{2}(u_{i,kj}+u_{k,ij}-u_{j,ki}-u_{k,ji})\\ &= \tilde{e}_{ki,j}-\tilde{e}_{jk,i}\end{aligned} \tag{2-8-4}$$

Putting Eqs. (2-8-4), (2-8-3), and (2-8-2) into Eq. (2-8-1) gives

$$u_i(B)-u_i(A) = \tilde{R}_{ij}(A)[x_j(A)-x_j(B)] + \int_A^B \{\tilde{e}_{ik}+[x_j-x_j(B)]\,(\tilde{e}_{ki,j}-\tilde{e}_{jk,i})\}\,dx_k \tag{2-8-5}$$

If the path L forms a closed circuit C so that the points A and B coincide,

$$u_i(B)-u_i(A) = \oint \{\tilde{e}_{ik}+[x_j-x_j(B)]\,(\tilde{e}_{ki,j}-\tilde{e}_{jk,i})\}\,dx_k \tag{2-8-6}$$

Let

$$Q_{ik} = \tilde{e}_{ik}+[x_j-x_j(B)]\,(\tilde{e}_{ki,j}-\tilde{e}_{jk,i}) \tag{2-8-7}$$

If the circuit C is reducible then there must exist a surface S lying wholly within the region and bounded by the curve C.

Applying the Stokes theorem,

$$\int_A^B Q_{ik}\,dx_k = \int_S (Q_{il,k}-Q_{ik,l})\,dx_k\,dx_l \tag{2-8-8}$$

Differentiating Eq. (2-8-7), interchanging the roles of k, l to obtain $Q_{il,k}$, and subtracting gives

$$Q_{il,k}-Q_{ik,l} = [x_j-x_j(B)]\,(\tilde{e}_{li,jk}+\tilde{e}_{jk,il}-\tilde{e}_{jl,ik}-\tilde{e}_{ki,jl}) \tag{2-8-9}$$

But, using Eq. (2-7-19), this can be written

$$Q_{il,k}-Q_{ik,l}=[x_j-x_j(B)]\tilde{c}_{jkil} \tag{2-8-10}$$

If the compatibility conditions are satisfied, that is, $\tilde{c}_{jkil}=0$, then in a simply connected region $u_i(B)=u_i(A)$ or the displacement is a single-valued function of position. In a multiply-connected region the vanishing of $\tilde{c}_{ijkl}$ is no longer sufficient to ensure that the displacement is single valued. To show this reduce the multiply-connected region to a simply-connected one by a system of barriers and take the circuit C to be nonreducible and cutting one and only one of the barriers at a point A. The change in u_i across the barrier will be denoted by Δu_i and will be given by a formula such as Eq. (2-8-6). All circuits such as C that pass through A have the same value for Δu_i. This is because two different paths can be combined into one circuit which is reducible and it has been shown above that the overall change of u_i round the whole circuit is then zero (note that line integrals evaluated in different directions have different signs). The compatibility conditions thus only ensure that Δu_i is independent of the circuit C round which the change is evaluated. To ensure that $\boldsymbol{u}$ is unique in an n-ply-connected system one requires $(n-1)$ supplementary conditions of the form

$$\oint Q_{ik}\,dx_k=0 \tag{2-8-11}$$

which is one set of three equations for each barrier.

To find how the displacement varies over a barrier consider two points A and B situated on the barrier. Let the points adjacent to A and B on one side of the barrier be denoted by A^+ and B^+ and adjacent points on the other side of the barrier be denoted by A^- and B^-. By Eq. (2-8-5),

$$u_i(B^+)-u_i(A^+)=\tilde{R}_{ij}(A^+)[x_j(A^+)-x_j(B^+)]+\int_{A^+}^{B^+}Q_{ik}\,dx_k \tag{2-8-12a}$$

$$u_i(B^-)-u_i(A^-)=\tilde{R}_{ij}(A^-)[x_j(A^-)-x_j(B^-)]+\int_{A^-}^{B^-}Q_{ik}\,dx_k \tag{2-8-12b}$$

As $\tilde{e}_{ij}$ is continuous so is Q_{ij}, and

$$\int_{A^+}^{B^+}Q_{ik}\,dx_k=\int_{A^-}^{B^-}Q_{ik}\,dx_k \tag{2-8-13}$$

Also,

$$x_j(A^+)=x_j(A^-)=x_j(A) \tag{2-8-14}$$

Subtracting Eq. (2-8-12b) from (2-8-12a) gives

$$u_i(B^+)-u_i(B^-)=u_i(A^+)-u_i(A^-)+[\tilde{R}_{ij}(A^+)-\tilde{R}_{ij}(A^-)] \times[x_j(A)-x_j(B)] \tag{2-8-15}$$

Equation (2-8-15) is a mathematical statement of the important result that the displacement on one side of the barrier relative to the displacement on the other side is one which would be possible in a rigid body. Places where such discontinuities in the displacement occur are called dislocations. They were originally investigated by Volterra (1907) and by Somigliana (1913, 1915) and were named "dislocations" by Love (1920).

2-9 JAUMANN AND CONVECTIVE TIME DERIVATIVES

When considering time derivatives it is desirable that they should not include an unwanted dependence on the translational and rotational movement of a material element. It is therefore convenient to estimate time derivatives in a reference frame in which these effects are absent. Following Oldroyd (1958), consider an arbitrarily moving reference frame κ' and let the coordinates of a material point P at time t be X' in such a system. Let the coordinates of P at time t in a fixed frame of reference κ be $\mathbf{x}$. Let the rate of change of any quantity for a given value of X' be denoted by d/dt and define

$$v' = \frac{d\mathbf{x}}{dt} \qquad v'^i = \frac{dx^i}{dt} \tag{2-9-1}$$

As with the material derivative,

$$\frac{d\psi}{dt} = \frac{\partial\psi}{\partial t} + v'^i \psi_{,i} \tag{2-9-2}$$

Consider a tensor $\mathbf{T}$ having components $T^{K...L}{}_{M...N}$ in κ' and $T^{k...l}{}_{m...n}$ in κ. By the definition of tensors

$$\frac{\partial x^k}{\partial X'^K} \cdots \frac{\partial x^l}{\partial X'^L} T^{K...L}{}_{M...N} = \frac{\partial x^m}{\partial X'^M} \cdots \frac{\partial x^n}{\partial X'^N} T^{k...l}{}_{m...n} \tag{2-9-3}$$

Differentiate with respect to d/dt, remembering that d/dt commutes with $\partial/\partial X'$, and using Eqs. (2-9-1) and (2-9-2) there follows the relation

$$\frac{d}{dt} T^{K...L}{}_{M...N} = \frac{\partial X'^K}{\partial x^k} \cdots \frac{\partial X'^L}{\partial x^l} \frac{\partial x^m}{\partial X'^M} \cdots \frac{\partial x^n}{\partial X'^N} \left(\frac{d}{dt} T^{k...l}{}_{m...n} \right) \tag{2-9-4}$$

where

$$\frac{d}{dt} T^{k\cdot s\cdot l}{}_{m\cdot r\cdot n} = \frac{\partial}{\partial t} T^{k\cdot s\cdot l}{}_{m\cdot r\cdot n} + v'^p T^{k\cdot s\cdot l}{}_{m\cdot r\cdot n,p} + \sum_{r=m}^{n} v'^p{}_{,r} T^{k\cdot s\cdot l}{}_{m\cdot p\cdot n} - \sum_{s=k}^{l} v'^s{}_{,q} T^{k\cdot q\cdot l}{}_{m\cdot r\cdot n} \tag{2-9-5}$$

Equivalently this can be written

$$\frac{d}{dt} T^{k\cdot s\cdot l}{}_{m\cdot r\cdot n} = \frac{\partial}{\partial t} T^{k\cdot s\cdot l}{}_{m\cdot r\cdot n} + v'^p T^{k\cdot s\cdot l}{}_{m\cdot r\cdot n,p} + \sum_{r=m}^{n} L'^p{}_r T^{k\cdot s\cdot l}{}_{m\cdot p\cdot n} - \sum_{s=k}^{l} L'^s{}_q T^{k\cdot q\cdot l}{}_{m\cdot r\cdot n} \tag{2-9-6}$$

$$L'^i{}_j = D'^i{}_j + W'^i{}_j \tag{2-9-7}$$

Formulas for particular types of total derivative follow directly from Eq. (2-9-6):

1. If the reference frame is at rest relative to the spatial frame, then $\boldsymbol{v}' = \boldsymbol{o}$, $\mathbf{L}' = \mathbf{O}$, and

$$\frac{d}{dt} \equiv \frac{\partial}{\partial t} \qquad \text{(the local derivative)} \tag{2-9-8}$$

2. If the reference frame is only translated with the material velocity but is rigid and nonrotating, then $v' = v$, $\mathbf{L}' = \mathbf{O}$, and

$$\begin{aligned}\frac{d}{dt} T^{k\ldots l}{}_{m\ldots n} &= \frac{\partial}{\partial t} T^{k\ldots l}{}_{m\ldots n} + v^p T^{k\ldots l}{}_{m\ldots n,p} \\ &\equiv \frac{D}{Dt} T^{k\ldots l}{}_{m\ldots n} \qquad \text{(the material derivative)}\end{aligned} \tag{2-9-9}$$

3. If the reference frame is instantaneously rigid and moving and rotating with the material, then $v' = v$, $\mathbf{L}' = \mathbf{W}$, $\mathbf{D}' = \mathbf{O}$, and

$$\frac{d}{dt} T^{k\ldots l}{}_{m\ldots n} = \frac{D}{Dt} T^{k\ldots l}{}_{m\ldots n} + \sum_{r=m}^{n} W^p{}_r T^{k\ldots l}{}_{m\cdot p\cdot n} - \sum_{s=k}^{l} W^s{}_q T^{k\cdot q\cdot l}{}_{m\ldots n} \tag{2-9-10}$$

   In conformity with Prager (1961) this type of time derivative will be called a Jaumann derivative and will be denoted either by $\mathscr{D}\mathbf{T}/\mathscr{D}t$ or $\overset{\circ}{\mathbf{T}}$.

4. If the reference frame moves and distorts with the material so that $\boldsymbol{v}' = v$ and $\mathbf{L}' = \mathbf{L}$, the corresponding time derivative is called a convective derivative and will be denoted by $\delta\mathbf{T}/\delta t$ or $\overset{\Delta}{\mathbf{T}}$.† Now the rate of change of the metric in the reference frame is given by

$$\frac{d}{dt} g'_{LM} = 2 \frac{\partial x^i}{\partial X'^L} \frac{\partial x^j}{\partial X'^M} D'_{ij} \tag{2-9-11}$$

For the first three types of reference frame $\mathbf{D}'$ is zero. Thus the operations of raising and lowering suffices commutes with the time differentiation in these reference frames. This is not true with the convective derivative and consequently, as noted by Oldroyd (1958), the convective derivative is not usually convenient to use in practice.

GENERAL REFERENCES

Lamb, H. (1932) "Hydrodynamics," 6th ed., Cambridge University Press, London.

Love, A. E. H. (1920) "A Treatise on the Mathematical Theory of Elasticity," Cambridge University Press, London.

Prager, W. (1961) "Introduction to Mechanics of Continua," Ginn, Boston.

Truesdell, C., and W. Noll (1965) The Non-linear Field Theories of Mechanics, in S. Flügge (ed.), "Handbuch der Physik," vol. III/3, Springer-Verlag, Berlin.

Truesdell, C., and R. A. Toupin (1960) The Classical Field Theories, in S. Flügge (ed.), "Handbuch der Physik," vol. III/1, pp. 226–858, Springer-Verlag, Berlin.

† It may be noted that $\mathbf{D} = \overset{\Delta}{\mathbf{e}}$.

REFERENCES

Almansi, E. (1911) *Rend. Lincei.* (*5A*), vol. 20, pp. 705–714.

Hamel, G. (1912) "Elementare Mechanik," Teubner, Leipzig and Berlin. (New York: Springer-Verlag, 1949.)

Oldroyd, J. G. (1958) *Proc. Roy. Soc.* (*London*), vol. A245, pp. 278–297.

Rivlin, R. S., and J. L. Ericksen (1955) *J. Rational Mech. Anal.*, vol. 4, pp. 323–425.

Somigliana, C. (1913) *Accad. Naz. Lincei. Rc.*, vol. 23, pp. 463–490.

Somigliana, C. (1915) *Accad. Naz. Lincei. Rc.*, vol. 24, pp. 655–675.

Volterra, V. (1907) *Arn. Ecole Norm.* (*3*), vol. 24, pp. 401–517.

CHAPTER

THREE

PHYSICAL PRINCIPLES OF CONTINUUM MECHANICS

3-1 CONSERVATION OF MASS

As the measurement of force depends on momentum and this in turn depends on mass it is appropriate to start by considering mass. It is known from experience that the ease with which movement of a body can be changed depends not on its size but on another quantity which is called its mass. Thus, in classical mechanics each material body has a characteristic mass which is a measure of its resistance to change of movement. The mass of a body is thus a measure of the amount of material contained in an arbitrary portion of a body. This concept of mass implies that the mass of a body $\mathscr{B}$ is a nonnegative, time-independent scalar quantity which is also generally independent of the size of the configuration occupied by the arbitrary portion of the body. Consider an arbitrary configuration B of a body $\mathscr{B}$. Let V be a set of points in B occupied by the particles in an arbitrary subset $\mathscr{V}$ of $\mathscr{B}$. Denoting the mass associated with $\mathscr{V}$ by the nonnegative real number M, the mass $M(\mathscr{V})$ possesses the following properties:

1. $M(\mathscr{V}_1 \cup \mathscr{V}_2) = M(\mathscr{V}_1) + M(\mathscr{V}_2)$ for all pairs $\mathscr{V}_1, \mathscr{V}_2$ of disjoint subsets of $\mathscr{B}$.
2. $M(\mathscr{V}) \to 0$ as the volume of V tends to zero.

In continuum physics, mass is regarded as being continuously distributed in space and hence the properties specified by properties 1 and 2 imply the existence of a scalar field ρ, defined on B. Thus,

$$M(\mathscr{V}) = M(V) = \int_V \rho \, dv \tag{3-1-1}$$

where ρ is variously referred to as the mass per unit volume, mass density, or simply as the density of the material of which the body $\mathscr{B}$ is composed.

With regard to the motion of a material body, the mass of a body $\mathscr{B}$ possesses one other property:

3. The mass $M(\mathscr{R})$ of the material $\mathscr{R}$ occupying an arbitrary material region R_t at time t in the current configuration B_t is independent of the motion of the body $\mathscr{B}$.

The time independence of the mass $M(\mathscr{R})$ gives, by way of Eq. (3-1-1), the equation of mass balance

$$\frac{d}{dt}\int_{R_t} \rho\, dv = 0 \tag{3-1-2}$$

which can be transformed, by way of the transport relation of Eq. (2-5-12), into the relation

$$\int_{R_t} \left(\frac{D\rho}{Dt} + \rho \operatorname{div} \boldsymbol{v}\right) dv = 0 \tag{3-1-3}$$

and hence, since Eq. (3-1-3) holds for all arbitrary material regions R_t, however small, it follows that

$$\frac{D\rho}{Dt} + \rho \operatorname{div} \boldsymbol{v} = 0 \qquad \frac{D\rho}{Dt} + \rho v^i_{,i} = 0 \tag{3-1-4}$$

which is the field equation form of the conservation of mass. From Eq. (2-1-21),

$$\frac{D\rho}{Dt} = \frac{\partial \rho}{\partial t} + (\operatorname{grad} \rho)\cdot \boldsymbol{v}$$

which can be used to rearrange Eq. (3-1-4) into the form

$$\frac{\partial \rho}{\partial t} + \operatorname{div}(\rho \boldsymbol{v}) = 0 \qquad \frac{\partial \rho}{\partial t} + (\rho v^i)_{,i} = 0 \tag{3-1-5}$$

Equations (3-1-4) and (3-1-5) are generally referred to as the equations of continuity.

The balance law of Eq. (3-1-2) and the field equation (3-1-4) are equivalent expressions of the principle of mass conservation. It is evident from Eq. (2-5-8) that the conservation of mass can be expressed in the form

$$\rho_r = J\rho \tag{3-1-6}$$

where ρ_r and ρ are the mass densities in the reference and current configurations respectively.

That J must be positive can be seen from the following physical considerations. For the condition that there are no empty spaces in R_r, the initial density ρ_r can be taken to be positive everywhere in R_r. It follows from Eq. (3-1-6) that, for this condition, J cannot vanish during the motion. Now if no displacement takes place, then $J = 1$. Thus, coupled with the requirement that the deformation be continuously possible, and since J cannot pass through the excluded zero value, it is concluded that J must be positive throughout the motion. A motion in which $\rho = \rho_r$ is called an isochoric, or volume-preserving, motion, while if the density is uniform in both space and time it is called a homochoric motion.

3-2 MOMENTUM, FORCE, AND TORQUE

The first precise formulation of the important concept of force was set out by Newton (1687) in the *Principia* in the form of the laws of motion. These laws laid down a prescription for the measurement of the force exerted on a particle in terms of the rate of change of momentum of that particle. The extension of these ideas to the continuum was finally achieved by Euler (1775).

Throughout this chapter, C_t, S_t, and R_t denote in turn a material curve, a material surface, and a material region in the configuration B_t assumed at time t by a moving material body $\mathscr{B}$.

The linear momentum $\boldsymbol{p}(R_t)$ of the material occupying R_t is defined by the relation

$$\boldsymbol{p}(R_t) = \int_{R_t} \rho \boldsymbol{v}\, dv \tag{3-2-1}$$

where $\boldsymbol{v}$ is the velocity of the material particle at the point $\boldsymbol{x}$.

Another important quantity is the angular momentum $\boldsymbol{H}(R_t)$ which is defined by the relation

$$\boldsymbol{H}(R_t) = \int_{R_t} (\boldsymbol{x} \times \boldsymbol{v})\rho\, dv \qquad H_i = \int_{R_t} \varepsilon_{ijk} x^j v^k \rho\, dv \tag{3-2-2}$$

where x is the position of a representative point of R_t relative to an origin o, $\boldsymbol{H}(R_t)$ being the angular momentum with respect to o.

Euler's equations of motion can now be stated in the form of two principles:

1. The rate of change of linear momentum $\boldsymbol{p}$ is equal to the total applied force $\boldsymbol{F}$ acting on the body; i.e.,

$$\dot{\boldsymbol{p}} = \boldsymbol{F} \tag{3-2-3}$$

2. The rate of change of angular momentum $\boldsymbol{H}$ is equal to the total applied torque $\boldsymbol{\Gamma}$; i.e.,

$$\boldsymbol{H} = \boldsymbol{\Gamma} \tag{3-2-4}$$

Two types of external force are assumed to act on a body $\mathscr{B}$: (1) body forces, such as gravity, which act on every material element throughout the body and can be described by a vector field $\boldsymbol{f}$ which is referred to as the body force per unit mass and which is defined on the configurations of $\mathscr{B}$; (2) surface forces, such as friction, which act on the surface elements of area and can be described by a vector $\boldsymbol{t}$ which is referred to as the surface traction per unit area and which is defined on the surface S of the material occupying that part of B_t which is acted upon by the surface traction.

The total force $\boldsymbol{F}$ is defined as

$$\boldsymbol{F} = \int_{R_t} \rho \boldsymbol{f} \, dv + \int_{\partial R_t} \boldsymbol{t} \, da \tag{3-2-5}$$

where the second integral on the right-hand side of Eq. (3-2-5) represents the contribution to $\boldsymbol{F}$ of the contact force acting on the boundary ∂R_t of the arbitrary material region R_t.

The total torque $\boldsymbol{\Gamma}$ about the origin o is defined as

$$\boldsymbol{\Gamma} = \int_{R_t} \rho(\boldsymbol{x} \times \boldsymbol{f}) \, dv + \int_{\partial R_t} (\boldsymbol{x} \times \boldsymbol{t}) \, da \tag{3-2-6}$$

where the second integral on the right-hand side of Eq. (3-2-6) represents the contribution to $\boldsymbol{\Gamma}$ of the contact forces acting on the boundary ∂R_t of the arbitrary material region R_t.

If the origin o is moved a distance $\boldsymbol{d}$ then the torque $\boldsymbol{\Gamma}'$ about the new origin o' can be expressed, by way of Eqs. (3-2-5) and (3-2-6), in the form

$$\begin{aligned} \boldsymbol{\Gamma}' &= \int_{R_t} (\boldsymbol{x} - \boldsymbol{d}) \times \boldsymbol{f} \rho \, dv + \int_{\partial R_t} (\boldsymbol{x} - \boldsymbol{d}) \times \boldsymbol{t} \, da \\ &= \boldsymbol{\Gamma} - \boldsymbol{d} \times \boldsymbol{F} \end{aligned} \tag{3-2-7}$$

The angular momentum will also change in such a change of origin to $\boldsymbol{H}'$ where

$$\boldsymbol{H}' = \int_{R_t} (\boldsymbol{x} - \boldsymbol{d}) \times \boldsymbol{v} \rho \, dv = \boldsymbol{H} - \boldsymbol{d} \times \boldsymbol{p} \tag{3-2-8}$$

Using Eq. (3-2-4) allows Eqs. (3-2-7) and (3-2-8) to be rearranged to give, by way of Eq. (3-2-3), the relation

$$\boldsymbol{\Gamma}' = \dot{\boldsymbol{H}}' - \boldsymbol{d} \times (\boldsymbol{F} - \dot{\boldsymbol{p}}) = \dot{\boldsymbol{H}}' \tag{3-2-9}$$

from which it is evident that if the principles 1 and 2 hold for one choice of origin they also hold for any other choice of origin, provided it is at rest relative to the first origin.

3-3 THE STRESS VECTOR

The cohesion of a body and the transmission throughout the body of the effects of forces acting on its surface depends on the interaction of the atoms and molecules of which the material is composed. Each atom interacts with every other atom of the material, but the strength of the interaction usually decays rapidly with increasing separation between them and consequently an atom or molecule only interacts strongly with its nearest neighbors. Thus any small volume element of a material is influenced by the rest of the material primarily through forces acting on its surface.

In 1822 Cauchy formulated the concept that the action of the rest of the material upon any volume element of it was of the same form as distributed surface forces. As has been pointed out, this is an approximation, albeit often a very good one. If the more detailed response is to be taken into account it is necessary to introduce, in addition to the surface forces, surface couples and possibly multipolar forces. However, returning to Cauchy's concept, this can be stated more mathematically in the form: the effect of the material on one side of a surface S, on the material on the other side, is equivalent to a distribution of force vectors $\boldsymbol{t}$ per unit area. The force vector per unit area $\boldsymbol{t}$ is also called the surface traction or stress vector.

It is important to note that the stress vector $\boldsymbol{t}$ at a point P is associated with an element of surface da having a unit normal $\boldsymbol{\nu}$. Thus, at each point in space there is a doubly infinite set of stress vectors corresponding to the different possible directions of the surface element da. When the dependence of the stress vector on direction is to be emphasized it will be written in the form $\boldsymbol{t}(\boldsymbol{\nu})$.

Consider the balance of linear momentum expressed in the form

$$\boldsymbol{t}(\boldsymbol{\nu})\,da+\boldsymbol{t}(-\boldsymbol{\nu})\,da = h(da\,\rho\ddot{\boldsymbol{x}} - \text{peripheral forces}) \tag{3-3-1}$$

for the disk-shaped body shown in Fig. 3-1. Letting h tend to zero in Eq. (3-3-1) gives

$$\boldsymbol{t}(\boldsymbol{\nu}) = -\boldsymbol{t}(-\boldsymbol{\nu}) \tag{3-3-2}$$

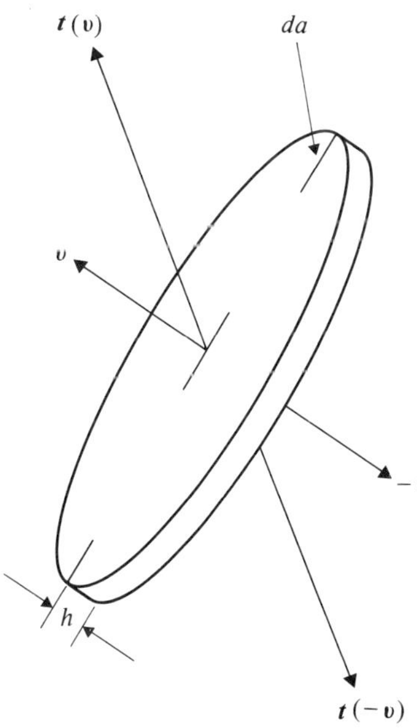

Figure 3-1 Surface tractions on a disk-shaped body.

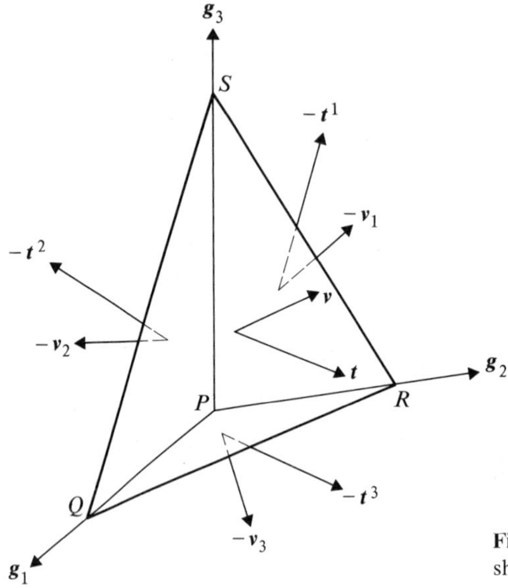

Figure 3-2 Forces acting on a small tetrahedron-shaped element.

This result has the same form as Newton's third law, namely, that action and reaction are equal and opposite.

Consider now a small tetrahedron as shown in Fig. 3-2. Let the sides of the tetrahedron be taken along the coordinate lines through the point P as shown with the vectorial elements of length being given by $\boldsymbol{PQ} = \boldsymbol{g}_1\,dx^1$, $\boldsymbol{PR} = \boldsymbol{g}_2\,dx^2$, $\boldsymbol{PS} = \boldsymbol{g}_3\,dx^3$, where $\boldsymbol{g}_i$ are the covariant base vectors.

Let the unit outward normals to the faces PRS, PSQ, PQR, and QRS be $-\boldsymbol{v}_1$, $-\boldsymbol{v}_2$, $-\boldsymbol{v}_3$, and $\boldsymbol{v}$, the tractions on the faces be $-\boldsymbol{t}^1$, $-\boldsymbol{t}^2$, $-\boldsymbol{t}^3$, and $\boldsymbol{t}$, and the areas of the faces be da_1, da_2, da_3, and da respectively. Now

$$\boldsymbol{PR}\times\boldsymbol{PS} = \boldsymbol{v}_1\,da_1 \qquad \boldsymbol{PS}\times\boldsymbol{PQ} = \boldsymbol{v}_2\,da_2$$
$$\boldsymbol{PQ}\times\boldsymbol{PR} = \boldsymbol{v}_3\,da_3 \qquad \boldsymbol{QR}\times\boldsymbol{QS} = \boldsymbol{v}\,da$$

Thus

$$\boldsymbol{v}\,da = (\boldsymbol{PR}-\boldsymbol{PQ})\times(\boldsymbol{PS}-\boldsymbol{PQ}) = \boldsymbol{v}_1\,da_1 + \boldsymbol{v}_2\,da_2 + \boldsymbol{v}_3\,da_3 \tag{3-3-3}$$

Now the contravariant base vectors $\boldsymbol{g}^i$ are normal to the coordinate planes. Thus the unit vectors are given by

$$\boldsymbol{v}_1 = \frac{\boldsymbol{g}^1}{\sqrt{g^{11}}} \qquad \boldsymbol{v}_2 = \frac{\boldsymbol{g}^2}{\sqrt{g^{22}}} \qquad \boldsymbol{v}_3 = \frac{\boldsymbol{g}^3}{\sqrt{g^{33}}}$$

where g_{ij} is the metric tensor. Substituting in Eq. (3-3-3) gives

$$\boldsymbol{v}\,da = \sum_{i=1}^{3}\left(\frac{da_i}{\sqrt{g^{ii}}}\right)\boldsymbol{g}^i \tag{3-3-4}$$

But resolving $\boldsymbol{v}$ with respect to the contravariant base vectors gives

$$\mathbf{v} = v_i \mathbf{g}^i \tag{3-3-5}$$

Thus

$$da_i = v_i \sqrt{g^{ii}}\, da \qquad (i \text{ not summed}) \tag{3-3-6}$$

Now the rate of change of linear momentum for the tetrahedron is given by

$$\mathbf{t}\, da - \mathbf{t}^i\, da_i = \rho \dot{\boldsymbol{v}} h\, da \tag{3-3-7}$$

where $\boldsymbol{v}$ is the velocity of the tetrahedron, ρ its density, and h the perpendicular distance from P to the surface QRS. Letting all the dimensions of the tetrahedron decrease in proportion, the right-hand side of Eq. (3-3-7) tends to zero faster than the left-hand side; using Eq. (3-3-6) to give da_i, it follows that in the limit Eq. (3-3-7) gives

$$\mathbf{t} = \sum_{i=1}^{3} \mathbf{t}^i v_i \sqrt{g^{ii}} \tag{3-3-8}$$

Having regard to the relations

$$\mathbf{t}^i \sqrt{g^{ii}} = T^{ij} \mathbf{g}_j \qquad \mathbf{t} = t^j \mathbf{g}_j \tag{3-3-9}$$

it is evident that Eq. (3-3-8) implies that $\mathbf{t}$ must be a linear transformation of $\mathbf{v}$; i.e.,

$$\mathbf{t} = \mathbf{T}^T \mathbf{v} \tag{3-3-10}$$

which relates the stress vector to the unit normal $\mathbf{v}$, it being noted that $\mathbf{T}$ is defined on the configurations of the material body and in particular does not depend upon $\mathbf{v}$.

3-4 BALANCE OF LINEAR MOMENTUM

Equation (3-2-5) can be expressed by way of Eqs. (3-2-1) and (3-2-3) in the form

$$\frac{d}{dt} \int_{R_t} \rho \boldsymbol{v}\, dv = \int_{R_t} \rho \boldsymbol{f}\, dv + \int_{\partial R_t} \mathbf{t}\, da \tag{3-4-1}$$

which is just the integral form of the equation of linear momentum balance. For this case it is evident that the body force $\boldsymbol{f}$ acts as an internal source of linear momentum while the surface traction promotes a flux of momentum across the surface on which it acts.

Substituting from Eq. (3-3-10) for $\mathbf{t}$ in Eq. (3-4-1) and rearranging by way of Eqs. (2-5-15) and (3-1-4) gives

$$\int_{R_t} \rho \left(\frac{D\boldsymbol{v}}{Dt} - \boldsymbol{f} \right) dv = \int_{\partial R_t} \mathbf{T}^T \mathbf{v}\, da \tag{3-4-2}$$

Using the divergence theorem expressed in the form of Eqs. (1-11-30) and (1-11-31), Eq. (3-4-2) can be rearranged into the form

$$\int_{R_t} \left[\rho \left(\frac{D\boldsymbol{v}}{Dt} - \boldsymbol{f} \right) - \operatorname{div} \mathbf{T} \right] dv = \boldsymbol{o} \tag{3-4-3}$$

The condition that Eq. (3-4-3) holds for all arbitrary material regions R_t, however small, leads to the field equation form of the balance of linear momentum:

$$\rho \frac{D\boldsymbol{v}}{Dt} = \operatorname{div}\mathbf{T} + \rho \boldsymbol{f} \qquad \rho \frac{Dv^i}{Dt} = T^{ij}{}_{,j} + \rho f^i \tag{3-4-4}$$

which is Cauchy's equation of motion. The tensor $\mathbf{T}$ is called Cauchy's stress tensor or simply the stress tensor.

Noting that

$$\dot{J} = J \operatorname{div} \boldsymbol{v} \tag{3-4-5}$$

Eq. (3-4-1) gives the alternative form

$$\frac{D}{Dt}(\rho \boldsymbol{v}) + \rho \boldsymbol{v} \operatorname{div} \boldsymbol{v} = \operatorname{div}\mathbf{T} + \rho \boldsymbol{f} \tag{3-4-6}$$

To express Eq. (3-4-4) in terms of the referential description, multiply it by J and use Eq. (3-1-6) to give

$$\rho_r \frac{D\boldsymbol{v}}{Dt} = J \operatorname{div}\mathbf{T} + \rho_r \boldsymbol{f} \tag{3-4-7}$$

where ρ_r is the mass density in the reference configuration.

3-5 BALANCE OF ANGULAR MOMENTUM

Equation (3-2-6) can be expressed, by way of Eqs. (3-2-2) and (3-2-4), in the form

$$\frac{d}{dt}\int_{R_t} \rho(\boldsymbol{x}\times\boldsymbol{v})\,dv = \int_{R_t} \rho(\boldsymbol{x}\times\boldsymbol{f})\,dv + \int_{\partial R_t} (\boldsymbol{x}\times\boldsymbol{t})\,da \tag{3-5-1}$$

which is just the integral form of the equation of angular momentum balance. For this case it is evident that $(\boldsymbol{x}\times\boldsymbol{f})$ is the supply of angular momentum per unit mass and $(\boldsymbol{x}\times\boldsymbol{t})$ is the influx of angular momentum per unit area in B_t.

Substituting from Eq. (3-3-10) for $\boldsymbol{t}$ in Eq. (3-5-1) and rearranging by way of Eqs. (2-5-15) and (3-1-4) gives

$$\int_{R_t} \rho\boldsymbol{x} \times \left(\frac{D\boldsymbol{v}}{Dt} - \boldsymbol{f}\right) dv = \int_{\partial R_t} \boldsymbol{x}\times(\mathbf{T}^T\boldsymbol{v})\,da \tag{3-5-2}$$

From Eq. (3-5-2) there can be obtained by way of the vector identity

$$\boldsymbol{a}\times(\boldsymbol{b}\times\boldsymbol{c}) = (\boldsymbol{b}\otimes\boldsymbol{c} - \boldsymbol{c}\otimes\boldsymbol{b})\boldsymbol{a} \tag{3-5-3}$$

the relation

$$\int_{R_t} \rho\left[\boldsymbol{x}\otimes\left(\frac{D\boldsymbol{v}}{Dt} - \boldsymbol{f}\right) - \left(\frac{D\boldsymbol{v}}{Dt} - \boldsymbol{f}\right)\otimes\boldsymbol{x}\right] dv = \int_{\partial R_t} [\boldsymbol{x}\otimes(\mathbf{T}^T\boldsymbol{v}) - (\mathbf{T}^T\boldsymbol{v})\otimes\boldsymbol{x}]\,da \tag{3-5-4}$$

Using the divergence theorem expressed in the form of Eqs. (1-11-30) and (1-11-31), Eq. (3-5-4) can be rearranged into the form

$$\int_{R_t} \left\{ \boldsymbol{x} \otimes \left[\rho \left(\frac{D\boldsymbol{v}}{Dt} - \boldsymbol{f} \right) - \operatorname{div} \mathbf{T} \right] - \left[\rho \left(\frac{D\boldsymbol{v}}{Dt} - \boldsymbol{f} \right) - \operatorname{div} \mathbf{T} \right] \otimes \boldsymbol{x} + \mathbf{T} - \mathbf{T}^T \right\} dv = \mathbf{O} \qquad (3\text{-}5\text{-}5)$$

The condition that Eq. (3-5-5) holds for all arbitrary material regions R_t, however small, gives, by way of Eq. (3-4-4), the further condition that

$$\mathbf{T} = \mathbf{T}^T \qquad (3\text{-}5\text{-}6a)$$

$$T^{ij} = T^{ji} \qquad (3\text{-}5\text{-}6b)$$

thus establishing the symmetry of the stress matrix in general without recourse to any assumption of equilibrium or of uniformity of the stress distribution.

If, following Dahler and Scriven (1963), allowance is made for a body couple density $\boldsymbol{G}$ and a surface couple density $\boldsymbol{c}(\boldsymbol{v})$ to act, then the rate of change of total angular momentum density $\boldsymbol{h}$ is obtained by extending Euler's laws to give

$$\frac{d}{dt} \int_{R_t} \rho \boldsymbol{h} \, dv = \int_{R_t} (\boldsymbol{x} \times \boldsymbol{f} + \boldsymbol{G}) \rho \, dv + \int_{\partial R_t} [\boldsymbol{x} \times \boldsymbol{t}(\boldsymbol{v}) + \boldsymbol{c}(\boldsymbol{v})] \, da \qquad (3\text{-}5\text{-}7)$$

In Eq. (3-5-7), the couple stress vector $\boldsymbol{c}$ at point x is a linear transformation of the unit normal $\boldsymbol{v}$; thus

$$\boldsymbol{c} = \mathbf{M}^T \boldsymbol{v} \qquad (3\text{-}5\text{-}8)$$

where $\mathbf{M}$ is referred to as the couple stress tensor, the concepts underlying the formulation of Eq. (3-5-8) paralleling those underlying the stress vector.

Substituting in Eq. (3-5-7) from Eqs. (3-3-10) and (3-5-8) for $\boldsymbol{t}$ and $\boldsymbol{c}$ respectively and setting

$$\boldsymbol{h} = \boldsymbol{m} + \boldsymbol{x} \times \boldsymbol{v}$$

allows Eq. (3-5-7) to be rearranged, by way of Eqs. (2-5-15) and (3-1-4), to give

$$\int_{R_t} \rho \boldsymbol{x} \times \left(\frac{D\boldsymbol{v}}{Dt} - \boldsymbol{f} \right) dv + \int_{R_t} \rho \left(\frac{D\boldsymbol{m}}{Dt} - \boldsymbol{G} \right) dv = \int_{\partial R_t} \boldsymbol{x} \times (\mathbf{T}^T \boldsymbol{v}) \, da + \int_{\partial R_t} \mathbf{M}^T \boldsymbol{v} \, da \qquad (3\text{-}5\text{-}9)$$

which is to be compared with Eq. (3-5-2). Equation (3-5-9) can be rearranged in the same way as Eq. (3-5-2), followed by using the divergence theorem, to give

$$\rho \frac{D\boldsymbol{m}}{Dt} = \operatorname{div} \mathbf{M} + \rho \boldsymbol{G} - 2\boldsymbol{\pi} \qquad (3\text{-}5\text{-}10)$$

where $\boldsymbol{\pi}$ is the axial vector of $\mathbf{T} - \mathbf{T}^T$ [see Eq. (1-7-37)], and

$$\boldsymbol{m} = \boldsymbol{h} - \boldsymbol{x} \times \boldsymbol{v} \qquad m^i = h^i - g^{il} \varepsilon_{ljk} x^j v^k \qquad (3\text{-}5\text{-}11)$$

is termed the internal angular momentum. As noted by Fung (1965), the terms **M** and $\boldsymbol{G}$ are usually of electromagnetic origin and in most engineering applications are very much smaller than the stress **T**. Thus, provided the coupling between the moment of linear momentum and the internal angular momentum is weak, the stress tensor will usually be very nearly symmetric.

3-6 PROPERTIES OF THE STRESS TENSOR

Let T^{ij} be the components of the stress tensor at some point x in the current configuration B_t; then it follows from Eqs. (2-1-10), (3-3-8), and (3-3-10) that

$$T^{ij} = \boldsymbol{g}^i \cdot (\mathbf{T}\boldsymbol{g}^j) = \boldsymbol{g}^j \cdot (\mathbf{T}^T \boldsymbol{g}^i) = \sqrt{g^{ii}}\,\boldsymbol{t}^i \cdot \boldsymbol{g}^j \tag{3-6-1}$$

It is evident from Eq. (3-6-1) that T^{ij} can be interpreted as the j component of the force per unit area in B_t acting on a surface segment whose outward normal at x is in the i direction.

The components of the stress tensor **T** do not necessarily have the dimensions of force per unit area. Expressing Eq. (3-3-8) in the form

$$\boldsymbol{t}^i = \sqrt{\frac{g_{jj}}{g^{ii}}}\, T^{ij} \frac{\boldsymbol{g}_j}{\sqrt{g_{jj}}} \qquad (i \text{ and } j \text{ not summed})$$

shows that, as $\boldsymbol{g}_j/\sqrt{g_{jj}}$ is a unit vector, the components, (for orthonormality)

$$T^{\langle ji \rangle} = \sqrt{\frac{g_{jj}}{g^{ii}}}\, T^{ji} \qquad (i \text{ and } j \text{ not summed})$$

have the correct physical dimensions, which are called the physical components of the tensor T^{ji}. Hence, having regard to Eq. (3-5-6b) it follows that

$$T^{\langle ij \rangle} = \sqrt{\frac{g_{ii}}{g^{jj}}}\, T^{ij} \tag{3-6-2}$$

are the physical components of the stress tensor T^{ij}. In a similar way it can be shown that

$$T\langle ij \rangle = \sqrt{\frac{g_{ii}}{g_{jj}}}\, T^i{}_j \tag{3-6-3}$$

It is to be noted that the physical components do not transform in a tensorial manner.

At a given point x, there exists in general three mutually perpendicular principal directions $\boldsymbol{n}_a$ $(a = 1, 2, 3)$, with components $(n_a)^i$ $(i = 1, 2, 3)$ which are the solutions of Eq. (1-10-10) (with $\mathbf{S} \equiv \mathbf{T}$, $\lambda \equiv \sigma$, $\boldsymbol{v} \equiv \boldsymbol{n}$), corresponding to the roots σ_a $(a = 1, 2, 3)$ which are called the *principal stresses*. The stress tensor and the stress vector at x now have the representations

$$\mathbf{T} = \sum_{a=1}^{3} \sigma_a \boldsymbol{n}^a \otimes \boldsymbol{n}_a \qquad t = \sum_{a=1}^{3} \sigma_a v_a \boldsymbol{n}_a \tag{3-6-4}$$

where here $\boldsymbol{v}$ is the unit normal vector of Eq. (3-3-10). Thus the principal stresses act along the normal to the surface elements and the magnitude of $\boldsymbol{t}(\boldsymbol{n}_1)$ is just σ_1, etc. The stress vector can be resolved into a component normal to the surface element $\boldsymbol{t}_{(n)}$ and a component $\boldsymbol{t}_{(s)}$ tangential to the surface element. The component $\boldsymbol{t}_{(n)}$ is called the normal stress and $\boldsymbol{t}_{(s)}$ the shearing stress.

The magnitude $\sigma_{(n)}$ of $\boldsymbol{t}_{(n)}$ is given by

$$\sigma_{(n)} = |\boldsymbol{t}_{(n)}| = \boldsymbol{t} \cdot \boldsymbol{v} = t^i v_i = T^{ji} v_i v_j \tag{3-6-5}$$

If the principal directions are taken as the coordinate axes it follows that they form a rectangular cartesian set and therefore there is no difference between covariant and contravariant components. Also the stress tensor takes a diagonal form, the diagonal elements being the principal stresses. Thus, using Eq. (3-6-5), $\sigma_{(n)}$ can be written in the form

$$\sigma_{(n)} = \sigma_1 v_1^2 + \sigma_2 v_2^2 + \sigma_3 v_3^2 \tag{3-6-6}$$

By its definition the shearing stress $\boldsymbol{t}_{(s)}$ can be written

$$\boldsymbol{t}_{(s)} = \boldsymbol{t} - \boldsymbol{t}_{(n)} \tag{3-6-7}$$

and its magnitude $\tau_{(s)}$ is given by

$$\begin{aligned} \tau_{(s)}^2 &= \boldsymbol{t}_{(s)} \cdot \boldsymbol{t}_{(s)} = t^2 - \sigma_{(n)}^2 \qquad (\boldsymbol{t}_{(s)} \text{ and } \boldsymbol{t}_{(n)} \text{ are perpendicular}) \\ &= \sigma_1^2 v_1^2 + \sigma_2^2 v_2^2 + \sigma_3^2 v_3^2 - (\sigma_1 v_1^2 + \sigma_2 v_2^2 + \sigma_3 v_3^2)^2 \end{aligned} \tag{3-6-8}$$

Noting that

$$v_1^2 + v_2^2 + v_3^2 = 1 \tag{3-6-9}$$

the stationary values of $\tau_{(s)}$ can be found by differentiating Eq. (3-6-8). When this is done it is found that $\tau_{(s)}$ possesses minimum values along the principal directions and maximum values on lines in the plane of two principal directions but bisecting the angle between them. For example,

$$v_1 = 0 \qquad v_2^2 = v_3^2 = \tfrac{1}{2} \tag{3-6-10}$$

Substituting in Eq. (3-6-8),

$$\tau_{(s)} = \tfrac{1}{2}|\sigma_2 - \sigma_3| \tag{3-6-11}$$

The values of $\tau_{(s)}$ satisfying such relations are called the principal shearing stresses. The maximum shearing stress is thus equal to half the difference between the maximum and minimum principal stresses and acts on a plane which makes an angle of 45° with the directions of these principal stresses.

The characteristic equation for $\mathbf{T}$ is derived in precisely the same way as the characteristic equation (2-3-7) for $\mathbf{U}$, and has the form

$$\sigma_a^3 - \sigma_a^2 \mathrm{I}_{\mathbf{T}} + \sigma_a \mathrm{II}_{\mathbf{T}} - \mathrm{III}_{\mathbf{T}} = 0 \qquad \text{for } a = 1, 2, 3 \tag{3-6-12}$$

Taking the coordinate axes as the principal directions it is easily shown that $\mathbf{T}$ takes the form

$$[T_{ij}] = \begin{bmatrix} \sigma_1 & 0 & 0 \\ 0 & \sigma_2 & 0 \\ 0 & 0 & \sigma_3 \end{bmatrix} \tag{3-6-13}$$

and the invariants can be written as

$$\begin{aligned} \mathrm{I_T} &= \sigma_1 + \sigma_2 + \sigma_3 \\ \mathrm{II_T} &= \sigma_1\sigma_2 + \sigma_2\sigma_3 + \sigma_3\sigma_1 \\ \mathrm{III_T} &= \sigma_1\sigma_2\sigma_3 \end{aligned} \tag{3-6-14}$$

Multiplying Eq. (3-6-13) n times gives

$$[T_{ij}]^n = \begin{bmatrix} \sigma_1^n & 0 & 0 \\ 0 & \sigma_2^n & 0 \\ 0 & 0 & \sigma_3^n \end{bmatrix} \tag{3-6-15}$$

and substituting in Eq. (3-6-12) gives the Cayley–Hamilton equation

$$\mathbf{T}^3 - \mathrm{I_T}\mathbf{T}^2 + \mathrm{II_T}\mathbf{T} - \mathrm{III_T}\mathbf{I} = 0 \tag{3-6-16}$$

Repeated application of Eq. (3-6-16) leads to the Lagrange–Sylvester equation for an arbitrary function f of $\mathbf{T}$:

$$f(\mathbf{T}) = f(\sigma_1)\frac{(\mathbf{T}-\sigma_2\mathbf{I})(\mathbf{T}-\sigma_3\mathbf{I})}{(\sigma_1-\sigma_2)(\sigma_1-\sigma_3)} + f(\sigma_2)\frac{(\mathbf{T}-\sigma_3\mathbf{I})(\mathbf{T}-\sigma_1\mathbf{I})}{(\sigma_2-\sigma_3)(\sigma_2-\sigma_1)} + f(\sigma_3)\frac{(\mathbf{T}-\sigma_1\mathbf{I})(\mathbf{T}-\sigma_2\mathbf{I})}{(\sigma_3-\sigma_1)(\sigma_3-\sigma_2)} \tag{3-6-17}$$

The deviator of the stress tensor $\mathbf{T}'$ is defined as

$$\mathbf{T}' = \mathbf{T} - \tfrac{1}{3}(\operatorname{tr}\mathbf{T})\mathbf{I} \tag{3-6-18}$$

The expression $T = \frac{1}{3}\operatorname{tr}\mathbf{T}$ is called the mean normal stress. For a discussion of geometrical representations of the stress tensor the reader is referred to Prager (1961).

3-7 NOMINAL AND PIOLA–KIRCHHOFF STRESS TENSORS

Cauchy's stress tensor, which is sometimes called the eulerian stress tensor, refers the stress to the instantaneous area of the surface element on which the stress vector acts. In conditions of finite strain, however, it is sometimes convenient to define the stress as the force per unit initial area. Such a procedure is used, for instance, to give the "engineering stress" in tensile tests.

Let the force on an element of area da in the present configuration $\chi(t)$ be denoted by dF. This is related to the Cauchy stress tensor by

$$dF = \mathbf{T}^T \nu \, da = \mathbf{T}^T \, d\mathbf{a} \tag{3-7-1}$$

By analogy with Eq. (3-7-1) a stress tensor $\mathbf{T}^0$ can be constructed which refers to the initial area dA by way of the relation

$$dF = (\mathbf{T}^0)^T \, dA \tag{3-7-2}$$

Substituting for da in Eq. (3-7-1) from Eq. (2-5-3) and for dA in the resulting relation from Eq. (3-7-2) gives

$$\mathbf{T}^0 = J\mathbf{F}^{-1}\mathbf{T} \tag{3-7-3}$$

where $\mathbf{T}^0$ is referred to as the nominal stress tensor and by its form it is clear that in general it is not a symmetric tensor. Because of its asymmetry it is difficult to incorporate into constitutive relations relating stress and strain. The first Piola–Kirchhoff stress tensor $\mathbf{T}_R$ is just $(\mathbf{T}^0)^T$ for $\mathbf{T} = \mathbf{T}^T$.

A more useful form is the second Piola–Kirchhoff stress tensor $\tilde{\mathbf{T}}$, defined by the relation

$$\tilde{\mathbf{T}} = J\mathbf{F}^{-1}\mathbf{T}(\mathbf{F}^{-1})^T = \mathbf{F}^{-1}\mathbf{T}_R = \mathbf{T}^0(\mathbf{F}^{-1})^T \tag{3-7-4a}$$

$$\tilde{T}^{\alpha\beta} = JX^{\alpha}{}_{,i}X^{\beta}{}_{,j}T^{ij} \tag{3-7-4b}$$

Interchanging i and j in Eq. (3-7-4b) gives

$$\begin{aligned}\tilde{T}^{\alpha\beta} &= JX^{\alpha}{}_{,j}X^{\beta}{}_{,i}T^{ji} = JX^{\beta}{}_{,i}X^{\alpha}{}_{,j}T^{ij} \\ &= \tilde{T}^{\beta\alpha} \qquad (T^{ij} \text{ symmetric})\end{aligned}$$

from which it follows that if $\mathbf{T}$ is symmetric then so also is $\tilde{\mathbf{T}}$, and thus it is suitable for the formulation of constitutive equations.

3-8 BALANCE OF ENERGY

Multiplying Eq. (3-4-4) by $\boldsymbol{v}$ and noting that

$$\operatorname{div}(\mathbf{T}\boldsymbol{v}) = \boldsymbol{v} \cdot \operatorname{div}\mathbf{T} + \operatorname{tr}(\mathbf{TL})$$

gives

$$\tfrac{1}{2}\rho \frac{D}{Dt}(\boldsymbol{v} \cdot \boldsymbol{v}) = \operatorname{div}(\mathbf{T}\boldsymbol{v}) - \operatorname{tr}(\mathbf{TL}) + \rho\boldsymbol{v} \cdot \boldsymbol{f} \tag{3-8-1}$$

Integrating Eq. (3-8-1) throughout an arbitrary material region R_t and noting that $D(\rho \, dv)/Dt$ is zero gives

$$\frac{d}{dt}K(R_t) = P(R_t) - \int_{R_t} \operatorname{tr}(\mathbf{TL})\, dv \tag{3-8-2}$$

where

$$K(R_t) = \int_{R_t} \tfrac{1}{2}\rho\boldsymbol{v} \cdot \boldsymbol{v}\, dv \tag{3-8-3}$$

defines the kinetic energy of the material occupying R_t and

$$P(R_t) = \int_{R_t} \rho\boldsymbol{v} \cdot \boldsymbol{f}\, dv + \int_{\partial R_t} \boldsymbol{v} \cdot \boldsymbol{t}\, da \tag{3-8-4}$$

is the mechanical power, i.e., the rate that work is being done on the element. If $\mathbf{T}$ is symmetric then tr $(\mathbf{TL})$ can be expressed in the form

$$\operatorname{tr}(\mathbf{TL}) = \operatorname{tr}(\mathbf{TD}) + \operatorname{tr}(\mathbf{TW}) = \operatorname{tr}(\mathbf{TD}) \tag{3-8-5}$$

and this term is called the stress power per unit volume in B_t. Equation (3-8-2) can now be expressed in terms of the stress power to give

$$\frac{d}{dt}K(R_t) = P(R_t) - M(R_t) \tag{3-8-6}$$

where

$$M(R_t) = \int_{R_t} \operatorname{tr}(\mathbf{TD})\, dv \tag{3-8-7}$$

Equation (3-8-6) is called the mechanical energy balance equation.

Energy can have many different forms such as nuclear, chemical, electromagnetic, potential, and thermal. As these forms can all be converted from one into the other it is desirable to have a balance equation to include all forms of energy. Such an equation, which is taken as a fundamental postulate, is called the first law of thermodynamics. The first law of thermodynamics takes the form

$$\frac{d}{dt}[K(R_t) + U(R_t)] = H(R_t) + P(R_t) \tag{3-8-8}$$

where $U(R_t)$ is the internal energy and $H(R_t)$ is the rate at which thermal energy is being fed into the element. The internal energy $U(R_t)$ is an additive set function like mass and therefore has the properties 1 and 2 given in Sec. 3-1. These conditions ensure the existence of a scalar field ε; hence $U(R_t)$ can be expressed in the form

$$U(R_t) = \int_{R_t} \rho\varepsilon\, dv \tag{3-8-9}$$

where ε is now seen to be the internal energy per unit mass. To find the rate at which thermal energy is being added to the material element it is assumed that there is a heat flux h which is the rate of flow of thermal energy per unit area into the material occupying the region R_t across the boundary ∂R_t and a heat supply or source r per unit mass and time. These assumptions give

$$H(R_t) = \int_{R_t} \rho r\, dv + \int_{\partial R_t} h\, da \tag{3-8-10}$$

Substituting in Eq. (3-8-8) for $K(R_t)$, $P(R_t)$, $U(R_t)$, and $H(R_t)$ from Eqs. (3-8-3), (3-8-4), (3-8-9), and (3-8-10) respectively gives, for the equation of total energy balance in integral form, the relation

$$\frac{d}{dt}\int_{R_t} \rho(\varepsilon + \tfrac{1}{2}\boldsymbol{v}\cdot\boldsymbol{v})\, dv = \int_{R_t} \rho(\boldsymbol{v}\cdot\boldsymbol{f} + r)\, dv + \int_{\partial R_t} (\boldsymbol{v}\cdot\boldsymbol{t} + h)\, da \tag{3-8-11}$$

The difference between the mechanical energy balance equation (3-8-6) and the total energy balance equation (3-8-8), i.e.,

$$\frac{d}{dt} U(R_t) = H(R_t) + M(R_t) \tag{3-8-12}$$

is referred to as the equation of thermal energy balance. Application of Eq. (3-8-12) of the thermal energy balance to a tetrahedral subregion of B_t, in a way similar to that described in Sec. 3-3, establishes the existence of a heat flux vector q, which relates to the unit normal $\mathbf{v}$ according to the relation

$$h = \boldsymbol{q} \cdot \mathbf{v} \tag{3-8-13}$$

The heat vector q is directed opposite to the flux of heat. Equation (3-8-13) can be regarded as the formulation of a concept which is as basic to heat flow as the concept of the stress vector is to the theory of stress. Substituting in Eq. (3-8-12) for $M(R_t)$, $U(R_t)$, and $H(R_t)$ from Eqs. (3-8-7), (3-8-9), and (3-8-10), with h replaced in Eq. (3-8-10) from Eq. (3-8-13), gives

$$\frac{d}{dt} \int_{R_t} \rho\varepsilon \, dv = \int_{R_t} [\rho r + \operatorname{tr}(\mathbf{TD})] \, dv + \int_{\partial R_t} \boldsymbol{q} \cdot \mathbf{v} \, da \tag{3-8-14}$$

which is the integral form of the equation of thermal energy balance. Equation (3-8-14) can be rearranged by way of Eqs. (2-5-12), (3-1-4), and the divergence theorem to give

$$\int_{R_t} \left[\rho \frac{D\varepsilon}{Dt} - \rho r - \operatorname{tr}(\mathbf{TD}) - \operatorname{div} \boldsymbol{q} \right] dv = 0 \tag{3-8-15}$$

The condition that Eq. (3-8-15) holds for all arbitrary material regions R_t, however small, leads to the relation

$$\rho\dot{\varepsilon} = \operatorname{tr}(\mathbf{TD}) + \operatorname{div} \boldsymbol{q} + \rho r \tag{3-8-16}$$

which is the field equation form of the balance of thermal energy. Equation (3-8-16) can be expressed in the referential description by way of Eqs. (2-6-11), (3-1-6), and (3-7-4) to give

$$\rho_r\dot{\varepsilon} = \operatorname{tr}(\tilde{\mathbf{T}}\dot{\mathbf{E}}) + \frac{\rho_r}{\rho} \operatorname{div} \boldsymbol{q} + \rho_r r \tag{3-8-17}$$

3-9 ENTROPY

When all the information necessary to characterize a material element, such as the quantity of each chemical substance, the stress, strain, velocity field, and some quantity which is a measure of hotness or coldness, are available it is said that the state of the system is known and the set of characterizing variables are called state variables. If the state variables of a homogeneous system are independent of time the system is said to be in thermodynamic equilibrium.

Two fundamental quantities are now introduced which characterize the thermal state of a system in thermodynamic equilibrium and which must be regarded as undefined concepts; these are absolute temperature θ and entropy S. The temperature θ is taken to be a positive quantity and practical ways of defining temperature have been given by Guggenheim (1949).

Entropy has the two properties:

1. The entropy of a system is the sum of the entropies of its parts (i.e., entropy, like mass, is an additive set function, the set being the material points).
2. The entropy of a system can change either by interacting with its surroundings or by changes within the system.

Thus,

$$\frac{dS}{dt} = \frac{dS_I}{dt} + \frac{dS_E}{dt} \tag{3-9-1}$$

where the term on the left-hand side is the rate of increase of the total entropy, the first term on the right-hand side is the rate of entropy increase due to internal causes, and the second term on the right-hand side is the rate of entropy increase due to interaction with the surroundings. The rate of entropy increase due to interaction with the surroundings is taken to be given by

$$\frac{dS_E}{dt} = \int_{R_t} \rho \frac{r}{\theta} dv + \int_{\partial R_t} \frac{(\mathbf{q} \cdot \mathbf{v})}{\theta} da \tag{3-9-2}$$

If η is the total entropy per unit mass and γ the entropy production per unit mass, then

$$S = \int_{R_t} \rho\eta \, dv \qquad \frac{dS_I}{dt} = \int_{R_t} \rho\gamma \, dv \tag{3-9-3}$$

Substituting in Eq. (3-9-1) from Eqs. (3-9-2) and (3-9-3) gives

$$\frac{d}{dt} \int_{R_t} \rho\eta \, dv = \int_{R_t} \rho\gamma \, dv + \int_{R_t} \rho \frac{r}{\theta} dv + \int_{\partial R_t} \frac{\mathbf{q} \cdot \mathbf{v}}{\theta} da \tag{3-9-4}$$

The second law of thermodynamics states that

$$\frac{dS_I}{dt} \geqslant 0 \qquad \text{and hence } \gamma \geqslant 0 \tag{3-9-5}$$

If dS_I/dt or γ are zero the flow process is said to be reversible, whereas if they are positive the process is said to be irreversible.

Equation (3-9-4) can be rearranged by way of Eqs. (2-5-12) and (3-1-4) to give

$$\int_{R_t} \left(\rho\dot{\eta} - \rho\frac{r}{\theta} - \rho\gamma \right) dv = \int_{\partial R_t} \frac{\mathbf{q} \cdot \mathbf{v}}{\theta} da \tag{3-9-6}$$

which can be further rearranged by way of the divergence theorem to give

$$\int_{R_t} \left(\rho\dot{\eta} - \rho\frac{r}{\theta} - \rho\gamma - \text{div}\frac{\boldsymbol{q}}{\theta} \right) dv = 0 \tag{3-9-7}$$

As the integral must be zero for any arbitrary material region R_t, the integrand must be zero, giving

$$\rho\gamma = \rho\dot{\eta} - \rho\frac{r}{\theta} - \text{div}\frac{\boldsymbol{q}}{\theta} \tag{3-9-8}$$

which is just the field equation form of Eq. (3-9-4). But by the second law of thermodynamics $\gamma \geqslant 0$, and thus

$$\rho\dot{\eta} - \rho\frac{r}{\theta} - \text{div}\frac{\boldsymbol{q}}{\theta} \geqslant 0 \tag{3-9-9}$$

This relation is called the Clausius–Duhem inequality after Clausius (1854) and Duhem (1901). It is to be noted, however, that reservations regarding the generality of the Clausius–Duhem inequality have been expressed by Green and Naghdi (1977).

Eliminating ρr between Eqs. (3-8-16) and (3-9-8) gives

$$\rho\gamma\theta \equiv \rho\theta\dot{\eta} - \rho\dot{\varepsilon} + \text{tr}\,(\mathbf{TD}) + \frac{1}{\theta}\boldsymbol{q}\cdot\text{grad}\,\theta \geqslant 0 \tag{3-9-10}$$

or, alternatively, r can be eliminated between Eqs. (3-8-17) and (3-9-8) to give

$$\rho_r\theta\gamma \equiv \rho_r\theta\dot{\eta} - \rho_r\dot{\varepsilon} + \text{tr}\,(\tilde{\mathbf{T}}\dot{\mathbf{E}}) + \frac{\rho_r}{\rho\theta}\boldsymbol{q}\cdot\text{grad}\,\theta \geqslant 0 \tag{3-9-11}$$

Often it is more convenient to use temperature rather than entropy as the independent variable, and in this case it is convenient to introduce a new thermodynamic potential called the Helmholtz free energy per unit mass ψ or sometimes just the free energy. The free energy ψ is defined by

$$\psi = \varepsilon - \theta\eta \tag{3-9-12}$$

Substituting for ε in Eq. (3-9-10) from Eq. (3-9-12) gives

$$\rho\theta\gamma \equiv -\rho\dot{\psi} - \rho\eta\dot{\theta} + \text{tr}\,(\mathbf{TD}) + \frac{1}{\theta}\boldsymbol{q}\cdot\text{grad}\,\theta \geqslant 0 \tag{3-9-13}$$

If the internal energy ε is a function of η and $\mathbf{E}$ only, that is,

$$\varepsilon = \varepsilon(\eta, \mathbf{E}) \tag{3-9-14}$$

then

$$\dot{\varepsilon} = \left(\frac{\partial\varepsilon}{\partial\eta}\right)_{\mathbf{E}}\dot{\eta} + \left(\frac{\partial\varepsilon}{\partial E_{\alpha\beta}}\right)_{\eta}\dot{E}_{\alpha\beta} \tag{3-9-15}$$

Substituting into Eq. (3-9-11) gives

$$\rho_r\left(\theta - \frac{\partial\varepsilon}{\partial\eta}\right)\dot{\eta} + \left(\tilde{T}^{\alpha\beta} - \rho_r\frac{\partial\varepsilon}{\partial E_{\alpha\beta}}\right)\dot{E}_{\alpha\beta} + \frac{\rho_r}{\rho\theta}q^i\theta_{,i} \geqslant 0 \tag{3-9-16}$$

If all the $E_{\alpha\beta}$ can be taken as independent variables then, for Eq. (3-9-16) to be true for all values of $\dot{\eta}$ and $\dot{\mathbf{E}}$, their coefficients must be zero, giving

$$\theta = \left(\frac{\partial \varepsilon}{\partial \eta}\right)_{\mathbf{E}} \tag{3-9-17}$$

$$\tilde{T}^{\alpha\beta} = \rho_r \left(\frac{\partial \varepsilon}{\partial E_{\alpha\beta}}\right)_{\eta} \tag{3-9-18}$$

Similarly, if the free energy depends only on θ and $\mathbf{E}$ and all the $E_{\alpha\beta}$ are assumed to be independent, then proceeding as above gives

$$\eta = -\left(\frac{\partial \psi}{\partial \theta}\right)_{\mathbf{E}} \tag{3-9-19}$$

$$\tilde{T}^{\alpha\beta} = \rho_r \left(\frac{\partial \psi}{\partial E_{\alpha\beta}}\right)_{\theta} \tag{3-9-20}$$

3-10 THE GENERAL BALANCE EQUATION

Reviewing those physical principles of continuum physics which have been considered so far, it is evident that the physical balance principles all take the form of integral equations. It has been shown that from these balance principles there can be derived the corresponding field equations by way of suitable conditions of continuity. It will be shown in the next section that the relation that must exist across a surface of discontinuity, i.e., the jump condition, can also be derived from the appropriate integral balance equation. Thus the integral balance equations include both the field equations and the jump equations, and hence the integral equation form of the physical balance principles is to be regarded as the fundamental form.

It is evident from Eqs. (3-1-2), (3-4-1), (3-5-1), (3-5-7), (3-8-6), (3-8-11), and (3-9-4) that all the physical balance principles can be accommodated within the general balance equation

$$\frac{d}{dt}\int_{R_t} \rho\psi \, dv = \int_{R_t} \rho s \, dv - \int_{\partial R_t} \boldsymbol{j}\cdot\boldsymbol{v}\, da \tag{3-10-1}$$

where ψ is some physical quantity per unit mass, which may be a scalar, vector, or tensor over the current configuration B_t of a body $\mathscr{B}$, s the supply of ψ per unit mass, and $-\boldsymbol{j}\cdot\boldsymbol{v}$ the influx of ψ per unit area in B_t. The various balance equations can be obtained using the forms of $-\boldsymbol{j}\cdot\boldsymbol{v}$ and s given in Table 3-1.

Taking the material region R_t which corresponds at time t to the configuration of the material region R_r and using Green's theorem gives Eq. (3-10-1) in the form

$$\int_{R_t} \frac{D}{Dt}(\rho\psi\, dv) = \int_{R_t} (\rho s - \operatorname{div}\boldsymbol{j})\, dv \tag{3-10-2}$$

Table 3-1 Particular forms of the flux and source terms appearing in the general balance equation

	ψ	s	$-j \cdot v$
Mass	1	0	0
Linear momentum	v	f	t
Angular momentum	$x \times v$	$x \times f$	$x \times t$
Total angular momentum	$x \times v + m$	$x \times f + G$	$x \times t(v) + c(v)$
Energy	$\varepsilon + \frac{1}{2} v \cdot v$	$v \cdot f + r$	$v \cdot t + h$
Entropy	η	$\gamma + \dfrac{r}{\theta}$	$\dfrac{1}{\theta}(q \cdot v)$

Noting that the conservation of mass requires

$$\frac{D}{Dt}(\rho \, dv) = 0$$

and that Eq. (3-10-2) holds for all material regions R_r, however small, leads to the field equation form of the general balance equation:

$$\rho \frac{D\psi}{Dt} = \rho s - \operatorname{div} j \tag{3-10-3}$$

or, using Eq. (3-1-4),

$$\frac{D}{Dt}(\rho\psi) + \rho\psi \operatorname{div} v = \rho s - \operatorname{div} j \tag{3-10-4}$$

3-11 JUMP CONDITIONS

Usually the displacement, velocity, and the fields representing properties of the material composing a body $\mathscr{B}$ vary smoothly throughout space. However, there may exist surfaces across which the variation occurs so rapidly that they can be regarded as being surfaces of discontinuity. Examples of such singular surfaces are the interfaces within a heterogeneous media, shock waves in gases, bores running up rivers, vortex sheets in fluids separating a main stream from the dead water, and cracks in solids. Moving surfaces of discontinuity are often referred to as waves.

Let $R_t^+, \partial R_t^+$ and $R_t^-, \partial R_t^-$ denote the parts of R_t and ∂R_t situated ahead and to the rear respectively of a singular surface S at time t. These parts of R_t and ∂R_t are shown schematically in Fig. 3-3 where $\hat{S} = S \cap R_t$. Consider Eqs. (2-1-5) and (2-1-6*b*), which are alternative representations of a motion of a body $\mathscr{B}$. Let S

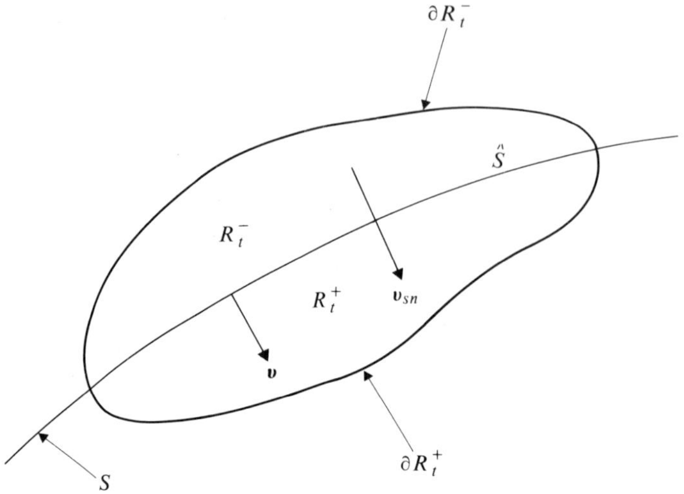

Figure 3-3 A surface of discontinuity S which is the common boundary of two regions R^+ and R^-.

denote an orientable smooth surface of order n ($\geqslant 1$). With regard to the motion of $\mathscr{B}$, the functions χ and χ^{-1}, together with their derivatives of orders $1, 2, \ldots, n-1$ with respect to position and time, are continuous on the singular surface S. The corresponding derivatives of χ and χ^{-1} of order n are almost everywhere discontinuous. Let ψ be a physical variable that is continuous and continuously differentiable throughout R_t^+ and R_t^-. The limit of $\psi(\boldsymbol{x})$ as $\boldsymbol{x}$ tends to a point $\boldsymbol{x}_s$ on S through any curve within the region R_t^+ will be denoted by $\psi^+(\boldsymbol{x}_s)$; $\psi^-(\boldsymbol{x}_s)$ is similarly defined. The jump in ψ across S will be denoted by

$$[\psi] = \psi^+ - \psi^- \tag{3-11-1}$$

The variable ψ may be a scalar, vector, or tensor over the current configuration B_t of a body $\mathscr{B}$. If ψ is a vector $\boldsymbol{a}$, then if $[\boldsymbol{a}]$ is normal to the singular surface S the discontinuity is said to be longitudinal, while if $[\boldsymbol{a}]$ is tangential to S it is said to be transversal.

Associated with each one of the balance laws, obtainable from Eq. (3-10-1) and Table 3-1, is a discontinuity relation which holds at a singular surface. With regard to Fig. 3-3, the regions $\hat{R}^\pm = R_t^\pm \cup \hat{S}$ are not, in general, material with respect to the motion of the body $\mathscr{B}$. However, the $\hat{R}^\pm$ may be regarded as material regions in a fictitious motion in which the velocity is equal to $\boldsymbol{v}$ on the portions $\partial R_t^\pm$ of the boundaries of $\hat{R}^\pm$ and equal to $V_n \boldsymbol{\nu}$ on the common part $\hat{S}$, where $\boldsymbol{\nu}$ is a unit normal vector field on S and V_n is the speed of propagation of S. The transport formulas of Eqs. (2-5-12) and (2-5-15) can be rearranged by way of the divergence theorem and then applied over $\hat{R}^+$ and $\hat{R}^-$ to give the relations

$$\frac{d}{dt}\int_{\hat{R}^+} \rho\psi\, dv = \int_{\hat{R}^+} \frac{\partial}{\partial t}(\rho\psi)\, dv + \int_{\partial R_t^+} \rho\psi \boldsymbol{v}\cdot\boldsymbol{\nu}\, da - \int_{\hat{S}} \rho\psi^+ V_n\, da \tag{3-11-2}$$

$$\frac{d}{dt}\int_{\hat{R}^-} \rho\psi\, dv = \int_{\hat{R}^-} \frac{\partial}{\partial t}(\rho\psi)\, dv + \int_{\partial R_t^-} \rho\psi v \cdot \mathbf{v}\, da + \int_{\hat{S}} \rho\psi^- V_n\, da \qquad (3\text{-}11\text{-}3)$$

it being here understood that the rates of change on the left-hand side of Eqs. (3-11-2) and (3-11-3) relate to the fictitious motion. Addition of these equations gives

$$\frac{d}{dt}\int_{R_t} \rho\psi\, dv = \int_{R_t} \frac{\partial}{\partial t}(\rho\psi)\, dv + \int_{\partial R_t} \rho\psi v \cdot \mathbf{v}\, da - \int_{\hat{S}} [\rho\psi] V_n\, da \qquad (3\text{-}11\text{-}4)$$

which is Thomas' form of the transport theorem (Truesdell and Toupin, 1960). Equations (3-10-1) and (3-11-4) can now be combined to give

$$\int_{R_t}\left(\frac{\partial}{\partial t}(\rho\psi) - \rho s\right) dv + \int_{\partial R_t} (\rho\psi v \cdot \mathbf{v} + \mathbf{j} \cdot \mathbf{v})\, da - \int_{\hat{S}} [\rho\psi] V_n\, da = 0 \qquad (3\text{-}11\text{-}5)$$

which, when followed by the collapse of ∂R_t onto the two sides of $\hat{S}$ together with the assumption that ρ, ψ, and s retain their usual smoothness properties away from S, further gives

$$\int_S \{[\rho\psi(v \cdot \mathbf{v}) + \mathbf{j} \cdot \mathbf{v}] - [\rho\psi] V_n\}\, da = 0 \qquad (3\text{-}11\text{-}6)$$

a relation which must hold for all segments of S formed by the intersection with regular regions of B_t. Subject to the condition that the integrand in Eq. (3-11-6) is continuous on S, the general jump condition is obtained from Eq. (3-11-6) in the form

$$[\rho\psi V - \mathbf{j} \cdot \mathbf{v}] = 0 \qquad (3\text{-}11\text{-}7)$$

where

$$V = V_n - v \cdot \mathbf{v} \qquad (3\text{-}11\text{-}8)$$

is referred to as the local speed of propagation of S and is the speed of travel of the singular surface relative to the material. The basic jump conditions can now be obtained from Eq. (3-11-7) by way of Table 3-1.

The theory of surfaces of discontinuity is almost entirely based upon Hadamard's lemma, which states that, if Ψ is a kinematical variable which is defined and continuously differentiable in the interior of a region R_t^+ with a smooth boundary S, and Ψ and $\Psi_{,i}$ approach finite limits Ψ^+ and $\Psi_{,i}^+$ as S is approached upon paths interior to R_t^+, then

$$\frac{d\Psi^+}{dl} = \Psi_{,i}^+ \frac{dx^i}{dl} \qquad (3\text{-}11\text{-}9)$$

where l is a parameter whose variation maps a smooth curve on the surface S, that is, $\mathbf{x} = \mathbf{x}(l)$ is a curve lying on S. To prove Eq. (3-11-9) it need only be noted that for two neighboring points P and Q as shown in Fig. 3-4 there exists a polygonal path lying entirely in R_t^+ which joins them. To find the total change in Ψ on passing from P to Q the contribution from each arm is added. Similarly, it can be

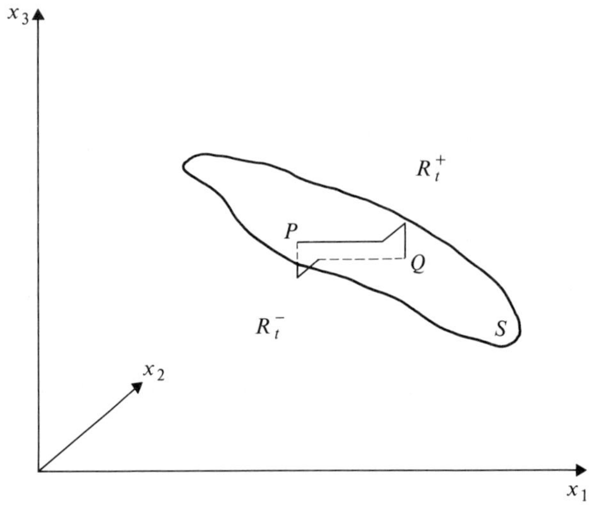

Figure 3-4 Two paths joining neighboring points on the surface S; one path lies entirely in the region R^+ and the other entirely in R^-.

shown that

$$\frac{d\Psi^-}{dl} = \Psi_{,i}^- \frac{dx^i}{dl} \tag{3-11-10}$$

Subtracting Eq. (3-11-10) from Eq. (3-11-9),

$$\frac{d[\Psi]}{dl} = [\Psi_{,i}] \frac{dx^i}{dl} \tag{3-11-11}$$

Thus the tangential derivative of the jump equals the jump of the tangential derivative (Hadamard, 1903).

Let the surface S be defined by the parametric equations

$$x^i = x^i(u^1, u^2) \qquad \text{for } i = 1, 2, 3 \tag{3-11-12}$$

Applying Hadamard's lemma to each coordinate system gives

$$\frac{\partial}{\partial u^a}[\Psi] = [\Psi_{,i}] \frac{\partial x^i}{\partial u^a} \qquad \text{for } a = 1, 2 \tag{3-11-13}$$

These conditions are known as the superficial conditions of compatibility. If Ψ is continuous across S or if $[\Psi]$ is constant on S,

$$[\Psi_{,i}] \frac{\partial x^i}{\partial u^a} = 0 \tag{3-11-14}$$

Multiplying by du^a,

$$[\Psi_{,i}]\, dx^i = 0 \tag{3-11-15}$$

where dx^i lies in the surface S. Thus $[\Psi_{,i}]$ must lie along the unit normal to S which will be denoted by n_i. This then gives Maxwell's theorem:

$$[\Psi_{,i}] = Bn_i \tag{3-11-16}$$

Multiplying by n^i,

$$B = [n^i\Psi_{,i}] \tag{3-11-17}$$

If Ψ stands for a vector $\boldsymbol{c}$ then Maxwell's theorem gives

$$[c_{k,i}] = n_i B_k \tag{3-11-18a}$$

$$B_k = [n^i c_{k,i}] \tag{3-11-18b}$$

Hence

$$[\text{curl}\, \boldsymbol{c}] = \boldsymbol{n} \times \boldsymbol{B} \tag{3-11-18c}$$

$$[\text{div}\, \boldsymbol{c}] = \boldsymbol{n} \cdot \boldsymbol{B} \tag{3-11-18d}$$

From Eq. (3-11-18c),

$$\boldsymbol{n} \times [\text{curl}\, \boldsymbol{c}] = \boldsymbol{n} \times (\boldsymbol{n} \times \boldsymbol{B}) = (\boldsymbol{n} \cdot \boldsymbol{B})\boldsymbol{n} - (\boldsymbol{n} \cdot \boldsymbol{n})\boldsymbol{B}$$

Then, using Eq. (3-11-18d),

$$\boldsymbol{n} \times [\text{curl}\, \boldsymbol{c}] = \boldsymbol{n}[\text{div}\, \boldsymbol{c}] - \boldsymbol{B}$$

Thus

$$\boldsymbol{B} = \boldsymbol{n}[\text{div}\, \boldsymbol{c}] - \boldsymbol{n} \times [\text{curl}\, \boldsymbol{c}] \tag{3-11-19}$$

These results comprise Weingarten's first theorem. The longitudinal and transversal jumps of the gradient of a continuous vector are the jumps of its divergence and curl respectively.

If $[\Psi]$ and $[\Psi_{,i}]$ are both zero, then replacing Ψ in Eq. (3-11-15) by $\Psi_{,k}$ gives

$$\begin{aligned} [\Psi_{,ik}] &= n_i C_k \\ &= n_k C_i \qquad \text{(because } \Psi_{,ik} = \Psi_{,ki}\text{)} \end{aligned} \tag{3-11-20}$$

Thus

$$\frac{C_i}{n_i} = \frac{C_k}{n_k} = C \qquad \text{say}$$

and

$$[\Psi_{,ik}] = n_i n_k C \tag{3-11-21}$$

The method is easily extended to higher orders.

In general such surfaces will be moving with respect to both the material and space. Let the parametric equations of such a surface be

$$x^i = x^i(u^1, u^2, t) \tag{3-11-22}$$

which can also be written

$$X^\alpha = X^\alpha(u^1, u^2, t) \tag{3-11-23}$$

Eliminating u^1 and u^2 from Eqs. (3-11-22) and (3-11-23) leads to the equivalent descriptions of S:

$$S(x^1, x^2, x^3, t) = 0 \tag{3-11-24a}$$

$$\Sigma(X^1, X^2, X^3, t) = 0 \tag{3-11-24b}$$

Differentiation with respect to time at a fixed point $R(u^1, u^2)$ on the surface will be denoted by D_t^S. The velocity of a point of the surface in space is given by

$$v_S^i = D_t^S x^i \tag{3-11-25}$$

Similarly, the intrinsic velocity or the velocity of the surface relative to the material is given by

$$V_S^\alpha = D_t^S X^\alpha \tag{3-11-26}$$

The relative velocity $\boldsymbol{v}_S^*$ in the neighborhood of a point on the surface is defined as

$$\boldsymbol{v}_S^* = \boldsymbol{v}_S - \boldsymbol{v} \tag{3-11-27}$$

Now the rate of change of properties on the moving surface S can be related to the rate of change at a fixed point in space by the relation

$$\begin{aligned} D_t^S &= \frac{\partial}{\partial t_{u\,\text{constant}}} = \frac{\partial}{\partial t_{x\,\text{constant}}} + \frac{\partial x^i}{\partial t_{u\,\text{constant}}} \frac{\partial}{\partial x^i} \\ \text{or} \qquad D_t^S &= \frac{\partial}{\partial t} + v_S^i \frac{\partial}{\partial x^i} \end{aligned} \tag{3-11-28}$$

Similarly, we find

$$D_t^S = \frac{D}{Dt} + V_S^\alpha \frac{\partial}{\partial X^\alpha} \tag{3-11-29}$$

Thus, from Eq. (3-11-25),

$$\begin{aligned} v_S^i &= D_t^S x^i = \frac{Dx^i}{Dt} + V_S^\alpha \frac{\partial x^i}{\partial X^\alpha} \\ v_S^i &= v^i + V_S^\alpha \frac{\partial x^i}{\partial X^\alpha} \qquad \text{or} \qquad \boldsymbol{v}_S^* = V_S \mathbf{F} \end{aligned} \tag{3-11-30}$$

In order that the moving discontinuity should not suddenly alter its character it will be assumed that the motion of the surface in space or in the body cannot be discontinuous. This expresses a condition of persistence.

Applying this condition to Eqs. (3-11-27) and (3-11-30),

$$[v^i] = -[v_S^{*i}] = -V_S^\alpha \left[\frac{\partial x^i}{\partial X^\alpha} \right] \tag{3-11-31}$$

Differentiating Eq. (3-11-24a) at a fixed time t,

$$\frac{\partial S}{\partial x^i} dx^i = 0$$

where the element $d\boldsymbol{x}$ lies in the surface S. Thus $\partial S/\partial x^i$ is normal to S. Letting $\boldsymbol{n}$ be a unit vector normal to S,

$$\frac{\partial S}{\partial x^i} = kn_i \qquad \text{and} \qquad k^2 = \frac{\partial S}{\partial x^i}\frac{\partial S}{\partial x^j}g^{ij} \tag{3-11-32}$$

Similarly,

$$\frac{\partial \Sigma}{\partial X^\alpha} = KN_\alpha \qquad \text{and} \qquad K^2 = \frac{\partial \Sigma}{\partial X^\alpha}\frac{\partial \Sigma}{\partial X^\beta}G^{\alpha\beta}$$

The speed of displacement of the surface is defined by the relation

$$v_{Sn} = \boldsymbol{v}_S \cdot \boldsymbol{n} \tag{3-11-33}$$

and the normal velocity by

$$\boldsymbol{v}_{Sn} = v_{Sn}\boldsymbol{n} \tag{3-11-34}$$

Operating on Eq. (3-11-24*a*) with Eqs. (3-11-28),

$$\frac{\partial S}{\partial t} + v_S^i \frac{\partial S}{\partial x^i} = 0 \tag{3-11-35}$$

Using Eqs. (3-11-32) and (3-11-33),

$$v_{Sn} = -\frac{1}{k}\frac{\partial S}{\partial t} \tag{3-11-36}$$

Similarly, we can define a speed of propagation V_{SN} by

$$V_{SN} = \boldsymbol{V}_S \cdot \boldsymbol{N} \tag{3-11-37}$$

From Eqs. (3-11-24*b*), (3-11-29), and (3-11-37),

$$V_{SN} = -\frac{1}{K}\frac{D\Sigma}{Dt} \tag{3-11-38}$$

The relative speed of propagation is defined as

$$v_{Sn}^* = (\boldsymbol{v}_S - \boldsymbol{v}) \cdot \boldsymbol{n} \tag{3-11-39}$$

From Eq. (3-11-30),

$$v_S^{*i} = V_S^\alpha \frac{\partial x^i}{\partial X^\alpha} \tag{3-11-40}$$

and

$$n_i = \frac{1}{k}\frac{\partial S}{\partial x^i} = \frac{1}{k}\frac{\partial \Sigma}{\partial X^\beta}\frac{\partial X^\beta}{\partial x^i}$$

$$= \frac{K}{k}N_\beta \frac{\partial X^\beta}{\partial x^i} \tag{3-11-41}$$

Substituting in Eqs. (3-11-39) and (3-11-40) gives

$$v_{Sn}^* = \frac{K}{k}V_{SN} \tag{3-11-42}$$

To proceed further, the "order" of the wave must be considered. Zero-order waves are difficult to handle, for each spatial position is occupied by two different material points, and will not be investigated here.

For first-order waves (that is, $[\dot{x}] \neq 0$) two principal types are distinguished. If $[\dot{x}_n] = 0$ the surface is called a vortex sheet while if $[\dot{x}_n] \neq 0$ the surface is called a shock wave. To find the equivalent compatibility conditions assume the surfaces given by Eqs. (3-11-24) are surfaces of discontinuity in a four-dimensional space and that the parameters characterizing the surface are (u^1, u^2, t).

Applying Hadamard's lemma gives the two forms:

Spatial *Referential*

$$\frac{\partial[\Psi]}{\partial u^a} = [\Psi_{,i}]\frac{\partial x^i}{\partial u^a} \qquad \frac{\partial[\Psi]}{\partial u^a} = [\Psi_{,\alpha}]\frac{\partial X^\alpha}{\partial u^a} \tag{3-11-43}$$

$$D_t^S[\Psi] = \left[\frac{\partial \Psi}{\partial t}\right] + v_S^i[\Psi_{,i}] \qquad D_t^S[\Psi] = \left[\frac{D\Psi}{Dt}\right] + V_S^\alpha[\Psi_{,\alpha}] \tag{3-11-44}$$

These relations are called the kinematical conditions of compatibility.

For the important case when Ψ is continuous, that is, $[\Psi] = 0$,

Spatial *Referential*

$$[\Psi_{,i}] = bn_i \qquad b = [n^i\Psi_{,i}] \qquad [\Psi_{,\alpha}] = BN_\alpha \qquad B = [N^\alpha\Psi_{,\alpha}] \tag{3-11-45}$$

$$\left[\frac{\partial \Psi}{\partial t}\right] = -bn_i v_S^i = -bv_{Sn} \qquad \left[\frac{D\Psi}{Dt}\right] = -BV_{SN} \tag{3-11-46}$$

Putting $\Psi = x^i$ in the referential representation gives

$$[\dot{x}^i] = -V_{SN}\left[N^\alpha \frac{\partial x^i}{\partial X^\alpha}\right] \tag{3-11-47}$$

For second-order waves Ψ is continuous and $[\Psi_{,i}]$ and $[\partial\Psi/\partial t]$ or $[\Psi_{,\alpha}]$ and $[D\Psi/Dt]$ are zero. Proceeding as for the static case (i.e., replacing Ψ by $\Psi_{,i}$ or $\partial\Psi/\partial t$, etc.) gives

Spatial *Referential*

$$[\Psi_{,ij}] = n_i n_j c \qquad [\Psi_{,\alpha\beta}] = N_\alpha N_\beta C \tag{3-11-48}$$

$$\left[\frac{\partial}{\partial t}(\Psi_{,i})\right] = -n_i v_{Sn} c \qquad \left[\frac{D}{Dt}(\Psi_{,\alpha})\right] = -N_\alpha V_{SN} C \tag{3-11-49}$$

$$\left[\frac{\partial^2 \Psi}{\partial t^2}\right] = cv_{Sn}^2 \qquad \left[\frac{D^2\Psi}{Dt^2}\right] = CV_{SN}^2 \tag{3-11-50}$$

Putting $\Psi = x^i$ in the referential representation gives the compatibility conditions for acceleration waves in the form

$$[\ddot{x}^i] = -V_{SN}\left[N^\alpha \frac{\partial \dot{x}^i}{\partial X^\alpha}\right] \tag{3-11-51}$$

$$[\ddot{x}^i] = C^i V_{SN}^2 \qquad C^i = N^\alpha N^\beta [x^i{}_{,\alpha\beta}] \tag{3-11-52}$$

The amplitude s^i of the acceleration wave is defined by

$$s^i = \frac{V_{SN}^2}{v_{Sn}^{*2}} C^i = \frac{k^2}{K^2} C^i \tag{3-11-53}$$

Thus

$$[\ddot{x}^i] = s^i v_{Sn}^{*2} \tag{3-11-54}$$

From Eq. (3-11-49),

$$\left[\frac{\partial \dot{x}^i}{\partial X^\alpha}\right] = -N_\alpha V_{SN} C^i$$

Using Eqs. (3-11-41), (3-11-42), and (3-11-53),

$$\left[\frac{\partial \dot{x}^i}{\partial X^\alpha}\right] = -v_{Sn}^* s^i n_j \frac{\partial x^j}{\partial X^\alpha} \tag{3-11-55}$$

Choosing $\boldsymbol{X}$ so that it instantaneously coincides with $\boldsymbol{x}$ gives

$$\left[\frac{\partial \dot{x}^i}{\partial x^j}\right] = -v_{Sn}^* s^i n_j \tag{3-11-56}$$

Hence

$$[\operatorname{div} \dot{\boldsymbol{x}}] = -v_{Sn}^* \boldsymbol{s} \cdot \boldsymbol{n} \qquad \text{and} \qquad [\operatorname{curl} \dot{\boldsymbol{x}}] = -v_{Sn}^* \boldsymbol{s} \times \boldsymbol{n} \tag{3-11-57}$$

GENERAL REFERENCES

Fung, Y. C. (1965) "Foundations of Solid Mechanics," Prentice-Hall, Englewood Cliffs, N.J.

Guggenheim, E. A. (1949) "Thermodynamics—An Advanced Treatise for Chemists and Physicists," Interscience, New York.

Prager, W. (1961) "Introduction to Mechanics of Continua," Ginn, New York.

Truesdell, C., and R. A. Toupin (1960) The Classical Field Theories, in S. Flügge (ed.), "Handbuch der Physik," vol. III/1, pp. 226–858, Springer-Verlag, Berlin.

REFERENCES

Dahler, J. S., and L. E. Scriven (1963) *Proc. Roy. Soc.* (*London*), vol. A275, pp. 504–527.

Duhem, P. (1901) *C.R. Acad. Sci.* (*Paris*), vol. 132, pp. 117–120.

Green, A. E., and P. M. Naghdi (1977) *Proc. Roy. Soc.* (*London*), vol. A357, pp. 253–270.

Hadamard, J. (1903) "Leçons sur la Propagation des Ondes et les Équations de l'Hydrodynamique," Hermann, Paris.

CHAPTER

FOUR

ELEMENTS OF STATISTICAL MECHANICS

4-1 STATISTICAL THERMODYNAMICS

If a physical system consists of a very large number of components, as with the atoms and molecules in a solid, liquid, or gas, it is not possible to determine the motion of each component and the information, even if it were available, would be too detailed to be of much interest. Instead, statistical methods are introduced to relate the macroscopic properties of the system to the properties of its components. Statistical physics (also called statistical mechanics) is concerned primarily with systems in thermodynamic equilibrium. Its scope is too broad to do more than extract a few results which will be of use in this text. The following books give a good presentation of the subject: Fowler (1936), Tolman (1938), Fowler and Guggenheim (1939), Schrödinger (1948), Rushbrooke (1949), ter Haar (1954), Landau and Lifshitz (1958), Rice (1967), Yvon (1969), and Mayer and Mayer (1977). The present outline follows the exposition of Landau and Lifshitz.

In equilibrium a finite system possesses a set of stationary quantum states which each have an associated energy. If several states have the same energy they are said to be degenerate. Now it can be shown that the number of levels in a given finite interval of the energy spectrum increases exponentially with the number of particles, and thus a macroscopic body which contains many particles has a very high density of states over most of the energy spectrum.

The following set of important results from statistical mechanics, which are of relevance to this book, will be stated without proof. The derivation of these relations can be found in the literature cited above.

Let w_n be the probability that a system is in a state n which has an associated energy U_n; w_n then satisfies the normalization condition

$$\sum_n w_n = 1 \tag{4-1-1}$$

and the entropy of the system S is given by

$$S = -k \sum_n w_n \ln w_n \tag{4-1-2}$$

where k is Boltzmann's constant ($1.38 \times 10^{-23}\,\mathrm{J\,K^{-1}}$). If the system consists of several noninteracting subsystems denoted by subscripts a, then

$$S = \sum_a S_a \tag{4-1-3}$$

i.e., the entropy of the compound system is the sum of the entropies of its parts.

Omitting for simplicity the index distinguishing a given subsystem, it can then be shown that w_n depends only on U_n, and in fact

$$w_n = C \exp\left(-\frac{U_n}{k\theta}\right) \tag{4-1-4}$$

where θ is the absolute temperature. Equation (4-1-4) is called the Gibbs distribution or canonical distribution. Taking the logarithm of Eq. (4-1-4) and substituting into Eq. (4-1-2) gives

$$S = -k \sum_n w_n \left(\ln C - \frac{U_n}{k\theta}\right) \tag{4-1-5}$$

Now, using Eq. (4-1-1) and noting that the mean value of the energy is given by

$$U = \sum_n w_n U_n \tag{4-1-6}$$

we find that

$$S = -k \ln C + \frac{U}{\theta} \tag{4-1-7}$$

The Helmholtz free energy A is given by $A = U - \theta S$ and thus

$$C = \exp\left(\frac{A}{k\theta}\right) \tag{4-1-8}$$

and the Gibbs distribution can be written

$$w_n = \exp\left(\frac{A - U_n}{k\theta}\right) \tag{4-1-9}$$

Using the normalization condition (4-1-1) and taking logarithms,

$$A = -k\theta \ln \sum_n \exp\left(\frac{-U_n}{k\theta}\right) \tag{4-1-10}$$

The quantity

$$Z = \sum_n \exp\left(\frac{-U_n}{k\theta}\right) \tag{4-1-11}$$

is called the *partition function.*

If the system consists of noninteracting subsystems then the energy of the whole system is the sum of the energies of its parts:

$$U = \sum_a U_a \tag{4-1-12}$$

Then, using Eq. (4-1-3) and the definition of free energy $A = U - \theta S$,

$$A = \sum_a A_a \tag{4-1-13}$$

Up to now it has been assumed that the number of particles in the system is constant. If the number of particles can vary the Gibbs distribution takes a different form. Let the probability that the body contains N particles and is at the same time in the nth quantum state be denoted by w_{nN}; then

$$w_{nN} = C\exp\left(\frac{\mu N - U_{nN}}{k\theta}\right) \tag{4-1-14}$$

where μ is called the chemical potential. Proceeding as above,

$$k\theta \ln C = U - \theta S - \mu N \tag{4-1-15}$$

Let

$$U - \theta S - \mu N = \Omega \tag{4-1-16}$$

be a generalized form of the Helmholtz free energy; then

$$w_{nN} = \exp\left(\frac{\Omega + \mu N - U_{nN}}{k\theta}\right) \tag{4-1-17}$$

From the normalization condition

$$\sum_N \sum_n w_{nN} = \exp\left(\frac{\Omega}{k\theta}\right) \sum_N \left[\exp\left(\frac{\mu N}{k\theta}\right) \sum_n \exp\left(-\frac{U_{nN}}{k\theta}\right)\right] = 1 \tag{4-1-18}$$

Hence

$$\Omega = -k\theta \ln \sum_N \left[\exp\left(\frac{\mu N}{k\theta}\right) \sum_n \exp\left(-\frac{U_{nN}}{k\theta}\right)\right] \tag{4-1-19}$$

Next consider the case of a perfect gas, i.e., one in which the interaction between the constituent particles is so weak that it can be neglected. The energy levels of the gas as a whole, U_n, now depend only on the energy levels e_i of each individual particle. If the gas is sufficiently rarefied that no more than one particle is in each quantum state then any exclusion due to quantum mechanical exchange will have no effect and can be ignored. Applying the Gibbs distribution to the gas molecules, the probability that a molecule is in the ith state and hence the average number of molecules $\bar{n}_i$ in this state is proportional to $\exp(-e_i/k\theta)$:

$$\bar{n}_i = a \exp\left(-\frac{e_i}{k\theta}\right) \tag{4-1-20}$$

where a is determined by the normalization condition

$$\sum_i \bar{n}_i = N \tag{4-1-21}$$

and N is the total number of molecules in the gas. The distribution given by Eq. (4-1-20) is called the Boltzmann distribution.

Taking the general form of the canonical distribution as given in Eq. (4-1-17) with $U_{nN} = n_i e_i$ and $N = n_i$ and denoting the Ω appertaining to the state i by Ω_i,

$$w_{ni} = \exp\left(\Omega_i + n_i \frac{\mu - e_i}{k\theta}\right) \tag{4-1-22}$$

Now $w_{oi} = \exp(\Omega_i/k\theta)$ is the probability of finding no particle in state i. As this must be near to unity, Ω_i must be nearly zero. To the same approximation, w_{ni} is zero for $n > 1$; thus

$$\bar{n}_i = \Sigma\, w_{ni} n = \exp\left(\frac{\mu - e_i}{k\theta}\right) \tag{4-1-23}$$

If, for a given density, the temperature is sufficiently low, then quantum effects become important. Equation (4-1-18) still applies, however, and so for particles in the ith state

$$\Omega_i = -k\theta \ln \sum_{n_i} \left[\exp n_i \left(\frac{\mu - e_i}{k\theta}\right)\right] \tag{4-1-24}$$

Now if the type of wave function describing the particles of the gas is antisymmetric, then no more than one particle can be in each quantum state and thus

$$\Omega_i = \quad k\theta \ln\left[1 + \exp\left(\frac{\mu - e_i}{k\theta}\right)\right] \tag{4-1-25}$$

The mean number of particles in the ith quantum state is given by

$$\bar{n}_i = -\frac{\partial \Omega_i}{\partial \mu} = \frac{1}{1 + \exp[(e_i - \mu)/k\theta]} \tag{4-1-26}$$

Statistics based on this distribution are called *Fermi–Dirac statistics*. The normalization condition

$$N = \sum_i \bar{n}_i = \sum_i \frac{1}{1 + \exp[(e_i - \mu)/k\theta]} \tag{4-1-27}$$

determines μ as a function of N and θ.

If the wave function describing the particles of the gas is symmetric the corresponding statistics is called *Bose–Einstein statistics*. As before, the particles in the ith state satisfy

$$\Omega_i = -k\theta \ln \sum_{n_i=0}^{\infty} \left(\exp n_i \frac{\mu - e_i}{k\theta} \right) \tag{4-1-28}$$

but now n_i can take any positive value. Performing the summation,

$$\Omega_i = k\theta \ln \left[1 - \exp\left(\frac{\mu - e_i}{k\theta} \right) \right] \tag{4-1-29}$$

and the mean occupation number $\bar{n}_i$ is given by

$$\bar{n}_i = -\frac{\partial \Omega_i}{\partial \mu} = \frac{1}{\exp[(e_i - \mu)/k\theta]} - 1 \tag{4-1-30}$$

The normalization condition gives

$$N = \sum_i \bar{n}_i = \sum_i \frac{1}{\exp[(e_i - \mu)/k\theta]} - 1 \tag{4-1-31}$$

and again determines μ as a function of N and θ. The degeneracy condition for a set of particles of mass m can be shown to be

$$k\theta \ll \left(\frac{\hbar^2}{m} \right) \left(\frac{N}{V} \right)^{2/3} \tag{4-1-32}$$

where $\hbar = h/2\pi$, h is Planck's constant (6.62×10^{-34} J s), and V is the volume of the N particles. If this condition holds quantum effects will be dominant.

From the above theory it can be seen that if all the energy levels available to the elements of the microstructure are known, then the thermodynamic potentials of the system can be determined when it is in thermal equilibrium. In this way the macroscopic thermal response of the system is related to the energetic characteristics of its microstructure.

Finally, if several chemical species $C_1, C_2, \ldots$ react according to the chemical equation

$$\sum_a \nu_a C_a = 0 \tag{4-1-33}$$

and if the chemical potentials of the species are $\mu_1, \mu_2, \ldots,$ if their partition functions per unit volume are $z_1, z_2, \ldots,$ and if their molecular number densities per unit volume are $n_1, n_2, \ldots,$ then

1. The equation of chemical equilibrium is

$$\sum_a \nu_a \mu_a = 0 \tag{4-1-34}$$

2. The law of mass action is

$$\prod_a n_a{}^{\nu_a} = \prod_a z_a{}^{\nu_a} = K_C \tag{4-1-35}$$

where K_C is called the chemical equilibrium constant.

4-2 MECHANICS OF SYSTEMS OF PARTICLES

The motion of a free particle is completely unrestricted, the particle being said to have three degrees of freedom since its position is uniquely determined at any instant by three independent coordinates. Consider a system containing N such particles, the instantaneous position of the ith particle of mass m_i being specified by the radius vector $\boldsymbol{r}_i$ in cartesian coordinates. Let $\boldsymbol{F}_i$ be the vector resultant of all the forces acting upon the particle, the dynamical behavior of the system being described by Newton's law of motion,

$$m_i\ddot{\boldsymbol{r}}_i = \boldsymbol{F}_i \qquad \text{for } i = 1, 2, \ldots, N \tag{4-2-1}$$

Equation (4-2-1) determines the complete dynamical behavior of the set of particles provided $\boldsymbol{r}_i$ and the velocity $\dot{\boldsymbol{r}}_i$ of each particle are known at some initial time t_0.

In general the particles are not free, being subjected to independent constraints, so that any one particle may have three degrees, two degrees, or one degree of freedom. Let the sum of the degrees of freedom of all N particles be $k < 3N$. Such a system can be regarded as a collection of free particles subjected to $l = 3N - k$ independent constraints so that only k coordinates are required to describe the motion of the system. These new coordinates $q_1, q_2, \ldots, q_k$, referred to as the generalized coordinates of Lagrange, are related to the cartesian coordinates

$$x_i = x_i(q_1, q_2, \ldots, q_k) \tag{4-2-2}$$

of the particles.

Let the cartesian components of $\boldsymbol{r}_1$ be x_1, x_2, x_3, of $\boldsymbol{r}_2$ be x_4, x_5, x_6, and similarly for all other particles; then the total kinetic energy K of the system is given by

$$\begin{aligned} K &= \tfrac{1}{2}\sum_{i=1}^{N} m_i\dot{\boldsymbol{r}}_i \cdot \dot{\boldsymbol{r}}_i \\ &= \tfrac{1}{2}\sum_{i=1}^{3N} m_i\left(\frac{\partial x_i}{\partial t}\right)^2 - \sum_{r=1}^{k}\sum_{s=1}^{k} \tfrac{1}{2}M_{rs}\dot{q}_r\dot{q}_s \end{aligned} \tag{4-2-3}$$

where

$$M_{rs} = \sum_{i=1}^{3N} m_i \frac{\partial x_i}{\partial q_r}\frac{\partial x_i}{\partial q_s} \tag{4-2-4}$$

it being noted that $m_1 = m_2 = m_3$, $m_4 = m_5 = m_6$, and similarly for all other particles. Generalized momenta p_r can be defined, by analogy to the cartesian components of momentum, in the form

$$p_r(q_1, q_2, \ldots; \dot{q}_1, \dot{q}_2, \ldots) = \frac{\partial K}{\partial \dot{q}_r} = \sum_{s=1}^{k} M_{rs}\dot{q}_s \tag{4-2-5}$$

For a conservative system, a potential function $U(q_1, q_2, \ldots, q_k)$ exists such that

$$Q_a = -\frac{\partial U}{\partial q_a} \qquad \text{for } a = 1, 2, \ldots, k \tag{4-2-6}$$

giving Lagrange's equations of motion in the form

$$\frac{d}{dt}\left(\frac{\partial K}{\partial \dot{q}_a}\right) - \frac{\partial K}{\partial q_a} = Q_a \qquad \text{for } a = 1, 2, \ldots, k \tag{4-2-7}$$

This is a set of k differential equations of second order with $q_1, q_2, \ldots, q_k$ as dependent variables and t as the independent variable. The physical significance of the generalized quantity Q_a depends upon the dimension of the generalized coordinate q_a which may be an angle.

Introducing the *lagrangian function*

$$L(q_a, \dot{q}_a) = K(q_a, \dot{q}_a) - U(q_a) \tag{4-2-8}$$

into Eq. (4-2-7) gives

$$\frac{d}{dt}\left(\frac{\partial L}{\partial \dot{q}_a}\right) - \frac{\partial L}{\partial q_a} = 0 \qquad \text{for } a = 1, 2, \ldots, k \tag{4-2-9}$$

The solution of Lagrange's equations of motion in either form, i.e., Eqs. (4-2-7) or (4-2-9), will result in an expression for each generalized coordinate q_a as a function of time and $2k$ constants of integration which are evaluated from the initial conditions of the N particles of the system. From Eqs. (4-2-5), (4-2-8), and (4-2-9), it follows that

$$p_a = \frac{\partial L}{\partial \dot{q}_a} \qquad \dot{p}_a = \frac{\partial L}{\partial q_a} \tag{4-2-10}$$

the generalized momenta p_a reducing to ordinary linear momenta and Lagrange's equations of motion reducing to Newton's equation of motion for the particular case that the generalized coordinates are cartesian coordinates.

The *hamiltonian function* H of a system is defined by the relation

$$H = \sum_{a=1}^{k} p_a \dot{q}_a - L \tag{4-2-11}$$

it being noted that in contrast to the lagrangian, which is a function of the coordinates and their time derivatives, the hamiltonian is a function of the coordinates and their conjugate momenta. Differentiation of Eq. (4-2-11) gives

$$dH = \sum_{a=1}^{k} p_a \, d\dot{q}_a + \sum_{a=1}^{k} \dot{q}_a \, dp_a - \sum_{a=1}^{k} \frac{\partial L}{\partial q_a} dq_a - \sum_{a=1}^{k} \frac{\partial L}{\partial \dot{q}_a} d\dot{q}_a$$

which reduces by way of Eq. (4-2-10) to the form

$$dH = \sum_{a=1}^{k} \dot{q}_a \, dp_a - \sum_{a=1}^{k} \frac{\partial L}{\partial q_a} dq_a \tag{4-2-12}$$

Since this equation depends only on dp_a and dq_a, but not on $d\dot{q}_a$, it follows that H is a function of q and p alone, and therefore the total differential dH may be expressed in the form

$$dH = \sum_{a=1}^{k} \frac{\partial H}{\partial p_a} dp_a + \sum_{a=1}^{k} \frac{\partial H}{\partial q_a} dq_a \tag{4-2-13}$$

where, by comparison with Eq. (4-2-12), it is evident that

$$\frac{\partial H}{\partial p_a} = \dot{q}_a \qquad \frac{\partial H}{\partial q_a} = -\frac{\partial L}{\partial q_a} = -\dot{p}_a \qquad \text{for } a = 1, 2, \ldots, k \tag{4-2-14}$$

The resulting $2k$ first-order equations (4-2-14) are *Hamilton's canonical equations of motion*; p_a and q_a are said to be *canonically conjugate variables*. Thus, the k second-order differential equations of Newton or Lagrange have been transformed into $2k$ first-order differential equations.

Euler's theorem for homogeneous polynomials states that if f_n is a homogeneous function of order n of a set of variables ξ_a, then

$$\sum_a \xi_a \frac{\partial f_n}{\partial \xi_a} = n f_n \tag{4-2-15}$$

and, consequently, since K is a homogeneous function of order two in the velocities it follows that

$$\sum_{a=1}^{k} \dot{q}_a \frac{\partial K}{\partial \dot{q}_a} = 2K \tag{4-2-16}$$

From the definition of the hamiltonian function given in Eq. (4-2-11) and that of the conjugate momenta in Eq. (4-2-10), H can be expressed in the form

$$H = \sum_{a=1}^{k} \dot{q}_a \frac{\partial L}{\partial \dot{q}_a} - L$$

and since the potential energy is not a function of the velocities it follows that

$$H = \sum_{a=1}^{k} \dot{q}_a \frac{\partial K}{\partial \dot{q}_a} - L$$

which, combined with Eqs. (4-2-8) and (4-2-16), gives

$$H = 2K - L = K + U \tag{4-2-17}$$

and the hamiltonian function is equal to the total energy for a conservative system.

4-3 MOLECULE CONFIGURATION, MOMENTUM, AND PHASE SPACE

The point location of a single monatomic molecule can be specified by three coordinates x_1, x_2, x_3 in a three-dimensional *molecule configuration space*. This concept of a particle configuration space approximates the N molecules to a system of N mass points. Thus the specification of the position vectors $\boldsymbol{r}_i$ $(i = 1, 2, \ldots, N)$ in molecule configuration space gives the configuration of the whole system. By considering a hyperspace characterized by $3N$ coordinate axes, it is possible for theoretical purposes to use only one point to define the configuration of the complete system, the molecule configuration space being now of $3N$ dimensions. In this concept of a molecule configuration space, the geometry of the system may be regarded as being given by a vector $\boldsymbol{r}^N$ ($\equiv \boldsymbol{r}_1, \boldsymbol{r}_2, \ldots, \boldsymbol{r}_N$), with $3N$ components.

The same construction can be made for the momenta, the momentum of a single monatomic molecule being described by a point in a three-dimensional *molecule momentum space*. Thus the vector moments of N molecules as a function of time can be regarded either as N points in a three-dimensional momentum space of a single molecule or as a single vector $\boldsymbol{p}^N$ ($\equiv \boldsymbol{p}_1, \boldsymbol{p}_2, \ldots, \boldsymbol{p}_N$), in a $3N$-dimensional momentum space for the entire system.

If all the momenta and coordinates of a system are known, then in principle the dynamical behavior of the system can be completely specified. For this reason the concept of *phase space* has been introduced. Molecule phase space, often referred to as μ space, is a combination of molecule configuration space and momentum configuration space and for a monatomic molecule is a six-dimensional subspace. A single point in this space gives both the configuration and momentum of the molecule. Thus the dynamical state of an N molecule system is given by N points in μ space, the motion of these points being determined by the classical equations of motion.

Instead of specifying N points in a six-dimensional phase space, the same information can be specified in terms of one point in a $6N$-dimensional phase space. This single point now represents the entire system, the phase space being often referred to as γ space and the single phase point as the γ point.

Since any given arrangement of the molecules in μ space is represented by its own point in γ space, it is important to note that for an assembly of N identical molecules there are $N!$ different points in γ space which correspond to the N points in μ space. This is because interchanging molecules gives indistinguishable arrangements in μ space and there are therefore $N!$ different arrangements of the molecules which correspond to the N points in μ space.

In the case of molecules with s degrees of freedom, the molecule configuration and momentum spaces are each of s dimensions, μ space therefore being of $2s$ dimensions and γ space of $2Ns$ dimensions.

The state of the system will change with time and the corresponding phase point will describe in phase space a line which is generally referred to as the phase trajectory. Newton's laws of motion or the equivalent relations of Hamilton determine the motion of the phase point. Denoting the N generalized coordinates of the system by $q_1, q_2, \ldots, q_N$ and the generalized momenta by $p_1, p_2, \ldots, p_N$, and since the system is assumed to obey classical mechanical laws, Hamilton's canonical equations (4-2-14), with k set equal to N, must be valid. Taking the components of p and q as cartesian coordinates an element $d\Omega$ of phase space can be constructed; i.e.,

$$d\Omega = dq_1\, dq_2 \cdots dq_N\, dp_1\, dp_2 \cdots dp_N \tag{4-3-1}$$

It is evident from Eqs. (4-2-14) and (4-3-1) that there is only one trajectory passing through every point in phase space since, when every p and q are given, Eq. (4-2-14) determines uniquely the rate of change of every coordinate in phase space. Hence the representative point can never cross its previous path.

4-4 DISTRIBUTION FUNCTIONS

A concept which finds application in statistical physics is that of a very large set of replica systems of the one real system of interest, the collection of these replica systems being referred to by Gibbs (see, for example, Gibbs, 1948) as an *ensemble of systems.* Each replica system, which is completely independent of the others, consists of the same number and type of molecules in the same type of enclosure. The replica systems represent the real system in various dynamical states resulting from different initial conditions, the instantaneous dynamical state of each replica system being described by its corresponding phase point in γ space.

Provided the number of replica systems in the ensemble is sufficiently large, the replica system points in γ space can be regarded as a cloud of points with a *distribution density* ρ. This density will in general be different in different parts of phase space, and it will change in time. Hence, whatever the functional form of the density

$$\rho = \rho(\boldsymbol{r}^N, \boldsymbol{p}^N)$$

the number of replica system points in the element $d\boldsymbol{r}^N\, d\boldsymbol{p}^N$ of γ space is $\rho\, d\boldsymbol{r}^N\, d\boldsymbol{p}^N$, such that

$$\int \rho(\boldsymbol{r}^N, \boldsymbol{p}^N)\, d\boldsymbol{r}^N\, d\boldsymbol{p}^N = \Gamma \qquad \text{for all time } t \tag{4-4-1}$$

where

$$d\boldsymbol{p}^N = \prod_{i=1}^{N} d\boldsymbol{p}_i \qquad d\boldsymbol{r}^N = \prod_{i=1}^{N} d\boldsymbol{r}_i$$

and Γ is the number of replica systems in the ensemble. A normalized function $f^{(N)}$ can be defined as

$$f^{(N)}(\boldsymbol{r}^N, \boldsymbol{p}^N) = \frac{\rho(\boldsymbol{r}^N, \boldsymbol{p}^N)}{\int \rho(\boldsymbol{r}^N, \boldsymbol{p}^N)\, d\boldsymbol{r}^N\, d\boldsymbol{p}^N} = \frac{\rho}{\Gamma} \tag{4-4-2}$$

subject to the normalization condition

$$\int f^{(N)}(\boldsymbol{r}^N, \boldsymbol{p}^N)\, d\boldsymbol{r}^N\, d\boldsymbol{p}^N = 1 \tag{4-4-3}$$

the quantity $f^{(N)}$ being referred to as the probability density function. Because

$$d\omega = f^{(N)}(\boldsymbol{r}^N, \boldsymbol{p}^N)\, d\boldsymbol{r}^N\, d\boldsymbol{p}^N \tag{4-4-4}$$

is the probability that at some instant of time t the particle 1 will be at location $\boldsymbol{r}_1$ and have momentum $\boldsymbol{p}_1$ within the respective limits $d\boldsymbol{r}_1$ and $d\boldsymbol{p}_1$ and simultaneously particle 2 will be at location $\boldsymbol{r}_2$ and have momentum $\boldsymbol{p}_2$ within the respective limits $d\boldsymbol{r}_2$ and $d\boldsymbol{p}_2$, and similarly for all other molecules, i.e., specified particles have specified phase, the function $f^{(N)}$ is called the *specific phase distribution* of order N for the system.

Various other distribution functions can be defined. For example, instead of defining the probability that each of the N particles shall be in a specified region of phase space, a distribution function $\hat{f}^{(N)}$ can be defined in such a way that

$$dW = \hat{f}^{(N)}(\boldsymbol{r}^N, \boldsymbol{p}^N)\, d\boldsymbol{r}^N\, d\boldsymbol{p}^N \tag{4-4-5}$$

is the probability that any one of the N particles will be found in the element $d\boldsymbol{r}_1\, d\boldsymbol{p}_1$ of phase space, while simultaneously any other particle will be found in the element $d\boldsymbol{r}_2\, d\boldsymbol{p}_2$ of phase space, and similarly for all other particles. Distribution functions which are defined without reference to specific particles are often referred to as *generic distribution functions*. The generic distribution function $\hat{f}^{(N)}$ is related to the *specific* distribution function $f^{(N)}$ by the equation

$$\hat{f}^{(N)}(\boldsymbol{r}^N, \boldsymbol{p}^N)\, d\boldsymbol{r}_1\, d\boldsymbol{r}_2 \cdots d\boldsymbol{r}_N\, d\boldsymbol{p}_1\, d\boldsymbol{p}_2 \cdots d\boldsymbol{p}_N = N!\, f^{(N)}(\boldsymbol{r}^N, \boldsymbol{p}^N)\, d\boldsymbol{r}^N\, d\boldsymbol{p}^N \tag{4-4-6}$$

which simply states that the probability of finding the points associated with the molecules of the system in any arbitrary order in the volume elements $d\boldsymbol{r}_1 d\boldsymbol{p}_1 \cdots d\boldsymbol{r}_N d\boldsymbol{p}_N$ around $\boldsymbol{r}_1 \boldsymbol{p}_1 \cdots \boldsymbol{r}_N \boldsymbol{p}_N$ in μ space is $N!$ times as large. The distribution function $f^{(N)}$ is said to be fine grained because it is associated with a phase volume element that becomes indefinitely small, it being noted that in classical mechanics there is no intrinsic limit to the fineness of the phase specification. Fine-grained distributions find application in the description of equilibrium systems. However, to describe nonequilibrium conditions it is usually necessary to proceed by way of coarse-grained distributions where the volume element of phase space cannot be made smaller than some lower limiting value.

Sufficient information can frequently be obtained by way of reduced distribution functions which can be deduced from $f^{(N)}$ by integrating away the unwanted information. Thus, for a subset of particles $h < N$, the specific phase distribution of order h for the system is obtained from $f^{(N)}$ by way of the relation

$$f^{(h)}(\boldsymbol{r}^h, \boldsymbol{p}^h) = \int f^{(N)}(\boldsymbol{r}^N, \boldsymbol{p}^N) \prod_{m=h+1}^{N} d\boldsymbol{r}_m\, d\boldsymbol{p}_m \tag{4-4-7}$$

An example of a reduced distribution function of use in the theory of liquids is the doublet distribution defined as

$$f^{(2)}(1, 2) = \int f^{(N)} \prod_{m=3}^{N} d\boldsymbol{r}_m\, d\boldsymbol{p}_m \tag{4-4-8}$$

the averaging process being with respect to the positions and momenta of $(N-2)$ particles. The lowest-order function of this type is the singlet distribution

$$f^{(1)} = \int f^{(N)} \prod_{m=2}^{N} d\boldsymbol{r}_m\, d\boldsymbol{p}_m \tag{4-4-9}$$

the averaging process being with respect to the position and momenta of $(N-1)$ particles. Similarly, reduced generic distribution functions can be obtained:

$$\begin{aligned} \hat{f}^{(h)}(\boldsymbol{r}^h, \boldsymbol{p}^h) &= \frac{N!}{(N-h)!} f^{(h)}(\boldsymbol{r}^h, \boldsymbol{p}^h) \\ &= \frac{1}{(N-h)!} \int \hat{f}^{(N)}(\boldsymbol{r}^N, \boldsymbol{p}^N) \prod_{m=h+1}^{N} d\boldsymbol{r}_m\, d\boldsymbol{p}_m \end{aligned} \tag{4-4-10}$$

The lowest-order generic function is the singlet distribution

$$\hat{f}^{(1)}(\boldsymbol{r}_1, \boldsymbol{p}_1) = N f^{(1)}(\boldsymbol{r}_1, \boldsymbol{p}_1)$$

which finds application in the statistical description of dilute gases.

For the theory of the transport of momentum in liquids, the probabilities of a particular configuration, i.e., generic configuration distribution functions from which all particle momentum information has been excluded, are found to have application. The generic configuration distribution function $n^{(N)}$ is defined in a way analogous to that used to define the phase distribution. Thus, the probabilities of a particular configuration are

$$n^{(N)}(\boldsymbol{r}^N) = N! f^{(N)}(\boldsymbol{r}^N) = N! \int f^{(N)}(\boldsymbol{r}^N, \boldsymbol{p}^N)\, d\boldsymbol{p}^N \tag{4-4-11}$$

and reduced functions

$$\begin{aligned} n^{(h)}(\boldsymbol{r}^h) &= \frac{N!}{(N-h)!} \int f^{(h)}(\boldsymbol{r}^h, \boldsymbol{p}^h) \prod_{i=1}^{h} d\boldsymbol{p}_i \\ &= \frac{1}{(N-h)!} \int n^{(N)}(\boldsymbol{r}^N)\, d\boldsymbol{r}^{N-h} \\ &= \int \hat{f}^{(h)}(\boldsymbol{r}^h, \boldsymbol{p}^h) \prod_{i=1}^{h} d\boldsymbol{p}_i \end{aligned} \tag{4-4-12}$$

From Eq. (4-4-12) it follows that the lowest-order function of this type is

$$n^{(1)}(\boldsymbol{r}_1) = N \int f^{(1)}\, d\boldsymbol{p}_1 = \frac{N}{V} = n \tag{4-4-13}$$

which is simply the number density.

The probability density for pairs, i.e., the function $n^{(2)}(\boldsymbol{r}_i, \boldsymbol{r}_{ij})$ is an important function in the theory of liquids, the quantity $n^{(2)}(\boldsymbol{r}_i, \boldsymbol{r}_{ij})\, d\boldsymbol{r}_i\, d\boldsymbol{r}_j$ being the probability that there is a molecule in the volume element $d\boldsymbol{r}_i$ about $\boldsymbol{r}_i$ and another in the volume element $d\boldsymbol{r}_j$ about $\boldsymbol{r}_j$. In the molecular theory of liquids it is useful to replace $n^{(N)}$ by a dimensionless distribution, or correlation function, denoted by $g^{(N)}(\boldsymbol{r}^N)$. If N is large compared to unity, the radial distribution function $g^{(2)}(\boldsymbol{r}_i, \boldsymbol{r}_{ij})$ can be defined by the relation

$$n^{(2)}(\boldsymbol{r}_i, \boldsymbol{r}_{ij}) = n(\boldsymbol{r}_i) n(\boldsymbol{r}_j) g^{(2)}(\boldsymbol{r}_i, \boldsymbol{r}_{ij}) \tag{4-4-14}$$

the function $g^{(2)}(\boldsymbol{r}_i, \boldsymbol{r}_{ij}) \to 1$ as $\boldsymbol{r}_{ij} = |\boldsymbol{r}_i - \boldsymbol{r}_j| \to \infty$, and deviations from unity are a measure of the correlations in the positions of pairs of molecules. The doublet distribution can be expressed in terms of the function $f^{(2)}(1, 2)$ of Eq. (4-4-8) in the form

$$n^{(2)}(1, 2) = N(N-1) \int f^{(2)}(1, 2)\, d\boldsymbol{p}_1\, d\boldsymbol{p}_2 \tag{4-4-15}$$

the averaging process being with respect to $(N-2)$ particles.

4-5 LIOUVILLE'S THEOREM

Consider a property of the system $\psi = \psi(q_a, p_a, t)$ which depends upon the dynamical state of the system and, in general, explicitly on time. The property ψ may be a constant of the motion such as the total energy, total linear momentum, or total angular momentum of the system. Alternatively, ψ may be a property which changes with time, such as the energy, momentum, or position of a single particle in the system. From the definition of the material, i.e., the substantial time derivative of Eq. (2-1-20), the change of ψ with time following a point along a trajectory is

$$\frac{D\psi}{Dt} = \frac{\partial\psi}{\partial t} + \sum_a \left(\frac{\partial\psi}{\partial q_a}\dot{q}_a + \frac{\partial\psi}{\partial p_a}\dot{p}_a\right) \tag{4-5-1}$$

which can be expressed, by way of Eq. (4-2-14), in the form

$$\frac{D\psi}{Dt} = \frac{\partial\psi}{\partial t} + \sum_a \left(\frac{\partial\psi}{\partial q_a}\frac{\partial H}{\partial p_a} - \frac{\partial\psi}{\partial p_a}\frac{\partial H}{\partial q_a}\right) \tag{4-5-2}$$

$$= \frac{\partial\psi}{\partial t} + [\psi, H] \tag{4-5-3}$$

where $[\psi, H]$ is referred to as the *Poisson bracket*.

In terms of the density distribution function $\rho(q_a, p_a, t)$ in phase space, generalization of the equation (3-1-5) of continuity gives

$$\frac{\partial\rho}{\partial t} + \sum_a \left[\frac{\partial}{\partial q_a}(\dot{q}_a\rho) + \frac{\partial}{\partial p_a}(\dot{p}_a\rho)\right] = 0$$

since there is no generation or decay of phase points. This generalization of the equation of continuity can be expressed by way of Hamilton's canonical equations (4-2-14) in the form

$$\frac{D\rho}{Dt} = \frac{\partial\rho}{\partial t} + \sum_a \left(\frac{\partial\rho}{\partial q_a}\frac{\partial H}{\partial p_a} - \frac{\partial\rho}{\partial p_a}\frac{\partial H}{\partial q_a}\right) = 0 \tag{4-5-4}$$

which is the celebrated *Liouville theorem*, and is the basis of all statistical mechanics. The Liouville equation (4-5-4) is an equation of continuity for the flow of representative points through phase space. Any group of points which initially fills a certain region of phase space will during its entire motion continue to occupy the same volume but with altered shape. That this is so can be shown as follows. In terms of the probability density $f^{(N)}(\mathbf{r}^N, \mathbf{p}^N, t)$ the Liouville equation has the form

$$\frac{\partial f^{(N)}}{\partial t} = \frac{\partial H}{\partial \mathbf{r}^N}\cdot\frac{\partial f^{(N)}}{\partial \mathbf{p}^N} - \frac{\partial H}{\partial \mathbf{p}^N}\cdot\frac{\partial f^{(N)}}{\partial \mathbf{r}^N} \tag{4-5-5}$$

or, from Eq. (4-5-4), it can be expressed in the form

$$\frac{Df^{(N)}}{Dt} = 0 \tag{4-5-6}$$

The vanishing of the substantial derivative indicates that the density of points in phase space remains constant in the neighborhood of a particle which is moving through phase space. For the small region in phase space, $\Delta\Omega$, the number of system points in the region is

$$\Delta\omega = f^{(N)}\Delta\Omega \tag{4-5-7}$$

the volume element $\Delta\Omega$ being sufficiently small for the density of the system points to approximate closely to a constant within this region. Differentiation of Eq. (4-5-7) gives

$$\frac{D(\Delta\omega)}{Dt} = \frac{Df^{(N)}}{Dt}\Delta\Omega + \frac{D(\Delta\Omega)}{Dt}f^{(N)}$$

which simplifies, by way of Eq. (4-5-6), to

$$\frac{D(\Delta\omega)}{Dt} = f^{(N)}\frac{D(\Delta\Omega)}{Dt} \tag{4-5-8}$$

Now, according to Hamilton's equations, $\Delta\omega$ does not change during the natural motion of the system, that is, $D(\Delta\omega)/Dt = 0$, because the uniqueness of the phase trajectories implies that no phase point originally inside $\Delta\Omega$ can cross the boundary. Hence

$$\frac{D}{Dt}(\Delta\Omega) = 0 \tag{4-5-9}$$

from which it follows that there is no change with time in the volume occupied by a given set of representative points, although the shape is continually changing. That this statement is true for any size of volume extension in phase may be shown by summing the results for a number of small elements of volume. The importance of phase space in statistical mechanics is primarily due to this particular property of conservation of volume in phase space during the motion. It is the fact that Hamilton's equations hold for the coordinates of phase space which gives it this unique property.

Liouville's equation (4-5-5) can be expressed in terms of $f^{(N)}$ in the form

$$-\frac{\partial f^{(N)}}{\partial t} = \sum_{i=1}^{N}\left[\frac{\partial f^{(N)}}{\partial \mathbf{r}_i}\cdot\frac{\mathbf{p}_i}{m} - \frac{\partial f^{(N)}}{\partial \mathbf{p}_i}\cdot\frac{\partial H}{\partial \mathbf{r}_i}\right] \tag{4-5-10}$$

it being noted that

$$\dot{\mathbf{p}}_i = -\frac{\partial H}{\partial \mathbf{r}_i} \qquad \dot{\mathbf{r}}_i = \frac{\partial H}{\partial \mathbf{p}_i}$$

where

$$H = \frac{1}{2m}\sum_{i=1}^{N}\mathbf{p}_i\cdot\mathbf{p}_i + U \tag{4-5-11}$$

Equation (4-5-10) is an important form of Liouville's equation since it has already been shown that it determines the time evolution of the system, in this case $f^{(N)}$, according to classical particle mechanics.

In terms of the reduced distribution function $f^{(h)}$, Liouville's equation (4-5-10) can be expressed in the form

$$\frac{\partial f^{(h)}}{\partial t} + \sum_{i=1}^{h} \left(\frac{\partial f^{(h)}}{\partial \boldsymbol{r}_i} \cdot \frac{\boldsymbol{p}_i}{m} \right) = \int \sum_{i=1}^{h} \left(\frac{\partial H}{\partial \boldsymbol{r}_i} \cdot \frac{\partial f^{(N)}}{\partial \boldsymbol{p}_i} \right) \prod_{m=h+1}^{N} d\boldsymbol{r}_m \, d\boldsymbol{p}_m \tag{4-5-12}$$

use being made of the property of $f^{(N)}$ to vanish if a momentum is infinite or any coordinate is larger than a characteristic dimension of the container.

4-6 TIME AVERAGES AND ENSEMBLE AVERAGES

At any instant of time, the system is represented by one point, the γ point in γ space, and in the course of time this representative point will move, describing a certain trajectory in phase space. From a knowledge of the instantaneous position of the representative point it is, in principle, possible to calculate its position at some other instant. Thus the macroscopic values of the thermodynamic quantities can now be regarded as time averages over the motion of the representative point of the system, i.e., over a segment of the trajectory in its own phase space. Hence, for a property of the system $\psi = \psi(t)$, its time average value $\bar{\psi}$ can be defined as

$$\bar{\psi} = \operatorname*{Lim}_{\tau \to \infty} \frac{1}{\tau} \int_0^{\tau} \psi(t)\, dt \tag{4-6-1}$$

where τ is some sufficiently large time interval.

The alternative approach of Gibbs is to replace time averages over a single system by ensemble averages over the ensemble at a fixed time. In terms of an ensemble average, the macroscopic value of a function $\psi(\boldsymbol{r}^N, \boldsymbol{p}^N, t)$, denoted as $\langle \psi(\boldsymbol{r}^N, \boldsymbol{p}^N, t) \rangle$, is derived by performing the ensemble phase average to give

$$\langle \psi(\boldsymbol{r}^N, \boldsymbol{p}^N, t) \rangle = \int \psi(\boldsymbol{r}^N, \boldsymbol{p}^N, t) f^{(N)}(\boldsymbol{r}^N, \boldsymbol{p}^N, t)\, d\boldsymbol{r}^N \, d\boldsymbol{p}^N \tag{4-6-2}$$

where, in general, $\langle \psi(\boldsymbol{r}^N, \boldsymbol{p}^N, t) \rangle$ is understood to be the expected mean value of the quantity ψ which would be obtained when ψ is actually measured at the time t.

The question which now arises is whether these two averages are equivalent. Boltzmann demonstrated the equivalence of time and ensemble averages by way of a postulate which he referred to as the *ergodic hypothesis*. The ergodic hypothesis states that the representative point of an isolated system visits in succession every point in phase space compatible with the energy of the system before returning to its original position. Any system satisfying this postulate is then said to be ergodic. It is, however, a simple matter to construct nonergodic systems. In the absence of ergodic systems, recourse has been made to the quasi-ergodic hypothesis which states that the trajectory would in time come arbitrarily close to any point in phase space whose energy was compatible with that of the system. However, there are difficulties associated with this approach (see, for example, Tolman, 1938; also see Caldirola, 1961; Farquhar, 1964).

With regard to the question of the equivalence of time averages and ensemble averages, Liouville's theorem suggests a fundamental postulate upon which statistical mechanics can be based. This fundamental postulate is adopted without proof and asserts that *all accessible regions of phase space have equal a priori probabilities for equal volumes.* That is, the representative point of a system is equally likely to be in any region of phase space which it visits. It is evident that the relation of Eq. (4-6-2) represents a formulation of the proposal by Gibbs to replace time averages by ensemble averages, which is in accord with this postulate.

The formulation of the function $f^{(N)}$ must be such as to give the required thermodynamic relations when thermodynamic quantities are computed by way of Eq. (4-6-2).

If the system is in thermodynamic equilibrium, the number of representative points in any given extension in phase, $\Delta\Omega$, must remain constant in time. The equilibrium condition can therefore be stated by way of Eq. (4-5-5) in the form

$$\frac{\partial}{\partial t} f^{(N)}(\boldsymbol{r}^N, \boldsymbol{p}^N, t) = 0 \qquad \text{(4-6-3)}$$

which is satisfied if

$$f^{(N)} = \text{constant}$$

with respect to the coordinates $\boldsymbol{r}^N, \boldsymbol{p}^N$ or if

$$f^{(N)} = f^{(N)}(\alpha)$$

where

$$\alpha = \alpha(\boldsymbol{r}^N, \boldsymbol{p}^N) \qquad \text{(4-6-4)}$$

is a constant of the motion. If the thermodynamic system is isolated, then the constants of motion are properties of the system which do not change with time, examples of which are energy, linear momentum, and angular momentum. If $f^{(N)} = f^{(N)}(\alpha)$, then the ensemble is said to represent a steady-state or equilibrium distribution.

There are three important steady-state distribution functions.

1. The *microcanonical ensemble*. Consider an isolated system composed of a large number N of molecules having total energy U. For an isolated system the energy U is a constant, independent of time. To obtain the macroscopic properties of the given system, it is necessary to identify an appropriate ensemble of similar systems. To satisfy the fundamental postulate of equal a priori probabilities to equal volumes of phase space, the representative points of such an ensemble must be uniformly distributed in the accessible regions of phase space. The conditions are satisfied by the ensemble whose probability density function is such that

$$f^{(N)} = \begin{cases} \text{constant} & U \leqslant H \leqslant U + \Delta U \\ 0 & \text{for all other values of } H \end{cases} \qquad \text{(4-6-5)}$$

   Such an ensemble is referred to as the microcanonical ensemble, or energy shell ensemble.

2. The *canonical ensemble*. This is defined as the probability density function

$$f^{(N)}(\mathbf{r}^N, \mathbf{p}^N) = \frac{1}{[Z_N N! h^{3N}]} \exp\left[-\frac{H(\mathbf{r}^N, \mathbf{p}^N)}{k\theta}\right]$$
$$Z_N = \frac{1}{N! h^{3N}} \int \exp(-\beta H)\, d\mathbf{r}^N\, d\mathbf{p}^N \tag{4-6-6}$$

where k is the Boltzmann constant and θ the absolute temperature, it being noted that Eq. (4-6-6) is simply Eq. (4-1-4) with C set equal to $[Z_N N! h^{3N}]^{-1}$, $U_n = H(\mathbf{r}^N, \mathbf{p}^N)$, and h^{3N} may be interpreted as the volume of a cell in γ space.

3. The *grand canonical ensemble*. The microcanonical and canonical ensembles are concerned with only those systems having a fixed number of molecules, i.e., closed systems. However, the most general case must take account of systems in which chemical reactions or physical changes of state are occurring, such systems being referred to as open systems. Consider a real system which consists of a large number n of subsystems, each of fixed volume V. The state of each subsystem is specified by its volume, its composition, and its energy. It can be shown that the ensemble which represents open systems at constant temperature θ in a given volume V is

$$f^{(n)}(\mathbf{r}^n, \mathbf{p}^n) = \frac{1}{[Z_{\mathrm{gr}} n! h^{3n}]} \exp\left(\frac{n\mu}{k\theta}\right) \exp\left[-\frac{H(\mathbf{r}^n, \mathbf{p}^n)}{k\theta}\right] \tag{4-6-7}$$

in which the normalizing constant Z_{gr} is the grand partition function for the system. Equation (4-6-7) is to be compared with Eq. (4-1-14) which relates to quantum mechanics.

GENERAL REFERENCES

Caldirola, P. (1961) In P. Caldirola (ed.), "Ergodic Theories," Academic Press, New York.

Farquhar, I. E. (1964) "Ergodic Theory in Statistical Mechanics," Interscience, New York.

Fowler, R. H. (1936) "Statistical Mechanics," Cambridge University Press, London.

Fowler, R. H., and E. A. Guggenheim (1939) "Statistical Thermodynamics," Cambridge University Press, London.

Gibbs, J. W. (1948) "Collected Works," Yale University Press, New Haven, Conn.

Haar, D. ter (1954) "Elements of Statistical Mechanics," Reinhardt, New York.

Landau, L. D., and E. M. Lifshitz (1958) "Statistical Physics," Pergamon Press, London.

Mayer, J. E., and M. G. Mayer (1977) "Statistical Mechanics," 2d ed., Wiley, Chichester.

Rice, O. K. (1967) "Statistical Mechanics, Thermodynamics and Kinetics," Freeman, San Francisco, Calif.

Rushbrooke, G. S. (1949) "Introduction to Statistical Mechanics," Oxford University Press, Oxford.

Schrödinger, E. (1948) "Statistical Thermodynamics," Cambridge University Press, Cambridge.

Tolman, R. C. (1938) "Principles of Statistical Mechanics," Clarendon Press, Oxford, 1938.

Yvon, J. (1969) "Correlations and Entropy in Classical Statistical Mechanics," Pergamon Press, Oxford.

CHAPTER

FIVE

PRINCIPLES OF CONSTITUTIVE THEORY

5-1 PRINCIPLES OF DETERMINISM AND LOCAL ACTION

For a given body there must exist some relation between the dynamical state of the body at some instant, the kinematical state of the body at the same instant, and possibly the kinematical state of the body at all instants of the body's past history. The equations which express the relation between the kinematical and dynamical variables are called constitutive equations. Continuum physics is concerned with establishing the equation of state for a given material and with showing how values can be assigned to the material coefficients appearing in the equation of state. In general, these coefficients are functions of thermodynamic variables such as pressure, volume, temperature, and composition, and therefore the constitutive equation of a body is influenced by the thermodynamic state of that body. Constitutive equations are mathematical models defining the ideal behavior of a particular class of real materials. The way in which a real material behaves, and in particular which equation of state describes it most satisfactorily, can only be determined by experiment.

In formulating constitutive equations certain general principles must be satisfied. The most important of them are the principles of determinism, local action, material objectivity, and material invariance.

Implicit in the statement that a constitutive equation expresses the relation between the kinematical and dynamical variables is the idea of determinism. In general, determinism is understood to imply that the past determines the future. However, in continuum mechanics it is sufficient to require that the past history of a body shall determine all the conditions in the body at the present time. Thus the causality of natural processes requires the time dependence of mechanical interactions to be specified if all mechanical influences are to be accounted for.

Regarding the body force f and the heat supply r as known, the field equations (3-1-4), (3-4-4), (3-5-6), and (3-8-16) provide eight scalar relations between the seventeen scalar fields consisting of the density ρ, internal energy ε, and the components of the velocity v, stress $\mathbf{T}$, and heat flux q. Hence, in general, the field equations do not by themselves determine the behavior of a material body. This conclusion implies some a priori knowledge of the way in which the materials respond to external fields. Without this experience, based on observation, it would be impossible to construct constitutive theories since in modern continuum mechanics there is no well-defined systematic rule whereby the basic constitutive variables can be identified. To see how constitutive equations can be arrived at by this essentially intuitive approach, consider the simple thermomechanical process described by Wang and Bowen (1966).

Consider a motion of a material body $\mathscr{B}$ to be described by Eqs. (2-1-6). Let the density in some reference configuration be prescribed and consider only a symmetric state of stress $\mathbf{T}$. In this way the spatial continuity equation (3-1-4) and the symmetry condition of Eq. (3-5-6) are satisfied. By appropriate choice of the body force f, it can be ensured that the Eq. (3-4-4) of motion also holds. There remains only the energy equation (3-8-16) to be satisfied. This observation implies that the way to obtain the further relations which must be adjoined to the field equations in order to obtain a complete system of governing equations is to specify $\mathbf{T}$, ε, and q. However, if the prescription of these fields depends only on a knowledge of the motion of $\mathscr{B}$ then the field equations would eventually furnish five scalar relations for only four scalar fields identified as the density and the components of v. Hence, it is evident that $\mathbf{T}$, ε, and q are not, in general, fully determined by the motion of $\mathscr{B}$ but depend also upon a scalar field which is not present as a dependent variable in the field equations. Having regard to the physical significance of ε and q, the additional scalar field is expected to measure the heat content and is identified as the temperature θ defined on the configurations of $\mathscr{B}$. Thus, for a simple thermomechanical process, experience identifies the motion of the material points of a body and their temperatures

$$\mathbf{x} = \chi(X, t) \qquad \theta = \theta(X, t)$$

as the independent constitutive variables, and the energy equation, expressed by way of the Clausius–Duhem entropy inequality in the form

$$\rho\gamma\theta = \rho(\theta\dot{\eta} - \dot{\varepsilon}) + \operatorname{tr}(\mathbf{TD}) + \frac{1}{\theta} q \cdot \operatorname{grad} \theta \geqslant 0$$

identifies $\mathbf{T}$, q, ε, and η as the dependent constitutive variables, where η is the entropy density and γ is the local entropy production.

The general continuum mechanics behavior of a material may be defined by a system of constitutive equations of the form

$$\Phi(t) = \underset{s=0}{\overset{\infty}{\mathscr{G}}} [\Psi^t(s)] \tag{5-1-1}$$

where Φ is some state function such as the stress tensor, the internal energy, entropy, etc., and

$$\Psi^t(s) = \Psi(t-s) \tag{5-1-2}$$

is a function which represents the history of the configuration of the material up to the present time t. In Eq. (5-1-1) the function $\mathscr{G}$ is generally referred to as a functional whenever some of the independent variables comprising the function are fields. The explicit notation of Eq. (5-1-1) due to Volterra (1930) will not, in general, be used unless it is essential to indicate the domain ($0 \leqslant s < \infty$) of s. Since the response of the material is characterized by the functional $\mathscr{G}$, called the constitutive functional of the material, the mechanical behavior of the special classes of materials, such as purely elastic materials, purely viscous materials, etc., is obtained by imposing special restrictions on the constitutive functional.

These considerations give rise to the following basic constitutive statement: for a simple thermomechanical process, the principle of determinism asserts that at the point x occupied at time t by an arbitrary particle of a body $\mathscr{B}$ the thermomechanical constitutive functions $\mathbf{T}$, q, ε, and η are uniquely determined by the history of the motion and temperature of the body $\mathscr{B}$.

The principle of determinism applied to a simple thermomechanical process gives the four constitutive equations:

$$\mathbf{T}(X,t) = \mathbf{G}[\chi^t(\cdot), \theta^t(\cdot), X, t] \tag{5-1-3}$$

$$q(X,t) = \mathbf{L}[\chi^t(\cdot), \theta^t(\cdot), X, t] \tag{5-1-4}$$

$$\varepsilon(X,t) = \mathscr{E}[\chi^t(\cdot), \theta^t(\cdot), X, t] \tag{5-1-5}$$

$$\eta(X,t) = \mathscr{H}[\chi^t(\cdot), \theta^t(\cdot), X, t] \tag{5-1-6}$$

where $\mathbf{G}$ and $\mathbf{L}$ are tensor-valued and vector-valued functionals respectively and $\mathscr{E}$ and $\mathscr{H}$ are scalar-valued functionals defined over the fields of real functions $\chi^t(\cdot)$ and $\theta^t(\cdot)$. For the present time t, the function $\chi^t(\cdot)$ describes the fixed history of the entire motion of the body and $\theta^t(\cdot)$ is the function describing the fixed history of the entire temperature field of the body. It is evident that determinism is effectively a principle of exclusion, insofar as it excludes the dependence of the material behavior at X, on any point outside the body and any future events. That is, Eqs. (5-1-3) to (5-1-6) are only applicable to all $X \in \mathscr{B}$ and only valid for all $\tau \leqslant t$. However, the very existence of a constitutive equation implies that all future phenomena concerning the behavior of the body can, in principle, be fully predicted when all past motions of the body are known and provided no chemical changes occur.

Continuum mechanics is essentially concerned with contact forces which are determined by the conditions in the immediate neighborhood of the point of application, and therefore to exclude possible actions at a distance from constitutive equations it is proposed to consider the response at a particle X. This procedure gives rise to the principle of local action, according to which the response at a particle X is determined if the conditions are known in an arbitrarily small neighborhood of X, the motion outside being disregarded. Consider a small neighborhood $\mathcal{N}(X)$ of X and let Z denote a typical particle in $\mathcal{N}(X)$. In a neighborhood $|\mathbf{Z}-X| < \delta$ it will be assumed that $\chi^t_X(\mathbf{Z}, s)$ can be approximated by N terms of a Taylor series expansion, provided δ is sufficiently small, to give

$$\begin{aligned}\chi^t_X(\mathbf{Z}, s) &= \chi^t(\mathbf{Z}, s) - \chi^t(X, s)\\ &= (Z_{\alpha_1} - X_{\alpha_1})\chi^t_{,\alpha_1}(X, s) + \frac{1}{2!}(Z_{\alpha_1} - X_{\alpha_1})(Z_{\alpha_2} - X_{\alpha_2})\chi^t_{,\alpha_1\alpha_2}(X, s) + \cdots\\ &\quad + \frac{1}{N!}(Z_{\alpha_1} - X_{\alpha_1})\cdots(Z_{\alpha_N} - X_{\alpha_N})\chi^t_{,\alpha_1\alpha_2\ldots\alpha_N}(X, s)\end{aligned} \tag{5-1-7}$$

where χ_X, defined to be the localization in $\mathcal{N}(X)$ of the motion χ at time τ, describes the motion of the neighborhood of X as it appears to an observer moving with the particle X. Similarly, for the fixed history of the temperature $\theta^t(\cdot)$ at a particle X, a Taylor series expansion gives

$$\begin{aligned}\theta^t(Z, s) &= \theta^t(X, s) + (Z_{\alpha_1} - X_{\alpha_1})\theta^t_{,\alpha_1}(X, s)\\ &\quad + \frac{1}{2!}(Z_{\alpha_1} - X_{\alpha_1})(Z_{\alpha_2} - X_{\alpha_2})\theta^t_{,\alpha_1\alpha_2}(X, s) + \cdots\\ &\quad + \frac{1}{N!}(Z_{\alpha_1} - X_{\alpha_1})\cdots(Z_{\alpha_N} - X_{\alpha_N})\theta^t_{,\alpha_1\alpha_2\ldots\alpha_N}(X, s)\end{aligned} \tag{5-1-8}$$

Provided the constitutive functionals are sufficiently smooth, the constitutive equation can be expressed in the form

$$\begin{aligned}\mathbf{T}(X, t) = \mathbf{G}[&\chi^t(X, s), \chi^t_{,\alpha_1}(X, s), \ldots, \chi^t_{,\alpha_1\alpha_2\ldots\alpha_N}(X, s),\\ &\theta^t(X, s), \theta^t_{,\alpha_1}(X, s), \ldots, \theta^t_{,\alpha_1\alpha_2\ldots\alpha_N}(X, s), X, t, dX]\end{aligned} \tag{5-1-9}$$

and the material at X for all $\tau \leqslant t$ is said to satisfy the smooth neighborhood hypothesis. If this is the case, it is evident that the behavior of the material point X is not affected by the histories of the distant points from X. That is, the stress depends functionally only on s, the history parameter, and therefore the dependence of the arguments of **G** on X can for simplicity be suppressed. Materials of this type, which depend on the first N gradients of the separate variables, are referred to as materials of grade N.

Simple materials are those corresponding to approximations of grade one, and are characterized by the absence of a microstructure. Many real materials tend to be approximated to these simple materials of grade one. Thus, for simple materials the constitutive equation (5-1-9) has the reduced form

$$\mathbf{T}(t) = \mathbf{G}[\chi^t(s), \mathbf{F}^t(s), \theta^t(s), \boldsymbol{k}^t(s), X, t, dX] \tag{5-1-10}$$

the response functional being assumed to be continuous. In Eq. (5-1-10),

$$k = \operatorname{Grad} \theta \tag{5-1-11}$$

and the differentials dX arise from approximations of the type

$$\chi^t(Z,s) \approx \chi^t(X,s) + \mathbf{F}^t(X,s)\,dX \tag{5-1-12}$$

where $\mathbf{F}^t(X,s)$ is the history up to time t of the deformation gradient $\mathbf{F}$ at X. All directional vectors $dX = Z - X$ originating from X in three dimensions can be decomposed along three vectors $\boldsymbol{H}_\alpha$ $(\alpha = 1,2,3)$ where $\boldsymbol{H}_\alpha$ is a material frame characterizing the local reference configuration B_r. Since the deformation gradient $\mathbf{F}^t(X,s)$ is independent of the neighborhood $\mathcal{N}(X)$ in contrast to $dX = Z - X$ which depends solely on the neighborhood $\mathcal{N}(X)$ in the reference configuration B_r and is independent of the motion χ, it follows that in the case of Eq. (5-1-12) there is no need to retain an explicit dependence of the fundamental $\mathbf{G}$ in Eq. (5-1-10) on Z. Hence, the constitutive equation (5-1-10) can be expressed in the general form

$$\mathbf{T}(t) = \mathbf{G}[\chi^t(s), \mathbf{F}^t(s), \theta^t(s), k^t(s), X, t, \boldsymbol{H}_\alpha] \tag{5-1-13}$$

where the appearance of $\boldsymbol{H}_\alpha$ in the constitutive equation serves to stress the possible directional dependence of the material properties at a material point X. This aspect of constitutive theory, i.e., anisotropy, is further considered in Sec. 5-3.

In the event that no confusion can arise in relation to the intended reference configuration, the explicit dependence of $\mathbf{G}$ upon $\boldsymbol{H}_\alpha$ is often replaced by an implicit dependence, the functional being replaced, for example, by $\mathbf{G}_\kappa$ such that

$$\mathbf{T}(t) = \mathbf{G}_\kappa[\chi^t(s), \mathbf{F}^t(s), \theta^t(s), k^t(s), X, t] \tag{5-1-14}$$

where the subscript κ denotes the reference configuration B_r spanned by $\boldsymbol{H}_\alpha$.

Inhomogeneity is a property of the body as a whole insofar as it relates to the way in which the material neighborhoods are pieced together. In the absence of any inhomogeneous aspects of the body, the explicit dependence of $\mathbf{G}$ on X and t can be suppressed to give

$$\mathbf{T}(t) = \mathbf{G}[\chi^t(s), \mathbf{F}^t(s), \theta^t(s), k^t(s)] \tag{5-1-15}$$

Materials of grade N, of which the simple materials described by constitutive equations of the form given in Eq. (5-1-13) are a limiting case, are said to be memory dependent since the stress at X, at the present time t, depends on the past history of the deformation gradients, the temperature, and the temperature gradients.

In Eq. (5-1-13) the response functional $\mathbf{G}$ is a material property that determines the stress for any choice of histories. For the case in which the domain of the response functionals contain no histories, the constitutive equation (5-1-13) can be expressed in the form

$$\mathbf{T} = \mathfrak{g}(\chi, \mathbf{F}, \theta, k, X, t, \boldsymbol{H}_\alpha) \tag{5-1-16}$$

where the response function $\mathfrak{g}$ is a property of a particle X having experienced a particular history. Since the dependence of $\mathfrak{g}$ upon history is implicit, it follows that if a particular history is replaced by another then $\mathfrak{g}$ must also be replaced by another response function. Materials that have an instantaneously viscous response can be described by generalizing Eq. (5-1-16) to include $\dot{\mathbf{F}}$ among the independent variables; thus

$$\mathbf{T}(t) = \mathfrak{g}(\chi, \mathbf{F}, \dot{\mathbf{F}}, \theta, \boldsymbol{k}, X, t, \boldsymbol{H}_\alpha) \qquad (5\text{-}1\text{-}17)$$

This aspect of constitutive theory is further considered in Sec. 5-5.

5-2 PRINCIPLE OF MATERIAL OBJECTIVITY

Measurements of distances and time intervals can only be made relative to a frame of reference which is given for all time. Material properties are intrinsic properties of a body and therefore effects produced in a material by a given set of causes must be independent of the observer, i.e., the frame of reference to which the effects are referred. These physical considerations give rise to the principle of material objectivity according to which material properties are invariant to observer transformations.

A point $\boldsymbol{x}$ in space and a time t, taken together as a pair $(\boldsymbol{x}, t)$, is called an event, the totality of all events being referred to as space-time. Two observers are said to be equivalent if they agree about:

1. The distance between an arbitrary pair of points in space
2. The orientation in space
3. The time elapsed between an arbitrary pair of events
4. The order in which two distinct events occur

An observer transformation, i.e., a change of frame, is a one-to-one mapping of space-time onto itself in which distances and time intervals are preserved. By selecting an arbitrary origin in space and identifying points with their position vectors, an event $(\boldsymbol{x}, t)$ and its image $(\boldsymbol{x}^*, t^*)$ under a change of frame are related by rigid transformations which can be expressed in terms of an arbitrary orthogonal tensor $\mathbf{Q}(t)$ and a point $\boldsymbol{c}(t)$ in the form

$$\boldsymbol{x}^* = \boldsymbol{c}(t) + \mathbf{Q}(t)\boldsymbol{x} \qquad (5\text{-}2\text{-}1)$$

$$t^* = t - a \qquad (5\text{-}2\text{-}2)$$

where a is a real, arbitrary number. It is assumed that $\boldsymbol{c}(t)$ and $\mathbf{Q}(t)$ are smooth functions of t, the point $\boldsymbol{c}(t)$ depending on both the change of frame and the choice of origin, whereas the transformation $\mathbf{Q}(t)$ is uniquely determined by the change of frame. Thus it is evident from Eqs. (5-2-1) and (5-2-2) that a change of frame, i.e., an observer transformation, can also be regarded as a time-dependent coordinate transformation.

The principle of material objectivity can be stated as follows. For reference frames $(o^*, \boldsymbol{x}^*)$ and $(o, \boldsymbol{x})$ related by Eq. (5-2-1) the corresponding stress tensors $\mathbf{T}^*$ and $\mathbf{T}$ are related by

$$\mathbf{T}^* = \mathbf{Q}\mathbf{T}\mathbf{Q}^T \qquad (5\text{-}2\text{-}3)$$

and the constitutive equation is invariant. Quantities which depend only on the orientation of the reference frame, that is, $\mathbf{Q}$, and not on the other aspects of the motion of the reference frame, such as its translational displacement, velocity and acceleration, angular velocity and angular accelerations, are said to be objective. Objective scalars f, vectors $\boldsymbol{v}$, and tensors $\mathbf{T}$ transform, according to the relations

$$f^* = f \qquad \boldsymbol{v}^* = \mathbf{Q}\boldsymbol{v} \qquad \mathbf{T}^* = \mathbf{Q}\mathbf{T}\mathbf{Q}^T \tag{5-2-4}$$

Objective quantities are, with the exception of rotation, independent of the reference frame. Thus the stress tensor, whose value and functional form depends on the material, is an objective quantity in the sense of Eq. (5-2-4). Consider the constitutive equation (5-1-3). The principle of material objectivity requires that the form of the functional G should be the same as any two objectively equivalent motions. Thus for a process $(\chi^*, \mathbf{T}^*)$ equivalent to $(\chi, \mathbf{T})$,

$$\mathbf{T}(t) = \mathsf{G}[\chi(Z,\tau), \theta(Z,\tau), X, t] \tag{5-2-5}$$

and

$$\mathbf{T}^*(t^*) = \mathsf{G}[\chi^*(Z,\tau^*), \theta(Z,\tau^*), X, t^*] \tag{5-2-6}$$

for all $\mathbf{Q}(t)$, $\boldsymbol{c}(t)$, and a such that

$$\boldsymbol{x}^*(Z,\tau^*) = \boldsymbol{c}(\tau) + \mathbf{Q}(\tau)\boldsymbol{x}(Z,\tau) \tag{5-2-7}$$

$$\tau^* = \tau - a \tag{5-2-8}$$

where τ is now being used as the time variable and all variable particles Z lie within the body $\mathscr{B}$.

The restrictions on the form of the functional G imposed by the principle of material objectivity can be examined by way of three special transformations of the frame $\boldsymbol{x}$. It will be shown that these special transformations enable the position, the time, and the orientation of a moving observer to be transformed away.

(a) Rigid translation of the spatial frame The particular conditions for which $\mathbf{Q}(\tau) = \mathbf{I}$, $a = 0$, $\boldsymbol{c}(\tau) = -\chi(X,\tau)$ determine a rigid translation of the spatial frame of reference in which the fixed material point X at time t remains at the origin. Inserting these conditions into Eqs. (5-2-7) and (5-2-8) gives

$$\chi_X^*(Z,\tau) = \chi(Z,\tau) - \chi(X,\tau) \qquad \tau^* = \tau \tag{5-2-9}$$

$$\mathbf{T}^*(t) = \mathbf{T}(t)$$

Insertion of these conditions into Eq. (5-2-6) gives the constitutive equation (5-2-5) in the reduced form

$$\mathbf{T}(t) = \mathsf{G}[\chi(Z,\tau) - \chi(X,\tau), \theta(Z,\tau), X, t] \tag{5-2-10}$$

(b) Shift of time Select $\mathbf{Q}(\tau) = \mathbf{I}$, $\boldsymbol{c}(\tau) = \boldsymbol{o}$, and $a = t$. This particular shift of time scale makes the present time t the reference time after the change of frame; i.e., setting $t = a$ in Eqs. (5-2-2) and (5-2-8) gives

$$\begin{gathered} \tau^* = \tau - t = -s \qquad t^* = t - t = 0 \qquad \mathbf{T}^*(0) = \mathbf{T}(t) \\ \chi^*(Z,\tau^*) = \chi(Z, t+\tau^*) \end{gathered} \tag{5-2-11}$$

Insertion of these conditions into Eq. (5-2-6) gives

$$\mathbf{T}(t) = \mathbf{T}^*(0) = \mathbf{G}[\chi(Z, t+\tau^*), \theta(Z, t+\tau^*), X, 0] \tag{5-2-12}$$

from which it is evident that $\mathbf{G}$ is independent of t. Combining this result with that of Eq. (5-2-10) gives

$$\mathbf{T}(t) = \mathbf{G}[\chi(Z, t+\tau^*) - \chi(X, t+\tau^*), \theta(Z, t+\tau^*), X] \tag{5-2-13}$$

where it is to be noted that, for all $s \geqslant 0$,

$$\chi_X^*(Z, -s) = \chi_X(Z, t-s) = \chi_X^t(Z, s)$$

From Eq. (5-2-13) it can be concluded that $\mathbf{G}$ is a functional in the history of the relative motions from the present and the temperatures of all particles $Z \in \mathscr{B}$.

Having regard to the principle of local action, for the two equivalent motions χ and χ^* to coincide in some neighborhood $\mathscr{N}(X)$ for all times $\tau \leqslant t$, it is evident from Eq. (5-2-13) that $\mathbf{G}$ is a functional in the history of the localization χ_X^t in $\mathscr{N}(X)$ of the motion χ for $s \geqslant 0$. Hence, from Eq. (5-2-13), there can be obtained the approximation [already given in Eq. (5-1-12)]

$$\chi_X^t(Z, s) \approx \mathbf{F}^t(s)\, dX$$

where the approximation can be made as precise as is required by confining dX to a small enough neighborhood of the origin. Thus, application of the principle of material objectivity reduces the constitutive equation (5-1-13) for a simple material to the form

$$\mathbf{T}(t) = \mathbf{G}[\mathbf{F}^t(s), \theta^t(s), k^t(s), X, H_\alpha] \tag{5-2-14}$$

from which it is evident that dependence on χ^t has been eliminated.

It is important to note that in observer transformations the deformation gradients transform as vectors. This can be shown by assuming that the two frames had the same orientation at some time t' when the neighborhood of X was in the reference configuration so that $dX^* = dX$ at t' giving

$$d\mathbf{x}^* = \mathbf{F}^*\, dX^* = \mathbf{F}^*\, dX \qquad d\mathbf{x} = \mathbf{F}\, dX$$

From Eq. (5-2-4),

$$d\mathbf{x}^* = \mathbf{Q}(t)\, d\mathbf{x} = \mathbf{Q}(t)\mathbf{F}\, dX = \mathbf{F}^*\, dX$$

from which it follows that

$$(\mathbf{F}^* - \mathbf{Q}(t)\mathbf{F})\, dX = 0$$

for all dX in the infinitesimal reference neighborhood of X. Hence,

$$\mathbf{F}^* = \mathbf{QF} \tag{5-2-15}$$

showing that the deformation gradient (which is a tensor) transforms like a vector under a change of frame at time t.

(c) Rigid rotations of the spatial frame Let $c(\tau) = \mathbf{o}$, $a = 0$, and take $\mathbf{Q}(\tau)$ to be arbitrary. This represents an arbitrary time-dependent rotation of the spatial frame of reference. In this rotation, the stress transforms as

$$\mathbf{T}^*(X, t) = \mathbf{Q}(t)\mathbf{T}(X, t)\mathbf{Q}(t)^T \tag{5-2-16}$$

which when applied to Eq. (5-2-13) gives the identity

$$\mathbf{Q}(0)\mathbf{G}[\chi_X^t(Z, s), \theta^t(Z, s), X]\mathbf{Q}(0)^T = \mathbf{G}[\mathbf{Q}(s)\chi_X^t(Z, s), \theta^t(Z, s), X] \tag{5-2-17}$$

The condition represented by Eq. (5-2-17) is the restriction that must be imposed on the functional $\mathbf{G}$ in order that it be objective.

Thus, the restrictions on the form of the functionals imposed by the principle of material objectivity gives Eq. (5-2-14), subject to the condition of Eq. (5-2-17), as the most general constitutive equation for the stress in the thermomechanical theories of continuous media. Equation (5-2-14) must of course satisfy the remaining principles of constitutive theory.

In the above formulation of the principle of material objectivity $\mathbf{Q}(t)$ is an arbitrary orthogonal tensor. The question which now arises is whether the full orthogonal group of transformations should be employed as the basis of the principle of material objectivity. If $\mathbf{Q}(t)$ is defined to be the proper orthogonal group of transformations, the response functionals are said to be spatially hemitropic in their tensor variables. Spatial isotropy is obtained by including reflection, for which case $\mathbf{Q}(t)$ is the full orthogonal group of transformations, the observer transformations being a rotoreflection combining a rotation and a reflection. However, as to whether rotoreflections should be admitted as observer transformations is at present an open question, since intrinsic orientations may be one of the properties of a material. Thus, for the present discussion the proper orthogonal group will be employed as the basis of the principle of material objectivity.

5-3 PRINCIPLE OF MATERIAL INVARIANCE

Structural ordering of the material points in a body gives rise to certain crystallographic symmetries in the material properties. Constitutive equations must be form invariant with respect to material symmetry transformations.

As a basis for developing the concepts underlying the symmetry properties of a material, Noll (1958) considered the effect of a change of reference configuration B_r on the deformation gradient $\mathbf{F}$ and hence upon the general constitutive equation (see also Truesdell and Noll, 1965). Consider the constitutive equation for a simple material in the absence of thermal effects. For this condition Eq. (5-2-14) can be expressed in the form

$$\mathbf{T}(t) = \mathbf{G}_\kappa(\mathbf{F}^t(s)) \tag{5-3-1}$$

$$\mathbf{T}(t) = \mathbf{G}_{\hat{\kappa}}(\hat{\mathbf{F}}^t(s)) \tag{5-3-2}$$

where κ and $\hat{\kappa}$ imply an arbitrary pair of reference configurations B_r and $\hat{B}_r$ of a body $\mathscr{B}$. The deformation gradients $\mathbf{F}$ and $\hat{\mathbf{F}}$ appearing in Eqs. (5-3-1) and (5-3-2) respectively relate the current configurations B_t and $\hat{B}_t$ of $\mathscr{B}$ to B_r and $\hat{B}_r$ respectively. Equation (2-2-2) gives

$$\mathbf{F} = F^i{}_\alpha g_i \otimes G^\alpha \qquad \hat{\mathbf{F}} = \hat{F}^i{}_\alpha g_i \otimes \hat{G}^\alpha \tag{5-3-3}$$

and hence

$$\begin{aligned}\mathbf{F} &= \hat{F}^i{}_\beta \frac{\partial \hat{X}^\beta}{\partial X^\alpha} g_i \otimes G^\alpha \\ &= (\hat{F}^i{}_\beta g_i \otimes \hat{G}^\beta)\left(\frac{\partial \hat{X}^\gamma}{\partial X^\alpha} \hat{G}_\gamma \otimes G^\alpha\right) = \hat{\mathbf{F}}\mathbf{P}\end{aligned} \tag{5-3-4}$$

and similarly

$$\hat{\mathbf{F}} = \mathbf{F}\mathbf{P}^{-1} \tag{5-3-5}$$

where the tensor

$$\mathbf{P} = \frac{\partial \hat{X}}{\partial X} = \frac{\partial \lambda(X)}{\partial X} \qquad \hat{X} = \lambda(X) \tag{5-3-6}$$

is the deformation gradient at X for the deformation λ mapping $B_r \to \hat{B}_r$. It follows from Eqs. (5-3-4) and (5-3-5) that the constitutive equations (5-3-1) and (5-3-2) can be expressed in the form

$$\mathbf{T}(t) = \mathbf{G}_\kappa(\mathbf{F}^t(s)) = \mathbf{G}_{\hat{\kappa}}(\mathbf{F}^t(s)\mathbf{P}^{-1}) \tag{5-3-7}$$

Equation (5-3-7) is the relation which must hold for all $\mathbf{F}$ between functionals for different reference configurations. It is evident from Eq. (5-3-7) that the response due to a given deformation gradient history $\mathbf{F}$ is the same as the response to the deformation gradient which first transforms $\hat{\kappa}$ into κ and is then followed by $\mathbf{F}$.

If for any two neighborhoods $\mathscr{N}(X)$ and $\mathscr{N}(\hat{X})$ there exist reference configurations B_r and $\hat{B}_r$ respectively in which the properties of the neighborhoods are identical, i.e.,

$$\rho_r = \hat{\rho}_r \qquad \mathbf{G}_\kappa = \mathbf{G}_{\hat{\kappa}} \tag{5-3-8}$$

then the two particles X and $\hat{X}$ are said to be materially isomorphic. For the body as a whole to be homogeneous there must exist a compatible reference configuration B_r for the whole body such that the response functional $\mathbf{G}_\kappa$ is the same at each point. Inhomogeneity is a property of the body as a whole relating specifically to the way in which the material neighborhoods are pieced together, in contrast to ρ and $\mathbf{G}$ which characterize properties of the material. The requirement that the response functional $\mathbf{G}$ with respect to B_r shall coincide with the response functional with respect to $\hat{B}_r$ implies that $\mathbf{G}_{\hat{\kappa}}$ is the same functional of the argument $\hat{\mathbf{F}}^t(s) = \mathbf{F}^t(s)\mathbf{P}^{-1}$ as $\mathbf{G}_\kappa$ is of $\mathbf{F}^t(s)$; i.e.,

$$\mathbf{G}_\kappa[\mathbf{F}^t(s)\mathbf{P}^{-1}] = \mathbf{G}_\kappa[\mathbf{F}^t(s)] \tag{5-3-9}$$

In Eq. (5-3-9), $\mathbf{P}$ is a unimodular tensor (that is, $\det \mathbf{P} = \pm 1$), since B_r and $\hat{B}_r$ have the same density. For convenience, set $\mathbf{P}^{-1} = \mathbf{H}$, giving

$$\mathbf{G}(\mathbf{F}^t(s)) = \mathbf{G}(\mathbf{F}^t(s)\mathbf{H}) \tag{5-3-10}$$

where $\mathbf{H}$ represents a unimodular mapping that leaves a response functional $\mathbf{G}$ invariant. Equation (5-3-10) states that the response functional $\mathbf{G}$ gives the same value for any deformation gradient history $\mathbf{F}^t(s)$ whether or not $\mathbf{F}$ is first preceded by the $\mathbf{H}$ which transforms from B_r to $\hat{B}_r$. This is to be compared with the statement concerning change of reference configurations as related to Eq. (5-3-7).

The property represented by Eq. (5-3-10) does not hold for all materials for all unimodular $\mathbf{H}$. For a given class of materials, the set of all unimodular tensors $\mathbf{H}$ for which Eq. (5-3-10) does hold forms a group called the isotropy group or material symmetry group g_κ of the material at the particle whose place in the reference configuration B_r is X. It is a subgroup of the group u of all unimodular transformations.

Since $\mathbf{H}$ belongs to g_κ, $\mathbf{F}^t(s)$ in Eq. (5-3-10) can be replaced by $\mathbf{F}^t(s)\mathbf{P}$ to give

$$\mathbf{G}_{\hat{\kappa}}(\mathbf{F}^t(s)\mathbf{P}) = \mathbf{G}_{\hat{\kappa}}(\mathbf{F}^t(s)\mathbf{P}\mathbf{H}) \tag{5-3-11}$$

Use of Eq. (5-3-7) now gives

$$\mathbf{G}_{\hat{\kappa}}(\mathbf{F}^t(s)\mathbf{P}\mathbf{H}\mathbf{P}^{-1}) = \mathbf{G}_{\hat{\kappa}}(\mathbf{F}^t(s)) \tag{5-3-12}$$

and hence, since $\mathbf{H} \in g_\kappa$, $\mathbf{P}\mathbf{H}\mathbf{P}^{-1}$ is an element of the symmetry group $g_{\hat{\kappa}}$ of the material relative to $\hat{B}_r$, from which it follows that

$$g_{\hat{\kappa}} = \mathbf{P} g_\kappa \mathbf{P}^{-1} \tag{5-3-13}$$

which is due to Noll (1958).

An orthogonal transformation $\mathbf{Q}$ belongs to the isotropy group g_κ if and only if

$$\mathbf{Q}\mathbf{G}(\mathbf{F}^t(s))\mathbf{Q}^T - \mathbf{G}(\mathbf{Q}\mathbf{F}^t(s)\mathbf{Q}^T) \tag{5-3-14}$$

If $\mathbf{Q}$ belongs to g_κ, then so does its inverse $\mathbf{Q}^T = \mathbf{Q}^{-1}$. Equation (5-3-14) can be established by replacing $\mathbf{F}$ by $\mathbf{QF}$ and $\mathbf{H}$ by $\mathbf{Q}^T$ in Eq. (5-3-10) to give

$$\mathbf{G}[\mathbf{Q}\mathbf{F}^t(s)\mathbf{Q}^T] - \mathbf{G}[\mathbf{Q}\mathbf{F}^t(s)] \tag{5-3-15}$$

where $\mathbf{Q}$ is time independent. The restriction imposed by the principle of material objectivity on the functional $\mathbf{G}$ can be expressed by way of Eq. (5-2-17) in the form

$$\mathbf{G}[\mathbf{Q}^t(s)\mathbf{F}^t(s)] = \mathbf{Q}(t)\mathbf{G}[\mathbf{F}^t(s)]\mathbf{Q}(t)^T \tag{5-3-16}$$

for all orthogonal tensor histories $\mathbf{Q}^t(s)$. Setting $\mathbf{Q}^t(s) \equiv \mathbf{Q}$, independent of s, gives

$$\mathbf{G}(\mathbf{Q}\mathbf{F}^t(s)) = \mathbf{Q}\mathbf{G}(\mathbf{F}^t(s))\mathbf{Q}^T \tag{5-3-17}$$

which combined with Eq. (5-3-15) gives Eq. (5-3-14). Thus Eq. (5-3-14) is the functional equation to be solved to determine the orthogonal members $\mathbf{Q}$ of g_κ.

It is evident from Eq. (5-3-14) that since g_κ is a group it contains both the identity transformation $\mathbf{I}$ and the inversion transformation $-\mathbf{I}$. Hence, by the definition of a group, if g_κ contains a certain proper orthogonal transformation $\mathbf{Q}$ it also contains the improper transformation $-\mathbf{Q}$.

For the condition that the material at a particle X is indistinguishable in its response both before and after it has been deformed into a new configuration, the group of all the possible static density-preserving deformations of the reference configuration B_r constitute the isotropy group g_K. If $\mathbf{H} = \text{Grad}\,\lambda$ where $\mathbf{H}$ is a member of the isotropy group, then λ is the motion which carries one configuration into another in which the material is indistinguishable, and $\mathbf{H}$ which is unimodular may, or may not, be orthogonal. A material which possesses a local reference configuration B_r such that the isotropy group contains the full orthogonal group o is said to be isotropic, and B_r is called an undistorted state of the material. For an isotropic material the response in any test to a deformation from an undistorted state is the same irrespective of whether the material is given an arbitrary rotation before the test. This says no more than that there are no preferred orientations of the material relative to the directions associated with the test.

A material is said to be anisotropic (or aelotropic) if its isotropy group relative to an undistorted state is a proper subgroup of the full orthogonal group o.

In principle, the mechanical response of the various idealized materials may now be classified according to the types of isotropy groups. It is of interest to note that out of the infinitely many members of the full orthogonal group, 12 distinct members suffice to describe the symmetries of 32 classes of crystalline materials.

5-4 INTERNAL CONSTRAINTS

Materials may be subjected to geometrical and internal constraints arising from other than material symmetries. For example, a body may be in a state of self-stress that cannot be relieved by a compatible deformation of the whole body. These internal constraints must be regarded as part of the constitutive equations.

The particular internal constraints of present interest can be represented by equations of the form

$$\eta(\mathbf{F}) = 0 \tag{5-4-1}$$

where η is a scalar-valued function. The principle of material objectivity leads, by way of Eq. (5-2-15), to the condition

$$\eta(\mathbf{QF}) = \eta(\mathbf{F}) = 0 \tag{5-4-2}$$

for all proper orthogonal $\mathbf{Q}$. Choosing $\mathbf{Q} = \mathbf{R}^T$ and using the right polar decomposition of Eq. (2-3-2) enables Eq. (5-4-2) to be expressed in the form

$$\eta(\mathbf{R}^T\mathbf{RU}) = \eta(\mathbf{F}) = 0$$

from which it is evident that η depends upon the deformation gradient through $\mathbf{U}$ and hence through the right Cauchy–Green deformation tensor $\mathbf{C}$ defined by Eq. (2-3-12). Hence, application of the principle of material objectivity shows that the internal constraints can be represented by the condition

$$\lambda(\mathbf{C}) = 0 \tag{5-4-3}$$

where $\lambda(\mathbf{C}) = \lambda(\mathbf{U}^2) \equiv \eta(\mathbf{U})$.

A material body $\mathscr{B}$ must be able to accommodate its motion to an internal constraint by way of suitably introduced contact forces. Hence the general constitutive equation governing the stress response of the body $\mathscr{B}$ must be modified to allow these forces to act. This is achieved by amplifying the statement of the principle of determinism given in Sec. 5-1. The required modification centers on the assumption that the part of the stress arising from the effect of the constraint does no work and is not determined by the motion.

For simple materials subject to internal constraints, the required amplification of the basic constitutive statement can be expressed as follows: the stress at time t is uniquely determined by the history of the deformation gradient only to within a stress $\mathbf{N}$ that does no work in any motion compatible with the constraints.

Differentiation of Eq. (5-4-3) with respect to τ gives

$$\dot{\lambda} = \frac{\partial \lambda}{\partial C^{\alpha\beta}} \dot{C}^{\alpha\beta} = \operatorname{tr}(\lambda_{\mathbf{C}} \dot{\mathbf{C}}) \tag{5-4-4}$$

where

$$\lambda_{\mathbf{C}} = \frac{\partial \lambda}{\partial C^{\alpha\beta}} \mathbf{G}^{\alpha} \otimes \mathbf{G}^{\beta} \tag{5-4-5}$$

From Eqs. (2-6-2) and (2-6-10),

$$\dot{\mathbf{C}}(t) = 2\mathbf{F}(t)^T \mathbf{D}(t) \mathbf{F}(t) \tag{5-4-6}$$

Substituting for $\dot{\mathbf{C}}$ in Eq. (5-4-4) from Eq. (5-4-6) gives

$$\operatorname{tr}(\lambda_{\mathbf{C}} \mathbf{F}^T \mathbf{D} \mathbf{F}) = \operatorname{tr}(\mathbf{F} \lambda_{\mathbf{C}} \mathbf{F}^T \mathbf{D}) = 0 \tag{5-4-7}$$

Having regard to Eq. (3-8-5) defining the stress power, the condition that Eq. (5-4-3) shall be a workless constraint can be expressed in the form

$$\operatorname{tr}(\mathbf{N}\mathbf{D}) = 0 \tag{5-4-8}$$

for all tensors $\mathbf{D}$ satisfying Eq. (5-4-7).

From Eqs. (5-4-7) and (5-4-8) it is evident that $\mathbf{N}$ is a scalar multiple of $\mathbf{F}\lambda_{\mathbf{C}}\mathbf{F}^T$:

$$\mathbf{N} = q\mathbf{F}\lambda_{\mathbf{C}}\mathbf{F}^T \tag{5-4-9}$$

from which it is evident that the stress $\mathbf{N}$ is determined, to within a scalar multiplier, by the expression of Eq. (5-4-3) for the constraint.

If there are several constraints represented by the equation

$$\lambda_m(\mathbf{C}) = 0 \qquad \text{for } m = 1, 2, \ldots, p$$

then

$$\mathbf{N} = \sum_{m=1}^{p} q_m \mathbf{F} (\lambda_m)_{\mathbf{C}} \mathbf{F}^T \tag{5-4-10}$$

The tensor $\mathbf{T}+\mathbf{N}$ is called the extra stress $\mathbf{T}_E$; i.e.,

$$\mathbf{T}_E = \mathbf{T} + \sum_{m=1}^{p} q_m \mathbf{F}(\lambda_m)_{\mathbf{C}} \mathbf{F}^T \tag{5-4-11}$$

An example of particular interest is that of an incompressible material. A material which can only undergo isochoric motions is said to be incompressible. For isochoric motions, $J(\tau) = |\det \mathbf{F}(\tau)| = 1$ and hence $\det \mathbf{C}(\tau) = 1$, from which it is evident that an appropriate form of Eq. (5-4-3) is

$$\lambda(\mathbf{C}) = \det \mathbf{C} - 1 = 0 \tag{5-4-12}$$

Now

$$\dot{\lambda} = \det \mathbf{C} \operatorname{tr}(\dot{\mathbf{C}}\mathbf{C}^{-1}) = \operatorname{tr}(\mathbf{C}^{-1}\dot{\mathbf{C}}) \tag{5-4-13}$$

which relation, when compared with Eq. (5-4-4), gives

$$\lambda_{\mathbf{C}} = \mathbf{C}^{-1} = \mathbf{F}^{-1}(\mathbf{F}^{-1})^T$$

from which it follows, by way of Eq. (5-4-9), that the constraint stress

$$\mathbf{N} = P\mathbf{I} \tag{5-4-14}$$

and hence the extra stress

$$\mathbf{T}_E = \mathbf{T} + P\mathbf{I} \tag{5-4-15}$$

the disposable scalar P $(\equiv q)$ here being identified as pressure.

5-5 SMOOTH MEMORY: RATE TYPE MATERIALS

The concept of smooth memory asserts that the value of the constitutive functional at distant past relative to the present does not appreciably affect its present value. This concept says no more than that only a very short part of the history of past motions and temperatures determines the response of the system; i.e., the memory of past motions and temperatures of any material point decays rapidly. Thus, the concept of smooth memory can be regarded as the counterpart in the domain of time to the concept of local action.

To formalize the concept of smooth memory, assume that the histories of $\boldsymbol{\chi}^t(s)$ and $\theta^t(s)$ can be approximated in the interval $|t-\tau_0| < \delta$ by N terms of a Taylor series expansion; i.e.,

$$\chi(\tau) = \chi(t) + (t-\tau)\dot{\chi}(t) + \cdots + \frac{(t-\tau)^N}{N!}\dot{\chi}_{(N)}(t) \tag{5-5-1}$$

and

$$\theta(\tau) = \theta(t) + (t-\tau)\dot{\theta}(t) + \cdots + \frac{(t-\tau)^N}{N!}\dot{\theta}_{(N)}(t) \tag{5-5-2}$$

where $\tau \in (\tau_0, t]$, the time constant τ_0 being characteristic of each material, and

$$\dot{\theta}_{(N)}(t) = \frac{\partial^N \theta}{\partial t^N} \qquad \dot{\chi}_{(N)}(t) = \frac{\partial^N \boldsymbol{\chi}}{\partial t^N}$$

The constitutive functionals are said to satisfy the concept of smooth memory, provided they are sufficiently smooth to permit χ^t and θ^t to be replaced by the real functions

$$\chi^t, \dot{\chi}^t, \ldots, \dot{\chi}^t_{(N)}$$
$$\theta^t, \dot{\theta}^t, \ldots, \dot{\theta}^t_{(N)}$$

such materials being referred to as materials of the differential type, and have constitutive equations of the form

$$\mathbf{T}(t) = \mathfrak{g}[\mathbf{F}^t, \dot{\mathbf{F}}^t_{(L)}, \theta^t, \dot{\theta}^t_{(M)}, k^t, k^t_{(P)}, X, t, \boldsymbol{H}_\alpha]$$
$$\text{for } L = 1, 2, \ldots, N;\ M = 1, 2, \ldots, N;\ P = 1, 2, \ldots, N \qquad (5\text{-}5\text{-}3)$$

where $\mathfrak{g}$, which is no longer a functional, is a tensor-valued function of the arguments.

A further generalization gives, for example, for a simple material in the absence of thermal effects, a differential equation of the form

$$\dot{\mathbf{T}}_{(p)} = \mathfrak{g}[\mathbf{T}, \dot{\mathbf{T}}, \ldots, \dot{\mathbf{T}}_{(p-1)}; \mathbf{F}, \dot{\mathbf{F}}, \ldots, \dot{\mathbf{F}}_{(N)}] \qquad (5\text{-}5\text{-}4)$$

where

$$\dot{\mathbf{T}}_{(p)} = \frac{\partial^p \mathbf{T}}{\partial t^p} \qquad \dot{\mathbf{F}}_{(N)} = \frac{\partial^N \mathbf{F}}{\partial t^N}$$

Materials conforming to Eq. (5-5-4) are referred to as materials of the rate type.

5-6 REDUCED CONSTITUTIVE EQUATIONS

From the definition of a simple material given in Sec. 5-1 and from Eq. (5-2-14) it is evident that, in the absence of thermal effects, the functional definition of a simple material takes the form

$$\mathbf{T}(X, t) = \mathbf{G}(\mathbf{F}^t(s), X, t) \qquad (5\text{-}6\text{-}1)$$

Under arbitrary rotations, the principle of material objectivity imposes the restriction that $\mathbf{G}$ must satisfy the relation

$$\mathbf{Q}(0)\mathbf{G}(\mathbf{F}^t(s), X, t)\mathbf{Q}(0)^T = \mathbf{G}(\mathbf{Q}(s)\mathbf{F}^t(s), X, t) \qquad (5\text{-}6\text{-}2)$$

According to Eq. (2-3-2), the deformation gradient $\mathbf{F}$ has the decomposition

$$\mathbf{F}^t(s) = \mathbf{R}^t(s)\mathbf{U}^t(s) \qquad (5\text{-}6\text{-}3)$$

where $\mathbf{R}^t(s)$ is the rotation tensor and $\mathbf{U}^t(s)$ is the right stretch tensor. By making the special choice

$$\mathbf{Q}(s) = [\mathbf{R}^t(s)]^T = \mathbf{R}(\tau)^T \qquad (5\text{-}6\text{-}4)$$

Eq. (5-6-2) becomes

$$\mathbf{G}(\mathbf{R}(t)^T\mathbf{F}^t(s), X, t) = \mathbf{R}(t)^T\mathbf{G}(\mathbf{F}^t(s), X, t)\mathbf{R}(t)$$

and since

$$\mathbf{R}^T\mathbf{F} = (\mathbf{R}^T\mathbf{R})\mathbf{U} = \mathbf{U}$$

it follows that

$$\mathbf{G}(\mathbf{U}^t(s), X, t) = \mathbf{R}(t)^T\mathbf{G}(\mathbf{F}^t(s), X, t)\mathbf{R}(t)$$

which in turn, by noting that $\mathbf{R}^T\mathbf{R} = \mathbf{I}$, can be transformed to

$$\mathbf{G}(\mathbf{F}^t(s), X, t) = \mathbf{R}(t)\mathbf{G}(\mathbf{U}^t(s), X, t)\mathbf{R}(t)^T$$

Hence the functional equation (5-6-1) for a simple material admits the reduced representation

$$\mathbf{T}(X, t) = \mathbf{R}(t)\mathbf{G}(\mathbf{U}^t(s), X, t)\mathbf{R}(t)^T \tag{5-6-5}$$

Although $\mathbf{R}$ and $\mathbf{U}$ have direct physical significance, neither can be simply calculated from the motion $\chi(X, t)$. However, $\mathbf{F}$ and therefore the deformation tensors $\mathbf{C}$ and $\mathbf{B}$, defined by Eqs. (2-3-12a) and (2-3-12b) respectively, can be readily calculated. Hence, defining another response function $\mathfrak{p}$ by way of the relation

$$\mathfrak{p}(\mathbf{U}^2) = \mathbf{G}(\mathbf{U}) = \mathfrak{p}(\mathbf{C})$$

it follows that Eq. (5-6-5) can be expressed in the alternative form

$$\mathbf{T}(X, t) = \mathbf{R}(t)\mathfrak{p}(\mathbf{C}^t(s), X, t)\mathbf{R}(t)^T \tag{5-6-6}$$

This is an important result which shows that:

1. The stress at time t is affected only by the present value of the rotation $\mathbf{R}(t)$, and not by the past history of the rotation $\mathbf{R}^t(s)$ for $s > 0$.
2. The stress, while initially assumed to depend on the history $\mathbf{F}^t(s)$ of the deformation gradient, can only depend on the history of one of the strain measures, such as the right stretch $\mathbf{U}$, the deformation tensor $\mathbf{C}$, or the strain $\mathbf{E} = \frac{1}{2}(\mathbf{C} - \mathbf{I})$.

Equation (5-6-5) states that the stress at a particle X at time t is the stress at time t that would result from a pure stretch history $\mathbf{U}^t(s)$ relative to a reference configuration B_r, followed by a rotation $\mathbf{R}(t)$ relative to B_r at time t. It is, however, possible to obtain a form of the general constitutive equation in which the configuration at the present time t is taken as the reference configuration, a form useful for describing the behavior of a fluid since fluids have no preferred reference configuration. Equations (2-2-15) and (2-3-2) can be combined to give

$$\mathbf{F}(\tau) = \mathbf{R}_{(t)}(\tau)\mathbf{R}(t)\mathbf{U}^*_{(t)}(\tau)\mathbf{U}(t) \qquad \tau = t - s \tag{5-6-7}$$

where

$$\mathbf{U}^*_{(t)}(\tau) = \mathbf{R}(t)^T\mathbf{U}_{(t)}(\tau)\mathbf{R}(t) \tag{5-6-8}$$

the subscript (t) denoting the use of the present configuration at time t as the reference configuration. By making the special choice

$$\mathbf{Q}(s) = [\mathbf{R}^t_{(t)}(s)\mathbf{R}(t)]^T$$

it follows that

$$\mathbf{Q}(s)\mathbf{F}^t(s) = \mathbf{U}^*_{(t)}(\tau)\mathbf{U}(t)$$

and for arbitrary rotations this gives, by way of Eq. (5-6-2),

$$\mathbf{T}(t) = \mathbf{G}(\mathbf{F}^t(s)) = \mathbf{R}(t)\mathbf{G}(\mathbf{U}^*_{(t)}(\tau)\mathbf{U}(t))\mathbf{R}(t)^T$$

or

$$\mathbf{T}(t) = \mathbf{R}(t)\mathbf{H}(\mathbf{U}^*_{(t)}(\tau), \mathbf{U}(t))\mathbf{R}(t)^T$$

The right Cauchy–Green deformation tensor can be introduced by noting that

$$\mathbf{C}^*_{(t)}(\tau) = [\mathbf{U}^*_{(t)}(\tau)]^2 = \mathbf{R}(t)^T\mathbf{C}_{(t)}(\tau)\mathbf{R}(t) \tag{5-6-9}$$

and defining a functional **L** by the relation

$$\mathbf{L}(\mathbf{U}^*_{(t)}(\tau)^2, \mathbf{U}(t)^2) \equiv \mathbf{H}(\mathbf{U}^*_{(t)}(\tau), \mathbf{U}(t))$$

to give

$$\mathbf{T}(t) = \mathbf{R}(t)\mathbf{L}(\mathbf{C}^*_{(t)}(\tau), \mathbf{C}(t))\mathbf{R}(t)^T \tag{5-6-10}$$

For a body held in the rest state for a sufficient length of time, the stress will depend only on the configuration of the rest state. This situation can be recognized by separating the response functional **L** into an equilibrium term $\mathfrak{p}(\mathbf{C})$ and a term that vanishes when the material has always been at rest, giving, by way of the strain history tensor

$$\mathbf{G}^*(s) = \mathbf{C}^*_{(t)}(\tau) - \mathbf{I} \qquad \text{for } \mathbf{G}^*(0) = \mathbf{O} \tag{5-6-11}$$

and Eq. (5-6-10), the relation

$$\mathbf{R}^T\mathbf{T}\mathbf{R} = \mathfrak{p}(\mathbf{C}) + \mathfrak{q}(\mathbf{G}^*(s), \mathbf{C}) \tag{5-6-12}$$

where the dependence on the present time has been suppressed. For the rest history the functional $\mathfrak{q}$ satisfies the condition

$$\mathfrak{q}(\mathbf{G}^*(s), \mathbf{C}) = 0 \tag{5-6-13}$$

for $\mathbf{G}^*(s) = \mathbf{O}$, Eq. (5-6-12) giving the relation

$$\mathbf{T} = \mathbf{R}\mathfrak{p}(\mathbf{C})\mathbf{R}^T \tag{5-6-14}$$

for a static deformation [cf. Eq. (5-6-6)].

5-7 SIMPLE MATERIALS

A simple material is defined to be a *simple fluid* if there is no detectable difference in its response functionals as arising from different configurations having the same density. Thus, for a simple fluid, the isotropy group g is the full unimodular group, that is, $g = u$ for all density-preserving deformations. For a simple material to be isotropic g must contain the full orthogonal group o, that is, $o \subset g$, and since $o \subset u$ it follows that all simple fluids, as defined, are isotropic. In Sec. 5-3, the quantity **H** was introduced as the unimodular transformation gradient from the reference configuration B_r to the reference configuration $\hat{B}_r$. If $\mathbf{H} \in g_\kappa$, then Eq. (5-3-12) states that the transformation

$$\hat{\mathbf{H}} = \mathbf{PHP}^{-1} \tag{5-7-1}$$

belongs to the isotropy group $g_{\hat{\kappa}}$ of the response functional $\mathbf{G}_{\hat{\kappa}}$. But using $g_\kappa = u$ it follows from Eq. (5-7-1) that

$$g_{\hat{\kappa}} = \mathbf{P}u\mathbf{P}^{-1} \tag{5-7-2}$$

from which it is evident that $g_{\hat{\kappa}} = u$, which implies that every configuration of a fluid is one for which the isotropy group is the full unimodular group, there being no preferred configurations in a fluid. Thus Eq. (5-3-10) is valid for all **H** in u for all B_r, from which it follows that Eq. (5-3-7) has the form

$$\mathbf{G}_\kappa(\mathbf{F}^t(s)) = \mathbf{G}_{\hat{\kappa}}(\mathbf{F}^t(s)) \tag{5-7-3}$$

subject to Eq. (5-3-6) and the condition det $\boldsymbol{P} = \pm 1$. Since, from Sec. 3-1,

$$\rho_r = |\det \mathbf{F}|\rho \tag{5-7-4}$$

it follows that the densities ρ_r and $\hat{\rho}_r$ in the reference configurations B_r and $\hat{B}_r$ must be related by the equation

$$\rho_r = |\det \mathbf{P}|\hat{\rho}_r \tag{5-7-5}$$

from which it follows that Eq. (5-7-3) applies, subject to the condition that

$$\rho_r = \hat{\rho}_r \tag{5-7-6}$$

Thus, for a fluid the condition that the form of the response functional $\mathbf{G}_\kappa$ depends only on the density ρ_r and not on any other properties of the reference configuration B_r at $\boldsymbol{X}$, in particular its orientation and shape, reduces the general constitutive equation (5-6-5) to the form

$$\mathbf{T}(t) = \mathbf{R}(t)\mathfrak{f}(\mathbf{U}^t(s), \rho_r)\mathbf{R}(t)^T \tag{5-7-7}$$

where ρ_r is the reference density in the reference configuration with respect to which $\mathbf{U}^t(s)$ is determined. Alternatively, taking the configuration at the present time t as the reference configuration, Eq. (5-7-7) becomes

$$\mathbf{T}(t) = \mathfrak{f}(\mathbf{U}^t_{(t)}(s), \rho(t)) \tag{5-7-8}$$

The Eq. (5-3-10) can be expressed in the form

$$\mathbf{G}(\mathbf{F}^t(s)) = \mathbf{G}(\mathbf{F}^t(s)\mathbf{H})$$

and replacing **F** by **QF** and **H** by $\mathbf{Q}^T$ and applying Eq. (5-2-17) with $\mathbf{Q}(s)$ a constant gives

$$\mathbf{QG}(\mathbf{F}^t(s))\mathbf{Q}^T = \mathbf{G}(\mathbf{QF}^t(s)\mathbf{Q}^T) \tag{5-7-9}$$

as the condition that an orthogonal transformation $\mathbf{Q}$ (and therefore also $\mathbf{Q}^T$) belongs to the isotropy group g. Since for a simple fluid the isotropy group u contains the orthogonal group o, Eq. (5-7-9) can be applied to Eq. (5-7-8) giving

$$\mathbf{Q}\mathfrak{f}(\mathbf{U}_{(t)}^t(s), \rho(t))\mathbf{Q}^T = \mathfrak{f}(\mathbf{Q}\mathbf{U}_{(t)}^t(s)\mathbf{Q}^T, \rho(t)) \tag{5-7-10}$$

for all orthogonal transformations $\mathbf{Q}$.

In Eq. (5-6-12), if the present configuration is used as the reference configuration, the argument in ρ_r is $\mathbf{C}_{(t)}(t) \equiv \mathbf{I}$; hence ρ_r must depend only on the density of B_r, so that $\mathfrak{p}_\kappa$ reduces to $\mathfrak{p}(\rho(t))$, to which in turn can be applied Eq. (5-7-10) giving

$$\mathbf{Q}\mathfrak{p}(\rho(t))\mathbf{Q}^T = \mathfrak{p}(\rho(t))$$

from which it is evident that $\mathfrak{p}$ is an isotropic tensor such that $\mathfrak{p}(\rho) = -P(\rho)\mathbf{I}$. Thus, equivalent reasoning applied to Eq. (5-6-12) gives, for the constitutive equation of a simple fluid, the form

$$\mathbf{T}(t) = -P(\rho(t))\mathbf{I} + \mathfrak{q}(\mathbf{G}(s), \rho(t)) \tag{5-7-11}$$

subject to the condition

$$\mathbf{Q}\mathfrak{q}(\mathbf{G}(s), \rho(t))\mathbf{Q}^T = \mathfrak{q}(\mathbf{Q}\mathbf{G}(s)\mathbf{Q}^T, \rho(t)) \tag{5-7-12}$$

where

$$\mathbf{G}(s) = \mathbf{C}_{(t)}^t(s) - \mathbf{I} \equiv \mathbf{R}\mathbf{G}^*(s)\mathbf{R}^T \tag{5-7-13}$$

From the discussion given in Sec. 5-4 of an incompressible material, in particular, Eq. (5-4-14), it is evident that, for an incompressible material, Eq. (5-7-11) takes the form

$$\mathbf{T}(t) = -P\mathbf{I} + \mathfrak{q}(\mathbf{G}(s)) \tag{5-7-14}$$

where, since the function $\mathfrak{q}$ is only determined to within an arbitrary isotropic functional, the first term on the right-hand side of Eq. (5-7-11) has been absorbed into $\mathfrak{q}$ and $\rho(t)$ is omitted since ρ is no longer time dependent.

If, as a result of deforming a material from one reference configuration to another, it is observed that, in general, there is a difference in the mechanical response from the two reference configurations, then the material is said to be a solid. This may be illustrated as follows. Consider a solid deformed from a stress-free state to a state in which it is necessary to maintain an applied stress in order to retain the new configuration. Nonzero stress is required to maintain zero strain relative to this new configuration, whereas zero stress is required for zero strain relative to the initial configuration. These considerations lead to the following definition of a simple solid.

A *simple solid* is a simple material for which at X there is a reference configuration B_r such that the associated isotropy group g is a subgroup of the orthogonal group o; i.e.,

$$g \subset o \qquad \text{(simple solid)}$$

The preferred reference configurations satisfying this condition are identified as the undistorted states of the solid.

It has already been noted in Sec. 5-3 that a simple material is called *isotropic* at a particle X if it possesses at least one local configuration such that for the response functional from that configuration the isotropy group g contains the full orthogonal group o; i.e.,

$$o \subset g \qquad \text{(isotropic material)}$$

The preferred reference configurations satisfying this condition are identified as the undistorted states of the isotropic material.

Thus, the definitions of a simple solid and an isotropic material may be combined to give a definition of an undistorted state of a simple material: an undistorted state of a simple material is a reference configuration for which the isotropy group contains the full orthogonal group or is a subgroup of it.

A theorem of group theory states that there is no group g which is a proper subgroup of the full unimodular group u and which in turn contains the full orthogonal group o as a proper subgroup (Brauer, 1965; Noll, 1965). Thus, if $o \subset g \subset u$, then either $o = g$ or else $g = u$; i.e., the orthogonal group is maximal in the unimodular group. It can therefore be concluded that isotropic simple materials must be either simple fluids or isotropic simple solids. Thus, for any undistorted state, if

$$g = u \qquad \text{(isotropic simple solid)}$$

then the simple material is said to be an isotropic simple solid.

If a simple material possesses a local reference configuration such that the isotropy group g is a proper subgroup of the full orthogonal group o, then the material is said to be a simple anisotropic (or aelotropic) solid. The concept of material symmetry is of great importance in obtaining a reduction in the number of distinct elastic coefficients required to specify the mechanical response of a solid. In solids, symmetry operations are rotations (and rotoreflections). This follows from Sec. 5-3 where it was noted that since g is a group it contains both the identity transformation $\mathbf{I}$ and the inversion transformation $-\mathbf{I}$, from which it is evident that g can be generated by a subgroup g_R, i.e., rotations, and $-\mathbf{I}$. Transverse isotropy is a continuous group of material symmetries defined by a rotation group g_R in an undistorted state for which any rotations about the axis of a given direction is a symmetry operation. That is, a material exhibits transverse isotropy if it has a single axis of rotational symmetry, every plane containing this axis being a plane of reflection symmetry (this applies to materials with a laminated structure). A material for which g consists of only $\mathbf{I}$ and $-\mathbf{I}$ is a solid corresponding to the triclinic system.

For the condition that $\mathbf{H}$ is in the isotropy group g_κ corresponding to the reference configuration B_r there can be obtained, from Eqs. (5-3-10) to (5-3-12), the relations

$$\mathbf{G}_{\hat{\kappa}}(\hat{\mathbf{F}}^t(s)) = \mathbf{G}_{\hat{\kappa}}(\hat{\mathbf{F}}^t(s)\hat{\mathbf{H}}) \tag{5-7-15}$$

where

$$\hat{\mathbf{H}} = \mathbf{PHP}^{-1} \tag{5-7-16}$$

and $\hat{\mathbf{H}}$ is in the isotropy group $g_{\hat{\kappa}}$ corresponding to the reference configuration $\hat{B}_r$. That is, Eq. (5-7-16) describes the transformation of the elements of an isotropy group in a change of local reference configuration (this has already been considered within the particular context of the simple fluid). If $\mathbf{P}$ is orthogonal, then so is $\hat{\mathbf{H}}$, and $\hat{B}_r$ is also an undistorted state. Thus it may be concluded that, in general, any orthogonal transformation of an undistorted state is an undistorted state, and the isotropy groups are similar. To investigate the undistorted configurations of simple solids, replace $\mathbf{H}$ in Eq. (5-7-16) by an orthogonal tensor $\mathbf{S}$, subject to

$$\hat{\mathbf{S}}\mathbf{P} = \mathbf{PS}$$

Now $\mathbf{P}$ admits the polar decomposition

$$\mathbf{P} = \mathbf{RU}$$

where $\mathbf{R}$ is orthogonal and $\mathbf{U}$ is positive and symmetric; hence

$$(\hat{\mathbf{S}}\mathbf{R})\mathbf{U} = (\mathbf{RS})(\mathbf{S}^T\mathbf{US})$$

The uniqueness of this polar decomposition now implies that

$$\hat{\mathbf{S}} = \mathbf{RSR}^T \tag{5-7-17}$$

and

$$\mathbf{U} = \mathbf{SUS}^T \tag{5-7-18}$$

Equation (5-7-18) is interpreted as a condition that the stretch tensor $\mathbf{U}$ of $\mathbf{P}$ must commute with $\mathbf{S}$; i.e.,

$$\mathbf{US} = \mathbf{SU} \tag{5-7-19}$$

The symmetry operations restrict, through Eq. (5-7-19), the stretch tensor $\mathbf{U}$ relating the undistorted configurations of a solid. Consider the stress on undistorted configurations, $\mathbf{T}_0 = \mathbf{G}(\mathbf{I})$. Material objectivity requires [Eq. (5-2-17)]

$$\mathbf{Q}(0)\mathbf{G}(\mathbf{F}^t(s))\mathbf{Q}(0)^T = \mathbf{G}(\mathbf{Q}(s)\mathbf{F}^t(s))$$

to hold for all orthogonal tensors $\mathbf{Q}$ and all deformation gradients $\mathbf{F}$. Now identify $\mathbf{Q}$ with a symmetry operation $\mathbf{S}$ and replace $\mathbf{F}$ by $\mathbf{FS}^T$ to give

$$\mathbf{SG}(\mathbf{F}^t(s))\mathbf{S}^T = \mathbf{G}(\mathbf{SF}^t(s)\mathbf{S}^T) \tag{5-7-20}$$

which is simply Eq. (5-3-14); i.e., the response functional is isotropic to all symmetry operations. For an undistorted configuration, $\mathbf{F}^t(s) = \mathbf{I}$, and Eq. (5-7-20) gives

$$\mathbf{SG}(\mathbf{I})\mathbf{S}^T = \mathbf{ST}_0\mathbf{S}^T = \mathbf{G}(\mathbf{I}) = \mathbf{T}_0$$

which can be interpreted as a condition that $\mathbf{T}_0$ must commute with $\mathbf{S}$; i.e.,

$$\mathbf{ST}_0 = \mathbf{T}_0\mathbf{S} \tag{5-7-21}$$

As already noted, for the triclinic class $\mathbf{S} = \mathbf{I}$, and since $\mathbf{I}$ commutes with every tensor, every configuration is undistorted. An undistorted state which corresponds to zero stress is referred to as a natural state.

For simple materials for which the stress depends only on the rest configuration and not on the strain history prior to the start of the static deformation, it has been shown in Sec. 5-6 that Eq. (5-6-14) in the form

$$\mathbf{T} = \mathbf{R}\mathfrak{p}(\mathbf{C})\mathbf{R}^T \tag{5-7-22}$$

is the constitutive equation for such materials. Thus Eq. (5-7-22) is the constitutive equation for simple solids which have been at rest for a long time, and is the starting point for the study of elasticity.

GENERAL REFERENCES

Guggenheim, E. A. (1949) "Thermodynamics—An Advanced Treatise for Chemists and Physicists," Interscience, New York.

Guggenheim, E. A. (1959) Thermodynamics, Classical and Statistical, in S. Flügge (ed.), "Handbuch der Physik," vol. III/2, pp. 1–118, Springer-Verlag, Berlin.

Truesdell, C., and W. Noll (1965) The Non-linear Field Theories of Mechanics, in S. Flügge (ed.), "Handbuch der Physik," vol. III/3, Springer-Verlag, Berlin.

Volterra, V. (1930) "Theory of Functionals and Integral and Integrodifferential Equations," Blackie Glasgow.

REFERENCES

Brauer, R. (1965) *Arch. Rational Mech. Anal.*, vol. 18, pp. 97–99.

Duhem, P. (1901) *C.R. Acad. Sci. (Paris)*, vol. 132, pp. 117–120.

Noll, W. (1958) *Arch. Rational Mech. Anal.*, vol. 2, pp. 197–226.

Noll, W. (1965) *Arch. Rational Mech. Anal.*, vol. 18, pp. 100–102.

Wang, C. C., and R. M. Bowen (1966) *Arch. Rational Mech. Anal.*, vol. 22, pp. 79–99.

CHAPTER

SIX

LINEAR ELASTICITY

6-1 CLASSICAL ELASTICITY

6-1-1 Generalized Hooke's Law

The foundation of the classical theory of purely elastic materials is Hooke's law, which states that the normal stress is proportional to the elongation per unit length. Hooke's law is the constitutive law for a hookean or linear elastic material for the limit of small strains. In addition to Hooke's law, the classical theory of elasticity involves two concepts which can be stated as follows. A purely elastic body is continuous and remains continuous under the action of external forces. There exists a unique unstressed state of the body, i.e., a natural state to which the body returns whenever all the external forces are removed.

Following the introduction of the concepts of stress and strain, Hooke's law can be generalized by specifying that the components of stress are to be linearly related to the components of strain. The classical elastic constitutive equations expressed in this form are often referred to as the generalized Hooke's law and comprise nine equations expressing the stress components as linear homogeneous functions of the nine strain components. For example, consider the finite strain tensor $\mathbf{E}$ and the second Piola–Kirchhoff stress tensor $\tilde{\mathbf{T}}$ to be taken as the constitutive field variables referred to the natural state of the material. Under triaxial loading, classical elasticity theory now assumes a generalized Hooke's law of the form

$$\tilde{\mathbf{T}} = \tilde{\mathbf{C}}\mathbf{E} \qquad \tilde{T}^{\alpha\beta} = \tilde{C}^{\alpha\beta\gamma\delta}E_{\gamma\delta} \tag{6-1-1}$$

which expresses each stress component as a linear combination of all the strains, $\tilde{\mathbf{C}}$ being the tensor of the elastic constants, or moduli of the material.

A condition of particular interest is that for which the displacement gradients are everywhere small compared to unity, such that the displacements are sufficiently small for there to be no need to distinguish between the initial referential coordinates X_α and the deformed spatial position x_i of the same particle in formulating the field equations. For these special conditions the constitutive equation equivalent to Eq. (6-1-1) is

$$\mathbf{T} = \mathbf{C}\tilde{\mathbf{e}} \tag{6-1-2}$$

where $\tilde{\mathbf{e}}$ is the infinitesimal strain tensor and $\mathbf{T}$ is the Cauchy stress.

The nine equations represented by Eq. (6-1-2) contain 81 constants. However, for a symmetric state of stress $\mathbf{T} = \mathbf{T}^T$ and so

$$C^{ijkl} = C^{jikl}$$

and since $\tilde{\mathbf{e}} = \tilde{\mathbf{e}}^T$, C^{ijkl} can be symmetrized with respect to k and l without altering the sum. Hence, without loss of generality it may be assumed that

$$C^{ijkl} = C^{ijlk}$$

It follows from these symmetry properties that, in general, the maximum number of independent elastic constants is 36.

Purely elastic materials differ in their rheological behavior from that of fluids in two important respects. The classical theory of elasticity is based on the concept of a natural state to which the material returns when stresses are removed. Consequently, the work expended in deforming a purely elastic material is entirely recovered when the material reverts to the natural state. That is, the deformation of a purely elastic material is a reversible process, which implies that the work of deformation may be conserved in solids, from which it follows that the stress-strain relations should be derivable from a potential function. Such a function W was introduced by Green into the classical theory of elasticity by defining it to be a function of the infinitesimal strain components $\tilde{\mathbf{e}}$, with the property that

$$T^i{}_j = \frac{\partial W}{\partial \tilde{e}_i{}^j} = C^i{}_j{}^k{}_l \tilde{e}_k{}^l \tag{6-1-3}$$

Substituting in Eq. (6-1-3) for $\mathbf{T}$ from Eq. (6-1-2) gives

$$W = \tfrac{1}{2} C^i{}_j{}^k{}_l \tilde{e}_k{}^l \tilde{e}_i{}^j \tag{6-1-4}$$

for the condition that W vanishes in the unstrained state. For the condition that the quadratic form of Eq. (6-1-4) is symmetric it follows that

$$C^i{}_j{}^k{}_l = C^k{}_l{}^i{}_j$$

This symmetry condition further reduces the number of independent elastic constants to 21, which is now generally accepted to be the number of independent elastic constants in the generalized Hooke's law in the most general form of anisotropy.

For particular types of material a further reduction of the number of coefficients follows from the symmetry properties of the material. Considerations of material symmetry, which have been generally discussed in Sec. 5-3, relate to rotations

$$d\hat{X} = \mathbf{P}\,dX \qquad \det \mathbf{P} = 1 \tag{6-1-5}$$

of a small specimen and whether this specimen is being viewed directly or in a mirror for which

$$d\hat{X} = \mathbf{P}\,dX \qquad \det \mathbf{P} = -1 \tag{6-1-6}$$

Material structures are identified in terms of local reference frames that are intrinsic to the material. A symmetry operation **P** transforms one reference frame into another, equivalent, reference frame. The class of symmetry operations defines the anisotropy of a body.

The matrices of coefficients **C** for the various crystal classes take their simplest form when referred to rectangular cartesian axes. Thus for the remainder of this section the discussion will be restricted to rectangular cartesian reference frames, but it should be remembered that the corresponding coefficients for use in general curvilinear coordinates can be obtained from the usual tensor transformation laws.

It is usual to simplify the notation by setting

$$T_{11} = T_1, T_{22} = T_2, T_{33} = T_3, T_{23} = T_4, T_{31} = T_5, T_{12} = T_6$$

$$\tilde{e}_{11} = \tilde{e}_1, \tilde{e}_{22} = \tilde{e}_2, \tilde{e}_{33} = \tilde{e}_3, 2\tilde{e}_{23} = \tilde{e}_4, 2\tilde{e}_{31} = \tilde{e}_5, 2\tilde{e}_{12} = \tilde{e}_6$$

Eq. (6-1-2) now taking the form

$$T_i = C_{ij}\tilde{e}_j \tag{6-1-7}$$

where C_{ij} will be referred to as the stiffness matrix. In the most general form of anisotropy discussed above, which corresponds to triclinic crystal symmetry, the stiffness matrix takes the form

$$\begin{bmatrix} C_{11} & C_{12} & C_{13} & C_{14} & C_{15} & C_{16} \\ C_{21} & C_{22} & C_{23} & C_{24} & C_{25} & C_{26} \\ C_{31} & C_{32} & C_{33} & C_{34} & C_{35} & C_{36} \\ C_{41} & C_{42} & C_{43} & C_{44} & C_{45} & C_{46} \\ C_{51} & C_{52} & C_{53} & C_{54} & C_{55} & C_{56} \\ C_{61} & C_{62} & C_{63} & C_{64} & C_{65} & C_{66} \end{bmatrix} \tag{6-1-8}$$

and remembering that $C_{ij} = C_{ji}$ this corresponds to 21 independent coefficients. If the elastic constants are invariant after a transformation **P** then

$$C_{ijkl} = P_{ia}P_{jb}P_{kc}P_{ld}C_{abcd} \tag{6-1-9}$$

This set of equations provides further relations between the components of **C** and thus reduces the number of independent components, as will now be illustrated by a few specific examples.

(a) Monoclinic crystal symmetry: symmetry with respect to a plane When the elastic constants at a point have the same value for every pair of coordinate systems which are the reflected images of one another with respect to a certain plane, that plane is called a plane of elastic symmetry. Suppose $X_3 = 0$ is the plane of elastic symmetry; then

$$[\mathbf{P}] = \begin{bmatrix} 1 & 0 & 0 \\ 0 & 1 & 0 \\ 0 & 0 & -1 \end{bmatrix} \tag{6-1-10}$$

Substituting in Eq. (6-1-9) it is clear that $a = i, b = j, c = k, d = l$ and that if there are an odd number of three suffixes then that component must be zero. Thus the stiffness matrix takes the form

$$\begin{bmatrix} C_{11} & C_{12} & C_{13} & 0 & 0 & C_{16} \\ C_{21} & C_{22} & C_{23} & 0 & 0 & C_{26} \\ C_{31} & C_{32} & C_{33} & 0 & 0 & C_{36} \\ 0 & 0 & 0 & C_{44} & C_{45} & 0 \\ 0 & 0 & 0 & C_{54} & C_{55} & 0 \\ C_{61} & C_{62} & C_{63} & 0 & 0 & C_{66} \end{bmatrix} \tag{6-1-11}$$

and remembering $C_{ij} = C_{ji}$ there are thus 13 independent components.

(b) Orthorhombic crystal symmetry or orthotropic symmetry These have three orthogonal planes of elastic symmetry which can be taken as the coordinate axes. Thus **C** must be invariant under the three transformations

$$[\mathbf{P}_1] = \begin{bmatrix} -1 & 0 & 0 \\ 0 & 1 & 0 \\ 0 & 0 & 1 \end{bmatrix} \quad [\mathbf{P}_2] = \begin{bmatrix} 1 & 0 & 0 \\ 0 & -1 & 0 \\ 0 & 0 & 1 \end{bmatrix} \quad [\mathbf{P}_3] = \begin{bmatrix} 1 & 0 & 0 \\ 0 & 1 & 0 \\ 0 & 0 & -1 \end{bmatrix} \tag{6-1-12}$$

and proceeding as in example (a) it is clear that coefficients which have an odd number of one, two, or three suffixes must be zero and hence the stiffness matrix becomes

$$\begin{bmatrix} C_{11} & C_{12} & C_{13} & 0 & 0 & 0 \\ C_{21} & C_{22} & C_{23} & 0 & 0 & 0 \\ C_{31} & C_{32} & C_{33} & 0 & 0 & 0 \\ 0 & 0 & 0 & C_{44} & 0 & 0 \\ 0 & 0 & 0 & 0 & C_{55} & 0 \\ 0 & 0 & 0 & 0 & 0 & C_{66} \end{bmatrix} \tag{6-1-13}$$

Taking account of the symmetry of C_{ij} there are thus nine independent components.

(c) Tetragonal symmetry Tetragonal symmetry has an extra axis of symmetry which can be taken as rotation through 90° about the X_3 axis. This ensures that **C** has the form given above with the extra conditions $C_{11} = C_{22}$, $C_{13} = C_{12}$, $C_{44} = C_{55}$, thus reducing the number of independent constants to six.

(d) Cubic symmetry Cubic symmetry applies the same symmetry requirements about the X_1 and X_2 axes and thus **C** has the above form with the extra conditions $C_{11} = C_{22} = C_{33}$, $C_{12} = C_{13} = C_{23}$, $C_{44} = C_{55} = C_{66}$, thus reducing the number of independent components to three.

(e) Hexagonal symmetry This corresponds to symmetry about a plane and rotation through 120° about an axis normal to this plane. If the axis is taken as the X_3 axis, **C** must be invariant to the transformations

$$[\mathbf{P}_1] = \begin{bmatrix} \cos\alpha & \sin\alpha & 0 \\ -\sin\alpha & \cos\alpha & 0 \\ 0 & 0 & 1 \end{bmatrix} \qquad [\mathbf{P}_2] = \begin{bmatrix} \cos 2\alpha & \sin 2\alpha & 0 \\ -\sin 2\alpha & \cos 2\alpha & 0 \\ 0 & 0 & 1 \end{bmatrix}$$

$$[\mathbf{P}_3] = \begin{bmatrix} 1 & 0 & 0 \\ 0 & 1 & 0 \\ 0 & 0 & -1 \end{bmatrix}$$

where $\alpha = 2\pi/3$. **C** then takes the form of Eq. (6-1-13) with the extra conditions $C_{11} = C_{22}$, $C_{13} = C_{23}$, $C_{44} = C_{55}$, $C_{66} = \frac{1}{2}(C_{11} - C_{12})$, which thus reduce the number of independent constants to five.

(f) Transverse isotropy This type of symmetry is not found among the crystal classes, which correspond only to finite deformations, for it is defined as invariant with respect to an arbitrary rotation about a given axis and as such is applicable to fibrous materials. If the axis is taken as $X_3 = 0$, then **C** must be invariant under the transformations

$$[\mathbf{P}] = \begin{bmatrix} \cos\theta & \sin\theta & 0 \\ -\sin\theta & \cos\theta & 0 \\ 0 & 0 & 1 \end{bmatrix} \tag{6-1-14}$$

for arbitrary θ. Clearly the stiffness coefficients must at least satisfy the orthotropic symmetry conditions and there can be no difference between the axes 1 and 2. Applying Eq. (6-1-14), the stiffness matrix takes the form

$$\begin{bmatrix} C_{11} & C_{12} & C_{13} & 0 & 0 & 0 \\ C_{21} & C_{22} & C_{23} & 0 & 0 & 0 \\ C_{31} & C_{32} & C_{33} & 0 & 0 & 0 \\ 0 & 0 & 0 & C_{44} & 0 & 0 \\ 0 & 0 & 0 & 0 & C_{55} & 0 \\ 0 & 0 & 0 & 0 & 0 & C_{66} \end{bmatrix} \tag{6-1-15}$$

Together with the relations $C_{ij} = C_{ji}$, $C_{11} = C_{22}$, $C_{13} = C_{23}$, $C_{44} = C_{55}$, and

$$C_{11} = C_{12} + 2C_{66} \tag{6-1-16}$$

There are thus five independent material constants characterizing transverse isotropy.

(g) Isotropy If the stiffness matrix is invariant to any arbitrary rotation it is said to be isotropic. It must therefore satisfy Eq. (6-1-15), but as there is no difference between any coordinate axis

$$\begin{array}{ll} C_{11} = C_{12} + 2C_{66} & C_{11} = C_{22} = C_{33} \\ C_{44} = C_{55} = C_{66} & C_{12} = C_{23} = C_{31} \end{array} \tag{6-1-17}$$

There are thus only two independent constants characterizing isotropy and it is usual to write the stiffness matrix in the form

$$\begin{bmatrix} \lambda+2G & \lambda & \lambda & 0 & 0 & 0 \\ \lambda & \lambda+2G & \lambda & 0 & 0 & 0 \\ \lambda & \lambda & \lambda+2G & 0 & 0 & 0 \\ 0 & 0 & 0 & G & 0 & 0 \\ 0 & 0 & 0 & 0 & G & 0 \\ 0 & 0 & 0 & 0 & 0 & G \end{bmatrix} \tag{6-1-18}$$

where λ and G are the Lamé elastic constants (Lamé, 1852). It can be shown by inspection that the above matrix can be written in the form

$$C_{ijkl} = \lambda\delta_{ij}\delta_{kl} + G(\delta_{ik}\delta_{jl} + \delta_{il}\delta_{jk}) \tag{6-1-19}$$

Equation (6-1-2) can be expressed in the form

$$T_{ij} = [\lambda\delta_{ij}\delta_{kl} + G(\delta_{ik}\delta_{jl} + \delta_{il}\delta_{kj})]\tilde{e}_{kl} \tag{6-1-20}$$

which in turn gives the general isotropic Hooke's law:

$$\mathbf{T} = \lambda(\mathrm{tr}\,\tilde{\mathbf{e}})\mathbf{I} + 2G\tilde{\mathbf{e}} \tag{6-1-21}$$

Hooke's law may be stated for an isotropic elastic material in the form

$$\mathrm{tr}\,\mathbf{T} = 3\kappa\,\mathrm{tr}\,\tilde{\mathbf{e}} \tag{6-1-22}$$

$$\mathbf{T}' = 2G\tilde{\mathbf{e}}' \tag{6-1-23}$$

where κ and G are constants and

$$\mathbf{T}' = \mathbf{T} - \tfrac{1}{3}(\mathrm{tr}\,\mathbf{T})\mathbf{I} \tag{6-1-24}$$

$$\tilde{\mathbf{e}}' = \tilde{\mathbf{e}} - \tfrac{1}{3}(\mathrm{tr}\,\tilde{\mathbf{e}})\mathbf{I} \tag{6-1-25}$$

are the deviators of the stress tensor $\mathbf{T}$ and strain tensor $\tilde{\mathbf{e}}$ respectively (Reiner, 1960) [see Eq. (3-6-18)].

For the case in which the strain is infinitesimal, $\mathrm{tr}\,\tilde{\mathbf{e}}$ is the change of volume per unit volume and $\frac{1}{3}\,\mathrm{tr}\,\mathbf{T}$ is the mean stress at a point, such that for the special case of hydrostatic compression $\operatorname{tr}\mathbf{T} = -3P$ where P is the pressure. Thus, Eq. (6-1-22), which states that the change in volume of the material is proportional to the mean stress $-P$, may be expressed in the form

$$\frac{\Delta v}{v} = -\frac{P}{\kappa} \tag{6-1-26}$$

where v and Δv denote the volume and change of volume respectively. Equation (6-1-26) relates the bulk changes in the volume of the material, and therefore the coefficient κ is appropriately called the bulk modulus of the material.

Equation (6-1-23) states that the stress deviation is simply proportional to the strain deviation $\tilde{\mathbf{e}}'$, it being noted that $\tilde{\mathbf{e}}'$ describes a deformation without volume change. The constant G is therefore called the shear modulus of rigidity. The coefficient 2 appears in Eq. (6-1-23) because, prior to the introduction of the tensor concept, it was customary to define the shear strain as $\gamma = 2\tilde{\mathbf{e}}'$.

Substituting Eqs. (6-1-24) and (6-1-25) into Eq. (6-1-23) gives, by way of Eq. (6-1-22), the stress-strain relations

$$\mathbf{T} = \lambda(\operatorname{tr}\tilde{\mathbf{e}})\mathbf{I} + 2G\tilde{\mathbf{e}} \tag{6-1-27}$$

or

$$\tilde{\mathbf{e}} = \frac{1+\nu}{E}\mathbf{T} - \frac{\nu}{E}(\operatorname{tr}\mathbf{T})\mathbf{I} \tag{6-1-28}$$

where E is called the modulus of elasticity or Young's modulus (Young, 1807) and the constant ν is called Poisson's ratio.

Several relationships between the constants λ, G, κ, E, and ν can be derived. These are

$$\begin{aligned}\lambda &= \frac{E\nu}{(1+\nu)(1-2\nu)} = \kappa - \tfrac{2}{3}G = \frac{2G\nu}{1-2\nu} \\ &= \frac{G(E-2G)}{3G-E} = \frac{3\kappa\nu}{1+\nu} = \frac{3\kappa(3\kappa-E)}{9\kappa-E}\end{aligned} \tag{6-1-29}$$

$$\begin{aligned}G &= \tfrac{3}{2}\kappa\frac{1-2\nu}{1+\nu} = 3\frac{\kappa E}{9\kappa-E} = \frac{\lambda}{2\nu}(1-2\nu) \\ &= \frac{E}{2(1+\nu)} = \tfrac{3}{2}(\kappa-\lambda)\end{aligned} \tag{6-1-30}$$

$$\begin{aligned}\kappa &= \lambda + \tfrac{2}{3}G = \frac{E}{3(1-2\nu)} = \frac{\lambda}{3\nu}(1+\nu) \\ &= \frac{2}{3}\frac{G(1+\nu)}{1-2\nu} = \frac{GE}{3(3G-E)}\end{aligned} \tag{6-1-31}$$

$$\begin{aligned}E &= \frac{G(3\lambda+2G)}{\lambda+G} = \frac{\lambda(1+\nu)(1-2\nu)}{\nu} = \frac{9\kappa(\kappa-\lambda)}{3\kappa-\lambda} \\ &= 2G(1+\nu) = \frac{9\kappa G}{3\kappa+G} = 3\kappa(1-2\nu)\end{aligned} \tag{6-1-32}$$

$$\nu = \frac{\lambda}{2(\lambda+G)} = \frac{\lambda}{3\kappa-\lambda} = \frac{E}{2G} - 1$$

$$= \frac{3\kappa-2G}{2(3\kappa+G)} = \frac{3\kappa-E}{6\kappa} \tag{6-1-33}$$

The constants λ, G, and ν can also be combined to give the useful relationships

$$\frac{G}{\lambda+G} = 1-2\nu \qquad \frac{\lambda}{\lambda+2G} = \frac{\nu}{1-\nu} \tag{6-1-34}$$

It follows from the second relation of Eq. (6-1-34) and the third relation of Eq. (6-1-33) that

$$-1 < \nu < \tfrac{1}{2} \tag{6-1-35}$$

the upper limiting value of $\nu = \frac{1}{2}$ corresponding to an incompressible solid for which $\lambda = \kappa = \infty$ and $G = \frac{1}{3}E$. Materials with negative values of ν are unknown. The condition that $\lambda = G$, known as Poisson's relation, is approximately true for many solids, and is often used, particularly in geophysics, to simplify calculations. For $\lambda = G$ it follows from Eq. (6-1-31) that $\kappa = \frac{5}{3}G$, from Eq. (6-1-32) that $E = \frac{5}{2}G$, and from Eq. (6-1-33) that $\nu = \frac{1}{4}$.

6-1-2 Thermoelasticity

Of the many types of deformation which may occur, there are two which are well known. These are the isothermal and the adiabatic deformation of a material.

For the condition that the state of stress in an elastic body is a unique function of the strain, and similarly for strain as a function of stress, it is sufficient to choose one of the two tensors $\tilde{\mathbf{e}}, \mathbf{T}$ as an independent state variable. The equation

$$d\varepsilon = \frac{1}{\rho_r} T^i{}_j \, d\tilde{e}_i{}^j + \theta \, d\eta \tag{6-1-36}$$

shows that the internal energy ε is a state function $\varepsilon(\tilde{\mathbf{e}}, \eta)$ of the strain $\tilde{\mathbf{e}}$ and the entropy η. Hence

$$d\varepsilon = \left(\frac{\partial \varepsilon}{\partial \eta}\right)_{\tilde{\mathbf{e}}} d\eta + \left(\frac{\partial \varepsilon}{\partial \tilde{e}_i{}^j}\right)_\eta d\tilde{e}_i{}^j \tag{6-1-37}$$

which, compared with Eq. (6-1-36), gives the relations

$$\rho_r \left(\frac{\partial \varepsilon}{\partial \tilde{e}_i{}^j}\right)_\eta = T^i{}_j \tag{6-1-38a}$$

$$\left(\frac{\partial \varepsilon}{\partial \eta}\right)_{\tilde{\mathbf{e}}} = \theta \tag{6-1-38b}$$

From a comparison of Eq. (6-1-3) with Eq. (6-1-38a) it is evident that in a reversible adiabatic process there exists a scalar function

$$W = \rho_r \varepsilon \tag{6-1-39}$$

whose partial derivative with respect to a strain component gives the corresponding stress component.

If the process is isothermal, the Helmholtz free energy [see Eq. (3-9-12)]

$$\psi = \varepsilon - \theta\eta \tag{6-1-40}$$

combined with Eq. (6-1-36) gives

$$d\psi = -\eta\, d\theta + \frac{1}{\rho_r} \operatorname{tr}(\mathbf{T}\, d\tilde{\mathbf{e}}) \tag{6-1-41}$$

from which is obtained the relations

$$\rho_r \left(\frac{\partial\psi}{\partial\tilde{e}_i{}^j}\right)_\theta = T^i{}_j \tag{6-1-42a}$$

$$\left(\frac{\partial\psi}{\partial\theta}\right)_{\tilde{\mathbf{e}}} = -\eta \tag{6-1-42b}$$

From a comparison of Eq. (6-1-3) and Eq. (6-1-42*a*) it is evident that in an isothermal process there also exists a scalar function

$$W = \rho_r \psi \tag{6-1-43}$$

whose partial derivative with respect to a strain component gives the corresponding component in an elastic body.

Equations (6-1-39) and (6-1-43) identify the strain energy function W, i.e., the strain energy density, for the two well-known thermodynamic processes—adiabatic and isothermal. It is evident from these two processes that the stress-strain relation can be written without reference to temperature. However, for other thermomechanical processes, W is in general a function of $\tilde{\mathbf{e}}$ and θ.

Substitution of Eq. (6-1-21) into Eq. (6-1-3) gives

$$\begin{aligned} dW &= \lambda(\operatorname{tr}\tilde{\mathbf{e}}) \operatorname{tr}(d\tilde{\mathbf{e}}) + 2G \operatorname{tr}(\tilde{\mathbf{e}}\, d\tilde{\mathbf{e}}) \\ &= (\lambda \tilde{e}^k{}_k \delta^i{}_j + 2G\tilde{e}^i{}_j)\, d\tilde{e}_i{}^j \end{aligned} \tag{6-1-44}$$

Neglecting the slight difference in λ and G for adiabatic and isothermal conditions, Eq. (6-1-44) can be integrated from the natural to the strained state to give

$$\begin{aligned} W &= \lambda \int_0^{\tilde{e}^k{}_k} \tilde{e}^k{}_k\, d\tilde{e}^k{}_k + 2G \int_0^{\tilde{e}^j{}_i} \tilde{e}^i{}_j\, d\tilde{e}_i{}^j \\ &= \tfrac{1}{2}\lambda(\operatorname{tr}\tilde{\mathbf{e}})^2 + G \operatorname{tr}\tilde{\mathbf{e}}^2 \end{aligned} \tag{6-1-45}$$

The interdependence of the two terms $\operatorname{tr}\tilde{\mathbf{e}}^2$ and $\operatorname{tr}\tilde{\mathbf{e}}$ tends to restrict the usefulness of Eq. (6-1-45). This restriction can be removed by introducing the strain deviator of Eq. (6-1-25), whose first invariant vanishes, to give W in the form

$$W = \tfrac{1}{2}(\lambda + \tfrac{2}{3}G)(\operatorname{tr}\tilde{\mathbf{e}})^2 + G(\operatorname{tr}\tilde{\mathbf{e}}'^2)$$

which, by way of the first relation of Eq. (6-1-31), can be expressed in the form

$$W = \tfrac{1}{2}\kappa(\operatorname{tr}\tilde{\mathbf{e}})^2 + G(\operatorname{tr}\tilde{\mathbf{e}}'^2) \tag{6-1-46}$$

where, since $\frac{1}{2}(\mathrm{tr}\,\tilde{\mathbf{e}}'^2)$ is the second invariant of the strain deviator, $\mathrm{tr}\,\tilde{\mathbf{e}}$ and the term $\mathrm{tr}\,(\tilde{\mathbf{e}}'^2)$ are independent. Hence, the necessary and sufficient condition for W to be positive and finite as required by thermodynamic considerations is

$$\kappa = \lambda + \tfrac{2}{3}G > 0 \qquad \text{for } G > 0 \tag{6-1-47}$$

or, alternatively, using the second relation of Eq. (6-1-31) and the fourth relation of Eq. (6-1-32), W is finite and positive if

$$E > 0 \qquad -1 < \nu < \tfrac{1}{2} \tag{6-1-48}$$

which is to be compared with the condition of Eq. (6-1-35).

6-2 FIELD EQUATIONS

6-2-1 Basic Equations

An elastic body has a unique natural state to which the body returns when all external loads are removed. All stresses, strains, and particle displacements are measured from this natural state, their values being taken to be zero in that state. To obtain a completely linear theory it is necessary to make certain assumptions. These are that:

1. The displacements and displacement gradient components are small compared to unity.
2. The velocities shall be sufficiently small so that all nonlinear terms involving the velocities can be neglected.
3. The constitutive equation shall be the generalized form of Hooke's law.

Two ways of describing a deformed body can be distinguished. These are the spatial and referential descriptions. Thus, there are two possible approaches to the formulation of the field equations of linearized isotropic elasticity.

The general equations in referential coordinates can be summarized as follows.

General equations: referential coordinates

Equation of continuity:

$$J = \frac{\rho_r}{\rho} \tag{6-2-1}$$

Kinematical relations:

$$v(X,t) = \frac{\partial}{\partial t}\mathbf{x}(X,t) \tag{6-2-2}$$

$$a(X,t) = \frac{\partial^2}{\partial t^2}\mathbf{x}(X,t) \tag{6-2-3}$$

Equations of motion:

$$\rho_r \frac{Dv}{Dt} = \operatorname{Div} \mathbf{T}^0 + \rho_r f \tag{6-2-4}$$

Geometric equations:

$$\mathbf{E} = \tfrac{1}{2}(\mathbf{C} - \mathbf{I}) \tag{6-2-5}$$

For classical infinitesimal linearized elasticity, assumption 1 above implies the use of the same reference axes for both $\boldsymbol{x}$ and X, such that

$$\boldsymbol{u} = \boldsymbol{x} - X = \boldsymbol{u}(X, t) \tag{6-2-6}$$

This approximation allows the introduction of the displacement gradient

$$\tilde{\mathbf{S}} = \mathbf{F} - \mathbf{I} \tag{6-2-7}$$

from which it follows that

$$\mathbf{C} = (\mathbf{I} + \tilde{\mathbf{S}}^T)(\mathbf{I} + \tilde{\mathbf{S}}) \tag{6-2-8}$$

Substituting in Eq. (6-2-5) for **C** from Eq. (6-2-8) gives

$$\mathbf{E} = \tfrac{1}{2}(\tilde{\mathbf{S}} + \tilde{\mathbf{S}}^T + \tilde{\mathbf{S}}^T\tilde{\mathbf{S}}) \tag{6-2-9}$$

Applying assumption 1 and neglecting second-order terms in the displacement gradients reduces **E** to the small strain tensor

$$\tilde{\mathbf{E}} = \tfrac{1}{2}(\tilde{\mathbf{S}} + \tilde{\mathbf{S}}^T) \tag{6-2-10}$$

Using Eq. (6-2-7) gives

$$\begin{aligned} J &= \det \mathbf{F} = \det(\mathbf{I} + \tilde{\mathbf{S}}) \\ &\approx 1 + \operatorname{tr} \tilde{\mathbf{S}}^T \end{aligned} \tag{6-2-11}$$

and

$$\begin{aligned} \mathbf{F}^{-1} &= (\mathbf{I} + \tilde{\mathbf{S}})^{-1} \\ &\approx \mathbf{I} - \tilde{\mathbf{S}} \end{aligned} \tag{6-2-12}$$

Substituting in Eq. (3-7-3) for J and $\mathbf{F}^{-1}$ from Eqs. (6-2-11) and (6-2-12) respectively gives

$$\mathbf{T}^0 \approx (1 + \operatorname{tr} \tilde{\mathbf{S}}^T)(\mathbf{I} - \tilde{\mathbf{S}})\mathbf{T} \tag{6-2-13}$$

from which it is evident that, provided **T** is of order one or more in $\tilde{\mathbf{S}}$, then to a first approximation

$$\mathbf{T}^0 \approx \mathbf{T} \tag{6-2-14}$$

To the proposed degree of accuracy the field equations of linearized isotropic isothermal elasticity can now be summarized in referential coordinates as follows:

Equations of motion:

$$\rho \frac{Dv}{Dt} = \operatorname{Div} \mathbf{T} + \rho f \tag{6-2-15}$$

Constitutive equation:

$$\mathbf{T} = \lambda(\text{tr}\,\tilde{\mathbf{E}})\mathbf{I} + 2G\tilde{\mathbf{E}} \tag{6-2-16}$$

Geometric equation:

$$\tilde{\mathbf{E}} = \tfrac{1}{2}(\tilde{\mathbf{S}} + \tilde{\mathbf{S}}^T) \tag{6-2-17}$$

which system of equations is seen to provide 15 equations for the 15 unknowns comprising six stresses $\mathbf{T}$, six strains $\tilde{\mathbf{E}}$, and three displacements $\boldsymbol{u}$.

The general equations in spatial coordinates can be summarized as follows.

General equations: spatial coordinates

Equation of continuity:

$$\frac{D\rho}{Dt} + \rho\,\text{div}\,\boldsymbol{v} = 0 \tag{6-2-18}$$

Kinematical relations:

$$\boldsymbol{v}(\mathbf{x}, t) = \frac{D\mathbf{x}}{Dt} \tag{6-2-19}$$

$$\boldsymbol{a}(\mathbf{x}, t) = \frac{D\boldsymbol{v}}{Dt} \tag{6-2-20}$$

Equations of motion:

$$\rho\frac{D\boldsymbol{v}}{Dt} = \text{div}\,\mathbf{T} + \rho\boldsymbol{f} \tag{6-2-21}$$

Geometric equations:

$$\mathbf{e} = \tfrac{1}{2}(\mathbf{I} - \mathbf{B}^{-1}) \tag{6-2-22}$$

Substituting for $\mathbf{F}$ in the first equation of (2-2-20) from Eq. (6-2-7) gives

$$\mathbf{B}^{-1} = (\mathbf{I} + \tilde{\mathbf{S}}^T)^{-1}(\mathbf{I} + \tilde{\mathbf{S}})^{-1} \tag{6-2-23}$$

To a first order in $\tilde{\mathbf{S}}$, Eq. (6-2-23) gives

$$\mathbf{B}^{-1} = \mathbf{I} - (\tilde{\mathbf{S}} + \tilde{\mathbf{S}}^T) \tag{6-2-24}$$

Substituting for $\mathbf{B}^{-1}$ in Eq. (6-2-22) from Eq. (6-2-24) reduces $\mathbf{e}$ to the small strain tensor

$$\tilde{\mathbf{e}} = \tfrac{1}{2}(\tilde{\mathbf{S}} + \tilde{\mathbf{S}}^T) \tag{6-2-25}$$

The assumption of small displacements and small velocities permits Eqs. (6-2-19) and (6-2-20) to be simplified to

$$\boldsymbol{v} = \frac{\partial \boldsymbol{u}}{\partial t} \tag{6-2-26}$$

and

$$\boldsymbol{a} = \frac{\partial \boldsymbol{v}}{\partial t} \tag{6-2-27}$$

To the proposed degree of accuracy the field equations of linearized isothermal elasticity can now be summarized as follows:

Equations of motion:

$$\rho \frac{\partial^2 \boldsymbol{u}}{\partial t^2} = \operatorname{div} \mathbf{T} + \rho f \tag{6-2-28}$$

Constitutive equation:

$$\mathbf{T} = \lambda(\operatorname{tr} \tilde{\mathbf{e}})\mathbf{I} + 2G\tilde{\mathbf{e}} \tag{6-2-29}$$

Geometric equations:

$$\tilde{\mathbf{e}} = \tfrac{1}{2}(\tilde{\mathbf{S}} + \tilde{\mathbf{S}}^T) \tag{6-2-30}$$

which system of equations is seen to provide 15 equations for the 15 unknowns comprising six stresses $\mathbf{T}$, six strains $\tilde{\mathbf{e}}$, and three displacements $\boldsymbol{u}$.

From an inspection of Eqs. (6-2-2), (6-2-3), (6-2-4) and Eqs. (6-2-19), (6-2-20), (6-2-21), it is evident that the kinematical relations appear simpler in the referential description, but the equations of motion appear more involved. However, with regard to the geometric equations it is to be noted that when the displacements and displacement gradients are sufficiently small, as assumed in the classical linearized theory of elasticity, the distinction between the two small strain definitions of Eqs. (6-2-17) and (6-2-30) is usually ignored. For this latter condition it is evident from a comparison of Eqs. (6-2-15), (6-2-16), (6-2-17) with Eqs. (6-2-28), (6-2-29), and (6-2-30) that to the proposed degree of accuracy there is no significant difference between the field equations when expressed in either the referential or spatial descriptions.

The number of unknowns can be simply reduced by substituting Eq. (6-2-30) into Eq. (6-2-29), so obtaining the stress in terms of the displacement gradients, and then by substituting the result into Eq. (6-2-28) to give three second-order partial differential equations for the three displacement components; these can be expressed in the form

$$G\nabla^2 \boldsymbol{u} + (\lambda + G)\boldsymbol{\nabla}(\boldsymbol{\nabla} \cdot \boldsymbol{u}) + \rho f = \rho \frac{\partial^2 \boldsymbol{u}}{\partial t^2} \tag{6-2-31}$$

which is the field equation of Navier (1827). Equation (6-2-31) can be expressed in an alternative form by setting

$$g = \boldsymbol{\nabla} \cdot \boldsymbol{u} \qquad \omega = \tfrac{1}{2}\boldsymbol{\nabla} \times \boldsymbol{u} \tag{6-2-32}$$

and using the vector identity

$$\nabla^2 \boldsymbol{u} = \boldsymbol{\nabla}(\boldsymbol{\nabla} \cdot \boldsymbol{u}) - \boldsymbol{\nabla} \times \boldsymbol{\nabla} \times \boldsymbol{u} \tag{6-2-33}$$

to give

$$(\lambda + 2G)\boldsymbol{\nabla} g - 2G\boldsymbol{\nabla} \times \omega + \rho f = \rho \frac{\partial^2 \boldsymbol{u}}{\partial t^2} \tag{6-2-34}$$

which is a form more convenient for evaluation in orthogonal curvilinear coordinates.

6-2-2 Equilibrium of an Elastic Body under Zero Body Force

In elastostatics the acceleration $\partial^2 \boldsymbol{u}/\partial t^2$ is zero in Eqs. (6-2-15), (6-2-28), and (6-2-31), giving the equations of equilibrium in terms of stress

$$\operatorname{div} \mathbf{T} + \rho f = 0 \tag{6-2-35}$$

or displacements

$$(\lambda + G)\nabla(\nabla \cdot \boldsymbol{u}) + G\nabla^2 \boldsymbol{u} + \rho f = 0 \tag{6-2-36}$$

It is evident from Eq. (6-2-35) that in the absence of displacements the elastostatic problem can, as an alternative, be formulated in terms of stresses alone.

If the body force vanishes, $f = 0$, taking the divergence of Eq. (6-2-36) gives

$$G\nabla \cdot (\nabla^2 \boldsymbol{u}) + (\lambda + G)\nabla^2(\nabla \cdot \boldsymbol{u}) = 0$$

or

$$\nabla^2 \operatorname{div} \boldsymbol{u} = 0 \qquad \nabla^2 \operatorname{tr} \tilde{\mathbf{e}} = 0 \tag{6-2-37}$$

which is a Laplace equation. A function satisfying Eq. (6-2-37) is called a harmonic function. Thus the dilation $\operatorname{tr} \tilde{\mathbf{e}}$ is a harmonic function under zero body force.

From Eq. (6-2-29),

$$(3\lambda + 2G)(\operatorname{tr} \tilde{\mathbf{e}}) = \operatorname{tr} \mathbf{T} = 3\bar{T}$$

where $\bar{T}$ is the mean stress. Hence, the mean stress is also a harmonic function; i.e.,

$$\nabla^2 \bar{T} = 0 \tag{6-2-38}$$

Setting $f = 0$ in Eq. (6-2-36) and operating on the resulting equation with the laplacian ∇^2 gives

$$(\lambda + G)\nabla(\nabla^2 \operatorname{tr} \tilde{\mathbf{e}}) + G\nabla^4 \boldsymbol{u} = 0$$

which, having regard to Eq. (6-2-37), gives

$$\nabla^4 \boldsymbol{u} = 0 \tag{6-2-39}$$

Equation (6-2-39) is a biharmonic equation and its solution is called a biharmonic function. Hence, the displacement $\boldsymbol{u}$ is biharmonic. It follows that when the body force is zero, each of the strain components and each of the stress components, being a linear combination of the first derivatives of the displacement, are all biharmonic functions; i.e.,

$$\nabla^4 \mathbf{T} = 0 \tag{6-2-40}$$

$$\nabla^4 \tilde{\mathbf{e}} = 0 \tag{6-2-41}$$

6-2-3 Boundary Conditions for the Field Equations

In problems involving the linear theory of elasticity, Navier's equation (6-2-31), which combines Hooke's law and the equation of motion, has to be solved for appropriate boundary and initial conditions.

Three ways in which the boundary conditions may be prescribed can be distinguished. These are displacement, traction, and mixed boundary conditions:

1. *Specified displacements.* The three components of $\boldsymbol{u}$ are prescribed on the boundary.
2. *Specified surface tractions.* The three traction components

$$\boldsymbol{t} = \mathbf{T}^T \boldsymbol{n}$$

   are prescribed at a boundary point where the boundary unit normal is $\boldsymbol{n}$.
3. *Mixed boundary conditions.* Two examples are where $\boldsymbol{t}$ is the surface traction prescribed on a part S_t of the surface of the body $S = S_t + S_u$ and $\boldsymbol{u}$ is the displacement field prescribed on the remaining part S_u not overlapping S_t, the second example being where, at each point of the boundary, $\boldsymbol{u}$ or $\boldsymbol{t}$, but not both, are prescribed along each coordinate axis.

It is physically impossible to specify the boundary conditions from outside for all the components of the stress tensor at a boundary point, only the traction components being accessible.

One important, simplifying, feature of a linear field theory is the possible superposition of solutions. For example, if fields $\mathbf{T}^{(1)}, \tilde{\mathbf{e}}^{(1)}, \boldsymbol{u}^{(1)}$ are a solution of Eqs. (6-2-28) to (6-2-31) for given body forces $\boldsymbol{f}^{(1)}$, and $\mathbf{T}^{(2)}, \tilde{\mathbf{e}}^{(2)}, \boldsymbol{u}^{(2)}$ are a solution for given body forces $\boldsymbol{f}^{(2)}$, then $\alpha\mathbf{T}^{(1)} + \beta\mathbf{T}^{(2)}$, $\alpha\tilde{\mathbf{e}}^{(1)} + \beta\tilde{\mathbf{e}}^{(2)}$, $\alpha\boldsymbol{u}^{(1)} + \beta\boldsymbol{u}^{(2)}$ are solutions corresponding to given $\boldsymbol{f} = \alpha\boldsymbol{f}^{(1)} + \beta\boldsymbol{f}^{(2)}$ where α and β are arbitrary, dimensionless, constants which must be small enough to ensure that the resultant solution satisfies the small-displacement condition. The boundary conditions are superposed in the same way.

The traction boundary condition for $\boldsymbol{t} = \mathbf{T}^T\boldsymbol{n}$ may be expressed in terms of the displacements by way of Eqs. (6-2-29) and (6-2-30), the relation having the form

$$\lambda(\operatorname{div}\boldsymbol{u})\boldsymbol{n} + G[\operatorname{grad}\boldsymbol{u} + (\operatorname{grad}\boldsymbol{u})^T]\boldsymbol{n} = \text{prescribed function} \tag{6-2-42}$$

In dynamic problems, a set of initial conditions on $\boldsymbol{u}$ or $\mathbf{T}$ must be specified in the region occupied by the elastic body and on the boundary surface S.

Thus far the displacements $\boldsymbol{u}$ have been taken to be the basic unknown variables. However, in problems of static equilibrium, the equations of equilibrium are usually solved for the stresses, Hooke's law being then used to obtain the strain. Such a solution will not necessarily be unique since, at least in principle, an infinite set of solutions will exist. The correct solution is identified by way of the conditions of compatibility. It is only this one solution, that satisfies both the equation of equilibrium and the equations of compatibility, which corresponds to a continuous displacement field.

The compatibility equation (2-7-19) can be expressed in the form

$$\tilde{e}_{ij,km} + \tilde{e}_{km,ij} - \tilde{e}_{ik,jm} - \tilde{e}_{jm,ik} = 0 \tag{6-2-43}$$

representing only six linearly independent equations. From Eq. (6-1-28), Hooke's law can be expressed in the form

$$\tilde{\mathbf{e}} = \frac{1+\nu}{E}\mathbf{T} - \frac{\nu}{E}(\operatorname{tr}\mathbf{T})\mathbf{I} \tag{6-2-44}$$

which, substituted into Eq. (6-2-43) gives the compatibility equation directly in terms of stress components, having the form

$$T_{ij,km}+T_{km,ij}-T_{ik,jm}-T_{jm,ik}=\frac{v}{1+v}(g_{ij}T^{p}_{p,km}+g_{km}T^{p}_{p,ij}-g_{ik}T^{p}_{p,jm}-g_{jm}T^{p}_{p,ik}) \quad (6\text{-}2\text{-}45)$$

which represents a total of 81 equations. However, since only six of these 81 equations are linearly independent, they are equivalent to the six linearly independent equations obtained by putting $k=m$ and summing to obtain

$$\nabla^2 T_{ij}+T^{k}_{k,ij}-T^{k}_{i,jk}-T^{k}_{j,ik}=\frac{v}{1+v}(g_{ij}\nabla^2 T^{p}_{p}+3T^{p}_{p,ij}-T^{p}_{p,ji}-T^{p}_{p,ij}) \quad (6\text{-}2\text{-}46)$$

which, although comprising a set of nine equations, amounts in effect to only six distinct equations because of the symmetry in the free indices i,j. These equations can be simplified by using the equations of equilibrium to yield

$$-T_{ik,jk}\equiv-(T_{ki,k})_{,j}=(\rho f_i)_{,j} \quad (6\text{-}2\text{-}47)$$

and

$$-T_{jk,ik}\equiv-(T_{kj,k})_{,i}=(\rho f_j)_{,i} \quad (6\text{-}2\text{-}48)$$

Eq. (6-2-46) now being expressed in the form

$$\nabla^2 T_{ij}+\frac{1}{1+v}T_{pp,ij}-\frac{v}{1+v}\delta_{ij}\nabla^2 T_{pp}=-(\rho f_i)_{,j}-(\rho f_j)_{,i} \quad (6\text{-}2\text{-}49)$$

With a contraction $k=i$ and $m=j$, Eq. (6-2-45) gives

$$T_{ij,ij}=\frac{1-v}{1+v}\nabla^2 T_{pp} \quad (6\text{-}2\text{-}50)$$

but, from Eq. (6-2-28) for $\partial u_j/\partial t=0$,

$$T_{ij,ij}=-(\rho f_j)_{,j} \quad (6\text{-}2\text{-}51)$$

which, combined with Eq. (6-2-50), gives

$$\nabla^2 T_{pp}=-\frac{1+v}{1-v}\nabla\cdot(\rho f) \quad (6\text{-}2\text{-}52)$$

Hence, the compatibility equations (6-2-49) take the simplified form

$$\nabla^2 T_{ij}+\frac{1}{1+v}T_{pp,ij}=-\frac{v}{1-v}\delta_{ij}\nabla\cdot(\rho f) \quad (6\text{-}2\text{-}53)$$

These equations were obtained by Michell (1900) and, for the case in which body forces are absent, by Beltrami (1892). They are known as the Beltrami–Michell compatibility equations. However, since Eqs. (6-2-53) actually represent only three functionally independent conditions it is necessary to add the three equilibrium equations

$$T_{ij,j}+\rho f_i=0 \quad (6\text{-}2\text{-}54)$$

to obtain a complete formulation.

6-2-4 Equilibrium and the Uniqueness of Solutions

An acceptable theory requires that the boundary conditions together with the initial conditions shall give a unique resultant motion. Thus the question arises of whether a condition can be formulated which, if a given solution satisfies it, would give a unique solution. There are two aspects of the problem. First, and of immediate interest, can a unique solution be expected from physically defensible considerations? Second, if so, has the mathematical formulation of the problem a unique solution? In practice, situations exist where unique solutions, are not expected, as, for example, certain buckling situations. Examples are buckling of thin-walled spherical shells subjected to a uniform external pressure and the buckling of thin columns or plates under axial compression. In these examples the possibility exists that the bodies will assume different forms of deformation, i.e., elastic states, of which only some will be stable. However, experience indicates that there exists a very wide range of mechanical phenomena whose characteristic response can be regarded as unique. Thus, ideally, any useful theory must correlate the physical aspects of a phenomenon with its mathematical formulation in such a way as to give a uniquely predictable system.

The starting point to establish that the linearized problem has a unique solution for $\mathbf{T}$ and $\boldsymbol{u}$ is the strain energy function W defined by Eq. (6-1-3); i.e.,

$$T^i{}_j = \frac{\partial W}{\partial \tilde{e}_i{}^j} \tag{6-2-55}$$

from which is obtained

$$W = \tfrac{1}{2}\operatorname{tr}(\mathbf{T}\tilde{\mathbf{e}}) \tag{6-2-56}$$

It has been shown [Eq. (6-1-45)] that for the isotropic Hooke's law there always exists such a function

$$W = \tfrac{1}{2}\lambda(\operatorname{tr}\tilde{\mathbf{e}})^2 + G(\operatorname{tr}\tilde{\mathbf{e}}^2) \tag{6-2-57}$$

which is positive definite provided the elastic modulus E and Poisson's ratio ν satisfy the conditions of Eq. (6-1-48); that is, $E > 0$ and $-1 < \nu < \frac{1}{2}$, where ν and E are related to G and λ by the relations

$$\lambda = G\frac{E-2G}{3G-E} \qquad \nu = \frac{E}{2G} - 1$$

Since Eq. (6-2-54) is linear, the remarks already made about the superposition of solutions, as relating to the field equations, applies here. Thus, if there were to exist two solutions $\mathbf{T}^{(1)}$, $\tilde{\mathbf{e}}^{(1)}$, $\boldsymbol{u}^{(1)}$ and $\mathbf{T}^{(2)}$, $\tilde{\mathbf{e}}^{(2)}$, $\boldsymbol{u}^{(2)}$ corresponding to the same boundary conditions and the same body forces, then

$$\mathbf{T} = \mathbf{T}^{(2)} - \mathbf{T}^{(1)} \tag{6-2-58}$$

$$\tilde{\mathbf{e}} = \tilde{\mathbf{e}}^{(2)} - \tilde{\mathbf{e}}^{(1)} \tag{6-2-59}$$

$$\boldsymbol{u} = \boldsymbol{u}^{(2)} - \boldsymbol{u}^{(1)} \tag{6-2-60}$$

would also be a solution of the equations for $f = 0$ satisfying boundary conditions such that $\boldsymbol{u}\cdot\boldsymbol{t} = 0$ at every boundary point, since at each point either $\boldsymbol{u}$ or $\boldsymbol{t} = 0$.

Hence, integrating over the boundary ∂R gives

$$\int_{\partial R} (\boldsymbol{u} \cdot \boldsymbol{t})\, da = 0 \tag{6-2-61}$$

or, substituting for $\boldsymbol{t} = \mathbf{T}^T \boldsymbol{n}$, gives

$$\int_{\partial R} \boldsymbol{u} \cdot \mathbf{T}^T \boldsymbol{n}\, da = 0 \tag{6-2-62}$$

which can be transformed to a volume integral to give

$$\int_V \operatorname{div} (\mathbf{T}\boldsymbol{u})\, dv = 0 \tag{6-2-63}$$

Now,

$$\operatorname{div} (\mathbf{T}\boldsymbol{u}) = \boldsymbol{u} \cdot \operatorname{div} \mathbf{T} + \operatorname{tr} (\mathbf{T}\tilde{\mathbf{S}})$$

For $f = 0$ it follows from the equilibrium equation (6-2-35) that

$$\operatorname{div} \mathbf{T} = 0$$

and from Eq. (6-2-56) it follows that

$$\operatorname{tr} (\mathbf{T}\tilde{\mathbf{e}}) = 2W = \operatorname{tr} (\mathbf{T}\tilde{\mathbf{S}})$$

since $\tilde{\mathbf{S}} = \tilde{\mathbf{e}} + \tilde{\boldsymbol{\Omega}}$ and $\operatorname{tr} (\mathbf{T}\tilde{\boldsymbol{\Omega}}) = 0$. Hence, it follows that Eq. (6-2-63) can be expressed in the form

$$\int_V 2W\, dv = 0 \tag{6-2-64}$$

However, it follows that if it is assumed that W is a positive-definitive homogeneous quadratic form in the $\tilde{\mathbf{e}}$, then the integral can only vanish if $W = 0$ everywhere, which in turn is only possible if all $\tilde{\mathbf{e}} = 0$ everywhere, a requirement which gives, by way of Eq. (6-2-59), the condition

$$\tilde{\mathbf{e}}^{(2)} = \tilde{\mathbf{e}}^{(1)}$$

and therefore, from Eq. (6-2-58) and Hooke's law,

$$\mathbf{T}^{(2)} = \mathbf{T}^{(1)}$$

everywhere. Thus it may be concluded that there cannot be two different stress and strain fields corresponding to the same externally imposed body forces and boundary conditions satisfying the linearized theory of elastostatics. It is important to note that this uniqueness proof, basically due to Kirchhoff (1859), only applies under the assumption that W is a positive-definite homogeneous quadratic form in the $\tilde{\mathbf{e}}$.

6-3 PLANE ELASTICITY

6-3-1 Plane Strain in Cartesian Coordinates

A state of plane strain is defined by the condition that the displacement normal to the plane vanishes everywhere. Taking the base vector g_3 to be a unit vector lying in the direction of this normal, it is clear that $g_1 \cdot g_3 = g_{13} = 0$, $g_{23} = g_2 \cdot g_3 = 0$ while $g_{33} = 1$. The displacement u_3 vanishes everywhere and both the displacement and the components of the metric tensor are independent of x_3. It follows that

$$u_3 = 0 \qquad u_{1,3} = u_{2,3} = 0 \tag{6-3-1}$$

for which conditions the isotropic Hooke's law [Eq. (6-1-27)] reduces to

$$\begin{aligned} T_{11} &= \lambda(\tilde{e}_{11} + \tilde{e}_{22}) + 2G\tilde{e}_{11} \\ T_{22} &= \lambda(\tilde{e}_{11} + \tilde{e}_{22}) + 2G\tilde{e}_{22} \\ T_{33} &= \lambda(\tilde{e}_{11} + \tilde{e}_{22}) \\ T_{12} &= 2G\tilde{e}_{12} \end{aligned} \tag{6-3-2}$$

or, inversely, from Eq. (6-1-28),

$$\begin{aligned} \tilde{e}_{11} &= \frac{1+\nu}{E}[(1-\nu)T_{11} - \nu T_{22}] \\ \tilde{e}_{22} &= \frac{1+\nu}{E}[(1-\nu)T_{22} - \nu T_{11}] \\ \tilde{e}_{33} &= \frac{1}{E}[T_{33} - \nu(T_{11} + T_{22})] = 0 \\ \tilde{e}_{12} &= \frac{1}{2G}T_{12} \\ \tilde{e}_{13} &= \tilde{e}_{23} = 0 \end{aligned} \tag{6-3-3}$$

The condition $\tilde{e}_{33} = 0$ gives, from Eqs. (6-3-2) and (6-3-3), the identity

$$T_{33} = \lambda(\tilde{e}_{11} + \tilde{e}_{22}) = \nu(T_{11} + T_{22}) \tag{6-3-4}$$

From Eq. (6-2-21), two equations of motion are obtained:

$$\begin{aligned} \frac{\partial T_{11}}{\partial x_1} + \frac{\partial T_{12}}{\partial x_2} + \rho f_1 &= \rho \frac{\partial^2 u_1}{\partial t^2} \\ \frac{\partial T_{12}}{\partial x_1} + \frac{\partial T_{22}}{\partial x_2} + \rho f_2 &= \rho \frac{\partial^2 u_2}{\partial t^2} \end{aligned} \tag{6-3-5}$$

to which must be added the compatibility equation

$$\frac{\partial^2 \tilde{e}_{11}}{\partial x_2^2} + \frac{\partial^2 \tilde{e}_{22}}{\partial x_1^2} = 2\frac{\partial^2 \tilde{e}_{12}}{\partial x_1 \partial x_2} \tag{6-3-6}$$

Equations (6-3-1) to (6-3-6) are the basic equations for plane strain, provided that for small displacements the difference between referential and spatial coordinates can be neglected.

In elastostatics the acceleration term $\partial^2\boldsymbol{u}/\partial t^2$ is zero, as in Eq. (6-3-5), the Eqs. (6-3-2), (6-3-3), (6-3-5), and (6-3-6) being now simply six equations for the six independent unknowns T_{11}, T_{22}, T_{12}, $\tilde{e}_{11}$, $\tilde{e}_{22}$, $\tilde{e}_{12}$ as functions of x_1 and x_2 which can, in principle, be solved provided the form of the boundary conditions and body forces are known. For zero acceleration terms, Eq. (6-3-5) gives, on differentiation of the first relation with respect to x_1 and the second relation with respect to x_2 followed by addition, the relation

$$2\frac{\partial^2 T_{12}}{\partial x_1 \partial x_2}+\frac{\partial^2 T_{11}}{\partial x_1^2}+\frac{\partial^2 T_{22}}{\partial x_2^2}+\frac{\partial(\rho f_1)}{\partial x_1}+\frac{\partial(\rho f_2)}{\partial x_2}=0$$

which, on substituting Eq. (6-3-3) into Eq. (6-3-6), can be used to give the relation

$$\frac{\partial^2 T_{11}}{\partial x_1^2}+\frac{\partial^2 T_{11}}{\partial x_2^2}+\frac{\partial^2 T_{22}}{\partial x_1^2}+\frac{\partial^2 T_{22}}{\partial x_2^2}+\frac{1}{1-\nu}\left[\frac{\partial(\rho f_1)}{\partial x_1}+\frac{\partial(\rho f_2)}{\partial x_2}\right]=0 \qquad \text{(plane strain)} \qquad (6\text{-}3\text{-}7)$$

which is the compatibility equation stated in terms of stresses for plane strain.

6-3-2 Plane Stress in Cartesian Coordinates

If the stress components T_{33}, T_{31}, T_{32} vanish everywhere, i.e.,

$$T_{33}=T_{31}=T_{32}=0 \qquad (6\text{-}3\text{-}8)$$

such that the only nonzero stresses in a body are T_{11}, T_{22}, T_{12} $(=T_{21})$, the state of stress is said to be plane stress parallel to the x_1x_2 plane. For these conditions the isotropic Hooke's law [Eq. (6-1-28)] reduces to

$$\begin{aligned}
\tilde{e}_{11}&=\frac{1}{E}(T_{11}-\nu T_{22})\\
\tilde{e}_{22}&=\frac{1}{E}(T_{22}-\nu T_{11})\\
\tilde{e}_{33}&=-\frac{\nu}{E}(T_{11}+T_{22}) \qquad (6\text{-}3\text{-}9)\\
\tilde{e}_{12}&=\frac{1}{2G}T_{12}\\
\tilde{e}_{13}&=\tilde{e}_{23}=0
\end{aligned}$$

or, inversely, from Eq. (6-1-27),

$$
\begin{aligned}
T_{11} &= \frac{E}{1-\nu^2}(\tilde{e}_{11}+\nu\tilde{e}_{22}) \\
T_{22} &= \frac{E}{1-\nu^2}(\tilde{e}_{22}+\nu\tilde{e}_{11}) \\
T_{12} &= 2G\tilde{e}_{12}
\end{aligned}
\tag{6-3-10}
$$

from which can be obtained the relation

$$T_{11}+T_{22} = \frac{E}{1-\nu}(\tilde{e}_{11}+\tilde{e}_{22}) \tag{6-3-11}$$

where

$$\tilde{e}_{11}+\tilde{e}_{22} = \frac{\partial u_1}{\partial x_1}+\frac{\partial u_2}{\partial x_2} \tag{6-3-12}$$

To these must be added the equations of motion

$$\frac{\partial T_{11}}{\partial x_1}+\frac{\partial T_{12}}{\partial x_2}+\rho f_1 = \rho\frac{\partial^2 u_1}{\partial t^2} \tag{6-3-13}$$

$$\frac{\partial T_{12}}{\partial x_1}+\frac{\partial T_{22}}{\partial x_2}+\rho f_2 = \rho\frac{\partial^2 u_2}{\partial t^2} \tag{6-3-14}$$

and the compatibility equation

$$\frac{\partial^2 \tilde{e}_{11}}{\partial x_2^2}+\frac{\partial^2 \tilde{e}_{22}}{\partial x_1^2} = 2\frac{\partial^2 \tilde{e}_{12}}{\partial x_1 \partial x_2} \tag{6-3-15}$$

which are the same as for plane strain. However, the compatibility equation stated in terms of the stresses now takes the form

$$\frac{\partial^2 T_{11}}{\partial x_1^2}+\frac{\partial^2 T_{11}}{\partial x_2^2}+\frac{\partial^2 T_{22}}{\partial x_1^2}+\frac{\partial^2 T_{22}}{\partial x_2^2} + (1+\nu)\left[\frac{\partial(\rho f_1)}{\partial x_1}+\frac{\partial(\rho f_2)}{\partial x_2}\right] = 0 \qquad \text{(plane stress)} \tag{6-3-16}$$

which is to be compared with Eq. (6-3-7).

6-3-3 Antiplane Strain in Cartesian Coordinates

A state of antiplane strain is defined by $u_1 = u_2 = 0$ and $u_3 = u_3(x_1, x_2)$. Hence, the only nonvanishing components of strain are

$$
\begin{aligned}
\tilde{e}_{13} &= \frac{1}{2}\frac{\partial u_3}{\partial x_1} \\
\tilde{e}_{23} &= \frac{1}{2}\frac{\partial u_3}{\partial x_2}
\end{aligned}
\tag{6-3-17}
$$

and thus, from Eq. (6-1-27), the only nonvanishing components of stress are

$$T_{13} = G\frac{\partial u_3}{\partial x_1} \qquad T_{23} = G\frac{\partial u_3}{\partial x_2} \tag{6-3-18}$$

The equations of motion reduce to only one:

$$\frac{\partial T_{13}}{\partial x_1} + \frac{\partial T_{23}}{\partial x_2} + \rho f_3 = \rho\frac{\partial^2 u_3}{\partial t^2} \tag{6-3-19}$$

This leads, by way of the one equation of equilibrium, to

$$\nabla^2 u_3 = 0 \tag{6-3-20}$$

from which it is evident that problems involving a state of antiplane strain reduce to solving Laplace's equation for u_3 together with the relevant boundary conditions. Antiplane strain finds application in fracture mechanics (see Chap. 14).

6-3-4 Strain Energy Function in Cartesian Coordinates

In Sec. 6-1-2 it was shown for an isotropic material that the strain energy function W can be expressed, by way of the isotropic Hooke's law, in the form

$$W = \tfrac{1}{2}\lambda(\operatorname{tr}\tilde{\mathbf{e}})^2 + G(\operatorname{tr}\tilde{\mathbf{e}}^2) \tag{6-3-21}$$

where W is the stored strain energy per unit volume.

For a state of plane strain this simplifies to

$$W = \tfrac{1}{2}\lambda(\tilde{e}_{11} + \tilde{e}_{22})^2 + G(\tilde{e}_{11}^2 + \tilde{e}_{22}^2 + 2\tilde{e}_{12}^2) \tag{6-3-22}$$

The total strain energy W_{tot} in a slab of material of unit thickness is

$$W_{\text{tot}} = \int_A W\,da$$

where the integration is taken over the area of cross section of the body perpendicular to the x_3 axis.

It is evident from the form of the isotropic Hooke's law given in Eqs. (6-1-27) and (6-3-21) that W can be expressed in the form

$$W = \tfrac{1}{2}\operatorname{tr}(\mathbf{T}\tilde{\mathbf{e}}) \tag{6-3-23}$$

which for both plane strain and plane stress becomes

$$W = \tfrac{1}{2}(T_{11}\tilde{e}_{11} + T_{22}\tilde{e}_{22} + 2T_{12}\tilde{e}_{12}) \tag{6-3-24}$$

Using the plane strain and plane stress forms of Hooke's law, i.e., Eqs. (6-3-3) and (6-3-9), Eq. (6-3-24) can be expressed in terms of stresses to give the relations:

Plane strain:

$$W = \frac{1-\nu^2}{2E}(T_{11}^2 + T_{22}^2) - \frac{\nu(1+\nu)}{E}T_{11}T_{22} + \frac{1}{2G}T_{12}^2 \tag{6-3-25}$$

Plane stress:

$$W = \frac{1}{2E}(T_{11}^2 + T_{22}^2) - \frac{\nu}{E} T_{11} T_{22} + \frac{1}{2G} T_{12}^2 \tag{6-3-26}$$

Alternatively, for a state of plane stress, W can be expressed, by way of Eq. (6-3-9), in terms of strains in the form

$$W = G\left[\frac{1}{1-\nu}(\tilde{e}_{11}^2 + \tilde{e}_{22}^2 + 2\nu\tilde{e}_{11}\tilde{e}_{22}) + 2\tilde{e}_{12}^2\right] \tag{6-3-27}$$

6-3-5 Navier's Displacement Equations of Motion in Cartesian Coordinates

For plane strain the Navier equations and traction boundary conditions can be obtained directly from the three-dimensional case by setting $u_3 = 0$ and noting that u_1 and u_2 are now independent of x_3. Thus Eq. (6-2-31) reduces to

$$G\nabla^2 u_1 + \frac{G}{1-2\nu}\frac{\partial}{\partial x_1}\left(\frac{\partial u_1}{\partial x_1} + \frac{\partial u_2}{\partial x_2}\right) + \rho f_1 = \rho \frac{\partial^2 u_1}{\partial t^2} \tag{6-3-28}$$

$$G\nabla^2 u_2 + \frac{G}{1-2\nu}\frac{\partial}{\partial x_2}\left(\frac{\partial u_1}{\partial x_1} + \frac{\partial u_2}{\partial x_2}\right) + \rho f_2 = \rho \frac{\partial^2 u_2}{\partial t^2} \tag{6-3-29}$$

Alternatively, Eqs. (6-3-28) and (6-3-29) can be obtained by substituting Eqs. (6-3-2) and (6-3-4) into the equations of motion (6-3-5). The traction boundary conditions take the form

$$t_1 = \frac{2G}{1-2\nu}\left[(1-\nu)\frac{\partial u_1}{\partial x_1} + \nu\frac{\partial u_2}{\partial x_2}\right] n_1 + G\left(\frac{\partial u_1}{\partial x_2} + \frac{\partial u_2}{\partial x_1}\right) n_2 \tag{6-3-30}$$

$$t_2 = \frac{2G}{1-2\nu}\left[\nu\frac{\partial u_1}{\partial x_1} + (1-\nu)\frac{\partial u_2}{\partial x_2}\right] n_2 + G\left(\frac{\partial u_1}{\partial x_2} + \frac{\partial u_2}{\partial x_1}\right) n_1 \tag{6-3-31}$$

$$t_3 = \frac{2\nu G}{1-2\nu}\left(\frac{\partial u_1}{\partial x_1} + \frac{\partial u_2}{\partial x_2}\right) n_3 \tag{6-3-32}$$

The Navier equations for plane stress can be obtained in the form

$$G\nabla^2 u_1 + G\frac{1+\nu}{1-\nu}\frac{\partial}{\partial x_1}\left(\frac{\partial u_1}{\partial x_1} + \frac{\partial u_2}{\partial x_2}\right) + \rho f_1 = \rho \frac{\partial^2 u_1}{\partial t^2} \tag{6-3-33}$$

$$G\nabla^2 u_2 + G\frac{1+\nu}{1-\nu}\frac{\partial}{\partial x_2}\left(\frac{\partial u_1}{\partial x_1} + \frac{\partial u_2}{\partial x_2}\right) + \rho f_2 = \rho \frac{\partial^2 u_2}{\partial t^2} \tag{6-3-34}$$

and the traction boundary conditions take the form

$$t_1 = \frac{2G}{1-\nu}\left(\frac{\partial u_1}{\partial x_1} + \nu\frac{\partial u_2}{\partial x_2}\right) n_1 + G\left(\frac{\partial u_1}{\partial x_2} + \frac{\partial u_2}{\partial x_1}\right) n_2 \tag{6-3-35}$$

$$t_2 = \frac{2G}{1-\nu}\left(\frac{\partial u_2}{\partial x_2} + \nu\frac{\partial u_1}{\partial x_1}\right) n_2 + G\left(\frac{\partial u_1}{\partial x_2} + \frac{\partial u_2}{\partial x_1}\right) n_1 \tag{6-3-36}$$

6-3-6 Curvilinear Coordinates

The development of plane and antiplane deformations of materials in general curvilinear coordinates parallels that given above for cartesian coordinates. As such, the present discussion will be restricted to the plane strain deformation of a circular cylinder, since this is required in Chap. 11.

For the plane strain deformation of a circular cylinder, it is convenient to use cylindrical polar coordinates (r, θ, z). The stress can be expressed in terms of the Airy stress function in the following way:

$$T^{ij} = \varepsilon^{ik}\varepsilon^{jl}\psi_{,kl} \tag{6-3-37}$$

where ε^{ik} is the permutation tensor in two dimensions. For cylindrical coordinates,

$$\varepsilon^{rr} = \varepsilon^{\theta\theta} = 0 \qquad \varepsilon^{r\theta} = -\varepsilon^{\theta r} = \frac{1}{r} \tag{6-3-38}$$

and hence

$$T^{rr} = \frac{1}{r^2}\psi_{,\theta\theta} \qquad T^{\theta\theta} = \frac{1}{r^2}\psi_{,rr} \qquad T^{r\theta} = -\frac{1}{r^2}\psi_{,r\theta} \tag{6-3-39}$$

The nonzero Christoffel symbols (see Sec. 1-8) for cylindrical coordinates are

$$\Gamma^r_{\theta\theta} = -r \qquad \Gamma^\theta_{r\theta} = \frac{1}{r} \tag{6-3-40}$$

so that

$$\begin{aligned} \psi_{,rr} &= \frac{\partial^2\psi}{\partial r^2} \\ \psi_{,\theta\theta} &= r\frac{\partial\psi}{\partial r} + \frac{\partial^2\psi}{\partial\theta^2} \\ \psi_{,r\theta} &= \frac{\partial^2\psi}{\partial r\partial\theta} - \frac{1}{r}\frac{\partial\psi}{\partial\theta} = r\frac{\partial}{\partial r}\left(\frac{1}{r}\frac{\partial\psi}{\partial\theta}\right) \end{aligned} \tag{6-3-41}$$

The physical components are obtained by noting that

$$T\langle rr\rangle = T^{rr} \qquad T\langle\theta\theta\rangle = r^2T^{\theta\theta} \qquad T\langle r\theta\rangle = rT^{r\theta} \tag{6-3-42}$$

Thus, in terms of the stress function the physical components of stress are

$$\begin{aligned} T\langle rr\rangle &= \frac{1}{r}\frac{\partial\psi}{\partial r} + \frac{1}{r^2}\frac{\partial^2\psi}{\partial\theta^2} \\ T\langle\theta\theta\rangle &= \frac{\partial^2\psi}{\partial r^2} \\ T\langle r\theta\rangle &= -\frac{\partial}{\partial r}\left(\frac{1}{r}\frac{\partial\psi}{\partial\theta}\right) \end{aligned} \tag{6-3-43}$$

The stress function ψ must generate only those stress fields that satisfy the conditions of compatibility; they will do so provided

$$\nabla^4\psi = 0 \tag{6-3-44}$$

The biharmonic equation (6-3-44) can be expressed in the form

$$\left[\frac{1}{r}\frac{\partial}{\partial r}\left(r\frac{\partial}{\partial r}\right)+\frac{1}{r^2}\frac{\partial^2}{\partial \theta^2}\right]^2\psi=0 \tag{6-3-45}$$

6-4 PROPAGATION OF ELASTIC WAVES IN ISOTROPIC SOLIDS

6-4-1 Wave Equations

A particularly useful diagnostic technique for studying the mechanical response of real materials is based on the propagation of waves, specifically in the form of a pressure, or stress, pulse. This provides information about the way solids behave when the forces acting on them are no longer in static equilibrium. For simplicity the present discussion will be limited to the propagation of waves in solids that extend to infinity.

The dynamical theory of isotropic elastic solids is governed by Navier's equation, whose vector form

$$(\lambda+2G)\boldsymbol{\nabla}(\boldsymbol{\nabla}\cdot\boldsymbol{u})-G\boldsymbol{\nabla}\times\boldsymbol{\nabla}\times\boldsymbol{u}+\rho\left(\boldsymbol{f}-\frac{\partial^2\boldsymbol{u}}{\partial t^2}\right)=0 \tag{6-4-1}$$

has been obtained from Eqs. (6-2-32) and (6-2-34). Any continuous differentiable vector field $\boldsymbol{v}$ can be represented as the sum of an irrotational vector field which is the gradient of a scalar potential φ plus a solenoidal vector field which is the curl of a vector potential $\boldsymbol{\psi}$. Specifically,

$$\boldsymbol{u}=\boldsymbol{\nabla}\varphi+\boldsymbol{\nabla}\times\boldsymbol{\psi} \qquad \boldsymbol{\nabla}\cdot\boldsymbol{\psi}=0 \tag{6-4-2}$$

$$\boldsymbol{f}=-\boldsymbol{\nabla}g-\boldsymbol{\nabla}\times\boldsymbol{h} \qquad \boldsymbol{\nabla}\cdot\boldsymbol{h}=0 \tag{6-4-3}$$

which is known as the Helmholtz representation. Substituting for $\boldsymbol{u}$ and $\boldsymbol{f}$ in Eq. (6-4-1) from Eqs. (6-4-2) and (6-4-3) gives

$$\boldsymbol{\nabla}(c_L^2\nabla^2\varphi-g-\ddot{\varphi})-\boldsymbol{\nabla}\times(c_T^2\boldsymbol{\nabla}\times\boldsymbol{\nabla}\times\boldsymbol{\psi}+\boldsymbol{h}+\ddot{\boldsymbol{\psi}})=\boldsymbol{o} \tag{6-4-4}$$

where

$$c_L=\left(\frac{\lambda+2G}{\rho}\right)^{1/2}=\sqrt{\frac{E(1-\nu)}{(1+\nu)(1-2\nu)\rho}} \tag{6-4-5a}$$

$$c_T=\left(\frac{G}{\rho}\right)^{1/2} \tag{6-4-5b}$$

Equation (6-4-4) is satisfied if

$$\frac{\partial^2\varphi}{\partial t^2}=c_L^2\nabla^2\varphi-g \tag{6-4-6}$$

and

$$\frac{\partial^2\boldsymbol{\psi}}{\partial t^2}=-c_T^2\boldsymbol{\nabla}\times\boldsymbol{\nabla}\times\boldsymbol{\psi}-\boldsymbol{h} \tag{6-4-7}$$

Eqs. (6-4-6) and (6-4-7) being the scalar and vector wave equations respectively, which are valid also in curvilinear coordinates. Thus, for this special case where the potentials have been uncoupled, Eqs. (6-4-6) and (6-4-7) indicate that a possible dynamic solution is one where two types of waves propagate separately, a dilatational wave governed by the scalar potential φ and propagating at speed c_L and an equivoluminal wave, which is also called a shear wave, a distortional wave, or a rotational wave, governed by the vector potential $\boldsymbol{\psi}$ and propagating at speed c_T. Since Young's modulus $E > 0$, it is evident that the velocities c_L and c_T are real and nonnegative in the region $-1 \leqslant \nu \leqslant \frac{1}{2}$ of Poisson's ratio ν.

6-4-2 Plane Waves in an Infinite Solid

Consider the solutions of Navier's equation in the form of plane waves propagating in an infinite solid, the boundary and initial conditions being initially unspecified.

A plane wave solution of Eq. (6-4-6) with $g = 0$ is a solution of the form

$$\varphi = \varphi(\mathbf{v} \cdot \mathbf{x} \mp ct) \tag{6-4-8}$$

where ν and c are constants. Substituting for φ from Eq. (6-4-8) into Eq. (6-4-6) gives a relation which is only satisfied if $c = c_L$ and

$$\mathbf{v} \cdot \mathbf{v} = 1 \tag{6-4-9}$$

from which it follows that in Eq. (6-4-8) the $\mathbf{v}$ are the cosine directors of a line l

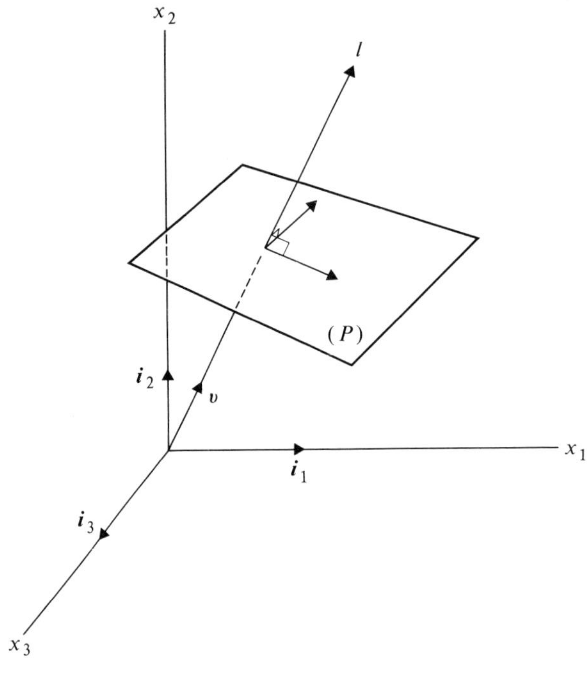

Figure 6-1 Propagation of a plane elastic wave in the direction of $\mathbf{v}$.

perpendicular to a plane P having the equation

$$\mathbf{v} \cdot \mathbf{x} \mp c_L t = \text{constant}$$

for a fixed time (see Fig. 6-1), the plane moving with a speed $\pm c_L$ along l. The plane is perpendicular to the line l for all time t.

Similarly, a vector plane wave solution of Eq. (6-4-7), in rectangular coordinates, is a solution of the form

$$\boldsymbol{\psi} = \boldsymbol{\psi}(\mathbf{v} \cdot \mathbf{x} \mp ct) \tag{6-4-10}$$

provided

$$\nabla \cdot \boldsymbol{\psi} = 0 \qquad h = o \tag{6-4-11}$$

Substituting for $\boldsymbol{\psi}$ in Eq. (6-4-7) from Eq. (6-4-10) gives a relation which is only satisfied if $c = c_T$ and $\mathbf{v}$ is a unit vector.

It is evident that the scalar plane waves $\varphi = \varphi(\mathbf{v} \cdot \mathbf{x} \mp c_L t)$ propagate along the direction $\mathbf{v}$ and are associated with the component of the displacement in this, the longitudinal, direction. In contrast, the vector waves $\boldsymbol{\psi} = \boldsymbol{\psi}(\mathbf{v} \cdot \mathbf{x} \mp c_T t)$ propagate in a plane perpendicular to the direction $\mathbf{v}$ and are associated with the transverse components of the displacement field in this plane. Thus, for a plane wave propagating along a direction $\mathbf{v}$, three independent components can be identified corresponding to a dilatational wave, which is also called an irrotational wave, and two plane polarized equivoluminal waves.

6-4-3 Spherical Waves

In spherical coordinates (r, θ, ϕ),

$$\nabla^2 \varphi = \frac{1}{r^2} \frac{\partial}{\partial r}\left(r^2 \frac{\partial \varphi}{\partial r}\right) + \frac{1}{r^2 \sin\theta} \frac{\partial}{\partial \theta}\left(\sin\theta \frac{\partial \varphi}{\partial \theta}\right) + \frac{1}{r^2 \sin^2\theta} \frac{\partial^2 \varphi}{\partial \phi^2} \tag{6-4-12}$$

Restricting the discussion to waves propagating with spherical fronts, i.e., the waves have central symmetry for which there is no dependence on the angles θ and ϕ, then φ may be taken to be of the form

$$\varphi = \varphi(r, t)$$

such that in the absence of body forces Eq. (6-4-6) reduces, by way of Eq. (6-4-12), to

$$\frac{\partial^2 \varphi}{\partial r^2} + \frac{2}{r} \frac{\partial \varphi}{\partial r} = \frac{1}{c_L^2} \frac{\partial^2 \varphi}{\partial t^2} \tag{6-4-13}$$

Assuming that φ has a solution of the form

$$\varphi = \frac{1}{r} F(r, t) \tag{6-4-14}$$

Eq. (6-4-13) reduces to

$$\frac{\partial^2 F}{\partial r^2} = \frac{1}{c_L^2} \frac{\partial^2 F}{\partial t^2} \tag{6-4-15}$$

which has a general solution of the form

$$F = F_1(r - c_L t) + F_2(r + c_L t) \tag{6-4-16}$$

where F_1 and F_2 are two arbitrary functions. Thus the general solution of Eq. (6-4-13) is

$$\varphi = \frac{1}{r} F_1(r - c_L t) + \frac{1}{r} F_2(r + c_L t) \tag{6-4-17}$$

where F_1 represents a spherical wave front $(r - c_L t) =$ constant advancing in the radial direction from the origin $r = 0$ and F_2 represents a system of spherical waves advancing toward the origin.

Similarly, a solution of Eq. (6-4-7), for the particular condition $\boldsymbol{h} = \boldsymbol{o}$, is given by

$$\psi = \frac{1}{r} G_1(r - c_T t) + \frac{1}{r} G_2(r + c_T t) \tag{6-4-18}$$

where G_1 and G_2 are two arbitrary functions.

From Eqs. (6-4-2), (6-4-17), and (6-4-18) the displacement field can now be obtained, the functions F_1, F_2, G_1, and G_2 being determined by the initial conditions.

The factor $1/r$, without which φ and ψ would not be solutions of Eqs. (6-4-6) and (6-4-7) respectively and therefore not a wave, accounts for the attenuation of a spherical wave as it moves out from its source.

6-4-4 Rayleigh Surface Waves

For an unbounded isotropic solid it has been shown that there are only two types of elastic wave which can be propagated. These are longitudinal, or irrotational, and transverse, or equivoluminal, waves. However, where there is a bounding surface there is the possibility of a third type of elastic wave which propagates over the surface and which penetrates only a little into the interior of the body. The surface waves of present interest are the Rayleigh (1887) waves, which are similar to gravitational surface waves in liquids, the velocity of propagation of Rayleigh waves being smaller than that of body waves.

The problem of surface waves can be formulated as follows. It is required to find a solution of the equations of motion which corresponds to a disturbance whose displacement amplitude is attenuated rapidly as the distance is increased from the stress-free boundary.

For simplicity, take the boundary to be the xy plane of an elastic half space $z \geqslant 0$, the xy plane (that is, $z = 0$) being stress free. Let the plane wave travel in the x direction. For the particular conditions $g = 0$, $\boldsymbol{h} = \boldsymbol{o}$, that is zero body force, Eqs. (6-4-6) and (6-4-7) can be expressed in rectangular coordinates in the form

$$\frac{\partial^2 \varphi}{\partial t^2} = c_L^2 \nabla^2 \varphi = c_L^2 \left(\frac{\partial^2 \varphi}{\partial x^2} + \frac{\partial^2 \varphi}{\partial y^2} + \frac{\partial^2 \varphi}{\partial z^2} \right) \tag{6-4-19}$$

and $$\frac{\partial^2 \psi_i}{\partial t^2} = c_T^2 \nabla^2 \psi_i = c_T^2 \left(\frac{\partial^2 \psi_i}{\partial x^2} + \frac{\partial^2 \psi_i}{\partial y^2} + \frac{\partial^2 \psi_i}{\partial z^2} \right) \qquad \text{for } i = 1, 2, 3 \tag{6-4-20}$$

As a solution of Eq. (6-4-19) try

$$\varphi = F(z) \exp [j(\omega t - kx)] \qquad \text{for } j = \sqrt{-1} \tag{6-4-21}$$

which represents a sinusoidal wave of frequency ω propagated in the x direction with velocity $c = \omega/k = \lambda_s \omega / 2\pi$ where λ_s is the wavelength and F is a function which determines the z dependence of the amplitude of the wave.

Substituting for φ in Eq. (6-4-19) from Eq. (6-4-21) gives

$$\frac{d^2 F}{dz^2} - b^2 F = 0 \tag{6-4-22}$$

which has the general solution

$$F(z) = A \exp(-bz) + B \exp(bz) \tag{6-4-23}$$

where

$$b^2 = k^2 - \frac{\omega^2}{c_L^2} = k^2 \left(1 - \frac{c^2}{c_L^2} \right) \tag{6-4-24}$$

Now the amplitude of the disturbance has been specified to decrease with increasing z; hence B must be zero, Eq. (6-4-23) reducing to

$$F(z) = A \exp(-bz) \tag{6-4-25}$$

where A is a constant.

Similarly, it can be shown that

$$\psi_y \equiv \psi = G(z) \exp [j(\omega t - kx)] \tag{6-4-26}$$

is a solution of Eq. (6-4-20) provided

$$G(z) = C \exp(-qz) \tag{6-4-27}$$

where

$$q^2 = k^2 - \frac{\omega^2}{c_T^2} = k^2 \left(1 - \frac{c^2}{c_T^2} \right) \tag{6-4-28}$$

and C is a constant.

Substituting for $F(z)$ and $G(z)$ in Eqs. (6-4-21) and (6-4-26) from Eqs. (6-4-25) and (6-4-27) gives for φ and ψ the relations

$$\varphi = A \exp(-bz) \exp [j(\omega t - kx)] \tag{6-4-29}$$

and $$\psi = C \exp(-qz) \exp [j(\omega t - kx)] \tag{6-4-30}$$

where $\omega = kc$.

Consider now the condition of a stress-free boundary which requires that $T\langle zz\rangle = T\langle zy\rangle = T\langle zx\rangle = 0$ at the surface where $z = 0$. Now, from the generalized Hooke's law [Eq. (6-1-21)],

$$T\langle zz\rangle = \lambda\left(\frac{\partial u}{\partial x} + \frac{\partial w}{\partial z}\right) + 2G\frac{\partial w}{\partial z}$$

and since $$\mathbf{u} = \nabla\varphi + \nabla\times\boldsymbol{\psi} \qquad \nabla\cdot\boldsymbol{\psi} = 0$$

the $T\langle zz\rangle$ component can be obtained in the form

$$T\langle zz\rangle = (\lambda+2G)\frac{\partial^2\varphi}{\partial z^2} + \lambda\frac{\partial^2\varphi}{\partial x^2} + 2G\frac{\partial^2\psi}{\partial x\partial z}$$

Substituting in this equation for φ and ψ from Eqs. (6-4-29) and (6-4-30), and noting that $T\langle zz\rangle = 0$ at $z = 0$, gives

$$T\langle zz\rangle = A[(\lambda+2G)b^2 - \lambda k^2] + 2jqkGC = 0 \tag{6-4-31}$$

Similarly, since

$$\begin{aligned} T\langle zx\rangle &= G\left(\frac{\partial u}{\partial z} + \frac{\partial w}{\partial x}\right) \\ &= G\left(2\frac{\partial^2\varphi}{\partial x\partial z} - \frac{\partial^2\psi}{\partial z^2} + \frac{\partial^2\psi}{\partial x^2}\right) \end{aligned}$$

substitution in this equation for φ and ψ from Eqs. (6-4-29) and (6-4-30), and noting that $T\langle zx\rangle = 0$ at $z = 0$, gives

$$T\langle zx\rangle = G[2jbkA - (q^2+k^2)C] = 0 \tag{6-4-32}$$

Equations (6-4-31) and (6-4-32) can be rearranged to give the identity

$$\frac{A}{C} = \frac{(q^2+k^2)}{2jbk} = -\frac{2jkqG}{(\lambda+2G)b^2 - \lambda k^2}$$

which can in turn be rearranged using Eqs. (6-4-24) and (6-4-28) to eliminate b and q, together with the relation

$$\frac{\lambda+2G}{G} = \frac{c_L^2}{c_T^2} = \frac{2(1-\nu)}{1-2\nu}$$

obtained from Eq. (6-4-5), to give

$$\left(2 - \frac{c^2}{c_T^2}\right)^2 = 4\left(1 - \frac{c^2}{c_L^2}\right)^{1/2}\left(1 - \frac{c^2}{c_T^2}\right)^{1/2}$$

further simplification giving the Rayleigh equation

$$\frac{c^2}{c_T^2}\left[\left(\frac{c}{c_T}\right)^6 - 8\left(\frac{c}{c_T}\right)^4 + \left(\frac{c}{c_T}\right)^2\left(24 - 16\frac{c_T^2}{c_L^2}\right) - 16\left(1 - \frac{c_T^2}{c_L^2}\right)\right] = 0 \tag{6-4-33}$$

which is a cubic in $(c/c_T)^2$.

For $c = 0$, Eq. (6-4-32) gives $A = -jC$, for which condition it can be shown that the displacement components are both zero—a solution of no interest. From Eq. (6-4-33), for $c = 0$ and $c_T < c_L$ the second factor is negative and for $c = c_T$ it is positive. Hence, since there is always a root of Eq. (6-4-33) in the range $(0 < c < c_T)$, surface waves can exist with a speed less than c_T.

Now A and C are independent of the frequency ω; hence, since

$$j\frac{A}{C} = \frac{2 - c^2/c_T^2}{2(1 - c^2/c_L^2)^{1/2}}$$

it is evident that the velocity of propagation of surface waves is independent of the frequency ω and depends only on the elastic constants of the material. These waves are therefore nondispersive, a plane surface wave traveling without change in form.

The displacements can be obtained in the form

$$u = Ak\left[\exp(-bz) - \tfrac{1}{2}\left(2 - \frac{c^2}{c_T^2}\right)\exp(-qz)\right]\sin(\omega t - kx) \tag{6-4-34}$$

and

$$w = Ab\left[2\left(2 - \frac{c^2}{c_T^2}\right)^{-1}\exp(-qz) - \exp(-bz)\right]\cos(\omega t - kx) \tag{6-4-35}$$

From Eqs. (6-4-24) and (6-4-28),

$$b = \omega\left(\frac{1}{c^2} - \frac{1}{c_L^2}\right)^{1/2} \qquad q = \omega\left(\frac{1}{c^2} - \frac{1}{c_T^2}\right)^{1/2}$$

and since c, c_L, and c_T are all independent of the frequency it is evident from an inspection of Eqs. (6-4-34) and (6-4-35) that ωz is the relevant parameter in the attenuation with depth of the vibrations both parallel to the surface and normal to it. Thus Rayleigh waves of high frequency will be attenuated more rapidly with depth than those of low frequency.

6-4-5 Vibrations of a Circular Cylinder

The vibration of a finite body under arbitrary boundary forces can in principle also be determined from Navier's equation (6-4-1), but the solution is in general very complex. However, a case of particular practical interest is that of a long isotropic circular cylinder, the theory of which is used in the split Hopkinson bar which will be discussed in detail in Chap. 13. The theory was first developed by Pochhammer (1876) and Chree (1889) and is presented by Love (1927). It is assumed that there are no traction forces at the surface of the cylinder and that the displacement is a simple harmonic function of z and t where the z axis coincides with the axis of the cylinder. These two conditions are not exactly compatible, but when the radius of the cylinder is small compared to its length they are very nearly reconciled.

It can easily be shown by taking the divergence and curl of Eq. (6-4-1) that

$$\frac{\partial^2 \Delta}{\partial t^2} = \left(\frac{\lambda + 2G}{\rho}\right)\nabla^2 \Delta \tag{6-4-36a}$$

$$\frac{\partial^2 \omega}{\partial t^2} = \frac{G}{\rho}\nabla^2\omega = -\frac{G}{\rho}\,\text{curl}\,\text{curl}\,\omega \tag{6-4-36b}$$

$$\frac{\partial^2 \boldsymbol{u}}{\partial t^2} = \frac{(\lambda+2G)}{\rho}\,\text{grad}\,\Delta - \frac{2G}{\rho}\,\text{curl}\,\omega \tag{6-4-37}$$

where

$$\Delta = \text{div}\,\boldsymbol{u} \qquad \omega = \tfrac{1}{2}\,\text{curl}\,\boldsymbol{u} \qquad \text{and hence} \qquad \text{div}\,\omega = 0 \tag{6-4-38}$$

In order that the boundary conditions should take a simple form it is convenient to write the equations in terms of cylindrical polar coordinates for which the above equations can be expressed as follows:

$$\nabla^2 = \boldsymbol{\nabla}\cdot\boldsymbol{\nabla} = \frac{\partial^2}{\partial r^2} + \frac{1}{r^2}\frac{\partial^2}{\partial \theta^2} + \frac{\partial^2}{\partial z^2} + \frac{1}{r}\frac{\partial}{\partial r} \tag{6-4-39}$$

$$\Delta = \text{div}\,\boldsymbol{u} = \frac{1}{r}\frac{\partial(ru\langle r\rangle)}{\partial r} + \frac{1}{r}\frac{\partial u\langle\theta\rangle}{\partial\theta} + \frac{\partial u\langle z\rangle}{\partial z} \tag{6-4-40}$$

$$2\omega\langle r\rangle = \frac{1}{r}\frac{\partial u\langle z\rangle}{\partial\theta} - \frac{\partial u\langle\theta\rangle}{\partial z} \qquad 2\omega\langle\theta\rangle = \frac{\partial u\langle r\rangle}{\partial z} - \frac{\partial u\langle z\rangle}{\partial r}$$
$$2\omega\langle z\rangle = \frac{1}{r}\left(\frac{\partial(ru\langle\theta\rangle)}{\partial r} - \frac{\partial u\langle r\rangle}{\partial\theta}\right) \tag{6-4-41}$$

$$\text{div}\,\omega = 0 = \frac{1}{r}\frac{\partial(r\omega\langle r\rangle)}{\partial r} + \frac{1}{r}\frac{\partial\omega\langle\theta\rangle}{\partial\theta} + \frac{\partial\omega\langle z\rangle}{\partial z} \tag{6-4-42}$$

The physical components of stress $T\langle rr\rangle$, $T\langle r\theta\rangle$, and $T\langle rz\rangle$ vanish at the surface of the cylinder $r = a$. Using Hooke's law these can be expressed in the form

$$T\langle rr\rangle = \lambda\Delta + 2G\frac{\partial u\langle r\rangle}{\partial r}$$
$$T\langle r\theta\rangle = G\left[\frac{1}{r}\frac{\partial u\langle r\rangle}{\partial\theta} + r\frac{\partial}{\partial r}\left(\frac{u\langle\theta\rangle}{r}\right)\right] \tag{6-4-43}$$
$$T\langle rz\rangle = G\left(\frac{\partial u\langle r\rangle}{\partial z} + \frac{\partial u\langle z\rangle}{\partial r}\right)$$

It is then assumed that $u\langle r\rangle$, $u\langle\theta\rangle$, and $u\langle z\rangle$ can be expressed as

$$u\langle r\rangle = U\exp[j(kz+\omega t)] \qquad u\langle\theta\rangle = V\exp[j(kz+\omega t)]$$
$$u\langle z\rangle = W\exp[j(kz+\omega t)] \tag{6-4-44}$$

where U, V, and W are functions of r, θ. These equations will now be used to examine the three basic modes of vibration of a long thin cylindrical bar which are termed torsional, longitudinal, and flexural vibrations.

(a) Torsional vibrations For such vibrations $u\langle r\rangle$ and $u\langle z\rangle$ are both zero and V is independent of θ so that Δ is also zero by Eq. (6-4-40). Substituting these conditions, and those of Eq. (6-4-44) into Eq. (6-4-41) gives

$$
\begin{aligned}
2\omega\langle r\rangle &= -jkV\exp[j(kz+\omega t)] \\
\omega\langle\theta\rangle &= 0 \\
2\omega\langle z\rangle &= \frac{1}{r}\frac{\partial(rV)}{\partial r}\exp[j(kz+\omega t)]
\end{aligned}
\tag{6-4-45}
$$

and Eq. (6-4-42) is thus also satisfied. Only the θ component of Eq. (6-4-37) is not identically zero, and this becomes

$$
\frac{\partial^2 \ddot{u}\langle\theta\rangle}{\partial t^2} = \frac{2G}{\rho}\left(\frac{\partial\omega\langle z\rangle}{\partial r} - \frac{\partial\omega\langle r\rangle}{\partial z}\right) \tag{6-4-46}
$$

Using Eqs. (6-4-44) and (6-4-45), this can be written in the form

$$
\frac{\partial^2 V}{\partial r^2} + \frac{1}{r}\frac{\partial V}{\partial r} - \frac{V}{r^2} + \varkappa'^2 V = 0 \tag{6-4-47}
$$

where

$$
\varkappa'^2 = \omega^2\frac{\rho}{G} - k^2 \tag{6-4-48}
$$

The solution of the above Bessel equation which is finite on the axis is

$$
V = BJ_1(\varkappa' r) \tag{6-4-49}
$$

where B is a constant and J_1 is Bessel's function of order unity.

From Eq. (6-4-43) it can be seen that $T\langle rr\rangle$ and $T\langle rz\rangle$ are zero everywhere and the condition that $T\langle r\theta\rangle$ should be zero on the surface of the cylinder gives

$$
\left[\frac{\partial}{\partial r}\left(\frac{J_1(\varkappa' r)}{r}\right)\right]_{r=a} = 0 \tag{6-4-50}
$$

One solution of this equation is $\varkappa' = 0$, which when inserted in Eq. (6-4-47) gives the solution

$$
V = B'r \tag{6-4-51}
$$

where B' is a constant. This torsional solution of Navier's equation is the only simple solution which satisfies the boundary conditions exactly. All waves propagate with the same phase velocity

$$
\frac{\omega}{k} = \left(\frac{G}{\rho}\right)^{1/2} = c_T \tag{6-4-52}
$$

and this type of wave suffers no dispersion. The other roots of Eq. (6-4-50) can be shown to correspond to more complex modes of torsional vibration involving nodal cylinders. As these waves are dispersive in character and difficult to excite experimentally they will not be considered further.

(b) Longitudinal vibrations For such vibrations V is zero while U and W are independent of θ. It follows from Eq. (6-4-41) that $\omega\langle r\rangle$ and $\omega\langle z\rangle$ are both zero and substituting in Eqs. (6-4-36) gives, by way of Eq. (6-4-40), the relations

$$\frac{\partial^2 \Delta}{\partial r^2} + \frac{1}{r}\frac{\partial \Delta}{\partial r} + h'^2 \Delta = 0 \tag{6-4-53}$$

$$\frac{\partial^2 \omega\langle\theta\rangle}{\partial r^2} + \frac{1}{r}\frac{\partial \omega\langle\theta\rangle}{\partial r} - \frac{\omega\langle\theta\rangle}{r^2} + \varkappa'^2 \omega\langle\theta\rangle = 0 \tag{6-4-54}$$

where

$$h'^2 = \frac{\omega^2 \rho}{\lambda + 2G} - k^2 \qquad \varkappa'^2 = \frac{\omega^2 \rho}{G} - k^2 \tag{6-4-55}$$

The solution of Eq. (6-4-53) which is finite on the axis is

$$\Delta = A' J_0(h'r) \exp[j(kz + \omega t)] \tag{6-4-56}$$

while the solution of Eq. (6-4-54) is

$$\omega\langle\theta\rangle = B' J_1(\varkappa' r) \exp[j(kz + \omega t)] \tag{6-4-57}$$

Now, from Eqs. (6-4-37), (6-4-56), and (6-4-57),

$$\omega^2 U = A \frac{\partial J_0(h'r)}{\partial r} + CkJ_1(\varkappa' r) \tag{6-4-58}$$

$$\omega^2 W = AjkJ_0(h'r) + \frac{1}{r} Cj \frac{\partial}{\partial r}[rJ_1(\varkappa' r)] \tag{6-4-59}$$

where A and C are constants. Substituting in Eq. (6-4-43) and setting $T\langle rr\rangle$ and $T\langle rz\rangle$ to zero when $r = a$, the radius of the cylinder, gives

$$A\left[2G\frac{\partial^2}{\partial r^2} J_0(h'r) - \frac{\lambda}{\lambda + 2G}\omega^2 \rho J_0(h'r)\right]_{r=a} + 2CGk\left[\frac{\partial}{\partial r} J_1(\varkappa' r)\right]_{r=a} = 0 \tag{6-4-60}$$

and

$$2Ak\left[\frac{\partial}{\partial r} J_0(h'r)\right]_{r=a} + C\left(2k^2 - \frac{\omega^2 \rho}{G}\right)[J_1(\varkappa' r)]_{r=a} = 0 \tag{6-4-61}$$

Eliminating A/C from these last two equations leaves the equation relating the frequency and wavelength of the waves; this is known as the frequency or dispersion equation. The condition that the ends are free from traction is not exactly satisfied but the residual stresses are very small if the length of the cylinder is large compared to a. J_0 and J_1 can be expanded in a power series to give

$$J_0(h'a) = 1 - \tfrac{1}{4}(h'a)^2 + \tfrac{1}{64}(h'a)^4 - \cdots \tag{6-4-62}$$

and

$$J_1(\varkappa' a) = \tfrac{1}{2}(\varkappa' a) - \tfrac{1}{16}(\varkappa' a)^3 + \cdots \tag{6-4-63}$$

Substituting these expressions in Eqs. (6-4-60) and (6-4-61), the frequency equation becomes

$$\left(2k^2 - \frac{\omega^2 \rho}{G}\right)\left[h'^2 + \frac{\lambda \omega^2 \rho}{G(\lambda + 2G)}\right] = 2k^2 h'^2 \tag{6-4-64}$$

to first order in a, which has the solution

$$\frac{\omega^2}{k^2} = \frac{G(3\lambda + 2G)}{(\lambda + G)\rho} = \frac{E}{\rho} \tag{6-4-65}$$

If second-order terms in a are retained the frequency equation becomes

$$\frac{\omega}{k} = \left(\frac{E}{\rho}\right)^{1/2}(1 - \tfrac{1}{4}\nu^2 k^2 a^2) \tag{6-4-66}$$

where ν is Poisson's ratio. Putting $(E/\rho)^{1/2} = c_0$, $\omega/k = c$, the phase velocity of the wave, and the wavelength $2\pi/k = \Lambda$, Eq. (6-4-66) can be written as

$$\frac{c}{c_0} = 1 - \pi^2 \nu^2 \frac{a^2}{\Lambda^2} \tag{6-4-67}$$

These waves are therefore dispersive and the original wave form will become distorted in propagating down the bar. Davies (1948) has calculated the exact variation of (c/c_0) as a function of a/Λ for $\nu = 0.29$ and then used this information to determine the distortion of an impact pulse by analyzing it into its Fourier components. He has shown that in traveling down a bar, a longitudinal pulse whose original length is comparable with the radius of the bar will develop a train of high-frequency oscillations superimposed on the rear of the pulse.

(c) Flexural waves For longitudinal and torsional waves the motion is symmetrical about the cylinder axis and so the displacements are independent of θ. In the case of flexural waves this is no longer true and U, V, and W will be functions of both r and θ. If it is assumed that $u\langle r\rangle$ and $u\langle z\rangle$ are proportional to $\cos\theta$ while $u\langle\theta\rangle$ is proportional to $\sin\theta$, then, from Eq. (6-4-41), $\omega\langle r\rangle$ and $\omega\langle z\rangle$ are proportional to $\sin\theta$ while $\omega\langle\theta\rangle$ is proportional to $\cos\theta$, as is also Δ by Eq. (6-4-40). From Eq. (6-4-36),

$$\frac{\partial^2 \Delta}{\partial r^2} + \frac{1}{r}\frac{\partial \Delta}{\partial r} - \frac{\Delta}{r^2} + h'^2 \Delta = 0 \tag{6-4-68}$$

whose solution may be written

$$\Delta = -\frac{\omega^2 \rho}{\lambda + 2G} A J_1(h'r)\cos\theta \exp[j(kz + \omega t)] \tag{6-4-69}$$

Using Eqs. (6-4-36b) and (6-4-37) gives relations of the form

$$\begin{aligned} U &= A\frac{\partial J_1(h'r)}{\partial r} + Bk\frac{\partial J_1(\varkappa' r)}{\partial r} + C\frac{J_1(\varkappa' r)}{r} \\ V &= -A\frac{J_1(h'r)}{r} - Bk\frac{J_1(\varkappa' r)}{r} - C\frac{\partial J_1(\varkappa' r)}{\partial r} \\ W &= jAkJ_1(h'r) - jB\varkappa'^2 J_1(\varkappa' r) \end{aligned} \tag{6-4-70}$$

Equating the stress on the surface to zero gives the frequency equation, but this leads to a very complicated equation which has been examined numerically by

Hudson (1943). If the Bessel functions in the complete solution are expanded to the second order in a then

$$\omega^2 = \tfrac{1}{4}a^2k^4c_0^2 \qquad \text{or} \qquad c'^2 = \tfrac{1}{4}a^2k^2c_0^2 \tag{6-4-71}$$

which agrees with the elementary theory according to which

$$\frac{\partial^2 u\langle y\rangle}{\partial t^2} = -c_0^2K^2\frac{\partial^4 u\langle y\rangle}{\partial z^4} \tag{6-4-72}$$

where $u\langle y\rangle$ is the displacement in the direction y which is orthogonal to z, and in the plane of which it is assumed all the flexural vibrations occur, and K is the radius of gyration of the cross section of the bar about an axis through the center of gravity perpendicular to the yz plane. For a circular rod $K = a/2$. The elementary theory is only valid for waves having a wavelength greater than $10a$ and gives very poor results for shorter wavelengths, as shown in Fig. 6-2 which is

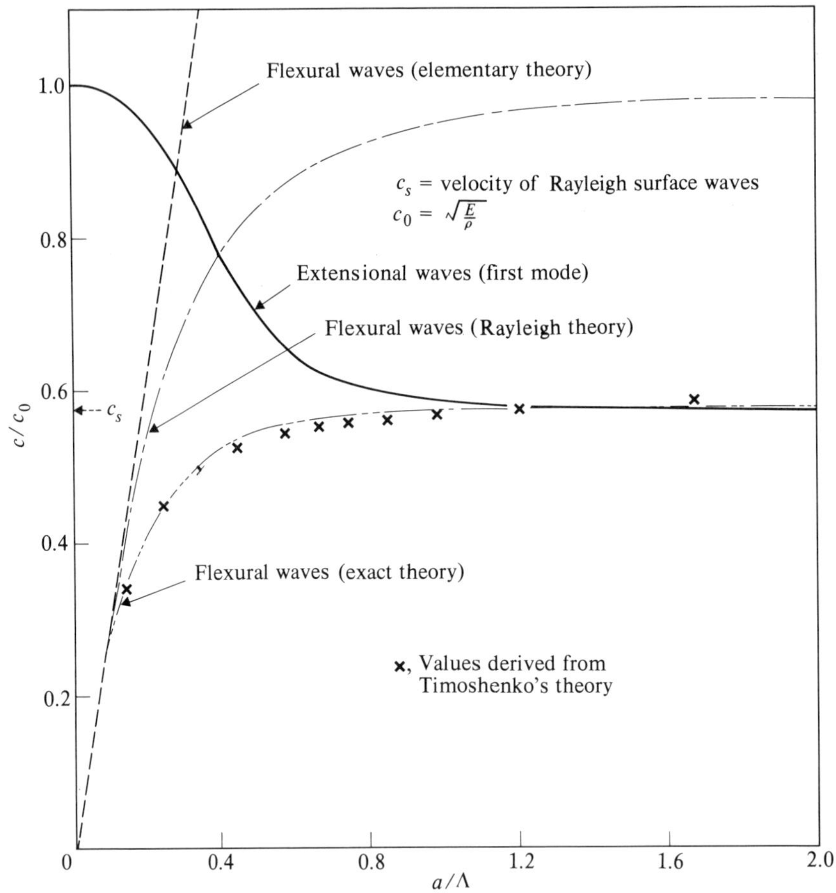

Figure 6-2 Phase velocity, c, c' of extensional and flexural waves of wavelength Λ in cylindrical bars of radius a (for $\nu = 0.29$). (*Davies, 1948.*)

taken from Davies (1948). A much better approximative equation which provides a good fit to the exact results of Hudson (1943) has been given by Timoshenko (1921), who proposed the flexural equation

$$c_0^2 K^2 \frac{\partial^4 u\langle y\rangle}{\partial z^4} + \frac{\partial^2 u\langle y\rangle}{\partial t^2} - K^2(1+\varepsilon)\frac{\partial^4 u\langle y\rangle}{\partial z^2 \partial t^2} + \frac{\varepsilon K^2}{c_0^2}\frac{\partial^4 u\langle y\rangle}{\partial t^4} = 0 \qquad (6\text{-}4\text{-}73)$$

where $\varepsilon = 2R'(1+\nu)$, R' being a numerical constant depending on the bar shape and ν being Poisson's ratio. For a given rod of circular cross section $R' = 10/9$.

6-5 EXPERIMENTAL APPLICATIONS

6-5-1 Uniaxial Test

External forces are often applied nonuniformly through various types of fastenings (e.g., bolts, rivets, etc.), the stresses varying sharply across such loading points, as also occurs across sharp discontinuities in the contours of a body such as at a notch. However, in practice this aspect of engineering structures usually has negligible effect on the overall elastic deformation of bodies, since these regions of the stress field, though complex, are often highly localized. This situation finds expression by way of St. Venant's principle (1855) which states that, in elastostatics, if the forces acting on a small part ∂R_1 of the boundary ∂R of a body are replaced by statically equivalent forces, i.e., they have the same resultant force and couple acting on the same part ∂R_1, the stress state is negligibly changed at distances from ∂R_1 which are large compared with the dimensions of ∂R_1. For example, in Fig. 6-3 the two tensile forces F act symmetrically about the compressive force $2F$, and since their resultant force and couple are zero, by St. Venant's principle they must produce negligible stresses in regions well beyond the radius r. This principle is of great importance in applied elasticity, since the long-range effects of applied forces can be determined by replacing the actual conditions at the loading points by the simplest statically equivalent conditions. This

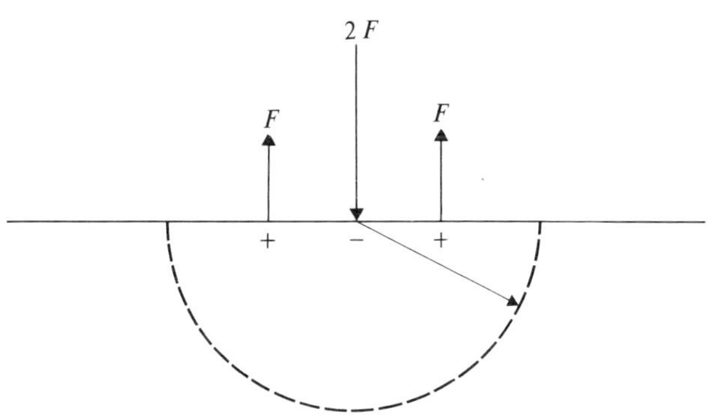

Figure 6-3 Example of St. Venant's principle.

principle is frequently invoked to justify solutions in long slender structural members, where the end traction boundary conditions are satisfied only in an average sense, so that the correct stress resultants act on the ends. In such solutions the actual stress distribution near the ends may differ markedly from the calculated stress distribution. When applying St. Venant's principle caution is required, particularly in the case of anisotropic materials such as fiber-reinforced materials. For a discussion of St. Venant's principle, see Fung (1965).

The strain state in a long cylindrical body acted on by loads that are normal to the axis of the cylinder (taken to be the z axis) and which are uniform in the axial direction, i.e., the boundary conditions involve only displacements or tractions parallel to the xy plane and are independent of z, is an example of a system which often can be approximated by a plane strain state. On the ends of the cylinder, the boundary conditions are $u\langle z\rangle = 0$ and $T\langle zx\rangle = T\langle zy\rangle = 0$. However, the z component of Eqs. (6-3-4) and (6-3-32) shows that the normal traction on the ends is not zero, except in special cases where $\tilde{e}\langle xx\rangle + \tilde{e}\langle yy\rangle = 0$ or, equivalently, $T\langle xx\rangle + T\langle yy\rangle = 0$. In the event that the plane strain solution gives $T\langle xx\rangle + T\langle yy\rangle < 0$, the required compressive loading on the ends can be considered to arise from the reaction with a frictionless rigid wall.

Since the boundary-value problem is linear, the actual strain can be determined by way of a superposition of a plane strain solution and another solution. For example, first determine the plane strain solution corresponding to the situation in which the lateral boundary conditions are all traction boundary conditions while the ends are free. If for this plane strain solution

$$T\langle xx\rangle + T\langle yy\rangle = f(x, y)$$

then, from Eq. (6-3-4),

$$T\langle zz\rangle = \nu f(x, y)$$

on the ends. Thus, it is required to find a second solution with

$$\begin{aligned} T\langle zz\rangle &= -\nu f(x, y) \\ T\langle zx\rangle &= T\langle zy\rangle = 0 \end{aligned}$$

on the ends and with traction-free lateral boundaries. Such a solution can be found by applying St. Venant's principle, providing the cylinder is long in comparison with its cross-sectional dimensions. Except near the ends, a good approximation is obtained by superposing a uniform uniaxial stress equal to the average value of the required $T\langle zz\rangle$ over the cross section, together with a bending solution corresponding to a linear distribution of $T\langle zz\rangle$.

The linear theory of elasticity of immediate interest has found extensive application in the deformation of engineering structures. Many real materials exhibit to a high degree of approximation a characteristic, linear, stress-strain response to deformation over a wide range of stress. For isotropic materials, the stress-strain relations are those of the generalized Hooke's law expressed in the form of Eq. (6-1-21), which is seen to contain only the two constants λ and G. In practice two other constants are frequently determined. These are the Young modulus E and the Poisson ratio ν, related to λ and G by the first relations of Eqs. (6-1-29) and (6-1-30). Since the four constants λ, G, E, and ν are interrelated by Eqs. (6-1-29) and (6-1-30), only two experiments are required, the experiment to determine E being referred to as the tension test and the experiment to determine G as the simple shear test.

To determine E and ν, a circular, cylindrical, test specimen is subjected to a uniform axial tension

$$T\langle 11\rangle \equiv T_0 = \frac{F}{A_0} \tag{6-5-1}$$

where F is the force applied and A_0 is the original, undeformed, cross-sectional area of the specimen. In Eq. (6-1-21), take the trace of $\mathbf{T}$ to give

$$\operatorname{tr}\mathbf{T} = I_{\mathbf{T}} = (3\lambda + 2G)I_{\tilde{\mathbf{e}}} = 3\kappa I_{\tilde{\mathbf{e}}} \tag{6-5-2}$$

which can now be used to eliminate $I_{\tilde{\mathbf{e}}}$ from Eq. (6-1-21) to give

$$\tilde{\mathbf{e}} = -\frac{\lambda}{2G(3\lambda + 2G)}(\operatorname{tr}\mathbf{T})\mathbf{I} + \frac{1}{2G}\mathbf{T} \tag{6-5-3}$$

Note that for $\tilde{\mathbf{e}}$ to be uniquely determined by $\mathbf{T}$ it is necessary that

$$G \neq 0 \qquad (3\lambda + 2G) \neq 0 \tag{6-5-4}$$

and in order not to have zero strain for a finite stress it is necessary that

$$|G| < \infty \qquad |3\lambda + 2G| < \infty \tag{6-5-5}$$

For the test conditions of immediate interest, Eqs. (6-5-3) and (6-1-32) give

$$\tilde{e}\langle 11\rangle = \frac{T_0}{E} = \tilde{e} = \frac{l - L}{L} \tag{6-5-6}$$

with

$$\tilde{e}\langle 22\rangle = \tilde{e}\langle 33\rangle = -\frac{\nu}{E}T_0 = \nu\tilde{e} \tag{6-5-7}$$

and

$$\tilde{e}\langle 23\rangle = \tilde{e}\langle 31\rangle = \tilde{e}\langle 12\rangle = 0 \tag{6-5-8}$$

where L is the initial, undeformed length of the cylindrical test specimen and l its current deformed length corresponding to an applied stress T_0.

From Eq. (6-5-6) it is evident that for $T_0 > 0$ a cylinder, initially of unit length and of unit diameter, will elongate by an amount T_0/E, and from Eq. (6-5-7) its diameter will contract by an amount $\nu T_0/E$. Measurement of this elongation and contraction determines E and ν and, by way of the first relations of Eqs. (6-1-29) and (6-1-30), also λ and G. It is important to note that the stress and strain values defined by Eqs. (6-5-1) and (6-5-6) respectively do not include the stress-induced changes in specimen dimensions, but are based on the original undeformed values of A_0 and L. For this reason, the stress T_0 and strain $\tilde{e}$ are called the nominal stress and strain, though they are also referred to as standard or conventional engineering stress and strain. Some typical values for E and ν are given in Table 6-1.

Table 6-1 Typical values of the elastic constants of various metals

Metal	*E*, TPa	*G*, TPa	ν
Magnesium	0.0448	0.0172	0.33
Aluminum	0.0689	0.0262	0.34
Iron	0.1965	0.0793	0.28
Copper	0.1241	0.04619	0.35
Zinc	0.0965	0.03447	0.35
Tin	0.04688	0.01724	0.36
Tungsten	0.3447	0.1517	0.17
Lead	0.0179	0.0055	0.45
Uranium	0.1723	0.0758	0.24

Some schematic stress-strain curves which are typical of the mechanical response of some common engineering materials are shown in Fig. 6-4. A ductile material is one in which a large deformation is produced before the specimen either ruptures or, more generally, the cross section of the specimen begins to decrease rapidly for only small increments in the stress. This behavior is in contrast to a brittle material for which only a small deformation occurs before rupture. A

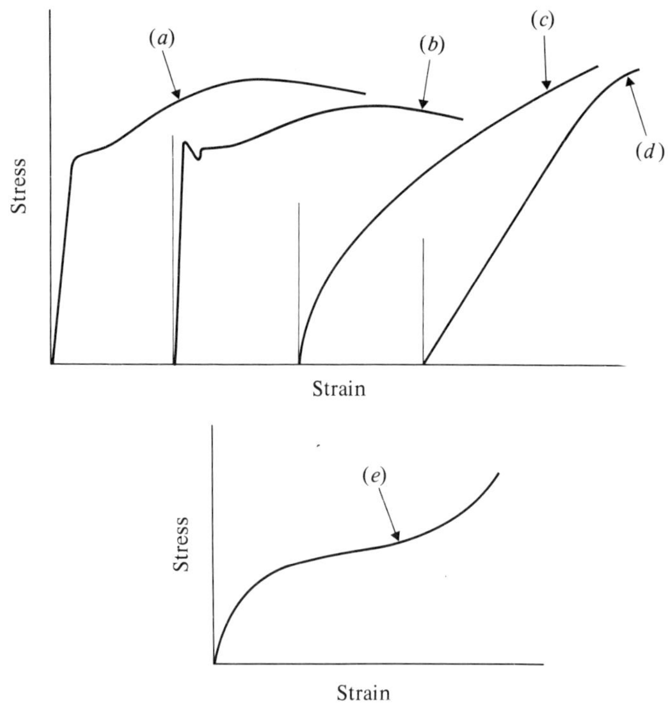

Figure 6-4 Typical nominal stress-strain curves. (*a*) Ductile material—I. (*b*) Ductile material—II. (*c*) Brittle material—I. (*d*) Brittle material—II. (*e*) Nonmetallic solid.

stress-strain curve characteristic of a ductile material such as an aluminum alloy is shown in Fig. 6-4*a* and for a mild or structural steel in Fig. 6-4*b*. These stress-strain curves for ductile materials are to be compared with the stress-strain curve characteristic of the response of a brittle material, such as concrete or malleable cast iron, shown in Fig. 6-4*c*. For some nonmetallic brittle materials as, for example, various types of plastic, stress-strain curves of the form shown in Fig. 6-4*d* are typical. It is evident from an inspection of Fig. 6-4 that there exists a considerable variation in the characteristic mechanical response of engineering materials. However, there is an initial region of the stress-strain curve which is substantially linear for three of the materials depicted in Fig. 6-4*a*, *b*, and *d*. It is this linear region of the characteristic stress-strain curve which is regarded as being described by Eq. (6-5-6), i.e., the materials conform to Hooke's law, this initial stage of the loading where the stress-strain relation is linear being referred to as the linear elastic range. Beyond the linear elastic range, where the strains become large and Hooke's law no longer applies, the materials are said to be in the nonlinear elastic range.

Various quality factors have been introduced in the mechanical testing of solids. Examples of these are as follows. A material has a high elastic strength if it resists loads without yielding and without being permanently deformed. The proportional limit is the elastic strength as represented by point *a* in Fig. 6-5*a* and is defined as the value of the stress beyond which the stress is no longer proportional to the strain. The elastic limit, denoted by point *b* in Fig. 6-5*a*, is defined as the maximum stress that can be applied to a material without producing a permanent plastic deformation when the load is removed. Elastic limit and proportional limit, which are alternative measures of the elastic strength, are rarely measured because of the practical difficulty of determining the points *a* and *b* in Fig. 6-5*a*. In Fig. 6-5*b*, an upper yield point *c* or a lower yield point *d* may be used to define the elastic strength or transition from the elastic to the plastic range

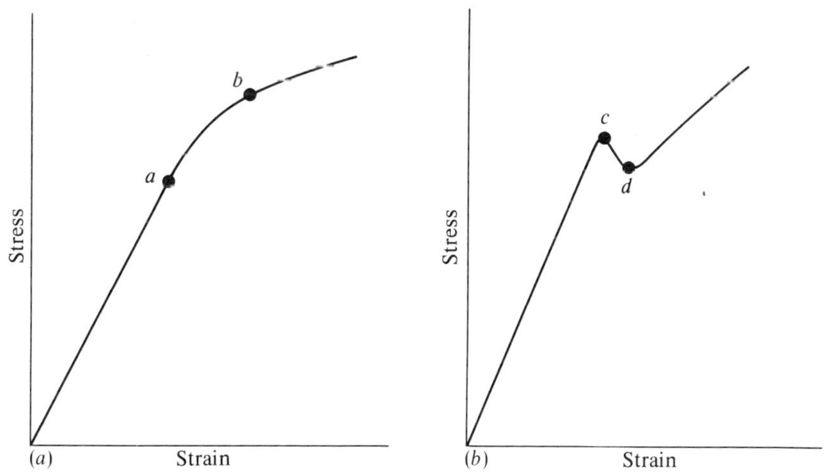

Figure 6-5 Typical uniaxial stress-strain curves illustrating elastic and plastic response.

where yielding takes place. A material is said to have a high resilience if its ability to absorb internal work or energy in the elastic range is high. Resilience is measured by the modulus of elastic resilience U_p, which is the internal work or strain energy per unit volume of the material required to stress the material to the proportional limit. Thus, by way of Eq. (6-5-6),

$$U_p = \int_0^{\tilde{e}_p} T_0 \, d\tilde{e} = \tfrac{1}{2} E \tilde{e}_p^2 \tag{6-5-9}$$

or, alternatively,

$$U_p = \tfrac{1}{2} \frac{T_p^2}{E} \tag{6-5-10}$$

where $(T_p, \tilde{e}_p)$ is the point a in Fig. 6-5a.

The stress-strain properties ascribed to materials in tension can, in principle, also be distinguished and similarly defined for materials in compression. Difficulties arise in the determination of the accurate mechanical response of materials in compression because of irregularities of alignment which are accentuated with an increase in loading and result in lateral deflections and bending stresses. There is also the question of how to allow for the lateral restraining forces at the ends of the specimen owing to the friction forces between the specimen ends and bearing plates. This effect gives rise to the phenomenon of "barreling" in compression which is the counterpart to "necking down" of a specimen in tension. Failure of the test specimen in compression may also be produced by lateral buckling if the specimen is too long.

6-5-2 Simple Torsion Test

In the case of a square block subjected to simple shear, as in Fig. 6-6, the stress components reduce to

$$T\langle 12 \rangle = T\langle 21 \rangle \tag{6-5-11}$$

with

$$T\langle 11 \rangle = T\langle 22 \rangle = T\langle 33 \rangle = T\langle 23 \rangle = T\langle 31 \rangle = 0 \tag{6-5-12}$$

For these conditions Eq. (6-5-3) gives

$$T\langle 12 \rangle = 2G\tilde{e}\langle 12 \rangle \tag{6-5-13}$$

or, since $2\tilde{e}\langle 12 \rangle = \gamma\langle 12 \rangle$, the shear modulus of rigidity

$$G = \frac{T\langle 12 \rangle}{\gamma\langle 12 \rangle} \tag{6-5-14}$$

is obtained by measuring the angle change $\gamma\langle 12 \rangle$. The practical difficulties involved in subjecting a square block to simple shear are such that in practice the shear modulus of rigidity G is generally determined by way of the torsion test.

The torsion test involves subjecting a specimen of solid circular cross section to a twisting moment Γ. The measurements recorded are the angle of twist θ,

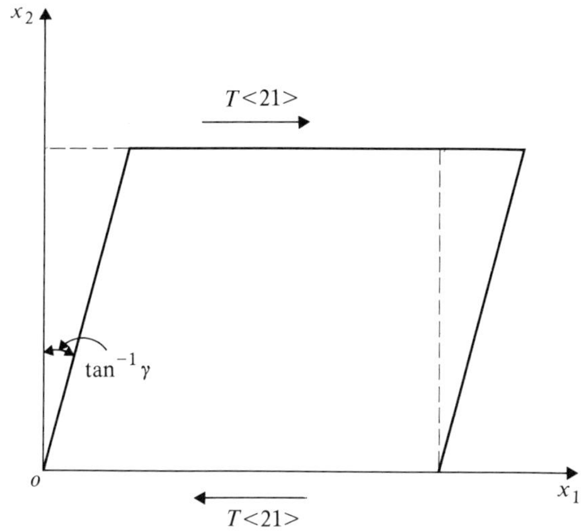

Figure 6-6 Simple shear test.

corresponding to a given value of Γ, for a specimen of length L, the mechanical response being evaluated from the (Γ, θ) curves in the form of a shear stress–shear strain curve as follows.

The two-dimensional shear behavior will be referred to cylindrical polar coordinates (r, θ, z) with the z axis coincident with the major axis of the cylindrical specimen. For a torque Γ, the shear stress

$$T\langle z\theta\rangle = \frac{r\Gamma}{\bar{I}} = \gamma\langle z\theta\rangle \; (\equiv \gamma) \tag{6-5-15}$$

where r is the radius of the circular cross section and

$$\bar{I} = \frac{\pi r^4}{2} \tag{6-5-16}$$

is the polar moment of area of the cross section. The shear strain $\tilde{e}\langle z\theta\rangle$ can be evaluated from the relation

$$2\tilde{e}\langle z\theta\rangle = \frac{r\theta}{L} = \gamma\langle z\theta\rangle \tag{6-5-17}$$

the modulus of rigidity being obtained from the relation

$$T\langle z\theta\rangle = 2G\tilde{e}\langle z\theta\rangle = G\frac{r\theta}{L} \tag{6-5-18}$$

in the form

$$G = \frac{\Gamma L}{\bar{I}\theta} \tag{6-5-19}$$

where the angle θ can be identified by reference to Fig. 6-7. This type of measurement is subject to an error arising from an apparent "strengthening" by

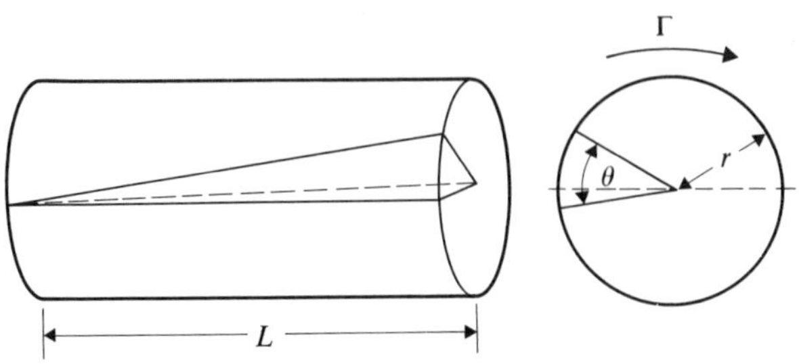

(a)

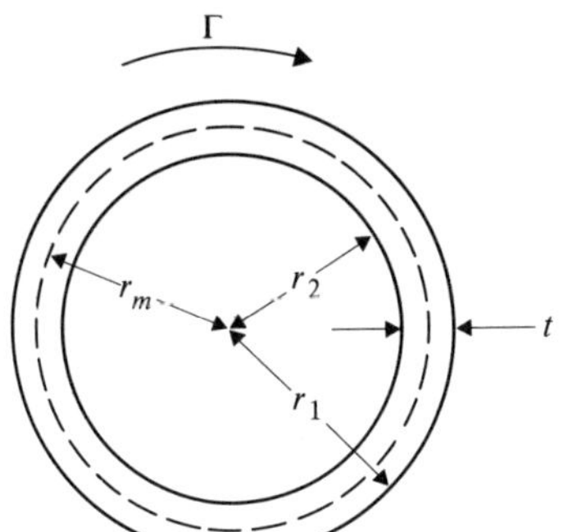

(b)

Figure 6-7 Shear stress and strain for cylindrical specimens as used in the torsion test: (a) solid rod, (b) thin-walled tube.

the material nearer the longitudinal axis of the specimen which is stressed to a lower level.

Greater accuracy of the mechanical properties of a material measured in shear can be obtained by using a thin-walled tubular specimen of circular cross section. For this type of specimen, the shear stress can be regarded as being substantially uniform across the specimen's annular cross section. Consider a thin-walled tubular specimen of circular cross section, of radii r_2 and r_1 where r_2 is the inner radius and r_1 is the outer radius. For a sufficiently thin-walled specimen the shear stress $T\langle\theta z\rangle$ can be assumed to be uniformly distributed, for which condition

$$T\langle\theta z\rangle = \frac{\Gamma}{2\pi r_m^2 t} = \frac{2\Gamma}{\pi(r_1+r_2)(r_1^2-r_2^2)} = T\langle z\theta\rangle \tag{6-5-20}$$

where $r_m = (r_1+r_2)/2$ and $t = r_1 - r_2$. For small strains

$$T\langle z\theta\rangle = \frac{2\Gamma}{\pi(r_1+r_2)(r_1^2-r_2^2)} = G\frac{(r_1+r_2)}{2}\frac{\theta}{L} \tag{6-5-21}$$

giving

$$G = \frac{1}{(r_1+r_2)^2(r_1^2-r_2^2)}\frac{4L\Gamma}{\pi\theta} \tag{6-5-22}$$

Some typical values of G are given in Table 6-1.

6-5-3 Hydrostatic Pressure Test

For an elastic body subject to a purely hydrostatic state of stress, the volume of the body decreases with increasing pressure. If the stress tensor has the form

$$\mathbf{T} = -P\mathbf{I} \qquad \text{for } P > 0 \tag{6-5-23}$$

where P is the hydrostatic pressure, then the state of stress at a point of the body is said to be hydrostatic. From Eqs. (6-5-2) and (6-5-23) it follows that

$$P = -\kappa I_{\tilde{\mathbf{e}}} \tag{6-5-24}$$

where

$$\kappa = \lambda + \tfrac{2}{3}G \tag{6-5-25}$$

is the modulus of compression or the bulk modulus. For an infinitesimal deformation, the volume change is given by $I_{\tilde{\mathbf{e}}}$, and for the volume to decrease with increasing hydrostatic pressure,

$$\kappa \geqslant 0 \tag{6-5-26}$$

that is,

$$\lambda + \tfrac{2}{3}G \geqslant 0 \tag{6-5-27}$$

The strain can be obtained from Eq. (6-5-3) in the form

$$\tilde{\mathbf{e}} = -\frac{P}{3\lambda + 2G}\mathbf{I} = -\frac{1-2\nu}{E}P\mathbf{I} \tag{6-5-28}$$

it being noted that κ can also be expressed in the alternative form

$$\kappa = \frac{E}{3(1-2\nu)} \tag{6-5-29}$$

from which, since $\kappa \geqslant 0$, it follows that $\nu \leqslant \frac{1}{2}$. For a discussion of the effects of subjecting materials to hydrostatic pressures, the reader is referred to Bridgman (1949, 1952).

6-5-4 Bend Test

For the curved rod or beam, the material filaments on the inside of the curve are compressed and the filaments on the outside are stretched. The transition from compression to tension occurs at a surface which is parallel to the principal axis of the beam and which experiences no change in length. This is called the neutral filament, neutral surface, or neutral axis. For small bending conditions in which the beam is not subject to additional compression or tension, i.e., the beam is subject only to pure bending, the neutral surface passes through the center of gravity of the cross section of simple beams.

In pure bending, a thin transverse element of the bar is distorted in such a way that the material filaments below the neutral surface exhibit a compressional strain which is proportional to the distance from the neutral surface, and the material filaments above are stretched also in proportion to their distance from the neutral axis. That is, the longitudinal stretch ΔL is proportional to the height y of the reference plane from the neutral surface, such that

$$\Delta L = \frac{Ly}{R} \tag{6-5-30}$$

where L is the length of the neutral filament, R the radius of the neutral filament, and L/R is the constant of proportionality. But $\Delta L/L$ is just the strain, and hence the stress is

$$T_0 = E\frac{y}{R} \tag{6-5-31}$$

Now consider the torque or bending moment M about the neutral plane. If the area of cross section is dA, the force across the area is

$$dF = \frac{Ey}{R}dA \tag{6-5-32}$$

The couple due to these forces is thus $Ey^2\,dA/R$ and the total couple or bending moment, due to all the filaments in the beam, which must be equal to the external couple M when the beam is in equilibrium, will be

$$M = \frac{E}{R}\int y^2\,dA = \frac{E\hat{I}}{R} = \frac{n}{R} \tag{6-5-33}$$

where

$$\hat{I} = \int y^2\,dA \tag{6-5-34}$$

is called the moment of inertia of the geometric cross section about the principal axis ($y = 0$) through its center of mass and is identical to the moment of inertia of an element of the beam having unit mass per unit area. The quantity

$$n = E\hat{I} \tag{6-5-35}$$

appearing in Eq. (6-5-33) is termed the flexural rigidity, or stiffness. Equation (6-5-33) can be made the basis of a mechanical test in a manner similar to that described for tension and shear.

6-6 STRUCTURAL THEORIES OF LINEAR ELASTICITY

In the preceding sections the macroscopic mechanical response of materials, as embodied in their constitutive equations, was examined. However, no account was taken of the ultimate particulate structure of matter or the way that this microstructure affects the macroscopic response. The classical concepts relating certain aspects of elasticity to atomic interactions are briefly reviewed, the origins of the main kinds of chemical forces in the atomic structure are examined, and the formation of the atoms at relatively low temperatures into the orderliness of the crystal state together with the relation of the crystal elastic constants to the crystal structure are discussed.

6-6-1 Atomic Structure

To understand the origins of the interaction forces between atoms it is necessary to consider the structure of the atom itself. Most of the mass of an atom is concentrated within an exceedingly small central volume called the nucleus whose diameter is of the order of 10^{-14} m. The nucleus itself is composed of protons and neutrons. The proton has a unit positive electrostatic charge $+e$ ($e = 1.6 \times 10^{-19}$ C) and a mass of 1.67×10^{-27} kg, while the neutron has no charge and is slightly heavier. The mass of the atom depends on the mass number which is the total number of protons and neutrons in its nucleus, while the chemical properties of the atom depend mainly on its atomic number Z, which is the number of protons only. The atomic number is important because the corresponding positive charge of the nucleus must, in the atom's normal electrically neutral state, be balanced by a corresponding number of electrons which each carry a unit negative electrostatic charge and are 1836 times lighter than a proton. Each chemical element is characterized by a different atomic number and the isotopes of an element have the same number of protons but different numbers of neutrons.

Each nucleus, which is positively charged, is surrounded by an electrically neutralizing cloud of electrons, and the forces between atoms are governed almost entirely by the electron configuration. The electron motion is determined by Schrödinger's equation for an electron moving in the electrostatic field of the nuclei and other electrons. For a single atom it can be shown that the quantum states which the electrons can occupy correspond to a discrete set of energy levels. Four quantum numbers are required to specify completely the electron's state, namely, the principal quantum number n, the total angular momentum quantum number l, the magnetic quantum number m, and the spin quantum number s. The quantum numbers n, l, and m must all be integers, and for a given principal quantum number l and m can take the values

$$l = 0, 1, 2, \ldots, n-1$$
$$m = 0, \pm 1, \pm 2, \ldots, \pm l$$

while s can only take the values $\pm\frac{1}{2}$.

Electronic states for which $l = 0, 1, 2$, and 3 are denoted by the letters s, p, d, and f. Pauli (1925) stated the following important principle: there cannot exist an atom in such a quantum state that two electrons within it have the same set of quantum numbers. A feature of the larger atoms is the formation of shells of eight electrons. This arises because two electrons having $s = \pm\frac{1}{2}$ can exist in each orbital shell and thus when the s and d orbitals are completely filled they contain eight electrons. Atoms whose outermost shells are of this type are particularly stable and they form the inert series of elements.

If an atom has too many electrons for a complete shell the surplus electrons,

Table 6-2 Electron configuration of atoms in their normal states

		He	Neon		Argon		Krypton			Xenon			Radon				Eka-radon			
		1*s*	2*s*	2*p*	3*s*	3*p*	3*d*	4*s*	4*p*	4*d*	5*s*	5*p*	4*f*	5*d*	6*s*	6*p*	5*f*	6*d*	7*s*	7*p*
H	1	1																		
He	2	2																		
Li	3	2	1																	
Be	4	2	2																	
B	5	2	2	1																
C	6	2	2	2																
N	7	2	2	3																
O	8	2	2	4																
F	9	2	2	5																
Ne	10	2	2	6																
Na	11	10 Neon core			1															
Mg	12				2															
Al	13				2	1														
Si	14				2	2														
P	15				2	3														
S	16				2	4														
Cl	17				2	5														
Ar	18	2	2	6	2	6														
K	19	18 Argon core						1												
Ca	20							2												
Sc	21						1	2												
Ti	22						2	2												
V	23						3	2												
Cr	24						5	1												
Mn	25						5	2												
Fe	26						6	2												
Co	27						7	2												
Ni	28						8	2												
Cu	29						10	1												
Zn	30						10	2												
Ga	31						10	2	1											
Ge	32						10	2	2											
As	33						10	2	3											
Se	34						10	2	4											
Br	35						10	2	5											
Kr	36	2	2	6	2	6	10	2	6											
Rb	37										1									
Sr	38										2									
Y	39									1	2									
Zr	40									2	2									
Nb	41									4	1									
Mo	42									5	1									
Tc	43									5	2									
Ru	44									7	1									
Rh	45									8	1									

Table 6-2 *(continued)*

		He	Neon		Argon		Krypton			Xenon			Radon				Eka-radon			
		1*s*	2*s*	2*p*	3*s*	3*p*	3*d*	4*s*	4*p*	4*d*	5*s*	5*p*	4*f*	5*d*	6*s*	6*p*	5*f*	6*d*	7*s*	7*p*
Pd	46					36				10										
Ag	47					Krypton core				10	1									
Cd	48									10	2									
In	49									10	2	1								
Sn	50									10	2	2								
Sb	51									10	2	3								
Te	52									10	2	4								
I	53									10	2	5								
Xe	54	2	2	6	2	6	10	2	6	10	2	6								
Cs	55														1					
Ba	56														2					
La	57													1	2					
Ce	58												1	1	2					
Pr	59												2	1	2					
Nd	60												3	1	2					
Pm	61												4	1	2					
Sm	62												5	1	2					
Eu	63												6	1	2					
Gd	64												7	1	2					
Tb	65												8	1	2					
Dy	66												9	1	2					
Ho	67												10	1	2					
Er	68												11	1	2					
Tm	69						54						12	1	2					
Yb	70						Xenon core						13	1	2					
Lu	71												14	1	2					
Hf	72												14	2	2					
Ta	73												14	3	2					
W	74												14	4	2					
Re	75												14	5	2					
Os	76												14	6	2					
Ir	77												14	7	2					
Pt	78												14	9	1					
Au	79												14	10	1					
Hg	80												14	10	2					
Tl	81												14	10	2	1				
Pb	82												14	10	2	2				
Bi	83												14	10	2	3				
Po	84												14	10	2	4				
At	85												14	10	2	5				
Rn	86	2	2	6	2	6	10	2	6	10	2	6	14	10	2	6				
Fr	87																		1	
Ra	88																		2	
Ac	89									86								1	2	
Th	90									Radon core								2	2	
Pa	91																	3	2	
U	92																	4	2	

being screened from the nucleus by the completed inner shells, are only weakly attracted to it and thus they are easily dissociated from the atom and move in large orbits round the nucleus, making the overall atomic size large. Atoms having too few electrons can be regarded as having "electron holes," in that they readily accept further electrons to complete their shells. Atoms of the first type are called electropositive elements or electron donors and they easily give up electrons to become positive ions. Atoms of the second type are called electronegative elements or electron acceptors and they accept extra electrons to become negative ions. A table showing the electron configurations of atoms in their normal states, which is taken from Pauling (1960), is given in Table 6-2.

6-6-2 Interatomic Forces

All atoms, even the inert gas atoms, are weakly attracted to one another by the fluctuating electron field due to the movement of electrons in the atom. This type of attractive force is called a van der Waals force and although it is important for condensation problems it will not be pursued further in the present discussion. The stronger types of atomic bonding can be divided into three main types: ionic bonds, covalent bonds, and metallic bonds.

(a) The ionic bond The atoms of metallic elements are electropositive and so lose their outer electrons easily whereas those of nonmetallic elements are electronegative and readily accept electrons; in this state positive and negative ions may be formed which retain their electronic structures as they approach one another to form a stable molecule or crystal. Sodium chloride crystals, for instance, are composed of positive sodium ions and negative ions of chlorine; both atoms then have completed shells and are held together by electrostatic forces. A crude estimate for the order of magnitude of bond energy may be obtained by taking the distance between nuclei to be 3×10^{-10} m. The energy is then given by the classical electrostatic formula e^2/r and, using the data in Sec. 6-6-1,

$$\frac{e^2}{r} = 4.8\ \text{eV}$$

where eV stands for electronvolt and is the energy acquired by an electron in dropping through a potential difference of one volt ($1\ \text{eV} = 1.6 \times 10^{-19}$ J).

(b) The covalent or homopolar bond In the covalent bond a pair of electrons of opposite spin are shared between two atoms, each electron helping to complete the shell on the other. Because the quantum states associated with this bond often extend in preferred directions from the atom it is called a directed bond. It is also a bond which can be saturated because only a limited number of bonding quantum states can be formed between an atom and its neighbors. Also, because of the limited number of places available in a partly filled outer electron shell, an atom with N electrons in that shell can bond with only $8-N$ neighbors by sharing electrons with them.

When $N = 7$ the atom has one bonding electron and a valence of one. When $N = 6$ two bonds can be formed with neighbors, giving the possibility of the formulation of long-chain molecules. $N = 5$ leads to the formation of sheets or layers of atoms, while if $N = 4$ one gets the solid tetrahedral bonding found between the carbon atoms in diamond. The average energy per bond is usually on the order of 1 eV.

(c) The metallic bond Many substances having $N < 4$ form metallic bonds. For such substances the bonding electrons, instead of being localized between two atoms, move freely through the whole group of positive ions as a free-electron gas. This mobility of the bonding electrons gives rise to the high electric and thermal conductivity of metals. For light atoms the metallic character arises because the bonding electron is only very weakly held to the nucleus and moves in a large orbit; it is therefore easy for this electron to migrate between different nuclei. For heavy atoms there are more bonding electron states than electrons to fill them and thus a valence electron can pass from one atom to another without being hindered by the quantum condition of filled quantum states. The bonds are thus unsaturated and the rule that no more than $8-N$ bonds can be formed no longer applies, which allows the number of neighbors surrounding an atom to be as numerous as their geometry will allow. Close-packed structures are therefore common. The average energy of the metallic bond is on the same order as the covalent bond, that is, 1 eV.

As the temperature is lowered the various bonding forces discussed above begin to dominate the random thermal motion of the atoms and at temperatures below the melting point an aggregate of atoms cohere into a solid structure. Solids can be divided into two broad classes: crystals which have a long-range order apparent in the regularity of the lattice and amorphous solids which do not possess this long-range order.

6-6-3 Crystals

The basic symmetry of a crystal lattice is its spatial periodicity which can be linked with the existence of a parallelepiped of repetition called the unit cell. The cell's concurrent edges will be denoted by the *basic* or *translation* vectors $\boldsymbol{a}_1, \boldsymbol{a}_2, \boldsymbol{a}_3$. Relative to any chosen lattice point as the origin, any other lattice point $\boldsymbol{x}$ is then given by

$$\boldsymbol{x} = l_1 \boldsymbol{a}_1 + l_2 \boldsymbol{a}_2 + l_3 \boldsymbol{a}_3 \tag{6-6-1}$$

where l_1, l_2, and l_3 are integers. The set of all such points forms the Bravais lattice of the crystal. The Bravais lattice does not necessarily include all the atoms in a crystal, however, for within the unit cell there may be other atoms which are not displaced by a basic vector. However, because of the crystal's periodicity such atoms must form part of an identical, though slightly displaced, Bravais lattice. Thus the entire crystal is composed of a set of interpenetrating Bravais lattices.

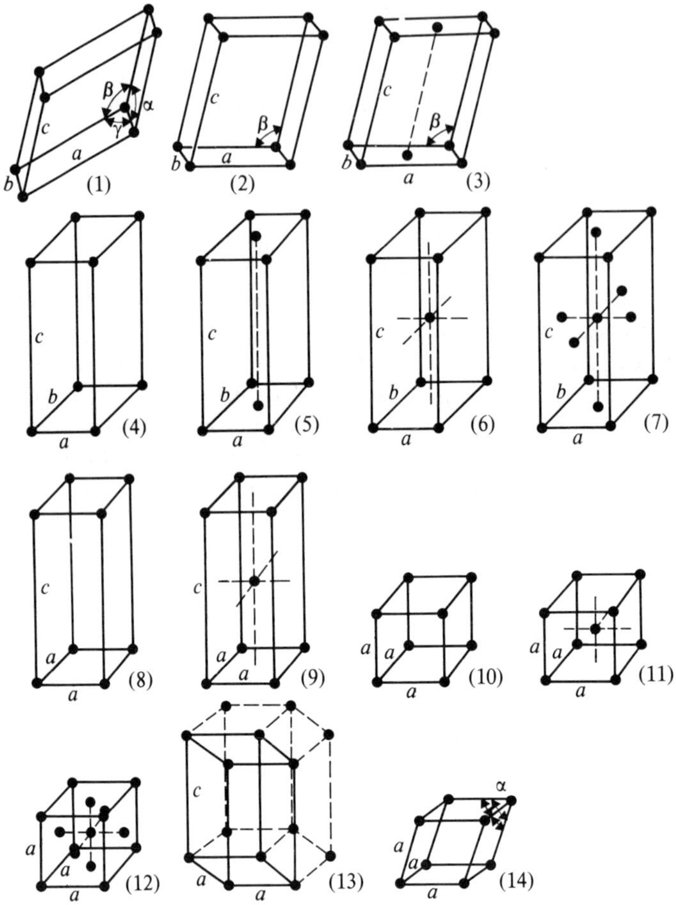

Figure 6-8 The 14 Bravais space lattices.

Lattices are classified according to their symmetry, where the symmetry of a lattice is determined by the assembly of all those rearrangements, after which the lattice is unaltered; these rearrangements are called *symmetry transformations*. Any symmetry transformation is composed of a combination of three fundamental types of transformation. These are *translations* by whole numbers of lattice vectors, *rotation* of the lattice through a definite angle about some axis, and *reflection* of it in some plane. If a lattice is unaltered on rotation through an angle $2\pi/n$ about an axis, then that axis is said to be an axis of symmetry of the nth order. An important symmetry transformation is *inversion*, in which each lattice point is taken through a point which is called a *center of symmetry* to an equidistant position on the opposite side of this center. It can easily be seen that a rotation through an angle π followed by reflection in a plane perpendicular to the axis of rotation is equivalent to an inversion transformation, so these three symmetry elements are mutually dependent. The set of all symmetry transformations for a given lattice is called its

Table 6-3 Types of Bravais lattice in each crystal system

Crystal system	Features	Bravais lattices
Triclinic	Three axes of arbitrary length at arbitrary angle to one another	(1) Simple
Monoclinic	Three axes, one pair not at right angles, of arbitrary length	(2) Simple (3) Base-centered
Orthorhombic	Rectangular parallelepiped with axes of arbitrary length	(4) Simple (5) Base-centered (6) Body-centered (7) Face-centered
Tetragonal	Three axes forming right square prism	(8) Simple (9) Body-centered
Cubic	Three equal orthogonal axes	(10) Simple (11) Body-centered (12) Face-centered
Hexagonal	Regular hexagonal prism	(13) Simple
Rhombohedral	Three axes equally inclined, not orthogonal but all equal in length	(14) Simple

symmetry group. There exist only 14 types of Bravais lattice, shown in Fig. 6-8, which it is usual to group into seven crystal systems as shown in Table 6-3.

The symmetry of the crystal leads to certain directions being equivalent. All macroscopic properties of the crystal will therefore be exactly the same along these equivalent directions. The translational symmetry of the lattice plays no part in such restrictions because parallel displacements do not change directions at all. The directional symmetry and hence the macroscopic properties of a crystal are determined by its axes and planes of symmetry. These sets of symmetry elements are called crystal classes and it can be shown that there are only 32 different crystal classes. The deduction of the possible symmetrical arrangements of particles in space so that they will give rise to the symmetry shown by crystals involves not only the directional symmetry of the crystal classes but also the translational symmetry of the Bravais lattice. The set of all these symmetry elements is called the space group of the crystal and it can be shown that there are in all 230 space groups. The space groups are important in determining the microscopic atomic structure of crystals, but it is the crystal class which determines the macroscopic response and, in particular, the symmetry properties of the crystals elastic moduli. Consider, for instance, the elastic constants $\tilde{C}_{ijkl}$ of a linear elastic or hookean material which relate the stress $\mathbf{T}$ and strain $\tilde{\mathbf{e}}$ by the relation $\mathbf{T} = \tilde{\mathbf{C}}\tilde{\mathbf{e}}$. The symmetry of the stress and strain tensors reduce the number of independent constants from $3^4 = 81$ to $36 = 6 \times 6$ and strain energy considerations introduce a further symmetry which reduces the number of constants to 21. To simplify the notation further it is usual to write one suffix for a pair of suffixes according to the rule $11 \to 1$, $22 \to 2$, $33 \to 3$, $23 \to 4$, $31 \to 5$, $12 \to 6$, for example, $\tilde{C}_{2231} \to \tilde{C}_{25}$. The

Table 6-4 Elastic constants in various crystal systems

Triclinic (21)	Orthorhombic (9)	Rhombohedral (6)	Tetragonal (6)	Hexagonal (5)	Cubic (3)
11	11	11	11	11	11
12	12	12	12	12	12
13	13	13	13	13	12
14	0	14	0	0	0
15	0	0	0	0	0
16	0	0	0	0	0
22	22	11	11	11	11
23	23	13	13	13	12
24	0	−14	0	0	0
25	0	0	0	0	0
26	0	0	0	0	0
33	33	33	33	33	11
34	0	0	0	0	0
35	0	0	0	0	0
36	0	0	0	0	0
44	44	44	44	44	44
45	0	0	0	0	0
46	0	0	0	0	0
55	55	44	44	44	44
56	0	14	0	0	0
66	66	$\frac{1}{2}$(11–12)	66	$\frac{1}{2}$(11–12)	44

Bracketed quantities denote the number of independent components while, in accord with Hearmon (1961), the *C*'s have been omitted to avoid undue repetition.

crystal symmetries then reduce the number of independent elastic constants even further. The independent constants associated with the main crystal systems are given in Table 6-4; for a more detailed discussion, see Hearmon (1961).

In principle the elastic constants can be obtained from a knowledge of the interatomic forces, which in turn may be derived from quantum mechanical considerations. In practice such derivations are usually very involved. Two approaches are followed, one dealing with generalized interatomic forces deriving the symmetry properties and interrelation between macroscopic parameters, such as the elastic, piezoelectric, and dielectric constants, and the other investigating particular materials from a quantum mechanical standpoint. In the following discussion only the first viewpoint is considered and even that must, for reasons of space, be of necessity cursory. For a detailed appraisal of this approach the reader is referred to Born and Huang (1954) and the comprehensive review article by Huntington (1958).

To arrive at such an atomistic derivation of the elastic constants it is recognized that the crystal is a mechanical system in a position of stable equilibrium and the elastic constants are determined by comparing the change in energy due to a small homogeneous deformation, as calculated on the atomic scale, with that which would be predicted by the macroscopic theory. For a homogeneous deformation the components of the strain tensor are constant and the work of deformation per unit volume W has been given in Eq. (6-1-4). For present purposes it is convenient to formulate the problem in cartesian coordinates, W now having the form

$$W = \tfrac{1}{2}\tilde{C}_{\alpha\beta\gamma\delta}\tilde{e}_{\delta\gamma}\tilde{e}_{\beta\alpha} \tag{6-6-2}$$

To calculate the change in energy due to atomic displacements it is noted that in a homogeneous deformation the deformed lattice is still a perfect lattice. In the undeformed state let the position of the kth atom in the lth Bravais cell be denoted by $X\binom{k}{l}$ which can be decomposed into a vector locating the Bravais cell $l_1 \boldsymbol{a}_1$ and a vector $X^{(\boldsymbol{k})}$ indicating the position of the kth atom in the cell; thus

$$X\binom{k}{l} = X(l) + X^{(\boldsymbol{k})} \tag{6-6-3}$$

where

$$X(l) = l_1 \boldsymbol{a}_1 + l_2 \boldsymbol{a}_2 + l_3 \boldsymbol{a}_3 \tag{6-6-4}$$

and the $\boldsymbol{a}$'s are the basic vectors. The displacement of each particle can also be considered as being composed of two parts, one being the homogeneous transformation and the other a displacement of all particles of each type by the same vector $\boldsymbol{u}(k)$. Thus the displacement of the kth particle in the lth cell is given by

$$u_\alpha\binom{l}{\boldsymbol{k}} = u_\alpha(k) + \tilde{u}_{\alpha\beta} X_\beta\binom{l}{\boldsymbol{k}} \tag{6-6-5}$$

Considering the case when the lattice particles interact with central forces and denoting the potential energy by $\psi(r^2)$, Born has shown that expanding the potential energy to second order in $u_\alpha(k)$ and $u_{\alpha\beta}$ the energy of the zeroth Bravais cell (that is, $l = 0$) is

$$U = \tfrac{1}{2}\left\{\begin{smallmatrix}\boldsymbol{kk'}\\ \alpha\beta\end{smallmatrix}\right\} u_\alpha(k) u_\beta(k') + \left\{\begin{smallmatrix}\boldsymbol{k}\\ \alpha,\beta\gamma\end{smallmatrix}\right\} u_\alpha(k)\tilde{u}_{\beta\gamma} + \tfrac{1}{2}\{\alpha\beta\gamma\delta\}\tilde{u}_{\alpha\beta}\tilde{u}_{\gamma\delta} \tag{6-6-6}$$

where

$$\left\{\begin{smallmatrix}\boldsymbol{kk'}\\ \alpha\beta\end{smallmatrix}\right\} = 2\left\{\delta_{\alpha\beta}\delta_{\boldsymbol{kk'}} \sum_{\boldsymbol{l'k''}} \psi'\binom{\boldsymbol{l'}}{\boldsymbol{k''k}} - \delta_{\alpha\beta} \sum_{\boldsymbol{l'}} \psi'\binom{\boldsymbol{l'}}{\boldsymbol{k'k}} \right.$$
$$\left. + 2\delta_{\boldsymbol{kk'}} \sum_{\boldsymbol{l'k''}} [\psi'' X_\alpha X_\beta]\binom{\boldsymbol{l'}}{\boldsymbol{k''k}} - 2\sum_{\boldsymbol{l'}} [\psi'' X_\alpha X_\beta]\binom{\boldsymbol{l'}}{\boldsymbol{k'k}}\right\} \tag{6-6-7†}$$

$$\left\{\begin{smallmatrix}\boldsymbol{k}\\ \alpha,\beta\gamma\end{smallmatrix}\right\} = -4\sum_{\boldsymbol{l'k'}} [\psi'' X_\alpha X_\beta X_\gamma]\binom{\boldsymbol{l'}}{\boldsymbol{k'k}} \tag{6-6-8}$$

$$\{\alpha\beta\gamma\delta\} = 2\sum_{\boldsymbol{l'k'k}} [\psi'' X_\alpha X_\beta X_\gamma X_\delta]\binom{\boldsymbol{l'}}{\boldsymbol{k'k}} \tag{6-6-9}$$

The terms in parentheses such as $\binom{\boldsymbol{l'}}{\boldsymbol{k'k}}$ indicate the particle coordinates to which the previous expression applies, that is, $\psi'\binom{\boldsymbol{l'}}{\boldsymbol{k'k}}$ is the derivative of the potential between the k'th particle and the l'th cell and the kth particle in the

† The summation convention has been used for both the atomic (k) and spatial (α) indices over their respective ranges of values.

zeroth cell. From the above definitions it follows that the brackets satisfy the symmetry conditions

$$\left\{\begin{smallmatrix}\boldsymbol{kk'}\\ \alpha\beta\end{smallmatrix}\right\} = \left\{\begin{smallmatrix}\boldsymbol{k'k}\\ \alpha\beta\end{smallmatrix}\right\} = \left\{\begin{smallmatrix}\boldsymbol{kk'}\\ \beta\alpha\end{smallmatrix}\right\} \tag{6-6-10}$$

$\left\{\begin{smallmatrix}\boldsymbol{k}\\ \alpha,\beta\gamma\end{smallmatrix}\right\}$ is completely symmetric in all the cartesian indices (6-6-11)

$\{\alpha\beta\gamma\delta\}$ is completely symmetric in all four cartesian indices (6-6-12)

From these symmetry conditions the energy depends only on the strain parameters

$$\left.\begin{aligned} 2\tilde{e}_{\alpha\beta} &= \tilde{u}_{\alpha\beta}+\tilde{u}_{\beta\alpha} \qquad \alpha \neq \beta \\ \tilde{e}_{\alpha\beta} &= \tilde{u}_{\alpha\beta} \qquad \alpha = \beta \end{aligned}\right\} \equiv \tilde{e}_\rho \qquad \text{for } \rho = 1, 2, \ldots, 6 \tag{6-6-13}$$

Dividing U by the Bravais cell volume v_a one obtains the energy density W. Redefining the terms in parentheses in Eqs. (6-6-7) to (6-6-9) by dividing by v_a leads to

$$W = \tfrac{1}{2}\left\{\begin{smallmatrix}\boldsymbol{kk'}\\ \alpha\beta\end{smallmatrix}\right\}u_\alpha(k)u_\beta(k') + \left\{\begin{smallmatrix}\boldsymbol{k}\\ \alpha,\rho\end{smallmatrix}\right\}u_\alpha(k)\tilde{e}_\rho + \tfrac{1}{2}\{\rho,\sigma\}\tilde{e}_\rho\tilde{e}_\sigma \tag{6-6-14}$$

The deformations due to $\boldsymbol{u}(k)$ and $\tilde{e}_\rho$ are called the internal and external strain respectively. The internal strains do not affect the dimensions of the specimen and are such as to make the energy density a minimum for a given elastic strain $\tilde{e}_\rho$. Thus the $\boldsymbol{u}(k)$ are determined by the condition

$$0 = \frac{\partial W}{\partial u_\alpha(k)} = \left\{\begin{smallmatrix}\boldsymbol{kk'}\\ \alpha\beta\end{smallmatrix}\right\}u_\beta(k') + \left\{\begin{smallmatrix}\boldsymbol{k}\\ \alpha,\rho\end{smallmatrix}\right\}\tilde{e}_\rho \tag{6-6-15}$$

If Eqs. (6-6-15) are used to eliminate the $\boldsymbol{u}(k)$ terms from Eq. (6-6-14) one is left with a quadratic expression in $\tilde{e}_{\alpha\beta} \equiv \tilde{e}_\rho$ which, when compared with Eq. (6-6-2), leads to expressions for $\tilde{C}_{\alpha\beta\gamma\delta}$.

If every atom lies at a center of symmetry then it cannot experience a net force from the totality of the other particles and thus no internal strains are induced. From Eqs. (6-6-14) and (6-6-9) it follows that

$$\tilde{C}_{\rho\sigma} \equiv \tilde{C}_{\alpha\beta\gamma\delta} = 2\sum_{\boldsymbol{l'k'k}}[\psi''X_\alpha X_\beta X_\gamma X_\delta]\left(\begin{smallmatrix}\boldsymbol{l'}\\ \boldsymbol{k'k}\end{smallmatrix}\right) \tag{6-6-16}$$

and as this is completely symmetric in all the cartesian indices so are the indices of $\tilde{C}_{\rho\sigma}$, and it follows that $\tilde{C}_{\rho\sigma}$ has at most 15 independent values. In general there are 21 independent elastic constants and the six relations

$$\begin{matrix} \tilde{C}_{23} = \tilde{C}_{44} & \tilde{C}_{31} = \tilde{C}_{55} & \tilde{C}_{12} = \tilde{C}_{66} \\ \tilde{C}_{14} = \tilde{C}_{56} & \tilde{C}_{25} = \tilde{C}_{64} & \tilde{C}_{36} = \tilde{C}_{45} \end{matrix} \tag{6-6-17}$$

which Eq. (6-6-16) implies, and which reduce the number of independent constants to 15, are called the Cauchy relations. Examples of structure possessing such a symmetry condition are the body-centered lattices of Li, Na, K, Mo, W, etc., and the face-centered cubic lattices of Cu, Ag, Au, Al, Pb, Ni, etc., but not the hexagonal close packing as in Be, Mg, Zn, Cd. The Cauchy relations can be expected to be satisfied if the particles interact with central forces and if every lattice particle occupies a center of symmetry, conditions most nearly satisfied for the ionic cubic crystals of the alkalihalides such as NaCl, LiF, and KCl.

At first it would seem the above theory could easily be extended to more general cases where the forces had a noncentral character, but this is not so because both in this case and if the interatomic forces are of long range leading to cooperative phenomena, the sums over lattice points are divergent. To avoid such divergencies Begbie and Born (1947) introduced the "method of long waves." In this approach the internal energy of an infinite crystal in its lowest electronic state is expressed as a function ϕ of the individual atomic coordinates and the atomic equations of motion are expressed in terms of the derivatives of ϕ and solved formally in terms of plane waves. The equation for the frequencies of the plane waves is then expanded about the long wavelength limit. Zero-, first-, and second-order equations are then obtained using the magnitude of the wave-number vector as the perturbation parameter. The first-order equations give the relative displacement of the various atoms in any unit cell, namely, the inner strains. The elastic constants are then obtained by comparing the second-order equations with the macroscopic equations.

A detailed discussion of further refinements of the theory can be found in the article by Liebfried and Ludwig (1961) dealing with the extension of the above harmonic analysis to include anharmonic effects (see also Martin, 1975; Johnson, 1977).

6-6-4 Polycrystalline and Heterogeneous Media

Because of the random nature of their formation the grains within a metal are usually oriented at random and most theories of polycrystalline response assume this to be true; however, mechanical working of the metal such as extrusion or rolling may orient the grains in a particular direction as well as severely deforming them. One would thus expect any theory relating the overall response of the metal to the single crystal response to be of a statistical nature. Most of the work to date is based on methods of continuum mechanics in which the postulates of homogeneity and isotropy imply a statistical averaging process. Polycrystalline response is treated as a special case of the theory of the mechanical behavior of heterogeneous media and a review of this general field has been given by Hashin (1967). A heterogeneous medium is a mixture of homogeneous phases which form regions large enough to be regarded as continua. A type of heterogeneous medium which is of particular interest is the composite material which comprises a "matrix" of homogeneous material in which inclusions of other materials are firmly embedded. In order to determine an overall response for the system it is assumed to be macroscopically homogeneous, which is defined by Hill (1967) as follows.

Let d be an order-of-magnitude linear measure of the spatial heterogeneity, e.g., mean grain size in a polycrystal or inclusion spacing in a dispersion composite. Take a representative sample of material in the form of a cube with side l and suppose that the distribution of the heterogeneity is such that the following property obtains for all sufficiently large l/d. Over each face the displacement under any uniform loading, or traction under any constraint, can be broadly described as a wavelike fluctuation about a mean with a period of order d and an amplitude independent of l. This may, indeed, be regarded as a minimal prescription for a representative macroelement, suitable for observing or calculating an overall constitutive law. Any such system will be called macroscopically homogeneous, and any loading or constraint with similar wavelike character will be called macroscopically uniform.

Two approaches are used to estimate overall elastic moduli. One approach provides precise bounds on the moduli depending on the relative volume of the constituents. If the difference between the individual moduli is small then the difference between the bounds can be made very small. Hill (1952) showed that the previous results of Voigt (1910), who assumed that the displacements of the crystals would all be of the same form, and Reuss (1929), who assumed the stress distribution to be of the same form for all the polycrystals, provided rigorous upper and lower bounds respectively for the overall elastic moduli. Improved bounds have been given by Hashin and Shtrikman (1962, 1963), Hill (1963, 1964, 1965), Walpole (1966) and by Lakkad, Miatt, and Parsons (1973). Walpole (1969) gives the following overall elastic bounds for aligned phases. If the average stress $\bar{\mathbf{T}}$ and strain $\bar{\mathbf{e}}$ are related by the fourth-order tensor $\bar{\mathbf{C}}$,

$$\bar{\mathbf{T}} = \bar{\mathbf{C}}\bar{\mathbf{e}} \tag{6-6-18}$$

and $\bar{\mathbf{C}}$ has the components

$$\begin{aligned}
&\bar{C}_{1111} = \bar{C}_{2222} = k+m && \bar{C}_{3333} = n \\
&\bar{C}_{1122} = \bar{C}_{2211} = k-m && \bar{C}_{1133} = \bar{C}_{2233} = \bar{C}_{3311} = \bar{C}_{3322} = 1 \\
&\bar{C}_{1212} = m && \bar{C}_{2323} = \bar{C}_{3131} = p \qquad \text{all other } \bar{C} \text{ zero}
\end{aligned}$$

Let the fractional concentration by volume of the rth phase be c_r. Define a set of overall moduli as follows:

$$\begin{aligned}
\bar{k} &= \left(\frac{\Sigma c_r}{k_r + m_0}\right)^{-1} - m_0 \\
\bar{l} &= \frac{\Sigma c_r l_r/(k_r + m_0)}{\Sigma c_r/(k_r + m_0)} \\
\bar{n} &= \frac{\Sigma c_r n_r + \Sigma c_r l_r [\Sigma c_s (l_s - l_r)/(k_s + m_0)]/(k_r + m_0)}{\Sigma c_r/(k_r + m_0)} \\
\bar{m} &= \left[\frac{\Sigma c_r}{(m_r + m_0 k_0)/(k_0 + 2m_0)}\right]^{-1} - \frac{m_0 k_0}{k_0 + 2m_0} \\
\bar{p} &= \left(\frac{\Sigma c_r}{p_r + p_0}\right)^{-1} - p_0
\end{aligned}$$

If k_0, m_0, and p_0 are equated to the greatest or least of k_r, m_r, and p_r, then we obtain the upper and lower bounds of the moduli respectively, that is, $\bar{k} - k$, $\bar{m} - m$, $\bar{n} - n$, $\bar{p} - p$, and $\bar{n} - n - (\bar{l} - l)^2/(\bar{k} - k)$ are either positive or negative.

The second approach is by way of a self-consistent theory which takes account of the shape of the inclusion and presumes that either the average stress or average strain within an inclusion can be evaluated by replacing the surrounding of the inclusion by a matrix whose elastic moduli coincides with the overall moduli. Inclusions of spherical or ellipsoidal shape have been studied by Hershey (1954), Eshelby (1957, 1961), Budiansky (1965), Hill (1965), Wu (1966), and Walpole (1969).

6-6-5 The Elastic Field of an Ellipsoidal Inclusion

This basic problem, on which much of the theory of heterogeneous media rests, can be applied directly to many different physical situations such as the formation of twins, differential thermal expansion of small internal regions, martensitic transformation, or the precipitation of a new phase. It has been solved for the ellipsoidal inclusion by Eshelby (1957), and his method of solution will be followed here.

It is assumed that the matrix is a hookean solid whose stress and strain satisfy the relation

$$\mathbf{T} = \lambda(\operatorname{tr}\tilde{\mathbf{e}})\mathbf{I} + 2G\tilde{\mathbf{e}} \tag{6-6-19}$$

The deviator of a tensor **f** will be denoted by $\mathbf{f}'$ and is defined by the relation

$$\mathbf{f}' = \mathbf{f} - \tfrac{1}{3}f\mathbf{I} \tag{6-6-20}$$

where

$$f = \operatorname{tr}\mathbf{f}$$

Thus, from Eq. (6-6-19),

$$T = 3\kappa\tilde{e} \qquad \text{and} \qquad \mathbf{T}' = 2G\tilde{\mathbf{e}}' \tag{6-6-21}$$

and $\kappa = \lambda + \frac{2}{3}G$ is termed the bulk modulus. The elastic energy density is thus given by

$$\tfrac{1}{2}\operatorname{tr}(\mathbf{T}\tilde{\mathbf{e}}) = \tfrac{1}{2}(\kappa\tilde{e}^2 + 2G\operatorname{tr}\tilde{\mathbf{e}}'^2) = \tfrac{1}{2}\left(\frac{1}{9\kappa}T^2 + \frac{1}{2G}\operatorname{tr}\mathbf{T}'^2\right) \tag{6-6-22}$$

Suppose that, apart from the constraint imposed by the surrounding matrix, the inclusion would undergo an arbitrary homogeneous stress-free strain $\tilde{\mathbf{e}}^F$; then the problem is to find the "constrained strain" $\tilde{\mathbf{e}}^c$. Let S be the surface separating the matrix and the inclusion and let $\boldsymbol{n}$ be a unit normal pointing into the matrix. The following set of simple imaginary cutting and welding operations are now performed:

1. Remove the inclusion and allow the unconstrained transformation to occur giving rise to the stress-free strain $\tilde{\mathbf{e}}^F$ without altering its elastic constants. Let the stress derived from $\tilde{\mathbf{e}}^F$ by Hooke's law be

$$\mathbf{T}^F = \lambda\tilde{e}^F\mathbf{I} + 2G\tilde{\mathbf{e}}^F \tag{6-6-23}$$

The stress in the inclusion and matrix is now zero.

2. Apply surface tractions $-\mathbf{T}^F \boldsymbol{n}$ to the inclusion. This brings it back to its original shape and size. Put it back in the matrix and reweld across S. The surface forces have now become a layer of body force spread over S.
3. Let the body forces relax or equivalently apply a stress distribution $+\mathbf{T}^F \boldsymbol{n}$ to the matrix through the surface S. The whole body, matrix plus inclusion, is now free from any external force, but in a state of self-stress due to the transformation.

According to Love (1927), the displacement $\boldsymbol{u}(\boldsymbol{r}-\boldsymbol{r}')$ at a point $\boldsymbol{r}$ due to a point force $\boldsymbol{F}$ at $\boldsymbol{r}'$ is given by

$$u_j(\boldsymbol{r}-\boldsymbol{r}') = F_k G_{kj} = F_k \left[\frac{1}{4\pi G} \frac{\delta_{jk}}{|\boldsymbol{r}-\boldsymbol{r}'|} - \frac{1}{16\pi G(1-\nu)} \frac{\partial^2}{\partial x_k \partial x_j} |\boldsymbol{r}-\boldsymbol{r}'| \right] \tag{6-6-24}$$

Thus the displacement impressed on the material in stage 3 is just

$$u_i^c(\boldsymbol{r}) = \int_S G_{ij} T_{jk}^F \, dS_k \tag{6-6-25}$$

As stage 2 has the same geometrical form as the initial state of the inclusion before the transformation occurred it is convenient to take the displacements at this stage as zero and thus Eq. (6-6-25) gives the actual displacements in the inclusion and matrix. The strain in the matrix or inclusion is

$$\tilde{e}_{ij}^c = \tfrac{1}{2}(u_{i,j}^c + u_{j,i}^c) \tag{6-6-26}$$

and the stress in the matrix is thus

$$\mathbf{T}^c = \lambda \tilde{e}^c \mathbf{I} + 2G\tilde{\mathbf{e}}^c \tag{6-6-27}$$

As the inclusion acquired a stress $-\mathbf{T}^F$ in stage 2 the stress in it is now

$$\mathbf{T}^I = \mathbf{T}^C - \mathbf{T}^F = \lambda(\tilde{e}^C - \tilde{e}^F)\mathbf{I} + 2G(\tilde{\mathbf{e}}^C - \tilde{e}^F) \tag{6-6-28}$$

where $\mathbf{T}^C$ is the stress derived by Hooke's law from the strain in the inclusion.

Making use of Gauss' theorem and the equivalence of $\partial/\partial x_i$ and $\partial/\partial x_i'$ when acting on $|\boldsymbol{r}-\boldsymbol{r}'|$, Eq. (6-6-25) may be written in the form

$$u_i^c = \frac{1}{16\pi G(1-\nu)} T_{jk}^F \psi_{,ijk} - \frac{1}{4\pi G} T_{ik}^F \phi_{,k} \tag{6-6-29}$$

where

$$\phi = \int_V \frac{dv}{|\boldsymbol{r}-\boldsymbol{r}'|} \qquad \text{and} \qquad \psi = \int_V |\boldsymbol{r}-\boldsymbol{r}'| \, dv \tag{6-6-30}$$

and V is the volume of the inclusion:

$$\begin{aligned} \nabla^2 \psi &= 2\phi \\ \nabla^4 \psi &= 2\nabla^2 \phi = \begin{cases} -8\pi & \text{inside } S \\ 0 & \text{outside } S \end{cases} \end{aligned} \tag{6-6-31}$$

Alternatively, noting that

$$16\pi G(1-\nu)G_{ij} = \frac{(3-4\nu)}{|\boldsymbol{r}-\boldsymbol{r}'|}\delta_{ij} + \frac{(x_i-x_i')(x_j-x_j')}{|\boldsymbol{r}-\boldsymbol{r}'|^3} \tag{6-6-32}$$

we can write Eq. (6-6-25) in the form

$$u_i^c(\boldsymbol{x}) = \frac{T_{jk}^F}{16\pi G(1-\nu)}\int_V \frac{dv}{r^2} f_{ijk}(\boldsymbol{l}) = \frac{e_{jk}^F}{8\pi(1-\nu)}\int_V \frac{dv}{r^2} g_{ijk}(\boldsymbol{l}) \tag{6-6-33}$$

where r and $\boldsymbol{l}$ are the length and direction of the line drawn from the volume element dv to the point of observation $\boldsymbol{x}$ and

$$f_{ijk} = (1-2\nu)(\delta_{ij}l_k + \delta_{ik}l_j) - \delta_{jk}l_i + 3l_il_jl_k \tag{6-6-34}$$

$$g_{ijk} = (1-2\nu)(\delta_{ij}l_k + \delta_{ik}l_j - \delta_{jk}l_i) + 3l_il_jl_k \tag{6-6-35}$$

For points remote from this inclusion, the terms in the integrand of Eq. (6-6-33) are constant within V and u_i^c has the simple form

$$u_i^c = \frac{VT_{jk}^F f_{ijk}}{16\pi G(1-\nu)r^2} = \frac{V\tilde{e}_{jk}^F g_{ijk}}{8\pi(1-\nu)r^2} \tag{6-6-36}$$

To discuss the elastic field inside an inclusion it is convenient to redefine the $\boldsymbol{l}$ to be the direction cosines of a line drawn from the point of observation to the volume element; thus the integrals of equation (6-6-33) will change sign. Integrating over a cone of solid angle $d\omega(\boldsymbol{l})$ centered on the direction $\boldsymbol{l}$ with its vertex at $\boldsymbol{x}$, Eq. (6-6-33) can be written

$$8\pi(1-\nu)u_i^c = -\tilde{e}_{jk}^F \int r(\boldsymbol{l})\, d\omega(\boldsymbol{l}) g_{ijk}(\boldsymbol{l}) \tag{6-6-37}$$

For the ellipsoid

$$\frac{X^2}{a^2} + \frac{Y^2}{b^2} + \frac{Z^2}{c^2} = 1 \tag{6-6-38}$$

$r(\boldsymbol{l})$ is the positive root of

$$\frac{(x+rl)^2}{a^2} + \frac{(y+rm)^2}{b^2} + \frac{(z+rn)^2}{c^2} = 1 \tag{6-6-39}$$

where l, m, n are the components of the vector $\boldsymbol{l}$. Thus

$$r(\boldsymbol{l}) = -\frac{f}{g} + \left(\frac{f^2}{g^2} + \frac{e}{g}\right)^{1/2} \tag{6-6-40}$$

where

$$g = \left(\frac{l}{a}\right)^2 + \left(\frac{m}{b}\right)^2 + \left(\frac{n}{c}\right)^2$$

$$f = \frac{lx}{a^2} + \frac{my}{b^2} + \frac{nz}{c^2}$$

$$e = 1 - \left(\frac{x}{a}\right)^2 - \left(\frac{y}{b}\right)^2 - \left(\frac{z}{c}\right)^2$$

Writing $\lambda_1 = l/a^2$, $\lambda_2 = m/b^2$, $\lambda_3 = n/c^2$ and noting that the square root term contributes nothing to the integral because it is an even function of l while g_{ijk} is an odd function, Eq. (6-6-37) is easily integrated to give

$$u_i^c(\mathbf{x}) = \frac{x_m \tilde{e}_{jk}^F}{8\pi(1-\nu)} \int_{4\pi} \frac{\lambda_m g_{ijk}}{g} \, d\omega \qquad (6\text{-}6\text{-}41)$$

and the strain

$$\tilde{e}_{il}^c(\mathbf{x}) = \frac{\tilde{e}_{jk}^F}{16\pi(1-\nu)} \int_{4\pi} \frac{\lambda_i g_{ljk} + \lambda_l g_{ijk}}{g} \, d\omega \qquad (6\text{-}6\text{-}42)$$

which are uniform and depend only on the shape of the ellipsoid.

Because stress and strain are homogeneous in the case of an ellipsoidal inclusion one can proceed to solve the further problem of an inclusion whose elastic constants differ from those of the matrix. To do this a uniform strain $\tilde{\mathbf{e}}^A$ is superimposed on the whole solid. The deformation of the surface of the inclusion is then given by $\tilde{\mathbf{e}}^C + \tilde{\mathbf{e}}^A$. Because a part of the stress arises from the transformation the uniform stress in the inclusion is given by applying Hooke's law not to $\tilde{\mathbf{e}}^C + \tilde{\mathbf{e}}^A$ but to $\tilde{\mathbf{e}}^C + \tilde{\mathbf{e}}^A - \tilde{\mathbf{e}}^F$. Thus although the strain in the inclusion is given by

$$\tilde{e} = \tilde{e}^C + \tilde{e}^A \qquad \tilde{\mathbf{e}}' = \tilde{\mathbf{e}}'^C + \tilde{\mathbf{e}}'^A \qquad (6\text{-}6\text{-}43)$$

the stress is given by

$$T = 3\kappa(\tilde{e}^C + \tilde{e}^A - \tilde{e}^F) \qquad \mathbf{T}' = 2G(\tilde{\mathbf{e}}'^C + \tilde{\mathbf{e}}'^A - \tilde{\mathbf{e}}'^F) \qquad (6\text{-}6\text{-}44)$$

If an inclusion having the same size and shape as the original inclusion before transformation but differing elastic constants λ_1, G_1, κ_1, when subjected to the strain given by Eq. (6-6-43) produces the stress equation (6-6-44), then it may be used to replace the inclusion, maintaining, however, continuity of displacement and surface traction. This may always be done by ensuring that

$$\kappa_1(\tilde{e}^C + \tilde{e}^A) = \kappa(\tilde{e}^C + \tilde{e}^A - \tilde{e}^F) \qquad (6\text{-}6\text{-}45)$$

$$G_1(\tilde{\mathbf{e}}'^C + \tilde{\mathbf{e}}'^A) = G(\tilde{\mathbf{e}}'^C + \tilde{\mathbf{e}}'^A - \tilde{\mathbf{e}}'^F) \qquad (6\text{-}6\text{-}46)$$

For a given applied uniform field $\tilde{\mathbf{e}}^A$ and prescribed values of κ_1 and G_1, Eqs. (6-6-45) and (6-6-46) can be solved, using Eq. (6-6-42), to find $\tilde{\mathbf{e}}^F$ and $\tilde{\mathbf{e}}^C$. The strain $\tilde{\mathbf{e}}$ inside the inclusion is just the sum of $\tilde{\mathbf{e}}^C$ and $\tilde{\mathbf{e}}^A$, and from Eqs. (6-6-45) and (6-6-46) it follows that

$$\tilde{e} = \frac{\kappa}{\kappa - \kappa_1} \tilde{e}^F \qquad \tilde{\mathbf{e}}' = \frac{G}{G - G_1} \tilde{\mathbf{e}}'^F \qquad (6\text{-}6\text{-}47)$$

It has also been shown by Eshelby (1957) that the interaction energy U_{int} between the uniform stress and the inclusion is given by

$$U_{\text{int}} = -\tfrac{1}{2} V[\kappa \tilde{e}^A \tilde{e}^F + 2G \operatorname{tr}(\tilde{\mathbf{e}}'^A \tilde{\mathbf{e}}'^F)] \qquad (6\text{-}6\text{-}48)$$

For a spherical inclusion the above formulas take the explicit form

$$\tilde{e}^C = \alpha \tilde{e}^F \qquad \tilde{\mathbf{e}}'^C = \beta \tilde{\mathbf{e}}'^F \qquad (6\text{-}6\text{-}49)$$

where

$$\alpha = \frac{1}{3}\frac{1+\nu}{1-\nu} \qquad \beta = \frac{2}{15}\frac{4-5\nu}{1-\nu} \tag{6-6-50}$$

For a spherical inhomogeneity with elastic constants κ_1, G_1 in an applied field $\tilde{\mathbf{e}}^A$, the equivalent $\tilde{\mathbf{e}}^F$ are given by

$$\tilde{e}^F = A\tilde{e}^A \qquad \tilde{\mathbf{e}}'^F = B\tilde{\mathbf{e}}'^A \tag{6-6-51}$$

where

$$A = \frac{\kappa - \kappa_1}{\kappa}\frac{4G+3\kappa}{4G+3\kappa_1} \tag{6-6-52}$$

and

$$B = \frac{G_1 - G}{(G-G_1)\beta - G} \tag{6-6-53}$$

The interaction energy U_{int} is

$$U_{\text{int}} = \tfrac{1}{2}V\left[\frac{A}{9\kappa}T^A T^A + \frac{B}{2G}\operatorname{tr}(\mathbf{T}'^A)^2\right] \tag{6-6-54}$$

Thus the energy density of a body containing a volume fraction v of such spheres is

$$\frac{1}{2}\left[\frac{1}{9\kappa}(1+Av)T^A T^A + \frac{1}{2G}(1+Bv)\operatorname{tr}(\mathbf{T}'^A)^2\right]$$

where v has been assumed small enough for the interaction between spheres to be neglected. The effective bulk elastic constants are thus

$$\kappa_{\text{eff}} = \frac{\kappa}{1+Av} \qquad G_{\text{eff}} = \frac{G}{1+Bv} \tag{6-6-55}$$

which, as v is assumed small, can also be written as

$$\kappa_{\text{eff}} = \kappa(1-Av) \qquad G_{\text{eff}} = G(1-Bv) \tag{6-6-56}$$

in agreement with Dewey (1947). Setting κ_1 and G_1 equal to zero gives the effective bulk elastic constants for a material with a uniform distribution of equal spherical cavities, in agreement with the results of Mackenzie (1950):

$$\begin{aligned} \kappa_{\text{eff}} &= \kappa\left[1 - \frac{3(1-\nu)}{2(1-2\nu)}v\right] \\ G_{\text{eff}} &= G\left[1 - \frac{15(1-\nu)}{7-5\nu}v\right] \end{aligned} \tag{6-6-57}$$

Mackenzie's theory is based on the general self-consistancy method of Fröhlich and Sack (1946), in which a "unit cell" is chosen consisting of one sphere surrounded by a spherical shell of the matrix material in the correct volumetric ratio for the whole dispersion. This spherical shell is embedded within a uniform medium having the macroscopic properties of the dispersion as a whole. It is then required that the macroscopic motion remains unchanged if the unit cell and shell be replaced by the macroscopic medium.

If κ_1 and G_1 are taken to be infinite, the effective bulk elastic constants of a material containing rigid inclusions are obtained which are in agreement with the results of Hashin (1955):

$$\kappa_{\text{eff}} = \kappa\left[1 + \frac{3(1-\nu)}{1+\nu}v\right]$$
$$G_{\text{eff}} = G\left[1 + \frac{15(1-\nu)}{2(4-5\nu)}v\right] \tag{6-6-58}$$

Equations having the same form as (6-6-52), (6-6-53), and (6-6-55) for small values of v can be obtained from Oldroyd (1956) for the case in which the spherical inclusions are filled with a viscous liquid having a viscosity η. The equation of state of the liquid is then taken to be of the form

$$T = \kappa_1\tilde{e} \qquad \mathbf{T}' = 2\eta\Delta\tilde{\mathbf{e}}' \tag{6-6-59}$$

where Δ is the differential operator D/Dt. If $\eta\Delta$ is represented by G_1 then the previous analysis is unaltered, but B must now be interpreted as an operator:

$$B = -\frac{\eta\Delta - G}{\beta\eta\Delta + (1-\beta)G} \tag{6-6-60}$$

Thus the equation of state for the bulk deviatoric components has the form

$$\left[\beta\eta\frac{D}{Dt} + (1-\beta)G\right]\mathbf{T}' = 2G\left[\eta(\beta+v)\frac{D}{Dt} + (1-\beta-v)G\right]\tilde{\mathbf{e}}' \tag{6-6-61}$$

GENERAL REFERENCES

Fung, Y. C. (1965) "Foundations of Solid Mechanics," Prentice-Hall, Englewood Cliffs, N.J.
Green, A. E., and W. Zerna (1954) "Theoretical Elasticity," Clarendon Press, Oxford.
Hearmon, R. F. S. (1961) "An Introduction to Applied Anisotropic Elasticity," University Press, Oxford.
Landau, L. D., and E. M. Lifshitz (1959) "Theory of Elasticity," Pergamon Press, Oxford.
Love, A. E. H. (1927) "The Mathematical Theory of Elasticity," Cambridge University Press, Cambridge.
Pauling, L. (1960) "The Nature of the Chemical Bond," 3d ed., Cornell University Press, New York.
Sokolnikoff, I. S. (1956) "Mathematical Theory of Elasticity," McGraw-Hill, New York.
Timoshenko, S., and J. N. Goodier (1951) "Theory of Elasticity," 2d ed., McGraw-Hill, New York.
Truesdell, C., and R. Toupin (1960) The Classical Field Theories, in S. Flügge (ed.), "Handbuch der Physik," vol. III/1, pp. 226–858, Springer-Verlag, Berlin.

REFERENCES

Begbie, G. H., and M. Born (1947) *Proc. Roy. Soc. (London)*, vol. A188, pp. 179–188.
Born, M., and K. Huang (1954) "Dynamical Theory of Crystal Lattices," Clarendon Press, Oxford.
Bridgman, P. W. (1949) "The Physics of High Pressure," Bell, London.

Bridgman, P. W. (1952) "Studies in Large Plastic Flow and Fracture," McGraw-Hill, New York.
Budiansky, B. (1965) *J. Mech. Phys. Solids*, vol. 13, pp. 223–227.
Davies, R. M. (1948) *Phil. Trans. Roy. Soc.*, vol. A240, pp. 375–457.
Dewey, J. M. (1947) *J. Appl. Phys.*, vol. 18, pp. 578–581.
Eshelby, J. D. (1957) *Proc. Roy. Soc. (London)*, vol. A241, pp. 376–396.
Eshelby, J. D. (1961) *Phil. Mag.*, vol. 6, pp. 953–963.
Fröhlich, H., and R. Sack (1946) *Proc. Roy. Soc. (London)*, vol. A185, pp. 415–430.
Hashin, Z. (1955) *Bull. Res. Coun. Israel*, vol. C5, pp. 46–59.
Hashin, Z. (1964) *Appl. Mech. Rev.*, vol. 17, pp. 1–9.
Hashin, Z. (1967) *Int. J. Engng. Sci.*, vol. 5, pp. 213–223.
Hashin, Z., and S. Shtrikman (1962) *J. Mech. Phys. Solids*, vol. 10, pp. 343–352.
Hashin, Z., and S. Shtrikman (1963) *J. Mech. Phys. Solids*, vol. 11, pp. 127–140.
Hershey, A. V. (1954) *J. Appl. Mech.*, vol. 21, pp. 236–240.
Hill, R. (1952) *Proc. Phys. Soc.*, vol. A65, pp. 349–354.
Hill, R. (1963) *J. Mech. Phys. Solids*, vol. 11, pp. 357–372.
Hill, R. (1964) *J. Mech. Phys. Solids.*, vol. 12, pp. 199–212.
Hill, R. (1965) *J. Mech. Phys. Solids*, vol. 13, pp. 89–101, 213–222.
Hill, R. (1967) *J. Mech. Phys. Solids*, vol. 15, pp. 79–95.
Hudson, G. E. (1943) *Phys. Rev.*, vol. 63, pp. 46–51.
Huntington, H. B. (1958) In F. Seitz and D. Turnbull (eds.), "Solid State Physics," vol. 7, pp. 213–351, Academic Press, New York.
Johnson, F. A. (1977) *Proc. Roy. Soc. (London)*, vol. A357, pp. 289–295, 297–307, 309–321.
Lakkad, S. C., B. B. Miatt, and B. Parsons (1973) *J. Phys. (D)*, vol. 6, pp. 1566–1584.
Leibfried, G., and W. Ludwig (1961) In F. Seitz and D. Turnbull (eds.), "Solid State Physics," vol. 12, pp. 275–444, Academic Press, New York.
Mackenzie, J. K. (1950) *Proc. Phys. Soc.*, vol. B63, pp. 2–11.
Martin, J. W. (1975) *J. Phys. C: Solid State Phys.*, vol. 8, pp. 2837–2857, 2858–2868.
Michell, J. H. (1900) *Proc. Lond. Math. Soc.*, vol. 31, pp. 100–124, 130–146.
Oldroyd, J. G. (1956) In R. Grammel (ed.), "Deformation and Flow of Solids," pp. 304–313, Springer-Verlag, Berlin.
Pauli, W. (1925) *Z. Physik.*, vol. 31, pp. 765–783.
Reiner, M. (1960) *J. Mech. Phys. Solids.*, vol. 8, pp. 255–261.
Reuss, A. (1929) *Z. f. angew Math. u. Mechanik*, vol. 9, pp. 49–58.
Timoshenko, S. (1921) *Phil. Mag.*, vol. 41, pp. 744–746.
Voigt, W. (1910) "Lehrbuch der Kristallphysik," Teubner, Leipzig and Berlin.
Walpole, L. J. (1966) *J. Mech. Phys. Solids*, vol. 14, pp. 151–162, 289–301.
Walpole, L. J. (1969) *J. Mech. Phys. Solids*, vol. 17, pp. 235–251.
Wu, T. T. (1966) *Int. J. Solids Struct.*, vol. 2, pp. 1–8.

CHAPTER

SEVEN

PURELY VISCOUS FLUIDS

7-1 LINEARLY VISCOUS FLUIDS

7-1-1 The Stokesian Fluid

A material is said to flow if under the action of finite forces the deformation of the body increases continuously and indefinitely with time. Consider the effect of energy dissipation, occurring during the motion of a fluid, on that motion itself. This process is the result of the thermodynamic irreversibility of the motion which always occurs to some extent, and is due to internal friction, i.e., viscosity, and thermal conduction. In the case of a purely viscous material, a stress is only generated if the amount of strain is changing, the stress generated by strain being considered to relax instantaneously. Thus, a purely viscous fluid can be recognized by noting that in any of its distorted shapes which leave the mass density unchanged it has no memory of its past states.

Euler's equation of motion of an ideal fluid, i.e., the inviscid (pascalian) fluid, can be written in the form

$$\frac{\partial}{\partial t}(\rho \boldsymbol{v}) = -\operatorname{div} \boldsymbol{\Pi} \tag{7-1-1}$$

where

$$\boldsymbol{\Pi} = P\mathbf{I} + \rho \boldsymbol{v} \otimes \boldsymbol{v} \qquad \text{(inviscid fluid)} \tag{7-1-2}$$

is the momentum flux density tensor and P is the thermodynamic pressure. The momentum flux given by Eq. (7-1-2) represents a completely reversible transfer of momentum, due simply to the mechanical transport of the different particles of fluid from one point to another and to the pressure forces acting on the fluid. The viscosity (i.e., internal friction) is due to another, irreversible, transfer of momentum from points where the velocity is large to those where it is small. The equation of motion of a viscous fluid may therefore be obtained by adding to the ideal momentum flux of Eq. (7-1-2) a term $-\mathbf{T}_E$ which gives the irreversible transfer of momentum in the fluid. Thus the momentum flux density tensor in a viscous fluid is written in the form

$$\begin{aligned}\Pi &= P\mathbf{I}+\rho v\otimes v-\mathbf{T}_E \\ &= -\mathbf{T}+\rho v\otimes v \qquad \text{(viscous fluid)}\end{aligned} \tag{7-1-3}$$

The tensor

$$\mathbf{T} = -P\mathbf{I}+\mathbf{T}_E \tag{7-1-4}$$

is the stress tensor and the tensor $\mathbf{T}_E$ is called the extra stress. For the no-flow condition the stress must be of the form

$$[\mathbf{T}]_0 = -\bar{P}_0\mathbf{I} \tag{7-1-5}$$

where $\bar{P}_0$ is the hydrostatic pressure. For an ideal nonviscous fluid, $P = \bar{P}_0$ where P is the thermodynamic pressure.

The first to propose a definition of a purely viscous fluid was Stokes (1845) who based the development of the general fluid on the assumption that the difference between the stress in a deforming fluid and the static equilibrium stress given by Eq. (7-1-5), i.e., the extra stress $\mathbf{T}_E$, depends only on the relative motion of the fluid, this difference in stress $\mathbf{T}_E$ being unaffected by eliminating the relative motion arising from any motion of rotation. It has been shown in Sec. 2-4 that relative motion from which rotational motion has been removed is determined by the rate of deformation, i.e., the rate of strain, or stretching tensor $\mathbf{D}$. Thus, the constitutive assumption of Stokes may be expressed in the form

$$\mathbf{T}_E = \mathfrak{f}(\mathbf{D}) \tag{7-1-6}$$

that is,

$$\mathbf{T} = -P\mathbf{I}+\mathfrak{f}(\mathbf{D}) \qquad \mathfrak{f}(\mathbf{O}) = \mathbf{O} \tag{7-1-7}$$

where P is the thermodynamic pressure.

That Stokes' intuitive definition of a purely viscous fluid is indeed correct can be seen as follows. Boussinesq (1868) proposed that the stress depended on the spin tensor $\mathbf{W}$ as well as the rate of stretch tensor $\mathbf{D}$. A further generalization is to assume that the stress depends on the velocity $\dot{x}$ as well as the velocity gradient $\partial\dot{\mathbf{x}}/\partial\mathbf{x}$ through $\mathbf{D}$ and $\mathbf{W}$. These assumptions lead to a more general constitutive equation which may be expressed in the form

$$\mathbf{T} = -P(\rho)\mathbf{I}+\mathfrak{f}(\rho, \mathbf{D}, \mathbf{W}, \dot{x}) \tag{7-1-8}$$

with the condition

$$\mathfrak{f}(\rho, \mathbf{O}, \mathbf{O}, o) = \mathbf{O} \tag{7-1-9}$$

From Sec. 5-2 the principle of material objectivity requires that

$$\mathbf{T}^* = -P(\rho^*)\mathbf{I} + \mathfrak{f}(\rho^*, \mathbf{D}^*, \mathbf{W}^*, \dot{\overline{\boldsymbol{x}^*}}) \tag{7-1-10}$$

must hold for all $\boldsymbol{c}(t)$ and orthogonal $\mathbf{Q}(t)$, where $\mathbf{T}^*$ is given by Eq. (5-2-3). Use of Eq. (5-2-4) gives

$$\begin{aligned} \rho^* &= \rho \\ \mathbf{D}^* &= \mathbf{QDQ}^T \\ \mathbf{W}^* &= \mathbf{QWQ}^T + \Omega \qquad \Omega = \dot{\mathbf{Q}}\mathbf{Q}^T \\ \dot{\overline{\boldsymbol{x}^*}} &= \mathbf{Q}\dot{x} + \dot{c} + \dot{\mathbf{Q}}\boldsymbol{x} \end{aligned} \tag{7-1-11}$$

Equations (5-2-3), (7-1-8), (7-1-10), and (7-1-11) give

$$\mathbf{Q}\mathfrak{f}(\rho, \mathbf{D}, \mathbf{W}, \dot{x})\mathbf{Q}^T = \mathfrak{f}(\rho, \mathbf{QDQ}^T, \mathbf{QWQ}^T + \Omega, \mathbf{Q}\dot{x} + \dot{c} + \dot{\mathbf{Q}}\boldsymbol{x}) \tag{7-1-12}$$

an equation which must hold for arbitrary $\dot{c}$, $\mathbf{Q}$, and $\dot{\mathbf{Q}}$.

(*a*) Set $\mathbf{Q} = \mathbf{I}$, $\dot{\mathbf{Q}} = \mathbf{O}$; then Eq. (7-1-12) reduces to

$$\mathfrak{f}(\rho, \mathbf{D}, \mathbf{W}, \dot{x}) = \mathfrak{f}(\rho, \mathbf{D}, \mathbf{W}, \dot{x} + \dot{c})$$

and since $\dot{\boldsymbol{c}}$ is an arbitrary vector it is evident that $\mathfrak{f}$ does not depend on the velocity $\dot{\boldsymbol{x}}$.

(*b*) At an arbitrary instant of time take $\mathbf{Q} = \mathbf{I}$ and $\dot{\mathbf{Q}} \neq \mathbf{O}$, thus giving

$$\mathfrak{f}(\rho, \mathbf{D}, \mathbf{W}) = \mathfrak{f}(\rho, \mathbf{D}, \mathbf{W} + \dot{\mathbf{Q}})$$

from which it is evident that $\mathfrak{f}$ cannot depend on the spin $\mathbf{W}$. Thus Eq. (7-1-12) reduces to

$$\mathbf{Q}\mathfrak{f}(\rho, \mathbf{D})\mathbf{Q}^T = \mathfrak{f}(\rho, \mathbf{QDQ}^T)$$

from which it is evident that $\mathfrak{f}(\mathbf{D})$ is an isotropic tensor function of the symmetric tensor $\mathbf{D}$, confirming Stokes' intuitive definition of a purely viscous fluid. That Stokes' hypothesis defines only isotropic fluids is consistent with the observation already made that a fluid has no preferred reference configuration, from which it follows that their isotropy group must be the full unimodular group for which all mappings $\mathbf{H}$ satisfy $\det \mathbf{H} = 1$ (see Sec. 5-3).

This is an important application of the principle of material objectivity, since its use has obviated the need to make a special physical assumption in order that a viscous fluid be isotropic.

At this stage an important distinction can be made between materials with a perfect memory, i.e., solids, and materials having no memory of past states, i.e., fluids. An observer transformation of deformation gradients is given by

$$\mathbf{F}^*(\tau) = \mathbf{Q}(\tau)\mathbf{F}(\tau)$$

which is to be compared with an observer transformation of the rate tensor which is given by

$$\mathbf{D}^* = \mathbf{QDQ}^T$$

The distinction recognizes that in observer transformations the deformation gradients, which span the reference and current configurations, transform as vectors, and hence response functions for solids are not, of necessity, required by the principal of material objectivity to be isotropic functions, whereas in observer transformations objective rate tensors transform as tensors wholly within the current configuration, and hence response functions for purely viscous fluids must, of necessity, be isotropic functions.

7-1-2 The Newtonian Fluid

When the function $\mathfrak{f}$ is linear, the fluid is referred to as a newtonian fluid and the constitutive equation can be expressed in the form

$$\mathbf{T} = -P\mathbf{I} + \mathbf{AD} \tag{7-1-13}$$

Both $\mathbf{T}$ and $\mathbf{D}$ are symmetric, and hence the property tensor A^{ij}_{kl} is symmetric in ij and also in kl. Since by definition stokesian fluids are isotropic, the property tensor $\mathbf{A}$ must be isotropic. Hence,

$$\begin{aligned} A^{ij}_{kl}D^{kl} &= [A_{(1)}g^{ij}g_{kl} + A_{(2)}(\delta^i{}_k\delta^j{}_l + \delta^i{}_l\delta^j{}_k)]D^{kl} \\ &= A_{(1)}g^{ij}(\text{tr}\,\mathbf{D}) + 2A_{(2)}D^{ij} \end{aligned} \tag{7-1-14}$$

where $A_{(1)}$ and $A_{(2)}$ are scalars and $(\text{tr}\,\mathbf{D})$ is the first invariant of the rate of strain tensor. The scalars $A_{(1)}$ and $A_{(2)}$ are not altered by flow and therefore Eq. (7-1-14) is valid for all instants of time. Substitution for $A^{ij}_{kl}D^{kl}$ in Eq. (7-1-13) from Eq. (7-1-14) gives

$$\mathbf{T} = -P\mathbf{I} + A_{(1)}(\text{tr}\,\mathbf{D})\mathbf{I} + 2A_{(2)}\mathbf{D} \tag{7-1-15}$$

Equation (7-1-15) can be expressed in the form

$$\mathbf{T} = -P\mathbf{I} + \mu(\text{tr}\,\mathbf{D})\mathbf{I} + 2\eta\mathbf{D} \tag{7-1-16}$$

where

$$A_{(1)} - \mu = \text{constant} \qquad A_{(2)} = \eta = \text{constant} \tag{7-1-17}$$

Equation (7-1-16) is the Navier–Poisson law; however, fluids described by Eq. (7-1-16) are generally referred to as newtonian fluids. It is evident from Eq. (7-1-16) that two constants η and μ, characteristic of material properties, are required to define the properties of the newtonian fluid.

Since the shearing components of $\mathbf{T}_E$ are given by the term $2\eta D^{ij}$ $(i \neq j)$, the coefficient η is called the shear, or dynamic viscosity.

Equation (7-1-16) can be expressed in terms of the deviators

$$\mathbf{T}' = \mathbf{T} - \tfrac{1}{3}(\text{tr}\,\mathbf{T})\mathbf{I} \tag{7-1-18}$$

$$\mathbf{D}' = \mathbf{D} - \tfrac{1}{3}(\text{tr}\,\mathbf{D})\mathbf{I} \tag{7-1-19}$$

to give the constitutive equation for a newtonian fluid in the form

$$\mathbf{T}' = (\bar{P} - P)\mathbf{I} + (\mu + \tfrac{2}{3}\eta)(\text{tr}\,\mathbf{D})\mathbf{I} + 2\eta\mathbf{D}' \tag{7-1-20}$$

where

$$\bar{P} = -\tfrac{1}{3}(\mathrm{tr}\,\mathbf{T}) \tag{7-1-21}$$

is the mean pressure [see Eq. (3-6-18)]. Since $\mathrm{tr}\,\mathbf{T}' = 0$ and $\mathrm{tr}\,\mathbf{D}' = 0$, Eq. (7-1-20) gives

$$\bar{P} - P + (\mu + \tfrac{2}{3}\eta)\,(\mathrm{tr}\,\mathbf{D}) = 0 \tag{7-1-22}$$

The condition of Eq. (7-1-22), together with the equation of continuity expressed in the form

$$\mathrm{tr}\,\mathbf{D} \equiv \mathrm{div}\,v = -\frac{1}{\rho}\frac{D\rho}{Dt} \tag{7-1-23}$$

reduces Eq. (7-1-20) to the relations

$$\mathbf{T}' = 2\eta\mathbf{D}' \tag{7-1-24}$$

$$\begin{aligned}\bar{P} &= P - \zeta(\mathrm{tr}\,\mathbf{D}) \\ &= P + \zeta\frac{1}{\rho}\frac{D\rho}{Dt}\end{aligned} \tag{7-1-25}$$

where

$$\zeta = \mu + \tfrac{2}{3}\eta \tag{7-1-26}$$

is the bulk viscosity. From Eq. (7-1-25) it is evident that the mean pressure $\bar{P}$ equals the thermodynamic pressure P if and only if one of the following two conditions is satisfied: either

$$\mathrm{tr}\,\mathbf{D} = 0 \tag{7-1-27}$$

or the bulk viscosity

$$\zeta = \mu + \tfrac{2}{3}\eta = 0 \tag{7-1-28}$$

which is called the Stokes condition.

For purely viscous fluids which are compressible, the pressure is taken to be the thermodynamic pressure which is determined by the density ρ and temperature θ of the fluid. Consider a reference state, such as a state of rest, and let $\bar{P}_0$ and ρ_0 be the corresponding pressure and density respectively. Taking a Taylor expansion about ρ_0 gives

$$P - \bar{P}_0 = \frac{\partial P}{\partial \rho}(\rho - \rho_0) + \cdots = \kappa\frac{(\rho - \rho_0)}{\rho_0} \tag{7-1-29}$$

provided density changes are not too large. The bulk modulus κ is a measure of the volume elasticity of the material. Substitution of Eq. (7-1-29) into Eq. (7-1-16) and rearranging using Eqs. (7-1-19) and (7-1-26) gives

$$\mathbf{T} = [\mathbf{T}]_0 - \kappa\frac{(\rho - \rho_0)}{\rho_0}\mathbf{I} + \zeta(\mathrm{tr}\,\mathbf{D})\mathbf{I} + 2\eta\mathbf{D}' \tag{7-1-30}$$

where $[\mathbf{T}]_0$ is given by Eq. (7-1-5) and is the stress prevailing if the material is at rest. The term involving the bulk modulus in Eq. (7-1-30) is the elastic stress which the material generates to oppose volume changes. Any change in volume is also opposed by the viscous stress term $\zeta(\mathrm{tr}\,\mathbf{D})\mathbf{I}$ which is generated by the material. Thus, it is evident that real materials with a nonvanishing bulk viscosity exhibit both viscous and elastic behavior under isotropic pressure.

For an incompressible fluid, its motions are always isochoric, i.e., the condition of Eq. (7-1-27) is always satisfied in an incompressible fluid. Hence, for an incompressible fluid, Eq. (7-1-16) reduces to

$$\mathbf{T} = -P\mathbf{I} + 2\eta\mathbf{D} \tag{7-1-31}$$

It is evident from Eqs. (7-1-4) and (7-1-31) that for an incompressible fluid the stress is determined by the deformation history only to within an indeterminate pressure or hydrostatic pressure P. Because of the indeterminate pressure P, Eq. (7-1-31) does not suffice to assign a unique functional to each material. This nonuniqueness can be removed by a normalization such as

$$\mathrm{tr}\,\mathbf{T}_E = \mathrm{tr}[\mathfrak{f}(\mathbf{D})] = 0 \tag{7-1-32}$$

When Eq. (7-1-32) holds, the extra stress $\mathbf{T}_E$ becomes the stress deviator $\mathbf{T}'$ and the indeterminate pressure $P = \bar{P}$ at all times.

Expressed in terms of $\mathbf{D}'$ defined by Eq. (7-1-19), the constitutive Eq. (7-1-16) gives the stress power

$$\mathrm{tr}(\mathbf{TD}) = -P\,\mathrm{tr}\,\mathbf{D} + \zeta(\mathrm{tr}\,\mathbf{D})^2 + 2\eta\,\mathrm{tr}\,(\mathbf{D}'^2) \tag{7-1-33}$$

The first term on the right-hand side of Eq. (7-1-33) can be expressed by way of Eq. (7-1-23) in the form

$$-P\,\mathrm{tr}\,\mathbf{D} = \frac{P}{\rho}\frac{D\rho}{Dt} = -\frac{P}{V}\frac{DV}{Dt}$$

where $V = 1/\rho$. The term $-P\,\mathrm{tr}\,\mathbf{D}$ can be positive or negative and may therefore represent a recoverable contribution to the internal energy, i.e., it may represent a recoverable power not contributing to the internal entropy production. The Clausius–Duhem inequality of Sec. 3-9 requires the rate of viscous dissipation to be nonnegative, and hence the dissipation function

$$\begin{aligned}\Phi_v &= \mathrm{tr}\,(\mathbf{TD}) + P\,\mathrm{tr}\,\mathbf{D} \\ &- \zeta(\mathrm{tr}\,\mathbf{D})^2 + 2\eta\,\mathrm{tr}\,(\mathbf{D}'^2) \geqslant 0\end{aligned} \tag{7-1-34}$$

which implies the inequalities

$$\eta \geqslant 0 \qquad \zeta \geqslant 0 \qquad \text{or} \qquad \mu \geqslant -\tfrac{2}{3}\eta \tag{7-1-35}$$

7-2 FIELD EQUATIONS OF NEWTONIAN FLUID

7-2-1 The Equations of Continuity and Motion

The field equations of a fluid obeying Fourier's law of heat conduction and a caloric equation of state can be summarized as follows.

General equations

Continuity equation (Sec. 3-1):

$$\frac{D\rho}{Dt} + \rho \operatorname{div} v = 0 \tag{7-2-1}$$

Equations of motion (Sec. 3-4):

$$\rho \frac{Dv}{Dt} = \operatorname{div} \mathbf{T} + \rho f \tag{7-2-2}$$

Energy equation (Sec. 3-8):

$$\rho \frac{D\varepsilon}{Dt} = \operatorname{tr}(\mathbf{TD}) + \rho r + \operatorname{div} q \tag{7-2-3}$$

Constitutive equations

Navier–Poisson law of a newtonian fluid (Sec. 7-1-2):

$$\mathbf{T} = -P\mathbf{I} + \mu(\operatorname{tr}\mathbf{D})\mathbf{I} + 2\eta\mathbf{D} \tag{7-2-4}$$

Kinetic equation of state:

$$F(P, \rho, \theta) = 0 \tag{7-2-5}$$

Fourier's law of heat conduction:

$$q = k\nabla\theta \tag{7-2-6}$$

Caloric equation of state:

$$\varepsilon = \varepsilon(\theta, \rho) \tag{7-2-7}$$

In these equations the rate of deformation $\mathbf{D}$ is related to the velocity v by the relation

$$\mathbf{D} = \tfrac{1}{2}(\mathbf{L} + \mathbf{L}^T) \tag{7-2-8}$$

The general equations and constitutive equations taken together total 16 equations. These 16 equations arise from the existence of six stress components $\mathbf{T}$, three velocity components v, three heat components q, one thermodynamic pressure P, one density ρ, one internal energy ε, and one temperature θ.

The number of equations can, however, be reduced as follows. From Eq. (7-2-8),

$$2\nabla\mathbf{D} = \nabla(\nabla \cdot v) + \nabla^2 v \tag{7-2-9}$$

which combined with Eq. (7-2-4) gives

$$\begin{aligned} \nabla\mathbf{T} &= -\nabla P\mathbf{I} + \mu\nabla(\nabla \cdot v)\mathbf{I} + 2\eta\nabla\mathbf{D} \\ &= -\nabla P + \mu\nabla(\nabla \cdot v) + \eta\nabla(\nabla \cdot v) + \eta\nabla^2 v \end{aligned} \tag{7-2-10}$$

Substitution of Eq. (7-2-10) into the equations of motion, i.e., Eq. (7-2-2), gives

$$\rho \frac{D\boldsymbol{v}}{Dt} = -\boldsymbol{\nabla} P + (\mu + \eta)\boldsymbol{\nabla}(\boldsymbol{\nabla} \cdot \boldsymbol{v}) + \eta \nabla^2 \boldsymbol{v} + \rho \boldsymbol{f} \tag{7-2-11}$$

which is known as the generalized Navier–Stokes equations of motion for fluids with bulk viscosity. Eliminating $\boldsymbol{q}$ from Eq. (7-2-3) by way of Eq. (7-2-6) gives the energy equation in the form

$$\rho \frac{D\varepsilon}{Dt} = \boldsymbol{\nabla} \cdot (k\boldsymbol{\nabla}\theta) + \rho r + \operatorname{tr}(\mathbf{TD}) \tag{7-2-12}$$

The term $\operatorname{tr}(\mathbf{TD})$ can, in turn, be expressed in terms of the dissipation function Φ_v by way of Eq. (7-1-34).

Thus the general equations reduce to seven equations in the seven unknowns $\boldsymbol{v}$, P, ρ, ε, and θ. These general equations can be expressed in the form:

Continuity equation:

$$\frac{D\rho}{Dt} + \rho \boldsymbol{\nabla} \cdot \boldsymbol{v} = 0 \tag{7-2-13}$$

Generalized Navier–Stokes equations for fluids with bulk viscosity:

$$\rho \frac{D\boldsymbol{v}}{Dt} = -\boldsymbol{\nabla} P + (\eta + \mu)\boldsymbol{\nabla}(\boldsymbol{\nabla} \cdot \boldsymbol{v}) + \eta \nabla^2 \boldsymbol{v} + \rho \boldsymbol{f} \tag{7-2-14}$$

Energy equations:

$$\rho \frac{D\varepsilon}{Dt} = \boldsymbol{\nabla} \cdot (k\boldsymbol{\nabla}\theta) + \rho r + \operatorname{tr}(\mathbf{TD}) \tag{7-2-15}$$

Equations of state:

$$F(P, \rho, \theta) = 0 \tag{7-2-16}$$

$$\varepsilon = \varepsilon(\theta, \rho) \tag{7-2-17}$$

In formulating flow problems it is generally assumed that there is no slip at a boundary in viscous flow, the fluid adhering to a rigid surface. Thus at a stationary surface $\boldsymbol{v} = 0$ for viscous flow, in contrast to ideal frictionless fluid flow for which only the normal component is required to vanish at a rigid surface. The consequences of slippage have been considered by Oldroyd (1949).

Two particular states of flow can be distinguished, depending on whether the fluid is incompressible or whether there is no bulk viscosity ($\zeta = 0$), the latter state being described by the Stokes condition $\mu = -2\eta/3$. These conditions reduce the generalized Navier–Stokes equations to the particular forms (Navier, 1822; Stokes, 1845):

Incompressible fluid ($\boldsymbol{\nabla} \cdot \boldsymbol{v} = 0$):

$$\frac{D\boldsymbol{v}}{Dt} = -\frac{1}{\rho}\boldsymbol{\nabla} P + \nu \nabla^2 \boldsymbol{v} + \boldsymbol{f} \tag{7-2-18}$$

Compressible fluid with no bulk viscosity ($\zeta = 0$):

$$\frac{Dv}{Dt} = -\frac{1}{\rho}\nabla P + \tfrac{1}{3}\nu\nabla(\nabla \cdot v) + \nu\nabla^2 v + f \qquad (\mu = -\tfrac{2}{3}\eta) \tag{7-2-19}$$

where

$$\nu = \frac{\eta}{\rho} \tag{7-2-20}$$

is the kinematic viscosity.

Introducing the vorticity vector ω defined by the relation

$$\omega = \nabla \times v = 2w \tag{7-2-21}$$

where w is the axial vector of the spin tensor $\mathbf{W}$. Using the vector identity

$$\nabla^2 v = \nabla(\nabla \cdot v) - \nabla \times (\nabla \times v)$$

enables Eq. (7-2-18) to be expressed in the form

$$\frac{Dv}{Dt} = -\frac{1}{\rho}\nabla P - 2\nu(\nabla \times w) + f \tag{7-2-22}$$

for an incompressible fluid. For irrotational flow, that is, $w = 0$, Eq (7-2-22) reduces to the equation of motion for a perfect fluid:

$$\frac{Dv}{Dt} = -\frac{1}{\rho}\nabla P + f \tag{7-2-23}$$

which, by comparison with Eq. (7-2-18), is seen to correspond to an inviscid fluid characterized by having $\nu = 0$.

7-2-2 Effects Arising from the Inertia of the Fluid

The field equations are nonlinear, and as such an exact analysis of any form of unsteady flow is, in general, of considerable mathematical complexity unless special, simplifying, assumptions are made. The nonlinearity arises from the inertial acceleration terms

$$\rho\frac{Dv}{Dt} = \rho\left[\frac{\partial v}{\partial t} + (v \cdot \nabla)v\right] \tag{7-2-24}$$

of the Navier–Stokes equations, the energy equation, and the equations of state.

Simplification of the mathematical treatment can be achieved, however, if the inertia of the fluid can be neglected. Such a simplification is of particular interest in nonsteady experiments which are directed solely at studying the rheological equation of state in relation to the physical and chemical properties of the fluids, and as such the question arises whether a simple experimental test can be devised which will indicate whether the inertia of the sample can be neglected.

Consider an incompressible fluid bounded by an infinite plane surface which executes a simple harmonic oscillation in its own plane with frequency ω. Take the solid surface as the YZ plane and the fluid region as $x > 0$; the Y axis is taken in the direction of the oscillation. The velocity u_y of the oscillating surface is a function of time, and as such can be expressed in the form $v_0 \cos(\omega t+\delta)$. At $x = 0$ the fluid velocity $\mathbf{v}$ must satisfy the boundary conditions $v_x(0) = v_z = 0$ and $v_y = u_y$.

It is evident from symmetry that all quantities will depend only on the coordinate x and the time t. From the equation of continuity it follows that, for all x, $\partial v_x/\partial x = 0$ which gives $v_x(x) =$ constant, and in order to satisfy the boundary condition at the solid surface $[x = 0, v_x(0) = 0]$ it follows that $v_x(x) = 0$ for all x. Since all quantities are independent of y and z it follows that the general equations of motion for the flow of an incompressible viscous fluid reduces to

$$\rho \frac{\partial v_y}{\partial t} = \eta \frac{\partial^2 v_y}{\partial x^2} \tag{7-2-25}$$

and

$$\frac{\partial P}{\partial x} = \frac{\partial P}{\partial y} = \frac{\partial P}{\partial z} = 0$$

which implies that P is everywhere a constant.

A solution of Eq. (7-2-25), subject to the boundary conditions $[x = 0, v_y = u_y = v_0 \cos(\omega t+\delta)]$ and $(x = \infty, v_y = 0)$ can be obtained in the form

$$v_y = v_0 \exp(-ax)\cos(\omega t - ax + \delta) \tag{7-2-26}$$

where v_0 is the velocity amplitude of the oscillating plate and x is the distance from the plate. Substituting Eq. (7-2-26) into Eq. (7-2-25) gives, in terms of the kinematic viscosity ν $(=\eta/\rho)$, the relation

$$2\nu a^2 = \omega$$

so that the velocity

$$v_y = v_0 \exp\left[-\left(\frac{\omega}{2\nu}\right)^{1/2} x\right]\cos\left[\omega t - \left(\frac{\omega}{2\nu}\right)^{1/2} x + \delta\right] \tag{7-2-27}$$

The quantity a is taken to have a positive imaginary part since otherwise the velocity would increase without limit in the interior of the fluid, which is physically impossible.

The amplitude of the transverse wave represented by Eq. (7-2-27) is rapidly damped in the interior of the fluid, the amplitude being attenuated exponentially with increasing distance x from the solid surface. Thus transverse waves can be generated in a viscous fluid. However, such waves are rapidly damped away from the solid surface whose motion generates the waves. The distance x_0 over which the amplitude falls off by a factor e is referred to as the depth of penetration of the wave, and, from Eq. (7-2-27),

$$x_0 = \left(\frac{2\nu}{\omega}\right)^{1/2} \tag{7-2-28}$$

from which it is evident that x_0 diminishes with increasing frequency but increases with the kinematic viscosity ν of the fluid.

Consider now the general case of a body of arbitrary shape executing oscillatory motion. Unlike the case of the oscillating plane, the convective term $(\boldsymbol{v} \cdot \boldsymbol{\nabla})\boldsymbol{v}$ occurring in the equation of motion of the fluid is not necessarily identically zero for a body of arbitrary shape. The present discussion will, however, be restricted to the case for which the convective term can be neglected in comparison with the other terms, the necessary conditions for this procedure to be valid being considered below.

Setting the convective term equal to zero in Eq. (7-2-18) and taking the curl of both sides of the resulting relation gives

$$\frac{\partial}{\partial t}(\text{curl}\, \boldsymbol{v}) = \nu \nabla^2 \,\text{curl}\, \boldsymbol{v} \tag{7-2-29}$$

since $\boldsymbol{\nabla} \times \boldsymbol{\nabla} P$ vanishes identically. It has been shown for the case of an oscillating plane surface that such an equation gives an exponential decrease of the quantity which satisfies it. Thus it is concluded that the vorticity curl $\boldsymbol{v}$ decreases toward the interior of the fluid. That is, the motion of the fluid caused by the oscillation of the body is rotational in a given boundary layer round the body, while at large distances it rapidly changes to potential flow. The depth of the boundary layer associated with the rotational flow is on the order of $x_0 \sim (\nu/\omega)^{1/2}$.

Two important cases are to be distinguished, depending upon whether x_0 is either large or small compared with the dimensions of the oscillating body.

Let l be a characteristic dimension of the oscillating body. The condition $x_0 \gg l$ implies that $l^2\omega \ll \nu$. In addition to this condition the assumption of a small Reynolds number

$$\text{Re} = \frac{ul\rho}{\eta} \tag{7-2-30}$$

will be made, where u is the velocity of the main stream. For an amplitude a_0 of oscillation the velocity of the body is on the order of $a_0\omega$, and the Reynolds number for the motion is therefore $\omega a_0 l/\nu$. Thus two conditions exist:

$$l^2\omega \ll \nu \qquad \frac{\omega a_0 l}{\nu} \ll 1 \tag{7-2-31}$$

This corresponds to the case of low frequencies of oscillation, which in turn implies that the velocity varies only slowly with time; as such the derivative $\partial \boldsymbol{v}/\partial t$ can be neglected in the general equation of motion. The convective term is neglected because the Reynolds number is small.

The absence of the term $\partial \boldsymbol{v}/\partial t$ from the equation of motion implies that the flow can be regarded as steady at any given instant, provided $x_0 \gg l$. This means that the flow at any given instant is what it would be if the body were moving uniformly with its instantaneous velocity. This condition is of considerable importance in developing an experimental technique for studying the transient flow behavior of fluids which exhibit the phenomenon of thixotropy.

Consider now the opposite case for which $l \gg x_0$. For the condition that the convective term $(\boldsymbol{v} \cdot \boldsymbol{\nabla})\boldsymbol{v}$ can again be neglected it is necessary that the amplitude of oscillations should be small compared with the dimensions of the body, a requirement which gives the inequalities

$$l^2\omega \gg \nu \qquad a_0 \ll l \tag{7-2-32}$$

for which, in contrast to the condition associated with $x_0 \ll l$, the Reynolds number need not be small. The above inequalities are verified by estimating the magnitude of $(\boldsymbol{v} \cdot \boldsymbol{\nabla})\boldsymbol{v}$ relative to the derivative $\partial \boldsymbol{v}/\partial t$. The operator $(\boldsymbol{v} \cdot \boldsymbol{\nabla})$ denotes differentiation in the direction of the velocity, the latter being nearly tangential in the vicinity of the surface of the body. In the tangential direction the velocity changes appreciably only over distances comparable with the dimensions of the body, from which it follows that

$$(\boldsymbol{v} \cdot \boldsymbol{\nabla})\boldsymbol{v} \sim \frac{v^2}{l} \sim \frac{a_0^2\omega^2}{l} \tag{7-2-33}$$

where the velocity has again been taken to be on the order of $a_0\omega$. Now the derivative

$$\frac{\partial \boldsymbol{v}}{\partial t} \sim v\omega \sim a_0\omega^2 \tag{7-2-34}$$

Comparing Eqs. (7-2-33) and (7-2-34) gives

$$(\boldsymbol{v} \cdot \boldsymbol{\nabla})\boldsymbol{v} \ll \frac{\partial \boldsymbol{v}}{\partial t} \tag{7-2-35}$$

if $a_0 \ll l$; i.e., the terms $\partial \boldsymbol{v}/\partial t$ and $\nu\nabla^2\boldsymbol{v}$ are of the same order.

Thus the conditions of Eq. (7-2-32) imply the existence of a thin layer near the surface of the body in which the flow is rotational; in the rest of the fluid there exists potential flow. This situation is encountered in the case of the free oscillation of a vessel enclosing a very low viscosity fluid, such as water or a liquid metal, and in the case of a freely oscillating disk suspended in an infinite fluid, such as liquid helium (Hollis-Hallett, 1951).

7-3 SPECIAL VISCOMETRIC FLOWS: I. STEADY STATE

7-3-1 Introduction

The characterization of the flow properties of fluids provides an example of the formulation of a complete particular theory in continuum mechanics by the combination of the general principles of Chap. 3, valid for all continua, with a particular constitutive equation. In principle it is required to observe directly the shear stress and rate of flow of an individual element of a flowing material. This, however, is not usually possible and as such the "shear stress–rate of flow" relation is in general inferred from observation of the motion of the fluid in bulk. The material in bulk is usually confined between bounding surfaces having a well-defined, relatively simple geometry, the flow behavior being obtained in terms of the shear stress and flow rate at one of the bounding surfaces. Thus the characterization of the flow behavior of a real fluid, by way of the appropriate constitutive equation (i.e., the rheological equation of state), reduces to a boundary-value problem, the required solution being obtained from the general equations of motion and continuity.

Among the earliest systematic studies of the steady flow behavior of liquids were those of Poiseuille, who was concerned with the capillary flow of blood, in 1840 and 1846. There now exists a large body of observational information relating to the flow properties of liquids in tubes and channels in which steady-state flow results from the presence of a pressure gradient, the motion of the liquid being referred to as Poiseuille-type flow. A significant advance was the study of flow in the absence of pressure gradients, as, for example, the steady flow of a liquid confined between semi-infinite rotating coaxial cylinders, the motion of the liquid being referred to as Couette-type flow (Couette, 1890). In the case of Poiseuille flow any given element of fluid can only be observed for a finite time insofar as pipes and channels are of finite length. This is in contrast to the flow between rotating coaxial cylinders, where an element of liquid can, in principle, be indefinitely maintained in some form of continuous orbital motion. Prolonged shearing under the latter conditions of flow can give rise to viscous heating of the test material, an undesirable feature of any method designed to determine the isothermal flow properties of the fluid.

Steady shearing flows of the Poiseuille and Couette types are viscometric flows which satisfy the dynamical equations exactly. However, from the discussion given in Sec. 7-2-2 of the effects arising from the inertia of the fluid, it is evident that exact solutions cannot, in general, be obtained due to the presence of the convective term $(\boldsymbol{v} \cdot \boldsymbol{\nabla})\boldsymbol{v}$ in the general equations of motion. For the purposes of discussing the various types of viscometric flow which are of immediate interest the convective term will be omitted, either by suitably specifying the nature of the flow or by limiting the discussion to the case of infinitesimally slow motion. With this simplification, Eq. (7-2-18) reduces to

$$\frac{\partial \boldsymbol{v}}{\partial t} = -\frac{1}{\rho}\boldsymbol{\nabla} P + \nu\nabla^2 \boldsymbol{v} + \boldsymbol{f} \tag{7-3-1}$$

for the flow of an incompressible newtonian fluid.

For the flow of newtonian fluids it will be assumed throughout that for any type of motion there is no slip at the interface between the fluid and bounding surface.

7-3-2 Flow through a Pipe

The flow through a pipe of radius a will be referred to cylindrical polar coordinates (r, θ, z) such that $r = 0$ is the axis of the pipe and z is measured downstream. Consideration of the flow properties of real fluids will be restricted to a particular type of slow viscous flow for which two-dimensional, axially symmetric, steady rectilinear flow is prescribed with velocity components $(0, 0, v\langle z \rangle)$ which satisfy the equation of continuity for an incompressible fluid. For these conditions the components of the rate of deformation at any point (r, θ, z) are

$$[\mathbf{D}] = \begin{bmatrix} 0 & 0 & \dfrac{1}{2}\dfrac{dv\langle z\rangle}{dr} \\ 0 & 0 & 0 \\ \dfrac{1}{2}\dfrac{dv\langle z\rangle}{dr} & 0 & 0 \end{bmatrix}$$

and the corresponding components of the stress are

$$[\mathbf{T}] = \begin{bmatrix} -P & 0 & T\langle rz\rangle \\ 0 & -P & 0 \\ T\langle zr\rangle & 0 & -P \end{bmatrix}$$

all such quantities being independent of θ.

For steady flow the general equations of motion reduce, for flow in the z direction, to

$$\frac{1}{r}\frac{\partial}{\partial r}(rT\langle rz\rangle) = \frac{\partial P}{\partial z} \qquad (T\langle rz\rangle = T\langle zr\rangle) \tag{7-3-2}$$

it being noted that $dP/dz < 0$ for all $v\langle z\rangle > 0$. The laws of motion also require that

$$\frac{\partial P}{\partial r} = 0 \qquad \frac{\partial P}{\partial \theta} = 0$$

from which it follows that the pressure gradient dP/dz is uniform over the cross section of the pipe such that Eq. (7-3-2) can be integrated, subject to the boundary conditions ($r = a$, $T\langle rz\rangle \equiv \tau_a$) and ($r = 0$, $T\langle rz\rangle = 0$) to give the relation

$$\frac{T\langle rz\rangle}{r} = \frac{\tau_a}{a} = \frac{1}{2}\frac{dP}{dz} \tag{7-3-3}$$

For the flow of a newtonian fluid it is evident from Eq. (7-2-4), or alternatively Eq. (7-1-24), that the shear stress $T\langle rz\rangle$ is related to the velocity $v\langle z\rangle$ at some intermediate radius r by the relation

$$T\langle rz\rangle = \eta\frac{dv\langle z\rangle}{dr} \qquad (T\langle rz\rangle = T\langle zr\rangle) \tag{7-3-4}$$

Equation (7-3-4) is the rheological equation of state for the flow of a newtonian fluid, relative to the flow through a circular pipe.

The velocity field is obtained by integration of Eq. (7-3-4) subject to the boundary condition ($r = a$, $v\langle z\rangle = 0$) to give, by way of Eq. (7-3-3),

$$v\langle z\rangle = -\frac{a^2 - r^2}{4\eta}\frac{dP}{dz} \tag{7-3-5}$$

from which it is evident that the velocity distribution for the laminar incompressible flow of a newtonian fluid in a tube is parabolic.

The observable volume flow rate is

$$Q = \int_0^a 2\pi r v\langle z\rangle \, dr \tag{7-3-6}$$

and substituting for $v\langle z\rangle$ from Eq. (7-3-5) gives Poiseuille's formula for the viscosity

$$\eta = -\frac{\pi a^3 \tau_a}{4Q} = -\frac{\pi a^4}{8Q}\frac{dP}{dz} \tag{7-3-7}$$

from which it follows, by comparison with Eq. (7-3-4), that

$$\left[\frac{dv\langle z\rangle}{dr}\right]_{r=a} = -\frac{4Q}{\pi a^3} \tag{7-3-8}$$

is the velocity gradient at the wall of the pipe.

The average velocity $\bar{v}\langle z\rangle$ is calculated by summing up all the velocities over a cross section and then dividing by the cross-sectional area. Thus

$$\frac{\displaystyle\int_0^{2\pi}\int_0^a v\langle z\rangle r \, dr \, d\theta}{\displaystyle\int_0^{2\pi}\int_0^a r \, dr \, d\theta} = -\frac{a^2}{8\eta}\frac{dP}{dz} = \frac{Q}{\pi a^2} = \bar{v}\langle z\rangle \tag{7-3-9}$$

which is just half the maximum velocity

$$v_{\max}\langle z\rangle = -\frac{a^2}{4\eta}\frac{dP}{dz} = \frac{2Q}{\pi a^2}$$

which occurs along the axis of the tube.

Poiseuille's formula is derived for the condition that the velocity is constant over the whole length of the tube. In practical applications the applied pressure is partly absorbed in giving kinetic energy to the liquid, the remaining pressure overcoming the viscous resistance. Since the kinetic energy given to the liquid per unit time is $\rho Q^3/\pi^2 a^4$ and the total work done per unit time by the applied pressure is PQ, it follows that the effective pressure P_e overcoming the viscous resistance is

$$P_e = P - \frac{\rho Q^2}{\pi^2 a^4} \tag{7-3-10}$$

For a capillary which is connected at each end to a reservoir it has been found experimentally that a correction of the same form is required. Applying the correction, the effective pressure becomes

$$P_e = P - m\frac{\rho Q^2}{\pi^2 a^4} \tag{7-3-11}$$

where m is a factor which tends to increase with an increasing Reynolds number.

Associated with the ends of the capillary tube is the Couette effect, which is concerned with an effective reduction in the pressure drop across the capillary tube. An amount of energy is expended in overcoming the viscous forces between the converging and diverging steamlines at the entrance and exit respectively of the capillary tube. The correction for this effect is made in the form of a small increase na in the actual length of the capillary tube.

Introducing the kinetic energy correction and the Couette correction into Poiseuille's equation (7-3-7) gives the relation

$$\eta = \frac{\pi a^4 P}{8Q(L+na)} - m\frac{\rho Q}{8\pi(L+na)} \tag{7-3-12}$$

for a capillary tube of actual length L and an applied pressure P.

It was shown by Reynolds (1883) that the observed deviation from Poiseiulle's equation (7-3-7), occurring at high rates of flow, was due to a change from streamline to turbulent flow. Turbulent flow normally occurs when the Reynolds number, defined by Eq. (7-2-30), exceeds a value of 2000. To avoid turbulence effects it is customary to work at values of $\mathrm{Re} \ll 2000$.

7-3-3 Flow between Rotating Coaxial Cylinders

For flow between coaxial cylinders the motion will be referred to cylindrical polar coordinates (r, θ, z) such that $r = 0$ is the common axis of the cylinders and z is measured vertically upward. The fluid is confined between boundary surfaces which are taken to be at $r = r_1, r = r_2$ where r_2 and r_1 are the radii of the inner and outer cylinders respectively. Consideration of the unsteady flow properties of fluids will be restricted to a particular type of slow viscous flow which is two dimensional and axially symmetric. For an incompressible fluid, the velocity components which satisfy the equation of continuity are $(0, v\langle\theta\rangle(r), 0)$. Each elementary cylindrical shell of the fluid rotates as a rigid body about the z axis. For these conditions the components of the rate of deformation at any point (r, θ, z) are

$$[\mathbf{D}] = \begin{bmatrix} 0 & \dfrac{r}{2}\dfrac{d}{dr}\left(\dfrac{v\langle\theta\rangle}{r}\right) & 0 \\ \dfrac{r}{2}\dfrac{d}{dr}\left(\dfrac{v\langle\theta\rangle}{r}\right) & 0 & 0 \\ 0 & 0 & 0 \end{bmatrix}$$

and the corresponding components of the stress are

$$[\mathbf{T}] = \begin{bmatrix} -P & T\langle r\theta\rangle & 0 \\ T\langle\theta r\rangle & -P & 0 \\ 0 & 0 & -P \end{bmatrix}$$

where all quantities involving $v\langle\theta\rangle(r, t)$ and $T\langle r\theta\rangle$ are independent of z.

In terms of the shear stress $T\langle r\theta\rangle$ the general equations of motion for the flow of an incompressible viscous fluid in the θ direction reduce to

$$\rho\frac{\partial v\langle\theta\rangle}{\partial t} = \frac{1}{r^2}\frac{\partial}{\partial r}(r^2 T\langle r\theta\rangle) \qquad (T\langle r\theta\rangle = T\langle\theta r\rangle) \tag{7-3-13}$$

The laws of motion also require that

$$\rho \frac{v\langle\theta\rangle^2}{r} = \frac{\partial P}{\partial r} \tag{7-3-14}$$

and

$$\rho g = \frac{\partial P}{\partial z} \tag{7-3-15}$$

For the flow of a newtonian fluid the shear stress is related to the velocity $v\langle\theta\rangle$ at some intermediate radius r by the relation

$$T\langle\theta r\rangle = \eta\left(\frac{\partial v\langle\theta\rangle}{\partial r} - \frac{v\langle\theta\rangle}{r}\right) = \eta r \frac{\partial \omega\langle\theta\rangle}{\partial r} \tag{7-3-16}$$

where $\omega\langle\theta\rangle$ is the angular velocity of an element of the fluid at some intermediate radius r. Equation (7-3-16) is the rheological equation of state for the flow of a newtonian fluid confined between coaxial cylinders and is considered to define the dynamic coefficient of viscosity η for all time t of shearing. For a newtonian fluid the equation (7-3-13) of motion in terms of $T\langle r\theta\rangle$, combined with the equation (7-3-16) of state, gives the Navier–Stokes form of the equation of motion

$$\frac{1}{\nu}\frac{\partial v\langle\theta\rangle}{\partial t} = \frac{\partial^2 v\langle\theta\rangle}{\partial r^2} + \frac{1}{r}\frac{\partial v\langle\theta\rangle}{\partial r} - \frac{v\langle\theta\rangle}{r^2} \tag{7-3-17}$$

for flow in the θ direction.

Flow between rotating coaxial cylinders of the form of immediate interest is thus seen to be a helical flow for which there is no driving force on the fluid in the axial direction.

A practical system for use in characterizing the flow properties of non-newtonian liquids consists of two vertical coaxial cylinders, the outer cylinder rotating at a given angular velocity $\omega_1(t)$, corresponding to some time t of shearing. The motion of the inner cylinder is resisted by some form of torque-measuring device such as a flat spiral spring or a torsion suspension wire. Examples of this type of coaxial cylinder viscometer have been described by Oldroyd, Strawbridge, and Toms (1951) and Billington (1965). In this type of instrument there is an upper free surface of the material under test, but there must be some change in the type of flow toward the bottom of the apparatus, though this end effect can be reduced to negligible proportions subject to the limitation imposed by the inertia of the liquid. Consider initially that all end effects can be neglected, so that any given element of the liquid can be taken to rotate in a circular orbit about the common axis of the cylinders, the angular velocity of any given element depending upon its radial distance from the common axis.

The damping of the inner cylinder can be considered to arise from the viscous property of the fluid and from such effects as the resistance to motion of the inner cylinder arising from the suspension unit carrying the inner cylinder. For the condition that all damping effects are simply additive and proportional to the angular velocity $\dot{\theta}_2$ of the inner cylinder, the total torque is given by

$$I\ddot{\theta}_2 = \Gamma_\eta(t) - \beta_e\dot{\theta}_2 - \Gamma_0\theta_2(t) \tag{7-3-18}$$

where I is the moment of inertia of the inner cylinder, $\theta_2(t)$ is the angular displacement of the inner cylinder at time t, Γ_0 is the restoring couple per unit angular displacement, β_e is the instrument damping factor, and $\Gamma_\eta(t)$ arises from the viscous effect.

For steady conditions of flow, the outer cylinder rotates at a constant angular velocity ω_1, the inner cylinder remaining stationary. Equation (7-3-17), expressed in terms of the angular velocity $\omega\langle\theta\rangle$ of an element of the liquid at some intermediate radius r after a long time of shearing ($t = \infty$), reduces to

$$\frac{1}{r^2}\frac{\partial}{\partial r}\left(r^3\frac{\partial\omega\langle\theta\rangle}{\partial r}\right) = 0 \tag{7-3-19}$$

which can be integrated, subject to the boundary conditions ($r = r_1, \omega\langle\theta\rangle = \omega_1$) and ($r = r_2, \omega\langle\theta\rangle = 0$), to give

$$\omega\langle\theta\rangle = \frac{r_1^2}{r^2}\frac{r^2 - r_2^2}{r_1^2 - r_2^2}\omega_1 = g(r)\omega_1 \tag{7-3-20}$$

The flow rate $D\langle r\theta\rangle$ is obtained from Eqs. (7-3-16) and (7-3-20) in the form

$$2D\langle r\theta\rangle = \frac{2r_1^2r_2^2}{r^2(r_1^2 - r_2^2)}\omega_1 = h_r\omega_1 = \dot{\gamma}\langle r\theta\rangle \tag{7-3-21}$$

where for rotational motion $\dot{\gamma}\langle r\theta\rangle$ is referred to as the rate of shear at some intermediate radius r.

With the inner cylinder stationary ($t = \infty$), the shear stress can be expressed in the form

$$T\langle r\theta\rangle(\infty) = \frac{\Gamma_\eta(\infty)}{2\pi r^2 H} \qquad \text{at } t = \infty \tag{7-3-22}$$

where H is the height to which the inner cylinder is immersed and $T\langle r\theta\rangle(\infty)$ is the shear stress at some intermediate radius r. The relation of Eq. (7-3-22) follows from the vanishing of the total couple on the material at any intermediate radius r and at $r = r_1, r = r_2$.

In practice the working equations for a newtonian liquid are obtained from Eqs. (7-3-21) and (7-3-22) in the form

$$2[D\langle r\theta\rangle(\infty)]_{r=r_2} = \frac{2r_1^2}{r_1^2 - r_2^2}\omega_1 - h_2\omega_1 = \dot{\gamma}_2(\infty) \tag{7-3-23}$$

and

$$[T\langle r\theta\rangle]_{r=r_2} = \frac{\Gamma_\eta(\infty)}{2\pi r_2^2 H} \equiv \tau_2(\infty) \tag{7-3-24}$$

where $\tau_2(\infty)$ and $\dot{\gamma}_2(\infty)$ are the shear stress and the rate of shear respectively, both quantities being measured at the wall of the inner cylinder, i.e., for $r = r_2$.

For the type of instrument in which the inner cylinder rotates through a measurable angle $\theta_2(\infty)$, such that $\Gamma_\eta(\infty) = \Gamma_0\theta_2(\infty)$, the deflection $\theta_2(\infty)$ varies with ω_1 according to the relation

$$\theta_2(\infty) = \lambda\omega_1 \tag{7-3-25}$$

where

$$\lambda = \eta\xi \tag{7-3-26}$$

and

$$\xi = \frac{4\pi r_1^2 r_2^2 H}{\Gamma_0(r_1^2 - r_2^2)} \tag{7-3-27}$$

is an instrument factor.

A similar analysis can be performed for the condition in which the outer cylinder is stationary and the inner cylinder rotates with an angular velocity $\omega_2(\infty)$.

The laminar shear flow may become unstable at high rotational speeds, the effect being enhanced for the case of the inner cylinder rotating and the outer cylinder fixed. This arises because laminar flow in the system considered above is strongly stabilized by centrifugal forces. Motion of an element of the fluid from an outer layer toward the axis of rotation is opposed because the centrifugal force on it is greater than on an element of the fluid nearer the axis of rotation. The outward movement of such an element of fluid is also resisted because of the higher centrifugal force on particles it would have to replace. For the system in which the outer cylinder is fixed and the inner cylinder is rotating the centrifugal force tends to introduce instability. The system in which the outer cylinder rotates, the inner cylinder being fixed, has been studied by Schlichting (1955); the transition Reynolds number, defined as $(\omega_1 r_1^2 \rho/\eta)_{\text{trans}}$, reaches a minimum of about 50,000 when $(1 - r_2/r_1)$ is about 0.05. Systems with rotating inner cylinders have been extensively studied by Taylor (1923, 1935, 1936), who predicted that when the inner cylinder is rotated and the outer cylinder is fixed, there should be a critical rotation speed, which can be calculated, above which the flow breaks down into a series of toroidal vortices filling the annular gap. These predictions were experimentally confirmed by Taylor.

For flow in the coaxial cylinder viscometer, the temperature rise $\theta - \theta_0$ due to viscous heating may be estimated as follows. The energy equation can be obtained in the form

$$0 = k\frac{1}{r}\frac{d}{dr}\left(r\frac{d\theta}{dr}\right) + \eta\left(r\frac{d}{dr}\omega\langle\theta\rangle\right)^2 \tag{7-3-28}$$

where k is the thermal conductivity of the fluid which, though it may vary with the rate of shear, will be regarded as constant for the present discussion and r, the radius, is here to be distinguished from the energy production (r), which is taken to be zero. Substituting for $\omega\langle\theta\rangle$ in Eq. (7-3-28) from Eq. (7-3-20) gives

$$0 = k\frac{1}{r}\frac{d}{dr}\left(r\frac{d\theta}{dr}\right) + 4\eta\omega_1^2\frac{r_1^4 r_2^4}{r^4(r_1^2 - r_2^2)^2}$$

Integrating twice with respect to r gives

$$\frac{d\theta}{dr} = \frac{2\eta\omega_1^2}{k}\frac{r_1^4 r_2^4}{(r_1^2 - r_2^2)^2}\frac{1}{r^3} + \frac{C}{r}$$

and
$$\theta = -\frac{\eta}{k}\frac{r_1^4 r_2^4}{(r_1^2-r_2^2)^2}\frac{\omega_1^2}{r^2} + C\ln r + D$$

where the constants of integration C and D are determined by the boundary conditions ($r = r_2, d\theta/dr = 0$) and ($r = r_1, \theta = \theta_0$) to give

$$\theta - \theta_0 = -\frac{\eta\omega_1^2}{k}\left[\frac{r_1^2-r^2}{r^2 r_1^2} - \frac{2}{r_2^2}\ln\left(\frac{r_1}{r}\right)\right]\frac{r_1^4 r_2^4}{(r_1^2-r_2^2)^2} \tag{7-3-29}$$

Hence the maximum temperature rise is

$$\theta_{\max} - \theta_0 = -\eta\frac{\omega_1^2}{2k}h_2 r_2^2\left[1 - h_2\ln\left(\frac{r_1}{r_2}\right)\right] \tag{7-3-30}$$

which occurs at the surface of the inner cylinder.

The preceding discussion of the steady-state flow properties of a liquid using a coaxial cylinder viscometer assumed characteristics of an infinitely long cylinder in that no account was taken of the viscous drag on the end of the inner cylinder or of the viscous drag over the inner surface of the base of the outer cylinder. Various methods of reducing the end effects have been employed.

Attempts have been made to eliminate end effects by the use of guard rings, as described by Couette (1890) and by Hatschek (1913). They consist of stationary continuations of the inner cylinder of the same diameter which are in close proximity to its ends, the object being to prevent movement of the liquid adjacent to the ends of the cylinder. In principle, they reduce the end effects to negligible proportions but introduce severe restrictions into the construction and operation of such instruments. Cylinder end effects can be eliminated by way of two completely separate measurements of the deflection of the inner cylinder, one with the inner cylinder immersed to a height H_1 and a repeat experiment with the inner cylinder immersed to a height H_2. This method of eliminating end effects has been employed, for example, by Oldroyd, Strawbridge, and Toms (1951). An arrangement to overcome end effects which is of particular interest is that devised by Mooney and Ewart (1934) in which both the inner and outer cylinders have conical extensions designed to give the same shearing conditions in the space between the cones as in the annulus. End effects are considerably reduced using coaxial cylinders which are sealed by mercury at their lower edges (Billington, 1965).

7-3-4 Flow between a Cone and Plate

The geometry of the cone and plate is shown in Fig. 7-1, the fluid being contained between the cone, which has a large apical angle, and the flat surface of the plate. For flow between the cone and plate the motion will be referred to the spherical polar coordinates (r, θ, ϕ) with the z axis measured vertically upward and coincident with the common axis of the plate and cone. The flow between the cone and plate will be assumed to have physical components of velocity $(0, 0, v\langle\phi\rangle(r, \theta))$

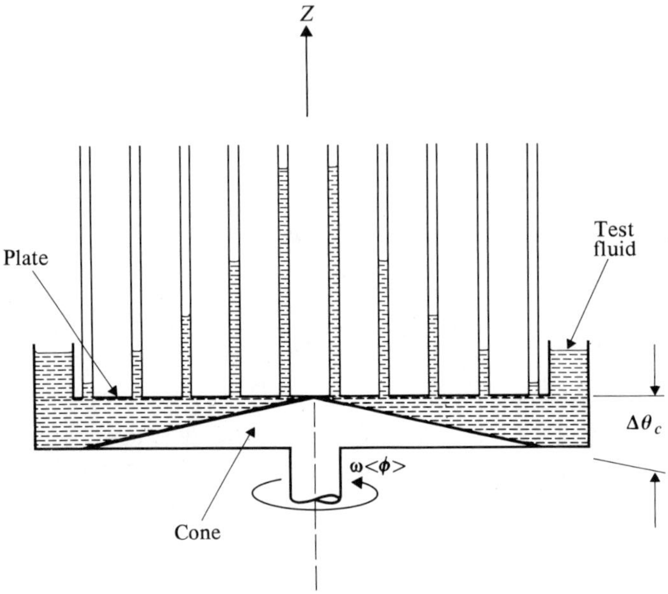

Figure 7-1 Cone and plate system.

which satisfy the equation of continuity for an incompressible fluid. The only nonvanishing components of **T** are $T\langle r\phi\rangle$ and $T\langle\theta\phi\rangle$. However, the system can be simplified by specifying the angular velocity

$$\omega\langle\phi\rangle = \frac{v\langle\phi\rangle}{r\sin\theta}$$

of an element of the fluid about the common axis of the cone and plate to be independent of r. Subject to this condition it is evident that $T\langle r\phi\rangle = 0$, such that the only components of the stress are

$$[\mathbf{T}] = \begin{bmatrix} -P & 0 & 0 \\ 0 & -P & T\langle\theta\phi\rangle \\ 0 & T\langle\phi\theta\rangle & -P \end{bmatrix}$$

where the $T\langle\theta\phi\rangle$ are functions only of θ. The corresponding components of the rate of deformation are

$$[\mathbf{D}] = \begin{bmatrix} 0 & 0 & 0 \\ 0 & 0 & \dfrac{1}{2}\dfrac{\sin\theta}{r}\dfrac{\partial}{\partial\theta}\left(\dfrac{v\langle\phi\rangle}{\sin\theta}\right) \\ 0 & \dfrac{1}{2}\dfrac{\sin\theta}{r}\dfrac{\partial}{\partial\theta}\left(\dfrac{v\langle\phi\rangle}{\sin\theta}\right) & 0 \end{bmatrix}$$

This particular state of flow is a unidirectional shear flow in which the shearing surfaces are the cones θ = constant, each such fluid cone rotating rigidly about the cone axis $\theta = 0$ with the angular velocity $\omega\langle\phi\rangle$.

In terms of the shear stress the general equations of motion for the flow of an incompressible viscous fluid reduce to

$$-\rho r(\omega\langle\phi\rangle)^2 \sin^2\theta = -\frac{\partial P}{\partial r} - \rho g \cos\theta \tag{7-3-31}$$

$$-\rho r(\omega\langle\phi\rangle)^2 \sin\theta\cos\theta = -\frac{1}{r}\frac{\partial P}{\partial\theta} + \rho g \sin\theta \tag{7-3-32}$$

$$\rho r \sin\theta \frac{\partial\omega\langle\phi\rangle}{\partial t} = \frac{1}{r}\frac{\partial T\langle\theta\phi\rangle}{\partial\theta} + 2\frac{\cot\theta}{r}T\langle\theta\phi\rangle \qquad (T\langle\theta\phi\rangle = T\langle\phi\theta\rangle) \tag{7-3-33}$$

For steady-state conditions of flow, Eq. (7-3-33) can be integrated to give

$$T\langle\theta\phi\rangle = \frac{B}{\sin^2\theta} \tag{7-3-34}$$

where B is the constant of integration. From Eq. (7-3-34) the difference in shear stress $\Delta T\langle\theta\phi\rangle$ between the boundaries of the cone $(\theta = \frac{1}{2}\pi - \Delta\theta_c)$ and plate $(\theta = \frac{1}{2}\pi)$ can be expressed in the form

$$\Delta T\langle\theta\phi\rangle = T\langle\tfrac{\pi}{2}\phi\rangle \tan^2\Delta\theta_c \tag{7-3-35}$$

from which can be obtained the following table:

$\Delta\theta_c$, degrees	$(\Delta T\langle\theta\phi\rangle / T\langle\frac{\pi}{2}\phi\rangle)$, %
$\frac{1}{2}$	0.0076
1	0.0306
2	0.1218
4	0.4886

Thus, for small angles $\Delta\theta_c$ ($< \sim 4°$) the variation in shear stress is less than $\frac{1}{2}$ percent, which is the feature of particular interest in the cone and plate principle.

For the flow of a newtonian fluid the shear stress $T\langle\theta\phi\rangle$ is related to the velocity $v\langle\phi\rangle$ at some intermediate angle θ by the relation

$$T\langle\theta\phi\rangle = \eta \sin\theta \frac{d\omega\langle\phi\rangle}{d\theta} \tag{7-3-36}$$

Substituting for $T\langle\theta\phi\rangle$ in Eq. (7-3-33) from Eq. (7-3-36) gives for a newtonian fluid the Navier–Stokes form of the equation of motion:

$$\frac{1}{\nu}\frac{\partial\omega\langle\phi\rangle}{\partial t} = \frac{1}{r^2\sin\theta}\frac{\partial}{\partial\theta}\left(\sin\theta\frac{\partial\omega\langle\phi\rangle}{\partial\theta}\right) + 2\frac{\cot\theta}{r^2}\frac{\partial\omega\langle\phi\rangle}{\partial\theta}$$

The simplest form of the cone and plate viscometer is that shown in Fig. 7-1. In operation the cone can rotate, the resulting viscous torque on the cone being recorded. Examples of this type of instrument have been described by McKennell (1956) and Jobling and Roberts (1959).

Introducing the condition associated with a small angle $\Delta\theta_c$, the equations (7-3-31) to (7-3-33) of motion for steady flow of an incompressible viscous fluid reduce to

$$-\rho r\omega\langle\phi\rangle^2 = -\frac{\partial P}{\partial r} - \rho g \cos\theta \tag{7-3-37}$$

$$\frac{1}{r}\frac{\partial P}{\partial \theta} = \rho g \sin\theta \tag{7-3-38}$$

and

$$\frac{\partial}{\partial \theta} T\langle\theta\phi\rangle(\infty) = 0 \tag{7-3-39}$$

From the form of Eq. (7-3-39) it is concluded that the shear stress is uniform.

Thus, from Eq. (7-3-39), it follows for the flow of a newtonian-type fluid that Eq. (7-3-39) reduces to

$$T\langle\theta\phi\rangle = \eta \sin\theta \frac{d\omega\langle\phi\rangle}{d\theta} = \text{constant} \tag{7-3-40}$$

which can be integrated, subject to the boundary conditions ($\theta = \frac{1}{2}\pi$, $\omega\langle\phi\rangle = 0$) and ($\theta = \frac{1}{2}\pi - \Delta\theta_c$, $\omega\langle\phi\rangle = \omega_c$) where ω_c is the rotational speed of the cone, to give

$$\frac{\omega\langle\phi\rangle}{\omega_c} = \frac{\frac{1}{2}\pi - \theta}{\Delta\theta_c} \tag{7-3-41}$$

from which it is evident that, at constant radius, the angular velocity varies linearly with the angular position θ. From Eqs. (7-3-40) and (7-3-41) it follows that the rate of shear has the form

$$\dot{\gamma}\langle\theta\phi\rangle = \sin\theta \frac{d\omega\langle\phi\rangle}{d\theta} \approx -\frac{\omega_c}{\Delta\theta_c} \tag{7-3-42}$$

Except near the edge of the cone, the velocity field has been shown by Jobling and Roberts (1959) to be well approximated by one for which the physical components of velocity are $(0, 0, v\langle\phi\rangle)$ and for which Eq. (7-3-41) is valid, provided that $\Delta\theta_c$ is $< \sim 4°$.

The viscous torque on the cone is

$$\begin{aligned}\Gamma_\eta(\infty) &= \int_0^{r_c} 2\pi r^2 T\langle\theta\phi\rangle \cos\Delta\theta_c \sin\left(\tfrac{1}{2}\pi - \Delta\theta_c\right) dr \\ &= \tfrac{2}{3}\pi r_c{}^3 B = \Gamma_0 \phi_c(\infty)\end{aligned}$$

where the relation of Eq. (7-3-34) has been used for the shear stress, $\phi_c(\infty)$ is the steady displacement of the cone, and Γ_0 is the restoring couple per unit angular displacement; r_c is the value of r for $\theta = \frac{1}{2}\pi - \Delta\theta_c$. Hence the shear stress $T\langle\theta\phi\rangle \equiv \tau_c(\infty)$ measured at the surface of the cone is given by

$$\tau_c(\infty) = \frac{3\Gamma_\eta(\infty)}{2\pi r_c^{\ 3}} = \frac{3\Gamma_0}{2\pi r_c^{\ 3}}\phi_c(\infty) \tag{7-3-43}$$

Eqs. (7-3-42) and (7-3-43) being the working equations for a newtonian fluid.

For the flow of a newtonian fluid

$$\tau_c(\infty) = \eta\dot{\gamma}_c(\infty)$$

and substituting for $\dot{\gamma}_c(\infty)$ and $\tau_c(\infty)$ from Eqs. (7-3-42) and (7-3-43) gives

$$\phi_c(\infty) = \lambda_c\omega_c \tag{7-3-44}$$

where

$$\lambda_c = \eta\xi_c \tag{7-3-45}$$

and

$$\xi_c = -\frac{2\pi r_c^{\ 3}}{3\Gamma_0\Delta\theta_c} \tag{7-3-46}$$

is an instrument factor, to be compared with Eq. (7-3-27).

As is to be expected, the steady-state flow of a newtonian fluid between a cone and plate is described by a set of equations, that is, Eqs. (7-3-44) to (7-3-46), which are of the same form as Eqs. (7-3-25) to (7-3-27) which characterize the flow of a newtonian fluid between rotating coaxial cylinders.

In addition to the uniformity of shearing conditions, advantages of this type of viscometer are the relative simplicity of operation, ease of filling and cleaning, together with the use of a small sample. One serious disadvantage is that some materials, particularly those which exhibit the flow properties associated with dilatant fluids, tend to be ejected out of the gap (Higginbotham, 1950; McKennell, 1953; Pollett, 1955; Bantoft, 1959; Cox, 1962; Bauer and Finkelstein, 1963; Hutton, 1963). At high speeds of rotation a *secondary flow* has been demonstrated, even for the flow of newtonian fluids, by Cox (1962) using a dye technique. The direction of the *secondary flow* is occasionally reversed in the case of the flow of nonnewtonian polymer solutions (Hoppmann and Miller, 1963; Hoppmann and Baronet, 1964).

For the cone and plate viscometer the problem of temperature rise due to viscous heating differs from that occurring in the coaxial cylinder viscometer insofar as the temperature in the cone and plate viscometer is described by a partial differential equation with r and θ as independent variables. There is also the additional complication of three boundaries to consider, consisting of the surface of the cone, the surface of the plate, and the surface of the test fluid exposed at the rim of the instrument.

Considering only the steady state and retaining the condition of small $\Delta\theta_c$, the energy equation for this case can be obtained in the form

$$k\nabla^2\theta_{cp} + \text{tr}\,(\mathbf{TD}) = 0 \tag{7-3-47}$$

where k is the thermal conductivity of the fluid and θ_{cp} is the temperature. The boundary conditions are such that the cone and plate surfaces are held at a fixed temperature $\theta_{cp}(0)$. Also, for all temperatures θ_{cp}, $\partial\theta_{cp}/\partial r = 0$ for $r = r_c$; i.e., no heat is transferred across the exposed surface of the test fluid. This last condition is difficult to satisfy experimentally, and as such there will be a small rate of heat transfer across the exposed surface of the fluid; thus the calculated temperature will tend to be a little too high.

Bird and Turian (1962), using variational methods to find an approximate solution of Eq. (7-3-47) subject to the above boundary conditions, have obtained for the maximum temperature rise θ_{max} the relation

$$(\theta_{\text{max}} - \theta(0))_{cp} = \frac{3\Gamma_\eta(\infty)\omega_c^2}{16\pi k r_c \dot{\gamma}\langle\theta\phi\rangle} \tag{7-3-48}$$

where the rate of shear $\dot{\gamma}\langle\theta\phi\rangle$ is given by Eq. (7-3-42).

7-3-5 Torsional Flow—The Parallel Plate System

For torsional flow between parallel circular plates the motion will be referred to cylindrical polar coordinates (r, θ, z), the origin of the coordinate system being taken at the center of the lower plate and the z axis being measured vertically upwards (Fig. 7-2). The fluid is confined between boundary surfaces which are taken to be at $z = 0$ and $z = z_0$. Consideration of the flow properties of fluids will be restricted to a particular type of slow viscous flow for which two-dimensional, axially symmetric flow is prescribed with velocity components $(0, v\langle\theta\rangle(z, r, t), 0)$ which satisfy the equation of continuity for an incompressible fluid. Thus the fluid

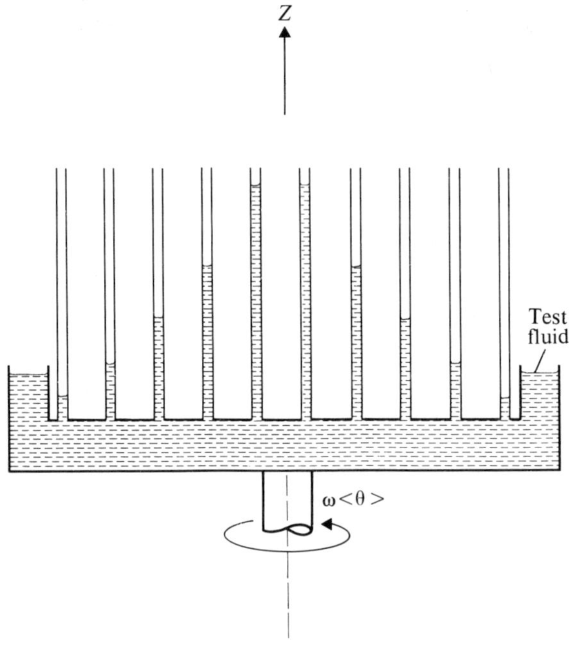

Figure 7-2 Torsional flow between parallel plates in relative motion.

is prescribed to be in a state of unidirectional shear flow in which the shearing surfaces are horizontal planes z = constant, each fluid plane rotating rigidly about the z axis with an angular velocity $\omega\langle\theta\rangle$. That is, the fluid moves in the form of elementary, coaxial disk-like laminae with axes on the axis of z, all edge effects being neglected.

With these restrictions as to the nature of the flow, in particular the neglect of edge effects at the rims of the plates, the components of the rate of deformation at any point (r, θ, z) are

$$[\mathbf{D}] = \begin{bmatrix} 0 & 0 & 0 \\ 0 & 0 & \dfrac{1}{2}\dfrac{\partial v\langle\theta\rangle}{\partial z} \\ 0 & \dfrac{1}{2}\dfrac{\partial v\langle\theta\rangle}{\partial z} & 0 \end{bmatrix}$$

and the corresponding components of the stress are

$$[\mathbf{T}] = \begin{bmatrix} -P & 0 & 0 \\ 0 & -P & T\langle\theta z\rangle \\ 0 & T\langle z\theta\rangle & -P \end{bmatrix}$$

All quantities involving $v\langle\theta\rangle(z, r, t)$ and $T\langle\theta z\rangle(t)$ are independent of θ.

In terms of the shear stress $T\langle\theta z\rangle$ the general equations of motion for the flow of an incompressible viscous fluid in the θ direction reduce to

$$\rho\frac{\partial v\langle\theta\rangle}{\partial t} = \frac{\partial T\langle\theta z\rangle}{\partial z} \qquad (T\langle\theta z\rangle = T\langle z\theta\rangle) \tag{7-3-49}$$

The laws of motion also require that

$$\rho\frac{v\langle\theta\rangle^2}{r} = \frac{\partial P}{\partial r} \tag{7-3-50}$$

and

$$-\rho g = \frac{\partial P}{\partial z} \tag{7-3-51}$$

For the flow of a newtonian fluid the shear stress $T\langle\theta z\rangle(t)$ is related to the velocity $v\langle\theta\rangle$ by the relation

$$T\langle\theta z\rangle(t) = \eta\frac{\partial v\langle\theta\rangle}{\partial z} = \eta\dot{\gamma}\langle\theta z\rangle(t) \tag{7-3-52}$$

where $\dot{\gamma}\langle\theta z\rangle(t)$ is the rate of shear. Equation (7-3-52) is the rheological equation of state for the flow of a newtonian fluid between parallel circular plates in relative angular rotation, and is considered to define the dynamic coefficient of viscosity η for all time t of shearing for the particular condition of flow in which the edge effects are neglected.

Substituting Eq. (7-3-52) into Eq. (7-3-49) gives the Navier–Stokes form of the equation of motion

$$\frac{1}{\nu}\frac{\partial v\langle\theta\rangle}{\partial t} = \frac{\partial^2 v\langle\theta\rangle}{\partial z^2} \tag{7-3-53}$$

for flow in the θ direction, neglecting edge effects.

For steady torsional flow, Eq. (7-3-53) reduces to

$$\frac{\partial^2 v\langle\theta\rangle}{\partial z^2} = 0 \tag{7-3-54}$$

which can be integrated, subject to the boundary conditions ($z = 0, \omega\langle\theta\rangle = 0$) and ($z = z_0, \omega\langle\theta\rangle = \omega_p$) where $\omega\langle\theta\rangle$ is the angular velocity of an element of the fluid about the z axis at a distance z from the lower plate and z_0 is the separation of the plates, the resulting velocity field being obtained in the form

$$v\langle\theta\rangle = r\omega\langle\theta\rangle = r\omega_p\frac{z}{z_0} \tag{7-3-55}$$

from which is obtained the rate of flow

$$\dot{\gamma}_p(\infty) = r\frac{\omega_p}{z_0} \tag{7-3-56}$$

It is evident from Eq. (7-3-56) that the rate of shear is independent of z but varies with r.

The viscous torque $\Gamma_\eta(\infty)$ acting at the upper plate is

$$\Gamma_\eta(\infty) = \int_0^{r_p} 2\pi r^2 \eta r\frac{\omega_p}{z_0}\,dr$$

Thus,

$$\Gamma_\eta(\infty) = \frac{\pi r_p{}^4}{2z_0}\eta\omega_p \tag{7-3-57}$$

where r_p is the radius of the plates.

7-4 SPECIAL VISCOMETRIC FLOWS: II. UNSTEADY MOTION

7-4-1 Forced Oscillatory Motion

The oscillatory motion is produced by subjecting the fluid to periodic shearing by maintaining, for example, the outer cylinder in forced harmonic oscillation, the resulting amplitude $\theta_2(t)$ of the inner cylinder together with its motion being recorded. At the surface of the outer cylinder the motion can be represented by the relation $\theta_1(t) = \theta_1 \cos\omega t$ since the fluid is following the forced motion of the cylinder which oscillates with frequency ω and has amplitude θ_1. The inner cylinder oscillates with the same frequency, but with different phase, since it is subject to a torque due to the viscous drag of the liquid. Thus the motion of the inner cylinder can be represented by the relation $\theta_2(t) = \theta_2 \cos(\omega t - \delta)$ where θ_2 is the amplitude of the inner cylinder for a given frequency of oscillation of the outer cylinder. It is convenient to express $\theta_1(t)$ and $\theta_2(t)$ as the real parts of complex quantities, such that

$$\theta_1(t) = re\theta_1 \exp j\omega t \tag{7-4-1}$$

and

$$\theta_2(t) = re\theta_2 \exp j(\omega t - \delta) \tag{7-4-2}$$

Provided the calculations involve only linear operations on the displacements, the sign *re* may be omitted, thus proceeding as if θ were complex. With this simplification the angular displacement of an element of a newtonian fluid, at some intermediate radius r, can be expressed in the form

$$\theta_r(t) = F(r) \exp(j\omega t) \tag{7-4-3}$$

where $F(r)$ incorporates an appropriate phase factor. Combination of Eqs. (7-3-17) and (7-4-3) gives the relation

$$\frac{d^2F}{dr^2} + \frac{3}{r}\frac{dF}{dr} + \alpha^2 F = 0 \tag{7-4-4}$$

which has a solution of the form (see, for example, Sneddon, 1951)

$$F(r) = \frac{1}{r}[AJ_1(\alpha r) + BG_1(\alpha r)] \tag{7-4-5}$$

where $J_1(\alpha r)$ and $G_1(\alpha r)$ are Bessel functions of complex argument, such that

$$\alpha^2 = -j\frac{\omega}{\nu} \tag{7-4-6}$$

and A and B are constants to be determined from the boundary conditions.

Equations (7-3-18), (7-4-2), and (7-4-3) can be combined to give the relation

$$-j\frac{\chi}{\lambda} h_2 \theta_2 \exp(-j\delta) = \left[r\frac{dF}{dr}\right]_{r=r_2} \qquad (\beta_e = 0) \tag{7-4-7}$$

where

$$\Gamma_\eta(t) = 2\pi r_2^2 H\eta \left[r\frac{\partial\omega\langle\theta\rangle}{\partial r}\right]_{r=r_2} \tag{7-4-8}$$

$$\chi = \frac{f_N^2 - f^2}{2\pi f f_N^2} \qquad (\omega = 2\pi f) \tag{7-4-9}$$

and

$$f_N = \frac{1}{2\pi}\left(\frac{\Gamma_0}{I}\right)^{1/2} \tag{7-4-10}$$

is the natural undamped frequency of oscillation of the inner cylinder. The quantity h_2 is defined by Eq. (7-3-23) and

$$\lambda = \eta\xi = \frac{4\pi r_1^2 r_2^2 H}{\Gamma_0(r_1^2 - r_2^2)}\eta \tag{7-4-11}$$

where λ and ξ are the same quantities as appear in Eqs. (7-3-26) and (7-3-27) respectively.

Inserting the boundary conditions into Eq. (7-4-5) and combining with Eq. (7-4-7) gives the relation for the amplitude ratio ψ $(=\theta_2/\theta_1)$ in the form

$$\frac{\exp(j\delta)}{\psi} = \frac{(\lambda - j\chi h_2)[J_1(\alpha r_1)G_1(\alpha r_2) - J_1(\alpha r_2)G_1(\alpha r_1)]}{\alpha\lambda r_1[J_1'(\alpha r_2)G_1(\alpha r_2) - J_1(\alpha r_2)G_1'(\alpha r_2)]} - \frac{r_2}{r_1}\frac{[J_1(\alpha r_1)G_1'(\alpha r_2) - J_1'(\alpha r_2)G_1(\alpha r_1)]}{[J_1'(\alpha r_2)G_1(\alpha r_2) - J_1(\alpha r_2)G_1'(\alpha r_2)]} \tag{7-4-12}$$

where $J_1'(\alpha r_2)$ is the value of $dJ_1(\alpha r)/d(\alpha r)$ for $r = r_2$, and similarly for $G_1'(\alpha r_2)$. It follows from Eq. (7-4-12) that the slope of the (f, ψ) curve at the origin is given by

$$\lim_{f\to 0}\frac{\psi(f)}{f} = \frac{8\pi^2 r_1^2 r_2^2 H}{\Gamma_0(r_1^2 - r_2^2)}\eta = 2\pi\lambda \tag{7-4-13}$$

where λ is given by Eq. (7-4-11).

The relation of Eq. (7-4-12) is an exact solution for the amplitude ratio; to utilize the relation it is necessary to recognize three arbitrary regions defined in terms of the magnitude of the Bessel functions. The three regions correspond to small, intermediate, and large arguments of the Bessel functions; the intermediate region is defined by Markovitz (1952) as being that for which neither the power series valid for small arguments nor the asymptotic expansion valid for large arguments can be employed.

For small arguments, Markovitz (1952) has obtained the expression

$$\frac{\exp(j\delta)}{\psi} = \left(1 - j\frac{\chi}{\lambda}\right) - \alpha^2\frac{r_1^2}{2h_2^2} + \alpha^4\left\{\frac{\chi h_2 r_2^2}{8\rho\xi\omega}\left[2\ln\left(\frac{r_2}{r_1}\right) + \frac{r_1^2 + r_2^2}{r_2^2 h_2}\right] + \frac{1}{192 r_1^2}\left[(r_1^2 - r_2^2)(r_1^4 - 5r_1^2 r_2^2 - 2r_2^4) + 12 r_1^2 r_2^4 \ln\left(\frac{r_1}{r_2}\right)\right]\right\} \tag{7-4-14}$$

where terms in α^6 and all higher terms have been neglected.

For large arguments of the Bessel functions Oldroyd (1951) has obtained the relation

$$\frac{\exp(j\delta)}{\psi} = \left[\left(\frac{r_2}{r_1}\right)^{3/2} - \frac{3}{8\pi(r_1 + r_2)\sqrt{r_1 r_2}}\left(\frac{\chi}{\xi\rho f}\right)\right]\cos\alpha(r_1 - r_2) + \left(\frac{r_2}{r_1}\right)^{3/2}\left[\frac{15r_1 - 3r_2}{8r_1 r_2\alpha} + \frac{15}{16 r_2^{3/2} r_1^{3/2}\alpha^3} + \left(\alpha + \frac{3}{8r_1 r_2\alpha} - \frac{45}{128 r_1^2 r_2^2\alpha^3}\right)\frac{\Gamma_0\chi}{4\pi^2\rho r_2^3 H f}\right]\sin\alpha(r_1 - r_2) \tag{7-4-15}$$

which, using the appropriate form of α, is the relation used by Oldroyd, Strawbridge, and Toms (1951) in the interpretation of their results for polymethylmethacralate.

It is evident from Eqs. (7-4-14) and (7-4-15) that considerable simplification results if both amplitude ratio and phase lag are measured in a given experiment, since in this case it would be possible to derive α as a direct function of the frequency ω (Walters, 1960, 1961). Instruments have been described by Markovitz et al. (1952), Cox, Nielsen, and Keeney (1957), and Billington (1965) which include facilities for measuring both amplitude ratio and phase lag.

7-4-2 Forced Oscillatory Motion—Fluid Inertia Neglected

Neglecting the inertia of the fluid leads to a considerable simplification of the treatment of forced oscillatory motion.

For the condition that the derivative $\partial v\langle\theta\rangle/\partial t$ can be neglected, Eq. (7-3-17) reduces to Eq. (7-3-19) which can be solved, subject to the boundary conditions $[r = r_1, \omega\langle\theta\rangle = \omega_1(t)]$ and $[r = r_2, \omega\langle\theta\rangle = \omega_2(t)]$, to give the relation

$$\omega\langle\theta\rangle = g(r)\omega_1(t) + f(r)\omega_2(t) \tag{7-4-16}$$

where

$$g(r) = 1 - f(r)$$

is given by Eq. (7-3-20).

The damping of the inner cylinder can be considered to arise from the viscous property of the fluid and from such effects as the resistance to motion of the inner cylinder arising from the suspension unit carrying the inner cylinder. For the condition that all damping effects are simply additive, and assuming that the contribution to the damping of the motion of the inner cylinder, due to all effects other than the viscous damping of the fluid, is to be proportional to the angular velocity $\dot{\theta}_2$ of the inner cylinder, the total torque can be expressed in the form

$$I\ddot{\theta}_2 = \Gamma_\eta(t) - \beta_e\dot{\theta}_2 - \Gamma_0\theta_2(t) \tag{7-4-17}$$

where β_e is the instrument damping factor. For the flow of a newtonian-type fluid Eq. (7-4-8) can be expressed by way of Eq. (7-4-16) in the form

$$\Gamma_\eta(t) = \beta(\omega_1(t) - \dot{\theta}_2) \tag{7-4-18}$$

where

$$\beta = \frac{4\pi r_1^2 r_2^2 H}{r_1^2 - r_2^2}\eta = \lambda\Gamma_0 \tag{7-4-19}$$

the choice of sign for $\dot{\theta}_2$ appearing in Eq. (7-4-18) depending upon whether, at any given instant, the cylinders are rotating in the same or opposite directions. Thus the equation of motion of the inner cylinder can be expressed in the form

$$I\ddot{\theta}_2 + (\beta_e + \beta)\dot{\theta}_2 + \Gamma_0\theta_2(t) = \beta\omega_1(t) \tag{7-4-20}$$

In connection with the shear torque $\Gamma_\eta(t)$, rotation of the cylinders in opposite directions gives rise to a neutral shear plane at some intermediate radius $r_n(t)$, for which the angular velocity $\omega_n(t) = 0$. That Eq. (7-4-20) holds for this condition of flow can be verified for a newtonian fluid by expressing Eq. (7-4-8) in the form

$$\Gamma_\eta(t) = 2\pi r_n^2 H\eta\dot{\gamma}_n(t)$$

where

$$r_n^2(t) = \frac{r_1^2 r_2^2[\omega_1(t) - \omega_2(t)]}{r_1^2\omega_1(t) - r_2^2\omega_2(t)} \tag{7-4-21}$$

and

$$\dot{\gamma}_n(t) = \left[r\frac{\partial\omega\langle\theta\rangle}{\partial r}\right]_{r=r_n} \tag{7-4-22}$$

From Eq. (7-4-1),

$$\omega_1(t) = j\omega\theta_1 \exp(j\omega t)$$

which, combined with Eq. (7-4-20), gives the equation of motion of the inner cylinder in the form

$$I\ddot{\theta}_2 + (\beta_e + \beta)\dot{\theta}_2 + \Gamma_0\theta_2(t) = j\beta\omega\theta_1 \exp(j\omega t) \tag{7-4-23}$$

for which a solution can be obtained by assuming that it is of the form of Eq. (7-4-2), the required solution being

$$\frac{\exp(j\delta)}{\psi} = \left(1 - j\frac{\chi}{\lambda}\right) + \frac{\beta_e}{\beta} \tag{7-4-24}$$

that is,

$$\theta_2^2 = \frac{\lambda^2}{\chi^2 + \lambda_0^2}\theta_1^2 \tag{7-4-25}$$

where

$$\lambda_0 = \frac{\beta + \beta_e}{\Gamma_0} = \lambda + \lambda_e \tag{7-4-26}$$

and λ and χ are given by Eqs. (7-4-11) and (7-4-9) respectively. The form of Eq. (7-4-25) indicates that the motion of the inner cylinder exhibits the phenomenon of resonance, the latter occurring for the particular condition that the applied frequency of oscillation f of the outer cylinder equals the natural undamped frequency of oscillation of the inner cylinder f_N.

The slope of the (f, θ_2) curve at the origin, obtained from Eq. (7-4-25) in the form

$$\operatorname*{Lim}_{f\to 0} \frac{\theta_2}{f\theta_1} = \frac{8\pi^2 r_2^2 r_1^2 H}{\Gamma_0(r_1^2 - r_2^2)}\eta = 2\pi\lambda \tag{7-4-27}$$

is the same as Eq. (7-4-13) which was obtained from Eq. (7-4-12).

7-4-3 Transient Motion—Fluid Inertia Neglected

For the transient motion the outer cylinder is taken to rotate at constant angular velocity ω_1, for all time t of shearing. The discussion is restricted to coaxial cylinder viscometers for which the moment of inertia I of the inner cylinder is made very small. Provided I can be made sufficiently small and the fluid inertia can be neglected, the inner cylinder will be expected to begin to rotate from the instant at which the outer cylinder is set in motion, there being no detectable time lag. That is, at $t \simeq 0$ both cylinders, together with the fluid, are prescribed to move instantaneously as a composite unit having the same angular velocity ω_1 at every point in the interval $(r_2 \leqslant r \leqslant r_1)$. With continued rotation of the outer cylinder the inner cylinder slows down; concomitant with the deceleration of the inner cylinder is the transient increase in the shear stress.

For these conditions Eq. (7-4-18) reduces to

$$\Gamma_\eta(t) = \beta(\omega_1 - \dot{\theta}_2) \tag{7-4-28}$$

Eq. (7-4-20) reducing in turn to

$$I\ddot{\theta}_2 + (\beta_e + \beta)\dot{\theta}_2 + \Gamma_0\theta_2(t) = \beta\omega_1 \tag{7-4-29}$$

which has a solution of the form

$$\Gamma_0\theta_2(t) - \beta\omega_1 = A\exp(\kappa t) \tag{7-4-30}$$

where A and κ are constants. For a small moment of inertia $(\beta_e + \beta)^2 \gg 4I\Gamma_0$ such that the only valid solution for κ is

$$\kappa \simeq -\frac{\Gamma_0}{\beta + \beta_e} = -\frac{1}{\lambda_0} \tag{7-4-31}$$

since the alternate value for κ does not satisfy the limiting conditions ($t = 0$, $\theta_2 = 0$, $\dot{\theta}_2 = \omega_1$). It also follows from the initial condition at $t = 0$ that $A = -\beta\omega_1$ which, together with the condition $[t = \infty, \Gamma_0\theta_2(\infty) = \beta\omega_1]$, leads to the relation

$$\theta_2(t) = \theta_2(\infty)\left[1 - \exp\left(\frac{-t}{\lambda_0}\right)\right] \tag{7-4-32}$$

for the dependence of $\theta_2(t)$ on the time of shearing, where

$$\lambda_0 = \lambda_e + \lambda \tag{7-4-33}$$

and λ is given by Eq. (7-4-11); λ_0 and λ_e have the same significance as the time constants appearing in Eq. (7-4-26).

7-4-4 Oscillation of a Vessel Enclosing a Liquid

For the flow of a liquid confined inside a spherical oscillating vessel, the motion will be referred to spherical polar coordinates (r, θ, ϕ), with the origin at the center of the sphere. Consideration of the unsteady flow properties of liquids will be restricted to a particular type of slow viscous flow for which three-dimensional, spherically symmetric flow is prescribed, each elementary spherical shell of the fluid rotating about the z axis. For this type of flow the velocity components are $(0, 0, v\langle\phi\rangle(r, \theta, t))$, which satisfy the equation of continuity for an incompressible fluid. The components of the rate of deformation at any point (r, θ, ϕ) are

$$[\mathbf{D}] = \begin{bmatrix} 0 & 0 & r\dfrac{d}{dr}\left(\dfrac{v\langle\phi\rangle}{r}\right) \\ 0 & 0 & 0 \\ r\dfrac{d}{dr}\left(\dfrac{v\langle\phi\rangle}{r}\right) & 0 & 0 \end{bmatrix}$$

and the corresponding components of the stress are

$$[\mathbf{T}] = \begin{bmatrix} -P & 0 & T\langle r\phi\rangle \\ 0 & -P & 0 \\ T\langle\phi r\rangle & 0 & -P \end{bmatrix}$$

All quantities involving $v\langle\phi\rangle(r, \theta, t)$ and $T\langle r\phi\rangle$ are independent of z.

In terms of the shear stress $T\langle r\phi\rangle$ the general equations of motion for the flow of an incompressible viscous fluid in the ϕ direction reduce to

$$\rho\frac{\partial v\langle\phi\rangle}{\partial t} = \frac{1}{r^2}\frac{\partial}{\partial r}(r^2 T\langle r\phi\rangle) + \frac{T\langle r\phi\rangle}{r} \tag{7-4-34}$$

The laws of motion also require that

$$-\rho\frac{v\langle\phi\rangle^2}{r} = -\frac{\partial P}{\partial r} - \rho g\cos\theta \tag{7-4-35}$$

and

$$-\rho\frac{v\langle\phi\rangle^2}{r}\cot\theta = -\frac{1}{r}\frac{\partial P}{\partial\theta} + \rho g\sin\theta \tag{7-4-36}$$

Consider a spherical vessel completely filled with liquid and which is freely oscillating about the z axis. For the flow of a newtonian liquid, the shear stress is related to the velocity $v\langle\phi\rangle$ at some intermediate radius r by the relation

$$T\langle r\phi\rangle = \eta r\frac{\partial}{\partial r}\left(\frac{v\langle\phi\rangle}{r}\right) = \eta r\frac{\partial\omega\langle\phi\rangle}{\partial r} \tag{7-4-37}$$

Substituting for $T\langle r\phi\rangle$ from Eq. (7-4-37) into Eq. (7-4-34) gives the Navier–Stokes form of the equation of motion

$$\frac{1}{\nu}\frac{\partial\omega\langle\phi\rangle}{\partial t} = \frac{\partial^2\omega\langle\phi\rangle}{\partial r^2} + \frac{4}{r}\frac{\partial\omega\langle\phi\rangle}{\partial r} \tag{7-4-38}$$

it being noted that in this form the convective term is absent. Let the periodic motion be of the form

$$\omega\langle\phi\rangle = \phi(r)\exp(\alpha t) \tag{7-4-39}$$

where

$$\alpha = -\beta + j\gamma \tag{7-4-40}$$

and

$$\beta = 2\pi\frac{\Delta}{T} \qquad \gamma = \frac{2\pi}{T} \tag{7-4-41}$$

where $2\pi\Delta$ is the logarithmic decrement between consecutive swings in the same sense and T is the period of oscillation. Substituting Eq. (7-4-39) into Eq. (7-4-38) gives

$$\frac{d^2\phi}{dr^2} + \frac{4}{r}\frac{d\phi}{dr} = b^2\phi \tag{7-4-42}$$

where

$$b = \left(\frac{\alpha}{v}\right)^{1/2} = \left(\frac{-\beta+j\gamma}{v}\right)^{1/2} \tag{7-4-43}$$

Equation (7-4-42) has a solution of the form

$$\phi = \frac{A}{r^3}[(br-1)\exp(br) + (br+1)\exp(-br)] \tag{7-4-44}$$

where the constant A is evaluated from the boundary conditions.

The torque on the surface of the sphere of radius R can be expressed in the form

$$\Gamma_\eta(t) = \tfrac{8}{3}\pi\eta R^4\left(\frac{d\phi}{dr}\right)_{r=R}\exp(\alpha t)$$

and since at $[r = R, \omega\langle\phi\rangle = \omega_0(t) = d\phi/dt]$ the torque $\Gamma_\eta(t)$ can be expressed in the form

$$\Gamma_\eta(t) = \tfrac{8}{3}\pi\eta\frac{R^4}{\Phi}\left(\frac{d\phi}{dr}\right)_{r=R}\frac{d\phi}{dt}$$

where $[r = R, \phi(R) = \Phi]$. From Eq. (7-4-44),

$$\Gamma_\eta(t) = L\frac{d\phi}{dt} \tag{7-4-45}$$

where

$$L - \tfrac{8}{3}\pi\eta R^4\left[\frac{(bR)^2 - 3bR\coth bR + 3}{bR\coth bR - 1}\right] \tag{7-4-46}$$

By comparison with Eq. (7-3-18) it is evident that the equation of motion of the oscillating sphere is

$$I\ddot{\phi} + L\dot{\phi} + \Gamma_0\phi = 0 \tag{7-4-47}$$

where I is the moment of inertia of the filled sphere, Γ_0 is the restoring couple per unit angular displacement of the sphere, and L is given by Eq. (7-4-46). From Eq. (7-4-47) it follows that the possible values of α are given by

$$I\alpha^2 + L\alpha + \Gamma_0 = 0 \tag{7-4-48}$$

Two relations are obtained by inserting Eq. (7-4-40) into Eq. (7-4-48) and separating real and imaginary parts to give

$$-2I\beta\gamma - \beta L_2 + \gamma L_1 = 0 \tag{7-4-49}$$

and

$$I(\beta^2 - \gamma^2) - \beta L_1 - \gamma L_2 + \Gamma_0 = 0 \tag{7-4-50}$$

where

$$L = L_1 + jL_2 \tag{7-4-51}$$

The viscosity η can in principle be calculated from the observed values of β and γ using either Eqs. (7-4-49) or (7-4-50). However, since L is a function of the complex quantity bR, where b is defined by Eq. (7-4-43), these formulas are difficult to handle. Hence, approximate methods of evaluating the viscosity η have been developed by Andrade and Chiong (1936) and Roscoe (1958). These approximate methods are considered in Sec. 7-5-2.

7-5 EXPERIMENTAL APPLICATIONS

7-5-1 The Capillary Tube Viscometer

Capillary tube viscometers, of which many variations exist, are the most commonly used type for determining the viscosity of newtonian liquids. They usually require small quantities of liquid, the temperature of which can be easily controlled. The liquid flows through a capillary tube, two modes of operation being distinguished. In one form of the instrument the liquid flows through the capillary tube under a known applied pressure difference, the resulting flow rate being measured. Alternatively, the liquid is forced through the capillary tube at a given flow rate, the resulting pressure drop produced across the capillary being measured. The driving pressure can be derived from the head of the liquid being tested, the flow of liquid through the capillary being due to gravity, thus allowing instruments of the type which involve a known pressure drop to be further classified as external pressure-type or gravity-type viscometers. A common form of the gravity-type viscometer is the U-tube instrument, the original form of which is ascribed to Oswald (1893). Accurate alignment of the capillary tube is essential, deviation of the instrument from the vertical leading to an error in measurement. A second source of error is the necessity of introducing an exact volume of liquid, variation in temperature of measurement involving adjustment of the working volume at each temperature. For a detailed account of capillary tube viscometry see, for example, Barr (1931), Merrington (1949), and Dinsdale and Moore (1962).

Introducing the time of flow t, Eq. (7-3-12) can be expressed in the form

$$Pt = \frac{m\rho V^2}{a^4\pi^2}\frac{1}{t} + \frac{8\eta V}{\pi a^4}(L + na) \tag{7-5-1}$$

where V is the volume of liquid flowing in time t. The product Pt should vary linearly with $1/t$, the factor m being obtained from the slope $m\rho V^2/a^4\pi^2$ (provided

m is a constant); n is determined from the intercept $8\eta V(L+na)/\pi a^4$, provided η is known with sufficient accuracy. If η is unknown n can be determined from a series of experiments with tubes of different lengths. A single capillary tube viscometer can be calibrated using a number of liquids, provided their viscosities are known with sufficient accuracy. Uncorrected values of the viscosities η' are determined from the time of flow for each liquid using Eq. (7-3-7). Defining a variable

$$e = \frac{\eta'}{\eta} - 1 \tag{7-5-2}$$

where η is the actual viscosity, Eq. (7-3-12) can be expressed in the form

$$e = \frac{ma\text{Re}}{16L} + \frac{na}{L} \tag{7-5-3}$$

where the Reynolds number

$$\text{Re} = \frac{2\rho V}{\pi a \eta t} \tag{7-5-4}$$

is the second variable. The factors m and n can be obtained from the slope and intercept respectively of Eq. (7-5-3).

For gravity-type instruments, Eq. (7-3-12) can be expressed in terms of the kinematic viscosity ν and the time of flow t in the form

$$\nu = A_0 t - \frac{B_0}{t} \tag{7-5-5}$$

where the constants A_0 and B_0 can be determined by observing the times of flow of two or more liquids of known kinematic viscosity.

Using the capillary tube method, Swindells, Coe, and Godfrey (1952) obtained the value

$$\eta = 1.0019 \pm 0.0003 \text{ mPa s}$$

for the viscosity of water at 20°C. The greatest inaccuracy in these measurements arises primarily in the determination of the end corrections.

7-5-2 Viscosity Determination by the Oscillation of a Vessel Enclosing a Liquid

A particularly simple viscometer consists of a vessel enclosing a liquid, the vessel being held by a torsional suspension. Free oscillation of the vessel about a vertical axis gives rise to a motion which is damped on account of the viscosity of the liquid η, the latter being calculated from measurements of the decrement and frequency of the oscillations. The practical advantages of the method lie in the comparative ease with which, to a high degree of accuracy, the decrement and frequency of the oscillations can be measured, together with good temperature control because the liquid is confined to a volume of small dimensions. There is the additional advantage that the method is suitable for molten metals and other liquids which must be kept out of contact with air. Examples of the use of this type of viscometer are the studies of Andrade and Chiong (1936), Andrade and Dobbs (1952), Roscoe (1958), and Roscoe and Bainbridge (1958).

The flow of a liquid confined inside a spherical oscillating vessel has been considered in Sec. 7-4-4. In principle the viscosity η can be evaluated from either Eq. (7-4-49) or Eq. (7-4-50). However, as noted in Sec. 7-4-4, these equations are difficult to handle.

Andrade and Chiong (1936) simplify their treatment by effectively setting $\coth bR$ equal to unity in Eq. (7-4-46), so giving the approximate relation

$$L = \tfrac{8}{3}\pi\eta R^4\left(bR - 2 + \frac{1}{bR-1}\right) \tag{7-5-6}$$

which can be justified as follows. Let $bR = A_1 + jA_2$; then

$$\coth bR = \frac{\tanh A_1 + \tanh A_1 \tan^2 A_2}{\tanh^2 A_1 + \tan^2 A_2} + j\left[\frac{(\tanh^2 A_1 - 1)\tan A_2}{\tanh^2 A_1 + \tan^2 A_2}\right]$$

from which it follows that the real part of $\coth bR$ lies between the bounds $\coth A_1$ and $\tanh A_1$ and the imaginary part between $\pm j(\coth A_1 - \tanh A_1)/2$. For measurements of the decrement to be accurate the experimental conditions must be such that a large number of oscillations can be observed; this condition is satisfied if $\beta/\gamma \ll 1$. From Eqs. (7-4-41) and (7-4-43), together with the condition $\beta/\gamma \ll 1$,

$$R\left(\frac{\pi}{\nu T}\right)^{1/2}(2j)^{1/2} \approx bR = A_1 + jA_2$$

Hence

$$A_1 \approx \left(\frac{\pi}{\nu T}\right)^{1/2} R \tag{7-5-7}$$

which is usually a large quantity. Hence $\coth A_1$ approximates closely to unity.

Thus far the treatment has followed that of Andrade and Chiong (1936). A simpler method of proceeding from this stage has been described by Roscoe (1958).

Roscoe proceeds to express L, as given by Eq. (7-5-6), in the form of a series of descending powers of bR which is convergent provided $|bR| > 1$. This enables Eq. (7-4-49) to be expressed in the form

$$I\Delta = \tfrac{2}{3}R^4(\pi\eta\rho T)^{1/2}\left(a_0 - \frac{2}{d} + \frac{a_2}{2d^2} + 0 - \frac{a_4}{4d^4} - \frac{a_5}{4d^5}\cdots\right) \tag{7-5-8}$$

where $d = (\pi/\nu T)^{1/2}R$ and the coefficients a_n, which are all close to unity, have the values

$$\begin{aligned} a_0 &= (1-\Delta)\left[\frac{(1+\Delta^2)^{1/2}+1}{2}\right]^{1/2} - (1+\Delta)\left[\frac{(1+\Delta^2)^{1/2}-1}{2}\right]^{1/2} \\ &= 1 - \tfrac{1}{2}\Delta - \tfrac{3}{8}\Delta^2 - \tfrac{1}{16}\Delta^3 \cdots \end{aligned} \tag{7-5-9}$$

$$a_2 = \left[\frac{(1+\Delta^2)^{1/2}+1}{2}\right]^{1/2} + \left[\frac{(1+\Delta^2)^{1/2}-1}{2}\right]^{1/2}$$
$$= 1+\tfrac{1}{2}\Delta+\tfrac{1}{8}\Delta^2-\tfrac{1}{16}\Delta^3 \cdots \qquad (7\text{-}5\text{-}10)$$

$$a_4 = \frac{a_2}{(1+\Delta^2)^{1/2}} = 1+\tfrac{1}{2}\Delta-\tfrac{3}{8}\Delta^2-\tfrac{5}{16}\Delta^3 \cdots \qquad (7\text{-}5\text{-}11)$$

$$a_5 = (1+\Delta^2)^{-1} = 1-\Delta^2+\Delta^4 \cdots \qquad (7\text{-}5\text{-}12)$$

The viscosity η can be evaluated from Eq. (7-5-8) by successive approximations.

Alternatively, the viscosity η can be evaluated from Eq. (7-4-50) expressed in the form

$$I\left[\left(\frac{T}{T_0}\right)^2 - (1-\Delta^2)\right] = \tfrac{4}{3}R^4(\pi\eta\rho T)^{1/2}$$
$$\times\left(b_0 - 2\frac{\Delta}{d} - \frac{b_2}{2d^2} - \frac{1}{2d^3} - \frac{b_4}{4d^4} + \frac{b_5\Delta}{d^5}\cdots\right) \qquad (7\text{-}5\text{-}13)$$

where $T_0 = 2\pi(I/\Gamma_0)^{1/2}$ is the periodic time of oscillation that the system would have in the absence of viscous damping, and the coefficients b_n are given by a similar set of expressions to the a_n coefficients. The expanded forms of the b_n coefficients are obtained by replacing Δ by $-\Delta$ in the corresponding expansions for the a_n. In Eq. (7-5-13) the square-bracket term on the left is difficult to determine accurately since it is the difference between two quantities which are obtained from measurements, both of which are close to unity. As such the accuracy of the results for η evaluated from Eq. (7-5-13) is relatively low. Also, Eq. (7-5-13) involves T_0 which is not simply determined with the vessel empty since it depends to some extent upon the weight of the suspended vessel.

The practical possibilities of the oscillating vessel method for accurate absolute measurements have been investigated by Roscoe and Bainbridge (1958). An improvement in apparatus was effected by using a capillary tube for the stem of the spherical bulb so that the liquid acted as its own thermometer. Large amplitudes of rotation were found to be accompanied by abnormally large logarithmic decrements. Roscoe and Bainbridge investigated statistical methods of calculating the decrement while avoiding this latter effect. A value of

$$\eta = 1.0025 \pm 0.0005 \text{ mPa s}$$

was obtained by Roscoe and Bainbridge for the viscosity of water at 20°C. This is to be compared with the value given at the end of Sec. 7-5-1, obtained using a capillary tube viscometer. It was concluded by Roscoe and Bainbridge from an analysis of the sources of error that the accuracy could be increased by further improvements well beyond that of the best capillary tube determination.

The use of a cylindrical vessel has been considered in detail by Roscoe (1958); in this connection Hopkins and Toye (1950) used a cylindrical vessel to determine the viscosity of zinc in the temperature range of 430 to 480°C.

7-5-3 Coaxial Cylinder Viscometry under Oscillatory Conditions of Flow

The way in which the (θ_2, f) curves for a newtonian liquid vary with the cylinder gap width for a given value of the viscosity η is shown in Fig. 7-3. The full-line curves shown in Fig. 7-3 have been calculated from Eq. (7-4-25) using the calculated values of λ obtained from Eq. (7-4-11) with $\eta = \eta_c$, where η_c is the viscosity of the test material obtained from a capillary tube viscometer, and the calculated values of λ_0 using the observed values of λ_e. Since the measured values of λ agree within the limits of experimental accuracy with the calculated values of $\eta_c\xi$ in the range of η_c from 0.32 to 4.23 Pa s, it follows that the values of η evaluated from the forced oscillatory measurements agree closely with the η_c values obtained with a capillary tube viscometer. Typical phase lag measurements, i.e., the variation of the displacement $\theta_2(t)$ of the inner cylinder and $\theta_1(t)$ of the outer cylinder with the time t of shearing, is shown in Fig. 7-4 for a fixed frequency $f = 0.92$ Hz.

For the condition $f = f_N$, $\chi = 0$, such that Eq. (7-4-25) reduces to

$$\theta_m = \theta_1 \frac{\lambda \Gamma_0}{\beta + \beta_e} = \theta_1 \frac{\lambda}{\lambda + \lambda_e} \qquad (7\text{-}5\text{-}14)$$

from which it is evident that a finite β_e leads to a maximum amplitude θ_m of the inner cylinder which is less than θ_1, the amplitude of the outer cylinder.

Equation (7-4-14) for general forced oscillatory motion was derived for zero instrument damping (that is, $\beta_e = 0$); in this connection it is of interest to note that

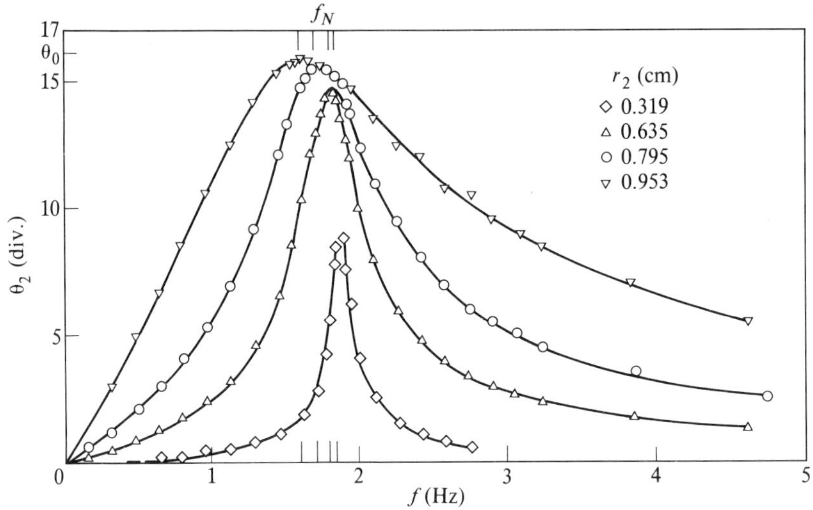

Figure 7-3 Variation of the amplitude of oscillation θ_2 of the inner cylinder with the frequency of oscillation f of the outer cylinder for a newtonian-type liquid having a viscosity $\eta = 8.0$ Pa s and, for various values of the cylinder gap width, $r_1 = 11.1$ mm, inner cylinder radius r_2: ◇, 3.19 mm; △, 6.35 mm; ○, 7.95 mm; ▽, 9.53 mm. (*Billington, 1965.*)

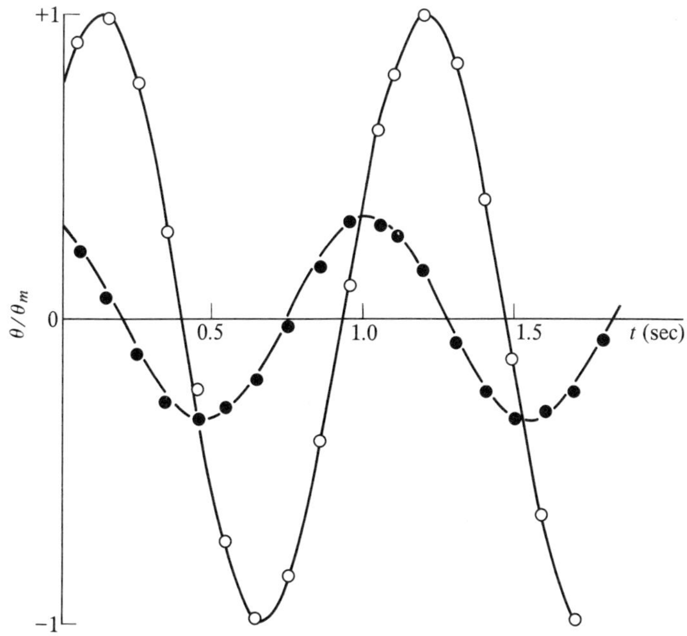

Figure 7-4 Variation of the displacement $\theta_2(t)$ of the inner cylinder and $\theta_1(t)$ of the outer cylinder with the time of shearing for a fixed frequency $f = 0.92$ Hz. Amplitude of the outer cylinder $\theta_1 = 0.70$ rad. Open circles: displacement of the outer cylinder $\theta_1(t)$; blocked circles: displacement of the inner cylinder $\theta_2(t)$. (*Billington, 1965.*)

the effect of setting α to zero in Eq. (7-4-14) is to give Eq. (7-4-24) for the condition $\beta_e = 0$. Thus, in Eq. (7-4-14) the terms involving α^2 and higher terms in α represent the effect of the inertia of the fluid. For $f = f_N$, $\chi = 0$ and Eq. (7-4-14) gives the relation

$$\theta_m = \frac{\theta_1}{(1 - 4\pi^2 C_r f_N^2/v^2) + j(\pi r_1^2/h_2^2)(f_N/v)} \tag{7-5-15}$$

where

$$C_r = \frac{1}{192 r_1^2}\left[(r_1^2 - r_2^2)(r_1^4 - 5r_1^2 r_2^2 - 2r_2^4) + 6r_1^2 r_2^4 \ln\left(\frac{r_1^2}{r_2^2}\right)\right]$$

from which it is evident that the effect of the inertia of the fluid is to lead to an amplitude θ_m of the inner cylinder which is also less than θ_1, the amplitude of the outer cylinder.

Whether mechanical damping of the motion of the inner cylinder or the inertia of the fluid is primarily responsible for the condition that at $f = f_N$ the amplitude θ_m of the inner cylinder is less than the amplitude θ_1 of the outer cylinder can be ascertained by reference to the two sets of conditions represented by Eqs. (7-2-31) and (7-2-32) for any particular instrument and test material.

7-5-4 Coaxial Cylinder Viscometry under Transient Conditions of Flow

For the newtonian-type liquids referred to in Sec. 7-5-3 and given values of the instrument factor ξ, the time constant λ can be calculated from Eqs. (7-4-11) and (7-4-31) using the value of λ_e determined from the oscillatory measurements, using the fluid having the smallest measured viscosity. In the case of the newtonian liquids of present interest this gave $\lambda/\lambda_e \simeq 400$, from which it is evident that the instrument damping factor has negligible effect on the transient measurements, Eq. (7-4-32) reducing to

$$\theta_2(t) = \theta_2(\infty)\left[1 - \exp\left(\frac{-t}{\lambda}\right)\right] \tag{7-5-16}$$

where λ is given by Eq. (7-4-11). For the coaxial cylinder viscometer described by

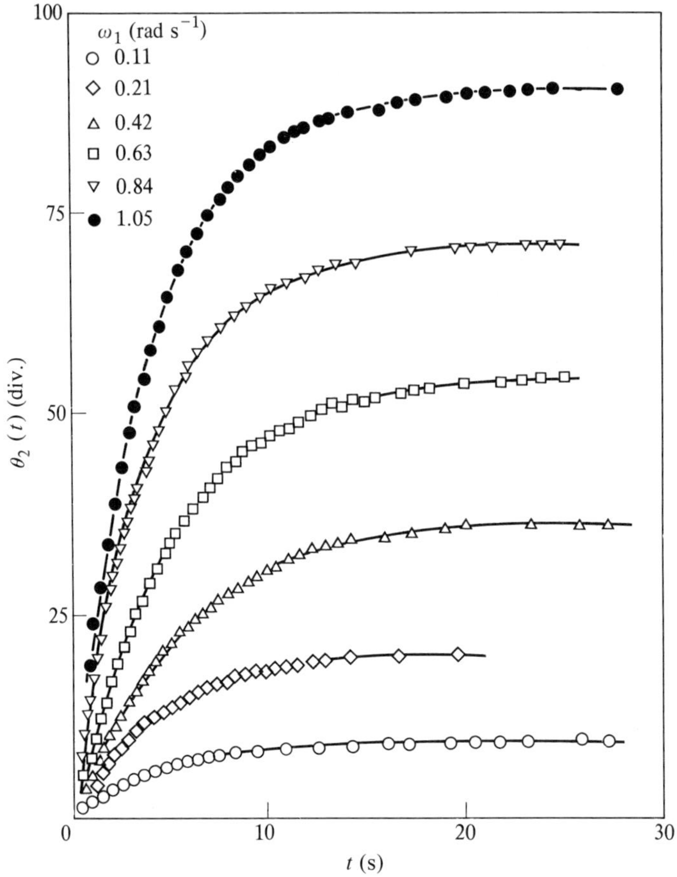

Figure 7-5 Typical displacement-time curves for a newtonian-type liquid having viscosity $\eta = 51.5$ Pa s. Radius of inner cylinder $r_2 = 19.0$ mm. ω_1: ○, 0.11 rad s^{-1}; ◇, 0.21 rad s^{-1}; △, 0.42 rad s^{-1}; □, 0.63 rad s^{-1}; ▽, 0.84 rad s^{-1}; ●, 1.05 rad s^{-1}. (*Billington, 1965.*)

Billington (1965) the quantity $4I\Gamma_0/\beta^2$ has the maximum value 0.04, which satisfies the required condition $4I\Gamma_0/\beta^2 \ll 1$ for a small moment of inertia I.

The way in which the angular displacement $\theta_2(t)$ of the inner cylinder varies with the time of shearing for several values of ω_1 is shown in Fig. 7-5 for a lubricating oil having a viscosity $\eta = 5.15$ Pa s. The (θ_2, t) curves shown in Fig. 7-5 are typical of the results obtained with newtonian liquids over the whole range of the rate of shear. For the (θ_2, t) curves of Fig. 7-5 the $(\ln[\theta_2(\infty)-\theta_2(t)], t)$ curves are linear, in agreement with Eq. (7-4-32), and have the same slope, from which it follows that for a given value of r_2 the time constant λ is independent of ω_1. Using the values of λ calculated from Eq. (7-4-11), where the viscosity $\eta = \eta_c$ has been determined from capillary tube viscometers, the calculated (θ_2, t) curves show good agreement with the experimental (θ_2, t) curves for ξ in the range from 0.68 to 2.68 Pa^{-1} and for η_c in the range from 0.23 to 5.15 Pa s. It is therefore concluded that, within the limits of experimental accuracy, fluid inertia effects can be neglected for the given range of cylinder dimensions and viscosity.

There are two experimental conditions which must be satisfied for this type of transient motion to be realized in practice. The time to accelerate the outer cylinder to its constant angular velocity ω_1 must be very much less than the time constant λ characterizing the motion of the inner cylinder. Similarly, the time taken to accelerate the inner cylinder to its initial angular velocity $\omega_2(0) = \omega_1$ must approximate to that taken to initially accelerate the outer cylinder to ω_1. This is achieved only if the moment of inertia I of the inner cylinder is made sufficiently small to satisfy the condition $4I\Gamma_0/\beta^2 \ll 1$. With these conditions satisfied both cylinders, together with the fluid, behave at $t \approx 0$ as if they were a solid unit, provided the further condition that the inertia of the fluid can be neglected is satisfied. These conditions at $t \approx 0$ provide one form of experimental test to verify that the conditions for transient motion are satisfied, the angular velocity $\omega_2(t)$ of the inner cylinder being determined by graphical methods from the (θ_2, t) curves and extrapolated to $t = 0$, for which condition $\omega_2(0)$ should equal ω_1. The time dependence of $\omega_2(t)$ for a newtonian fluid is of the form

$$\omega_2(t) = \omega_1 \exp\left(\frac{-t}{\lambda}\right) \tag{7-5-17}$$

The transient motion of immediate interest has been considered in detail since it has found a particular application in the study of the phenomenon of thixotropy which is discussed in Chap. 9.

7-6 IRREVERSIBILITY

7-6-1 The Problem of Irreversibility

Because the liquid state appears as an interpolated state, lying between the solid and gas states of matter, two approaches to the liquid state can be identified. These are concerned with liquids considered as dense gases and liquids considered as disordered solids. Of immediate interest are the theories of the liquid state which regard a liquid as a limiting case of a dense gas. The object of the approach to the liquid state regarded as a compressed gas, and assuming the motion to be given, is the calculation of the stresses defined as arising from the transport of momentum. This approach is effectively an attempt to correct the gas kinetic formulas by considering the dense packing and shortened mean free path, and therefore introduces the time between collisions directly. However, the mechanism of momentum transport in a liquid must be different from that in a gas. Instead of transporting momentum by their own motion from one layer to another, the atoms of a liquid are considered to transfer momentum to their neighbors by the interaction of interatomic forces. The interparticle force between two liquid atoms or molecules arises from the short-range attractive and repulsive forces between the particles. If the distance between particle centers is large the force is attractive but changes to a strong repulsion at short distances. Thus, the simple molecule regarded as a point mass must be replaced by a hard sphere surrounded by an attractive field of force. Before discussing viscosity in relation to molecular structure it is necessary, however, to consider a particular problem which is central to all kinetic theories of the transport of momentum.

Any macroscopic system which is not in equilibrium will, in the absence of any form of external restraining agencies, always return to equilibrium in the same direction in time. For example, if a flowing liquid is brought to rest, it will remain indefinitely at rest in the absence of any driving forces. That is, a liquid at rest will not spontaneously acquire any form of macroscopic motion. This biasing of the macroscopic system in one direction is expressed in thermodynamics by the law of increase of entropy, which states that the entropy of an isolated system increases to a maximum at equilibrium and that continued observation reveals no spontaneous decreases in entropy. For a macroscopic system, this irreversible return to equilibrium occupies a finite time, the time-dependent response being characterized by a relaxation time. The return to equilibrium is described in terms of the action of dissipative processes which work to destroy the nonequilibrium states through which the system must pass. In terms of continuum theories of the response of macroscopic systems, these dissipative processes appear in the form of empirical transport coefficients, examples of which are the coefficients of viscosity, thermal conductivity, and diffusion. These are examples of the material coefficients appearing in constitutive theory. Ideally it is required to define these material coefficients in terms of the collective behavior of the interacting particles comprising the material.

In contrast to the irreversible behavior of a macroscopic system, physical mechanisms at the molecular level are reversible in that the equations describing the behavior of a mechanism contain no information which can be used to decide in which direction in time the process will occur. For example, the laws of conservation of energy and momentum which describe the collision mechanism between two gas molecules are unable to distinguish the direction in time of the molecular paths and therefore apply equally well to the reversed molecular paths. It can be shown, at least in principle, that this observation applies equally to complex, multiparticle collision interactions of both gas molecules and liquid molecules. The problem arising from the incompatibility of the reversibility in time of microphysical mechanisms and the irreversible drift to equilibrium always observed macroscopically is in this form referred to as Loschmidt's (1876, 1877) reversibility paradox.

This particular problem has been stated in other ways. For example, there is Zermelo's (1896) recurrence paradox which is based on a theorem due to Poincaré. For a closed conservative dynamical system of interacting particles, where the force on each particle depends only on the relative configuration of the particles, Poincaré showed that any given initial state must recur to any arbitrarily specified degree of accuracy an indefinitely large number of times. The molecular paths representing a system of N interacting particles contained in a finite space volume V must, according to Poincaré, have a quasi-periodic structure. Thus, any initial state will reappear again if sufficient time is allowed to elapse. The time taken for this to happen is the Poincaré time period τ_p and the whole repetitive process is the Poincaré cycle. The existence of such a cycle is, in principle, contrary to the general observation of macroscopic irreversibility, this contradiction being Zermelo's paradox.

Thus, central to the whole problem of relating the macroscopic material coefficients appearing in the constitutive equations for liquids, i.e., the transport coefficients, to the microphysics of the material is the paradox of macroscopic irreversibility and the reversibility of microphysical mechanisms. It is evident, therefore, that in any attempt to describe macroscopic properties in terms of microphysical mechanisms, there must be inserted into the theory an assumption which excludes the reversed solutions, thus ensuring that the final equations embodying the collective behavior of the microphysical mechanisms shall describe the always-observed irreversible drift of a macroscopic system to equilibrium. This assumption relates directly to the averaging process involved in relating the statistical properties of the microphysical mechanisms to the macroscopic material property.

For the rigorous theories of the viscosity of a dilute gas the assumption takes the form of the hypothesis of molecular chaos. This states that the momentum of a molecule after a small number of collisions is statistically independent, in magnitude and direction, of its original momentum. Introducing this assumption precludes the possibility of precise prediction associated with the reversibility otherwise inherent in the equation of motion, the future behavior of a system now being a matter of probability rather than certainty. The observed irreversibility is now associated with the most probable evolution of the molecular distribution functions, calculated using the above hypothesis. For liquids the concept of collisions is not applicable and therefore an alternative hypothesis to that of molecular chaos is required.

The condition that a physical system shall be macroscopically irreversible requires that there shall exist some physical process which will smooth out the initial nonuniformities during the drift to equilibrium. It is this smoothing-out process which characterizes irreversibility. The smoothing-out process is identified with the particle interactions considered in statistical terms. Thus the physical aspects of any kinetic theory are concerned with specifying the interaction process in such a way as to ensure the required irreversible behavior of the macroscopic system.

Using the subjective terminology introduced in Chap. 5, irreversibility can be interpreted as the condition that at any instant a liquid shall have no memory of its past history insofar as it ceases to be aware of information which, at any instant in the course of its time evolution toward equilibrium, it can no longer recover.

At the present stage of development of the molecular theory of liquids, there does not appear to be any generally applicable way in which a nonuniform system can progressively discard information concerning the state of a system of interacting particles during the time evolution of the system toward equilibrium. The only methods presently available are for restricted situations.

7-6-2 The Particle-Mobility Approximation for Liquids

For liquids it is assumed that both the irreversible approach to equilibrium and the equilibrium condition itself are determined by the same microphysical mechanisms. As such, it is to be expected that the interparticle force will be as important in determining the nature of the transport coefficients for a nonuniform liquid as it is known to be in determining the thermodynamic functions describing the uniform state.

A liquid molecule is permanently in strong interaction with its nearest neighbors and the resultant force $\boldsymbol{F}(t)$ on it at some instant t will be the vector sum of the forces due to those neighbors. At time $(t+t')$, where t' is small, this force will differ both in magnitude and direction due to its rapid fluctuations arising from the molecular motion. For a molecule at rest, it is evident that the fluctuating force is completely random in direction. In a liquid, however, all molecules are in continual motion and if at some instant a particular molecule acquires a drift velocity, then the fluctuations will no longer be entirely random in direction. Thus, on the microphysical scale, the behavior of the molecules of a liquid can be regarded as being formally similar to the behavior of brownian-type particles.

The particle mobility–diffusion approximation approach to brownian theory is characterized by three general features.

First, the concept of probability is used to describe the particular system at time t, the previous history of the system being irrelevant. The basis of the theory is the concept of chance, and it is therefore a stochastic theory. Systems where the present states are not described in terms of the explicit history are said to be Markov systems, brownian theory being therefore an example of the Markov stochastic type.

The second general feature is the introduction of a time interval τ which cannot be indefinitely small nor indefinitely large. It is assumed that during this small, but finite, time interval τ the brownian-type particle suffers an incremental change in its momentum of amount Δp resulting from the competition between the fluctuating acceleration due to the molecular agitation and the systematic deceleration resulting from the dynamic internal friction, i.e., the viscosity of the continuous medium.

Third, the conditions relating to the increments Δp are represented by the equations

$$\langle \Delta p^2 \rangle = 2k\theta\left(\frac{\tau}{\mu}\right) \qquad \text{and} \qquad \langle \Delta p \rangle = -\frac{p}{m}\left(\frac{\tau}{\mu}\right)$$

where $\mu = 1/6\pi\eta a$ is the mobility of the brownian particle of radius a drifting with velocity $\boldsymbol{v}$ in a liquid of viscosity η. These conditions relating to $\langle \Delta p^2 \rangle$ and $\langle \Delta p \rangle$ apply only to those values of τ which are neither very large nor very small as compared with the magnitude of μ, and in particular the upper value of τ must be consistent with the condition $(\tau/m\mu) \ll 1$.

Kirkwood (1946) proceeded to develop a general theory of the nonequilibrium properties of liquids on the assumption that these three characteristic features of brownian theory would also form the basis of more general theories of irreversible behavior. The principal difference relates to the condition that $\mu = 1/6\pi\eta a$ for brownian movement whereas for molecular motion the mobility must be interpreted in a statistical sense.

In principle, provided the equations of motion for all the molecules can be solved, the force $\boldsymbol{F}(t+t')$ at time $(t+t')$ can be calculated from $\boldsymbol{F}(t)$ for any value of t'. In practice this is virtually impossible and therefore Kirkwood (1946) assumed that after a certain time interval τ the force on the molecule can be regarded as being statistically independent of its earlier value. It is important to note that τ must be less than the duration of any observation of the behavior of a liquid property.

The concept of statistical independence is given physical significance in statistical physics by way of a correlation coefficient. This approach can be extended to Kirkwood's hypothesis. Consider the scalar product

$$\phi(t') = \boldsymbol{F}(t) \cdot \boldsymbol{F}(t+t')$$

which can be either positive or negative. The average value of ϕ can now be obtained over an equilibrium ensemble of systems, each containing a representative molecule, this average being expressed in the form

$$\langle \phi(t') \rangle = \langle \boldsymbol{F}(t) \cdot \boldsymbol{F}(t+t') \rangle \tag{7-6-1}$$

The expression $\langle \phi(t') \rangle$ is called the autocorrelation function of the force $\boldsymbol{F}(t)$.

For a molecule at rest, the fluctuating force is completely random in direction. In a liquid, however, all molecules are in continual motion and if at some instant a particular molecule acquires a drift velocity $\boldsymbol{v}$ through the liquid, then the fluctuations will no longer be entirely random in direction, and on average these fluctuations will tend to oppose the motion. This effect can be interpreted in terms of a retarding, or frictional, force which becomes effective when any molecule is in motion. Kirkwood (1946) assumed the frictional force to depend only on the particle velocity $\boldsymbol{v}$, assuming in particular direct proportionality. Hence, for a given molecule, the total force

$$F(t) = -\beta mv + G(t) \tag{7-6-2}$$

where $G(t)$ is the always-present, directionally random, fluctuating force on a moving molecule. In Eq. (7-6-2), the constant β is referred to as the friction constant, and is proportional to the reciprocal of the mobility. Hence, Kirkwood's hypothesis for liquids can be regarded as giving rise to the particle-mobility approximation for liquids. In particular, as for brownian motion, the asymmetric part of the force is represented as a steady braking force which implies an averaging process to smooth the asymmetry. This property, which is introduced by way of Kirkwood's hypothesis in the form of Eq. (7-6-1), was formulated by Kirkwood who obtained the relation

$$\beta = \frac{1}{3mk\theta}\int_0^\tau \langle F(t)\cdot F(t+t')\rangle_{N-2}\,dt' \tag{7-6-3}$$

where the integrand is here understood to be the average of the force on particles 1 and 2 due to the remaining $(N-2)$ particles.

For β to be independent of τ, the autocorrelation function $\langle\phi(t')\rangle$ must vanish for values of τ greater than some time interval τ_1, which must itself be less than τ_p. Thus the formulation of Kirkwood's hypothesis requires that

$$\begin{aligned} &\text{for} && t' < \tau_1 : \langle\phi(t')\rangle \neq 0 \\ &\text{and for} && t' > \tau_1 : \langle\phi(t')\rangle = 0 \end{aligned} \tag{7-6-4}$$

these conditions always being satisfied for all values of t. To avoid difficulties with the Poincaré time period τ_p, it follows that $\tau_1 < \tau_p$. These conditions ensure that the integral in Eq. (7-6-3) will reach a constant value, referred to as the plateau value, extension of the range of τ beyond τ_1 giving no further contribution to β.

In an attempt to estimate β, Kirkwood (1946) assumed the force correlation of Eq. (7-6-3) to have the form of a gaussian function. For this condition,

$$\beta = \frac{(2\pi)^{1/2}\langle\Delta F^2\rangle}{6mk\theta\omega} \tag{7-6-5}$$

where $\langle\Delta F^2\rangle$ is the mean square of the force on the single particle and ω is some characteristic frequency describing the time changes of the force. Rice and Kirkwood (1959) obtain the alternative expression

$$\beta = \left[\frac{4\pi n}{3m}\int_0^\infty r^2\left(\frac{\partial^2\psi}{\partial r^2} + \frac{2}{r}\frac{\partial\psi}{\partial r}\right)g^{(2)}(r)\,dr\right]^{1/2} \tag{7-6-6}$$

where $g^{(2)}(r)$ is the equilibrium radial distribution function referring to the spherically symmetric pair interaction potential ψ. For liquid argon at 89 K, Eq. (7-6-6) gives $\beta = 7.3\times10^{12}\,\mathrm{s}^{-1}$. Numerical estimates of β have been made, for example, by Cole (1960) and Helfand (1961). Rather than attempt the difficult problem of approximating the force correlations in time in terms of actual particle trajectories, Turner (1960) and Eisenschitz and Wilford (1962) have made the approximation in terms of the Fourier components and have obtained the values $\beta = 0.56 \times 10^{11}\ s^{-1}$ and $\beta = 3.5 \times 10^{11}\ s^{-1}$ respectively for argon at 84 K. The accurate evaluation of β is one of the outstanding problems of nonequilibrium theory.

7-7 VISCOSITY AND MOLECULAR STRUCTURE

7-7-1 Liquids Considered as Dense Gases

The first theory to give physical insight into the structural basis of the transport of momentum in liquids is that of Kirkwood (1946), Kirkwood, Buff, and Green (1949), and Eisenschitz (1949). Since it would appear likely that the Kirkwood–Eisenschitz approach to the transport of momentum in liquids will form part of any complete theory, there follows an outline account of this particular approach.

A perfect single crystal is characterized by a regular geometrical arrangement of the elementary particles, i.e., the atoms or molecules, this structural pattern being referred to as a space lattice. The lattice structure extends throughout the crystal and crystalline solids are therefore said to exhibit long-range order. As the melting point is approached, the kinetic energy of the elementary particles increases, thus tending to destroy the long-range order; the interparticle forces tend to preserve the long-range order. Above the melting point most, but not all, of the order disappears. In the vicinity of some reference particle the immediate neighbors are grouped around it with some local degree of order, which to a limited extent resembles the arrangement in a solid. The next-nearest neighbors show only a very slight degree of order relative to the reference particle, and at greater distances there is no correlation at all between the position of any particle and the reference particle. Liquids are therefore said to exhibit short-range order only. This short-range order, which extends over only a few particle diameters, is here considered to account for the viscous property of a liquid. Effects arising from the short-range order diminish with increasing temperature.

The physical significance of the short-range order in liquids has been formulated by way of the concept of a radial distribution function. In general, the radial distribution function specifies the number of particles to be found at any distance r from an arbitrarily chosen reference particle. Place the center of a reference particle at the origin of a thin spherical shell of radius r and thickness dr comparable to a molecular diameter. Let this shell of volume $4\pi r^2\,dr$ contain dN particles. The number of particles per unit volume

$$\rho(r) = \frac{dN}{4\pi r^2\,dr}$$

or

$$dN = \rho(r) 4\pi r^2\,dr \tag{7-7-1}$$

In Eq. (7-7-1), $\rho(r)$ is independent of the angular molecular arrangement. Now consider the whole of the space surrounding the reference particle to be divided into a series of concentric spheres such that the volume interval between neighboring spheres is a constant. The average number of particles contained between each spherical shell, and therefore also $\rho(r)$, now varies with r. For large distances corresponding to the limit $dr \to 0$, $\rho(r)$ becomes continuous, being simply the average, i.e., the macroscopic, number density $n = N/V$, where N is the total number of particles in the volume V of the system. Since the problem of present interest is primarily concerned with pairs of particles, the local-order characteristic of the liquid state is described in terms of a pair correlation function $g^{(2)}(r)$. The radial distribution function $g^{(2)}(r)$ is defined as the ratio of the average particle density $\rho(r)$, at distance r from the reference particle, to the macroscopic density n of the liquid. This definition of $g^{(2)}(r)$ gives the relation

$$\rho(r) = ng^{(2)}(r) \tag{7-7-2}$$

which can be substituted into Eq. (7-7-1) to give

$$dN = 4\pi r^2 n g^{(2)}(r)\, dr \tag{7-7-3}$$

where dN, the average number of particles in the thin spherical shell of radius r and thickness dr, has now been expressed in terms of the radial distribution function. Setting $n = N/V$ in Eq. (7-7-3) and rearranging gives

$$\frac{dN}{N} = \frac{4\pi r^2\, dr}{V} g^{(2)}(r) \tag{7-7-4}$$

from which it is evident that Eq. (7-7-2) defines $g^{(2)}(r)$ as a dimensionless distribution function. For large radial distances, that is, $r \to R$, where R is a large characteristic dimension compared with the molecular separation,

$$\operatorname*{Lim}_{r \to R} g^{(2)}(r) = 1 \tag{7-7-5}$$

whereas, for $r = 0$,

$$g^{(2)}(0) = 0 \tag{7-7-6}$$

since the molecules are impenetrable. Thus $g^{(2)}(r)$ is the probability of finding a molecule at r if there is a molecule at the origin, and it is normalized to unit probability at large r. Information regarding these functions can be obtained from x-ray and neutron-diffraction studies (see, for example, Furakawa, 1962; Guinier, 1963). The radial distribution function for a liquid is shown in Fig. 7-6, from which it is evident that long-range order no longer exists. Also shown in Fig. 7-6 by a broken-line curve is the limiting condition corresponding to $r \to \infty$. Departures of the actual full-line curve from the broken line give the evidence for the existence of short-range order. The full-time curve starts from the point $(0, r_0)$ where r_0 is the radius of the reference molecule and rises to a maximum at a distance corresponding to the shell of the nearest neighbors. With increasing distance the departures of the full-line curve from the broken line decrease in magnitude and are spaced further apart, the short-range order finally disappearing. Thus, in liquids, there is a maximum structural resemblance to that of a solid over limited regions.

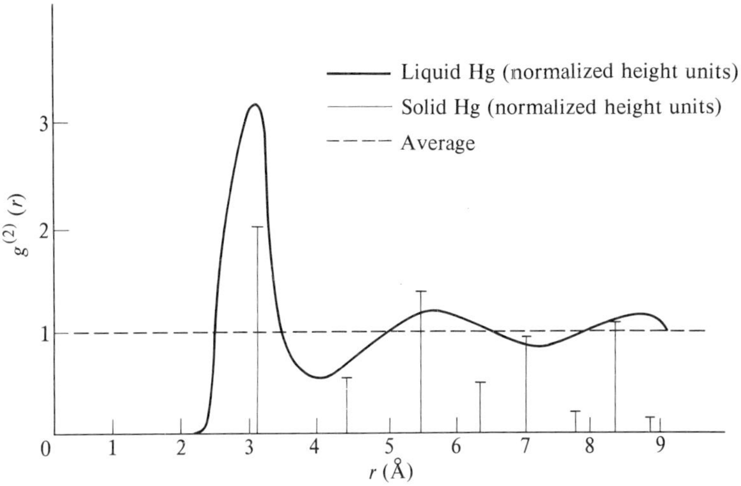

Figure 7-6 Variation of the radial distribution function with the radius measured from an arbitrary atom chosen as the origin. (*Guinier, 1963.*)

For the theory of present interest it is useful to define a potential of mean force $\phi(r)$ by the relation

$$g^{(2)}(r) = \exp\left[-\frac{\phi(r)}{k\theta}\right] \tag{7-7-7}$$

The formulation of Eq. (7-7-7) recognizes the radial distribution function for a liquid in equilibrium as a thermodynamic potential, and rearrangement gives

$$\phi(r) = -k\theta \ln g^{(2)}(r) \tag{7-7-8}$$

which can be differentiated to give a mean relative force

$$\langle F \rangle = -\frac{\partial \phi}{\partial r} \tag{7-7-9}$$

acting on each member of a pair of molecules. It is under the action of this force that molecular movement proceeds. In Eq. (7-7-9), the ensemble average $\langle F \rangle$ is understood to be the expected mean value of the force F which would be obtained when F is actually measured at some given time t. With regard to the theory of present interest, it is assumed that biased molecular movement, i.e., nonuniform liquid properties, are determined by the action of this same mean force, it being noted, however, that for nonequilibrium properties the equilibrium pair distribution function must be replaced by the appropriate nonequilibrium pair distribution. The very close approach of two molecules is strongly resisted because their electronic atmospheres overlap insofar as their positively charged nuclei are no longer fully screened and they can therefore interact to exert an electrostatic repulsion. It is this strong intermolecular repulsive force at small separation distances which ensures that the individual molecules always retain their identity, i.e., there is no aggregation. However, the existence of the condensed state of matter is evidence for cohesive effects which are attributed to intermolecular attractive forces at separation distances of a few molecular diameters. These attractive forces are referred to as the van der Waals or dispersion forces, and have their origin in the fluctuations in the electronic atmospheres of the individual molecules. The fluctuations instantaneously exhibit the properties of an electrostatic dipole which in turn polarizes a nearest-neighbor molecule to produce another electrostatic dipole to which the first molecule is now attracted. For symmetrical molecules the time average of the fluctuating dipole moment is zero. However, the average attractive force depends on the average of the square of the dipole moment which is positive. In the simplest case of symmetrical molecules it is sufficient to assume that the force between two molecules depends only on the distance between the centers of the two molecules, being in particular independent of a third molecule. This brief review of the properties of the intermolecular force suggests that it can be defined as the negative gradient of a simple potential function. With regard to this potential function, the total potential energy for a system of N molecules located at the vector positions $\boldsymbol{r}_i$ $(i = 1, 2, \ldots, N)$ will be expressed as $\psi(\boldsymbol{r}_i)$ $(i = 1, 2, \ldots, N)$ and for the particular conditions of interest the total potential

$$\Psi(\boldsymbol{r}_i) = \sum_{1 \leqslant i < j \leqslant N} \psi(\boldsymbol{r}_i, \boldsymbol{r}_j) \tag{7-7-10}$$

where ψ is the interaction potential for the ith and jth molecules taken as a pair, ψ being referred to as the pair potential. The summation in Eq. (7-7-10) extends over all $N(N-1)/2$ pairs of particles that can arise from the N particles of the system. Since the force has been taken to depend only on $r_{ij} = |\boldsymbol{r}_i - \boldsymbol{r}_j|$, it follows that

$$\psi(\boldsymbol{r}_i, \boldsymbol{r}_j) = \psi(r_{ij}) \tag{7-7-11}$$

and the intermolecular force

$$F(r) = -\frac{\partial \psi}{\partial r} \tag{7-7-12}$$

Unsymmetrical molecules may have permanent asymmetries of electric charge distribution which will give rise to additional forces whose magnitudes will depend on the orientation of these molecules.

7-7-2 The Stress Relation

Consider the flow of a system of N interacting molecules contained in a volume V relative to cartesian coordinates (x, y, z), the origin of coordinates being located at the center of a representative liquid molecule. The x axis is taken parallel to the direction of the mean macroscopic laminar flow direction of the liquid, the momentum flux in the liquid being along the y direction. In Fig. 7-7, let $\boldsymbol{i}$ and $\boldsymbol{j}$ be any pair of the N molecules in the liquid, the molecules being specified to have no internal degrees of freedom. The jth particle momentum is denoted by $\boldsymbol{p}_j$, with components (p_{jx}, p_{jy}, p_{jz}). Molecule j will also be subject to a force $F(r_{ij})$ due to

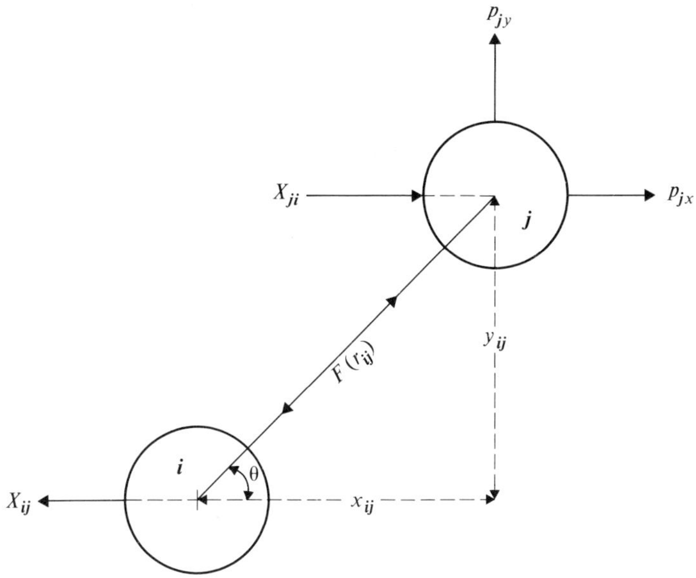

Figure 7-7 Momentum transfer in a liquid, arising from the motion and interaction of two molecules.

molecule ***i*** acting along the line of centers. Two ways in which momentum flow can occur are to be distinguished. The motion of molecule ***j*** is such that in a time δt it will move a distance $(p_{jy}/m)\delta t$ in the y direction. Molecule ***j*** will therefore transport its own x component of momentum p_{jx} in the y direction at a rate $[p_{jx}(p_{jy}/m)\delta t]/\delta t = p_{jx}p_{jy}/m$. A similar expression must apply to all other molecules in the liquid. Hence this mode of transfer, which is effectively a kinetic contribution to the momentum flow, is of total magnitude

$$\sum_{j=1}^{N} \frac{p_{jx}p_{jy}}{m} \tag{7-7-13}$$

Due to the action of intermolecular forces the resultant momentum of ***j*** is continually changing at a rate which is just the vector sum of all the forces acting on it. From Fig. 7-7, the x component of these forces, in particular X_{ji} which is due only to molecule ***i***, is

$$X_{ji} = F(r_{ij})\cos\theta$$

which in time δt will contribute an increment of momentum

$$\delta p_{jx} = F(r_{ij})\cos\theta\delta t$$

to molecule ***j***. Similarly, molecule ***i*** will incur a change in its x component of momentum due to the molecule ***j***. Now these changes are equal in magnitude but opposite in direction and are such that if $F(r_{ij})$ is positive molecule ***j*** gains x momentum while molecule ***i*** loses it. The net effect, therefore, is as though x momentum equal to $F(r_{ij})\cos\theta\delta t$ has been transferred during δt through space from molecule i to molecule j. Thus the rate of transfer is

$$F(r_{ij})\cos\theta\delta t\left(\frac{y_{ij}}{\delta t}\right) = F(r_{ij})\cos\theta y_{ij}$$

where y_{ij} is the distance in the y direction between molecules i and j. Since $\cos\theta = x_{ij}/r_{ij}$, the total flow is obtained by summing over all pairs of molecules in the volume V to give

$$\sum_{\text{pairs}} F(r_{ij})\frac{x_{ij}y_{ij}}{r_{ij}} \tag{7-7-14}$$

which is the contribution to the momentum flow due to the intermolecular forces. Hence, the sum of Eqs. (7-7-13) and (7-7-14) gives the total momentum flow

$$\sum_{j=1}^{N}\frac{p_{jx}p_{jy}}{m} + \sum_{\text{pairs}} F(r_{ij})\frac{x_{ij}y_{ij}}{r_{ij}}$$

which in turn gives for the stress tensor T_{xy} the relation

$$T_{xy} = \frac{1}{V}\left[\sum_{j=1}^{N}\frac{p_{jx}p_{jy}}{m} + \sum_{\text{pairs}} F(r_{ij})\frac{x_{ij}y_{ij}}{r_{ij}}\right] \tag{7-7-15}$$

This simple approach can be generalized by way of statistical theory. Define a tensor

$$\mathbf{S} = \sum_{i=1}^{N} \boldsymbol{r}_i \otimes \boldsymbol{p}_i \tag{7-7-16}$$

and differentiate $\mathbf{S}$ with respect to the time to give the relation

$$\frac{d\mathbf{S}}{dt} = \Sigma\left(\frac{\partial\mathbf{S}}{\partial\boldsymbol{r}_i}\frac{\boldsymbol{p}_i}{m} - \frac{\partial\mathbf{S}}{\partial\boldsymbol{p}_i}\frac{\partial H}{\partial\boldsymbol{r}_i}\right) \tag{7-7-17}$$

where

$$H = \frac{1}{2m}\sum_{i=1}^{N}\boldsymbol{p}_i\cdot\boldsymbol{p}_i + \Psi + \Phi \tag{7-7-18}$$

is the hamiltonian function. In Eq. (7-7-18), Ψ is the potential of Eq. (7-7-10) and Φ is a potential introduced to take account of interactions of particles with the walls of the container. Substituting for H in Eq. (7-7-17) from Eq. (7-7-18) gives

$$\frac{d\mathbf{S}}{dt} = \Sigma\left[\frac{\boldsymbol{p}_i\otimes\boldsymbol{p}_i}{m} - \boldsymbol{r}_i\otimes\frac{\partial\Phi}{\partial\boldsymbol{r}_i} - \sum_{i\neq j}\boldsymbol{r}_i\otimes\frac{\partial\psi(r_{ij})}{\partial\boldsymbol{r}_i}\right] \tag{7-7-19}$$

If $\mathbf{S}$ is time independent, then averaging Eq. (7-7-19) gives

$$\left\langle\frac{d\mathbf{S}}{dt}\right\rangle = \Sigma\left[\frac{\langle\boldsymbol{p}_i\otimes\boldsymbol{p}_i\rangle}{m} - \left\langle\boldsymbol{r}_i\otimes\frac{\partial\Phi}{\partial\boldsymbol{r}_i}\right\rangle - \tfrac{1}{2}\Sigma\left\langle(\boldsymbol{r}_i-\boldsymbol{r}_j)\otimes\frac{\partial\psi}{\partial\boldsymbol{r}_i}\right\rangle\right] = \mathbf{O} \tag{7-7-20}$$

where $\langle d\mathbf{S}/dt\rangle$ is here understood to be the expected mean value of the quantity $d\mathbf{S}/dt$ which would be obtained when $d\mathbf{S}/dt$ is actually measured at some given time t. At the surface of a body the stress tensor $\mathbf{T}$ may be assumed to be balanced at every point by the forces of interaction between the particles and the wall of the container, and it therefore follows from Eqs. (7-7-9) and (7-7-18) that

$$\sum_i \left\langle \boldsymbol{r}_i \otimes \frac{\partial \Phi}{\partial \boldsymbol{r}_i} \right\rangle = \int \boldsymbol{r} \otimes (\mathbf{T}^T \boldsymbol{n})\, da = V\mathbf{T} \tag{7-7-21}$$

where da is a surface element and $\boldsymbol{n}$ the unit normal to the surface. Hence, by using Eq. (7-7-21) in Eq. (7-7-20), the stress relation is obtained in the form

$$\mathbf{T} = n\left[\frac{1}{m}\langle \boldsymbol{p}_i \otimes \boldsymbol{p}_i \rangle - \frac{N}{2}\left\langle (\boldsymbol{r}_i - \boldsymbol{r}_j) \otimes \frac{\partial \psi}{\partial \boldsymbol{r}_i} \right\rangle\right] \tag{7-7-22}$$

In Eq. (7-7-22) the first term on the right-hand side refers to one particle and can therefore be reduced to a phase integration over the particle phase of one particle; this requires that the singlet distribution function be known. Thus, this first term can be regarded as applying specifically to a dilute gas where the interparticle force is negligible. The second term on the right-hand side of Eq. (7-7-22) involves two molecules through the intermolecular force potential ψ and will therefore involve the doublet distribution function. This second term applies to a dense gas or liquid. Hence, the two terms taken together allow for the calculation of the shear and bulk viscosities for the whole range of densities from that of a dilute gas up to that of a liquid. Since for liquids it is the second term on the right-hand side of Eq. (7-7-22) which provides the principal contribution to the flow of momentum, the contribution from the first term, i.e., the kinetic term, will here be neglected.

7-7-3 The Coefficient of Viscosity

The microphysical properties can now be related to the macrophysics of the transport of momentum in liquids by comparing the nondiagonal terms in Eq. (7-7-22) with the corresponding terms in Eq. (7-1-31).

Having regard to the dependence of the second term on the right-hand side of Eq. (7-7-22) on the relative separation of a pair of molecules, it is convenient to introduce relative coordinates. In this way the configuration of the pair is written in terms of the motion of the center of mass of the pair and the relative movement of the particles about the mass center. Thus, define

$$\boldsymbol{r}_+ = \tfrac{1}{2}(\boldsymbol{r}_2 + \boldsymbol{r}_1) \qquad \boldsymbol{r}_- = \tfrac{1}{2}(\boldsymbol{r}_2 - \boldsymbol{r}_1) \tag{7-7-23}$$

$$\boldsymbol{F}_+ = \tfrac{1}{2}(\boldsymbol{F}_2 + \boldsymbol{F}_1) \qquad \boldsymbol{F}_- = \tfrac{1}{2}(\boldsymbol{F}_2 - \boldsymbol{F}_1) \tag{7-7-24}$$

$$\boldsymbol{p}_+ = \boldsymbol{p}_2 + \boldsymbol{p}_1 \qquad \boldsymbol{p}_- = \boldsymbol{p}_2 - \boldsymbol{p}_1 \tag{7-7-25}$$

where the subscripts 1 and 2 refer to the two particles forming a pair.

In terms of relative coordinates, Eqs. (7-7-7) and (7-7-9) take the form

$$g^{(2)}(r_+, r_-) = A \exp\left[-\frac{\phi(\boldsymbol{r}_-)}{k\theta}\right] \tag{7-7-26}$$

and

$$2\langle \boldsymbol{F}_- \rangle = -\frac{\partial \phi}{\partial \boldsymbol{r}_-} \tag{7-7-27}$$

where ϕ is now the potential of the mean force associated with the difference coordinate r_- and A accounts for the averaged effects of the remainder.

Consider one particle of a pair to be located at the origin of a spherical polar coordinate system (r, θ, ϕ). In equilibrium the radial distribution function $g^{(2)}$ is spherically symmetrical, being only a function of r_-.

For nonequilibrium conditions, e.g., laminar shear flow, this symmetry will in general not be maintained. That is, the radial distribution function will be distorted, and for newtonian-type flow this distortion will be assumed to be in proportion to some unknown function $\psi_{\mathbf{D}}$. Thus, in nonequilibrium,

$$\frac{n^{(2)}(r_-, \theta, \phi, t) - g^{(2)}(r_-)}{g^{(2)}(r_-)} = \psi_{\mathbf{D}}(r_-) = \hat{\psi}_{\mathbf{D}}(r_-)\hat{R}(\theta, \phi) \tag{7-7-28}$$

where $n^{(2)}$ is the probability distribution function in coordinates which here determines the nonequilibrium behavior arising from momentum flow, and

$$\hat{R}(\theta, \phi) = \sin^2\theta \sin\phi \cos\phi \tag{7-7-29}$$

is an orientation factor, the unknown function $\hat{\psi}_{\mathbf{D}}(\boldsymbol{r}_-)$ now representing purely the radial part of the distortion.

The number of molecules whose centers lie within a volume element δV, which is centered a distance r from the origin and oriented at the angles θ and ϕ, is

$$\begin{aligned} \delta N &= \rho(r, \theta, \phi)\delta V \\ &= \frac{N}{V} n^{(2)} \delta V \end{aligned}$$

The contribution of these δN molecules to the shear stress tensor is

$$\frac{N}{V} n^{(2)}(\boldsymbol{r}_- \otimes \boldsymbol{F}_-)\delta V = \frac{N}{V}(\boldsymbol{r}_- \otimes \boldsymbol{F})r_-^2 n^{(2)} \sin\theta \hat{R}(\theta, \phi)\, dr_-\, d\theta\, d\phi$$

which can be integrated over the volume of the liquid for all N molecules to give, for the shear stress, the relation

$$\mathbf{T} = n^2 \int (\boldsymbol{r}_- \otimes \boldsymbol{F}_-) n^{(2)} r_-^2 \sin\theta \hat{R}(\theta, \phi)\, dr_-\, d\theta\, d\phi \tag{7-7-30}$$

Substituting for $n^{(2)}$ from Eq. (7-7-28) into Eq. (7-7-30) gives, by way of Eq. (7-7-27),

$$\mathbf{T} = \frac{2\pi}{15} n^2 \int \left(\boldsymbol{r}_- \otimes \frac{d\phi(r_-)}{d\boldsymbol{r}_-} \right) \hat{\psi}_{\mathbf{D}}(r_-) g^{(2)}(r_-) r_-^2\, dr_- \tag{7-7-31}$$

The simplest assumption regarding the dependence of the function $\psi_{\mathbf{D}}(\boldsymbol{r}_-)$ on the rate of deformation $\mathbf{D}$ is that of direct proportionality. This assumption enables $\psi_{\mathbf{D}}(\boldsymbol{r}_-)$ to be expressed in the form

$$\psi_{\mathbf{D}}(\boldsymbol{r}_-) = 2\frac{\beta m}{k\theta} u(\boldsymbol{r}_-) \frac{\boldsymbol{r}_- \cdot (\mathbf{D}\boldsymbol{r}_-)}{r_-^2} \tag{7-7-32}$$

For two-dimensional rectilinear shear flow, this simplifies to

$$\hat{\psi}_{\mathbf{D}}(\boldsymbol{r}_-) = 2\frac{\beta m}{k\theta} u(\boldsymbol{r}_-)\dot{\gamma} \tag{7-7-33}$$

where the constant of proportionality has been set equal to $2\beta m/k\theta$, $\dot{\gamma}$ is the rate of shear, and the unknown function $u(\boldsymbol{r}_-)$ represents the purely radial part of the distortion. Substituting for $\psi_{\mathbf{D}}(\boldsymbol{r}_-)$ in Eq. (7-7-31) from Eq. (7-7-33) gives

$$\eta = \frac{\pi}{15} \frac{\beta m n^2}{k\theta} \int_0^\infty \frac{d\phi(r_-)}{dr_-} g^{(2)}(r_-) u(r_-) r_-^3 \, dr_- \tag{7-7-34}$$

In Eq. (7-7-34), much is known about $\phi(\boldsymbol{r}_-)$ and $g^{(2)}(\boldsymbol{r}_-)$. For example, $\phi(\boldsymbol{r}_-)$ can be represented by the Lennard–Jones 6–12 potential and information about $g^{(2)}(\boldsymbol{r}_-)$ can be obtained from equilibrium theory and diffraction experiments.

Thus the problem reduces to evaluating the properties of the unknown function $u(\boldsymbol{r}_-)$.

7-7-4 Radial Distortion Function

In the molecular theory of liquids it is the distribution functions for molecular pairs which are important, in particular the now time-dependent radial distribution, i.e., the radial distortion function $n^{(2)}(r, \theta, \phi, t)$ of Eq. (7-7-28). The necessary reformulation of the brownian-motion-type theory for it to be applicable to molecular pairs has been given by Eisenschitz (1949) and by Kirkwood, Buff, and Green (1949).

Starting with Liouville's equation, the procedures outlined in Sec. 7-6-2 provide the necessary statistical foundation for the formulation of a Fokker–Planck equation for the pair. In terms of the time-smoothed, particle pair, specific distribution function $\bar{f}^{(2)}$ the Fokker–Planck equation can be expressed in the form

$$\frac{d\bar{f}^{(2)}}{dt} = \operatorname{tr}\,(\beta_{\mu\nu}^{(2)} \psi_{\mathrm{FP}}) \bar{f}^{(2)} \equiv M_{\mathrm{FP}} \tag{7-7-35}$$

where

$$\frac{d\bar{f}^{(2)}}{dt} = \frac{\partial \bar{f}^{(2)}}{\partial t} + \frac{1}{2m}\left[\left(\boldsymbol{p}_+ \cdot \frac{\partial}{\partial \boldsymbol{r}_+}\right) + \left(\boldsymbol{p}_- \cdot \frac{\partial}{\partial \boldsymbol{r}_-}\right)\right]\bar{f}^{(2)} + 2\left(\langle \boldsymbol{F}_- \rangle \cdot \frac{\partial}{\partial \boldsymbol{p}_-}\right)\bar{f}^{(2)} \tag{7-7-36}$$

and M_{FP} is the Fokker–Planck operator (see, for example, Eisenschitz, 1958; Cole, 1967). In Eq. (7-7-35), $\beta^{(2)}$ is the friction tensor (see, for example, Eisenschitz, 1958).

Since the second term on the right-hand side of Eq. (7-7-22) is regarded as the relevant contribution to the transport of momentum in liquids, it is important to note that this term depends on the relative coordinates of a pair of molecules. It is evident, therefore, that what is required is the probability distribution in relative coordinates. The required nonequilibrium doublet distribution $n^{(2)}$ can be expressed in relative coordinates in the form

$$n^{(2)}(\boldsymbol{r}, t) = N(N-1) \int f^{(2)}(\boldsymbol{r}, \boldsymbol{p}, t) \, d\boldsymbol{p}_+ \, d\boldsymbol{p}_- \tag{7-7-37}$$

where $f^{(2)}$ is the unsmoothed probability. Provided deviations from equilibrium are not too large, it is unnecessary to distinguish between the time-smoothed probability $\bar{f}^{(2)}$ and the unsmoothed probability $f^{(2)}$.

To simplify the derivation of $n^{(2)}$ from the Fokker–Planck equation it is assumed that the friction tensor $\beta^{(2)}$ of Eq. (7-7-35) can be expressed in the form

$$\beta^{(2)} = \beta \mathbf{I} \tag{7-7-38}$$

where the scalar β is just the friction constant of Eqs. (7-6-3) and (7-7-32). Regarding any physical basis for Eq. (7-7-38), it is to be noted that it has been shown in Sec. 7-1-1 that liquids are isotropic. Equation (7-7-35) now simplifies to the form

$$\frac{df^{(2)}}{dt} = \beta \psi_{\mathrm{FP}} f^{(2)} \tag{7-7-39}$$

where $\bar{f}^{(2)}$ has been replaced by the unsmoothed probability $f^{(2)}$. A condition on β is that it must be sufficiently large for the condition $\beta\tau \gg 1$ to hold.

With this simplification the required solution of the Fokker–Planck equation is now obtained by multiplying throughout Eq. (7-7-39) successively by $\boldsymbol{p}_+ \, d\boldsymbol{p}_+ \, d\boldsymbol{p}_-$ and $\boldsymbol{p}_- \, d\boldsymbol{p}_+ \, d\boldsymbol{p}_-$ and performing the integration. After applying certain approximations to simplify the integration, the Smoluchowski equation is obtained in the form

$$\frac{\partial n^{(2)}}{\partial t} + \left[\frac{\partial}{\partial \boldsymbol{r}_+} \cdot \frac{\partial}{\partial \boldsymbol{r}_+}\left(\frac{k\theta}{2m\beta} n^{(2)}\right)\right] + \left\{\frac{\partial}{\partial \boldsymbol{r}_-} \cdot \left[\frac{\partial}{\partial \boldsymbol{r}_-}\left(\frac{k\theta}{2m\beta} n^{(2)}\right) - \frac{1}{m\beta}\langle \boldsymbol{F}_- \rangle n^{(2)}\right]\right\} + O\left(\frac{1}{\beta^2}\right) = 0 \tag{7-7-40}$$

For laminar shear flow, terms must be added to the right-hand side of Eq. (7-7-40) to take account of the flow velocity $\boldsymbol{v}$. With this addition, substituting for $n^{(2)}$ in Eq. (7-7-40) from Eq. (7-7-28) gives, for the unknown function $\psi_{\mathbf{D}}(\boldsymbol{r}_-)$, the steady-state equation

$$\frac{k\theta}{2m\beta} \nabla^2 \psi_{\mathbf{D}} + \frac{1}{m\beta}\langle \boldsymbol{F}_- \rangle \cdot \boldsymbol{\nabla} \psi_{\mathbf{D}} = 2\frac{\langle \boldsymbol{F}_- \rangle \cdot (\mathbf{D}\boldsymbol{r}_-)}{k\theta} \tag{7-7-41}$$

This form of the Smoluchowski equation, due to Eisenschitz (1949) and Kirkwood, Buff, and Green (1949), is based on the concept of the particle mobility–forced diffusion approximation, the biasing of the free diffusion movement being due to the presence of an applied drift velocity in viscous flow.

Values of the viscosity can only be computed from Eq. (7-7-34) if the equation

$$\frac{d^2u(r_-)}{dr_-^2} + \left[\frac{2}{r_-} - \frac{1}{k\theta}\frac{d\phi(r_-)}{dr_-}\right]\frac{du(r_-)}{dr_-} - 6\frac{u(r_-)}{r_-^2} = -\frac{r_-}{k\theta}\frac{d\phi(r_-)}{dr_-} \tag{7-7-42}$$

obtained by substituting for $\psi(\boldsymbol{r}_-)$ in Eq. (7-7-41) from Eq. (7-7-32), can be solved to give a time-independent solution of $u(r_-)$.

Since Eq. (7-7-42) is a second-order equation, a solution can only be obtained by specifying two boundary conditions.

One condition, adopted by both Eisenschitz and Kirkwood, is that

$$u(r_-) = 0 \qquad \text{for } r_- \to \infty \tag{7-7-43}$$

a necessary condition to ensure that the distortion of equilibrium distributions does not increase with the distance between the molecules.

The second condition has been specified in two ways:

1. *Weak condition.* Kirkwood specifies that the function $u(r_-)$ must remain finite everywhere, and hence requires that

$$u(r_-) = 0 \qquad \text{for } r_- \to 0 \tag{7-7-44}$$

2. *Strong condition.* Eisenschitz drew attention to the fact that the condition of Eq. (7-7-44), although ensuring that $u(r_-)$ remains finite, permits a degree of anisotropy of the shear flow which does not appear to be in accord with experiment, and hence suggested the strong condition

$$r_-^3 u(r_-) = 0 \qquad \text{for } r_- \to \infty \tag{7-7-45}$$

   As r_- approaches zero, this condition implies that $u(r_-)$ can tend to infinity, a situation which is basically undesirable, although, as noted by Eisenschitz, this is not a serious objection provided the calculated viscosity remains finite, which it does in practice for both forms of the second condition.

Although the theory developed by Kirkwood and Eisenschitz provides an insight into the structure of liquids, it is, however, restricted in application by the hypothesis that, during the interval τ, only small momentum or energy transfers occur, thus excluding hard-core collisions and strong deflections. For this reason, the Kirkwood–Eisenschitz theory of molecular viscosity is generally referred to as the soft contribution to the transport coefficient. Various developments of the theory have been proposed; e.g., there is the Rice and Kirkwood (1959) small-step diffusion model which has also been extended to liquid mixtures by Rice and Allnatt (1961). The Rice–Allnatt theory (Allnatt and Rice, 1961; Rice and Allnatt, 1961) extends the basic concept to account for the contribution from hard-core collisions.

7-7-5 Hard-Core Collisions

The Rice–Allnatt theory is concerned with a potential which has the form of a rigid-core repulsion superimposed on an arbitrary soft potential. In this theory, an elementary dynamical event is characterized by an instantaneous hard-core collision, i.e., rigid-core repulsion, followed by quasi-brownian motion which is assumed to occur within the plateau time τ, and which has its origin in the long-range attractive part of the potential. Thus Rice and Allnatt consider the effective viscosity of a liquid to be the sum of two terms, such that

$$\eta = \eta_V(\sigma) + \eta_V(r > \sigma) \tag{7-7-46}$$

where $\eta_V(\sigma)$ represents the contribution due to rigid-core encounters and $\eta_V(r > \sigma)$ is the component arising from the intermolecular force contribution to the shear viscosity of liquids, being in particular simply that of Eq. (7-7-42).

Starting from the Liouville equation, the Rice–Allnatt kinetic equations for the distribution function are obtained in the form

$$\frac{d\bar{f}^{(2)}}{dt} = (\beta\psi_{FP} + \psi_H)\bar{f}^{(2)} \tag{7-7-47}$$

where $\psi_H \bar{f}^{(2)}$ describes the hard-core collisions. Equation (7-7-47) is to be compared with Eq. (7-7-39).

Rice and Allnatt proceed to express $\eta_V(\sigma)$ in the form

$$\eta_V(\sigma) = \sum_{i=1}^{3} \eta_V^{(i)}(\sigma) \tag{7-7-48}$$

where $\eta_V^{(1)}(\sigma)$ arises from the distortion of the momentum dependence of the local equilibrium distribution function $\bar{f}^{(2)}$, $\eta_V^{(2)}(\sigma)$ arises from the correction to the equilibrium value of $\bar{f}^{(2)}$, and $\eta_V^{(3)}(\sigma)$ arises from a distortion of the pair correlation function.

The component $\eta_V^{(1)}(\sigma)$ has the form

$$\eta_V^{(1)}(\sigma) = \frac{5k\theta}{8g^{(2)}(\sigma)}\left(\frac{2\pi n\sigma^3}{15}\right)\left[1 + \frac{4\pi}{15}n\sigma^3 g^{(2)}(\sigma)\right]\xi \tag{7-7-49}$$

where

$$\xi = \frac{1}{\Omega^{(2,2)} + 5\beta/8ng^{(2)}(\sigma)}\left[1 + \frac{4\Omega^{(2,2)}}{4\Omega^{(2,2)} + 5\beta/ng^{(2)}(\sigma)}\right] \tag{7-7-50}$$

$$\Omega^{(2,2)} = \left(\frac{4\pi k\theta}{m}\right)^{1/2}\sigma^2 \tag{7-7-51}$$

is the reduced collision cross section for the hard-core part of the potential, σ is the rigid-core diameter, $g^{(2)}(\sigma)$ is the pair correlation function evaluated at $r = \sigma$, and β is the appropriate friction constant determined by the intermolecular potential. The remaining two components have the form

$$\eta_V^{(2)}(\sigma) = \frac{5k\theta}{8g^{(2)}(\sigma)}\left(\frac{2\pi n\sigma^3}{15}\right)\left[\frac{4\pi}{15}n\sigma^3 g^{(2)}(\sigma)\right]\left[\frac{24}{\pi}\frac{g^{(2)}(\sigma)}{\Omega^{(2,2)}}\right] \tag{7-7-52}$$

$$\eta_V^{(3)}(\sigma) = -\frac{37}{70}\left(\frac{2\pi n\sigma^3}{15}\right)ng^{(2)}(\sigma)\beta m u(\sigma) \tag{7-7-53}$$

it being noted that the component $\eta_V^{(3)}(\sigma)$ does not appear in the theory of Kirkwood and Eisenschitz. Because of the appearance of the friction constant β in Eq. (7-7-53), the $\eta_V^{(3)}(\sigma)$ component is referred to by Rice and Allnatt as a cross-effect.

It is evident from Eqs. (7-7-49), (7-7-52), and (7-7-53) that the $\eta_V(\sigma)$ component of the viscosity is essentially a dense gas viscosity, its effectiveness

Table 7-1 Comparison of the calculated and observed values of the shear viscosity of liquid argon, in mPa s

θ, K	90	128	133.5	185.5
η_K	0.031	0.059	0.059	0.095
$\eta_V^{(1)}$	0.012	0.058	0.036	0.076
$\eta_V^{(2)}$	0.282	0.330	0.350	0.532
$\eta_V^{(3)}$	0.155	0.058	0.042	0.050
$\eta_V(r > \sigma)$	0.926	0.184	0.190	0.111
$\eta_{(\mathrm{calc})}$	1.406	0.689	0.677	0.864
$\eta_{(\mathrm{obs})}$	2.390	0.840	0.840	0.870

increasing with increasing temperature like the gas kinetic viscosity. For further details, the reader is referred to Rice and Gray (1965).

The predictions of the Rice–Allnatt theory of the shear viscosity have been compared with experimental results for liquid argon. The computations are based on the modified Lennard–Jones potential using the results given by Kirkwood, Lewinson, and Alder (1952). The problem of obtaining a reliable value for the friction constant β has been discussed in Sec. 7-6-2. Values of the calculated viscosity are given in Table 7-1, where they are compared with the experimental values given by Lowry, Rice, and Gray (1964) (see also Cook, 1961).

7-7-6 Concluding Remarks

Recent work has been directed toward obtaining general transport equations starting from the exact laws of mechanics. For example, there is the study of classical mechanical systems by Bogoliubov (1962) (see also Cohen, 1962; Green and Piccirelli, 1963) and the study of quantum mechanical systems by Hove (1961). Prigogine and coworkers (Prigogine, 1962; Balescu, 1963; Resibois, 1966) have developed a theory which can be applied both to classical and quantum mechanical systems.

The general approach by Prigogine (1962) has been applied by Prigogine, Nicolis, and Misguich (1965) to the study of transport processes in dense media. Development of the theory, which is based on the general kinetic equations of nonequilibrium mechanics, is by way of two approximations. The collisions between particles are considered to be efficient enough to relax the one-particle distribution function $f^{(1)}$ to its local equilibrium value $f_0^{(1)}$ and to build up very quickly a local equilibrium state for the two-particle distribution function. This approximation gives only the potential part of the transport coefficients. That is, the study of purely kinetic transport such as diffusion is excluded. The second approximation is concerned with the particular way of linearizing the kinetic equation with respect to the stresses. Values of the viscosity of argon, calculated by way of the Prigogine, Nicolis, Misguich (PNM) theory are compared with experiment in Table 7-2.

Table 7-2 Comparison of calculated and observed values of the shear viscosity of liquid argon, in mPa s

θ, K	90	128	133.5	185.5
$\eta_{\text{(observed)}}$	2.39	0.84	0.84	0.87
$\eta(\text{PNM})_{\text{calc}}$	1.31	0.82	0.83	0.86
$\eta(\text{RA})_{\text{calc}}$	1.41	0.69	0.70	0.86
$\eta(\text{WD})_{\text{calc}}$	1.76	0.80	0.81	1.01

Table 7-3 Comparison of the hard-core and soft force contributions to the viscosity

	θ, K	$\dfrac{\eta_V(\sigma)}{\hat{\eta}}$	$\dfrac{\eta_V(r > \sigma)}{\hat{\eta}}$
Argon	85.7	0.921	1.805
	104.3	0.998	1.389
	120.0	1.045	1.123
	141.2	1.073	0.975
Krypton	122.1	0.914	1.800
	148.7	0.994	1.359
	171.0	1.035	1.101
	201.2	1.062	0.855
Xenon	192.2	1.040	1.511
	221.0	1.088	1.228
	260.0	1.116	0.926

In Table 7-3 there is compared the relative contributions to the PNM shear viscosity of the inert gas liquids under their saturation vapour pressure. It is evident that for the PNM theory the hard and soft parts of the viscosity are of comparable magnitude for temperatures ranging from the melting points to the critical points.

A comparison has been made in Table 7-2 between the PNM predictions and those of the Rice–Allnatt (RA) theory. Also included in the table are predictions based on the Wei and Davis (WD) (1966, 1967) modification of the Rice–Allnatt theory. The lowest temperature (90 K) is the only point below the critical pressure. At this point all three theories seriously underestimate the viscosity. Above the critical pressure, the PNM predictions are all within the experimental limit of accuracy (2 percent), in contrast to 12 percent for RA and 8 percent for WD.

Having regard to the present state of development of the respective theories, it may be tentatively concluded that the PNM theory is in somewhat better agreement with experiment than the RA-type theory, when the predictions of both theories are computed with the same pair correlation function.

It is important to note, however, that the solution of the many-body problem inherent in the formal results of modern transport theory still remains an open question.

Both the approximate theories of liquid transport and those theories derived from more general kinetic equations are at present applicable mainly to monatomic liquids such as argon, krypton, and xenon. However, most liquids of technical interest are generally composed of complex groupings of the elementary particles, i.e., the elementary particles, be they atoms or molecules, form multiparticle aggregates. Attempts to describe the liquid state in such a way as to take account of the possibility of aggregation have been made without direct reference to either the gaseous or solid states of matter. The simplest method is that due to Bernal and coworkers (Bernal, 1958, 1959, 1960; Bernal and Mason, 1960) and of Scott (1960, 1962). This approach is at present concerned with the geometrical problem of the close packing of particles and the related problem of aggregation. Bernal (1965) has drawn attention to the importance in any theory of the liquid state of constructing a statistical geometry. Thus, Scott, Bernal, Mason, and Knight (1962) have studied the geometrical arrangements in the packing of some thousand steel balls. It is not immediately apparent as to how this approach will give a physically satisfying theory for the transport of momentum in complex liquids, though the structure of aggregates in relation to the molecular kinematics of the viscosity of a Bernal liquid has been considered by Smith (1964). Various semiempirical correlation schemes for relating viscosity, temperature, and pressure have been proposed. The general aim of this type of approach is to deduce in a mainly qualitative sense the effect of molecular structure on the viscosity of nonassociated and associated liquids, and to some extent mixtures and solutions (see, for example, Bondi, 1967).

GENERAL REFERENCES

Barr, G. (1931) "A Monograph of Viscometry," Oxford University Press, Oxford.

Barker, J. A. (1963) Lattice Theories of the Liquid State, in J. S. Rowlinson (ed.), "The International Encyl. Phys. Chem. and Chem. Phys.," Pergamon Press, Oxford.

Batchelor, G. K. (1970) "Fluid Dynamics," Cambridge University Press, London.

Born, M., and H. S. Green (1949) "A General Kinetic Theory of Liquids," Cambridge University Press, London.

Cole, G. H. A. (1967) "The Statistical Theory of Classical Simple Dense Fluids," Pergamon Press, Oxford.

Coleman, B. D., H. Markovitz, and W. Noll (1966) "Viscometric Flows of Non-Newtonian Fluids," Springer-Verlag, Berlin.

Croxton, C. A. (1974) "Liquid State Physics—A Statistical Mechanical Introduction," Cambridge University Press, London.

Croxton, C. A. (1975) "Introduction to Liquid State Physics," Wiley, London.

Dinsdale, A., and F. Moore (1962) "Viscosity and its Measurement," Chapman and Hall, London.

Eisenschitz, R. (1958) "Statistical Theory of Irreversible Processes," Oxford University Press, Oxford.

Green, H. S. (1952) "Molecular Theory of Fluids," Cambridge University Press, London.

Haken, H. (1977) "Synergetics," Springer-Verlag, Berlin.

Hausen, J. P., and I. R. McDonald (1976) "Theory of Simple Liquids," Academic Press, New York.

Hirschfelder, J. O., C. F. Curtiss, and R. B. Bird (1954) "Molecular Theory of Gases and Liquids," Wiley, New York.
Lamb, H. (1932) "Hydrodynamics," 6th ed., Cambridge University Press, London.
Landau, L. D., and E. M. Lifshitz (1959) "Fluid Mechanics," Pergamon Press, Oxford.
March, N. H., and M. P. Tosi (1976) "Atomic Dynamics in Liquids," Macmillan, London.
Merrington, A. C. (1949) "Viscometry," Edward Arnold, London.
Milne-Thomson, L. M. (1967) "Theoretical Hydrodynamics," 5th ed., Macmillan, London.
Prigogine, I. (1955) "Introduction to Thermodynamics of Irreversible Processes," Thomas, New York.
Prigogine, I. (1962) "Non-equilibrium Statistical Mechanics," Interscience, New York.
Rice, S. A., and P. Gray (1965) "The Statistical Mechanics of Simple Liquids," Interscience, New York.
Rowlinson, J. S. (1959) "Liquids and Liquid Mixtures," Academic Press, New York.
Truesdell, C., and R. Toupin (1960) The Classical Field Theories, in S. Flügge (ed.), "Handbuch der Physik," vol. III/1, pp. 226–858, Springer-Verlag, Berlin.

REFERENCES

Allnatt, A. R., and S. R. Rice (1961) *J. Chem. Phys.*, vol. 34, pp. 2156–2165.
Andrade, E. N. da C., and Y. S. Chiong (1936) *Proc. Phys. Soc.*, vol. 48, pp. 247–260.
Andrade, E. N. da C., and E. R. Dobbs (1952) *Proc. Roy. Soc. (London)*, vol. A211, pp. 12–30.
Balescu, R. (1963) "Statistical Mechanics of Charged Particles," Interscience, New York.
Bantoft, E. (1959) Some Measurements on Pigment-Plasticizer Dispersions, in C. C. Mill (ed.), "Rheology of Disperse Systems," pp. 105–126, Pergamon Press, New York.
Bauer, W. H., and A. P. Finkelstein (1963) *Trans. Soc. Rheol.*, vol. 7, pp. 424–425.
Bernal, J. D. (1958) *Acta. Phys. Hungary*, vol. 8, pp. 269–276.
Bernal, J. D. (1959) *Nature*, vol. 183, pp. 141–147.
Bernal, J. D. (1960) *Nature*, vol. 185, pp. 68–70.
Bernal, J. D. (1965) The Geometry of the Structure of Liquids, in Th. J. Hugel (ed.), "Liquids: Structure, Properties, Solid Interactions," pp. 25–47, Elsevier, Amsterdam.
Bernal, J. D., and J. Mason (1960) *Nature*, vol. 188, pp. 910–911.
Billington, E. W. (1965) *J. Sci. Instrum.*, vol. 42, pp. 569–575.
Bird, R. B., and R. M. Turian (1962) *Chem. Engng. Sci.*, vol. 17, pp. 331–334.
Bogoliubov, N. N. (1962) Problems of Dynamical Theory in Statistical Physics, in J. de Boer and G. E. Uhlenbeck (eds.), "Studies in Statistical Mechanics," vol. 1, pp. 1–118, North-Holland, Amsterdam.
Bondi, A. (1967) Viscosity and Molecular Structure, in F. R. Eirich (ed.), "Rheology," vol. 4, pp. 1–83, Academic Press, New York.
Cohen, E. G. D. (1962) *Physica*, vol. 28, pp. 1025–1044, 1045–1059, 1060–1073.
Cole, G. H. A. (1960) *Nature*, vol. 186, p. 301.
Cook, G. A. (1961) "Argon, Helium and Rare Gases," Interscience, New York.
Cox, D. B. (1962) *Nature*, vol. 193, p. 670.
Cox, W. P., L. E. Nielsen, and R. Keeney (1957) *J. Polym. Sci.*, vol. 26, pp. 365–374.
Eisenschitz, R. (1949) *Proc. Phys. Soc.*, vol. A62, pp. 41–49.
Eisenschitz, R., and M. J. Wilford (1962) *Proc. Phys. Soc.*, vol. 80, pp. 1078–1090.
Furakawa, K. (1962) *Rep. Prog. Phys.*, vol. 25, pp. 395–440.
Green, M. S., and H. Piccirelli (1963) *Phys. Rev.*, vol. 132, pp. 1388–1410.
Guinier, A. (1963) "X-ray Diffraction in Crystals, Imperfect Crystals and Amorphous Bodies," Freeman, San Francisco, Calif.
Hatschek, E. (1913) *Trans. Faraday Soc.*, vol. 9, pp. 80–92.
Helfand, E. (1961) *Phys. Fluids*, vol. 4, pp. 681–691.
Higginbotham, R. S. (1950) *J. Sci. Instrum.*, vol. 27, pp. 139–141.
Hollis-Hallett, A. C. (1951) *Proc. Roy. Soc. (London)*, vol. A210, pp. 404–426.
Hopkins, M. R., and T. C. Toye (1950) *Proc. Phys. Soc.*, vol. B63, pp. 773–782.

Hoppmann, W. H., and C. N. Baronet (1964) *Nature*, vol. 201, pp. 1205–1206.
Hoppmann, W. H., and C. E. Miller (1963) *Trans. Soc. Rheol.*, vol. 7, pp. 181–193.
Hove, L. van (1961) "Fundamental Problems in Statistical Mechanics," North-Holland, Amsterdam.
Hutton, J. F. (1963) *Nature*, vol. 200, pp. 646–648.
Jobling, A., and J. E. Roberts (1959) *J. Polym. Sci.*, vol. 36, pp. 421–431, 433–441.
Kirkwood, J. G. (1946) *J. Chem. Phys.*, vol. 14, pp. 180–201.
Kirkwood, J. G., F. P. Buff, and H. S. Green (1949) *J. Chem. Phys.*, vol. 17, pp. 988–994.
Kirkwood, J. G., V. A. Lewinson, and B. J. Alder (1952) *J. Chem. Phys.*, vol. 20, pp. 929–938.
Lowry, B. A., S. A. Rice, and P. Gray (1964) *J. Chem. Phys.*, vol. 40, pp. 3673–3683.
Markovitz, H. (1952) *J. Appl. Phys.*, vol. 23, pp. 1070–1077.
Markovitz, H., P. M. Yavorsky, R. C. Harper, L. J. Zapas, and T. W. De Witt (1952) *Rev. Sci. Instrum.*, vol. 23, pp. 430–437.
McKennell, R. (1953) In V. G. W. Harrison (ed.), *Proc. Second Int. Congr. Rheol.*, Oxford, pp. 350–359, Butterworths, London.
McKennell, R. (1956) *Anal. Chem.*, vol. 28, pp. 1710–1714.
Mooney, M., and R. G. Ewart (1934) *Physics*, vol. 5, pp. 350–354.
Oldroyd, J. G. (1949) *J. Colloid Sci.*, vol. 4, pp. 333–342.
Oldroyd, J. G. (1951) *Quart. J. Mech. Appl. Math.*, vol. 4, pp. 271–282.
Oldroyd, J. G., D. J. Strawbridge, and B. A. Toms (1951) *Proc. Phys. Soc.*, vol. B64, pp. 44–57.
Pollett, W. F. O. (1955) *Brit. J. Appl. Phys.*, vol. 6, pp. 199–206.
Prigogine, I., G. Nicolis, and J. Misguich (1965) *J. Chem. Phys.*, vol. 43, pp. 4516–4521.
Resibois, P. (1966) A Perturbative Approach to Irreversible Statistical Mechanics, in E. Meeron (ed.), "Physics of Many-Particle Systems," pp. 495–629, Gordon and Breach, New York.
Rice, S. A., and A. R. Allnatt (1961) *J. Chem. Phys.*, vol. 34, pp. 2144–2155.
Rice, S. A., and J. G. Kirkwood (1959) *J. Chem. Phys.*, vol. 31, pp. 901–908.
Roscoe, R. (1958) *Proc. Phys. Soc.*, vol. 72, pp. 576–584.
Roscoe, R., and W. Bainbridge (1958) *Proc. Phys. Soc.*, vol. 72, pp. 585–595.
Schlichting, H. (1955) "Boundary Layer Theory," McGraw-Hill, New York.
Scott, G. D. (1960) *Nature*, vol. 188, pp. 908–909.
Scott, G. D. (1962) *Nature*, vol. 194, pp. 956–957.
Scott, G. D., J. D. Bernal, J. Mason, and K. R. Knight (1962) *Nature*, vol. 194, pp. 957–958.
Smith, F. W. (1964) *Canadian J. Phys.*, vol. 42, pp. 304–320.
Sneddon, I. N. (1951) "Fourier Transforms," McGraw-Hill, New York.
Swindells, J. F., J. R. Coe, and T. B. Godfrey (1952) *J. Res. Nat. Bur. Stand.*, vol. 48, pp. 1–31.
Taylor, G. I. (1923) *Phil. Trans. Roy. Soc.*, vol. A223, pp. 289–343.
Taylor, G. I. (1935) *Proc. Roy. Soc. (London)*, vol. A151, pp. 421–478.
Taylor, G. I. (1936) *Proc. Roy. Soc. (London)*, vol. A157, pp. 546–578.
Turner, R. E. (1960) *Physica*, vol. 26, pp. 274–283.
Walters, K. (1960) *Quart. J. Mech. Appl. Math.*, vol. 13, pp. 444–461.
Walters, K. (1961) *Quart. J. Mech. Appl. Math.*, vol. 14, pp. 431–436.
Wei, C. C., and H. T. Davis (1966) *J. Chem. Phys.*, vol. 45, pp. 2533–2544.
Wei, C. C., and H. T. Davis (1967) *J. Chem. Phys.*, vol. 46, pp. 3456–3467.

CHAPTER

EIGHT

NONLINEAR ELASTICITY

8-1 PURELY ELASTIC MATERIALS

8-1-1 Finite Elasticity

If for a material maintained at rest for a long time in static deformation the stress depends only on the rest configuration and not on the strain history prior to the start of the static deformation, then the material is said to exhibit elastic response. In terms of a single particle the principles of determinism and local action imply for an elastic material that the stress at a particle X is determined by the present local configuration of an arbitrary small neighborhood $\mathcal{N}(X)$ of X relative to its initial configuration. That is, the stress $\mathbf{T}(X,t)$ at X is determined by the set of relative motions $\chi(Z,t)-\chi(X,t)$ of all particles Z in the neighborhood $\mathcal{N}(X)$ of X. For a sufficiently small neighborhood, assuming differentiability, it has been shown in Sec. 5-1 that the general constitutive equation has the form

$$\mathbf{T}(X,t) = \mathbf{G}[\chi^t(s), \mathbf{F}^t(s), X, t] \tag{8-1-1}$$

before the principle of material objectivity has been applied. The most general constitutive equation of an elastic material for a given X under static or equilibrium conditions can now be obtained, by application of the principle of material objectivity, in the form

$$\mathbf{T} = \mathfrak{g}(\mathbf{F}) \tag{8-1-2}$$

by assuming that $\mathbf{F}^t(s) = \mathbf{F}$, a constant for all s [cf. Eqs. (5-1-16) and (5-2-14)]. In Eq. (8-1-2), $\mathfrak{g}$ is the response function of the elastic material.

Alternatively the stress can be assumed to depend only on $\mathbf{F}^t(s) = \mathbf{F}(t)$, i.e., for the particular condition $s = 0$, giving the stress relation

$$\mathbf{T}(t) = \mathfrak{g}(\mathbf{F}(t)) \tag{8-1-3}$$

which is entirely independent of the history of the motion of the body before it arrived in the present configuration upon which it alone depends. The behavior of Eq. (8-1-3) when combined with Cauchy's laws of motion is referred to as dynamic elasticity. The present discussion will, however, be restricted to a study of elastic materials in static, or equilibrium, deformation.

It is important to note that in terms of purely elastic materials the concept of elasticity has been introduced by way of an idealized physical assumption, relating to the way in which the materials respond to deformation. However, for simple materials this physical assumption is quite general. Thus the concept of elasticity applies to a narrow idealized class of materials in all motions and to a broad class of simple materials at rest. In general the response function $\mathfrak{g}$ is a nonlinear tensor-valued function of the single-tensor argument $\mathbf{F}$.

As already noted in Sec. 5-3, the deformation gradient tensor depends on the reference configuration B_r selected. Application of the special condition relating to elasticity, that is, $\mathbf{F}^t(s) = \mathbf{F}$, a constant for all s [see Eq. (5-3-7)], gives

$$\mathbf{T} = \mathfrak{g}_{\varkappa}(\mathbf{F}) = \mathfrak{g}_{\hat{\varkappa}}(\mathbf{F}\mathbf{P}^{-1}) \tag{8-1-4}$$

for all $\mathbf{F}$. An elastic solid is an elastic material for which there exists a reference configuration, called an undistorted state, such that the isotropy group is a subgroup of the orthogonal group. For an elastic material, an undistorted configuration in which the stress is zero is called a natural state (see Sec. 5-3).

Under arbitrary rotations, the principle of material objectivity imposes the restriction that $\mathfrak{g}$ must satisfy the relation

$$\mathbf{Q}\mathfrak{g}(\mathbf{F})\mathbf{Q}^T = \mathfrak{g}(\mathbf{Q}\mathbf{F}) \tag{8-1-5}$$

for all $\mathbf{F}$ and all orthogonal $\mathbf{Q}$ in order that the response functions $\mathfrak{g}$ shall be admissible. This identity, for a special choice of $\mathbf{Q}$, gives further important information about the possible forms which acceptable constitutive equations can have.

Equation (8-1-5) states that in observer transformations, i.e., when $\mathbf{F}$ is transformed as a vector, the response function $\mathfrak{g}$ is an isotropic function. The physical significance of this condition, imposed by the principle of material objectivity, can be seen as follows. According to Eq. (2-3-2) the deformation gradient $\mathbf{F}$ has the decomposition

$$\mathbf{F} = \mathbf{R}\mathbf{U} \tag{8-1-6}$$

where $\mathbf{R}$ is the orthogonal rotation tensor and $\mathbf{U}$ is the positive, symmetric, right stretch tensor. Since $\mathbf{R}^T\mathbf{R} = \mathbf{I}$, Eq. (8-1-6) gives

$$\mathbf{R}^T\mathbf{F} = (\mathbf{R}^T\mathbf{R})\mathbf{U} = \mathbf{U} \tag{8-1-7}$$

By making the special choice

$$\mathbf{Q} = \mathbf{R}^T \tag{8-1-8}$$

Eq. (8-1-5) becomes

$$\mathfrak{g}(\mathbf{R}^T\mathbf{F}) = \mathbf{R}^T\mathfrak{g}(\mathbf{F})\mathbf{R}$$

which in turn can be expressed by way of Eq. (8-1-7) in the form

$$\mathfrak{g}(\mathbf{U}) = \mathbf{R}^T\mathfrak{g}(\mathbf{F})\mathbf{R}$$

and since $\mathbf{R}^T\mathbf{R} = \mathbf{I}$ this transforms to

$$\mathfrak{g}(\mathbf{F}) = \mathbf{R}\mathfrak{g}(\mathbf{U})\mathbf{R}^T \tag{8-1-9}$$

Hence the elastic stress relation of Eq. (8-1-2) admits the reduced representation

$$\mathbf{T} = \mathbf{R}\mathfrak{g}(\mathbf{U})\mathbf{R}^T \tag{8-1-10}$$

and since $\mathbf{U}$ is objective, $\mathfrak{g}$ as a function of $\mathbf{U}$ is an isotropic function in all observer transformations. Thus, according to Eq. (8-1-10), material objectivity requires that the dependence on $\mathbf{F}$ must take the form of an arbitrary function of $\mathbf{U}$ together with the additional explicit dependence on $\mathbf{R}$. From Eq. (8-1-10) the response function $\mathfrak{g}$ can be determined from observations involving pure deformations alone, relative to a reference configuration B_r, since the effect of rotation is merely to rotate the stress by means of the rotation $\mathbf{R}$ relative to B_r.

Although $\mathbf{R}$ and $\mathbf{U}$ have direct physical significance, it has, however, been noted in Sec. 5-6 that neither can be simply calculated from the motion $\chi(X, t)$. However, $\mathbf{F}$ and therefore $\mathbf{C} = \mathbf{U}^2 = \mathbf{F}^T\mathbf{F}$ can be readily calculated. Hence, it is sometimes convenient to introduce another response function $\mathfrak{f}$ by means of the definition

$$\mathfrak{f}(\mathbf{U}^2) = \mathbf{U}\mathfrak{g}(\mathbf{U})\mathbf{U} \tag{8-1-11}$$

from which Eq. (8-1-10) can be written in the alternative form

$$\bar{\mathbf{T}} = \mathfrak{f}(\mathbf{C}) \tag{8-1-12}$$

where

$$\bar{\mathbf{T}} = \mathbf{F}^T\mathbf{T}\mathbf{F} \tag{8-1-13}$$

is called the convective stress tensor. In Eq. (8-1-12), $\mathbf{C}$ is the right Cauchy–Green deformation tensor with respect to the reference configuration B_r.

Setting

$$\mathfrak{h}(\mathbf{C}) = \mathbf{C}^{-1}\mathfrak{f}(\mathbf{C})\mathbf{C}^{-1} \tag{8-1-14}$$

permits Eq. (8-1-12) to be expressed, by way of Eq. (8-1-13), in the form

$$\mathbf{T} = \mathbf{F}\mathfrak{h}(\mathbf{C})\mathbf{F}^T \tag{8-1-15}$$

Introducing a modified response function $\mathfrak{p}(\mathbf{C})$ by the relation

$$\mathfrak{p}(\mathbf{C}) = \mathbf{U}\mathfrak{h}(\mathbf{C})\mathbf{U} \tag{8-1-16}$$

permits Eq. (8-1-15) to be expressed in the form

$$\mathbf{T} = \mathbf{R}\mathfrak{p}(\mathbf{C})\mathbf{R}^T \tag{8-1-17}$$

Equations (8-1-10), (8-1-12), and (8-1-17) are reduced forms of Eq. (8-1-2); many other reduced forms are possible. For the particular condition that the reference configuration is a natural state, the various response functions must satisfy the condition

$$\mathfrak{g}(\mathbf{I}) = \mathfrak{f}(\mathbf{I}) = \mathfrak{p}(\mathbf{I}) = \mathbf{O}$$

Replacement of the Cauchy stress $\mathbf{T}$ in the constitutive equation by the nominal stress $\mathbf{T}^0$ defined in Sec. 3-7 by the relation

$$\mathbf{T}^0 = J\mathbf{F}^{-1}\mathbf{T} \tag{8-1-18}$$

where $J = \det \mathbf{F}$, gives further reduced forms. From Eq. (8-1-10),

$$J^{-1}\mathbf{F}\mathbf{T}^0 = \mathbf{R}\mathfrak{g}(\mathbf{U})\mathbf{R}^T$$

or

$$J^{-1}\mathbf{R}^T\mathbf{F}\mathbf{T}^0 = \mathfrak{g}(\mathbf{U})\mathbf{R}^T$$

which takes, by way of Eq. (8-1-7), the form

$$\mathbf{T}^0 = \mathfrak{g}^0(\mathbf{U})\mathbf{R}^T \tag{8-1-19}$$

where

$$\mathfrak{g}^0(\mathbf{U}) = J\mathbf{U}^{-1}\mathfrak{g}(\mathbf{U}) \tag{8-1-20}$$

is an arbitrary function of $\mathbf{U}$.

In Sec. 3-7 the symmetric second Piola–Kirchhoff tensor $\tilde{\mathbf{T}}$ is shown related to $\mathbf{T}^0$ by the relation

$$\mathbf{T}^0 = \tilde{\mathbf{T}}\mathbf{F}^T \tag{8-1-21}$$

Since $\tilde{\mathbf{T}}$ is symmetric, Eqs. (8-1-19) and (8-1-21) give

$$(\mathbf{T}^0)^T = \mathbf{F}\tilde{\mathbf{T}} = \mathbf{R}[\mathfrak{g}^0(\mathbf{U})]^T$$

or

$$\mathbf{R}^T\mathbf{F}\tilde{\mathbf{T}} = [\mathfrak{g}^0(\mathbf{U})]^T$$

and by Eq. (8-1-7) this gives

$$\mathbf{U}\tilde{\mathbf{T}} = [\mathfrak{g}^0(\mathbf{U})]^T$$

and hence

$$\tilde{\mathbf{T}} = \tilde{\mathfrak{g}}(\mathbf{U}) \tag{8-1-22}$$

but, since $\mathbf{C} = \mathbf{U}^2$,

$$\tilde{\mathbf{T}} = \tilde{\mathfrak{p}}(\mathbf{C}) \tag{8-1-23}$$

where

$$\tilde{\mathfrak{p}}(\mathbf{C}) = \tilde{\mathfrak{g}}(\mathbf{C}^{1/2})$$

Equations (8-1-22) and (8-1-23) are alternative reduced forms.

Thus, in contrast to the use of the Cauchy stress tensor **T**, the principle of material objectivity leads to the conclusion that in terms of the second Piola–Kirchhoff stress the response function of finite elasticity depends only on one of the deformation tensors and not on the rotation.

Equation (8-1-18) can be expressed, by way of Eq. (8-1-2), in the alternative form

$$\mathbf{T}^0 = (\det \mathbf{F})\mathbf{F}^{-1}\mathfrak{g}(\mathbf{F}) = \mathfrak{g}^+(\mathbf{F}) \tag{8-1-24}$$

from which follows the relation

$$\mathfrak{g}^+(\mathbf{QF}) = \det(\mathbf{QF})\,(\mathbf{QF})^{-1}\mathfrak{g}(\mathbf{QF})$$

which in turn takes, by way of Eq. (8-1-5), the form

$$\mathbf{QF}\mathfrak{g}^+(\mathbf{QF}) = \det(\mathbf{QF})\mathbf{Q}\mathfrak{g}(\mathbf{F})\mathbf{Q}^T$$

that is,

$$\mathfrak{g}^+(\mathbf{QF}) = (\det \mathbf{F})\mathbf{F}^{-1}\mathfrak{g}(\mathbf{F})\mathbf{Q}^T$$

and, using Eq. (8-1-24), this gives

$$\mathfrak{g}^+(\mathbf{QF}) = \mathfrak{g}^+(\mathbf{F})\mathbf{Q}^T \tag{8-1-25}$$

as the identity the function $\mathfrak{g}^+(\mathbf{F})$ must satisfy for arbitrary orthogonal **Q** if $\mathbf{T}^0$ is to be taken as a function of **F**. Similarly, setting

$$\tilde{\mathbf{T}} = \tilde{\mathfrak{g}}^+(\mathbf{F}) \tag{8-1-26}$$

then since $\tilde{\mathbf{T}}$ is symmetric, it follows from Eq. (8-1-21) that

$$\mathfrak{g}^+(\mathbf{F}) = \mathbf{F}^{-1}(\mathbf{T}^0)^T = \mathbf{F}^{-1}\tilde{\mathfrak{f}}(\mathbf{F})$$

and

$$\mathfrak{g}^+(\mathbf{QF}) = (\mathbf{QF})^{-1}\tilde{\mathfrak{f}}(\mathbf{QF})$$

or

$$\tilde{\mathfrak{g}}(\mathbf{QF}) = \mathbf{F}^{-1}\tilde{\mathfrak{f}}(\mathbf{F}) = \tilde{\mathfrak{g}}^+(\mathbf{F}) \tag{8-1-27}$$

which is just the identity to be satisfied by $\tilde{\mathfrak{g}}^+$ for arbitrary orthogonal **Q**.

8-1-2 Isotropic Elasticity

From Eq. (5-3-14), an elastic material can be defined as an isotropic material if and only if its response function [Eq. (8-1-2)], taken relative to an undistorted state as the reference configuration, is an isotropic tensor function of **F**, for which condition $\mathfrak{g}$ must satisfy the identity

$$\mathbf{Q}\mathfrak{g}(\mathbf{F})\mathbf{Q}^T = \mathfrak{g}(\mathbf{QFQ}^T) \tag{8-1-28}$$

for all **F** and all orthogonal **Q**.

Application of the isotropy property of Eq. (8-1-28) to Eqs. (8-1-10) and (8-1-17) gives the isotropic stress relations

$$\mathbf{T} = \mathfrak{g}(\mathbf{V}) \tag{8-1-29}$$

$$\mathbf{T} = \mathfrak{p}(\mathbf{B}) \tag{8-1-30}$$

where $\mathbf{V}$ is the left stretch tensor (see Sec. 2-3) and

$$\mathbf{B} = \mathbf{V}^2 = \mathbf{F}\mathbf{F}^T = \mathbf{R}\mathbf{C}\mathbf{R}^T$$

is the left Cauchy–Green strain tensor (see Sec. 2-3). In Eq. (8-1-30) the response function $\mathfrak{p}$ is defined by $\mathfrak{g}(\mathbf{V}) = \mathfrak{p}(\mathbf{V}^2)$. Since a material is isotropic with respect to certain properties only if these properties are the same in all directions, it is to be expected that isotropic elastic materials will be independent of an arbitrary rotation, a conclusion in agreement with Eqs. (8-1-29) and (8-1-30). It is also evident from Eqs. (8-1-29) and (8-1-30) that for the case of two applied stretches differing only by an orthogonal tensor, the resultant stress will differ by the same orthogonal tensor.

In the constitutive equations (8-1-29) and (8-1-30), both $\mathfrak{g}$ and $\mathfrak{p}$ are isotropic tensor functions. Hence, the representation theorems available for isotropic tensor functions of one tensor variable can be applied to these constitutive equations. For example, it is evident from Sec. 1-10 that the function $\mathfrak{p}(\mathbf{B})$ has a representation of the form [see Eq. (1-10-21)]

$$\mathbf{T} = \mathfrak{p}(\mathbf{B}) = \alpha_0\mathbf{I} + \alpha_1\mathbf{B} + \alpha_2\mathbf{B}^2 \tag{8-1-31}$$

where α_0, α_1, and α_2 are scalar invariant functions of the three principal invariants $I_{\mathbf{B}}$, $II_{\mathbf{B}}$, $III_{\mathbf{B}}$ of $\mathbf{B}$ or, equivalently, of the principal values of $\mathbf{B}$.

For the case that the argument tensor has an inverse, it follows from the Cayley–Hamilton theorem, that is, Eq. (1-10-19) expressed in the form

$$\mathbf{B}^3 = I_{\mathbf{B}}\mathbf{B}^2 - II_{\mathbf{B}}\mathbf{B} + III_{\mathbf{B}}\mathbf{I}$$

that

$$\mathbf{B}^2 = I_{\mathbf{B}}\mathbf{B} - II_{\mathbf{B}}\mathbf{I} + III_{\mathbf{B}}\mathbf{B}^{-1}$$

and hence that an equivalent form of Eq. (8-1-31) can be obtained in the form

$$\mathbf{T} = \mathfrak{p}(\mathbf{B}) = \beta_0\mathbf{I} + \beta_1\mathbf{B} + \beta_{-1}\mathbf{B}^{-1} \tag{8-1-32}$$

where again β_0, β_1, and β_{-1} are scalar invariant functions of $I_{\mathbf{B}}$, $II_{\mathbf{B}}$, $III_{\mathbf{B}}$ of $\mathbf{B}$ or, equivalently, of the principal values of $\mathbf{B}$.

The α_i $(i = 0, 1, 2)$ and the β_i $(i = 0, \pm 1)$ are in general connected by the relations

$$\beta_0 = \alpha_0 - II_{\mathbf{B}}\alpha_2 \qquad \beta_1 = \alpha_1 + I_{\mathbf{B}}\alpha_2 \qquad \beta_{-1} = III_{\mathbf{B}}\alpha_2 \tag{8-1-33}$$

$$\alpha_0 = \beta_0 + \frac{II_{\mathbf{B}}}{III_{\mathbf{B}}}\beta_{-1} \qquad \alpha_1 = \beta_1 - \frac{I_{\mathbf{B}}}{III_{\mathbf{B}}}\beta_{-1} \qquad \alpha_2 = \frac{1}{III_{\mathbf{B}}}\beta_{-1} \tag{8-1-34}$$

For incompressible isotropic elastic materials the stress is replaced by the extra stress to give, by way of Eq. (5-4-15), the relation

$$\mathbf{T} = -P\mathbf{I} + \mathfrak{p}(\mathbf{B}) \tag{8-1-35}$$

or

$$\mathbf{T} = -P_\alpha\mathbf{I} + \alpha_1\mathbf{B} + \alpha_2\mathbf{B}^2 \tag{8-1-36}$$

$$\mathbf{T} = -P_\beta\mathbf{I} + \beta_1\mathbf{B} + \beta_{-1}\mathbf{B}^{-1} \tag{8-1-37}$$

where, from Eqs. (8-1-33) and (8-1-34),

$$\beta_1 = \alpha_1 + \mathrm{I}_{\mathbf{B}}\alpha_2 \qquad \beta_{-1} = \alpha_2 \qquad \alpha_1 = \beta_1 - \mathrm{I}_{\mathbf{B}}\beta_{-1} \tag{8-1-38}$$

since for incompressible isotropic elastic materials only isochoric deformations are possible, characterized by $\mathrm{III}_{\mathbf{B}} = \det \mathbf{B} = 1$.

8-2 HYPERELASTICITY

In Sec. 5-1 the principle of determinism applied to a simple thermomechanical process was shown to intuitively yield four constitutive equations, one for each of the four quantities $\mathbf{T}$, q, ε, and η where q is the heat vector, ε is the internal energy density, and η is the entropy density. Replacing ε by the free energy ψ, application of the principle of material objectivity gives Eqs. (5-1-3) to (5-1-6) in the form

$$\mathbf{T}(X,t) = \mathbf{G}_{\varkappa}[\mathbf{F}^t(s), \theta^t(s), k^t(s)] \tag{8-2-1}$$

$$q(X,t) = \boldsymbol{K}_{\varkappa}[\mathbf{F}^t(s), \theta^t(s), k^t(s)] \tag{8-2-2}$$

$$\psi(X,t) = \Psi_{\varkappa}[\mathbf{F}^t(s), \theta^t(s), k^t(s)] \tag{8-2-3}$$

$$\eta(X,t) = H_{\varkappa}[\mathbf{F}^t(s), \theta^t(s), k^t(s)] \tag{8-2-4}$$

A thermoelastic material will be defined by the constitutive equations

$$\mathbf{T}(t) = \mathbf{G}[\mathbf{F}(t), \theta(t), k(t)] \tag{8-2-5}$$

$$q(t) = \boldsymbol{K}[\mathbf{F}(t), \theta(t), k(t)] \tag{8-2-6}$$

$$\psi(t) = \Psi[\mathbf{F}(t), \theta(t), k(t)] \tag{8-2-7}$$

$$\eta(t) = H[\mathbf{F}(t), \theta(t), k(t)] \tag{8-2-8}$$

for the particular condition $s = 0$ for which $\mathbf{F}^t(s) = \mathbf{F}(t)$, noting that, in general, since $\mathbf{F}(t)$ depends on the reference configuration B_r which is chosen, the response functions $\mathbf{G}$, $\boldsymbol{K}$, Ψ, and H will also depend on the reference configuration. For a thermoelastic material which is inhomogeneous, the constitutive functions will also depend on the material particle X. The principle of material objectivity further requires that

$$\mathbf{Q}\mathbf{G}(\mathbf{F}, \theta, k)\mathbf{Q}^T = \mathbf{G}(\mathbf{Q}\mathbf{F}, \theta, \mathbf{Q}k) \tag{8-2-9}$$

$$\mathbf{Q}\boldsymbol{K}(\mathbf{F}, \theta, k) = \boldsymbol{K}(\mathbf{Q}\mathbf{F}, \theta, \mathbf{Q}k) \tag{8-2-10}$$

$$\Psi(\mathbf{F}, \theta, k) = \Psi(\mathbf{Q}\mathbf{F}, \theta, \mathbf{Q}k) \tag{8-2-11}$$

$$H(\mathbf{F}, \theta, k) = H(\mathbf{Q}\mathbf{F}, \theta, \mathbf{Q}k) \tag{8-2-12}$$

for all $\mathbf{F}$, θ, k and all arbitrary orthogonal $\mathbf{Q}$.

The entropy production associated with a single thermomechanical process can be expressed in the form

$$\theta\rho\gamma = -\rho(\dot{\psi} + \eta\dot{\theta}) + \operatorname{tr}(\mathbf{TD}) + \frac{1}{\theta} q \cdot \operatorname{grad}\theta \geqslant 0 \tag{8-2-13}$$

where γ is the local entropy production and

$$\dot{\psi} = \operatorname{tr}\left[\left(\frac{\partial \Psi}{\partial \mathbf{F}}\right)^T \dot{\mathbf{F}}\right] + \frac{\partial \Psi}{\partial \theta}\dot{\theta} + \frac{\partial \Psi}{\partial k}\cdot \dot{k} \tag{8-2-14}$$

Substituting for $\dot{\psi}$ from Eq. (8-2-14) and $\mathbf{T}$, q, ψ, and η from Eqs. (8-2-5) to (8-2-8) into Eq. (8-2-13) gives the inequality

$$\operatorname{tr}\left[\mathbf{TD} - \rho \frac{\partial \Psi(\mathbf{F},\theta,k)^T}{\partial \mathbf{F}}\dot{\mathbf{F}}\right] - \rho\left[\frac{\partial \Psi(\mathbf{F},\theta,k)}{\partial \theta} + H(\mathbf{F},\theta,k)\right]\dot{\theta}$$
$$+ \frac{1}{\theta}K(\mathbf{F},\theta,k)\cdot k - \rho\frac{\partial \Psi(\mathbf{F},\theta,k)}{\partial k}\cdot \dot{k} \geqslant 0$$

which can be simplified further by noting that

$$\operatorname{tr}(\mathbf{TD}) = \operatorname{tr}(\mathbf{TL}) = \operatorname{tr}(\mathbf{F}^{-1}\mathbf{T}\dot{\mathbf{F}})$$

to give

$$\operatorname{tr}\left\{\left[\mathbf{G}(\mathbf{F},\theta,k) - \rho\mathbf{F}\frac{\partial \Psi(\mathbf{F},\theta,k)^T}{\partial \mathbf{F}}\right]\mathbf{L}\right\} - \rho\left[\frac{\partial \Psi(\mathbf{F},\theta,k)}{\partial \theta} + H(\mathbf{F},\theta,k)\right]\dot{\theta}$$
$$+ \frac{1}{\theta}\mathbf{K}(\mathbf{F},\theta,k)\cdot k - \rho\frac{\partial \Psi(\mathbf{F},\theta,k)}{\partial k}\cdot \dot{k} \geqslant 0 \tag{8-2-15}$$

This is an important inequality since, for any thermoelastic material, it must be valid for all functions $\mathbf{F}(t)$, $\theta(t)$, and $k(t)$ that satisfy the principle of admissibility, a requirement which can itself always be fulfilled because there are body forces and thermal sources which can be selected that will satisfy the balance equations. From an inspection of Eq. (8-2-15) it is evident that the presence of the term in $\dot{k}$ could possibly, for a given choice of $\dot{k}$, violate the inequality; hence

$$\frac{\partial \Psi(\mathbf{F},\theta,k)}{\partial k} = o$$

Thus Eq. (8-2-7) must simplify to $\psi = \Psi(\mathbf{F},\theta)$, the free energy being independent of the temperature gradient. Similar considerations applied to the coefficients of $\mathbf{L}$ and $\dot{\theta}$ give Eqs. (8-2-5), (8-2-7), and (8-2-8) in the form

$$\mathbf{T} = \mathbf{G}(\mathbf{F},\theta) = \rho\mathbf{F}\frac{\partial \Psi(\mathbf{F},\theta)^T}{\partial \mathbf{F}} \tag{8-2-16}$$

$$\psi = \Psi(\mathbf{F},\theta) \tag{8-2-17}$$

$$\eta = H(\mathbf{F},\theta) = -\frac{\partial \Psi(\mathbf{F},\theta)}{\partial \theta} \tag{8-2-18}$$

thus giving the condition that, in addition to ψ, $\mathbf{T}$ and η are also independent of the temperature gradient. This condition reduces the inequality of Eq. (8-2-15) to

$$q\cdot k = \mathbf{K}(\mathbf{F},\theta,k)\cdot k \geqslant 0 \tag{8-2-19}$$

which is referred to as the entropy production due to heat flow.

Replacing the free energy by the internal energy ε, interchanging the role of entropy and temperature, and pursuing a similar analysis to the above now gives, in general, the constitutive equations as functions of $\mathbf{F}(t)$, $\eta(t)$, and $k(t)$. However, the entropy inequality again gives

$$\varepsilon = E(\mathbf{F}, \eta) \tag{8-2-20}$$

$$\mathbf{T} = \rho\mathbf{F}\frac{\partial E(\mathbf{F}, \eta)^T}{\partial \mathbf{F}} \tag{8-2-21}$$

$$\theta = \frac{\partial E(\mathbf{F}, \eta)}{\partial \eta} \tag{8-2-22}$$

and

$$q \cdot k \geqslant 0 \tag{8-2-23}$$

that is, $\mathbf{T}$, ε, and θ are again independent of the temperature gradient.

From Eqs. (8-2-16) and (8-2-21) it is evident that, for the condition that either θ or η are regarded as constants, the stress $\mathbf{T}$ is determined by the derivative of a scalar function σ; i.e.,

$$\mathbf{T} = \rho\sigma_{\mathbf{F}}(\mathbf{F})\mathbf{F}^T \tag{8-2-24}$$

where

$$\sigma_{\mathbf{F}}(\mathbf{F}) \equiv \frac{\partial \sigma}{\partial \mathbf{F}}$$

and σ is referred to as the strain energy function. Green (1839, 1841) introduced this concept in discussing perfect elasticity. The term Green elasticity has recently, however, given way to the term hyperelasticity (see, for example, Truesdell and Toupin, 1960; Truesdell and Noll, 1965). For the case of an isothermal deformation with reversible heat conduction $\sigma = \psi$, and for an adiabatic isentropic deformation $\sigma = \varepsilon$.

The hyperelastic formulation of Truesdell and Noll (1965) is given in the form

$$\mathbf{T} = \rho\mathbf{F}\sigma_{\mathbf{F}}(\mathbf{F})^T = \rho\sigma_{\mathbf{F}}(\mathbf{F})\mathbf{F}^T \tag{8-2-25}$$

where $\sigma(\mathbf{F})$ is the strain energy per unit mass. If the scalar σ is objective, then

$$\sigma(\mathbf{QF}) = \sigma(\mathbf{F})$$

and, choosing the particular condition $\mathbf{Q} = \mathbf{R}^T$, this gives

$$\sigma(\mathbf{U}) = \sigma(\mathbf{F})$$

Thus it is possible to introduce a strain energy per unit mass defined by the relation

$$\hat{\sigma}(\mathbf{C}) = \sigma(\mathbf{C}^{1/2})$$

By noting that since $\mathbf{C}$ is a symmetric tensor

$$\hat{\sigma}_{\mathbf{F}}(\mathbf{F})^T = 2\hat{\sigma}_{\mathbf{C}}(\mathbf{C})\mathbf{F}^T$$

the relation

$$\mathbf{T} = 2\rho \mathbf{F} \hat{\sigma}_{\mathbf{C}}(\mathbf{C}) \mathbf{F}^T \tag{8-2-26}$$

can be obtained for the Cauchy stress **T**.

The Piola–Kirchhoff stresses $\mathbf{T}_R$ and $\tilde{\mathbf{T}}$ are related to **T** using Eqs. (3-7-3) and (3-7-4*a*), thus giving, by way of Eq. (8-2-26) and the relations

$$X_i^\alpha x_\beta^i = \delta_{\alpha\beta} \qquad X_j^\beta x_\gamma^j = \delta_{\beta\gamma}$$

the stress relations

$$\mathbf{T}_R = 2\rho_r \mathbf{F} \sigma_{\mathbf{C}}(\mathbf{C}) \tag{8-2-27}$$

$$\tilde{\mathbf{T}} = 2\rho_r \hat{\sigma}_{\mathbf{C}}(\mathbf{C}) \tag{8-2-28}$$

Setting $W = \rho_r \hat{\sigma}$, Eq. (8-2-27) can be expressed in the form

$$\mathbf{T}_R = 2\mathbf{F} W_{\mathbf{C}}(\mathbf{C})$$

which is completely determined by the simple scalar function $W(\mathbf{C})$ of the components of the deformation tensor **C**. For an isotropic hyperelastic material,

$$W = W(\mathrm{I}_{\mathbf{C}}, \mathrm{II}_{\mathbf{C}}, \mathrm{III}_{\mathbf{C}}) \tag{8-2-29}$$

and hence, for this type of material,

$$\mathbf{T}_R = 2\mathbf{F}\left(\frac{\partial W}{\partial \mathrm{I}_{\mathbf{C}}}\frac{\partial \mathrm{I}_{\mathbf{C}}}{\partial \mathbf{C}} + \frac{\partial W}{\partial \mathrm{II}_{\mathbf{C}}}\frac{\partial \mathrm{II}_{\mathbf{C}}}{\partial \mathbf{C}} + \frac{\partial W}{\partial \mathrm{III}_{\mathbf{C}}}\frac{\partial \mathrm{III}_{\mathbf{C}}}{\partial \mathbf{C}}\right)$$

which leads by way of the relations

$$\frac{\partial \mathrm{I}_{\mathbf{C}}}{\partial \mathbf{C}} = \mathbf{I}$$

$$\frac{\partial \mathrm{II}_{\mathbf{C}}}{\partial \mathbf{C}} = (\mathrm{tr}\,\mathbf{C})\mathbf{I} - \mathbf{C}^T = \mathrm{I}_{\mathbf{C}}\mathbf{I} - \mathbf{C}^T$$

$$\frac{\partial \mathrm{III}_{\mathbf{C}}}{\partial \mathbf{C}} = (\det \mathbf{C})\mathbf{C}^{-T} = \mathrm{III}_{\mathbf{C}}\mathbf{C}^{-T}$$

to the stress relation

$$\mathbf{T} = 2\frac{\rho}{\rho_r}\left[\left(\frac{\partial W}{\partial \mathrm{I}_{\mathbf{C}}} + \mathrm{I}_{\mathbf{C}}\frac{\partial W}{\partial \mathrm{II}_{\mathbf{C}}}\right)\mathbf{B} - \frac{\partial W}{\partial \mathrm{II}_{\mathbf{C}}}\mathbf{B}^2 + \frac{\partial W}{\partial \mathrm{III}_{\mathbf{C}}}\mathrm{III}_{\mathbf{C}}\mathbf{I}\right] \tag{8-2-30}$$

for the Cauchy stress. From the Cayley–Hamilton theorem can be obtained the relation

$$\mathbf{B}^2 = \mathrm{I}_{\mathbf{B}}\mathbf{B} - \mathrm{II}_{\mathbf{B}}\mathbf{I} + \mathrm{III}_{\mathbf{B}}\mathbf{B}^{-1}$$

which together with the relations

$$\mathrm{I}_{\mathbf{C}} = \mathrm{I}_{\mathbf{B}} \qquad \mathrm{II}_{\mathbf{C}} = \mathrm{II}_{\mathbf{B}} \qquad \mathrm{III}_{\mathbf{C}} = \mathrm{III}_{\mathbf{B}}$$

can be used to express Eq. (8-2-30) in the alternative form

$$\mathbf{T} = 2\frac{\rho}{\rho_r}\left[\left(\mathrm{II}_\mathbf{B}\frac{\partial W}{\partial \mathrm{II}_\mathbf{B}} + \mathrm{III}_\mathbf{B}\frac{\partial W}{\partial \mathrm{III}_\mathbf{B}}\right)\mathbf{I} + \frac{\partial W}{\partial \mathrm{I}_\mathbf{B}}\mathbf{B} - \mathrm{III}_\mathbf{B}\frac{\partial W}{\partial \mathrm{II}_\mathbf{B}}\mathbf{B}^{-1}\right]$$

which are the stress-strain relations of Finger (1894) for isotropic materials. In the natural state $x^i = X^\alpha$ and it therefore follows from the definition of $\mathbf{B}$ that for this condition $(\mathbf{B} = \mathbf{I})_{x^i = X^\alpha}$ which in turn gives the conditions that

$$\mathrm{I}_\mathbf{B} = \mathrm{II}_\mathbf{B} = 3 \qquad \mathrm{III}_\mathbf{B} = 1$$

Hence, for the natural state to be stress free ($\mathbf{T} = \mathbf{O}$), it is evident that these conditions reduce Finger's stress-strain relations to

$$\left(\frac{\partial W}{\partial \mathrm{I}_\mathbf{B}} + 2\frac{\partial W}{\partial \mathrm{II}_\mathbf{B}} + \frac{\partial W}{\partial \mathrm{III}_\mathbf{B}}\right)_{\mathbf{T}=\mathbf{O}} = 0$$

thus giving a condition which must be satisfied by W.

For infinitesimal strain $\tilde{\mathbf{e}}$,

$$\mathrm{II}_\mathbf{B} \approx (3 + 4\mathrm{I}_{\tilde{\mathbf{e}}}) \qquad \mathrm{III}_\mathbf{B} \approx (1 + 2\mathrm{I}_{\tilde{\mathbf{e}}}) \qquad \frac{\rho_r}{\rho} \approx 1 + \mathrm{I}_{\tilde{\mathbf{e}}}$$

and setting

$$\frac{\partial W}{\partial \mathrm{I}_\mathbf{B}} \approx G + \frac{\lambda + 2G}{2}\mathrm{I}_{\tilde{\mathbf{e}}}$$

$$\frac{\partial W}{\partial \mathrm{II}_\mathbf{B}} \approx -\tfrac{1}{2}G$$

$$\frac{\partial W}{\partial \mathrm{III}_\mathbf{B}} \approx 0$$

Finger's stress-strain relations reduce to

$$\mathbf{T} = \lambda \mathrm{I}_{\tilde{\mathbf{e}}}\mathbf{I} + 2G\tilde{\mathbf{e}}$$

which in Sec. 6-1-1 is shown to be the stress-strain relationship of the classical infinitesimal theory of the elasticity of an isotropic material, λ and G being material constants.

The form of the strain energy function depends on the reference configuration B_r, such that

$$\sigma_\varkappa(\mathbf{F}) = \sigma_{\hat{\varkappa}}(\mathbf{F}\mathbf{P}^{-1}) \tag{8-2-31}$$

for all $\mathbf{F}$, where $\mathbf{P}$ is the deformation gradient transformation from B_r to another configuration $\hat{B}_r$.

For an incompressible thermoelastic material, the constitutive assumption of Eq. (8-2-5) is replaced by

$$\mathbf{T}(t) = -P(t)\mathbf{I} + \mathbf{G}(\mathbf{F}(t), \theta(t), k(t)) \tag{8-2-32}$$

where the tensor function $\mathbf{G}$ is now determined only to within an arbitrary isotropic tensor and $\mathbf{F}(t)$ is restricted to the incompressibility condition $\det \mathbf{F} = 1$

or $\operatorname{tr}\mathbf{D} = 0$, from which it follows that the pressure term in Eq. (8-2-32) contributes nothing to the term $\operatorname{tr}(\mathbf{TD})$, and hence the foregoing thermodynamic considerations remain substantially unaltered.

Thus $\mathrm{III}_\mathbf{B} = 1$ and the stress-strain relationships of Finger reduce to those derived by Rivlin (1948a, 1948b):

$$\mathbf{T} = -P\mathbf{I} + 2\frac{\partial W}{\partial \mathrm{I}_\mathbf{B}}\mathbf{B} - 2\frac{\partial W}{\partial \mathrm{II}_\mathbf{B}}\mathbf{B}^{-1} \tag{8-2-33}$$

where P is an arbitrary hydrostatic pressure, and, from Eq. (8-1-43), it follows that

$$\beta_1 = 2\frac{\partial W}{\partial \mathrm{I}_\mathbf{B}} \qquad \beta_{-1} = -2\frac{\partial W}{\partial \mathrm{II}_\mathbf{B}} \tag{8-2-34}$$

8-3 HOMOGENEOUS STRAIN OF COMPRESSIBLE ISOTROPIC ELASTIC MATERIALS

8-3-1 Simple Extension

A homogeneous deformation has been defined in Sec. 2-6. Since in homogeneous deformations all measures of deformation and rotation are constant throughout the body, it follows that the stresses are also constant, for which condition

$$\operatorname{div}\mathbf{T} = o \tag{8-3-1}$$

That is, homogeneous deformations can be maintained in equilibrium by means of surface tractions alone, no body forces ($f = 0$) being required.

The present discussion is restricted to homogeneous isotropic materials, it being assumed that there exists a natural, stress-free state which is taken as the reference configuration. For present purposes, the stress relation of Eq. (8-1-38) will be used. The solutions of immediate interest describe static configurations.

Let the cartesian coordinates (X, Y, Z) be the referential coordinates of a body in the natural state and the cartesian coordinates (x, y, z) be the spatial coordinates in the deformed configuration. For simple extension with stretch in the Z direction defined by the principal stretches

$$x = \lambda_1 X \qquad y = \lambda_2 Y \qquad z = \lambda_3 Z \tag{8-3-2}$$

where

$$\lambda_1 = \lambda_2 = \alpha\lambda_3$$

the deformation tensor

$$\mathbf{B} = \mathbf{F}\mathbf{F}^T \tag{8-3-3}$$

is a diagonal tensor such that

$$[\mathbf{B}] = \operatorname{diag}(\alpha^2\lambda_3^2, \alpha^2\lambda_3^2, \lambda_3^2) \tag{8-3-4}$$

where the quantities $\lambda_1^2, \lambda_2^2, \lambda_3^2$ are the principal values of $\mathbf{B}$ and α is a constant to be determined. From Eq. (8-1-38) it follows that the principal axes of **B** and **T** coincide, so that using Eq. (8-3-4) gives, by way of Eq. (8-1-38), the stress relations

$$T\langle xx\rangle = T\langle yy\rangle = \beta_0 + \beta_1\alpha^2\lambda_3^2 + \beta_{-1}\frac{1}{\alpha^2\lambda_3^2} \tag{8-3-5}$$

$$T\langle zz\rangle = \beta_0 + \beta_1\lambda_3^2 + \beta_{-1}\frac{1}{\lambda_3^2} \tag{8-3-6}$$

$$T\langle xy\rangle = T\langle yz\rangle = T\langle zx\rangle = 0 \tag{8-3-7}$$

where the response functions β_0, β_1, and β_{-1} are functions of λ_3^2 and α^2 through the principal invariants

$$\mathrm{I}_\mathbf{B} = \lambda_3^2 + 2\alpha^2\lambda_3^2 \tag{8-3-8}$$

$$\mathrm{II}_\mathbf{B} = 2\alpha^2\lambda_3^4 + \alpha^4\lambda_3^4 \tag{8-3-9}$$

$$\mathrm{III}_\mathbf{B} = \alpha^4\lambda_3^6 \tag{8-3-10}$$

From Eq. (8-3-5), the lateral faces $x =$ constant and $y =$ constant of the body are stress free if

$$\beta_0 + \beta_1\alpha^2\lambda_3^2 + \beta_{-1}\frac{1}{\alpha^2\lambda_3^2} = 0 \tag{8-3-11}$$

is satisfied. It is evident that a given extension with stress-free cylindrical walls may be produced by more than one tensile stress.

Uniform dilatation, defined by

$$\lambda_1 = \lambda_2 = \lambda_3 = \lambda \tag{8-3-12}$$

is the special case of simple extension for which $\alpha = 1$. From the stress relations $T\langle xx\rangle = T\langle yy\rangle$ and $T\langle zz\rangle$ the isotropic stress tensor is of the form

$$\mathbf{T} = -P\mathbf{I} \qquad P = \kappa(\lambda) \tag{8-3-13}$$

where

$$\kappa(\lambda) = -\beta_0(3\lambda^2, 3\lambda^4, \lambda^6) - \lambda^2\beta_1(3\lambda^2, 3\lambda^4, \lambda^6) - \frac{1}{\lambda^2}\beta_{-1}(3\lambda^2, 3\lambda^4, \lambda^6) \tag{8-3-14}$$

from which it is evident that only hydrostatic stress can produce uniform dilatation. For the elastic response in compression and tension to have equal but opposite effects, it is necessary that $\kappa(\lambda) = -\kappa(1/\lambda)$. However, it is evident from Eqs. (8-3-13) and (8-3-14) that, in general, for compression and tension the condition of equality $\kappa(\lambda) = -\kappa(1/\lambda)$ will not necessarily be satisfied.

8-3-2 Simple Shear

Shown in Fig. 8-1 is the simple shear of a rectangular block aligned with the cartesian coordinate axes (X, Y, Z). A simple shear of amount γ is given by

$$x = X + \gamma Y \qquad y = Y \qquad z = Z \tag{8-3-15}$$

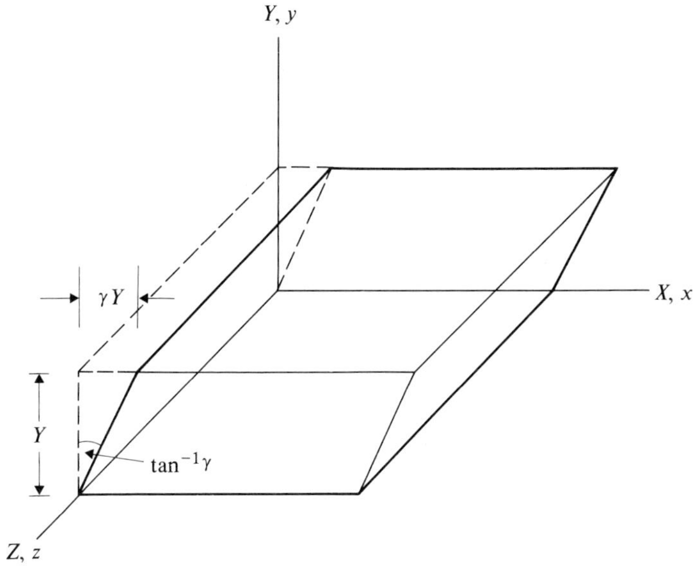

Figure 8-1 Simple shear.

Let the unit vectors $\boldsymbol{l}$, $\boldsymbol{m}$, $\boldsymbol{n}$ form an orthonormal set. The unit vector $\boldsymbol{l}$ defines the direction of shear and the planes orthogonal to $\boldsymbol{m}$ and $\boldsymbol{n}$ are referred to as the glide planes and the planes of shear respectively. From Eq. (8-3-15) the matrices of

$$\mathbf{F} = \mathbf{I} + \gamma \boldsymbol{l} \otimes \boldsymbol{m} \qquad \mathbf{B} = \mathbf{F}\mathbf{F}^T \tag{8-3-16}$$

and $\mathbf{B}^{-1}$ are

$$[\mathbf{F}] = \begin{bmatrix} 1 & \gamma & 0 \\ 0 & 1 & 0 \\ 0 & 0 & 1 \end{bmatrix} \tag{8-3-17}$$

$$[\mathbf{B}] = \begin{bmatrix} 1+\gamma^2 & \gamma & 0 \\ \gamma & 1 & 0 \\ 0 & 0 & 1 \end{bmatrix} \qquad [\mathbf{B}^{-1}] = \begin{bmatrix} 1 & -\gamma & 0 \\ -\gamma & 1+\gamma^2 & 0 \\ 0 & 0 & 1 \end{bmatrix} \tag{8-3-18}$$

The principal invariants are

$$\mathrm{I}_{\mathbf{B}} = \mathrm{II}_{\mathbf{B}} = 3 + \gamma^2 \qquad \mathrm{III}_{\mathbf{B}} = 1 \tag{8-3-19}$$

from which it follows that, since $\mathrm{III}_{\mathbf{B}} = 1$, simple shear is an isochoric deformation, in accord with Fig. 8-1.

From Eq. (8-3-16),

$$\mathbf{B} = (1+\gamma^2)\boldsymbol{l} \otimes \boldsymbol{l} + \boldsymbol{m} \otimes \boldsymbol{m} + \boldsymbol{n} \otimes \boldsymbol{n} + \gamma(\boldsymbol{l} \otimes \boldsymbol{m} + \boldsymbol{m} \otimes \boldsymbol{l}) \tag{8-3-20}$$

and similarly, from Eq. (2-3-12),

$$\mathbf{C} = \boldsymbol{l} \otimes \boldsymbol{l} + (1+\gamma^2)\boldsymbol{m} \otimes \boldsymbol{m} + \boldsymbol{n} \otimes \boldsymbol{n} + \gamma(\boldsymbol{l} \otimes \boldsymbol{m} + \boldsymbol{m} \otimes \boldsymbol{l}) \tag{8-3-21}$$

Hence,

$$\mathbf{B}\boldsymbol{l} = \mathbf{C}\boldsymbol{l} + \gamma^2 \boldsymbol{l} \qquad \mathbf{B}\boldsymbol{m} = \mathbf{C}\boldsymbol{m} - \gamma^2 \boldsymbol{m} \qquad \mathbf{B}\boldsymbol{n} = \mathbf{C}\boldsymbol{n} = \boldsymbol{n} \tag{8-3-22}$$

which shows that one of the principal stretches is unity and that the stretch axis is in the direction defined by $\boldsymbol{n}$ in both the reference configuration B_r and in the current configuration B_t, thus establishing that the deformation is two dimensional with respect to the planes of shear.

The deformation tensor $\mathbf{B}$ can be expressed, by way of Eq. (1-10-1), in the form

$$\begin{aligned}\mathbf{B} = \mathbf{V}^2 = \mathbf{BI} &= \mathbf{B}(\boldsymbol{v}_a \otimes \boldsymbol{v}_a) \\ &= (\mathbf{B}\boldsymbol{v}_a) \otimes \boldsymbol{v}_a = \sum_{a=1}^{3} \lambda_a^2 (\boldsymbol{v}_a \otimes \boldsymbol{v}_a)\end{aligned} \tag{8-3-23}$$

where the $\boldsymbol{v}_a$ $(a = 1, 2, 3)$ are an orthonormal set of proper vectors associated with the λ_a $(a = 1, 2, 3)$. Equation (8-3-23) is generally referred to as the spectral representation of a symmetric tensor. Since from Eq. (8-3-19), $\lambda_1 \lambda_2 \lambda_3 = 1$ and, from Eq. (8-3-22), $\lambda_3 = 1$, it follows that Eq. (8-3-23) can be rearranged into the form

$$\mathbf{B} = \lambda_1^2 \boldsymbol{v}_1 \otimes \boldsymbol{v}_1 + \frac{1}{\lambda_1^2} \boldsymbol{v}_2 \otimes \boldsymbol{v}_2 + \boldsymbol{v}_3 \otimes \boldsymbol{v}_3 \tag{8-3-24}$$

Since $\boldsymbol{v}_1$ and $\boldsymbol{v}_2$ are an orthogonal pair of unit vectors, orthogonal to $\boldsymbol{n}$, it follows that $\boldsymbol{v}_1$ and $\boldsymbol{v}_2$ can be expressed in the form

$$\boldsymbol{v}_1 = \cos\theta \boldsymbol{l} + \sin\theta \boldsymbol{m} \qquad \boldsymbol{v}_2 = -\sin\theta \boldsymbol{l} + \cos\theta \boldsymbol{m} \qquad \boldsymbol{v}_3 = \boldsymbol{m} \tag{8-3-25}$$

Substituting for the $\boldsymbol{v}_a$ $(a = 1, 2, 3)$ in Eq. (8-3-24) from Eq. (8-3-25) and comparing with Eq. (8-3-20) gives

$$\begin{aligned}\lambda_1^2 \cos^2\theta + \frac{1}{\lambda_1^2} \sin^2\theta &= 1 + \gamma^2 \\ \lambda_1^2 \sin^2\theta + \frac{1}{\lambda_1^2} \cos^2\theta &= 1 \\ \frac{\lambda_1^4 - 1}{\lambda_1^2} \sin 2\theta &= 2\gamma\end{aligned} \tag{8-3-26}$$

which in turn give

$$\lambda_1^2 = 1 + \tfrac{1}{2}\gamma^2 + \gamma(1 + \tfrac{1}{4}\gamma^2)^{1/2} = \frac{1}{\lambda_2^2} \tag{8-3-27a}$$

$$\lambda_2^2 = 1 + \tfrac{1}{2}\gamma^2 - \gamma(1 + \tfrac{1}{4}\gamma^2)^{1/2} = \frac{1}{\lambda_1^2} \tag{8-3-27b}$$

$$\lambda_3 = 1 \tag{8-3-27c}$$

and

$$
\begin{aligned}
v_1 &= \frac{(\frac{1}{2}\gamma + \sqrt{1+\frac{1}{4}\gamma^2})\boldsymbol{l} + \boldsymbol{m}}{\sqrt{2+\frac{1}{2}\gamma^2 + \gamma(1+\frac{1}{4}\gamma^2)^{1/2}}} \\
v_2 &= \frac{(\frac{1}{2}\gamma - \sqrt{1+\frac{1}{4}\gamma^2})\boldsymbol{l} + \boldsymbol{m}}{\sqrt{2+\frac{1}{2}\gamma^2 + \gamma(1+\frac{1}{4}\gamma^2)^{1/2}}} \\
v_3 &= \boldsymbol{n}
\end{aligned}
\tag{8-3-28}
$$

For $\gamma = 0$, Eqs. (8-3-28) reduce to

$$
v_1 = \frac{1}{\sqrt{2}}\boldsymbol{l} + \frac{1}{\sqrt{2}}\boldsymbol{m} \qquad v_2 = -\frac{1}{\sqrt{2}}\boldsymbol{l} + \frac{1}{\sqrt{2}}\boldsymbol{m} \qquad v_3 = \boldsymbol{n}
$$

while, for $\gamma \to \infty$,

$$
v_1 \to \boldsymbol{l} \qquad v_2 \to \boldsymbol{m} \qquad v_3 = \boldsymbol{n}
$$

Thus the principal axes $\boldsymbol{v}_1$ and $\boldsymbol{v}_2$ rotate about the z axis through the angle $\pi/4$ for $0 \leqslant \gamma \leqslant \infty$.

The constitutive equation (8-1-38) can now be expressed in the form

$$
\begin{aligned}
[\mathbf{T}] = (\beta_0 + \beta_1 + \beta_{-1}) \begin{bmatrix} 1 & 0 & 0 \\ 0 & 1 & 0 \\ 0 & 0 & 1 \end{bmatrix} + (\beta_1 - \beta_{-1})\gamma \begin{bmatrix} 0 & 1 & 0 \\ 1 & 0 & 0 \\ 0 & 0 & 0 \end{bmatrix} \\
+ \beta_1 \gamma^2 \begin{bmatrix} 1 & 0 & 0 \\ 0 & 0 & 0 \\ 0 & 0 & 0 \end{bmatrix} + \beta_{-1}\gamma^2 \begin{bmatrix} 0 & 0 & 0 \\ 0 & 1 & 0 \\ 0 & 0 & 0 \end{bmatrix}
\end{aligned}
\tag{8-3-29}
$$

where the response functions β_i $(i = 0, \pm 1)$ are functions of γ^2 through the principal invariants defined by Eq. (8-3-19).

Let

$$
\hat{G}(\gamma^2) = \beta_1(3+\gamma^2, 3+\gamma^2, 1) - \beta_{-1}(3+\gamma^2, 3+\gamma^2, 1) \tag{8-3-30}
$$

so that, from Eq. (8-3-26),

$$
T\langle xy \rangle = \hat{G}(\gamma^2)\gamma \tag{8-3-31}
$$

From Eqs. (8-3-15) and (6-2-13), for $\gamma \to 0$, $2E\langle xy \rangle \simeq \gamma$ and therefore, from Eq. (6-2-12),

$$
\hat{G}(0) = G \tag{8-3-32}
$$

where G is just the shear modulus of rigidity of the linearized theory. For this reason $\hat{G}(\gamma^2)$ is referred to as the generalized or secant shear modulus. Since $\hat{G}$ is an even function of γ, the shear stress is an odd function of γ, in accord with the physical condition that changing the direction of the shear stress changes the direction of the shear.

The principal stresses σ_a $(a = 1, 2, 3)$ are related to the principal values B_a and B_a^{-1} $(a = 1, 2, 3)$ of $\mathbf{B}$ and $\mathbf{B}^{-1}$ respectively by the relation

$$\sigma_a = \beta_0 + \beta_1 B_a + \beta_{-1} B_a^{-1} \qquad \text{for } a = 1, 2, 3 \tag{8-3-33}$$

from which there can be obtained, by way of Eqs. (8-3-27*a*, *b*),

$$\sigma_1 - \sigma_2 = 2(\lambda_1^2 - \lambda_2^2) \tag{8-3-34}$$

Equation (8-3-32) can be rearranged, by way of Eqs. (8-3-27*a*, *b*), to give

$$\hat{G}(\gamma^2) = (\sigma_1 - \sigma_2)(\lambda_1^2 - \lambda_2^2) \tag{8-3-35}$$

Since by definition the stretches are positive, $\lambda_1 + \lambda_2 > 0$ and hence the physical condition

$$\hat{G}(\gamma^2) > 0 \tag{8-3-36}$$

is satisfied provided

$$\frac{\sigma_1 - \sigma_2}{\lambda_1 - \lambda_2} > 0 \tag{8-3-37}$$

which are just the inequalities suggested by Baker and Ericksen (1954), who proposed that the greater principal stress occurs always in the direction of the greater principal stretch.

From Eq. (8-3-29),

$$T\langle xx \rangle = T\langle zz \rangle + \beta_1 \gamma^2 \tag{8-3-38a}$$

$$T\langle yy \rangle = T\langle zz \rangle + \beta_{-1} \gamma^2 \tag{8-3-38b}$$

$$T\langle zz \rangle = \beta_0 + \beta_1 + \beta_{-1} \tag{8-3-38c}$$

Hence, by way of Eq. (8-3-31),

$$T\langle xx \rangle - T\langle yy \rangle = \gamma T\langle xy \rangle \tag{8-3-39}$$

from which it is evident that the normal stresses cannot be equal.

The mean normal stress can be obtained from Eqs. (8-3-38) in the form

$$T_m = \tfrac{1}{3}(T\langle xx \rangle + T\langle yy \rangle + T\langle zz \rangle) = T\langle zz \rangle + \tfrac{1}{3}(\beta_1 + \beta_{-1})\gamma^2 \tag{8-3-40}$$

from which it follows that if in simple shear the necessary normal stresses are not provided, then a shearing stress will produce a dilatation or compression of a specimen according to whether T_m is positive or negative. The possibility of this result was considered by Kelvin and Tait (1867), the effects arising from the presence of a hydrostatic stress which vanishes in the corresponding deformation considered in terms of the infinitesimal theory being now referred to as Kelvin effects.

For the particular condition $T\langle xx \rangle = T\langle yy \rangle$, it follows from Eqs. (8-3-38) that $\beta_1 - \beta_{-1} = 0$ and hence from Eq. (8-3-30) it follows that $\hat{G} = 0$. Hence, the already noted inequality of the normal stresses is essential in order to effect a simple shear, it being further concluded that for the particular condition $T\langle xx \rangle = T\langle yy \rangle$ an initially cubical specimen when subjected to a shearing stress will tend to dilate unequally. This latter effect is now referred to as the Poynting (1909, 1912) effect, which relates to the observation by Poynting that the length of various steel, copper, and brass wires increased when twisted in the elastic range, the elongation being proportional to the square of the twist.

8-4 HOMOGENEOUS STRAIN OF INCOMPRESSIBLE ISOTROPIC ELASTIC MATERIALS

8-4-1 Simple Extension

The present discussion is restricted to homogeneous isotropic materials, it being assumed that there exists a natural, stress-free state which is taken as the reference configuration. For incompressible isotropic elastic materials, the stress relation of Eq. (8-1-43) will be used. In the case of an incompressible isotropic elastic material, Eq. (3-4-4) takes, by way of Eq. (8-1-31), the form

$$\operatorname{div} \mathbf{T}_E - \operatorname{grad} P + \rho_r f = \rho_r \boldsymbol{a} \tag{8-4-1}$$

where ρ_r is the uniform and constant density. For homogeneous strain fields, it is evident from Eqs. (8-1-32) and (8-4-1) that for $\boldsymbol{a} = \mathbf{o}$, $\boldsymbol{f} = \mathbf{o}$ the indeterminate pressure P appearing in Eq. (8-1-43) is a uniform constant.

For simple tension, Eq. (8-1-43) gives

$$T\langle xx\rangle = T\langle yy\rangle = -P + \beta_1 \alpha^2 \lambda_3^2 + \beta_{-1} \frac{1}{\alpha_2 \lambda_3^2} \tag{8-4-2}$$

$$T\langle zz\rangle = -P + \beta_1 \lambda_3^2 + \beta_{-1} \frac{1}{\lambda_3^2} \tag{8-4-3}$$

where the β_i $(i = \pm 1)$ are functions of the first two principal invariants $\mathrm{I}_{\mathbf{B}}$ and $\mathrm{II}_{\mathbf{B}}$. The condition of incompressibility, that is, $\mathrm{III}_{\mathbf{B}} = 1$, gives

$$\alpha = \lambda_3^{-3/2} \tag{8-4-4}$$

and hence

$$\mathrm{I}_{\mathbf{B}} = \frac{2}{\lambda_3} + \lambda_3^2 \qquad \mathrm{II}_{\mathbf{B}} = 2\lambda_3 + \frac{1}{\lambda_3^2} \tag{8-4-5}$$

From the above stress relations the condition that the lateral faces $x =$ constant and $y =$ constant of the body are stress free is satisfied by setting $T\langle xx\rangle = T\langle yy\rangle = 0$ in Eq. (8-4-2) to give, by way of Eq. (8-4-5), the condition

$$P = \beta_1 \frac{1}{\lambda_3} + \beta_{-1}\lambda_3 \tag{8-4-6}$$

Substituting for P in Eq. (8-4-3) from Eq. (8-4-6) gives

$$T\langle zz\rangle = \left(\beta_1 - \frac{1}{\lambda_3}\beta_{-1}\right)\left(\frac{\lambda_3^3 - 1}{\lambda_3}\right) \tag{8-4-7}$$

which, by comparison with the simple extension of a compressible isotropic elastic material, illustrates the general observation that it is a simpler problem to determine the response of an incompressible isotropic elastic material. This is because in a given problem involving stress boundary conditions the pressure P is not arbitrary, and hence there is effectively one less boundary condition.

8-4-2 Simple Shear

The results for the simple shear of an incompressible isotropic elastic material can be obtained from the simple shear of a compressible isotropic elastic material by simply replacing the response function β_0 by the arbitrary pressure $-P$.

As for simple extension, the number of boundary conditions can be reduced by defining P in such a way that $T\langle zz\rangle = 0$. From Eq. (8-3-38c) this gives the condition

$$P = \beta_1 + \beta_{-1} \tag{8-4-8}$$

and hence, from Eqs. (8-3-38a, b) and (8-3-7),

$$T\langle xx\rangle = \beta_1 \gamma^2 \tag{8-4-9a}$$

$$T\langle yy\rangle = \beta_{-1} \gamma^2 \tag{8-4-9b}$$

$$T\langle xy\rangle = (\beta_1 - \beta_{-1})\gamma = \hat{G}(\gamma^2)\gamma \tag{8-4-9c}$$

where Eq. (8-4-9c) is just Eq. (8-3-31); rearranging Eqs. (8-4-9) gives

$$T\langle xx\rangle - T\langle yy\rangle = \gamma T\langle xy\rangle \tag{8-4-10}$$

These relations were obtained by Rivlin (1948d) for incompressible hyperelastic materials.

Incompressible bodies may exhibit the Poynting effect, but do not exhibit Kelvin effects.

8-4-3 Biaxial Stretching of a Sheet

Let the sheet be of thickness $2h_0$ and assume a homogeneous deformation of the form

$$x = \lambda_1 X \qquad y = \lambda_2 Y \qquad z = \lambda_3 Z \tag{8-4-11}$$

the condition of incompressibility giving

$$\lambda_1 \lambda_2 \lambda_3 = 1 \tag{8-4-12}$$

The deformation tensor $\mathbf{B} = \mathbf{F}\mathbf{F}^T$ is of the form

$$[\mathbf{B}] = \operatorname{diag}(\lambda_1^2, \lambda_2^2, \lambda_3^2) \tag{8-4-13}$$

and hence Eq. (8-1-43) gives the stress relations

$$T\langle xx\rangle = -P + \beta_1 \lambda_1^2 + \beta_{-1} \frac{1}{\lambda_1^2} \tag{8-4-14}$$

$$T\langle yy\rangle = -P + \beta_1\lambda_2^2 + \beta_{-1}\frac{1}{\lambda_2^2} \tag{8-4-15}$$

$$T\langle zz\rangle = -P + \beta_1\lambda_3^2 + \beta_{-1}\frac{1}{\lambda_3^2} \tag{8-4-16}$$

where the β_i $(i = \pm 1)$ are functions of the principal invariants

$$\mathrm{I}_{\mathbf{B}} = \lambda_1^2 + \lambda_2^2 + \frac{1}{\lambda_1^2\lambda_2^2} \qquad \mathrm{II}_{\mathbf{B}} = \frac{1}{\lambda_1^2} + \frac{1}{\lambda_2^2} + \lambda_1^2\lambda_2^2 \tag{8-4-17}$$

Consider the deformation to be produced by means of edge tractions only, so that the major surfaces $Z = \pm h_0$ are free from applied stress. This condition can be satisfied by specifying P to be of such a form that $T\langle zz\rangle = 0$. Thus, from Eq. (8-4-16),

$$P = \beta_1\lambda_3^2 + \beta_{-1}\frac{1}{\lambda_3^2} \tag{8-4-18}$$

Hence, substituting for P in Eqs. (8-4-14) and (8-4-15) from Eq. (8-4-18) gives

$$T\langle xx\rangle = \left(\lambda_1^2 - \frac{1}{\lambda_1^2\lambda_2^2}\right)(\beta_1 - \lambda_2^2\beta_{-1}) \tag{8-4-19}$$

$$T\langle yy\rangle = \left(\lambda_2^2 - \frac{1}{\lambda_2^2\lambda_1^2}\right)(\beta_1 - \lambda_1^2\beta_{-1}) \tag{8-4-20}$$

These relations were obtained by Rivlin (1948d) for incompressible hyperelastic materials.

It is evident from Eqs. (8-4-19) and (8-4-20) that if $T\langle xx\rangle$ and $T\langle yy\rangle$ can be evaluated for any given values of λ_1 and λ_2, then the corresponding values of the β_i $(i = \pm 1)$ can be evaluated.

8-5 NONHOMOGENEOUS DEFORMATIONS

8-5-1 Extension, Inflation, and Torsion of an Incompressible Circular Cylindrical Tube

A classical problem of nonlinear elasticity theory is the combined extension and torsion of an incompressible solid-rod or cylindrical tube. This is a typical case of nonhomogeneous deformation.

Consider a right-circular cylindrical tube of incompressible material. Let (R, Θ, Z) be the cylindrical referential coordinates in the initial state of a particle that is located in the deformed configuration by the cylindrical current coordinates (r, θ, z). The principal axis of the tube is aligned with the Z axis. The simple deformations to be considered are

$$r = \left(\frac{K + R^2}{F}\right)^{1/2} \qquad \theta = \Theta + DZ \qquad z = FZ \tag{8-5-1}$$

where D, F, and K are constants subject to the condition that F and K have values such that $(K+R^2)/F > 0$ when R is in the interval $(R_1 \geqslant R \geqslant R_2)$; R_1 and R_2 are the outer and inner radii respectively of the undeformed tube.

The deformation tensor $\mathbf{B} = \mathbf{FF}^T$ is of the form

$$[B^{ij}] = \begin{bmatrix} \dfrac{1}{\eta F} & 0 & 0 \\ 0 & \dfrac{1+R^2D^2}{R^2} & DF \\ 0 & DF & F^2 \end{bmatrix} \tag{8-5-2}$$

and

$$[(B^{-1})_{ij}] = \begin{bmatrix} \eta F & 0 & 0 \\ 0 & R^2 & \dfrac{-R^2D}{F} \\ 0 & \dfrac{-R^2D}{F} & \dfrac{1+R^2D^2}{F^2} \end{bmatrix} \tag{8-5-3}$$

where

$$\eta = \frac{r^2}{R^2}F = 1 + \frac{K}{R^2} \tag{8-5-4}$$

The physical components of the stress can now be obtained from Eqs. (8-1-43), (8-5-2), and (8-5-3), it being evident that the resulting stresses are functions of r only. Hence, from considerations of equilibrium,

$$\frac{\partial P}{\partial z} = 0 \qquad \frac{\partial P}{\partial \theta} = 0 \tag{8-5-5}$$

that is, $P = P(r)$. The condition of equilibrium in the radial direction, i.e.,

$$\frac{\partial T\langle rr\rangle}{\partial r} + \frac{1}{r}(T\langle rr\rangle - T\langle\theta\theta\rangle) = 0 \tag{8-5-6}$$

gives the relation

$$T\langle rr\rangle = -\int \frac{T\langle rr\rangle - T\langle\theta\theta\rangle}{r}\,dr \tag{8-5-7}$$

where the constant of integration in Eq. (8-5-7) can be chosen to determine the condition at a boundary surface, as, for example, at the surface of a rod or at the outer surface of a tube. The stress relations can now be obtained in the form

$$T\langle rr\rangle = -\int_{r_2}^{r} \frac{1-\eta^2}{\eta F}(\beta_1 - F^2\beta_{-1})\frac{dr}{r} + D^2\int_{r_2}^{r} \beta_1 r\,dr + [T\langle rr\rangle]_{r=r_2} \tag{8-5-8}$$

$$T\langle\theta\theta\rangle = T\langle rr\rangle - \frac{1-\eta^2}{\eta F}(\beta_1 - F^2\beta_{-1}) + D^2\beta_1 r^2 \tag{8-5-9}$$

$$T\langle zz\rangle = T\langle rr\rangle - \left(\frac{1}{\eta F} - F^2\right)\beta_1 + \left(\frac{1}{F^2} - \eta F + \frac{R^2}{F^2}D^2\right)\beta_{-1} \tag{8-5-10}$$

$$T\langle\theta z\rangle = rF\left(\beta_1 - \frac{1}{\eta F}\beta_{-1}\right)D, \qquad T\langle\theta r\rangle = T\langle zr\rangle = 0 \tag{8-5-11}$$

where the material coefficients β_i $(i = \pm 1)$ are infinite power series in the invariants

$$\mathrm{I}_{\mathbf{B}} = F^2 + \frac{1}{\eta F} + \frac{\eta}{F} + \eta\frac{R^2}{F}D^2 \tag{8-5-12}$$

$$\mathrm{II}_{\mathbf{B}} = \frac{1}{F^2} + \eta F + \frac{F}{\eta} + \frac{R^2}{F^2}D^2 \tag{8-5-13}$$

of the deformation tensor $\mathbf{B}$, since for an incompressible solid $\mathrm{III}_{\mathbf{B}} = 1$.

The resultant torque

$$\begin{aligned}\Gamma &= 2\pi\int_{r_2}^{r_1} r^2 T\langle\theta z\rangle\, dr \\ &= 2\pi\frac{D}{F}\int_{R_2}^{R_1}\eta\left(\beta_1 - \frac{1}{\eta F}\beta_{-1}\right)R^3\, dR\end{aligned} \tag{8-5-14}$$

where r_1, r_2 are the outer and inner radii respectively of the deformed tube. The deformed radius r is related to the undeformed radius R by Eq. (8-5-1).

The resultant force acting on the ends of the cylinder is

$$N_z = 2\pi\int_{r_2}^{r_1} rT\langle zz\rangle\, dr \tag{8-5-15}$$

Substituting for $T\langle zz\rangle$ in Eq. (8-5-15) from Eq. (8-5-10) gives

$$\begin{aligned}N_z &= 2\pi\int_{r_2}^{r_1} rT\langle rr\rangle\, dr \\ &\quad + 2\pi\int_{r_2}^{r_1}\left[\left(F^2 - \frac{1}{\eta F}\right)\left(\beta_1 - \frac{\eta}{F}\beta_{-1}\right) + \frac{D^2}{F^2}\beta_{-1}R^2\right]r\, dr\end{aligned} \tag{8-5-16}$$

where η is defined by Eq. (8-5-4).

If the external surface is free from applied stress, so that $[T\langle rr\rangle]_{r=r_1} = 0$, then it follows from the condition of equilibrium in the radial direction that

$$\int_{r_2}^{r_1} rT\langle rr\rangle\, dr = -\tfrac{1}{2}r_2^2[T\langle rr\rangle]_{r=r_2} - \tfrac{1}{2}\int_{r_2}^{r_1} r(T\langle\theta\theta\rangle - T\langle rr\rangle)\, dr \tag{8-5-17}$$

where $[T\langle rr\rangle]_{r=r_2}$ is the value of $T\langle rr\rangle$ at $r = r_2$.

Substituting for the first term on the right-hand side of Eq. (8-5-16) from Eq.

(8-5-17) and rearranging gives

$$
\begin{aligned}
N_z = & -\pi r_2^2[T\langle rr\rangle]_{r=r_2} + \pi \int_{r_2}^{r_1} \frac{1-\eta^2}{\eta F}(\beta_1 - F^2\beta_{-1}) r\,dr \\
& + 2\pi \int_{r_2}^{r_1} \left(F^2 - \frac{1}{\eta F}\right)\left(\beta_1 - \frac{\eta}{F}\beta_{-1}\right) r\,dr \\
& - \pi \int_{r_2}^{r_1} D^2 \left(\beta_1 - \frac{2}{\eta F}\beta_{-1}\right) r^3\,dr
\end{aligned} \tag{8-5-18}
$$

where $[T\langle rr\rangle]_{r=r_2}$ can be obtained from Eq. (8-5-8) for the particular condition $[T\langle rr\rangle]_{r=r_1} = 0$. Hence substituting for $[T\langle rr\rangle]_{r=r_2}$ in Eq. (8-5-18) from Eq. (8-5-8) for the particular condition $[T\langle rr\rangle]_{r=r_1} = 0$ gives

$$
\begin{aligned}
N_z = & \; 2\pi \int_{R_2}^{R_1} \frac{\eta F^3 - 1}{\eta F^2}\left(\beta_1 - \frac{\eta}{F}\beta_{-1}\right) R\,dR \\
& + \pi \int_{R_2}^{R_1} \frac{\eta^2 - 1}{\eta^2 F^2}(\beta_1 - F^2\beta_{-1})\left(\frac{R_2^2}{R^2} - 1\right) R\,dR \\
& + \pi \frac{D^2}{F^2} \int_{R_2}^{R_1} \left[(R_2^2 - R^2)\beta_1 + 2\frac{R^2}{F}\beta_{-1}\right] R\,dR
\end{aligned} \tag{8-5-19}
$$

If the outer surface of the tube is free from surface traction, the inflating pressure P_i on the inner surface can be obtained from Eq. (8-5-8) in the form

$$
\begin{aligned}
P_i & = -[T\langle rr\rangle]_{r=r_2} \\
& = \int_{R_2}^{R_1} \frac{\eta^2 - 1}{\eta^2 F}(\beta_1 - F^2\beta_{-1}) \frac{dR}{R} + \frac{D^2}{F} \int_{R_2}^{R_1} \beta_1 R\,dR
\end{aligned} \tag{8-5-20}
$$

Provided Γ, N_z, and P_i have specified values, Eqs. (8-5-14), (8-5-19), and (8-5-20) can be used to determine F, D, and K.

8-5-2 Extension and Torsion of an Incompressible Solid Circular Cylinder

For a solid rod $R_2 = r_2 = 0$, $K = 0$, and $\eta = 1$ for all D and F. These limiting conditions reduce Eqs. (8-5-8), (8-5-9), (8-5-10), (8-5-11), (8-5-14), and (8-5-19) to the form

$$
T\langle rr\rangle = -D^2 \int_r^{r_1} \beta_1 r\,dr \tag{8-5-21}
$$

$$
T\langle \theta\theta\rangle = T\langle rr\rangle + \beta_1 D^2 r^2 \tag{8-5-22}
$$

$$
T\langle zz\rangle = T\langle rr\rangle + \frac{F^3 - 1}{F}\left(\beta_1 - \frac{1}{F}\beta_{-1}\right) + \beta_{-1}\frac{D^2}{F^2}R^2 \tag{8-5-23}
$$

$$
T\langle \theta z\rangle = \left(\beta_1 - \frac{1}{F}\beta_{-1}\right) FrD \qquad T\langle r\theta\rangle = T\langle rz\rangle = 0 \tag{8-5-24}
$$

$$\Gamma = 2\pi \frac{D}{F} \int_0^{R_1} \left(\beta_1 - \frac{1}{F}\beta_{-1}\right) R^3\, dR \tag{8-5-25}$$

and

$$N_z = 2\pi \int_0^{R_1} \frac{F^3 - 1}{F^2} \left(\beta_1 - \frac{1}{F}\beta_{-1}\right) R\, dR - \pi \frac{D^2}{F^2} \int_0^{R_1} \left(\beta_1 - \frac{2}{F}\beta_{-1}\right) R^3\, dR \tag{8-5-26}$$

Setting $D = 0$ and taking $F \equiv \lambda_3$ reduces Eq. (8-5-23) to the solution for simple extension, i.e., Eq. (8-4-7).

Pure torsion is obtained by setting $F = 1$. For this limiting condition, and for a solid rod, Eqs. (8-5-21), (8-5-22), (8-5-23), and (8-5-24) reduce to

$$T\langle rr\rangle = -D^2 \int_r^{r_1} \beta_1 r\, dr \tag{8-5-27a}$$

$$T\langle \theta\theta\rangle = T\langle rr\rangle + \beta_1 D^2 r^2 \tag{8-5-27b}$$

$$T\langle zz\rangle = T\langle rr\rangle + \beta_{-1} D^2 r^2 \tag{8-5-27c}$$

$$T\langle \theta z\rangle = r(\beta_1 - \beta_{-1})D \tag{8-5-27d}$$

With the substitution $\gamma = rD$, the difference between Eqs. (8-5-27*b*) and (8-5-27*c*), together with Eq. (8-5-27*d*) gives

$$T\langle \theta\theta\rangle = T\langle zz\rangle + (\beta_1 - \beta_{-1})\gamma^2$$

$$T\langle \theta z\rangle = (\beta_1 - \beta_{-1})\gamma$$

from which it follows, by comparison with Eqs. (8-4-9) and (8-4-10), that the shear response in pure torsion and in simple shear is equivalent.

Setting $N_z = 0$ in Eq. (8-5-26), the elongation of a twisted rod can be obtained in the form

$$F = \frac{1}{F^2} + \frac{1}{2}\frac{D^2}{F^2} \frac{\displaystyle\int_0^{R_1} (\beta_1 - (2/F)\beta_{-1}) R^3\, dR}{\displaystyle\int_0^{R_1} (\beta_1 - (1/F)\beta_{-1}) R\, dR} \tag{8-5-28}$$

and as $D \to 0$, $F \to 1$, such that the axial strain e_p can be obtained in the form

$$e_p = \tfrac{1}{12}(1 + \hat{q}) R_1^2 D^2 \tag{8-5-29}$$

where

$$\hat{q} = \frac{-\beta_{-1}}{\beta_1 - \beta_{-1}} \tag{8-5-30}$$

In Eq. (8-5-29) the first term on the right-hand side represents the component of the axial strain due to geometric nonlinearity alone and the second term on the right-hand side represents approximately the component of the axial strain due to tensorial nonlinearity alone. Equation (8-5-29) states that the elongation of a twisted rod is proportional to the square of the twist D, this being the classical statement of Poynting's (1909, 1912) effect. These exact solutions of isotropic elasticity were first discussed by Rivlin (1948a, 1949a, 1949b).

8-5-3 Bending of a Block into an Annular Wedge

Consider a rectangular block of incompressible material for which the initial configuration is described in the cartesian coordinates (X, Y, Z), the block being bounded by the planes $X = X_1$, $X = X_2$, $Y = \pm Y_0$, and $Z = \pm Z_0$. The loaded configuration is described in the cylindrical coordinates (r, θ, z). Of immediate interest are the deformations

$$r = \sqrt{2AX} \qquad \theta = BY \qquad z = \frac{Z}{AB} \tag{8-5-31}$$

where A and B are constants. This deformation carries the block into the annular wedge bounded by the cylinders $r = r_1 = \sqrt{2AX_1}$, $r = r_2 = \sqrt{2AX_2}$ and the planes $\theta = \pm\theta_0 = \pm BY_0$, $z = \pm z_0 = \pm Z_0/AB$. Assuming B as given, then from Eqs. (8-5-31) the stretch $1/AB$ can be regarded as being arbitrary, so allowing the radial stretch to be adjusted in order to render the deformation isochoric.

Substituting Eqs. (8-5-31) into the component form of $\mathbf{B} = \mathbf{F}\mathbf{F}^T$ gives

$$[B^{ij}] = \begin{bmatrix} \dfrac{A^2}{r^2} & 0 & 0 \\ 0 & B^2 & 0 \\ 0 & 0 & \dfrac{1}{A^2B^2} \end{bmatrix} \tag{8-5-32}$$

$$[(B^{-1})_{ij}] = \begin{bmatrix} \dfrac{r^2}{A^2} & 0 & 0 \\ 0 & \dfrac{1}{B^2} & 0 \\ 0 & 0 & A^2B^2 \end{bmatrix} \tag{8-5-33}$$

From Eq. (8-1-43), together with Eqs. (8-5-31), (8-5-32), and (8-5-33), there can be obtained the stress difference $T\langle\theta\theta\rangle - T\langle rr\rangle$ which can be substituted into Eq. (8-5-7) to give $T\langle rr\rangle$. In this way the stress relations can be obtained from Eqs. (8-1-43), (8-5-32), and (8-5-33) in the form

$$T\langle rr\rangle = \int_{r_2}^{r}\left[\left(B^2 r - \frac{A^2}{r^3}\right)\beta_1 + \left(\frac{1}{r^3B^2} - \frac{r}{A^2}\right)\beta_{-1}\right]dr + [T\langle rr\rangle]_{r=r_2} \tag{8-5-34}$$

$$T\langle\theta\theta\rangle = T\langle rr\rangle + \left(B^2r^2 - \frac{A^2}{r^2}\right)\beta_1 + \left(\frac{1}{r^2B^2} - \frac{r^2}{A^2}\right)\beta_{-1} \tag{8-5-35}$$

$$T\langle zz\rangle = T\langle rr\rangle + \left(\frac{1}{A^2B^2} - \frac{A^2}{r^2}\right)\beta_1 - \left(\frac{r^2}{A^2} - A^2B^2\right)\beta_{-1} \tag{8-5-36}$$

where the β_i $(i = \pm 1)$ are functions of the invariants

$$\mathrm{I}_{\mathbf{B}} = \frac{A^2}{r^2} + B^2r^2 + \frac{1}{A^2B^2} \tag{8-5-37}$$

$$\mathrm{II}_{\mathbf{B}} = \frac{r^2}{A^2} + \frac{1}{B^2 r^2} + A^2 B^2 \tag{8-5-38}$$

For the particular case that $T\langle rr\rangle$ is zero at $r = r_1$, that is, $[T\langle rr\rangle]_{r=r_1} = 0$, then $T\langle rr\rangle$ can be expressed by way of Eq. (8-5-7) in the form

$$T\langle rr\rangle = \int_{r_1}^{r} \left[\left(B^2 r - \frac{A^2}{r^3}\right)\beta_1 + \left(\frac{1}{B^2 r^3} - \frac{r}{A^2}\right)\beta_{-1}\right] dr \tag{8-5-39}$$

The condition that both surfaces are free of traction, that is, $[T\langle rr\rangle]_{r=r_1} = [T\langle rr\rangle]_{r=r_2} = 0$, can be obtained from Eq. (8-5-34) in the form

$$\int_{r_2}^{r_1} \left[\left(\frac{A^2}{r^3} - B^2 r\right)\beta_1 - \left(\frac{1}{r^3 B^2} - \frac{r}{A^2}\right)\beta_{-1}\right] dr = 0 \tag{8-5-40}$$

which, in principle, gives a relation between the constants A and B. However, it is evident that to use Eq. (8-5-40) requires some simplifying assumption regarding the form of the response functions. Thus, Eq. (8-5-40) would appear to imply that the bending of a block into a circular wedge cannot be accomplished by end loads and couples alone.

This special case of pure bending was first discussed by Rivlin (1949a, 1949b).

8-6 TRANSVERSE ISOTROPY

The concept of material symmetry has been discussed in Secs. 5-3 and 5-7. In Sec. 5-7 a material has been defined as having transverse isotropy if it has a single axis of rotational symmetry, every plane containing this axis being a plane of reflection symmetry.

Taking the X_3 axis as the axis of rotational symmetry, Rivlin (1966) considered the irreducible integrity basis of the deformation tensor $\mathbf{C}$. Such an integrity (or polynomial) basis will consist of the irreducible polynomials of $\mathbf{C}$ (i.e., polynomials which cannot be factorized into the products of polynomials of lower degree) which remain invariant under transformations of the reference frame $\mathbf{H}$ belonging to the relevant isotropy group. In the present case, if the reference frame is rectangular cartesian, $\mathbf{H}$ consists of the continuous groups of matrices

$$[\mathbf{H}] = \begin{bmatrix} \cos\theta & \sin\theta & 0 \\ -\sin\theta & \cos\theta & 0 \\ 0 & 0 & 1 \end{bmatrix}$$

for arbitrary θ. Thus, in addition to the usual invariants

$$\mathrm{I} = \mathrm{I}_{\mathbf{B}} = \mathrm{I}_{\mathbf{C}} \qquad \mathrm{II} = \mathrm{II}_{\mathbf{B}} = \mathrm{II}_{\mathbf{C}} \qquad \mathrm{III} = \mathrm{III}_{\mathbf{B}} = \mathrm{III}_{\mathbf{C}} \tag{8-6-1}$$

belonging to the full orthogonal group, we have the extra invariants

$$C_{3'3'} = H_{3'\alpha} H_{3'\beta} C_{\alpha\beta} = C_{33} \qquad [H_{3'1} = H_{3'2} = 0, H_{3'3} = 1]$$

and as

$$C_{1'3'} = \cos\theta C_{13} + \sin\theta C_{23}$$

and
$$C_{2'3'} = -\sin\theta C_{13} + \cos\theta C_{23}$$

this gives

$$(C_{1'3'})^2 + (C_{2'3'})^2 = (C_{13})^2 + (C_{23})^2$$

as stated above. More generally,

$$\mathrm{IV} = C^1{}_3 C_1{}^3 + C^2{}_3 C_2{}^3 \qquad \mathrm{V} = C^3{}_3 \tag{8-6-2}$$

Denoting by $\frac{1}{2}\beta_i$ $(i = \pm 1, 3, 4, 5)$ the derivatives of W with respect to this set of invariants, i.e.,

$$\beta_1 = 2\frac{\partial W}{\partial \mathrm{I}} \qquad \beta_{-1} = -2\frac{\partial W}{\partial \mathrm{II}} \qquad \beta_3 = 2\frac{\partial W}{\partial \mathrm{III}} \tag{8-6-3}$$

$$\beta_4 = 2\frac{\partial W}{\partial \mathrm{IV}} \qquad \beta_5 = 2\frac{\partial W}{\partial \mathrm{V}} \tag{8-6-4}$$

the stress relation can be expressed in the form

$$\begin{aligned}\mathbf{T} &= \frac{1}{J}\mathbf{F}\frac{\partial W}{\partial \mathbf{F}} \\ &= \frac{1}{2J}\mathbf{F}\left(\beta_1\frac{\partial \mathrm{I}}{\partial \mathbf{F}} - \beta_{-1}\frac{\partial \mathrm{II}}{\partial \mathbf{F}} + \beta_3\frac{\partial \mathrm{III}}{\partial \mathbf{F}} + \beta_4\frac{\partial \mathrm{IV}}{\partial \mathbf{F}} + \beta_5\frac{\partial \mathrm{V}}{\partial \mathbf{F}}\right)\end{aligned} \tag{8-6-5}$$

Use of the relations

$$\mathbf{F}\frac{\partial \mathrm{I}}{\partial \mathbf{F}} = 2\mathbf{B} \tag{8-6-6}$$

$$\mathbf{F}\frac{\partial \mathrm{II}}{\partial \mathbf{F}} = 2\mathrm{I}\mathbf{B} - 2\mathbf{B}^2 \tag{8-6-7}$$

$$\mathbf{F}\frac{\partial \mathrm{III}}{\partial \mathbf{F}} = 2\mathrm{III}\mathbf{I} \tag{8-6-8}$$

$$\mathbf{F}\frac{\partial \mathrm{IV}}{\partial \mathbf{F}} = 2\mathbf{M} \tag{8-6-9}$$

$$\mathbf{F}\frac{\partial \mathrm{V}}{\partial \mathbf{F}} = 2\mathbf{N} \tag{8-6-10}$$

$$N^i{}_j = F^i{}_3 F_j{}^3 \qquad M^i{}_j = C^\alpha{}_3(F^i{}_\alpha F_j{}^3 + F_{j\alpha}F^{i3}) \tag{8-6-11}$$

where α takes on the values 1 and 2, summation over repeated α being understood to allow the stress relation of Eq. (8-6-5) to be expressed in the form

$$\mathbf{T} = \frac{1}{J}[\beta_1\mathbf{B} + \beta_{-1}(\mathbf{B}^2 - \mathrm{I}\mathbf{B}) + \beta_3\mathrm{III}\mathbf{I} + \beta_4\mathbf{M} + \beta_5\mathbf{N}] \tag{8-6-12}$$

From the Cayley–Hamilton equation expressed in the form

$$\mathbf{B}^3 = \mathrm{I}\mathbf{B}^2 - \mathrm{II}\mathbf{B} + \mathrm{III}\mathbf{I} \tag{8-6-13}$$

there can be obtained the relation

$$\mathbf{B}^2 = \mathrm{I}\mathbf{B} - \mathrm{II}\mathbf{I} + \mathrm{III}\mathbf{B}^{-1} \tag{8-6-14}$$

which permits the stress relation of Eq. (8-6-12) to be expressed in the form

$$\mathbf{T} = \frac{1}{J}[(\mathrm{III}\beta_3 - \mathrm{II}\beta_{-1})\mathbf{I} + \beta_1\mathbf{B} + \beta_{-1}\mathrm{III}\mathbf{B}^{-1} + \beta_4\mathbf{M} + \beta_5\mathbf{N}] \tag{8-6-15}$$

If the medium is incompressible, $\mathrm{III} = 1$ and Eq. (8-6-15) takes the form [see Eq. (8-1-31)]

$$\mathbf{T} = -P\mathbf{I} + \beta_1\mathbf{B} + \beta_{-1}\mathbf{B}^{-1} + \beta_4\mathbf{M} + \beta_5\mathbf{N} \tag{8-6-16}$$

which is seen to differ from Eq. (8-1-43) only by the addition of the two terms involving β_4 and β_5 which are specific to the transversely isotropic case.

The solution of problems is similar to that for the isotropic case.

8-7 EXPERIMENTAL APPLICATIONS: I. METALS

The existence of unequal normal stresses in simple shear is now referred to as the Poynting effect, which relates to the observation by Poynting (1909, 1912) that the length of various steel, copper, and brass wires increased when twisted in the elastic range, the elongation being proportional to the square of the twist, an observation which is in general accord with Eq. (8-5-28). These length changes in twisted wires were accompanied by a resulting volume increase which was also approximately proportional to the square of the twist.

Of particular interest are the measurements of Poynting (1912) for two steel wires having diameters $2R_1 = 0.986$ and 1.21 mm and a copper wire of diameter $2R_1 = 1.1219$ mm. All three wires were of length $L = 1.605$ m. The observed axial strain $e_p = \Delta l/L$ for the first full turn was $e_p = 1.71 \times 10^{-6}$ and 2.90×10^{-6} for the steel wires and $e_p = 4.25 \times 10^{-6}$ for the copper wire. These length changes were accompanied by the observed contraction strains $\Delta R_1/R_1 = 3.2 \times 10^{-7}$ and 5.2×10^{-7} for the two steel wires and $\Delta R_1/R_1 = 1.75 \times 10^{-7}$ for the copper wire, the associated volume increases being $\Delta V/V = 1.07 \times 10^{-6}$ and 1.85×10^{-6} for the steel wires and $\Delta V/V = 0.75 \times 10^{-6}$ for the copper wire. Using the values $R_1 D = 1.93 \times 10^{-3}$ and 2.37×10^{-3} for one full turn of the steel wires and $R_1 D = 2.4 \times 10^{-3}$ for one full turn of the copper wire, the values of $\hat{q}$ can be evaluated from Eq. (8-5-30) to give $\hat{q} = 5.5$ and 6.2 for the steel wires and $\hat{q} = 7.9$ for the copper wire.

The measurements of Foux (1964) confirm the observations of Poynting. For this extensive series of tests on hard steel wires of similar length and diameter, Foux obtained a range of values of $\hat{q}$ extending from ~ 5.4 to ~ 6.2 which are to be compared with those evaluated from Poynting's (1912) results.

8-8 EXPERIMENTAL APPLICATIONS: II. RUBBER

8-8-1 The Form of the Strain Energy Function

The present discussion is mainly concerned with the investigation by Rivlin and Saunders (1951) of the form of the strain energy function W for vulcanized rubber. Basic to these studies are the assumptions that the body is isotropic, substantially homogeneous, and effectively incompressible. With regard to the assumption of incompressibility, small volume changes in rubber of the order of one part in 10^4 during stretching have, however, been measured by Gee, Stern, and Treloar (1950). For an incompressible material, $\mathrm{III}_{\mathbf{B}} = 1$ and the strain energy may be expressed in the form

$$W = W(\mathrm{I}_{\mathbf{B}}, \mathrm{II}_{\mathbf{B}}) \qquad (\mathrm{III}_{\mathbf{B}} = 1) \tag{8-8-1}$$

where $\mathrm{I}_{\mathbf{B}}$, $\mathrm{II}_{\mathbf{B}}$, and $\mathrm{III}_{\mathbf{B}}$ are first, second, and third invariants respectively of the deformation tensor $\mathbf{B}$. Restrictions on the functional form of W for an incompressible isotropic material have been discussed by Baker and Ericksen (1954).

A discussion of the form of the strain energy function and its relation to the molecular structure of rubber has been given by Treloar (1958, 1970). This is a statistical theory based upon the molecular concept of vulcanized rubber as a network of long-chain molecules. As a first approximation, this particular molecular theory of rubber-like materials gives a strain energy function of the form

$$W = C(\mathrm{I}_{\mathbf{B}} - 3) \tag{8-8-2}$$

where C is a constant.

The linear form

$$W = C_1(\mathrm{I}_{\mathbf{B}} - 3) + C_2(\mathrm{II}_{\mathbf{B}} - 3) \tag{8-8-3}$$

where C_1 and C_2 are constants, was suggested by Mooney (1940) on the basis of shearing experiments on rubber blocks. This is to be compared with the form

$$W = C_1(\mathrm{I}_{\mathbf{B}} - 3) + f(\mathrm{II}_{\mathbf{B}} - 3) \tag{8-8-4}$$

suggested by the experiments of Rivlin and Saunders (1951), where C_1 is a constant and f is a function whose general behavior can be evaluated from experiment. It is evident that any extension of the statistical theory which considers the structural properties in greater detail must give a form of W which is in accord with that deduced from the experiments of Rivlin and Saunders.

For a detailed discussion of the relation of experiment to the theory of finite elasticity, in particular the experiments of Rivlin and Saunders (1951), the reader is referred to Green and Adkins (1960).

In the following subsections

$$\beta_1 = 2\frac{\partial W}{\partial \mathrm{I}_{\mathbf{B}}} \qquad \beta_{-1} = -2\frac{\partial W}{\partial \mathrm{II}_{\mathbf{B}}} \tag{8-8-5}$$

the experiments of Rivlin and Saunders (1951) being concerned with the evaluation of $\beta_1/2$ and $-\beta_{-1}/2$.

8-8-2 Pure Homogeneous Strain of a Thin Rubber Sheet

For the pure homogeneous deformation of a thin sheet, the principal forces F_1 and F_2, measured per unit length of the edge in the undeformed configuration, are defined by

$$F_1 = 2h_0\lambda_2\lambda_3 T\langle xx\rangle = \frac{2h_0}{\lambda_1} T\langle xx\rangle \tag{8-8-6a}$$

$$F_2 = 2h_0\lambda_3\lambda_1 T\langle yy\rangle = \frac{2h_0}{\lambda_2} T\langle yy\rangle \tag{8-8-6b}$$

where the stresses $T\langle xx\rangle$ and $T\langle yy\rangle$ are given by Eqs. (8-4-19) and (8-4-20). From measurements of these forces for any given values of λ_1 and λ_2, $T\langle xx\rangle$ and $T\langle yy\rangle$ can be calculated, and hence the values of the β_i $(i = \pm 1)$ can be evaluated from Eqs. (8-4-19) and (8-4-20) expressed in the form

$$\begin{aligned} \beta_1 &= \frac{\lambda_1^2\lambda_2^2}{\lambda_1^2-\lambda_2^2}\left(\frac{\lambda_1^2 T\langle xx\rangle}{\lambda_1^4\lambda_2^2-1} - \frac{\lambda_2^2 T\langle yy\rangle}{\lambda_2^4\lambda_1^2-1}\right) \\ \beta_{-1} &= \frac{\lambda_1^2\lambda_2^2}{\lambda_1^2-\lambda_2^2}\left(\frac{T\langle xx\rangle}{\lambda_1^4\lambda_2^2-1} - \frac{T\langle yy\rangle}{\lambda_2^4\lambda_1^2-1}\right) \end{aligned} \tag{8-8-7}$$

From Eqs. (8-8-7) values of the β_i $(i = \pm 1)$ can be obtained for various values of $\mathbf{I_B}$, subject to the particular condition that λ_1 and λ_2 are varied in such a way that $\mathbf{II_B}$ remains constant. Similarly, by varying λ_1 and λ_2 in such a way that $\mathbf{I_B}$ remains constant while $\mathbf{II_B}$ varies, the dependence of the β_i $(i = \pm 1)$ upon $\mathbf{II_B}$ for various fixed values of $\mathbf{I_B}$ can be determined by way of Eqs. (8-8-7).

Equation (8-4-17) can be rearranged to give

$$\lambda_2^2 = \frac{1}{2}\left\{(\mathbf{I_B}-\lambda_1^2) \pm \left[(\mathbf{I_B}-\lambda_1^2)^2 - \frac{4}{\lambda_1^2}\right]^{1/2}\right\} \tag{8-8-8a}$$

$$\lambda_2^2 = \frac{1}{2\lambda_1^2}\left\{\left(\mathbf{II_B} - \frac{1}{\lambda_1^2}\right) \pm \left[\left(\mathbf{II_B} - \frac{1}{\lambda_1^2}\right)^2 - 4\lambda_1^2\right]^{1/2}\right\} \tag{8-8-8b}$$

The corresponding values of λ_1 and λ_2 for which $\mathbf{I_B}$ is a constant can be obtained from Eq. (8-8-8*a*) and similarly, for $\mathbf{II_B}$ a constant, the corresponding values of λ_1 and λ_2 are given by Eq. (8-8-8*b*). The variation of λ_2 with λ_1 for given constant values of $\mathbf{I_B}$ and $\mathbf{II_B}$ is shown in Fig. 8-2, where the broken-line curves represent

$$\lambda_2 = \lambda_1^{-1/2}(F_2 = 0) \qquad \lambda_1 = \lambda_2^{-1/2}(F_1 = 0) \tag{8-8-9}$$

which correspond to simple extensions parallel to the X and Y axes respectively.

Rivlin and Saunders (1951) developed an experimental technique which enabled them to obtain results for the pure homogeneous deformation of a thin sheet for which $\mathbf{I_B}$ or $\mathbf{II_B}$ was held constant. In Fig. 8-3 are shown the results obtained by Rivlin and Saunders for the approximate ranges $5 \leqslant \mathbf{I_B} \leqslant 12$ and $5 \leqslant \mathbf{II_B} \leqslant 30$. These results suggest the form of the strain energy function given in Eq. (8-8-4), subject to the particular conditions

$$\beta_1 = 2\frac{\partial W}{\partial \mathbf{I_B}} = 2C_1 = \text{constant} \tag{8-8-10a}$$

$$\beta_{-1} = -2\frac{\partial W}{\partial \mathrm{II}_{\mathbf{B}}} = -2\frac{df}{d\mathrm{II}_{\mathbf{B}}} \tag{8-8-10b}$$

The ratio

$$\frac{\partial W/\partial \mathrm{II}_{\mathbf{B}}}{\partial W/\partial \mathrm{I}_{\mathbf{B}}} = -\frac{\beta_{-1}}{\beta_1} \simeq \tfrac{1}{8} \tag{8-8-11}$$

for low values of $\mathrm{II}_{\mathbf{B}}$ and decreases steadily with increasing $\mathrm{II}_{\mathbf{B}}$. The experiment is unsuitable for very large deformations which are difficult to obtain in the manner described, and is also unsuitable for small deformations since, for values of $\mathrm{I}_{\mathbf{B}}$ and $\mathrm{II}_{\mathbf{B}}$ less than ~ 5, the results obtained are subject to large experimental errors. For these reasons, Rivlin and Saunders conclude that the absolute dependence on the β_i $(i = \pm 1)$ on $\mathrm{I}_{\mathbf{B}}$ and $\mathrm{II}_{\mathbf{B}}$ cannot be determined from this experiment alone. However, it is to be noted that the particular form of strain energy function W, defined by Eq. (8-8-4), enables the theory to predict results for other types of deformation which are in good accord with experiment.

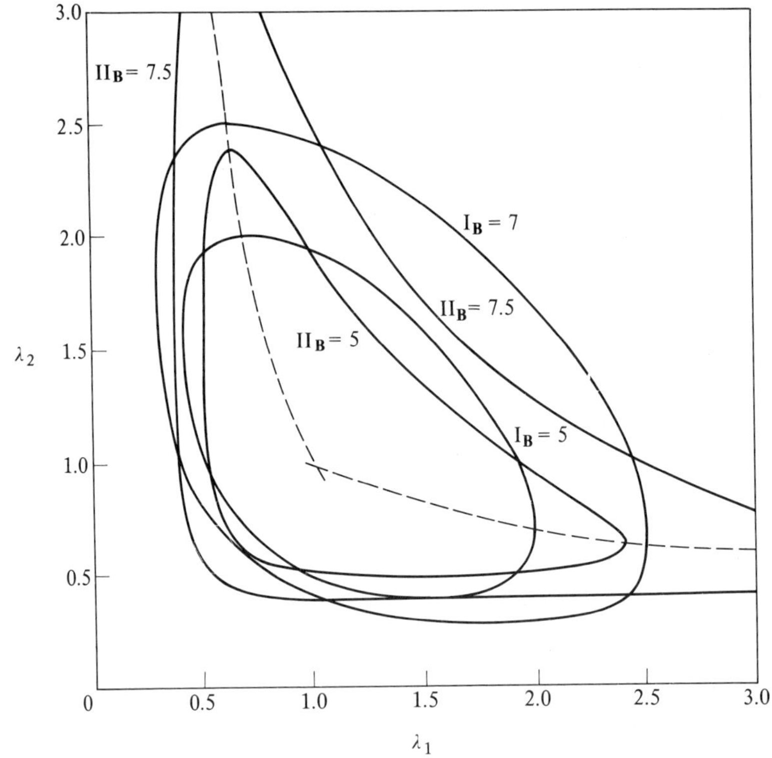

Figure 8-2 The relation between λ_1 and λ_2 at constant $\mathrm{I}_{\mathbf{B}}$ or $\mathrm{II}_{\mathbf{B}}$. (*Rivlin and Saunders, 1951.*)

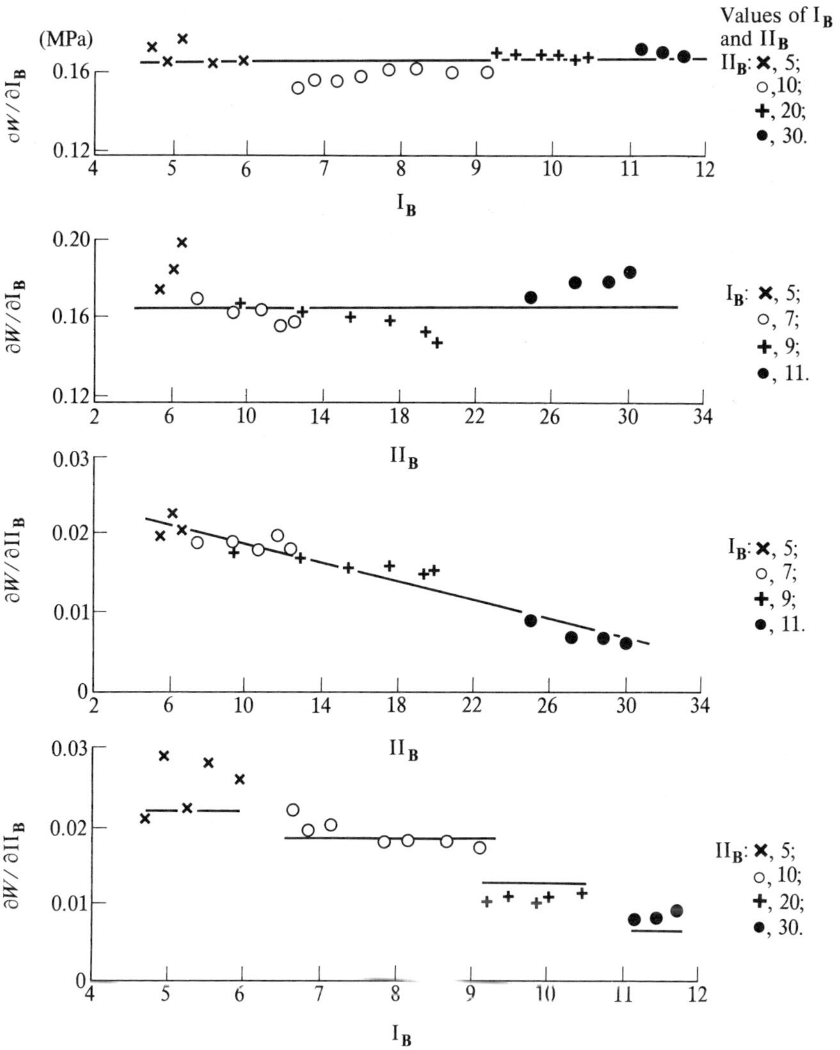

Figure 8-3 Dependence of $\partial W/\partial \mathrm{I_B}$ and $\partial W/\partial \mathrm{II_B}$ on $\mathrm{I_B}$ and $\mathrm{II_B}$. (*Rivlin and Saunders, 1951.*)

8-8-3 Pure Shear

From Eqs. (8-4-19) and (8-8-6*a*) there is obtained the relation

$$
\begin{aligned}
F_1 &= 2h_0\left(\lambda_1 - \frac{1}{\lambda_1^3\lambda_2^2}\right)(\beta_1 - \lambda_2^2\beta_{-1}) \\
&= 4h_0\left(\lambda_1 - \frac{1}{\lambda_1^3\lambda_2^2}\right)\left(\frac{\partial W}{\partial \mathrm{I_B}} + \lambda_2^2\frac{\partial W}{\partial \mathrm{II_B}}\right)
\end{aligned}
\tag{8-8-12}
$$

If it is assumed, by way of the experiments on the pure homogeneous deformation of a thin sheet, that the condition of Eq. (8-8-10*a*) remains valid for measurements in pure shear, then if the deformation is varied so that λ_2 remains constant in Eq. (8-8-12), β_{-1} can be evaluated from the variation of $\beta_1 - \lambda_2^2\beta_{-1}$ with $\mathrm{II}_\mathbf{B}$. These conditions can be obtained by stretching a strip of rubber in such a way that its width remains unchanged. This is approximated to by the distribution of strain in a wide sheet stretched between two clamped edges. Provided the width of the sheet is sufficiently large compared with its length, the nonuniformity of strain arising from the outer edges being free may be neglected, and the extension ratio λ_2 in the center, in a direction parallel to the edges, can then be regarded as being effectively unaltered in any deformation produced by moving these edges further apart.

The results obtained by Rivlin and Saunders (1951) are shown in Fig. 8-4, curve I corresponding to $\lambda_2 = 1$ and curve II corresponding to $\lambda_2 = 0.776$.

For $\lambda_2 = 1$, the ratio defined by Eq. (8-8-11) was assumed to have the value $\frac{1}{8}$ at $\mathrm{II}_\mathbf{B} = 5$. The measurements of $(\beta_1 - \beta_{-1})/2$, $(\lambda_2 = 1)$ at this point gave separate values of $\beta_1/2$ and $-\beta_{-1}/2$. Values of $-\beta_{-1}/2$ at other values of $\mathrm{II}_\mathbf{B}$ were found from the appropriate readings of $(\beta_1 - \beta_{-1})/2$, by way of Eqs. (8-8-4) and (8-8-10*a*).

The values of $(\beta_1 - \lambda_2^2\beta_{-1})/2$ for $\lambda_2 = 0.776$, calculated using the values of $\beta_1/2$ and $-\beta_{-1}/2$ thus obtained, were found to be in close accord with the experimental values derived directly from curve II.

8-8-4 Simple Extension

For a simple extension of ratio λ, the load N required to extend a test piece of rubber of initial uniform cross-sectional area A can be obtained from Eq. (8-4-7) in

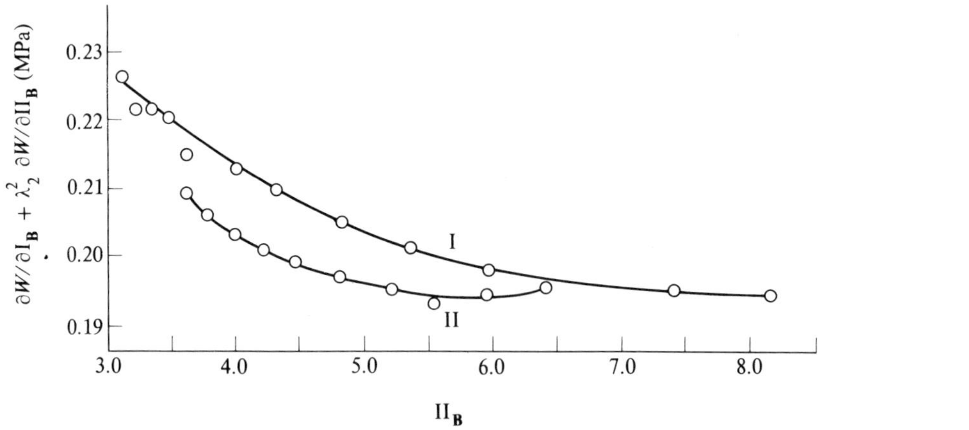

Figure 8-4 Variation of $(\partial W/\partial \mathrm{I}_\mathbf{B} + \lambda_2^2 \partial W/\partial \mathrm{II}_\mathbf{B})$ with $\mathrm{II}_\mathbf{B}$ from experiments on pure shear and pure shear superposed on simple extension. Curve I, pure shear ($\lambda_2 = 1$); curve II, pure shear superposed on simple extension ($\lambda_2 = 0.776$). (*Rivlin and Saunders, 1951.*)

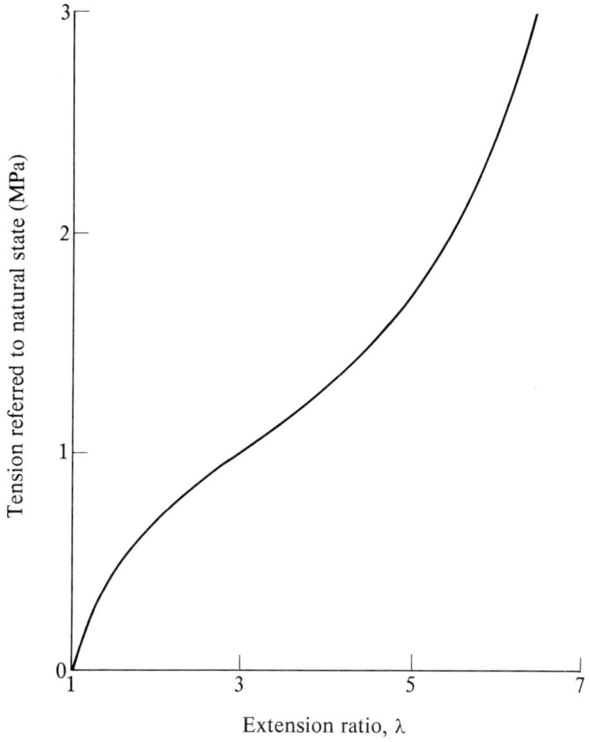

Figure 8-5 Typical force-extension ratio curve for rubber in simple extension. (*Treloar, 1958.*)

the form

$$N = \frac{A}{\lambda} T\langle zz\rangle = A\left(\frac{\lambda^3 - 1}{\lambda^2}\right)\left(\beta_1 - \frac{1}{\lambda}\beta_{-1}\right)$$
$$= 2A\left(\frac{\lambda^3 - 1}{\lambda^2}\right)\left(\frac{\partial W}{\partial \mathrm{I_B}} + \frac{1}{\lambda}\frac{\partial W}{\partial \mathrm{II_B}}\right) \tag{8-8-13}$$

a typical experimental load-deformation curve being shown in Fig. 8-5. Of particular interest is the variation with $1/\lambda$ of the quantity $[\beta_1 - (1/\lambda)\beta_{-1}]/2$, calculated from the results of Rivlin and Saunders (1951) by way of Eq. (8-8-13) and shown in Fig. 8-6. It is evident from Fig. 8-6 that there is an effectively linear relation for various values of $1/\lambda$ from 0.9 to 0.45, corresponding to values of $\mathrm{II_B}$ lying between 3.03 and 4.65. The values of $[\beta_1 - (1/\lambda)\beta_{-1}]/2$, obtained by Rivlin and Saunders from the simple extension experiment, agree closely with those calculated using the values of $\beta_1/2$ and $-\beta_{-1}/2$ derived from pure shear.

8-8-5 Uniform Two-Dimensional Extension

Setting $\lambda_1 = \lambda_2 > 1$ in Eq. (8-4-11) gives a uniform two-dimensional extension in the xy plane. Hence, setting

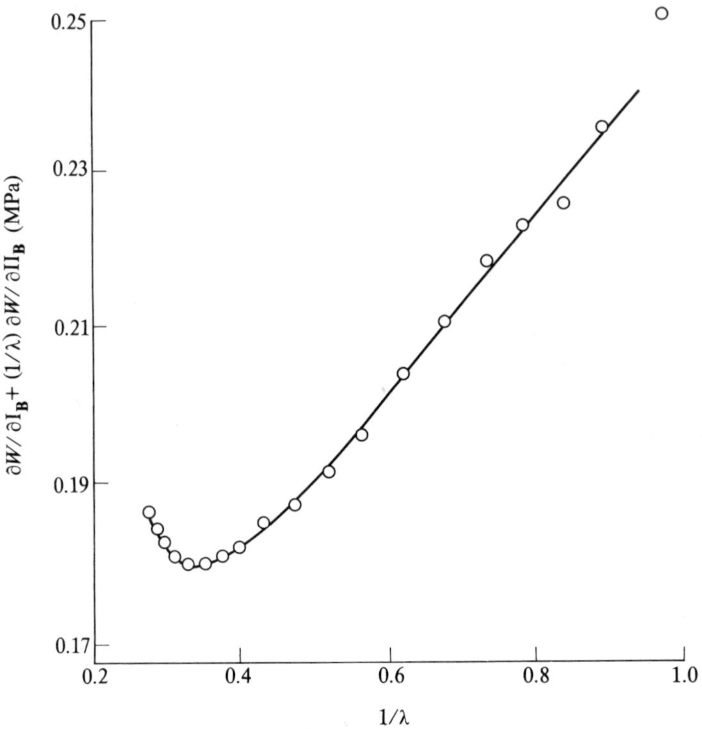

Figure 8-6 Variation of $[\partial W/\partial \mathrm{I_B} + (1/\lambda)\partial W/\partial \mathrm{II_B}]$ with $1/\lambda$ from experiment or simple extension. (*Rivlin and Saunders, 1951.*)

$$\lambda_1 = \lambda_2 = \lambda \qquad \lambda_3 = \frac{1}{\lambda^2} \qquad (\lambda_1\lambda_2\lambda_3 = 1) \tag{8-8-14}$$

in Eqs. (8-4-14), (8-4-15), (8-4-16), and (8-4-17) gives

$$T\langle xx\rangle = T\langle yy\rangle = -P + \beta_1\lambda^2 + \beta_{-1}\frac{1}{\lambda^2} \tag{8-8-15}$$

$$T\langle zz\rangle = -P + \beta_1\frac{1}{\lambda^4} + \beta_{-1}\lambda^4 \tag{8-8-16}$$

$$\mathrm{I_B} = 2\lambda^2 + \frac{1}{\lambda^4} \qquad \mathrm{II_B} = \frac{2}{\lambda^2} + \lambda^4 \tag{8-8-17}$$

The particular conditions of Eq. (8-4-18), corresponding to $T\langle zz\rangle = 0$, and Eq. (8-8-13) give, by way of Eq. (8-8-14),

$$T\langle xx\rangle = T\langle yy\rangle = \left(\lambda^2 - \frac{1}{\lambda^4}\right)(\beta_1 - \lambda^2\beta_{-1}) \tag{8-8-18}$$

Alternatively, setting $T\langle xx\rangle = T\langle yy\rangle = 0$ in Eq. (8-8-15) gives

$$P = \beta_1 \lambda^2 + \beta_{-1} \frac{1}{\lambda^2} \tag{8-8-19}$$

which substituted into Eq. (8-8-16) gives

$$T\langle zz \rangle = -\left(\lambda^2 - \frac{1}{\lambda^4}\right)(\beta_1 - \lambda^2 \beta_{-1}) \tag{8-8-20}$$

It is evident that the same state of strain can be maintained by the stress system $T\langle zz \rangle = 0$ and $T\langle xx \rangle = T\langle yy \rangle$ given by Eq. (8-8-18), or, alternatively, by the stress system $T\langle xx \rangle = T\langle yy \rangle = 0$ and the unidirectional compressive stress $T\langle zz \rangle$ parallel to the Z axis and given by Eq. (8-8-20). Thus, it follows from Eq. (8-8-20) that measurements of $T\langle xx \rangle$ or $T\langle yy \rangle$ in the case represented by Eq. (8-8-18) enables the stress associated with unidirectional compression of the material ratio $1/\lambda^2$ in the Z direction to be determined.

Treloar (1943) and Rivlin and Saunders (1951) have reproduced the conditions described by Eq. (8-8-14) to a good approximation by inflating a thin circular sheet of rubber of uniform initial thickness h_0 clamped around its edge, by applying an air pressure $\bar{P}_i$ to one face. The radial tension T_0 at the pole of the inflated sheet, measured per unit length of the major surface of the sheet in its deformed state, varies with the pressure $\bar{P}_i$ according to the relation

$$\bar{P}_i = \frac{2T_0}{r} \tag{8-8-21}$$

where r is now the radius of curvature at the pole. Provided that $h_0 \ll r$, the components of stress at the pole can be regarded as being very nearly uniform across the thickness of the sheet and are given by Eq. (8-8-18) with $T\langle zz \rangle = 0$, the Z axis being taken along the normal to the sheet at its pole. The thickness of the deformed sheet at the center is then h_0/λ^2 and hence, by way of Eqs. (8-8-18) and (8-8-21),

$$\begin{aligned}\bar{P}_i &= \frac{2}{r}\frac{h_0}{\lambda^2}\delta = \frac{2h_0}{r}\left(1 - \frac{1}{\lambda^6}\right)(\beta_1 - \lambda^2 \beta_{-1}) \\ &= \frac{4h_0}{r}\left(1 - \frac{1}{\lambda^6}\right)\left(\frac{\partial W}{\partial \mathrm{I}_{\mathbf{B}}} + \lambda^2 \frac{\partial W}{\partial \mathrm{II}_{\mathbf{B}}}\right)\end{aligned} \tag{8-8-22}$$

where $\delta = T\langle xx \rangle = T\langle yy \rangle$ as defined by Eq. (8-8-18). It is evident from Eq. (8-8-22) that the quantity $(\beta_1 - \lambda^2 \beta_{-1})/2$ can be determined, provided $\bar{P}_i$, r, and λ can be measured.

Setting $\lambda^2 = 1/\lambda'$ enables Eq. (8-8-22) to be rearranged and expressed in the form

$$\beta_1 - \frac{1}{\lambda'}\beta_{-1} = \frac{r\bar{P}_i \lambda^6}{2h_0(\lambda^6 - 1)} \tag{8-8-23}$$

The variation of the quantity $[\beta_1 - (1/\lambda')\beta_{-1}]/2$, as defined by Eq. (8-8-23), with $1/\lambda'$ is shown in Fig. 8-7 for the compression measurements of Rivlin and Saunders, it being noted that for these measurements $\lambda' < 1$. For $\lambda' < 1$ and λ

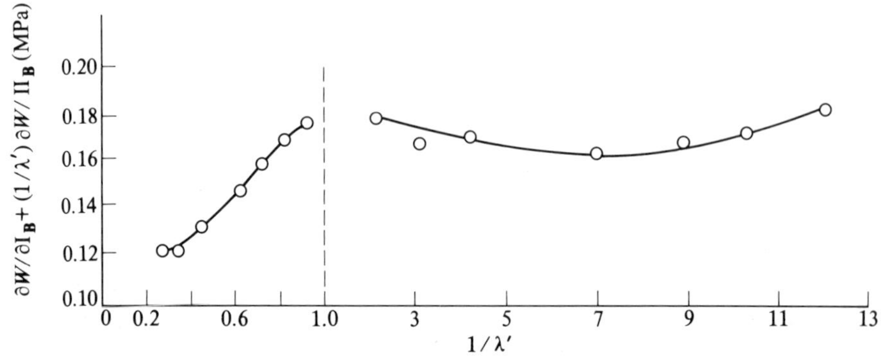

Figure 8-7 Dependence of $[\partial W/\partial I_B + (1/\lambda')\partial W/\partial II_B]$ on $1/\lambda'$ from experiments on simple compression and simple extension. (Note change of scale at $1/\lambda' = 1$.) (*Rivlin and Saunders, 1951.*)

having the significance ascribed to it in Eqs. (8-8-14) to (8-8-22), it follows that λ' represents a compression ratio.

For comparison purposes, in Fig. 8-7 there is also shown the variation of $[\beta_1 - (1/\lambda')\beta_{-1}]/2$ with $1/\lambda'$ for $\lambda' > 1$; this portion of the curve, which corresponds to simple extension, was obtained by Rivlin and Saunders from their measurements in simple elongation in the manner described in Sec. 8-8-4. In this region, λ' is simply the extension ratio λ of Sec. 8-8-4.

Values of $-\beta_{-1}/2$ over the entire range of deformation were obtained by Rivlin and Saunders using the procedure outlined in Sec. 8-8-3 for pure shear. The variation of $-\beta_{-1}/2$ with II_B is shown in Table 8-1, which has been reproduced from Rivlin and Saunders (1951).

8-8-6 Torsion

Measurements in the torsion mode of deformation have been made by Rivlin (1947) and by Rivlin and Saunders (1951). The behavior of rubber as obtained from experiments on torsion is in general accord with the characteristic behavior of rubber described in terms of the strain energy function W as inferred from the measurements described in Secs. 8-8-2 to 8-8-5.

Rivlin (1947), assuming the Mooney (1940) form of strain energy function, as defined by Eq. (8-8-3), obtained an estimate of the ratio C_1/C_2 from measurements of the normal end tractions on a twisted solid-rod cylinder of rubber. For a solid-rod specimen of rubber deformed in pure torsion, Eqs. (8-5-27) and (8-8-3) give

$$T\langle zz\rangle = -D^2[2R_1^2C_2 + (R_1^2 - R^2)(C_1 - 2C_2)] \qquad (8\text{-}8\text{-}24)$$

Values of $T\langle zz\rangle/D^2$ can be obtained from measurements of $T\langle zz\rangle$ over one of the plane ends of a solid-rod specimen for given values of D. From Eq. (8-8-24) it is evident that the variation of these values of $T\langle zz\rangle/D^2$ with $R_1^2 - R^2$ gives values of C_2 and $C_1 - 2C_2$. In the experiments of Rivlin a plane circular metal disk was bonded at each end of the solid-rod specimen. One disk contained five circular

Table 8-1

$1/\lambda'$	0.5	0.6	0.7	0.8	3	5	7	9	11
II_B	4.25	3.69	3.35	3.14	9.67	25.4	49.3	81.2	121
$-\beta_{-1}/\beta_1$	0.16	0.26	0.33	0.39	0.12	0.06	0.04	0.03	0.035

holes spaced at different distances from the axis of the rod which passed through the center of the disk. The torsion was produced by rotating this disk in its plane about its center relative to the other unperforated disk, the rubber being observed to bulge out into the holes. Assuming the height of each bulge to be proportional to the pressure which would have been exerted on the rubber at that point by an unperforated plate, Rivlin found that this pressure was approximately proportional to D^2, in accord with the relation of Eq. (8-8-24). The pressure distribution observed in these experiments yielded $C_1/C_2 = 7.1$ which is in satisfactory accord with the results described in Sec. 8-8-2.

In Fig. 8-8 are shown the $(\Gamma, D/F)$ curves obtained by Rivlin and Saunders (1951). For the limiting condition $D = 0$, Eq. (8-5-26) reduces to the form

$$[N_z]_{D=0} = \pi R_1^2 \frac{(F^3 - 1)}{F^2} \left(\beta_1 - \frac{1}{F} \beta_{-1} \right)_{D=0} \tag{8-8-25}$$

and for $D = 0$, Eq. (8-5-25) reduces to the form

$$\left[\frac{\Gamma F}{D} \right]_{D=0} = \tfrac{1}{2} \pi R_1^4 \left(\beta_1 - \frac{1}{F} \beta_{-1} \right)_{D=0} \tag{8-8-26}$$

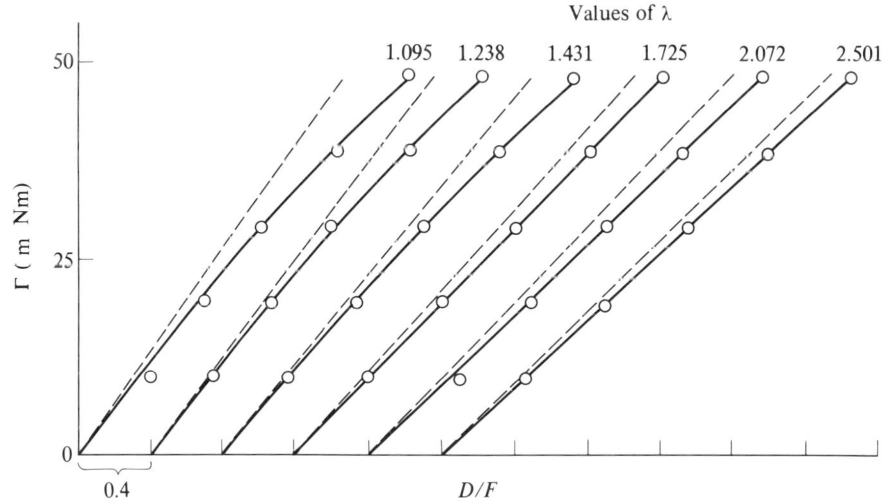

Figure 8-8 Variation of torsional couple Γ with amount of torsion D/F for various values of the extension ratio λ (large range of D/F; the origin is shifted by 0.4 units parallel to the abscissa for each increment in the value of λ). (*Rivlin and Saunders, 1951.*)

Equations (8-8-25) and (8-8-26) give the relation

$$\left[\frac{\Gamma F}{D}\right]_{D=0} = \frac{R_1^2 F^2}{2(F^3-1)}[N_z]_{D=0} \tag{8-8-27}$$

which can be rearranged to give

$$2\frac{(F^3-1)}{F^2}\frac{[\Gamma F/D]_{D=0}}{R_1^2[N_z]_{D=0}} = 1 \tag{8-8-28}$$

From direct measurements of Γ, N, and D at various values of F, Rivlin and Saunders found the left-hand side of Eq. (8-8-28) to lie in the range 0.98 to 1.00. Hence, values of $[\Gamma F/D]_{D=0}$ can be evaluated from Eq. (8-8-27) using measured values of $[N_z]_{D=0}$ at various values of F. Using these values of $[\Gamma F/D]_{D=0}$, the broken lines in Fig. 8-8 represent the linear relation

$$\Gamma = \left[\frac{\Gamma F}{D}\right]_{D=0}\frac{D}{F} \tag{8-8-29}$$

these straight lines being tangential to the observed $(\Gamma, D/F)$ curves at $D = 0$. The variation of $-\beta_{-1}/2$ with $\mathrm{II}_{\mathbf{B}}$ for the measurements in torsion are shown in Fig. 8-9. For these measurements, values of $-\beta_{-1}/2$ were evaluated assuming the functional form of Eq. (8-8-4) for W and taking the ratio $-\beta_{-1}/\beta_1$ to have the particular value 0.16 at a fixed reference point determined by $\mathrm{II}_{\mathbf{B}} = 4.25$.

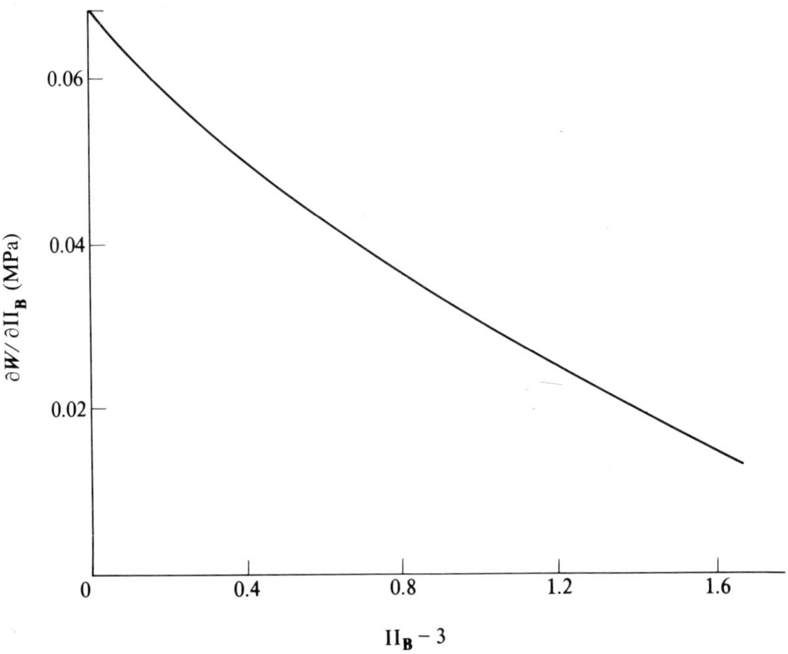

Figure 8-9 Dependence of $\partial W/\partial \mathrm{II}_{\mathrm{B}}$ on $(\mathrm{II}_{\mathrm{B}} - 3)$. (*Rivlin and Saunders, 1951.*)

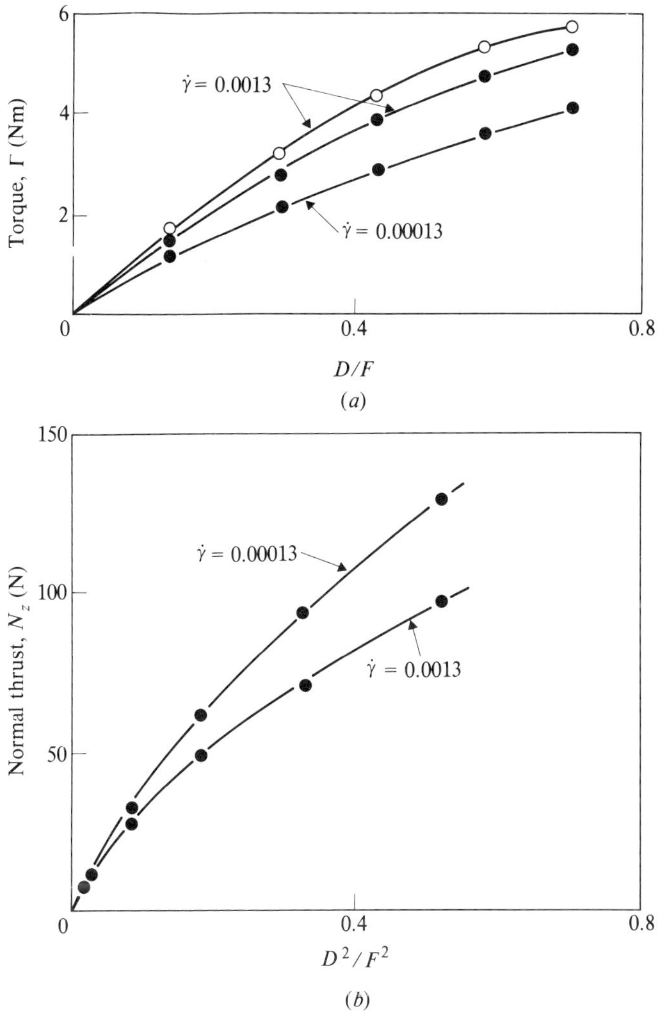

Figure 8-10 (*a*) Variation of torsional couple Γ with the amount of torsion D/F for solid cylindrical specimens of highly filled polyurethane rubber tested in torsion at constant rates of strain. ○, fixed length; ●, free end. (*b*) Relation between D^2/F^2 and normal thrust, i.e., the axial force N_z necessary to sustain constant length in the same test. (*Lenoe, Heller, and Freudenthal, 1965.*)

The nonlinear mechanical response of polyurethane rubbers with a high volume percentage of inorganic filler has been studied by Lenoe, Heller, and Freudenthal (1965). In Fig. 8-10*a* are shown the $(\Gamma, D/F)$ curves for solid cylindrical specimens of highly filled polyurethane rubber tested in torsion at constant rates of strain. These $(\Gamma, D/F)$ curves are to be compared with those shown in Fig. 8-8. The relation between D^2/F^2 and the axial force N_z necessary to sustain constant length in the same series of tests is shown in Fig. 8-10*b*. Although it is evident from Fig. 8-10*b* that the relation between N_z and D^2/F^2 is nonlinear, it

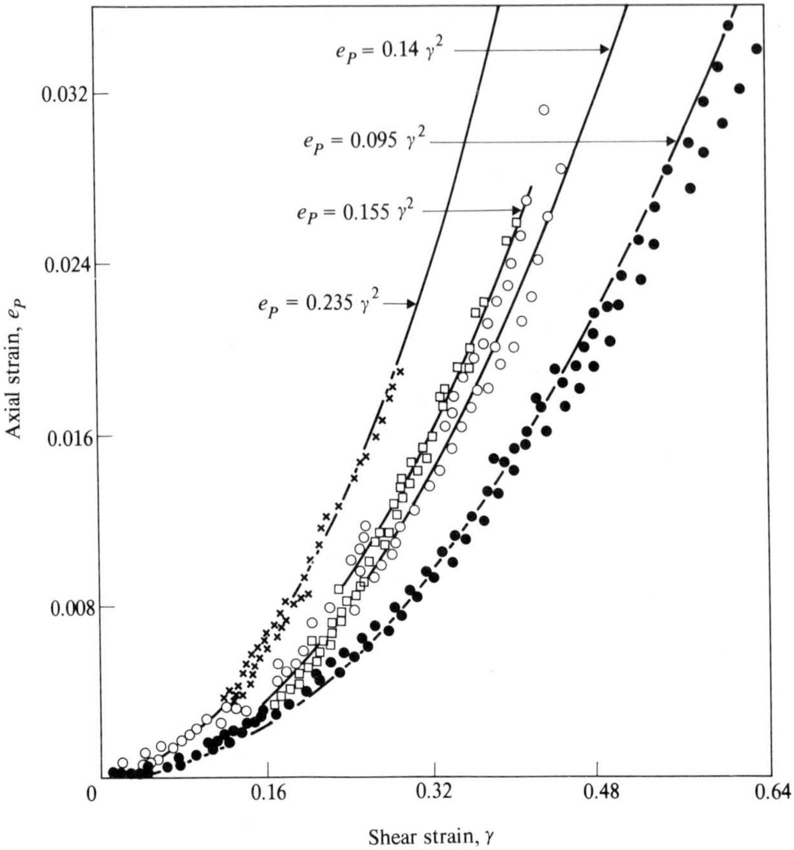

Figure 8-11 Second order axial extension as a function of applied shear strain in solid cylindrical specimens of highly filled polyurethane rubber in torsion at constant strain rates $\dot{\gamma}$ and in torsion under sustained shear stress. (*Lenoe, Heller, and Freudenthal, 1965.*)

is to be noted from Fig. 8-11 that for $N_z = 0$ the relation between the shear strain γ, as defined by Eq. (6-5-17), and the axial strain $e_P = (l-L)/L$ under torsion without axial restraint approximates closely to the quadratic relation of Eq. (8-5-29).

The observed departures from the statistical theory of rubber elasticity at moderate extension ratios has to some extent been accounted for by Gent and Thomas (1958). The mechanical response of natural and latex rubber has been shown by Hart-Smith (1966) to be adequately represented over a wide range of deformation by the relations

$$\beta_1 = 2G\exp[k_1(\mathrm{I}_\mathbf{B}-3)^2] \qquad \beta_{-1} = -2G\frac{k_2}{\mathrm{II}_\mathbf{B}} \tag{8-8-30}$$

where the k_i $(i = 1, 2)$ and G are constants. The relations of Eqs. (8-8-30) are primarily based on the results of Rivlin and Saunders (1951).

8-9 STRUCTURAL THEORY OF FINITE ELASTICITY

8-9-1 Polymers

To conform with the definition of elasticity a body must return to its original configuration when the deforming stresses are released; if it does not return to its original configuration the deformation is termed plastic. Finite strain measures do not differ significantly from the infinitesimal strain measure for strains less than about 4 percent and most materials based on a crystalline structure, such as metals, start to deform plastically before such a strain is reached. For these materials the theory of finite elasticity is inapplicable. There is, however, a class of materials, generally referred to as polymeric materials, which do exhibit large recoverable distortions, of which the best-known example is rubber.

Polymers are formed by the linking together of a large number of identical chemical molecular units. Polymers whose repeat units are linked by the bonding of two carbon atoms are called "addition polymers" and all other types are called "condensation polymers." The number of bonds which a free unit can acquire during synthesis of a polymer is termed the "functionality" of the unit. If the unit elements are difunctional they can only be joined together in the form of a linear or chain polymer, but if some of the unit elements are more than difunctional there is the possibility of cross-linking between the chains to form a three-dimensional network, as is found in resins. The structural order and symmetry of a polymer determine to a large extent its mechanical response to loading. Polymers having the same unit elements but different geometric or structural forms are said to be isomeric. If more than one type of unit element is used in forming a chain the resulting entity is called a copolymer.

The linking together of like units proceeds in a repetitive manner until the radicals allowing chain growth are destroyed, which usually is a random process. Thus the chains of a particular type of polymer do not in general have the same number of units, i.e., they are not homodisperse, but have a distribution of number of units, or are heterodisperse. Of the many different averages and moments that can be used to characterize the distribution of chain lengths two are of particular importance; these are the number average $\bar{M}_n$ and the weight average $\bar{M}_w$ which are defined as follows. If n_x is the number of polymers in the mixture having x meric units,

$$\bar{M}_n = \frac{\sum_{x=1}^{\infty} n_x x}{\sum_{x=1}^{\infty} n_x} \tag{8-9-1}$$

$$\bar{M}_w = \frac{\sum_{x=1}^{\infty} n_x x^2}{\sum_{x=1}^{\infty} n_x x} \tag{8-9-2}$$

For a homodisperse system of x meric chain length $\bar{M}_n = \bar{M}_w = x$ while if a variety of chain lengths exist $\bar{M}_w$ is always greater than $\bar{M}_n$ and the ratio $\bar{M}_w/\bar{M}_n$ is a measure of the width of the molecular-weight distribution (Flory, 1953).

The great length of the polymer chains, which can be 10^3 to 10^4 monomer units long, plays a major role in determining their mechanical response, because it allows each chain to assume many different formations and thus the entropy contribution to the free energy is large; also, the chains become entangled leading to enhanced resilience.

Polymers are found in three different solid states which are termed rubber-like, glassy, and crystalline. The rubber-like and glassy states are both amorphous and consist of randomly coiled and intertwined chains or possibly bundles of chains, while the crystalline state consists of an ordered aggregate of folded chains forming a regular crystal network.

8-9-2 Rubbery State

If the temperature of a liquid polymer is slowly decreased or if the chain length is increased there is a gradual change from a low viscosity liquid through a highly viscous state to a viscoelastic state and then to a rubbery state. Normally the process is reversible in that if the temperature is increased or the chain length decreased this sequence of states is traversed in the reverse order. In the rubbery state the polymer chains possess enough thermal energy for the monomer units to rotate about their connecting bonds and thus to assume a vast number of equal-energy configurations. Of these different configurations most correspond to a highly coiled and compact state whereas there are only a few greatly extended configurations, and indeed only one fully extended one. As each configuration is equally accessible there is a greater probability and hence a larger entropy associated with the more compact state. An extended rubbery polymer thus contracts not because its molecules thereby decrease their energy, which is the case in a metallic crystal lattice, but because the shorter length has a higher entropy. It has been found in experiments on rubber that the internal energy is almost independent of the extension of the rubber, which can be expressed mathematically in the form

$$\left(\frac{\partial U}{\partial r}\right)_{\theta, v} = 0 \tag{8-9-3}$$

Materials for which Eq. (8-9-3) is exact are called ideal elastomers. In equilibrium let the tensile force required to maintain a length r be denoted by F. The work done on the system if it is held at a constant volume is then given by

$$dW = F\,dr \qquad \text{or} \qquad F = \left(\frac{\partial W}{\partial r}\right)_{\theta, v} \tag{8-9-4}$$

But by the second law of thermodynamics

$$\left(\frac{\partial U}{\partial r}\right)_{\theta, v} = \theta\left(\frac{\partial S}{\partial r}\right)_{\theta, v} + \left(\frac{\partial W}{\partial r}\right)_{\theta, v} \tag{8-9-5}$$

Then, using Eqs. (8-9-3) and (8-9-4),

$$F = -\theta\left(\frac{\partial S}{\partial r}\right)_{\theta, v} \tag{8-9-6}$$

The detailed statistical description of the properties of a real molecular chain are very complicated and depend on such factors as the number and length of each link, the bond angles, and the activation energies between different stable rotational states and the influence of neighboring chains. However, the average properties characterizing a chain are not usually very sensitive to its detailed structure and it is sufficient to treat the chain as if each individual link could be randomly oriented. Now it has been shown by Kuhn (1934) and Guth and Mark (1934) that if one end of the polymer is situated at the origin of a coordinate system the probability $P(r)\,dv$ of finding the other end at a distance r and within a volume element dv is

$$P(r)\,dv = \frac{b^3}{\pi^{3/2}} \exp(-b^2 r^2)\,dv \tag{8-9-7}$$

where b is a parameter depending on the detailed chain geometry and the chain length. For an ideal chain consisting of n identical links of length l,

$$b^2 = \frac{3}{2nl^2} \tag{8-9-8}$$

The mean square value $(\overline{r^2})$ is thus given by

$$(\overline{r^2}) = \int_0^\infty r^2 P(r)\,dv = \frac{b^3}{\pi^{3/2}} \int_0^\infty r^2 \exp(-b^2 r^2) 4\pi r^2\,dr$$

or

$$(\overline{r^2}) = \frac{3}{2b^2} = nl^2 \tag{8-9-9}$$

Thus

$$\sqrt{(\overline{r^2})} = ln^{1/2} \tag{8-9-10}$$

Using the statistical mechanical relation $S = k \ln P$, the entropy S is given, by way of Eq. (8-9-7), in the form

$$S = k \ln P = c - kb^2 r^2 \tag{8-9-11}$$

where c is a constant. Using this in Eq. (8-9-6),

$$F = 2k\theta b^2 r \tag{8-9-12}$$

or, using Eq. (8-9-9),

$$F = \frac{3k\theta r}{(\overline{r^2})} \tag{8-9-13}$$

Before the above simple statistical theory for a single chain can be used as a model for rubber-like material behavior a few refinements are required:

1. It is assumed that the intermolecular forces between chains is very small and thus the work done by such forces during the material deformation is negligible.
2. In order that the network of chains should have some resilience and not merely flow when stressed, it is assumed that the chains are linked together at a few points. The number of linkages may be very small because the chains are long and each chain need only be linked at a few points. As the number of linkages is small the restriction of possible configurations and thus the entropy of any particular chain is very little different from the entropy of a free chain.
3. It is assumed that the vector distance between the ends of each chain in the network is deformed in the same ratio as the bulk material, i.e., the chain undergoes an affine deformation.

To find the entropy and Helmholtz free-energy change after a deformation and thus the appropriate constitutive equation for the rubber, we follow the approach of Treloar (1958).

Consider a cube of material whose edges are parallel to the principal axes. Let the extension ratios along the principal axes be λ_1, λ_2, λ_3, and let the rubber contain N chains per unit volume, where the different chains will in general have different lengths and therefore different values of b. Consider a particular chain having one end at the origin and the other at the point (X, Y, Z). After the deformation, taking axes (x, y, z) along the deformed principal axes, the coordinates of the other end now lie at

$$x = \lambda_1 X \qquad y = \lambda_2 Y \qquad z = \lambda_3 Z \tag{8-9-14}$$

The entropy before deformation is, by Eq. (8-9-11),

$$S = c - kb^2(X^2 + Y^2 + Z^2) \tag{8-9-15}$$

After deformation it will be

$$S' = \begin{cases} c - kb^2(x^2 + y^2 + z^2) \\ c - kb^2(\lambda_1^2 X^2 + \lambda_2^2 Y^2 + \lambda_3^2 Z^2) \end{cases} \tag{8-9-16}$$

The change in entropy ΔS is thus given by

$$\Delta S = S' - S = -kb^2[(\lambda_1^2 - 1)X^2 + (\lambda_2^2 - 1)Y^2 + (\lambda_3^2 - 1)Z^2] \tag{8-9-17}$$

The contribution ΔS_p of all chains, N_p in number, having nearly the same value of b, namely, b_p, is thus

$$\Delta S_p = \sum_{i=1}^{N_p} - kb_p^2[(\lambda_1^2 - 1)X_{ip}^2 + (\lambda_2^2 - 1)Y_{ip}^2 + (\lambda_3^2 - 1)Z_{ip}^2] \tag{8-9-18}$$

Averaged over many chains each direction is equally likely and as

$$\Sigma(X_{ip}^2 + Y_{ip}^2 + Z_{ip}^2) = N_p(\overline{r_p^2}) \tag{8-9-19}$$

then

$$\Sigma X_{ip}^2 = \Sigma Y_{ip}^2 = \Sigma Z_{ip}^2 = \tfrac{1}{3} N_p(\overline{r_p^2}) \tag{8-9-20}$$

Substituting in Eq. (8-9-18),

$$\Delta S_p = -\tfrac{1}{3}kN_p b_p^2 (\overline{r_p^2})(\lambda_1^2+\lambda_2^2+\lambda_3^2-3) \tag{8-9-21}$$

Using Eq. (8-9-9) this reduces to

$$\Delta S_p = -\tfrac{1}{2}kN_p(\lambda_1^2+\lambda_2^2+\lambda_3^2-3) \tag{8-9-22}$$

Finally, summing over all the groups of values of b_p and taking

$$\Sigma N_p = N \tag{8-9-23}$$

gives

$$\Delta S = -\tfrac{1}{2}kN(\lambda_1^2+\lambda_2^2+\lambda_3^2-3) \tag{8-9-24}$$

As there is no change in the internal energy the work of deformation per unit volume W is given by

$$W = -\theta\Delta S = \tfrac{1}{2}k\theta N(\lambda_1^2+\lambda_2^2+\lambda_3^2-3) = C(\mathrm{I}_{\mathbf{B}}-3) \tag{8-9-25}$$

which is simply Eq. (8-8-2). Now if E_1, E_2, E_3 are the principal values of the Green–St. Venant strain tensor, it follows from Sec. 2-6 that

$$E_1 = \tfrac{1}{2}(\lambda_1^2-1) \qquad E_2 = \tfrac{1}{2}(\lambda_2^2-1) \qquad E_3 = \tfrac{1}{2}(\lambda_3^2-1) \tag{8-9-26}$$

and thus Eq. (8-9-25) can also be written in the form

$$W = G(E_1+E_2+E_3) = G(\operatorname{tr}\mathbf{E}) \tag{8-9-27}$$

where

$$G = k\theta N \tag{8-9-28}$$

as has been shown by Rivlin (1948). For a more detailed study and further extensions of the theory the reader is referred to Treloar (1958, 1970).

It is to be noted that all the thermodynamic expressions considered so far apply to closed systems (in particular, see Sec. 8-2). By analogy, the open system functions can be treated in a similar manner, as discussed by Larché and Cahn (1978), who have developed a nonlinear theory of thermochemical equilibrium of solids under stress.

GENERAL REFERENCES

Flory, P. J. (1953) "Principles of Polymer Chemistry," Cornell University Press, New York.

Green, A. E., and J. E. Adkins (1960) "Large Elastic Deformations and Non-linear Continuum Mechanics," Clarendon Press, Oxford.

Murnaghan, F. D. (1951) "Finite Deformations of an Elastic Solid," Wiley, New York.

Treloar, L. R. G. (1958) "The Physics of Rubber Elasticity," 2d ed., Oxford University Press, Oxford.

Treloar, L. R. G. (1970) "Introduction to Polymer Science," Wykeham, London.

Truesdell, C., and W. Noll (1965) The Non-linear Field Theories of Mechanics, in S. Flügge (ed.), "Handbuch der Physik," vol. III/3, Springer-Verlag, Berlin.

Truesdell, C., and R. A. Toupin (1960) The Classical Field Theories, in S. Flügge (ed.), "Handbuch der Physik," vol. III/1, pp. 226–858, Springer-Verlag, Berlin.

REFERENCES

Baker, M., and J. L. Ericksen (1954) *J. Wash. Acad. Sci.*, vol. 44, pp. 33–35.

Foux, A. (1964) Second Order Effects, in M. Reiner and D. Abir (eds.), "Elasticity, Plasticity and Fluid Dynamics," pp. 228–251, Pergamon Press, London.

Gee, G., J. Stern, and L. R. G. Treloar (1950) *Trans. Faraday Soc.*, vol. 46, pp. 1101–1106.

Gent, A. N., and A. G. Thomas (1958) *J. Polym. Sci.*, vol. 28, pp. 625–628.

Guth, E., and H. Mark (1934) *Monatsh. Chem.*, vol. 65, pp. 93–121.

Hart-Smith, L. J. (1966) *Z. angew. Math. Phys.*, vol. 17, pp. 608–626.

Kuhn, W. (1934) *Kolloid. Z.*, vol. 68, pp. 2–15.

Larché, F., and J. W. Cahn (1978) *Acta Met.*, vol. 26, pp. 53–60.

Lenoe, E. M., R. A. Heller, and A. M. Freudenthal (1965) *Trans. Soc. Rheol.*, vol. 9, pp. 77–102.

Mooney, M. (1940) *J. Appl. Phys.*, vol. 11, pp. 582–592.

Poynting, J. H. (1909) *Proc. Roy. Soc. (London)*, vol. A82, pp. 546–559.

Poynting, J. H. (1912) *Proc. Roy. Soc. (London)*, vol. A86, pp. 534–561.

Reiner, M. (1960) *J. Mech. Phys. Solids*, vol. 8, pp. 255–261.

Rivlin, R. S. (1947) *J. Appl. Phys.*, vol. 18, pp. 444–449.

Rivlin, R. S. (1948a) *Phil. Trans. Roy. Soc.*, vol. A240, pp. 459–490, 491–508, 509–525.

Rivlin, R. S. (1948b) *Phil. Trans. Roy. Soc.*, vol. A241, pp. 379–397.

Rivlin, R. S. (1949a) *Proc. Roy. Soc. (London)*, vol. A195, pp. 463–473.

Rivlin, R. S. (1949b) *Phil. Trans. Roy. Soc.*, vol. A242, pp. 173–195.

Rivlin, R. S., and D. W. Saunders (1951) *Phil. Trans. Roy. Soc.*, vol. A243, pp. 251–288.

Rivlin, R. S. (1966) The Fundamental Equations of Non-linear Continuum Mechanics, in S-I Pai (ed.), "Dynamics of Fluid and Plasmas," pp. 83–126, Academic Press, New York.

Treloar, L. R. G. (1943) *Trans. Faraday Soc.*, vol. 39, pp. 241–246.

CHAPTER

NINE

NONLINEAR VISCOUS FLUIDS

9-1 NONLINEAR FLUIDS

9-1-1 The Generalized Stokesian Fluid

The concept of linear viscous response has been considered in Sec. 7-1-1 in relation to the principle of Stokes, the formulation of this principle being expressed as

$$\mathbf{T} = -P\mathbf{I} + \mathfrak{f}(\mathbf{D}) \qquad \mathfrak{f}(\mathbf{O}) = \mathbf{O} \tag{9-1-1}$$

where the pressure term $-P\mathbf{I}$ describes the state of stress in fluids at rest and $\mathbf{D}$ is the rate of deformation tensor (see Sec. 2-4). In Sec. 7-1-1 it was shown by way of the principle of material objectivity that the principle of Stokes defines only isotropic fluids, and hence $\mathfrak{f}(\mathbf{D})$ is an isotropic tensor function of the symmetric tensor $\mathbf{D}$.

To derive the form of the function in Eq. (9-1-1), Reiner (1945) assumed $\mathfrak{f}(\mathbf{D})$ to be a polynomial, this assumption being expressed in the form

$$\mathfrak{f}(\mathbf{D}) = \sum_{k=0}^{\infty} C_k \mathbf{D}^k \tag{9-1-2}$$

where the coefficients C_k are material properties of the fluid and hence depend on the thermodynamic state of the system. Flow of the material must alter the property tensors, from which it follows that Eq. (9-1-2) can only be valid at one particular instant of time, a limitation which could be removed by using a convected coordinate system (Oldroyd, 1950; see also Sec. 2-9). The Cayley–Hamilton equation (1-10-19) can now be used to eliminate all powers of $\mathbf{D}^k$ for $k \geqslant 3$. However, since it has been shown in Sec. 7-1-1 that $\mathfrak{f}(\mathbf{D})$ is an isotropic tensor function of the symmetric tensor $\mathbf{D}$, it follows that the representation theorem available for isotropic tensor functions of one tensor variable can be applied to the constitutive equation (9-1-1). Thus, it follows from Sec. 1-10, in particular Eq. (1-10-21), that the constitutive equation (9-1-1) can be expressed in the form

$$\mathbf{T} = -P\mathbf{I} + f_0\mathbf{I} + f_1\mathbf{D} + f_2\mathbf{D}^2 \tag{9-1-3}$$

where the f_i $(i = 0, 1, 2)$ are scalar invariants of the principal invariants $\mathrm{I}_\mathbf{D}$, $\mathrm{II}_\mathbf{D}$, $\mathrm{III}_\mathbf{D}$ of the rate of deformation tensor $\mathbf{D}$.

For a compressible fluid the pressure P is the thermodynamic pressure, and therefore the properties of a stokesian fluid which are assumed compressible can only be specified if the form of the three functions f_i $(i = 0, 1, 2)$ of the rate of deformation tensor are known.

In the case of an incompressible, stokesian fluid, for which $\mathrm{I}_\mathbf{D} = 0$, the dependence of f_0 on the rate of deformation can be transferred to the pressure, the latter quantity then being a dynamical variable. The constitutive equation then has the form

$$\mathbf{T} = -P\mathbf{I} + f_1\mathbf{D} + f_2\mathbf{D}^2 \tag{9-1-4}$$

the pressure P now being determined only to within an additive constant.

Generalized stokesian fluids, the equations for which were first derived by Reiner (1945), are sometimes referred to as Reiner–Rivlin fluids, Rivlin (1948, 1949) having arrived at a reduction from Eqs. (9-1-1) to (9-1-3) without making the assumptions associated with Eq. (9-1-2). Reiner noted that a fluid whose flow properties are described by Eq. (9-1-4) should exhibit both nonlinear viscosity and normal stress effects. Thus, Eq. (9-1-4) can be expressed in the form

$$\mathbf{T} = -P\mathbf{I} + 2\eta\mathbf{D} + \eta_c\mathbf{D}^2 \tag{9-1-5}$$

where η_c is the cross-viscosity, or viscous normal stress coefficient (Reiner, 1945) of the stokesian fluid and η is the shear viscosity.

9-1-2 Nonnewtonian Liquids

It is now widely accepted that there are no stokesian fluids save those that are purely viscous, i.e., fluids for which f_2 vanishes. A particularly restricted type of fluid is obtained from the stokesian fluid for the limiting conditions that the coefficient f_2 vanishes and f_1 depends only on the second invariant $\mathrm{II}_\mathbf{D}$ of the rate of deformation tensor $\mathbf{D}$. For these limiting conditions, Eq. (9-1-5) reduces to

$$\mathbf{T} = -P\mathbf{I} + 2\eta\mathbf{D} \tag{9-1-6}$$

which is simply Eq. (7-1-31) with, however, the difference that now

$$\eta = \eta(\mathrm{II}_\mathbf{D}) \tag{9-1-7}$$

thus implying the possibility of a nonlinear relation between shear stress and the rate of deformation. In particular, liquids whose flow properties can be described by Eq. (9-1-6), subject to the condition of Eq. (9-1-7), are generally referred to as nonnewtonian liquids, thus recognizing that in contrast to newtonian fluids for which η = constant, for all $\mathbf{D}$, the coefficient η of Eq. (9-1-6) may now vary with the rate of deformation.

Consider the characteristic equation

$$\lambda^3 - \mathrm{I}_\mathbf{D}\lambda^2 + \mathrm{II}_\mathbf{D}\lambda - \mathrm{III}_\mathbf{D} = 0 \tag{9-1-8}$$

of the rate of deformation tensor $\mathbf{D}$. For an incompressible fluid for which $\mathrm{I}_\mathbf{D} = 0$, Eq. (9-1-8) reduces to

$$\lambda^3 + \mathrm{II}_\mathbf{D}\lambda = 0 \tag{9-1-9}$$

for the particular condition $\mathrm{III}_\mathbf{D} = 0$. Hence, if λ_1, λ_2, and λ_3 are the roots of Eq. (9-1-9), then $\lambda_1 = 0$ and

$$\lambda_2 = -\lambda_3 = \sqrt{-\mathrm{II}_\mathbf{D}} = \sqrt{\tfrac{1}{2}\operatorname{tr}\mathbf{D}^2} \tag{9-1-10}$$

The invariant

$$Z^{1/2} = (\tfrac{1}{2}\operatorname{tr}\mathbf{D}^2)^{1/2} \tag{9-1-11}$$

is an important quantity called the intensity of the rate of deformation. It has been shown in Sec. 7-1-2 that the dissipation function Φ_v, appropriate to the constitutive equation (9-1-6), can be expressed in the form

$$\Phi_v = 2\eta \operatorname{tr}\mathbf{D}^2 \tag{9-1-12}$$

for an incompressible liquid. Equation (9-1-12) can be rearranged, by way of Eq. (9-1-11), to give

$$Z = \tfrac{1}{2}\operatorname{tr}\mathbf{D}^2 = \frac{\Phi_v}{4\eta} \tag{9-1-13}$$

from which it follows that the components of the intensity of the rate of deformation in rectangular, cylindrical, and spherical coordinates can be evaluated from the components of the dissipation function Φ_v.

A simple assumption to make is that f_1 in Eq. (9-1-4), or η in Eq. (9-1-5), is simply a function of Z. That is, the parameter η is, in general, strain rate dependent and hence the term nonnewtonian fluid. For incompressible nonnewtonian liquids having

$$f_1 = f(Z) \qquad \eta = \eta(Z) \tag{9-1-14}$$

an equation of state which has proved useful can be expressed in the form

$$\begin{aligned} \mathbf{T} &= -P\mathbf{I} + f(Z)\mathbf{D} \\ &= -P\mathbf{I} + 2\eta(Z)\mathbf{D} \end{aligned} \tag{9-1-15}$$

where Z is given by Eq. (9-1-13). Several approximate, analytical representations of the function $f(Z)$ have been proposed, of which the following are examples. The two-parameter equation

$$f(Z) = 2m^{1/n}|Z|^{(n-1)/2n} = 2\eta(Z) \tag{9-1-16}$$

which was proposed by Waele (1923) and Ostwald (1925, 1929), is also known as the power law. In Eq. (9-1-16), the constants m and n are temperature dependent, the equation of state reducing to Newton's law for $n = 1$ and $m = \eta$. If $n < 1$, then $\eta(Z)$ decreases with an increasing rate of deformation (pseudoplastic fluids), whereas if $n > 1$, then $\eta(Z)$ increases with an increasing rate of deformation (dilatant fluid). The reaction rate theory of viscosity, based on Eyring's (1936) hole theory of the liquid state, gives the relation

$$f(Z) = p_0 \frac{\sinh^{-1}(\beta_0\sqrt{Z})}{\beta_0\sqrt{Z}} = 2\eta(Z) \tag{9-1-17}$$

where p_0 and β_0 are material constants. This particular theory of viscosity has been particularly successful in explaining many of the results of experiment and has also had a great influence on theoretical considerations (see, for example, Eyring, Henderson, Stover, and Eyring, 1964). It has been shown (Billington, 1967) that the Goodeve (1938, 1949) impulse theory of flow can be extended to include the rotational motion of the particles of the minority component of certain types of suspension, which exhibit nonnewtonian flow behavior. This gives the relations

$$f(Z) = p_0 \frac{\ln(1 + Z^{1/2}/D_0)}{Z^{1/2}} \qquad (0 \leqslant \operatorname{tr}\mathbf{D}^2 \leqslant D_a^2) \tag{9-1-18a}$$

$$f(Z) = 2\eta_a + \frac{p_{e0}}{Z^{1/2}} \qquad (\operatorname{tr}\mathbf{D}^2 \geqslant D_a^2) \tag{9-1-18b}$$

which contain two adjustable parameters p_0, D_0 for the range of response $0 \leqslant \operatorname{tr}\mathbf{D}^2 \leqslant D_a^2$ and η_a, p_{e0} for $\operatorname{tr}\mathbf{D}^2 \geqslant D_a^2$ where D_a is the critical value of the rate of deformation $\mathbf{D}$ at which the response changes over from the nonlinear behavior of Eq. (9-1-18a) to the response of Eq. (9-1-18b).

An alternative approach is to define a quantity

$$X^{1/2} = (\tfrac{1}{2}\operatorname{tr}\mathbf{T}'^2)^{1/2} \tag{9-1-19}$$

which is called the stress intensity. The Ellis equation

$$\frac{2}{f(X)} = \frac{1}{\eta_0} + m^{-1/n}|X|^{(1-n)/2n} \tag{9-1-20}$$

contains three adjustable positive parameters m, n, and η_0. For $0 < n < 1$, this equation predicts $f \to 2\eta_0$ as $X \to 0$, and as such is useful if the stresses encountered in practice are small. This equation includes the power law ($1/\eta_0 = 0$) as the limiting case. Sisko (1958) proposed the equation

$$f(X) = 2\eta_\infty + 2m^{1/n}|X|^{(n-1)/2n} \qquad \text{for } n < 1 \tag{9-1-21}$$

which contains three adjustable parameters m, n, and η_∞. At high stresses this equation predicts that $f \to 2\eta_\infty$ as $X \to \infty$, whereas at low stresses $f \to \infty$ as $X \to 0$. Some essentially nonnewtonian fluids exhibit newtonian behavior at both low and very high shear stresses. This behavior can be accounted for by the relation

$$f(X) = 2\eta_\infty + \frac{2(\eta_0 - \eta_\infty)}{1 + X/p_0^2} \tag{9-1-22}$$

which contains three adjustable parameters η_0, η_∞, and p_0 and is seen to reduce to Newton's law for the limiting conditions $\eta = \eta_0$ at $X = 0$ and $\eta \to \eta_\infty$ as $X \to \infty$, the constant p_0 being determined from experiment (see Philippoff, 1935). Reiner (1960) has proposed the relation

$$\frac{2}{f(X)} = \frac{1}{\eta_\infty} - \left(\frac{1}{\eta_\infty} - \frac{1}{\eta_0}\right)\exp\left(-\frac{X}{\chi^2}\right) \tag{9-1-23}$$

which contains three adjustable parameters η_0, η_∞, and χ. In practice Reiner has observed that χ^2 is a weak function of X.

Constitutive equations obtained using the above expressions for the functions $f(Z)$ and $f(X)$ are empirical expressions designed to approximate the actual behavior of various materials. The essential condition which these empirical expressions satisfy is that they transform properly in going from one coordinate system to another. This ensures that the parameters, such as m and n, would be the same when determined in various geometrical arrangements.

9-1-3 Viscoplasticity

An ideal plastic material may exhibit simultaneously both viscous and elastic properties. The response of this type of material to shear is such that the viscous and elastic behavior occurs in different portions of the body of the material. Thus, in general, one part of a body may be a viscous liquid, whereas the remaining part of the body may exhibit the behavior characteristic of a solid. It is to be noted that this behavior does not imply that the material is rheologically inhomogeneous, the difference in behavior being attributed to a difference in stress prevailing between different parts of the body.

The rheological behavior of this type of material is characterized by a yield stress k. If the applied stress is less than the yield stress k the material behaves as a solid and if the applied stress exceeds the yield stress k then the material flows as if it were a viscous liquid characterized by a viscosity η_θ called the plastic viscosity or mobility.

A yield condition will be defined by the relation

$$f(\mathbf{T}) \equiv \tfrac{1}{2}\operatorname{tr}\mathbf{T}'^2 - k^2 = 0 \tag{9-1-24}$$

which is the von Mises' yield condition given by Eqs. (10-2-1) and (10-2-13). Oldroyd (1947, 1948, 1949) defines an ideal plastic material to be a material obeying von Mises' yield condition

$$\mathbf{D} = \mathbf{O} \qquad \text{if } \tfrac{1}{2}\operatorname{tr}\mathbf{T}'^2 \leqslant k^2 \tag{9-1-25}$$

and for which

$$\mathbf{T}' = \frac{2\eta_\theta}{1-k/(\frac{1}{2}\operatorname{tr}\mathbf{T}'^2)^{1/2}}\mathbf{D}' \qquad \tfrac{1}{2}\operatorname{tr}\mathbf{T}'^2 > k^2 \tag{9-1-26}$$

or, in its equivalent form,

$$\mathbf{T}' = \left[2\eta_\theta + \frac{k}{(2\operatorname{tr}\mathbf{D}'^2)^{1/2}}\right]\mathbf{D}' \qquad \tfrac{1}{2}\operatorname{tr}\mathbf{T}'^2 > k^2 \tag{9-1-27}$$

Materials whose rheological behavior is approximated to by the constitutive equations (9-1-26) and (9-1-27) are referred to as Bingham plastics (Bingham and Green, 1919).

It is to be noted that Oldroyd's theory of the Bingham plastic involves a discontinuity when $\frac{1}{2}\operatorname{tr}\mathbf{T}'^2 = k^2$.

The ideal Bingham plastic may be regarded as a limiting type of non-newtonian fluid, for which the coefficients $f(Z)$ and $f(X)$ of Sec. 9-1-2 have the form

$$f(Z) = 2\eta_\theta + \frac{k}{(2\operatorname{tr}\mathbf{D}'^2)^{1/2}} \tag{9-1-28}$$

and

$$\frac{2}{f(X)} = \frac{1}{\eta_\theta}\left[1 - \frac{k}{(\frac{1}{2}\operatorname{tr}\mathbf{T}'^2)^{1/2}}\right] \tag{9-1-29}$$

9-1-4 Rivlin–Ericksen Fluids

The two normal stress functions resulting from the Reiner–Rivlin constitutive equation (9-1-4) are the same, which implies that the normal stresses in simple shearing flow must coincide. There is a growing body of experimental evidence which is not in accord with this conclusion. Hence, it would appear that there are no stokesian fluids except those that are purely viscous.

Reference has been made in Sec. 5-5 to materials of the differential type. A theory of materials of the differential type has been proposed by Rivlin and Ericksen (1955). They start from the basic assumption that the stress at a given time depends only on the spatial gradients of the displacements, velocity, acceleration, second acceleration, and higher-order accelerations at that time. From certain invariance considerations, Rivlin and Ericksen arrived at a constitutive equation which for incompressible fluids reduces to

$$\mathbf{T} = -P\mathbf{I} + \mathfrak{f}(\mathbf{A}_1, \mathbf{A}_2, \ldots, \mathbf{A}_n) \tag{9-1-30}$$

where $\mathfrak{f}$ is an isotropic function of the Rivlin–Ericksen tensors $\mathbf{A}_n$ which have been defined in Sec. 2-4. Assuming that $\mathfrak{f}$ is a polynomial function in $\mathbf{A}_1$ and $\mathbf{A}_2$ only, then for incompressible fluids it can be shown by way of the Cayley–Hamilton theorem that (see, for example, the appendix of Green and Adkins, 1960)

$$\begin{aligned}\mathfrak{f}(\mathbf{A}_1,\mathbf{A}_2) = {} & a_1\mathbf{A}_1 + a_2\mathbf{A}_2 + a_3\mathbf{A}_1^2 + a_4\mathbf{A}_2^2 \\ & + a_5(\mathbf{A}_1\mathbf{A}_2 + \mathbf{A}_2\mathbf{A}_1) + a_6(\mathbf{A}_1^2\mathbf{A}_2 + \mathbf{A}_2\mathbf{A}_1^2) \\ & + a_7(\mathbf{A}_1\mathbf{A}_2^2 + \mathbf{A}_2^2\mathbf{A}_1) + a_8(\mathbf{A}_1^2\mathbf{A}_2^2 + \mathbf{A}_2^2\mathbf{A}_1^2)\end{aligned} \tag{9-1-31}$$

where $a_1, a_2, \ldots, a_8$ are functions, expressible as polynomials, of the ten invariants

$$\operatorname{tr} \mathbf{A}_1, \operatorname{tr} \mathbf{A}_1^2, \operatorname{tr} \mathbf{A}_1^3$$

$$\operatorname{tr} \mathbf{A}_2, \operatorname{tr} \mathbf{A}_2^2, \operatorname{tr} \mathbf{A}_2^3$$

$$\operatorname{tr} \mathbf{A}_1\mathbf{A}_2, \operatorname{tr} \mathbf{A}_1^2\mathbf{A}_2, \operatorname{tr} \mathbf{A}_1\mathbf{A}_2^2, \operatorname{tr} \mathbf{A}_1^2\mathbf{A}_2^2$$

it being noted by way of Eq. (2-4-18) that $\mathbf{A}_1 = 2\mathbf{D}$.

Since $\mathbf{A}_1 = 2\mathbf{D}$, it follows that if all dependence on $\mathbf{A}_2$ is suppressed, then Eqs. (9-1-30) and (9-1-31) give the Reiner–Rivlin constitutive equation (9-1-4).

Of particular interest are incompressible fluids of grade two defined by the constitutive equation

$$\mathbf{T} = -P\mathbf{I} + a_1\mathbf{A}_1 + a_2\mathbf{A}_2 + a_3\mathbf{A}_1^2 \tag{9-1-32}$$

which is obtained from Eqs. (9-1-30) and (9-1-31) by suppressing all dependence on coefficients higher than a_3. Rivlin (1956) observed that, for fluids whose flow properties are described by the constitutive equation (9-1-32), the normal stresses in simple shearing flow need not be the same. That this is so can be seen as follows.

9-1-5 The Normal Stress Differences

Simple fluids have been considered in Sec. 5-7, where it was shown that the constitutive equation (5-7-11) for a simple fluid can be expressed in the form

$$\mathbf{T}(t) = -P(\rho(t))\mathbf{I} + \mathfrak{q}(\mathbf{G}(s), \rho(t)) \tag{9-1-33a}$$

$$\mathfrak{q}(\mathbf{O}) = \mathbf{O} \tag{9-1-33b}$$

where

$$\mathbf{G}(s) = \mathbf{C}_{(t)}^t(s) - \mathbf{I} \tag{9-1-34}$$

it being noted that for compressible fluids both P and $\mathfrak{q}$ are functions of the density $\rho(t)$ at the present time t. For incompressible fluids, P is indeterminate and the density is a constant so that $\mathfrak{q}$ need not be regarded as a function of the density.

To introduce the concept of the normal stress difference, consider a motion of a body such that the velocity field $\dot{x}(x, t)$ is described in a cartesian coordinate system (x, y, z) by the components

$$\dot{x} = 0 \qquad \dot{y} = v(x, t) \qquad \dot{z} = 0 \tag{9-1-35}$$

such motions being called lineal flows by Truesdell and Noll (1965). The relative deformation function ξ of Eq. (2-2-8) can now be obtained by solving the differential equations

$$\frac{d\xi}{d\tau} = \dot{x}(\xi, \tau) \tag{9-1-36}$$

subject to the initial conditions

$$[\xi]_{\tau=t} = \mathbf{x} \tag{9-1-37}$$

For the velocity components of Eq. (9-1-35) the differential equations (9-1-36) reduce to

$$\frac{d\xi}{d\tau} = 0 \qquad \frac{d\eta}{d\tau} = v(\xi, \tau) \qquad \frac{d\zeta}{d\tau} = 0 \tag{9-1-38}$$

which can be solved, subject to the conditions of Eq. (9-1-37), to give

$$\xi = x \qquad \eta = y + \int_t^\tau v(x, \tau')\, d\tau' \qquad \zeta = z \tag{9-1-39}$$

It follows from Eq. (2-2-9), which defines the relative deformation gradient $\mathbf{F}_{(t)}(\tau)$, that

$$[\mathbf{F}_{(t)}(\tau)] = \begin{bmatrix} 1 & 0 & 0 \\ k(\tau) - k(t) & 1 & 0 \\ 0 & 0 & 1 \end{bmatrix} \tag{9-1-40}$$

where

$$k(\tau) = \int_0^\tau v'(x, \bar{\tau})\, d\bar{\tau} \tag{9-1-41}$$

the prime denoting differentiation with respect to x. It is evident from Eq. (9-1-40) that $\mathbf{F}_{(t)}(\tau)$ must be of the form

$$\mathbf{F}_{(t)}(\tau) = \mathbf{I} + [k(\tau) - k(t)]\mathbf{N} \tag{9-1-42}$$

where

$$[\mathbf{N}] = \begin{bmatrix} 0 & 0 & 0 \\ 1 & 0 & 0 \\ 0 & 0 & 0 \end{bmatrix} \tag{9-1-43}$$

Equation (9-1-42) can be expressed in the form

$$\mathbf{F}_{(t)}(t - s) = \mathbf{I} + g(s)\mathbf{N} \tag{9-1-44}$$

where

$$g(s) = k(\tau) - k(t) = k(t - s) - k(t) \tag{9-1-45}$$

Substitution of Eq. (9-1-44) into $\mathbf{C}_{(t)} = \mathbf{F}_{(t)}^T \mathbf{F}_{(t)}$ and the result into Eq. (9-1-34) gives

$$\mathbf{G}(s) = g(s)(\mathbf{N} + \mathbf{N}^T) + g(s)^2 \mathbf{N}^T \mathbf{N} \tag{9-1-46}$$

from which it is inferred that

$$\mathbf{A}_1 = -\dot{g}(0)\,(\mathbf{N} + \mathbf{N}^T) \tag{9-1-47}$$

$$\mathbf{A}_2 = \ddot{g}(0)\,(\mathbf{N} + \mathbf{N}^T) + 2\dot{g}(0)^2 \mathbf{N}^T \mathbf{N} \tag{9-1-48}$$

If a new function $\mathfrak{k}$ is defined by the relation

$$\mathfrak{k}(g(s), \mathbf{N}) = \mathfrak{q}(g(s)(\mathbf{N} + \mathbf{N}^T) + g(s)^2 \mathbf{N}^T \mathbf{N}) \tag{9-1-49}$$

where $\mathfrak{k}$ is a functional of the scalar function $g(s)$ and a function of the tensor $\mathbf{N}$, then the stress **T** can, by reference to Eq. (9-1-33), be expressed in the form

$$\mathbf{T} = -P\mathbf{I} + \mathfrak{f}(g(s), \mathbf{N}) \tag{9-1-50}$$

The property of isotropy can now be used to determine which response functions vanish.

It is evident from Eqs. (9-1-49) and (5-7-12) that $\mathfrak{f}$ satisfies the identity

$$\mathbf{Q}\mathfrak{f}(g(s), \mathbf{N})\mathbf{Q}^T = \mathfrak{f}\left(\alpha g(s), \frac{1}{\alpha}\mathbf{QNQ}^T\right) \tag{9-1-51}$$

for all orthogonal tensors **Q** and all real $\alpha \neq 0$. The constant factor α will now be used to test the parity of the response functions.

Selecting the orthogonal tensor **Q** such that

$$[\mathbf{Q}] = \begin{bmatrix} 1 & 0 & 0 \\ 0 & 1 & 0 \\ 0 & 0 & -1 \end{bmatrix} \tag{9-1-52}$$

gives, by way of Eq. (9-1-43), the condition

$$\mathbf{QNQ}^T = \mathbf{N} \tag{9-1-53}$$

Using Eq. (9-1-51) with $\alpha = 1$ and the choice of **Q** determined by Eq. (9-1-52), it is evident that Eq. (9-1-50) gives

$$\mathbf{QTQ}^T = \mathbf{T} \tag{9-1-54}$$

However, the conditions of immediate interest are such that

$$[\mathbf{QTQ}^T] = \begin{bmatrix} T\langle 11\rangle & T\langle 12\rangle & -T\langle 13\rangle \\ T\langle 21\rangle & T\langle 22\rangle & -T\langle 23\rangle \\ -T\langle 31\rangle & -T\langle 32\rangle & T\langle 33\rangle \end{bmatrix} \qquad (T\langle 12\rangle = T\langle 21\rangle)$$

which implies that the matrix of the stress **T** has the form

$$[\mathbf{T}] = \begin{bmatrix} T\langle 11\rangle & T\langle 12\rangle & 0 \\ T\langle 21\rangle & T\langle 22\rangle & 0 \\ 0 & 0 & T\langle 33\rangle \end{bmatrix} \qquad (T\langle 12\rangle = T\langle 21\rangle) \tag{9-1-55}$$

where the $T\langle ij\rangle$ are now understood to be the physical components of the stress. Equation (9-1-55) shows that application of the principle of material objectivity requires for a simple fluid in a lineal flow that two of the six elements of the stress tensor be identically zero. Since, by Eq. (9-1-43), **N** is a fixed constant, it follows from Eq. (9-1-50) that the six components of $\mathbf{T}_E$ are functionals of $g(s)$ only, and in consequence of Eq. (9-1-55) only four of these functionals can be nonzero. The shear stress $T\langle 12\rangle$ and the normal stress differences $T\langle 11\rangle - T\langle 33\rangle$ and $T\langle 22\rangle - T\langle 33\rangle$ are not affected by the pressure term $P\mathbf{I}$. The functional dependence of $T\langle 12\rangle$ and of the normal stress differences on $g(s)$ can be denoted by

$$T\langle 12\rangle = f(g(s)) \tag{9-1-56}$$

$$T\langle 11\rangle - T\langle 33\rangle = \beta_1(g(s)) \tag{9-1-57}$$

$$T\langle 22\rangle - T\langle 33\rangle = \beta_2(g(s)) \tag{9-1-58}$$

The material functionals f, β_1, and β_2, together with $P(\rho)$ characterize the mechanical behavior of simple fluids in lineal flows.

The parity of the response functionals can be established by choosing $\mathbf{Q}$ in Eq. (9-1-51) to be of the form

$$[\mathbf{Q}] = \begin{bmatrix} 1 & 0 & 0 \\ 0 & -1 & 0 \\ 0 & 0 & 1 \end{bmatrix} \tag{9-1-59}$$

together with $\alpha = -1$. This choice gives, by way of Eq. (9-1-43), the condition

$$\frac{1}{\alpha}\mathbf{QNQ}^T = -\mathbf{QNQ}^T = \mathbf{N} \tag{9-1-60}$$

From Eqs. (9-1-50) and (9-1-51) it is evident that the stress corresponding to $-g(s)$ is given by $\mathbf{QTQ}^T$, where $\mathbf{T}$ is the stress corresponding to $g(s)$. The conditions of immediate interest are such that

$$[\mathbf{QTQ}^T] = \begin{bmatrix} T\langle 11\rangle & -T\langle 12\rangle & 0 \\ -T\langle 21\rangle & T\langle 22\rangle & 0 \\ 0 & 0 & T\langle 33\rangle \end{bmatrix} \qquad (T\langle 12\rangle = T\langle 21\rangle) \tag{9-1-61}$$

from which it follows, by way of Eqs. (9-1-56), (9-1-57), and (9-1-58) that the material functionals f and β_i ($i = 1, 2$) obey the identities

$$f(-g(s)) = -f(g(s)) \qquad \beta_i(-g(s)) = \beta_i(g(s)) \qquad \text{for } i = 1, 2 \tag{9-1-62}$$

from which it is concluded that f must be an odd functional and the β_i ($i = 1, 2$) must be even functionals. The isotropy condition, i.e., Eq. (9-1-51), imposes no further restrictions on f and the β_i ($i = 1, 2$), while the condition of Eq. (9-1-33b) requires simply that

$$f(0) = 0 \qquad \beta_i(0) = 0 \qquad \text{for } i = 1, 2 \tag{9-1-63}$$

Consider steady-state flow. For this particular condition, Eqs. (9-1-41) and (9-1-45) give

$$g(s) = -\dot{\gamma}s \tag{9-1-64}$$

where $g(s)$ is a function of s which does not vary with t. Equation (9-1-50) now simplifies to the form

$$\mathbf{T} = -P\mathbf{I} + \mathfrak{k}(\dot{\gamma}, \mathbf{N}) \tag{9-1-65}$$

for steady-state flow and the material functionals $\beta_i(g(s))$ ($i = 1, 2$) reduce to material functions, Eqs. (9-1-56), (9-1-57), and (9-1-58) simplifying to

$$T\langle 12\rangle = f(\dot{\gamma}) \tag{9-1-66}$$

$$T\langle 11\rangle - T\langle 33\rangle = \beta_1(\dot{\gamma}) \tag{9-1-67}$$

$$T\langle 22\rangle - T\langle 33\rangle = \beta_2(\dot{\gamma}) \tag{9-1-68}$$

Since the state of stress is determined by the strain only to within an additive hydrostatic pressure, it follows that the absolute value of any one normal component of stress is of no rheological interest. Thus for an isotropic, incompressible liquid in steady rectilinear shear flow there are only three independent stress quantities of rheological interest. These are a tangential stress component $T\langle 12\rangle$ $(=T\langle 21\rangle)$ and two differences $(T\langle 11\rangle - T\langle 33\rangle$ and $T\langle 22\rangle - T\langle 33\rangle)$ of the normal components of stress which depend only on the rheological properties of the material. In practice the required values of the normal stress differences cannot be obtained by direct measurement of stress components in rectilinear shear flow. However, Weissenberg (1947) has shown that they can be obtained from measurements of certain pressure distributions in suitable curvilinear shear flows in which the streamlines are circular. For steady shear flow in the systems discussed in Secs. 9-3-2, 9-3-3, and 9-3-4, the assumptions are made that

$$T\langle 23\rangle = T\langle 32\rangle = T\langle 13\rangle = T\langle 31\rangle = 0 \tag{9-1-69}$$

and the quantities

$$T\langle 11\rangle - T\langle 33\rangle,\ T\langle 11\rangle - T\langle 22\rangle,\ T\langle 22\rangle - T\langle 33\rangle, \text{and } T\langle 21\rangle\ (=T\langle 12\rangle) \tag{9-1-70}$$

depend only on the local value of the rate of deformation **D**. This follows from the basic assumption that the stress is determined by the local shape which in turn is determined by **D**.

9-2 VISCOELASTICITY

9-2-1 Stress Relaxation

Shearing deformations in purely viscous fluids are dissipative processes, the work done on a fluid in changing its shape being entirely converted to heat. In contrast, the work expended in deforming a perfectly elastic solid is entirely recovered when the material reverts to the natural state.

In principle, all materials possess all rheological properties. Thus it is to be expected that there exist materials whose behavior is partly like a fluid and partly like a solid. Materials for which the work of shearing deformation is not completely dissipated, as in fluids, or is not completely conserved, as in solids, are called viscoelastic materials.

Two well-defined types of viscoelastic material can be distinguished. These are the viscoelastic fluid and the elasticoviscous solid. The elasticoviscous solid is distinguished from the purely elastic solid by the phenomenon of creep. Viscoelastic fluids are distinguished from purely viscous fluids by the phenomenon of strain recovery, or recoil, upon the release of stress.

A small strain imposed on a perfectly elastic solid leads to the generation of a stress which persists as long as the strain is maintained. That is, for perfectly elastic solids the stress generated by strain does not relax at all. This is in contrast to the behavior of purely viscous fluids. Consider a small strain to be imposed upon a purely viscous fluid. This small deformation will only generate a stress if the amount of strain is changing. Terminate the deformation process and this generated stress disappears. In particular, for a purely viscous fluid the stress generated by strain is considered to relax instantaneously.

Since viscoelastic materials are defined as materials which exhibit behavior characteristics of both purely viscous fluids and purely elastic solids, it is to be expected that for viscoelastic materials the stress produced by strain will relax at a finite rate.

One of the central problems in devising mechanical descriptions of materials which exhibit viscoelastic behavior is the question of how to describe the appearance of both elastic and viscous effects. One approach is to assume simple linear superposition of elastic and viscous responses.

As an introduction, consider first the deviatoric relationships for visco-elasticity, taking as a starting point Eqs. (7-1-24) and (6-1-23) expressed in the form

$$\mathbf{T}'_\eta = 2\eta\mathbf{D}'_\eta \tag{9-2-1}$$

and

$$\mathbf{T}'_e = 2G\tilde{\mathbf{e}}'_e \tag{9-2-2}$$

where Eq. (9-2-2) is the deviatoric relationship for an isotropic elastic solid, the discussion being initially restricted to infinitesimal deformations.

For the particular case when η and G can be regarded as constants, independent of the deformation, Eqs. (9-2-1) and (9-2-2) are linear and the response has been referred to by Zener (1948) as anelasticity. The viscous term is usually associated with internal friction in solids.

There are two possible ways in which viscous and elastic effects can be linearly superposed. The total velocity $\mathbf{v}$ of an element of a viscoelastic material can be regarded as the sum of the components $\mathbf{v}_e$, $\mathbf{v}_\eta$ of velocity arising independently from elastic and viscous effects respectively. That is, the total rate of deformation $\mathbf{D}$ of an element of the material is just

$$\mathbf{D} = \mathbf{D}_e + \mathbf{D}_\eta \tag{9-2-3}$$

where $\mathbf{D}_e$ and $\mathbf{D}_\eta$ arise from elastic and viscous effects respectively. For this system, both the elastic and viscous effects arise from the same force on the element.

Application of the particular condition represented by Eq. (9-2-3) to Eqs. (9-2-1) and (9-2-2) gives

$$\mathbf{T}' + \tau_0 \frac{\partial \mathbf{T}'}{\partial t} = 2\eta\mathbf{D}' \tag{9-2-4}$$

where

$$\tau_0 = \frac{\eta}{G} \tag{9-2-5}$$

is a characteristic time constant. Equation (9-2-4) forms the basis of Maxwell's (1867) relaxation theory of elasticity. The constitutive equation (9-2-4) can be regarded as an example of a differential constitutive equation. Two particular types of behavior can be identified, depending upon whether the material is subjected to a constant load for a long time of deformation, giving rise to the phenomenon of creep, or whether the specimen is kept at a constant strain for a long time of deformation, giving rise to the phenomenon of stress relaxation. For constant strain, $\mathbf{D} = \mathbf{0}$ and Eq. (9-2-4) reduces to

$$\left(1+\tau_0\frac{\partial}{\partial t}\right)\mathbf{T}' = \mathbf{0} \qquad (\tilde{\mathbf{e}}' = \tilde{\mathbf{e}}'_0 = \text{constant})$$

which has the solution

$$\mathbf{T}' = \mathbf{T}'_0 \exp\left(-\frac{t}{\tau_0}\right) \tag{9-2-6}$$

where $\mathbf{T}'_0 = 2G\tilde{\mathbf{e}}'_0$ is the value of the stress at time $t = 0$. Steady creep takes place under constant stress $\mathbf{T}'_0$ according to the relation

$$\tilde{\mathbf{e}}' = \tilde{\mathbf{e}}'_0\left(1+\frac{t}{\tau_0}\right) \tag{9-2-7}$$

where for $t = 0$, $\mathbf{T}'_0 = 2G\tilde{\mathbf{e}}'_0$. Alternatively, it may be assumed that the elastic and viscous deformations are equal $(\mathbf{D}_e = \mathbf{D}_\eta = \mathbf{D})$, the associated forces being additive; i.e.,

$$\mathbf{T}' = \mathbf{T}'_e + \mathbf{T}'_\eta \tag{9-2-8}$$

where $\mathbf{T}'_e$ and $\mathbf{T}'_\eta$ relate to Eqs. (9-2-1) and (9-2-2). Combination of Eqs. (9-2-1), (9-2-2), and (9-2-8) gives

$$\mathbf{T}' = 2G(\tilde{\mathbf{e}}' + \lambda_0\mathbf{D}') \tag{9-2-9}$$

where

$$\lambda_0 = \frac{\eta}{G} \tag{9-2-10}$$

It is important to note that the instantaneous response of a material to which Eq. (9-2-9) can be applied is essentially fluid-like because the viscous property inhibits any sudden displacements. Such a system is said to exhibit retarded elastic behavior and the time constant is referred to as the retardation time. For these two simple models, $\tau_0 = \lambda_0$; however, the possibility exists that for more complicated systems $\tau_0 \neq \lambda_0$. Equation (9-2-9) forms the basis of the Kelvin–Voigt (Kelvin, 1875; Voigt, 1889) theory of the retarded elasticity. For constant stress $\mathbf{T}'_0$ applied at $t = 0$, Eq. (9-2-9) gives

$$\tilde{\mathbf{e}}' = \tilde{\mathbf{e}}'_0\left[1-\exp\left(-\frac{t}{\lambda_0}\right)\right] \tag{9-2-11}$$

where $\mathbf{T}'_0 = 2G\tilde{\mathbf{e}}'_0$ and $\tilde{\mathbf{e}}'_0$ is the value which the strain attains for $t = \infty$. The quantity $J = 1/G$ is referred to as the equilibrium compliance.

An equation of state which has found application in the linear theory of viscoelasticity is the operator equation

$$\left(1 + \sum_{r=0}^{N} \tau_r \frac{\partial^{r+1}}{\partial t^{r+1}}\right)\mathbf{T}' = 2\eta_0\left(1 + \sum_{r=0}^{M} \lambda_r \frac{\partial^{r+1}}{\partial t^{r+1}}\right)\mathbf{D}' \tag{9-2-12}$$

where η_0 is the viscosity of the material at zero rate of deformation and τ_r and λ_r are parameters characteristic of the properties of the material and are taken to be independent of the deformation process. Setting $\eta_0 = \eta$ and all λ and τ parameters, except τ_0, to zero in Eq. (9-2-12) gives Eq. (9-2-4). Alternatively, the operator equation may be expressed in the form

$$\left(1 + \sum_{r=0}^{N} \tau_r \frac{\partial^{r+1}}{\partial t^{r+1}}\right)\mathbf{T}' = 2G\left(1 + \sum_{r=0}^{M} \lambda_r \frac{\partial^{r+1}}{\partial t^{r+1}}\right)\tilde{\mathbf{e}}' \tag{9-2-13}$$

Setting all λ and τ parameters, except λ_0, to zero in Eq. (9-2-13) gives Eq. (9-2-9). With all λ and τ parameters, except λ_0 and τ_0, set to zero, Eq. (9-2-13) reduces to

$$\mathbf{T}' + \tau_0\dot{\mathbf{T}}' = 2G(\tilde{\mathbf{e}}' + \lambda_0\mathbf{D}') \tag{9-2-14}$$

which is the constitutive equation for the standard linear solid. It is important to note that the constitutive equations (9-2-12), (9-2-13), and (9-2-14) only apply for the limiting condition of small velocities and small velocity gradients.

9-2-2 Spectrum of Relaxation Times

The Maxwell and Kelvin–Voigt models are limited in application by being characterized in terms of a single time constant. Where the relaxation extends across several decades of logarithmic time, the usefulness of the two models can be extended by introducing the concept of N such models, or elements, in series. The Maxwell model can be generalized to the form

$$\mathbf{T}' = 2\tilde{\mathbf{e}}'_0\left[G_0 + \sum_{n=1}^{N} G_n \exp\left(-\frac{t}{\tau_n}\right)\right] \tag{9-2-15}$$

the system relaxing to constant strain $\tilde{\mathbf{e}}'_0$, and the Kelvin–Voigt model to the form

$$\tilde{\mathbf{e}}' = \tfrac{1}{2}\mathbf{T}'_0\left\{J_0 + \frac{1}{\eta}t + \sum_{n=1}^{N} J_n\left[1 - \exp\left(-\frac{t}{\lambda_n}\right)\right]\right\} \tag{9-2-16}$$

where $J_n = 1/G_n$ and the presence of the compliance J_0 provides instantaneous response. Further generalization can be effected by introducing the concept of a continuous spectrum, or distribution, of relaxation times by replacing G_n in Eq. (9-2-15) by $g(\tau)\,d\tau$ for the relaxation times between τ and $\tau + d\tau$, giving

$$G(t) = G_0 + \int_0^\infty g(\tau)\exp\left(-\frac{t}{\tau}\right)d\tau \tag{9-2-17}$$

it being noted that $g(\tau)$ is really the density of the distribution of shear moduli. Equation (9-2-17) can be expressed in the sometimes more convenient form

$$G(t) = G_G + \int_{-\infty}^{\infty} H(\tau)\exp\left(-\frac{t}{\tau}\right)d(\ln\tau) \tag{9-2-18}$$

where $H(\tau) = \tau g(\tau)$, Eq. (9-2-18) being taken as the definition of the distribution function H, usually called the relaxation spectrum. Equation (9-2-16) can similarly be replaced by a continuous spectrum of retardation times to give a compliance

$$J(t) = J_0 + \frac{1}{\eta}t + \int_0^{\infty} f(\lambda)\left[1-\exp\left(-\frac{t}{\lambda}\right)\right]d\lambda \tag{9-2-19}$$

or, with $L(\lambda) = \lambda f(\lambda)$,

$$J(t) = J_0 + \frac{1}{\eta}t + \int_{-\infty}^{\infty} L(\lambda)\left[1-\exp\left(-\frac{t}{\lambda}\right)\right]d(\ln\lambda) \tag{9-2-20}$$

Gross (1953) has given the relationships between the relaxation spectrum H and the retardation spectrum L (see also Leaderman, 1958; Ferry, 1961).

9-2-3 Boltzmann Superposition Principle

It is evident from Eq. (9-2-4) that the Maxwell element is described by a differential equation that can equally well be applied to variable strain. In contrast, the stress relaxation equation (9-2-6) was obtained as a particular solution corresponding to a constant strain. Thus it is necessary to adopt a principle of superposition, introduced by Boltzmann (1874), according to which an increment of strain $d\tilde{\mathbf{e}}'(\tau)$ produces an increment of stress

$$d\mathbf{T}'(t) = 2d\tilde{\mathbf{e}}'(\tau)G(t-\tau) \tag{9-2-21}$$

that is independent of the values of strain outside of the interval $(\tau, \tau+d\tau)$.

For strain increments $\tilde{\mathbf{e}}'_k$ applied at times τ_k, Eq. (9-2-21) gives

$$\mathbf{T}' = 2\sum_{k=1}^{N} \tilde{\mathbf{e}}'_k G(t-\tau_k) \tag{9-2-22}$$

where $G(t)$ is given by Eqs. (9-2-17) and (9-2-18). A similar superposition for stress increments $\mathbf{T}'_k$ applied at times τ_k gives by way of the relation

$$d\tilde{\mathbf{e}}'(t) = \tfrac{1}{2}d\mathbf{T}'(\tau)J(t-\tau) \tag{9-2-23}$$

the equation

$$\tilde{\mathbf{e}}' = \frac{1}{2}\sum_{k=1}^{N} \mathbf{T}'_k J(t-\tau_k) \tag{9-2-24}$$

where $J(t)$ is given by Eqs. (9-2-19) and (9-2-20).

For a continuously varying load, with increments $d\mathbf{T}' = (\partial\mathbf{T}/\partial t)\,d\tau$ applied at time τ, or with increments $d\tilde{\mathbf{e}}'$ applied at time τ, the superposition of Eqs. (9-2-22) and (9-2-24) take the form

$$\mathbf{T}'(t) = 2\int_{-\infty}^{t} \frac{d\tilde{\mathbf{e}}'(\tau)}{d\tau}G(t-\tau)\,d\tau \tag{9-2-25}$$

and
$$\tilde{\mathbf{e}}'(t) = \frac{1}{2}\int_{-\infty}^{t} \frac{d\mathbf{T}'(\tau)}{d\tau} J(t-\tau)\,d\tau \tag{9-2-26}$$

The instantaneous elastic response is sometimes separated out by expressing $G(t)$ and $J(t)$ in the form

$$G(t) = G_0[1-\phi(t)] \tag{9-2-27}$$

$$J(t) = J_0[1+\psi(t)] \tag{9-2-28}$$

where $\phi(t)$ is the dimensionless relaxation function and $\psi(t)$ is the dimensionless, retarded elastic function, or creep function. Equations (9-2-25) and (9-2-26) now take the form

$$\tilde{\mathbf{e}}'(t) = \tfrac{1}{2}J_0\left[\mathbf{T}' + \int_{-\infty}^{t} \psi(t-\tau)\frac{d\mathbf{T}'}{d\tau}\,d\tau\right] \tag{9-2-29}$$

and
$$\mathbf{T}'(t) = 2G_0\left[\tilde{\mathbf{e}}' - \int_{-\infty}^{t} \phi(t-\tau)\frac{d\tilde{\mathbf{e}}'}{d\tau}\,d\tau\right] \tag{9-2-30}$$

assuming that stress and strain vanish at $\tau = -\infty$. In terms of the elapsed time $s = t-\tau$, Eqs. (9-2-25) and (9-2-26) take the form

$$\mathbf{T}'(t) = 2\int_{\infty}^{0} \frac{d\tilde{\mathbf{e}}'(t-s)}{ds} G(s)\,ds \tag{9-2-31}$$

and
$$\tilde{\mathbf{e}}'(t) = \frac{1}{2}\int_{\infty}^{0} \frac{d\mathbf{T}'(t-s)}{ds} J(s)\,ds \tag{9-2-32}$$

which on integration take the form

$$\mathbf{T}'(t) = 2G_0\tilde{\mathbf{e}}'(t) + 2\int_{0}^{\infty} \tilde{\mathbf{e}}'(t-s)\frac{dG(s)}{ds}\,ds \tag{9-2-33}$$

and
$$\tilde{\mathbf{e}}'(t) = \tfrac{1}{2}J_0\mathbf{T}'(t) + \frac{1}{2}\int_{0}^{\infty} \mathbf{T}'(t-s)\frac{dJ(s)}{ds}\,ds \tag{9-2-34}$$

it being assumed that stress and strain are zero at $\tau = -\infty$, and where, for $s = 0$, $J = J_0$ and $G = G_0$.

The constitutive equations (9-2-25) to (9-2-34) are examples of linear hereditary stress-strain laws, in which stress depends on the history of the strain instead of only on its present value. History-dependent materials are also referred to as materials with memory. A detailed study of this type of material can only be made by way of the principle of fading memory, which is beyond the scope of the present discussion.

9-2-4 Stress Rates and Material Objectivity

The concepts underlying the constitutive equations of viscoelasticity have so far been presented without regard to the question of whether they satisfy the principle of material objectivity. There is at least one aspect with regard to which they do not, in their present form, satisfy material objectivity and that is in the choice of stress rate. The correct choice of stress rate can be determined by first demonstrating that as a possible choice the material derivative is itself not objective.

From Eq. (5-2-16),

$$\mathbf{T}^* = \mathbf{Q}\mathbf{T}\mathbf{Q}^T \tag{9-2-35}$$

which gives in terms of the material derivative

$$\dot{\mathbf{T}}^* = \mathbf{Q}\dot{\mathbf{T}}\mathbf{Q}^T + \dot{\mathbf{Q}}\mathbf{T}\mathbf{Q}^T + \mathbf{Q}\mathbf{T}\dot{\mathbf{Q}}^T \tag{9-2-36}$$

which, because of the presence of the last two terms on the right-hand side, is not objective. To eliminate these terms write the spin tensor in the form

$$\mathbf{W}^* = \mathbf{A} + \mathbf{Q}\mathbf{W}\mathbf{Q}^T \tag{9-2-37}$$

where $\mathbf{A} = \dot{\mathbf{Q}}\mathbf{Q}^T = -\mathbf{Q}\dot{\mathbf{Q}}^T$ is the angular velocity tensor of the unstarred frame relative to the starred frame (see Sec. 2-1). Now form the product $\mathbf{W}^*\mathbf{Q}$ to give

$$\mathbf{W}^*\mathbf{Q} = \dot{\mathbf{Q}}\mathbf{Q}^T\mathbf{Q} + \mathbf{Q}\mathbf{W}\mathbf{Q}^T\mathbf{Q} \tag{9-2-38}$$

from which it follows that

$$\dot{\mathbf{Q}} = \mathbf{W}^*\mathbf{Q} - \mathbf{Q}\mathbf{W} \tag{9-2-39}$$

with the transpose

$$\dot{\mathbf{Q}}^T = -\mathbf{Q}^T\mathbf{W}^* + \mathbf{W}\mathbf{Q}^T \tag{9-2-40}$$

Substitute Eqs. (9-2-39) and (9-2-40) into Eq. (9-2-36), which, using Eq. (9-2-35), gives

$$\dot{\mathbf{T}}^* - \mathbf{W}^*\mathbf{T}^* + \mathbf{T}^*\mathbf{W}^* = \mathbf{Q}(\dot{\mathbf{T}} - \mathbf{W}\mathbf{T} + \mathbf{T}\mathbf{W})\mathbf{Q}^T$$

which can be expressed in the form

$$\mathring{\mathbf{T}}^* = \mathbf{Q}\mathring{\mathbf{T}}\mathbf{Q}^T \tag{9-2-41}$$

where

$$\mathring{\mathbf{T}} = \dot{\mathbf{T}} - \mathbf{W}\mathbf{T} + \mathbf{T}\mathbf{W} \tag{9-2-42}$$

is referred to as the corotational stress rate or Jaumann derivative. The Jaumann derivative is seen to be equal to the material derivative of the stress as it would appear to an observer in a frame of reference attached to the particle and rotating with it at an angular velocity equal to the instantaneous value of the angular velocity $\mathbf{W}$ of the material. It is evident from Eq. (9-2-41) that the Jaumann derivative satisfies the principle of material objectivity. The Jaumann derivative can be expressed in the form

$$\begin{aligned}\mathring{\mathbf{T}} = \frac{\mathscr{D}\mathbf{T}}{\mathscr{D}t} &= \dot{\mathbf{T}} - \mathbf{W}\mathbf{T} + \mathbf{T}\mathbf{W} \\ &= \frac{\partial \mathbf{T}}{\partial t} + (v \cdot \nabla)\mathbf{T} - \mathbf{W}\mathbf{T} + \mathbf{T}\mathbf{W}\end{aligned} \tag{9-2-43}$$

from which it follows that a constitutive equation can be made to satisfy the principle of material objectivity simply by adding to the stress rate appearing in the equation, be it $\partial\mathbf{T}/\partial t$ or $\dot{\mathbf{T}}$, the remaining terms in Eq. (9-2-43) which are necessary to modify the form of the constitutive equation in order that it shall finally contain only the Jaumann form of the stress derivative. This procedure is only valid if all other dependent and independent variables appearing in the constitutive equation are themselves objective.

The addition of the objective quantity

$$\mathbf{DT}+\mathbf{TD}$$

to $\mathring{\mathbf{T}}$ gives the convected stress rate, or stress flux

$$\overset{\Delta}{\mathbf{T}} = \mathring{\mathbf{T}}+\mathbf{DT}+\mathbf{TD} \tag{9-2-44}$$

or, alternatively, since $\mathbf{D}-\mathbf{W} = \mathbf{L}^T$ and $\mathbf{D}+\mathbf{W} = \mathbf{L}$,

$$\overset{\Delta}{\mathbf{T}} = \dot{\mathbf{T}}+\mathbf{L}^T\mathbf{T}+\mathbf{TL} \tag{9-2-45}$$

Higher-order corotational and convected stress rates may appear in constitutive equations of the rate type (see Sec. 5-5; also see Oldroyd, 1958; Truesdell and Noll, 1965).

9-2-5 Functional Constitutive Equations

Viscoelastic materials may be characterized by functional constitutive equations. Since a viscoelastic material need not have a preferred configuration, it is convenient to use the general constitutive equation for a simple material in the form in which the configuration at the present time t is taken as the reference configuration. This can be allowed for by considering the deformation gradient

$$\mathbf{F}(\tau) = \mathbf{F}_{(t)}(\tau)\mathbf{F} \tag{9-2-46}$$

and noting that a function of the product $\mathbf{F}_{(t)}(\tau)\mathbf{F}$ defines a function of $\mathbf{F}_{(t)}(\tau)$ and $\mathbf{F}$, from which it follows that the general definition of simple materials can be stated as follows:

$$\mathbf{T} = \mathbf{J}(\mathbf{F}_{(t)}(\tau), \mathbf{F}) \tag{9-2-47}$$

It has been shown in Sec. 5-6 that Eq. (9-2-47) can be reduced to the form [see Eq. (5-6-12)]

$$\mathbf{R}^T\mathbf{TR} = \mathfrak{p}(\mathbf{C})+\mathfrak{q}(\mathbf{G}^*(s), \mathbf{C}) \tag{9-2-48}$$

where for the rest history the functional $\mathfrak{q}$ satisfies the condition

$$\mathfrak{q}(\mathbf{G}^*(s), \mathbf{C}) = \mathbf{o}$$

for $\mathbf{G}^*(s) = \mathbf{O}$, Eq. (9-2-48) giving the relation

$$\mathbf{T} = \mathbf{R}\mathfrak{p}(\mathbf{C})\mathbf{R}^T$$

for a static deformation [see Eq. (5-6-14)].

Stress relaxation, interpreted as the fading of a material's memory of past deformations, has been discussed by Coleman and Noll (1961) (see also Truesdell and Noll, 1965), who have shown that the constitutive functional $\mathfrak{q}$ for such materials can be approximated by a sum of n linear functionals. The special case $n = 1$ can be represented by an integral, so that Eq. (9-2-48) becomes

$$\mathbf{T} = \mathfrak{h}(\mathbf{B}) + \int_0^\infty \mathfrak{g}(\mathbf{B}, s)\,[\mathbf{G}(s)]\,ds \tag{9-2-49}$$

where

$$\mathbf{G}(s) = \mathbf{C}_{(t)}(\tau) - \mathbf{I} = \mathbf{R}\mathbf{G}^*(s)\mathbf{R}^T$$

and $\mathbf{B} = \mathbf{RCR}^T$. Coleman and Noll define Eq. (9-2-49) as the basis of the theory of the linear viscoelastic response of an isotropic solid.

The equations of classical viscoelasticity can be obtained by way of the linearization $\boldsymbol{u} = \boldsymbol{x} - \boldsymbol{X}$, $\tilde{\mathbf{S}} = \mathbf{F} - \mathbf{I}$, and by introducing as a measure of smallness the quantity $\varepsilon = |\tilde{\mathbf{S}}| = \sqrt{\operatorname{tr} \tilde{\mathbf{S}}\tilde{\mathbf{S}}^T}$ where $\varepsilon \ll 1$. For these conditions

$$\begin{aligned}
\mathbf{F} &= \mathbf{I} + \tilde{\mathbf{S}} = \mathbf{I} + O(\varepsilon) \\
\mathbf{F}^{-1} &= \mathbf{I} - \tilde{\mathbf{S}} + O(\varepsilon^2) = \mathbf{I} + O(\varepsilon) \\
\mathbf{F}_{(t)}(\tau) &= \mathbf{F}(\tau)\mathbf{F}^{-1} = [\mathbf{I} + \tilde{\mathbf{S}}(\tau)]\,[\mathbf{I} - \tilde{\mathbf{S}}(t)] \\
&= \mathbf{I} + \tilde{\mathbf{S}}(\tau) - \tilde{\mathbf{S}}(t) \\
\mathbf{C}_{(t)}(\tau) &= \mathbf{F}_{(t)}(\tau)^T \mathbf{F}_{(t)}(\tau) = \mathbf{I} + 2[\tilde{\mathbf{e}}(\tau) - \tilde{\mathbf{e}}(t)]
\end{aligned}$$

Setting $\mathfrak{h}(\mathbf{B}) = \mathbf{K}(\tilde{\mathbf{e}})$, and since

$$\mathbf{B} = \mathbf{F}\mathbf{F}^T \approx \mathbf{I} + 2\tilde{\mathbf{e}}$$

Eq. (9-2-49) gives

$$\mathbf{T} = \mathbf{K}(\tilde{\mathbf{e}}) + 2\int_0^\infty \mathfrak{g}(\mathbf{I}, s)\,[\tilde{\mathbf{e}}(\tau) - \tilde{\mathbf{e}}(t)]\,ds \tag{9-2-50}$$

or, alternatively,

$$\begin{aligned}
\mathbf{T} = \left\{[\lambda + \bar{\lambda}(0)]\operatorname{tr}\tilde{\mathbf{e}} + \int_0^\infty \dot{\bar{\lambda}}(s)\operatorname{tr}\tilde{\mathbf{e}}(t-s)\,ds\right\}\mathbf{I} \\
+ 2[G + \bar{G}(0)]\tilde{\mathbf{e}} + 2\int_0^\infty \dot{\bar{G}}(s)\tilde{\mathbf{e}}(t - s)\,ds
\end{aligned} \tag{9-2-51}$$

where λ and G are Lamé's coefficients of elasticity, and $\bar{\lambda}$ and $\bar{G}$ are material functions.

Consider the operator equation (9-2-12) expressed in the form

$$\begin{aligned}
\left(1 + \tau_0 \frac{\partial}{\partial t} + \tau_1 \frac{\partial^2}{\partial t^2} + \cdots + \tau_N \frac{\partial^{N+1}}{\partial t^{N+1}}\right)\mathbf{T}' \\
= 2\eta_0 \left(1 + \lambda_0 \frac{\partial}{\partial t} + \lambda_1 \frac{\partial^2}{\partial t^2} + \cdots + \lambda_M \frac{\partial^{M+1}}{\partial t^{M+1}}\right)\mathbf{D}'
\end{aligned} \tag{9-2-52}$$

Under certain conditions, it is possible to solve this equation for $\mathbf{T}'$ in terms of $\mathbf{D}'$ and its derivatives, the resulting equation being expressed in the form

$$\mathbf{T}' = 2\left(A_1 + A_2\frac{\partial}{\partial t} + A_3\frac{\partial^2}{\partial t^2} + \cdots\right)\mathbf{D}' \tag{9-2-53}$$

where the coefficients $A_1, A_2, \ldots$ are related to the original coefficients $\tau_0, \tau_1, \ldots, \lambda_0, \lambda_1, \ldots$ by

$$\begin{aligned} A_1 &= \eta_0 \\ A_2 &= -\eta_0(\tau_0 - \lambda_0) \\ A_3 &= \eta_0(\tau_0^2 - \tau_1 - \tau_0\lambda_0 + \lambda_1) \\ &\cdots\cdots\cdots\cdots\cdots\cdots \end{aligned}$$

Equation (9-2-53) can be generalized by writing the partial time derivatives as Jaumann or convected time derivatives, but this procedure does not lead to predictions of a nonnewtonian viscosity. Having regard to the form of Eq. (9-2-53) and in the quest for nonlinear viscoelasticity, Rivlin and Ericksen (1955) proposed a theory of what are sometimes called materials of the differential type. From certain invariance considerations, Rivlin and Ericksen arrived at a constitutive equation which for incompressible fluids reduces to that of Eq. (9-1-30). Markovitz (1957) has suggested that three material coefficients are sufficient to describe simple flows of general fluids, which can be satisfied by suppressing all dependence on coefficients higher than a_3 in Eq. (9-1-31), for which particular limiting condition Eq. (9-1-30) reduces to

$$\mathbf{T} = -P\mathbf{I} + a_1\mathbf{A}_1 + a_2\mathbf{A}_2 + a_3\mathbf{A}_1^2 \tag{9-2-54}$$

which is simply Eq. (9-1-32). If all dependence on $\mathbf{A}_2$ is suppressed, then Eq. (9-2-54) reduces to the Reiner–Rivlin equation (9-1-4).

9-3 VISCOMETRIC FLOWS AND THE NORMAL STRESS DIFFERENCES

9-3-1 Flow through a Pipe

For the purposes of the present discussion the nature of the flow will be taken to be that described in Sec. 7-3-2, differing only in so far that now normal stress effects are considered. There is symmetry in the angular direction so that derivatives with respect to θ are zero. The components of the stress are

$$\begin{bmatrix} T\langle 11\rangle & T\langle 12\rangle & T\langle 13\rangle \\ T\langle 21\rangle & T\langle 22\rangle & T\langle 23\rangle \\ T\langle 31\rangle & T\langle 32\rangle & T\langle 33\rangle \end{bmatrix} = \begin{bmatrix} T\langle rr\rangle & 0 & T\langle rz\rangle \\ 0 & T\langle \theta\theta\rangle & 0 \\ T\langle zr\rangle & 0 & T\langle zz\rangle \end{bmatrix}$$

where the $T\langle ij\rangle$ are now understood to be the physical components of the stress.

In terms of the physical components of the stress the general equations of motion for steady-state flow in the z direction of an incompressible viscous liquid are

$$-\frac{\partial T\langle zz\rangle}{\partial z} = \frac{1}{r}\frac{\partial}{\partial r}(rT\langle zr\rangle) \tag{9-3-1}$$

$$r\frac{\partial T\langle rr\rangle}{\partial r} = T\langle\theta\theta\rangle - T\langle rr\rangle \tag{9-3-2}$$

provided that

$$\frac{\partial T\langle zr\rangle}{\partial z} = 0$$

Since it has been shown that $T\langle\theta\theta\rangle - T\langle rr\rangle$ and $T\langle rr\rangle - T\langle zz\rangle$ depend only on the local value of the rate of deformation $\dot{\gamma}$, which is dependent only on r [see Eqs. (9-1-66), (9-1-67), and (9-1-68)], it follows that

$$\frac{\partial}{\partial z}(T\langle zz\rangle - T\langle rr\rangle) = \frac{\partial}{\partial z}(T\langle rr\rangle - T\langle\theta\theta\rangle) = 0 \tag{9-3-3}$$

are satisfied and therefore

$$\frac{\partial^2 T\langle zz\rangle}{\partial r\partial z} = \frac{\partial^2 T\langle rr\rangle}{\partial r\partial z}$$

which gives, by way of Eq. (9-3-2), the relation

$$\frac{\partial^2 T\langle rr\rangle}{\partial r\partial z} = \frac{\partial}{\partial z}\left(\frac{T\langle\theta\theta\rangle - T\langle rr\rangle}{r}\right) = \frac{\partial^2 T\langle zz\rangle}{\partial r\partial z} = 0$$

Thus the pressure gradient is independent of r and θ, as for the flow of a newtonian-type liquid, Eq. (9-3-1) being integrated to give

$$-\tfrac{1}{2}r^2\frac{\partial T\langle zz\rangle}{\partial z} = rT\langle zr\rangle + \text{constant}$$

and it follows from setting $r = 0$ that the constant must be zero, such that

$$-\frac{1}{2}\frac{\partial T\langle zz\rangle}{\partial z} = \frac{T\langle zr\rangle}{r} = \frac{[T\langle zr\rangle]_a}{a} \tag{9-3-4}$$

which is to be compared with Eq. (7-3-3) for the flow of a newtonian-type liquid through a pipe, and the relevant treatment of pipe flow discussed in Sec. 7-3-2 can be applied.

Consider the total axial thrust

$$N_a = 2\pi\int_0^a T\langle zz\rangle r\,dr \tag{9-3-5}$$

For the condition that there is no contribution to the momentum flux parallel to the axis of the tube from the incoming liquid, the required momentum flux is ρQ, which is a determinable quantity, and hence N_a is a measurable quantity. Equation (9-3-5) can be expressed in the form

$$N_a = \pi a^2 T\langle zz\rangle(a) - \pi \int_0^a r^2 \frac{\partial T\langle zz\rangle}{\partial r} dr \tag{9-3-6}$$

Now from the identity

$$T\langle zz\rangle = T\langle rr\rangle + (T\langle zz\rangle - T\langle rr\rangle)$$
$$\frac{\partial T\langle zz\rangle}{\partial r} = \frac{\partial T\langle rr\rangle}{\partial r} + \frac{\partial}{\partial r}(T\langle zz\rangle - T\langle rr\rangle)$$

and from Eq. (9-3-2) it follows that

$$\frac{\partial T\langle zz\rangle}{\partial r} = \frac{(T\langle\theta\theta\rangle - T\langle rr\rangle)}{r} + \frac{\partial}{\partial r}(T\langle zz\rangle - T\langle rr\rangle) \tag{9-3-7}$$

which can be combined with Eq. (9-3-6) to give

$$N_a = \pi a^2 T\langle zz\rangle(a) - \pi \int_0^a r(T\langle\theta\theta\rangle - T\langle rr\rangle)\, dr - \pi \int_0^a r^2 d(T\langle zz\rangle - T\langle rr\rangle) \tag{9-3-8}$$

Equation (9-3-8) can be simplified to the form

$$N_a - \pi a^2 T\langle rr\rangle(a) = \pi \int_0^a (2T\langle zz\rangle - T\langle rr\rangle - T\langle\theta\theta\rangle) r\, dr$$

Changing the variable of integration from r to $T\langle zr\rangle$ by way of Eq. (9-3-4) and differentiating with respect to $[T\langle zr\rangle]_a \equiv \tau_a$ (i.e., the shear stress at the wall of the tube) gives

$$[2T\langle zz\rangle - T\langle rr\rangle - T\langle\theta\theta\rangle]_{\tau_a} = \frac{1}{\tau_a}\frac{\partial}{\partial \tau_a}\left[\frac{[N_a - \pi a^2 T\langle rr\rangle(a)]\tau_a^2}{\pi a^2}\right] \tag{9-3-9}$$

Since $T\langle rr\rangle(a)$ is the pressure on the tube wall at the exit, the right-hand side of Eq. (9-3-9) is a measurable quantity, and hence the left-hand side can be determined.

The presence of normal stress effects must also be manifested in an alteration of the radius of the jet issuing from the pipe. In particular the material may be in tension in the axial direction for flow through the pipe. Relaxation of the tensile stress in the jet will lead to a concentration in the axial direction and hence an expansion in the radial direction. A method of determining $T\langle zz\rangle - T\langle rr\rangle$ has been described by Gaskins and Philippoff (1959) which involves measurement of the expansion ratio, i.e., the ratio of the jet radius, at a number of flow velocities.

9-3-2 Flow between Rotating Coaxial Cylinders

The nature of the flow will be taken to be that described in Sec. 7-3-3, differing only in that now normal stress effects are considered to be present.

From the symmetry of the system with regard to rotation about the common axis of the cylinders, all stress components are assumed to be independent of θ. The components of the stress are

$$\begin{bmatrix} T\langle 11\rangle & T\langle 12\rangle & T\langle 13\rangle \\ T\langle 21\rangle & T\langle 22\rangle & T\langle 23\rangle \\ T\langle 31\rangle & T\langle 32\rangle & T\langle 33\rangle \end{bmatrix} = \begin{bmatrix} T\langle rr\rangle & T\langle r\theta\rangle & 0 \\ T\langle \theta r\rangle & T\langle \theta\theta\rangle & 0 \\ 0 & 0 & T\langle zz\rangle \end{bmatrix}$$

In terms of the physical components of the stress, the general equations of motion for the steady-state flow in the θ direction of an incompressible viscous liquid are

$$0 = \frac{1}{r^2}\frac{\partial}{\partial r}(r^2 T\langle r\theta\rangle) \qquad (T\langle r\theta\rangle = T\langle \theta r\rangle) \tag{9-3-10}$$

$$-\rho\frac{v\langle\theta\rangle^2}{r} = \frac{\partial T\langle rr\rangle}{\partial r} - \frac{T\langle\theta\theta\rangle - T\langle rr\rangle}{r} \tag{9-3-11}$$

$$0 = \frac{\partial T\langle zz\rangle}{\partial z} - \rho g \tag{9-3-12}$$

which are to be compared with Eqs. (7-3-13), (7-3-14), and (7-3-15) respectively.

Equation (9-3-11) is the fundamental equation for this method of determining the difference of normal stress components. In practice the pressures on the two cylinder walls can be measured, giving

$$\Delta T\langle rr\rangle = T\langle rr\rangle(r_1) - T\langle rr\rangle(r_2) \tag{9-3-13}$$

and for the particular condition that $\Delta r/r_1$ is small

$$r\frac{\partial T\langle rr\rangle}{\partial r} \approx r\frac{\Delta T\langle rr\rangle}{\Delta r} \tag{9-3-14}$$

where $\Delta r = r_1 - r_2$. Thus the difference $T\langle\theta\theta\rangle - T\langle rr\rangle$ can be obtained from Eq. (9-3-11) in the form

$$\begin{aligned} T\langle\theta\theta\rangle - T\langle rr\rangle &= \left(r\frac{\partial T\langle rr\rangle}{\partial r}\right)_{\text{cc}} + \rho v\langle\theta\rangle^2 \\ &\approx r\frac{\Delta T\langle rr\rangle}{\Delta r} + \rho v\langle\theta\rangle^2 \end{aligned} \tag{9-3-15}$$

and neglecting inertial forces, by working at low speeds,

$$\begin{aligned} T\langle\theta\theta\rangle - T\langle rr\rangle &\approx \left(r\frac{\partial T\langle rr\rangle}{\partial r}\right)_{\text{cc}} \\ &\approx r\frac{\Delta T\langle rr\rangle}{\Delta r} \end{aligned} \tag{9-3-16}$$

where the subscripts cc are used to distinguish flow between rotating coaxial cylinders.

Equation (9-3-12) is satisfied by a solution of the form

$$T\langle zz\rangle = \rho g z + A(r) \qquad (9\text{-}3\text{-}17)$$

where $A(r)$ determines, in part, the boundary condition at the upper free surface of the liquid. Differentiation of the identity

$$T\langle zz\rangle = T\langle rr\rangle - (T\langle rr\rangle - T\langle zz\rangle)$$

gives, by way of Eqs. (9-3-10) and (9-3-16), the relation

$$\left(r\frac{dT\langle zz\rangle}{dr}\right)_{cc} = (T\langle\theta\theta\rangle - T\langle rr\rangle) + b\dot{\gamma}\frac{d}{d\dot{\gamma}}(T\langle rr\rangle - T\langle zz\rangle) \qquad (9\text{-}3\text{-}18)$$

where

$$b = 2\frac{d\ln\dot{\gamma}}{d\ln T\langle r\theta\rangle} \qquad (9\text{-}3\text{-}19)$$

has been obtained from Eq. (9-3-10) expressed in the form

$$r\frac{d\dot{\gamma}}{dr}\frac{dT\langle r\theta\rangle}{d\dot{\gamma}} + 2T\langle r\theta\rangle = 0$$

and noting that this can in turn be expressed in the form

$$\frac{r}{\dot{\gamma}}\frac{d\dot{\gamma}}{dr} = -2\frac{T\langle r\theta\rangle}{\dot{\gamma}}\frac{d\dot{\gamma}}{dT\langle r\theta\rangle} = -b \qquad (9\text{-}3\text{-}20)$$

In steady shear flow, Eq. (9-1-66) can be expressed in the form

$$\dot{\gamma} = f(T\langle r\theta\rangle) \qquad (9\text{-}3\text{-}21)$$

where the form of the function f is assumed to be known. Equation (9-3-21) can, in principle, be combined with Eq. (9-3-19) to give b as a known function of $\dot{\gamma}$. Having established the dependence of b on $\dot{\gamma}$ it follows from Eq. (9-3-20) that $\dot{\gamma}$ can be determined, to within a constant of integration, as a function of r. Since it has been assumed that $T\langle\theta\theta\rangle - T\langle rr\rangle$ and $T\langle rr\rangle - T\langle zz\rangle$ depend only on the local value of the rate of shear $\dot{\gamma}$, it follows that the right-hand side of Eq. (9-3-18) must be a function of r, and therefore the dependence of $T\langle zz\rangle$ on r is determined to within a constant of integration by Eq. (9-3-20), which in turn determines the function $A(r)$ appearing in Eq. (9-3-17).

Let Λ be the excess of the atmospheric pressure P_0 over the thrust $-T\langle zz\rangle$ exerted by the fluid in the axial direction:

$$\Lambda = P_0 + T\langle zz\rangle$$

For the case that

$$T\langle\theta\theta\rangle - T\langle rr\rangle > 0 \qquad T\langle rr\rangle - T\langle zz\rangle = 0$$

it follows from Eq. (9-3-18) that

$$\frac{d}{dr}(-T\langle zz\rangle) < 0$$

so that, in general, Λ is not zero and hence the free surface at the upper end of the fluid mass cannot be a plane $z =$ constant.

The pressure will be uniform over the free upper surface of the liquid as long as it remains horizontal, and the above condition will not be satisfied. It is therefore to be expected that the liquid will tend to climb up the inner cylinder, thus producing on lower horizontal material planes a pressure distribution which approximates to that required in shear flow. Such an effect has been observed in many liquids (Weissenberg, 1947).

Thus, measurements involving flow between rotating coaxial cylinders can be made to give one difference $(T\langle\theta\theta\rangle - T\langle rr\rangle)$ of normal stress components, provided the measurements of pressure on the cylinder walls are made at points so sufficiently far removed from the top surface of the annular gap that the climbing effect can be neglected.

9-3-3 The Cone and Plate System

The flow will be taken to be that described in Sec. 7-3-4, differing only by the presence of normal stress effects.

Neglecting inertial body forces enables the general equations of motion for steady conditions of flow between a cone and plate to be expressed in the form

$$\frac{\partial T\langle rr\rangle}{\partial r} + \frac{1}{r}(2T\langle rr\rangle - T\langle\theta\theta\rangle - T\langle\phi\phi\rangle) = \rho g \cos\theta \tag{9-3-22}$$

$$\frac{1}{r}\frac{\partial T\langle\theta\theta\rangle}{\partial\theta} + \frac{1}{r}(T\langle\theta\theta\rangle - T\langle\phi\phi\rangle)\cot\theta = -\rho g \sin\theta \tag{9-3-23}$$

$$\frac{1}{r}\frac{\partial T\langle\theta\phi\rangle}{\partial\theta} + \frac{2}{r}\cot\theta\, T\langle\theta\phi\rangle = 0 \qquad (T\langle\theta\phi\rangle = T\langle\phi\theta\rangle) \tag{9-3-24}$$

when normal stress effects are present. Equations (9-3-22), (9-3-23), and (9-3-24) are to be compared with Eqs. (7-3-31), (7-3-32), and (7-3-33) respectively. It is evident from Eqs. (9-3-24) and (7-3-33) that the same analysis as developed in Sec. 7-3-4 can be applied to Eq. (9-3-24), from which it follows that this particular stress equation of motion can be satisfied.

Although inertial body forces are neglected, Eqs. (9-3-22) and (9-3-23) are, however, generally (though not exclusively) incompatible for liquids which exhibit normal stress effects in shear flow; this can be seen as follows. From Eq. (9-3-22),

$$\frac{\partial^2 T\langle rr\rangle}{\partial\theta\partial r} + \frac{1}{r}\frac{\partial}{\partial\theta}(2T\langle rr\rangle - T\langle\theta\theta\rangle - T\langle\phi\phi\rangle) = -\rho g \sin\theta \tag{9-3-25}$$

and, from Eq. (9-3-23),

$$\frac{\partial^2 T\langle\theta\theta\rangle}{\partial r\partial\theta} + \cot\theta \frac{\partial}{\partial r}(T\langle\theta\theta\rangle - T\langle\phi\phi\rangle) = -\rho g \sin\theta \tag{9-3-26}$$

For flow between a cone and plate the rate of shear $\dot{\gamma}$ is seen from Eq. (7-3-42) to be independent of r, which coupled with the condition relating to Eq. (9-1-70) gives

$$\frac{\partial^2 T\langle\theta\theta\rangle}{\partial r \partial\theta} = \frac{\partial^2 T\langle rr\rangle}{\partial r \partial\theta}$$

and
$$\frac{\partial}{\partial r}(T\langle\phi\phi\rangle - T\langle\theta\theta\rangle) = 0$$

which conditions combined with Eqs. (9-3-25) and (9-3-26) give the relation

$$\frac{\partial}{\partial\theta}(2T\langle rr\rangle - T\langle\theta\theta\rangle - T\langle\phi\phi\rangle) = 0 \qquad \text{(9-3-27)}$$

However, since

$$2T\langle rr\rangle - T\langle\theta\theta\rangle - T\langle\phi\phi\rangle = T\langle rr\rangle - T\langle\theta\theta\rangle + T\langle rr\rangle - T\langle\phi\phi\rangle$$

from which it follows that the quantity $2T\langle rr\rangle - T\langle\theta\theta\rangle - T\langle\phi\phi\rangle$ is therefore solely a function of $\dot{\gamma}$ by way of the condition associated with Eq. (9-1-70). For the shear conditions of flow commonly encountered, the expression

$$\frac{d\dot{\gamma}}{d\theta} = \frac{d\dot{\gamma}}{dT\langle\theta\phi\rangle}\frac{\partial T\langle\theta\phi\rangle}{\partial\theta}$$

can be used together with Eqs. (9-3-19) and (9-3-24) to give

$$\frac{1}{\dot{\gamma}}\frac{d\dot{\gamma}}{d\theta} = -b\cot\theta \qquad \text{(9-3-28)}$$

For the general condition

$$T\langle\theta\phi\rangle = 2\eta D\langle\theta\phi\rangle = \eta\dot{\gamma}$$

and so

$$\frac{2}{b} = \frac{d\ln T\langle\theta\phi\rangle}{d\ln\dot{\gamma}} = 1 + \frac{d\ln\eta}{d\ln\dot{\gamma}}$$

which, together with the observation that for many liquids which exhibit normal stress effects in shear flow the viscosity decreases with increasing $\dot{\gamma}$, gives the condition

$$\frac{d\ln\eta}{d\ln\dot{\gamma}} = \frac{\dot{\gamma}}{\eta}\frac{d\eta}{d\dot{\gamma}} \leqslant 0$$

which implies that $2/b \leqslant 1$; i.e.,

$$b \geqslant 2 \qquad \text{provided } \frac{d\eta}{d\dot{\gamma}} \leqslant 0 \qquad \text{(9-3-29)}$$

Equation (9-3-28) enables Eq. (9-3-27) to be expressed in the form

$$b\dot{\gamma}\cot\theta\frac{d}{d\dot{\gamma}}(2T\langle rr\rangle - T\langle\theta\theta\rangle - T\langle\phi\phi\rangle) = 0 \qquad \text{(9-3-30)}$$

from which it follows that for $b \geqslant 2$, Eq. (9-3-30) is not satisfied for points in the gap between the cone and plate unless the properties of the liquid are such that

$$\frac{d}{d\dot{\gamma}}(2T\langle rr\rangle - T\langle\theta\theta\rangle - T\langle\phi\phi\rangle) = 0 \tag{9-3-31}$$

Newtonian liquids and a certain viscoelastic liquid considered by Zaremba (1937), Fromm (1948), and De Witt (1955) do satisfy this condition. However, there is a large group of liquids which do not satisfy the condition of Eq. (9-3-31). Thus, in practice it is likely that the condition of Eq. (9-3-30) will not be satisfied and it is therefore to be expected that there will be some disturbance of the state of flow (Oldroyd, 1958; Ericksen, 1960) which has been assumed to exist for the purposes of the present discussion. It is important to note that the condition associated with Eq. (9-3-30) is quite distinct from the incompatibility of Eqs. (7-3-31) and (7-3-32) arising when inertial forces are neglected.

This situation can usually be resolved in practice to a sufficient approximation by assuming that the angle $\Delta\theta_c$ is sufficiently small for $\cot\theta$ and $\cos\theta$ to be neglected.

To this approximation Eq. (9-3-23) reduces to

$$\frac{\partial T\langle\theta\theta\rangle}{\partial\theta} = -\rho g r \sin\theta$$

which can be integrated to give

$$T\langle\theta\theta\rangle(r) = \rho g r \cos\theta + E(r) \tag{9-3-32}$$

and, for $\theta = \frac{1}{2}\pi$,

$$-T\langle\theta\theta\rangle(r) = -E(r)$$

where $-T\langle\theta\theta\rangle(r)$ is the pressure on the plate at a distance r from the cone apex.

Equation (9-3-22) may be expressed to the required approximation in the form

$$\left(r\frac{\partial T\langle rr\rangle}{\partial r}\right)_{\text{cp}} = T\langle\phi\phi\rangle + T\langle\theta\theta\rangle - 2T\langle rr\rangle \tag{9-3-33}$$

where $\cos\theta \approx 0$ for $\theta = \frac{1}{2}\pi$ (this is equivalent to neglecting the term $\rho r g \cos\theta$) and the subscripts cp are used to distinguish flow in a cone and plate system.

Consider a sphere of radius r whose center is at the cone apex. The quantity $-T\langle rr\rangle$ is normal to the surface of this sphere. In particular if the free surface of the liquid in the gap is of this spherical shape then, in the absence of surface tension effects, it follows that

$$-T\langle rr\rangle(r_p) = P_0 \tag{9-3-34}$$

where P_0 is the atmospheric pressure. Thus, in principle, $T\langle rr\rangle(r_p)$ can be determined. By changing the amount of test liquid in the gap, and hence the effective value of r_p, Weissenberg and Roberts (see, for example, Roberts, 1952) obtained the variation of $-T\langle rr\rangle(r_p)$ with r_p. The total quantity of liquid in the gap was maintained constant by replacing the fraction of test liquid removed by an immiscible newtonian liquid to facilitate measurement of $T\langle rr\rangle$. Also the variation of the pressure $-T\langle\theta\theta\rangle$ on the plate with distance r from the cone apex is obtained. These measurements taken together enable the derivative $\partial T\langle rr\rangle/\partial r$ and $T\langle\theta\theta\rangle$ to be evaluated, and hence enable the values of $T\langle\phi\phi\rangle$ to be obtained by calculation from Eq. (9-3-33).

Thus, measurements involving flow between a cone and plate in relative rotation can be made to give, at least in principle, all three normal components of stress in steady shear flow, from which in turn can be derived the differences of the normal components of stress. The successful application of this type of measurement is subject, however, to the condition that the free boundary of the test liquid is of a spherical shape.

From the condition associated with Eq. (9-1-70), it follows since $\dot{\gamma}$ is independent of r for the cone and plate system that

$$\frac{\partial T\langle\theta\theta\rangle}{\partial r}=\frac{\partial T\langle rr\rangle}{\partial r} \tag{9-3-35}$$

which condition can be used to express Eq. (9-3-33) in the form

$$\left(r\frac{\partial T\langle\theta\theta\rangle}{\partial r}\right)_{\text{cp}}=T\langle\phi\phi\rangle+T\langle\theta\theta\rangle-2T\langle rr\rangle \tag{9-3-36}$$

the pressure gradient appearing on the left-hand side of this equation being determined from measurements of pressure on the plate, and therefore the quantity $T\langle\phi\phi\rangle+T\langle\theta\theta\rangle-2T\langle rr\rangle$ can be evaluated. It is of interest to note, by way of the condition associated with Eq. (9-1-70), that the quantity $T\langle\phi\phi\rangle+T\langle\theta\theta\rangle-2T\langle rr\rangle$ is independent of r since $\dot{\gamma}$ is independent of r for the cone and plate system. Therefore Eq. (9-3-36) can be integrated to give

$$T\langle\theta\theta\rangle(r)=T\langle\theta\theta\rangle(r_p)+(T\langle\phi\phi\rangle+T\langle\theta\theta\rangle-2T\langle rr\rangle)\ln\left(\frac{r}{r_p}\right) \tag{9-3-37}$$

where

$$T\langle\theta\theta\rangle(r_p)-(T\langle\phi\phi\rangle+T\langle\theta\theta\rangle-2T\langle rr\rangle)\ln r_p$$

is the integration constant. Thus for $\theta=\frac{1}{2}\pi$, the function $E(r)$ appearing in Eq. (9-3-32) has been determined in the form of Eq. (9-3-37).

From Eq. (9-3-37) can be obtained the total thrust N_T on the plate due to the normal stress effect in the form

$$\begin{aligned}N_T&=-\int_0^{r_p}T\langle\theta\theta\rangle(r)\,dA-P_0\pi r_p^2\\&=-2\pi\int_0^{r_p}rT\langle\theta\theta\rangle(r)\,dr-P_0\pi r_p^2\end{aligned}$$

where the term $P_0\pi r_p^2$ represents the upthrust due to atmospheric pressure on the underside of the plate. Thus

$$N_T=-[T\langle\theta\theta\rangle(r_p)+P_0]\pi r_p^2+\tfrac{1}{2}\pi r_p^2(T\langle\phi\phi\rangle+T\langle\theta\theta\rangle-2T\langle rr\rangle) \tag{9-3-38}$$

If the free boundary is part of a sphere of radius r_p, the last term on the right-hand side of Eq. (9-3-38) can be evaluated at r_p, which together with the boundary condition of Eq. (9-3-34) gives

$$T\langle\phi\phi\rangle - T\langle\theta\theta\rangle = \frac{2N_T}{\pi r_p^2} \tag{9-3-39}$$

Thus a measurement of total thrust N_T should enable the normal stress difference $T\langle\phi\phi\rangle - T\langle\theta\theta\rangle$ to be evaluated, subject to the condition that the unidirectional state of shear flow persists up to the free boundary of the test material.

9-3-4 Torsional Flow—The Parallel Plate System

For steady conditions of flow with normal stress effects present, the general equations of motion for flow between parallel plates, i.e., Eqs. (7-3-49), (7-3-50), and (7-3-51), take the form

$$\frac{1}{r}\frac{\partial T\langle\theta\theta\rangle}{\partial\theta} + \frac{\partial T\langle\theta z\rangle}{\partial z} = 0 \qquad (T\langle\theta z\rangle = T\langle z\theta\rangle) \tag{9-3-40}$$

$$-\rho\frac{v\langle\theta\rangle^2}{r} = \frac{\partial T\langle rr\rangle}{\partial r} + \frac{T\langle rr\rangle - T\langle\theta\theta\rangle}{r} \tag{9-3-41}$$

and

$$\frac{\partial T\langle zz\rangle}{\partial z} = \rho g \tag{9-3-42}$$

Neglecting inertial body forces, and noting that from considerations of the symmetry of the flow the term $\partial T\langle\theta\theta\rangle/\partial\theta$ is set to zero, gives Eqs. (9-3-40), (9-3-41), and (9-3-42) in the form

$$\frac{\partial T\langle\theta z\rangle}{\partial z} = 0 \tag{9-3-43}$$

$$\left(r\frac{\partial T\langle rr\rangle}{\partial r}\right)_{\text{pp}} = T\langle\theta\theta\rangle - T\langle rr\rangle \tag{9-3-44}$$

and

$$\frac{\partial T\langle zz\rangle}{\partial z} = \rho g \tag{9-3-45}$$

where the subscripts pp are used to distinguish flow between parallel plates. Thus the stress equations of motion (9-3-43), (9-3-44), and (9-3-45) are compatible for the state of flow assumed to exist in Sec. 7-3-5, provided that the boundary conditions, particularly at the edges of the plates, can be satisfied.

Equation (9-3-45) is satisfied by a solution of the form

$$T\langle zz\rangle = \rho g z + G(r) \tag{9-3-46}$$

From Eq. (7-3-56),

$$\frac{\partial}{\partial\dot{\gamma}} = \frac{z_0}{\omega_p}\frac{\partial}{\partial r}$$

from which it follows that

$$\dot{\gamma}\frac{\partial}{\partial \dot{\gamma}} = r\frac{\partial}{\partial r} \tag{9-3-47}$$

Differentiation of the identity

$$T\langle zz\rangle = T\langle rr\rangle + (T\langle zz\rangle - T\langle rr\rangle)$$

gives, by way of Eqs. (9-3-44) and (9-3-47), the relation

$$\left(r\frac{\partial T\langle zz\rangle}{\partial r}\right)_{\text{pp}} = T\langle\theta\theta\rangle - T\langle rr\rangle + \dot{\gamma}\frac{\partial}{\partial\dot{\gamma}}(T\langle zz\rangle - T\langle rr\rangle) \tag{9-3-48}$$

Now it is evident from Eq. (7-3-56) that the stress components which depend upon $\dot{\gamma}$ may be regarded as depending on r, and as the right-hand side of Eq. (9-3-48) is a function of $\dot{\gamma}$ [by way of Eq. (9-1-67)] and therefore also of r, it follows that the variation of $T\langle zz\rangle$ with r is determined to within a constant of integration.

Equation (9-3-48), which is the basic equation for the parallel plate system, relates the distribution of pressure on either plate to both differences of normal stress components, the right-hand side of Eq. (9-3-48) being independent of z since, by Eq. (7-3-56), the rate of shear $\dot{\gamma}$ is independent of z.

9-4 EXPERIMENTAL APPLICATIONS: I. NORMAL STRESS MEASUREMENTS

9-4-1 The Cone and Plate Rheogoniometer

Weissenberg (1949) formulated the principles of a modified cone and plate instrument which would enable measurements to be made of the components of the stress and flow rate round the full solid angle in space at every point in the flowing material, a preliminary development of an instrument approximating to this type having been described by Russell (1946). In the Roberts–Weissenberg instrument, the construction and operation of which has been described by Jobling and Roberts (1959), shear flow between a cone and plate, operated on the cone and plate principle as discussed in Secs. 7-3-4 and 9-3-3, is employed, the radial distribution of pressure on the plate, measured by capillary tubes (Roberts, 1957), giving values of $T\langle\phi\phi\rangle + T\langle\theta\theta\rangle - 2T\langle rr\rangle$ determined from the pressure gradient and $T\langle\theta\theta\rangle - T\langle rr\rangle$ from the pressure at the rim of the cone. Measurements of the total thrust on the cone enables $T\langle\phi\phi\rangle - T\langle\theta\theta\rangle$ to be determined. This form of apparatus is shown in outline in Fig. 7-1 where the plate of a cone and plate viscometer has been modified to carry a manometer head consisting of a series of vertical capillary tubes. With the addition of the manometer the modified cone and plate instrument is referred to as the cone and plate rheogoniometer and the measurement of normal stress by this type of apparatus as rheogoniometry.

For the cone and plate system, Eqs. (9-1-67) and (9-1-68) reduce to

$$T\langle\theta\theta\rangle - T\langle rr\rangle = \beta_1(\dot{\gamma}) \tag{9-4-1}$$

$$T\langle\phi\phi\rangle - T\langle rr\rangle = \beta_2(\dot{\gamma}) \tag{9-4-2}$$

From Eq. (9-3-37) it is to be expected for flow in a cone and plate apparatus that the change in level $h(r)$ of the liquid in the manometer tubes will vary linearly with $\log r$. This has been experimentally verified by Markovitz and Brown (1964), who present their results for a 6.8 % solution of polyisobutylene in cetane in the form of $(h(r) - h_c(r), \log r)$ curves, where $h_c(r)$ is a centrifugal correction. Using these results, Markovitz and Brown determined the variation of $\beta_1 + \beta_2$ with $\dot{\gamma}$. Adams and Lodge (1964) present their basic results for a solution (I) of 5.76 gm of polyisobutylene of number-average molecular weight 10^6 in a 100 ml of dekalin and for a solution (II) of 2.44 gm of polymethylmethacrylate of number-average molecular weight 10^6 in 100 ml dimethylphthalate in the form of pressure distributions. The results obtained by Adams and Lodge for the pressure distributions in the two solutions give a pressure distribution which varies linearly with $\log r$ in agreement with Eq. (9-3-37). Using these results, Adams and Lodge (1964) determined the variation of $\beta_1 + \beta_2$ with $\dot{\gamma}$. The measurements of Markovitz and Brown and those of Adams and Lodge for polyisobutylene are in good general agreement, a result in accord with Eq. (9-3-37).

9-4-2 The Parallel Plate System

This instrument, a form of which has been described by Russell (1946), is shown in Fig. 7-2. Among the first detailed descriptions of such an instrument were those by Garner, Nissan, and Wood (1950) and Greensmith and Rivlin (1953). The instrument in a modified form has been used, for example, by Kotaka, Kurata, and Tamura (1959), Reiner (1952, 1957, 1958, 1960), and Adams and Lodge (1964).

Using a parallel plate apparatus, Markovitz and Brown (1964) have obtained normal stress measurements for a 6.9 % solution of polyisobutylene in cetane. They obtained the variation of $\bar{h}(r)$ $[\equiv h(r) - h_c(r)]$ with r for various speeds of rotation and a fixed-plate separation d. Using these results, Markovitz and Brown determined the variation of $\bar{h}(0) - \bar{h}(r)$ with $\dot{\gamma}$. These results were shown by Markovitz and Brown to be consistent with the equation

$$T\langle zz\rangle(r) - T\langle zz\rangle(0) = \beta_1(\dot{\gamma}) + \int_0^{\dot{\gamma}} \frac{1}{\dot{\gamma}'}\beta_2(\dot{\gamma}')\,d\dot{\gamma}' \tag{9-4-3}$$

obtained from Eq. (9-3-48) using the relations

$$T\langle zz\rangle - T\langle rr\rangle = \beta_1(\dot{\gamma}) \tag{9-4-4}$$

$$T\langle\theta\theta\rangle - T\langle rr\rangle = \beta_2(\dot{\gamma}) \tag{9-4-5}$$

which follows from Eqs. (9-1-67) and (9-1-68) for the parallel plate system. For these results, Markovitz and Brown obtained a single $[(r\,dT\langle zz\rangle/dr)_{pp}, \dot{\gamma}]$ curve in accord with Eq. (9-3-48). Having obtained an approximate value for the difference $\beta_2 - \beta_1$ from coaxial cylinder measurements and $\beta_1 + \beta_2$ from the cone and plate results, Markovitz (1965a, 1965b, 1965c) evaluated β_1 and β_2 and proceeded to use these values to calculate the difference $T\langle zz\rangle(r) - T\langle zz\rangle(0)$ as a function of $\dot{\gamma}$ for a

5.4% solution of polyisobutylene in cetane; he obtained good agreement between calculation and experiment.

Using a parallel plate apparatus, Adams and Lodge (1964) obtained the pressure distributions for their solutions (I) and (II) and the variation of $(r\,\partial T\langle zz\rangle/\partial r)_{\text{pp}}$ with $\dot{\gamma}$ for these results. The methods of measuring $T\langle\theta\theta\rangle - T\langle rr\rangle$ and $T\langle\phi\phi\rangle - T\langle\theta\theta\rangle$ using the cone and plate rheogoniometer are only valid if the required state of shear flow persists right up to the exposed surface of the liquid at the rim of the cone. It has been shown by Adams and Lodge (1964) that this can only hold if the free boundary is part of a sphere whose center is at the cone vertex, a condition which is unlikely to be accurately realized in practice for the cone and plate geometry. To overcome this restriction, and provide an alternative method of determining $T\langle\theta\theta\rangle - T\langle rr\rangle$ and $T\langle\phi\phi\rangle - T\langle\theta\theta\rangle$, Adams and Lodge describe a technique involving combined measurements from both a cone and plate instrument and a parallel plate instrument, a proposal to combine the information from two such instruments having also been made by Markovitz (1957, 1960). Adams and Lodge combine Eqs. (9-3-36) and (9-3-48) to give a relation for the variation of β_1 with the rate of shear. From their measurements, Adams and Lodge conclude that β_2 is significantly different from zero and increases with the rate of shear, a situation which conflicts with the predictions of a phenomenological theory (Weissenberg, 1947) and a molecular theory (Lodge, 1956) but is in agreement with the conclusion of Markovitz, who obtained a markedly similar dependence of β_2 on $\dot{\gamma}$.

The results of Markovitz and Brown (1964) and of Adams and Lodge (1964) are both in general accord with the overall physical concepts underlying the generation of normal stress differences and their evaluation from viscometric flow measurements. However, when the normal stress difference is simply taken to be the difference between the appropriate rim pressure and atmospheric pressure, Adams and Lodge (1964) have noted that certain discrepancies exist between the results of measuring normal stress differences in different ways, in particular as characterized by Eqs. (9-3-36) and (9-3-48). In this connection it has been suggested that the use of small capillary tubes, filled with static liquid, to measure normal stress differences in polymer solutions in steady shear flow may give rise to an unwanted pressure

$$P_H = P + T\langle 22\rangle \tag{9-4-6}$$

where P is the pressure recorded using a capillary tube and $-T\langle 22\rangle$ is the pressure which would act on the wall if the hole associated with the capillary tube were absent. Assuming that the hole pressure P_H is determined by $T\langle 21\rangle$ and that P_H is independent of the gap between the walls, Eq. (9-3-36) can be given in the form

$$\Delta_{\text{cp}} \equiv -\left(r\frac{dP}{dr}\right)_{\text{cp}} = T\langle\theta\theta\rangle + T\langle\phi\phi\rangle - 2T\langle rr\rangle \tag{9-4-7}$$

the equation for the rim pressure $P(r_p)$ taking the form

$$P(r_p) = T\langle rr\rangle - T\langle\theta\theta\rangle + P_H + P_0 \tag{9-4-8}$$

where r_p denotes the radius of the rotating member. Equation (9-3-48) for the parallel plate system becomes

$$\Delta_{\mathrm{pp}} \equiv -\left(r\frac{dP}{dr}\right)_{\mathrm{pp}} = T\langle\theta\theta\rangle - T\langle rr\rangle + \dot{\gamma}\frac{d}{d\dot{\gamma}}(T\langle zz\rangle - T\langle rr\rangle) - \left(\dot{\gamma}\frac{dP_H}{d\dot{\gamma}}\right)_{\mathrm{pp}} \tag{9-4-9}$$

and Eq. (9-3-18) for flow between rotating coaxial cylinders becomes

$$\Delta_{\mathrm{cp}} \equiv -\left(r\frac{dP}{dr}\right)_{\mathrm{cc}} = T\langle\theta\theta\rangle - T\langle rr\rangle + 2\left(T\langle r\theta\rangle\frac{dP_H}{dT\langle r\theta\rangle}\right)_{\mathrm{cc}} \tag{9-4-10}$$

Although the possible significance of the hole pressure P_H has been noted, e.g., by Schultz–Grünow (1955), it has generally been neglected. This is partly because estimates for newtonian liquids suggest that the hole pressure P_H will be small (see, for example, Thom and Apelt, 1958); typically P_H will be about $0.05T\langle 21\rangle$, where $T\langle 21\rangle$ denotes the value of the shear stress near to but unperturbed by the hole. The observation by Greensmith and Rivlin (1953) that holes of diameter 1.0, 1.7, and 2.6 mm gave indistinguishable results also provides evidence in favor of the conclusion that the hole pressure P_H can, for some systems, be neglected. Eliminating the normal stress differences between Eqs. (9-4-7), (9-4-9), and (9-4-10) gives

$$\Delta_{\mathrm{pp}} - \Delta_{\mathrm{cp}} + \tfrac{1}{2}\left(\dot{\gamma}\frac{d}{d\dot{\gamma}} - 1\right)(\Delta_{\mathrm{cc}} - \Delta_{\mathrm{cp}}) = \left(\dot{\gamma}\frac{d}{d\dot{\gamma}} - 1\right)\left(T\langle r\theta\rangle\frac{dP_H}{dT\langle r\theta\rangle}\right)_{\mathrm{cc}} - \left(\dot{\gamma}\frac{dP_H}{d\dot{\gamma}}\right)_{\mathrm{pp}} \tag{9-4-11}$$

Markovitz (1965a, 1965b, 1965c) has shown for a 5.4% solution of polyisobutylene in cetane that the left-hand side of Eq. (9-4-11) is small, a result which implies that either the hole pressure can be neglected or that for these particular hole sizes and liquid properties the difference on the right-hand side can be neglected. However, Broadbent, Kaye, Lodge, and Vale (1968) concluded from their studies that certain polymer solutions generate much higher values of the hole pressure P_H (typically $P_H \sim 2T\langle 21\rangle$) and suggested that by taking account of P_H it might be possible to resolve the discrepancies observed by Adams and Lodge (1964) between the results of measuring normal stress differences in different ways. For further details regarding hole-pressure errors arising in the measurement of normal stress differences in polymer solutions in steady shear flow, the reader is referred to the continuing studies of Lodge and his coworkers (see, for example, Lipson and Lodge, 1968; Kaye, Lodge, and Vale, 1968; Broadbent and Lodge, 1971; Higashitani and Lodge, 1975).

9-5 EXPERIMENTAL APPLICATIONS: II. NONNEWTONIAN FLOW

9-5-1 Flow through a Pipe

In general, for the flow of nonnewtonian fluids, a relationship of the form

$$-\frac{dv\langle z\rangle}{dr} = \dot{\gamma}(\tau) \qquad (\tau \equiv T\langle rz\rangle = T\langle zr\rangle) \tag{9-5-1}$$

will be assumed to exist between the shear stress τ and the rate of shear $\dot{\gamma}(\tau)$ (see Sec. 7-3-2).

For the general case, allowance must be made for an effective slip of the fluid relative to the wall of the tube, such that the fluid at $r = a$ appears to have a velocity $v_a(\tau_a)$ which varies with the stress at the wall. From Eqs. (7-3-3) and (9-5-1),

$$v\langle z\rangle = v_a(\tau_a) + \int_r^a \dot{\gamma}(\tau)\,dr = v_a(\tau_a) + \frac{a}{\tau_a}\int_\tau^{\tau_a} \dot{\gamma}(\bar{\tau})\,d\bar{\tau} \tag{9-5-2}$$

and substituting for $v\langle z\rangle$ from Eq. (9-5-2) into Eq. (7-3-6) and integrating by parts gives

$$Q = \pi a^2 v_a(\tau_a) + \frac{\pi a^3}{\tau_a^3}\int_0^{\tau_a} \tau^2\dot{\gamma}(\tau)\,d\tau \tag{9-5-3}$$

For a nonnewtonian fluid, an apparent viscosity

$$\eta_{ap} = \frac{\tau_a}{4Q/\pi a^3} \tag{9-5-4}$$

can be defined by way of Eq. (7-3-7) for a single experiment. This quantity is not, however, of fundamental significance in relation to the rheological properties of the material since it depends upon the flow rate and pipe diameter from one experiment to the next. From Eq. (9-5-3), the general relation (Reiner, 1930, 1931)

$$\frac{1}{\eta_{ap}} = \frac{4\zeta(\tau_a)}{a} + \phi(\tau_a) \tag{9-5-5}$$

is obtained where

$$\zeta(\tau_a) = \frac{v_a(\tau_a)}{\tau_a} \tag{9-5-6}$$

$$\phi(\tau_a) = \frac{4}{\tau_a^4}\int_0^{\tau_a} \tau^2\dot{\gamma}(\tau)\,d\tau \tag{9-5-7}$$

and $\zeta(\tau_a)$ is the effective slip coefficient (Oldroyd, 1949), being the effective velocity of slip per unit shear stress at the wall. Oldroyd (1956) envisages $\zeta(\tau_a)$ arising from a possible interaction of the wall with the motion of large asymmetric molecules, such as polymer molecules in the fluid, the presence of the wall exerting an orienting effect upon such molecules. This effect could only be present within a few molecular diameters of the wall, but nevertheless could, in principle, exert an appreciable influence on the measured rheological properties of the fluid. The quantities which characterize the rheological properties of the material are $\zeta(\tau_a)$ and $\dot{\gamma}(\tau)$. For each pair of observed values of Q and dP/dz, the values of η_{ap} and τ_a can be calculated and used to obtain the variation of the fluidity $1/\eta_{ap}$ with the stress at the wall τ_a, or as some function of τ_a, for each pipe diameter $2a$. The problem of obtaining the "shear stress rate of flow" characteristic for the material independent of the pipe diameter and wall effects has been discussed by Rabinowitch (1929), Schofield and Scott-Blair (1930, 1931), Reiner (1931), and Mooney (1931) who have pointed out that in the absence of any wall anomaly the curves for different pipe diameters should coincide and give a direct plot of

$$\frac{1}{\eta_{ap}} = \phi(\tau_a) \tag{9-5-8}$$

it being shown by Rabinowitch (1929) and Mooney (1931) that a formula for the function $\dot{\gamma}$ in terms of ϕ of the form

$$\dot{\gamma}(\tau) = \frac{1}{4\tau^2}\frac{d[\tau^4\phi(\tau)]}{d\tau} \tag{9-5-9}$$

can be obtained, the differentiation being carried out numerically.

It has been emphasized by Oldroyd (1949) that at large values of τ_a separation of the experimental $(1/\eta_{ap}, \tau_a)$ curves for different pipe diameters cannot be used as evidence of anomalous flow at the wall. For example, if the flow ceases to be purely laminar the observed values of $1/\eta_{ap}$ may fall below the expected values $\phi(\tau_a)$, and the experimental values for wide pipes should for these conditions fall below those for narrow pipes. Observations of this nature might be misinterpreted as evidence of a wall effect and, as such, must be treated with care when analyzing measurements at large τ_a.

In the absence of nonlaminar flow a separation of the experimental $(1/\eta_{ap}, \tau_a)$ curves for different values of a can be interpreted as evidence of a wall effect, the flow measurements being analyzed on the basis of the theoretical relation of Eq. (9-5-5). From the experimental $(1/\eta_{ap}, \tau_a)$ curves, corresponding to different values of a, a plot of $1/\eta_{ap}$ against $1/a$ for particular constant values of τ_a can be obtained which should give a straight line with the slope $4\zeta(\tau_a)$ and intercept $\phi(\tau_a)$ on the $1/\eta_{ap}$ axis. Thus the magnitude of a wall effect can, in principle, be assessed and using Eq. (9-5-9) the stress rate of flow characteristic can, in principle, be obtained.

If the form of the relationship between τ and the flow rate $\dot{\gamma}(\tau)$ is known, the relationship between P and Q can be derived in a manner analogous to the derivation of Poiseuille's equation. In the case of a Bingham plastic solid, defined by the rheological equation of state

$$\tau = k + \eta_\theta\dot{\gamma}(\tau)$$

where k is a constant [see Sec. 9-1-3, in particular Eqs. (9-1-24) and (9-1-27)],

$$\dot{\gamma}(\tau) = 0 \qquad \text{for } 0 \leqslant \tau < k$$

and

$$\dot{\gamma}(\tau) = \frac{\tau - k}{\eta_\theta} \qquad \text{for } \tau \geqslant k$$

and there is no anomaly at the wall, streamline flow being assumed to exist. If the stress at the wall is less than the yield value k [see Eq. (7-3-3)], i.e.,

$$\frac{dP}{dz} < \frac{2k}{a}$$

no flow occurs, the apparent fluidity being zero for $\tau_a < k$. From Eq. (9-5-7), if $\tau_a \geqslant k$,

$$\frac{1}{\eta_{ap}} = \phi(\tau_a) = \frac{1}{\eta_\theta}\left[1 - \frac{4}{3}\frac{k}{\tau_a} + \frac{1}{3}\left(\frac{k}{\tau_a}\right)^4\right] \tag{9-5-10}$$

the associated velocity distribution for this condition being obtained from Eq. (9-5-2), with v_a set to zero, in the form

$$v\langle z\rangle(r) = \frac{a(\tau_a - k)^2}{2\eta_\theta \tau_a} \qquad \text{for } 0 \leqslant r \leqslant \frac{ak}{\tau_a}$$

and
$$v\langle z\rangle(r) = \frac{(a^2 - r^2)\tau_a - 2a(a-r)k}{2a\eta_\theta} \qquad \text{for } \frac{ak}{\tau_a} \leqslant r \leqslant a$$

from which it is evident that, within a distance $r_y = ak/\tau_a$ from the axis, the material behaves as an elastic solid without flowing and moves through the tube as a rigid body (the so-called plug flow). In terms of Q and P, Eq. (9-5-10) can be expressed in the form

$$Q = \frac{\pi a^4 P}{8\eta_\theta L}\left[1 - \frac{4}{3}\frac{P_y}{P} + \frac{1}{3}\left(\frac{P_y}{P}\right)^4\right] \tag{9-5-11}$$

where P_y is the pressure corresponding to the yield value; Eq. (9-5-11) was derived by Buckingham (1921) and Reiner (1930).

For a further discussion of pipe-flow measurements with nonnewtonian fluids the reader is referred to the work of Toms (1949, 1958).

9-5-2 Flow between Rotating Coaxial Cylinders

The flow of nonnewtonian fluids between rotating coaxial cylinders may give rise to anomalous flow at the walls of a similar nature to that discussed for the flow of nonnewtonian fluids in a capillary, the effective velocities of slip being in general different at the inner and outer cylinders. Let the effective velocities of slip be $v_s(\tau_2)$ and $v_s(\tau_1)$ at the walls of the inner and outer cylinders respectively. The boundary conditions with slippage are $[r = r_2,\ v\langle\theta\rangle = v_s(\tau_2)]$ and $[r = r_1,\ v\langle\theta\rangle = r_1\omega_1 - v_s(\tau_1)]$.

In general, for the flow of nonnewtonian fluids, a relationship of the form

$$\frac{dv\langle\theta\rangle}{dr} - \frac{v\langle\theta\rangle}{r} = \dot{\gamma}(\tau) \qquad (\tau \equiv T\langle r\theta\rangle = T\langle\theta r\rangle) \tag{9-5-12}$$

will be assumed to exist between the shear stress τ and the rate of shear $\dot{\gamma}(\tau)$ (see Sec. 7-3-3).

For this condition, and from Eq. (7-3-22) for steady conditions of flow, the relation

$$r\frac{d\omega\langle\theta\rangle}{dr} = \dot{\gamma}\left(\frac{\Gamma_\eta(\infty)}{2\pi r^2 H}\right)$$

is obtained for the velocity distribution. Hence

$$\omega\langle\theta\rangle = \frac{v_s(\tau_2)}{r_2} + \int_{r_2}^{r} \dot{\gamma}\left(\frac{\Gamma_\eta(\infty)}{2\pi\bar{r}^2 H}\right)\frac{d\bar{r}}{\bar{r}} = \frac{v_s(\tau_2)}{r_2} + \int_{\tau(r)}^{\tau_2} \frac{\dot{\gamma}(\bar{\tau})}{2\bar{\tau}}\,d\bar{\tau} \tag{9-5-13}$$

Substituting $r = r_1$ gives an experimentally observable relationship between ω_1 and $\Gamma_\eta(\infty)$ in terms of $\dot{\gamma}$, of the form

$$\omega_1 = \omega_s(\tau_2) + \omega_s(\tau_1) + \int_{r_2}^{r_1} \dot{\gamma}\left(\frac{\Gamma_\eta(\infty)}{2\pi r^2 H}\right)\frac{dr}{r}$$

or
$$\omega_1 = \omega_s(\tau_2) + \omega_s(\tau_1) + \int_{\tau_1}^{\tau_2} \frac{\dot{\gamma}(\tau)}{2\tau}\, d\tau$$

where $r_2\omega_s(\tau_2) = v_s(\tau_2)$ and $r_1\omega_s(\tau_1) = v_s(\tau_1)$.

Anomalous flow at the walls can be detected by way of a series of experiments in which ω_1 is varied for various values of the cylinder gap width, r_1 being maintained constant. In this way it is possible to plot ω_1 against τ_2 for different selected values of τ_1, the latter being obtained using the various $(\omega_1, \Gamma_\eta(\infty))$ curves for each value of r_2. In the absence of a wall effect, ω_1 should tend to zero as τ_2 tends to τ_1; if this is not the case, then half the extrapolated value of ω_1 corresponding to $\tau_2 = \tau_1$ is a measure of $\omega_s(\tau_1)$. For the condition that there is no anomalous flow at the wall, the curves for constant τ_1 represent the functional relationship

$$\omega_1(\tau_2) = \int^{\tau_2} \frac{\dot{\gamma}(\tau)}{2\tau}\, d\tau + \text{constant}$$

from which $\dot{\gamma}(\tau)$ is obtained by differentiation in the form

$$\dot{\gamma}(\tau) = 2\tau\frac{d\omega_1(\tau)}{d\tau}$$

For a Bingham plastic, defined by the rheological equation of state

$$\tau = k + \eta_\theta\dot{\gamma} \tag{9-5-14}$$

where the yield stress k and the parameter η_θ are both constants, no flow will occur if $\tau_2(\infty) < k$. If $\tau_2(\infty) \geqslant k \geqslant \tau_1(\infty)$, so that the yield stress is attained on a surface at $r = r_y$ lying within the annular gap, flow will occur in that part of the material for which $r_2 \leqslant r \leqslant r_y$ and $\tau_2(\infty) \geqslant \tau(\infty) \geqslant k$ but the material in the region $r_y < r \leqslant r_1$ remains an elastic solid and moves as a rigid body attached to the outer cylinder.

Thus, for a Bingham plastic,

$$\dot{\gamma}(\tau) = \begin{cases} 0 & \text{for } 0 \leqslant \tau \leqslant k \\ \tau - k/\eta_\theta & \text{for } \tau \geqslant k \end{cases}$$

and there is no anomaly at the wall, streamline flow being assumed to exist; that is, $\omega_s = 0$. For the condition $\tau_2(\infty) \geqslant k$, the relation of Eq. (9-5-13), for $\omega_s = 0$, gives

$$\omega\langle\theta\rangle = \frac{\Gamma_\eta(\infty)}{4\pi\eta_\theta H}\left(\frac{1}{r_2^2} - \frac{1}{r^2}\right) - \frac{k}{\eta_\theta}\ln\left(\frac{r}{r_2}\right) \qquad \text{for } r_2 \leqslant r \leqslant r_y \tag{9-5-15}$$

where r_y is defined by

$$k = \frac{\Gamma_\eta(\infty)}{2\pi r_y^2 H}$$

and
$$\omega\langle\theta\rangle = \omega\langle\theta\rangle(r_y) \qquad \text{for } r \geqslant r_y$$

the latter equation implying rigid-body rotation in a region adjacent to the wall of the outer cylinder, together with continuity of velocity at the yield surface $r = r_y$. Two cases of interest can be distinguished:

1. $r_1 > r_y \geqslant r_2$: for which

$$\omega_1 = \omega\langle\theta\rangle(r_y)$$

such that Eq. (9-5-15) takes the form

$$\omega_1 = \frac{\Gamma_\eta(\infty)}{4\pi\eta_\theta r_2^2 H} - \frac{k}{2\eta_\theta}\left[1 + \ln\left(\frac{\Gamma_\eta(\infty)}{2\pi r_2^2 Hk}\right)\right]$$

$$(2\pi r_2^2 Hk \leqslant \Gamma_\eta(\infty) \leqslant 2\pi r_1^2 Hk) \qquad (9\text{-}5\text{-}16)$$

2. $r_1 \leqslant r_y$: for which the velocity distribution is defined by Eq. (9-5-15) everywhere in the annular gap, such that

$$\omega_1 = \frac{\Gamma_\eta(\infty)}{4\pi\eta_\theta H}\left(\frac{1}{r_2^2} - \frac{1}{r_1^2}\right) - \frac{k}{\eta_\theta}\ln\left(\frac{r_1}{r_2}\right)(\Gamma_\eta(\infty) \geqslant 2\pi r_1^2 Hk) \qquad (9\text{-}5\text{-}17)$$

For the limiting condition $2\pi r_2^2 Hk > \Gamma_\eta(\infty)$ the outer cylinder is unable to rotate, that is, $\omega_1 = 0$.

The material will be sheared throughout when $r_1 = r_y$ in Eq. (9-5-16), a condition which is only satisfied when the angular velocity of the outer cylinder attains the critical value

$$\omega_1 = \frac{k}{2\eta_\theta}\left[\frac{r_1^2}{r_2^2} - 1 - \ln\left(\frac{r_1^2}{r_2^2}\right)\right] \qquad \text{for } r_1 = r_y \qquad (9\text{-}5\text{-}18)$$

it being noted in this connection that for the flow of a Bingham plastic in a pipe the solid plug never disappears for a finite Q.

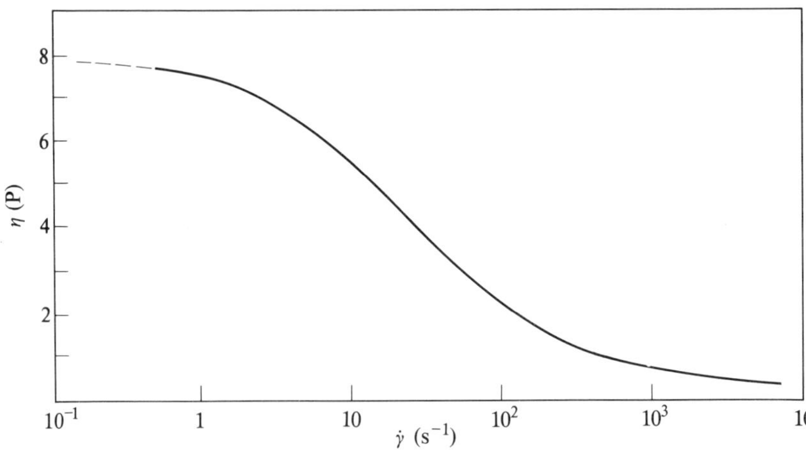

Figure 9-1 Variation of $\eta(\dot{\gamma})$ with $\dot{\gamma}$ of a mixture of polymethylmethacrylate and pyridine. (*Oldroyd, Strawbridge, and Toms, 1951.*)

In Fig. 9-1 is shown the variation of the viscosity $\eta(\dot{\gamma})$ with $\dot{\gamma}$ of a mixture of polymethylmethacrylate and pyridine, obtained by Oldroyd, Strawbridge, and Toms (1951) from measurements using a coaxial cylinder viscometer. In these measurements end effects were eliminated by way of two experiments in which the annular gap was filled to different heights H' and H''. It was assumed in these experiments that the annular gap was sufficiently small so that setting $r = r_a = \sqrt{r_1 r_2}$ the viscosity η, the rate of shear $\dot{\gamma}$, and the shear stress $T\langle r\theta\rangle$ could be taken to be defined by the relations

$$T\langle r\theta\rangle = \frac{\Gamma''_\eta(\omega_1) - \Gamma'_\eta(\omega_1)}{2\pi r_1 r_2 (H'' - H')} \equiv \tau \tag{9-5-19}$$

$$\dot{\gamma} = \frac{2r_1 r_2}{r_1^2 - r_2^2}\omega_1 = h_{r_a}\omega_1 \qquad \text{for } r_a = \sqrt{r_1 r_2} \tag{9-5-20}$$

$$\begin{aligned}\theta''_2(\infty) - \theta'_2(\infty) &= \eta(\dot{\gamma})\frac{4\pi r_1^2 r_2^2 (H'' - H')}{\Gamma_0(r_1^2 - r_2^2)}\omega_1 \\ &= \eta(\xi'' - \xi')\omega_1\end{aligned} \tag{9-5-21}$$

the rate of shear $\dot{\gamma}$ and the shear stress τ being evaluated at r_a $(=\sqrt{r_1 r_2})$. Equations (9-5-19), (9-5-20), and (9-5-21) follow directly from Eqs. (7-3-16), (7-3-21), (7-3-22), and (7-3-25) for the particular conditions of immediate interest. Since it is evident from Fig. 9-1 that the liquid exhibits marked nonnewtonian flow behavior, $\dot{\gamma}$ can only be regarded as a first-order approximation to the rate of shear evaluated at r_a $(=\sqrt{r_1 r_2})$.

In Fig. 9-2 is shown the variation of the shear stress τ_2 with $\dot{\gamma}_2$, obtained by Billington (1967) using a coaxial cylinder viscometer. The liquid was obtained by adding a given concentration c of aluminum laurate to a newtonian oil of viscosity 3.5 cP. In these measurements, end effects were eliminated by using coaxial cylinders which were sealed by mercury at their lower edges (Billington, 1965). The effect of varying r_2 and keeping r_1 = constant, i.e., varying the cylinder gap width $(r_1 - r_2)$ over a limited range, is also shown in Fig. 9-2. For a given value of the shear stress the three values of ω_1, corresponding to the three values of r_2, are seen from Fig. 9-2 to correlate with the radius of the inner cylinder according to the relation

$$\omega_1 = L + Kr_2^2 \tag{9-5-22}$$

where L and K are constants for the given value of the shear stress. This empirical correlation must satisfy the limiting condition that, for $r_1 = r_2$, $\omega_1 = 0$ such that $L_2 = -K_2 r_1^2$ which leads to the relation

$$L_2 = \tfrac{1}{2}h_2\omega_1 = \tfrac{1}{2}\dot{\gamma}_2(\infty) \tag{9-5-23}$$

where h_2 is given by Eq. (7-3-23). The correlation by way of $\dot{\gamma} = \dot{\gamma}_2(\infty)$, defined as for the flow of a newtonian liquid, does not imply that $\dot{\gamma}_2(\infty)$ can be identified as the rate of shear at the wall of the inner cylinder, the use of Eq. (9-5-23) being merely a convenient way of expressing the change in $\tau_2(\infty)$ with ω_1 for the three cylinder gap widths. In general, this type of measurement cannot be used to

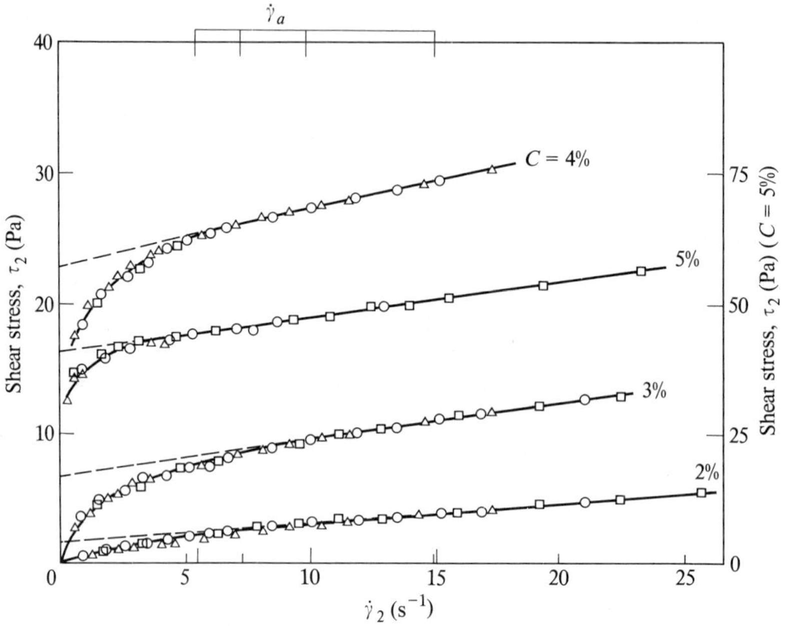

Figure 9-2 Variation of the shear stress τ_2 with $\dot{\gamma}_2$, for three values of r_2: △, 19.0 mm; ○, 20.6 mm; □, 22.2 mm. (*Billington, 1967.*)

establish the form of the rate of shear, as measured relative to the wall of the inner cylinder, for the flow of nonnewtonian liquids. To establish the significance of $\dot{\gamma}_2(\infty)$ in relation to the rate of shear at the wall of the inner cylinder would require direct measurement of the velocity at various points across the gap between the viscometer cylinders. The full-line curves shown in Fig. 9-2 have been calculated from the constitutive equations (9-1-18) where

$$p_0 = \eta_0 \dot{\gamma}_0 = \frac{\eta_a \dot{\gamma}_0}{1 + \dot{\gamma}_a / \dot{\gamma}_0} \tag{9-5-24}$$

and p_0, p_{e0}, γ_0, $\dot{\gamma}_a$, η_0, and η_a are material constants evaluated from the experimental $(\tau_2(\infty), \dot{\gamma}_2(\infty))$ curves. The $(\tau_2(\infty), \dot{\gamma}_2(\infty))$ measurements shown in Fig. 9-2 correlate with the full-line curves calculated from Eqs. (9-1-18) to within ± 2 percent. The useful range of the coaxial cylinder viscometer is limited by the observation that some nonnewtonian liquids are ejected from the gap between the cylinders at a critical rate of shear.

9-5-3 Time Dependence and the Phenomenon of Thixotropy

When considering the characterization of the mechanical response of real materials, a distinction must be made between materials which exhibit a time-independent response to an applied stress and those materials whose mechanical response is time dependent. The individual elements of the internal, i.e., the submacroscopic, structure of certain types of nonnewtonian liquid can be linked to produce a continuous structure throughout the bulk of the liquid, an example being the continuous structure developed in certain dilute carbon suspensions (McDowel and Usher, 1931). Materials of this type generally exhibit the behavior of a weak solid for small flow rates, the simplest example being the ideal Bingham plastic considered in Secs. 9-1-3 and 9-5-1. McDowel and Usher (1931) observed that the electrical conductivity of dilute carbon suspensions decreased with continuous shearing, this behavior being attributed to the breakdown of the continuously linked structure. The breakdown of the linked structure, which is initially continuous throughout the material, is also observed to produce a decrease in the resistance to flow of the carbon suspension which is dependent upon the period of time for which the fluid has been subjected to an applied shear stress. This time-dependent flow behavior is referred to as thixotropy. Time-dependent flow behavior, i.e., the phenomenon of thixotropy, is characteristic of many nonnewtonian liquids, the flow properties of which exhibit an isothermal, reversible time-dependent decrease of the resistance to flow accompanying the application of a shear stress. This definition of the phenomenon of thixotropy is here restricted to time-dependent systems for which the decrease in the resistance to flow is due solely to the shear-induced breakdown of some form of submacroscopic internal structure which is initially continuous throughout the material.

For liquids exhibiting a time-independent mechanical response it is sufficient to observe their behavior by way of steady-state measurements. Near-steady-state measurements of the flow properties of liquids which exhibit thixotropic behavior (Green and Weltman, 1943; Thornton and Rae, 1953; Goodeve, 1938, 1939a, 1939b, 1949; Goodeve and Whitfield, 1938; Arnold and Goodeve, 1940) do not take direct account of the time dependence of the system, though Weltman (1943) and Thornton (1953) have attempted to develop methods to relate breakdown to time of shearing. In the case of liquids exhibiting a time-dependent response it is essential that their flow behavior be observed by way of experiments in which the applied stress is programmed to change in a specified way with the time of deformation. The measurements of Oldroyd, Strawbridge, and Toms (1951), Markovitz et al. (1952), and Billington and Huxley (1965), in which the outer cylinder of a coaxial cylinder viscometer is made to execute sinusoidal oscillations, are typical examples of experiments in which the applied stress is programmed to change in a specified way with the time of deformation. For this type of unsteady motion it is important to note that the amplitude of oscillation of the inner cylinder, for a given frequency of oscillation of the outer cylinder, is independent of the time of shearing, and in this sense does not distinguish between liquids whose flow properties are time independent and those liquids which exhibit a time-dependent change in the resistance to flow arising from a shear-induced breakdown of the submacroscopic internal structure of the liquid.

The essentially transient aspect of thixotropy, emphasized by Billington (1960a, 1960b) implies that the phenomenon of thixotropy should be studied by way of transient measurements, as distinct from either steady-state or oscillatory-type flow. A coaxial cylinder viscometer has been described by Billington (1965) in which the test liquid can be subjected to impulsive loading, thus giving transient flow. Using this instrument it has been shown (Billington, 1967) that thixotropic liquids can exist in at least two distinguishable, limiting, mechanical states. The initial, or upper, instantaneous state has been identified as corresponding to maximum submacroscopic internal structure and the final, or lower, steady state to minimum submacroscopic internal structure, the transition from the initial to the final state being identified as a transient phenomenon.

9-5-4 Transient Flow Properties of Some Typical Liquids

The equations, basic to a description of the flow properties of a liquid confined between rotating coaxial cylinders, have been discussed in Secs. 7-4-3 and 7-5-4 for the particular condition that fluid inertia can be neglected. For transient measurements with fluid inertia neglected, the instantaneous value at some time t of shearing of the shear stress $T\langle\theta r\rangle$ $[\equiv\tau_2(t)]$ measured at the wall of the inner cylinder of radius r_2 can, by reference to Eq. (7-3-22), be expressed in the form

$$T\langle r\theta\rangle \equiv \tau_2(t) = \frac{\Gamma_\eta(t)}{2\pi r_2^2 H} = \frac{h_2}{\xi}\theta_2(t) \tag{9-5-25}$$

where h_2 and ξ are defined by Eqs. (7-3-23) and (7-3-27) respectively and $\theta_2(t)$ is

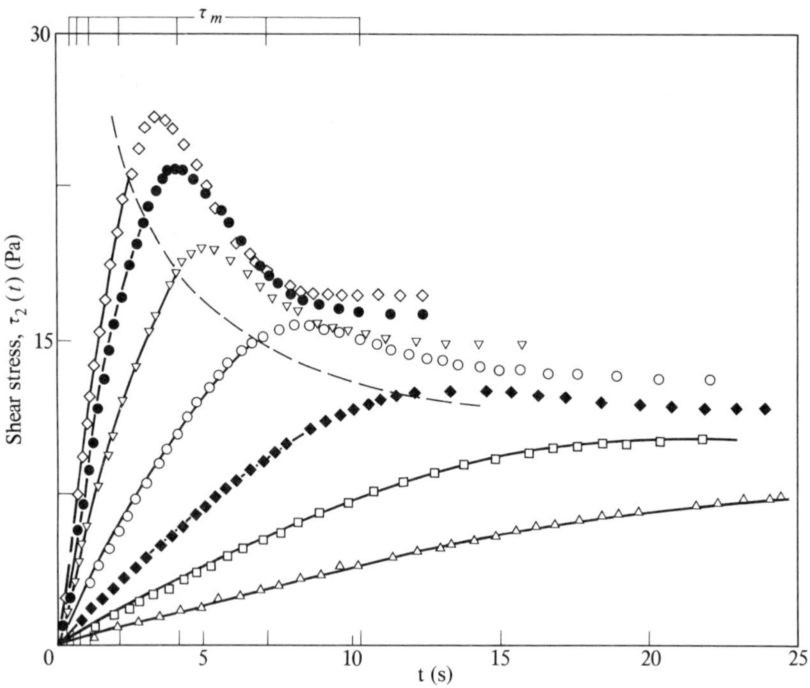

Figure 9-3 Typical examples of the variation of the shear stress $\tau_2(t)$ with the time of shearing for $r_2 = 20.6$ mm and several values of ω_1: △, 0.06 rad s^{-1}; □, 0.11 rad s^{-1}; ◆, 0.21 rad s^{-1}; ○, 0.42 rad s^{-1}; ▽, 0.84 rad s^{-1}; ●, 1.26 rad s^{-1}; ◇, 1.68 rad s^{-1}. (*Billington, 1967.*)

the instantaneous angular displacement of the inner cylinder. As for the static measurements described above, the instantaneous value of the shear stress has been observed to correlate with the effective rate of shear

$$\dot{\gamma}_2(t) = \frac{2r_1^2}{r_1^2 - r_2^2}\left(\omega_1 - \frac{d\theta_2}{dt}\right) = h_2[\omega_1 - \omega_2(t)] \tag{9-5-26}$$

where the instantaneous value of the angular velocity $\omega_2(t)$ of the inner cylinder is obtained from the experimental (θ_2, t) curves. It is to be noted that the individual dynamic $(\tau_2(t), \dot{\gamma}_2(t))$ curves, one for each value of ω_1 and a given value of the cylinder gap width, compound together to give a single dynamic $(\tau_2(t), \dot{\gamma}_2(t))$ curve which for each value of the concentration c is indistinguishable from that shown in Fig. 9-2. The observation that the steady-state $(\tau_2(\infty), \dot{\gamma}_2(\infty))$ and dynamic $(\tau_2(t), \dot{\gamma}_2(t))$ curves are indistinguishable is taken as evidence that the fluid inertia can be neglected for the investigated range of the effective rate of shear. That is, the essential experimental conditions discussed in Secs. 7-4-3 and 7-5-4 are satisfied by this particular liquid for transient conditions of flow between rotating coaxial cylinders.

The viscoelastic liquid obtained by adding a given concentration of aluminum laurate to a newtonian oil, referred to above as an example of the steady-state flow of a nonnewtonian liquid, can also be obtained in a form which exhibits the phenomenon of thixotropy. Some typical (τ_2, t) curves for this liquid are shown in Fig. 9-3. It is evident that only part of the transient motion, represented by the full-line curves which extend from $t = 0$ to $t = t_b$, can be correlated with $\omega_2(t)$ by way of Eq. (9-5-26). In Figs. 9-4 and 9-5 is shown the variation of $\tau_2(t)$ with $\dot{\gamma}_2(t)$ for

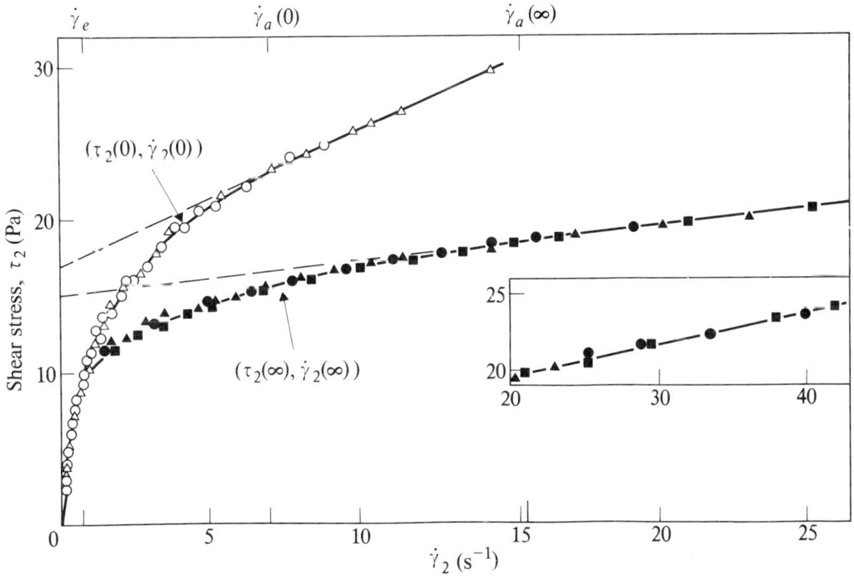

Figure 9-4 Variation of the shear stress τ_2 with $\dot{\gamma}_2$ for various values of r_2: ○, ●, 22.2 mm; △, ▲, 19.0 mm; ■, 20.6 mm. (*Billington, 1967.*)

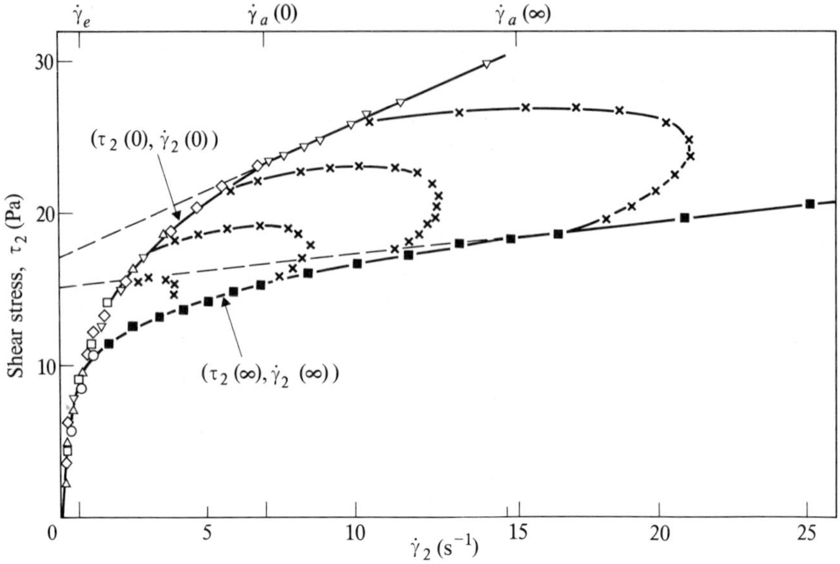

Figure 9-5 Variation of the shear stress τ_2 for the liquid with $\dot{\gamma}_2$ for $r_2 = 20.6$ mm and several values of ω_1: ○, 0.21 rad s^{-1}; □, 0.63 rad s^{-1}; △, 1.05 rad s^{-1}; ◇, 1.68 rad s^{-1}; ▽, 2.1 rad s^{-1}; ×, $\dot{\gamma}(t)$ for $t > t_b$. *(Billington, 1967.)*

several values of ω_1 and three values of r_2 for t in the range $(0 \leqslant t \leqslant t_b)$. The full-line curves shown in Figs. 9-4 and 9-5 are identical, and it therefore follows that the individual $(\tau_2(t), \dot{\gamma}_2(t))$ curves corresponding to each value of ω_1 and the different r_2 compound together, thus giving a single $(\tau_2(t), \dot{\gamma}_2(t))$ curve characteristic of the flow behavior of this particular liquid in the time interval $(0 \leqslant t \leqslant t_b)$ during which the material, i.e., the viscous property of the liquid, is independent of the time of shearing. For the range $(0 \leqslant t \leqslant t_b)$ the characteristic $(\tau_2(t), \dot{\gamma}_2(t))$ curve can be described by the relations

$$\tau_2(t) = \begin{cases} p_0(0)\ln\left(1 + \dfrac{\dot{\gamma}_2(t)}{\dot{\gamma}_0(0)}\right) & \text{for } 0 \leqslant \dot{\gamma}_2(t) \leqslant \dot{\gamma}_a(0) \\ p_{e0} + \eta_a(0)\dot{\gamma}_2(t) & \text{for } \dot{\gamma}_2(t) \geqslant \dot{\gamma}_a(0) \end{cases} \tag{9-5-27}$$

as for the steady-state conditions of flow considered in Sec. 9-5-2 for the same liquid in its nonthixotropic state. Equation (9-5-27) follows directly from the constitutive equation (9-1-18). The value of $\dot{\gamma}_2(t)$ having been obtained, and hence the flow parameters $p_0(0)$, $\dot{\gamma}_0(0)$, $\dot{\gamma}_a(0)$, and $\eta_a(0)$ by graphical methods, the correlation between experiment and the proposed scalar equation (9-5-27) of state can be further verified by calculating the value of $\dot{\gamma}_2(t)$ for a given value of the stress $\tau_2(t)$. For the transient motion, as discussed in Secs. 7-4-3 and 7-5-4, in the range $[0 \leqslant \dot{\gamma}_2(t) \leqslant \dot{\gamma}_a(0)]$ the relation

$$\dot{\gamma}_2(t) = \dot{\gamma}_2(\infty)\frac{1 - \exp(-t/\tau_m)}{1 + [\dot{\gamma}_2(\infty)/\dot{\gamma}_0(0)]\exp(-t/\tau_m)} \tag{9-5-28}$$

where

$$\tau_m = \frac{p_0(0)}{\dot{\gamma}_0(0)}\left[\frac{1}{1+(\dot{\gamma}_2(\infty)/\dot{\gamma}_0(0))}\right]\xi = \eta_e(0)\xi \tag{9-5-29}$$

can be obtained, ξ being defined by Eq. (7-3-27). For given values of t, the required value of $\dot{\gamma}_2(t)$ can be calculated from Eq. (9-5-28), the values of the parameters $p_0(0)$ and $\dot{\gamma}_0(0)$ having been obtained by graphical methods. Although the parameters $p_0(0)$ and $\dot{\gamma}_0(0)$ are determined by graphical methods it is the actual experimental values of the stress $\tau_2(t)$ which are shown in Figs. 9-4 and 9-5, the corresponding values of $\dot{\gamma}_2(t)$ having been calculated from Eq. (9-5-28) for given values of t and the graphically determined values of the parameters $p_0(0)$ and $\dot{\gamma}_0(0)$. For $\dot{\gamma}_2(t) > \dot{\gamma}_a(0)$ the effective rate of shear can be calculated from the relation

$$\dot{\gamma}_2(t) = \dot{\gamma}_2(\infty) - [\dot{\gamma}_2(\infty) - \dot{\gamma}_a(0)]\exp\left(-\frac{t-t_a}{\tau_a}\right) \tag{9-5-30}$$

where

$$\tau_a = \eta_a(0)\xi \tag{9-5-31}$$

the constant t_a being the time corresponding to which $\dot{\gamma}_2(t) = \dot{\gamma}_a(0)$ and ξ is defined by Eq. (7-3-27). Since $\theta_2(t)$ and t are the measurements basic to the experimental technique, it is of interest to calculate the (θ_2, t) curves or, since h_2/ξ can be determined to a high degree of accuracy, the (τ_2, t) curves. By substituting for $\dot{\gamma}_2(t)$ from Eqs. (9-5-28) and (9-5-30) into the constitutive relations of Eq. (9-5-27) the (τ_2, t) curves can be calculated for the range $(0 \leqslant t \leqslant t_b)$. The full-line curves shown in Fig. 9-3 have been calculated in this way and are seen to be in good agreement with experiment.

For $t \geqslant t_b$ the effective rate of shear $\dot{\gamma}_2(t)$ increases more rapidly than is consistent with the correlation over the initial range of time.

The change in flow behavior in the region of t_b is attributed to the onset of thixotropic behavior of the liquid.

With continued shearing the inner cylinder comes momentarily to rest and then begins to rotate in the opposite direction to that of the outer cylinder (see Sec. 7-4-2). From Sec. 7-4-2 it follows that for the contrarotation of the two cylinders, replacement of $\omega_2(t)$ by $-\omega_2(t)$ gives the effective rate of shear $\dot{\gamma}_2(t)$ for $t \geqslant t_b$. The transient change in $\tau_2(t)$ with $\dot{\gamma}_2(t)$ for $t \geqslant t_b$ is shown in Fig. 9-5 for several values of ω_1 and $r_2 = 20.6$ mm. These results are typical for all three values of r_2. It is to be noted that these $(\tau_2(t), \dot{\gamma}_2(t))$ curves for $t \geqslant t_b$ represent a true time-dependent change in the material properties of the liquid and are in accord with the generally accepted definition of thixotropic behavior.

After a long time of shearing the inner cylinder comes to rest, corresponding to the steady-state flow of the liquid. The variation of $\tau_2(\infty)$ with $\dot{\gamma}_2(\infty)$ is shown in Figs. 9-4 and 9-5 for the three values of r_2, the results being represented by a single curve as for the steady-state measurements shown in Fig. 9-2 for the same liquid in its nonthixotropic state.

Similar results have been obtained for a markedly different liquid consisting of a colloidal suspension of graphite particles dispersed in a newtonian-type liquid.

It is evident from Figs. 9-4 and 9-5 that this type of liquid can exist in at least two distinguishable rheological states. The rheological state prior to the onset of thixotropy is characterized by the $(\tau_2(0), \dot{\gamma}_2(0))$ curve and by the $(\tau_2(\infty), \dot{\gamma}_2(\infty))$ curve for the final steady state.

For the viscoelastic liquid, the time interval required to restore the flow properties, in particular to recreate the maximum observed stress for a given ω_1, is about 30 s for all $\dot{\gamma}_2$ compared with about 30 min for the colloidal suspension of graphite particles.

With regard to the phenomenon of thixotropy, it is of interest to note the marked, formal similarity between the shear stress–rate of strain hysteresis loops shown in Fig. 9-5 and the stress-strain loops characteristic of the dynamic plastic response of many solids as is shown in Chap. 13.

9-5-5 Oscillatory Flow between Rotating Coaxial Cylinders

The forced oscillatory flow of a liquid confined between coaxial cylinders is of particular interest in relation to the characterization of a liquid which exhibits viscoelastic behavior. The equations, basic to a description of the flow properties of a liquid confined between oscillating coaxial cylinders, have been discussed in Secs. 7-4-1 and 7-4-2. In Fig. 9-6 is shown the variation of the amplitude ratio ψ $(=\theta_2/\theta_1)$ with the frequency f of a mixture of polymethylmethacrylate and pyridine, obtained by Oldroyd, Strawbridge, and Toms (1951), from measurements using a coaxial cylinder viscometer. The cylinder and effects were neglected in these particular experiments, the full length of the annulus being filled with

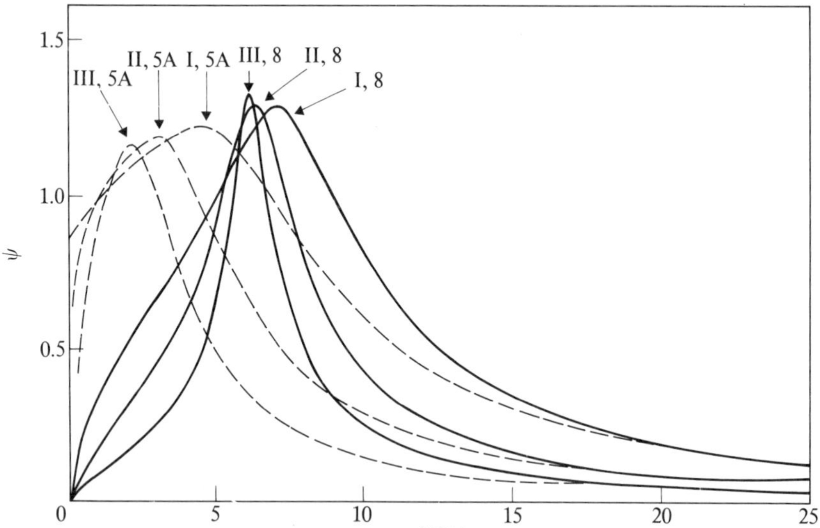

Figure 9-6 Variation of the amplitude ratio $\psi = \theta_2/\theta_1$ with the frequency f of a mixture of polymethylmethacrylate and pyridine. (*Oldroyd, Strawbridge, and Toms, 1951.*)

liquid. In each case the maximum observed value of the amplitude ratio $\psi > 1$ and occurs at a frequency $f > f_N$ where f_N is the natural undamped frequency of oscillation of the inner cylinder. This behavior is in marked contrast to that for newtonian liquids, as can be seen from Fig. 7-3. For newtonian-type liquids the maximum value of the amplitude ratio ψ is less than unity for all known measurements and is seen from Fig. 7-3 to occur at the natural undamped frequency f_N of oscillation of the inner cylinder. Observations of oscillatory flow with other polymer solutions have been described by Toms and Strawbridge (1953) using a coaxial cylinder viscometer, their results being formally similar to those shown in Fig. 9-6.

Oldroyd (1951) initially considered the observed (ψ, f) curves of Fig. 9-6 in terms of the scalar constitutive equation

$$T\langle\theta z\rangle + \tau_0 \frac{\partial T\langle\theta z\rangle}{\partial t} = \eta_0 \left(\dot{\gamma}\langle\theta z\rangle + \lambda_0 \frac{\partial \dot{\gamma}\langle\theta z\rangle}{\partial t} \right) \tag{9-5-32}$$

which can be obtained directly from Eq. (9-2-12) for $\tau_1 = \cdots = \tau_N = 0$ and $\lambda_1 = \cdots = \lambda_M = 0$. In Eq. (9-5-32) the shear stress $T\langle\theta z\rangle$ is defined by Eq. (9-5-19) with $\Gamma'_\eta(\omega_1)$ and H' both set to zero and the effective rate of shear $\dot{\gamma}$ is defined by Eq. (9-5-20). To enable theoretical predictions to be made, the amplitude of oscillation is supposed small, so that the amplitude ratio ψ is independent of the absolute value of the amplitude and η_0 can be identified with the viscosity observed in steady rotation at very low rates of shear. The experimental (ψ, f) curves shown in Fig. 9-6 differ only slightly from those predicted for an idealized material and shown in Fig. 9-7. Oldroyd (1951) therefore

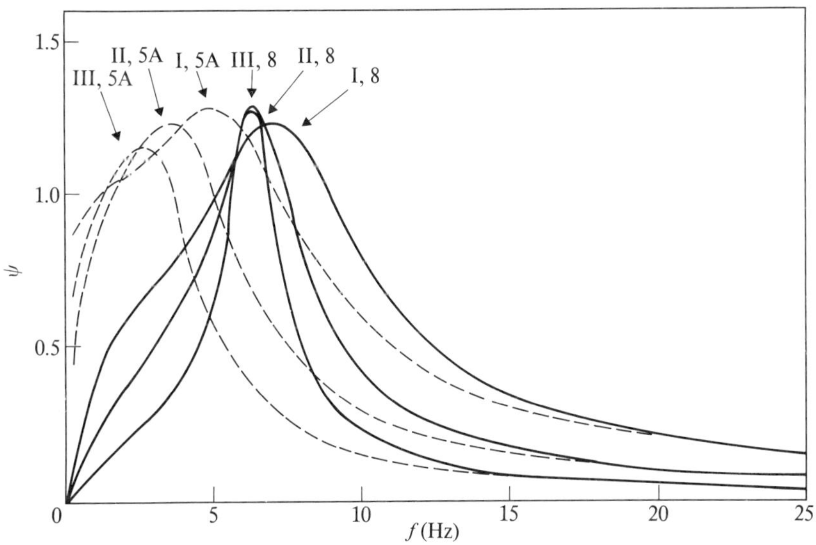

Figure 9-7 Predicted relations between ψ and f, in the case of an idealized elasticoviscous liquid. *(Oldroyd, Strawbridge, and Toms, 1951.)*

concludes that, to a first approximation, the mixture of polymethylmethacrylate and pyridine can be characterized by the scalar constitutive equation (9-5-32), subject to the limitation of small rates of shear, with relaxation times $\tau_0 = 0.0655$ s and $\lambda_0 = 0.015$ s.

9-5-6 Flow between a Cone and Plate

Although the use of cone and plate flow in viscometry is relatively recent, it is now a well-established method (see, for example, Mooney and Ewart, 1934; Weissenberg, 1949; Higginbotham, 1950; McKennell, 1954; Markovitz, Elyash, Padden, and De Witt, 1955; Kepes, 1956; Adams and Lodge, 1964). The equations, basic to a description of the flow properties of a viscous liquid in cone and plate geometry, have been discussed in Sec. 7-3-4. In Fig. 9-8 is shown the variation of the viscosity $\eta(\dot{\gamma})$ with $\dot{\gamma}$ for a 13% solution of polyisobutylene in decalin (Markovitz, Elyash, Padden, and De Witt, 1955). The variation of $\eta(\dot{\gamma})$ with $\dot{\gamma}$ is shown for two different values of the cone angle $\Delta\theta_c = \frac{1}{2}$ and 2°. It is evident from Fig. 9-8 that the two $(\eta, \dot{\gamma})$ curves, one for each value of $\Delta\theta_c$, are effectively indistinguishable, the existence of a single curve being in accord with the theory discussed in Sec. 7-3-4.

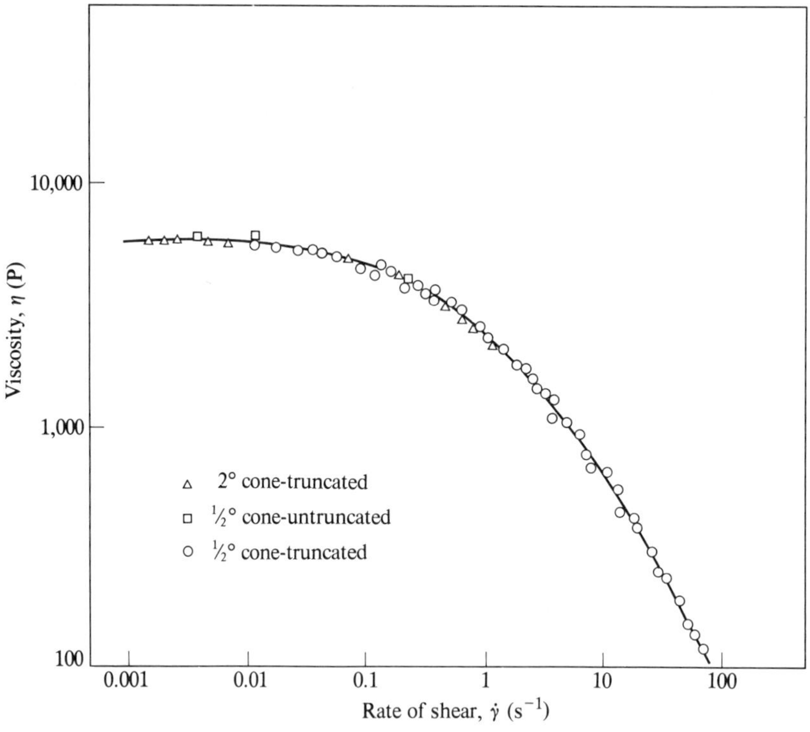

Figure 9-8 Variation of the viscosity $\eta(\dot{\gamma})$ with $\dot{\gamma}$ for a 13 percent solution of polyisobutylene in decalin. (*Markovitz, Elyash, Padden, and De Witt, 1955.*)

From Fig. 9-8, it is evident that the flow behavior is markedly nonnewtonian. The useful range of the cone and plate viscometer is limited by the observation that some nonnewtonian liquids are ejected from the gap at a critical rate of shear which, for the same liquid, is often much lower than for other geometries, such as flow between rotating coaxial cylinders.

9-5-7 Velocity Profile Measurements

The determination of the shape of the velocity profile, though experimentally difficult, is of considerable interest in relation to characterizing the flow properties of nonnewtonian liquids.

Velocity measurements can be made in large-diameter pipes using a free-running propeller system. The overall size of the propeller is limited by the need to produce as small a perturbation of the flow pattern as possible. The method of use is to measure the number of revolutions per second of the propeller at various points across the diameter of the pipe. The velocity at any point on the radius of the pipe is obtained from an independent calibration of the angular velocity ω of the propeller for a given constant speed V at which the propeller is drawn through a stationary liquid at a fixed depth of immersion. Velocity measurements using a small free-running propeller to determine the velocity profiles of a liquid flowing in a large-diameter pipe has been described by Billington, Jones, and Wyllie (1962). A set of typical velocity profiles are shown in the $(V, r/a)$ curves of Fig. 9-9 where V is the fluid velocity at some point r on the radius of a pipe of diameter $2a$. In these experiments the ratio of pipe diameter to propeller diameter was approximately 20 to 1. In Fig. 9-9 the full-line curve for the newtonian-type liquid, which represents a parabolic velocity profile calculated from Eq. (7-3-5), is seen to be in good agreement with the results of experiment. This apparatus has also been used for measurements in nonnewtonian liquids, some typical profiles being shown in Fig. 9-9.

The measurement of the point-to-point velocity in flow may be made in a coaxial cylinder viscometer using a form of the hot-wire anemometer described by Richardson and Tyler (1933). The essential condition, as for pipe-flow measurements, is that the ratio of the wire diameter to cylinder gap width must be as small as possible in order to minimize the disturbance of the flow pattern. Measurements of the velocity distribution in a fluid confined between concentric cylinders when the outer one is rotating and the inner one is at rest have also been described by Taylor (1936). The shape of the velocity profile across the gap of a coaxial cylinder viscometer can also be determined from the pattern produced by a thin streak of dyed liquid initially introduced into the stationary liquid and lying along a radius. The profile can be deduced from a series of photographs.

For the flow of certain types of liquid suspensions, the motion of the particles can be photographed under time exposure. The shape of the velocity profile can be deduced from the length of the streaks on the photograph. It is required to observe the light scattered by the solid phase and not the transmitted light. This is achieved by use of a narrow beam of light.

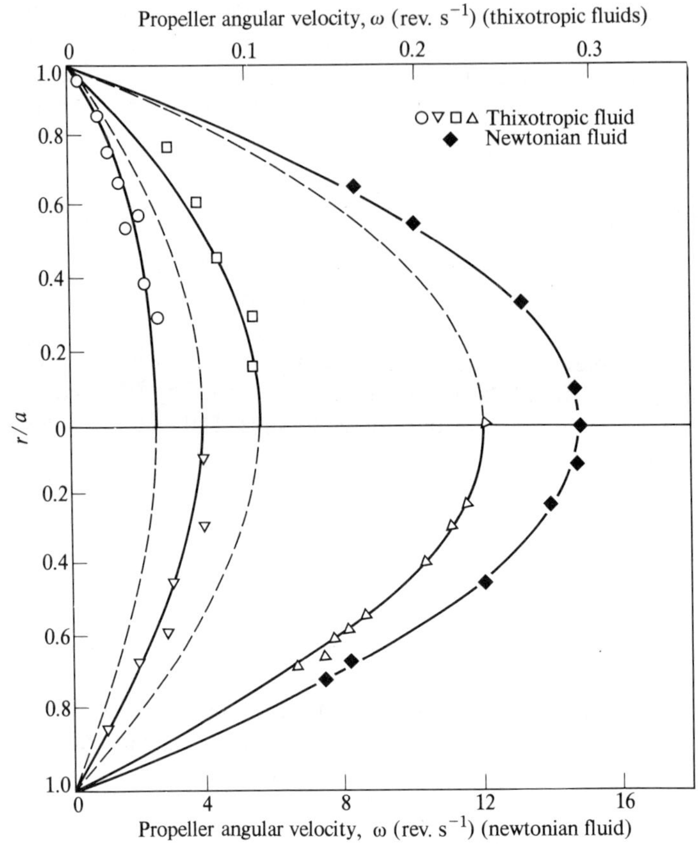

Figure 9-9 Velocity profiles. Open symbols: thixotropic liquids; blocked symbol: newtonian liquid. Flow rate Q: ○, 2.0 $dm^3 s^{-1}$; ▽, 3.0 $dm^3 s^{-1}$; □, 3.6 $dm^3 s^{-1}$; △, 6.2 $dm^3 s^{-1}$. (*Billington, Jones, and Wyllie, 1962.*)

An experimental mapping of the velocity field in cone and plate flow has been obtained by Jobling and Roberts (1959), who observed the motion of small suspended particles. For the fluids studied it was concluded that, except near the edge of the cone, the velocity field is well approximated by one for which the physical components of velocity are $(0, 0, v\langle\phi\rangle(\theta, t))$ and for which Eq. (7-3-41) is valid, provided $\Delta\theta_c$ is $< \sim 4°$ (see Sec. 7-3-4).

9-6 NONNEWTONIAN LIQUIDS AND MOLECULAR STRUCTURE

9-6-1 Liquids Considered as Disordered Solids

In the theories of monatomic liquids regarded as having a microstructure which approximates to a solid-like structure, each molecule is considered to be confined within a fairly small volume, or cage, and to interact directly only with a few neighbors. These theories generally fall into one of four types, classified as hole, vibration, phonon, and relaxation theories.

Vibration theories, of which that of Andrade (1934, 1952) is an example, regard the molecule as vibrating around an equilibrium position, occasionally transferring momentum to its neighbors. In phonon theories (see, for example, Bosworth, 1956), it is vibrations of the lattice as a whole, i.e., sound waves or phonons, which are regarded as transmitting momentum. The phonon concept has found application in recent theories of liquid helium. Relaxation theories regard viscous flow as being similar to the deformation of elastic solids. In the hole theory the liquid is regarded as a lattice with some sites occupied by molecules and others empty. Viscous flow is characterized by the jumping of molecules from one site to another under the influence of an applied force. Eyring's (1936) hole theory which is based on this concept of a momentum transport in a liquid and the Eyring theory of absolute reaction rates (see, for example, Eyring, Henderson, Stover, and Eyring, 1964) is probably the most widely discussed example of a theory of monatomic liquids regarded as having a solid-like structure. A general discussion of lattice theories of the liquid state has been given by Barker (1963).

The present discussion will be limited to an outline description of the physical concepts underlying the theory of absolute reaction rates as applied to the transport of momentum in flowing liquids.

When external forces are applied to a system or when energy in the form of radiation, thermal currents, chemical, or nuclear reactions are added to a system, it will be driven away from its old equilibrium state and will seek a new equilibrium relative to the imposed influence. Usually some parts of the system move toward the new equilibrium much faster than other parts and, as Eyring, Henderson, Stover, and Eyring (1964) have indicated, these parts form equilibrated pools but currents continue to flow from one pool to another until complete equilibrium is reached. The general evolution of systems which are far from equilibrium is very complex and involves kinematic theory which is outside the scope of this book. However, if a system remains sufficiently close to equilibrium that the entropy can be defined by classical thermodynamic variables, two theories can be used to give some estimation of the rate at which the equilibrating currents flow. The first theory is called the thermodynamics of irreversible processes and the second the theory of absolute reaction rates.

9-6-2 The Thermodynamics of Irreversible Processes

This is a rapidly growing subject of which it is only possible to indicate some of the more important assumptions and results. It is assumed that:

1. The system always remains sufficiently close to equilibrium for it to possess an entropy function which, together with the other state variables, is sufficient to define the internal energy exactly.
2. The rate of internal entropy production is given by an expression of the form

$$\dot{S}_i = \sum_k X_k J_k \tag{9-6-1}$$

where X_k are generalized forces (e.g., temperature gradient, electric field, or chemical affinity) and J_k are the corresponding fluxes (e.g., heat flow, electric current, and chemical diffusion).

3. The fluxes are linearly related to the forces

$$J_k = L_{km} X_m \tag{9-6-2}$$

where the coefficients L_{km} are called phenomenological or kinetic coefficients.

4. Microscopic reversibility applies, i.e., all the mechanical equations of motion of individual particles are symmetrical with respect to time reversal.

Using this last assumption Onsager (1931a, 1931b) showed, using a statistical mechanical argument, that the L_{km} satisfy the symmetry relation

$$L_{km} = L_{mk} \tag{9-6-3}$$

which is usually referred to as Onsager's reciprocal relation.

For slight disturbances from equilibrium of a body which is either rotating with angular velocity $\boldsymbol{\Omega}$ or in an external magnetic field $\boldsymbol{H}$, the reciprocal relations take the special form

$$\begin{aligned} L_{km}(\boldsymbol{\Omega}) &= L_{mk}(-\boldsymbol{\Omega}) \\ L_{km}(\boldsymbol{H}) &= L_{mk}(-\boldsymbol{H}) \end{aligned} \tag{9-6-4}$$

The derivation of these relations can be found in Groot and Mazur (1962) and Landau and Lifshitz (1958).

Substituting Eq. (9-6-2) into (9-6-1) gives the quadratic form

$$\dot{S}_i = \sum_m \sum_k L_{km} X_m X_k \tag{9-6-5}$$

and from the second law of thermodynamics $\dot{S}_i$ must always be positive. Thus the quadratic form must be positive definite which implies certain restrictive relations between the L_{km}. A further restriction in the form of L_{km} arises from Curie's principle (1908), which states that the macroscopic causes always have fewer elements of symmetry than the effects they produce, e.g., scalar "forces" cannot give rise to vector "fluxes."

This framework has been extended, in the works of Biot (1958) and Schapery (1964), for example, through the use of normal coordinates and hidden variables to establish the form of the relations between temperature, stress, and deformation in viscoelastic media.

9-6-3 The Theory of Absolute Reaction Rates

This theory is applicable to systems which can be much further from equilibrium than is permissible for the above theory of irreversible thermodynamics and reduces to the above theory as a limiting case.

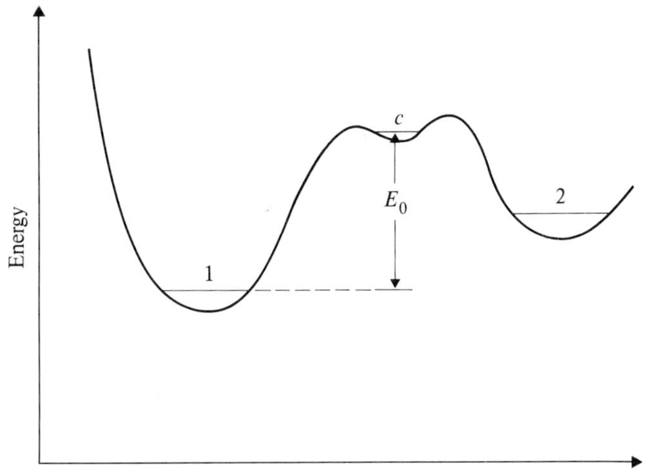

Figure 9-10 Diagram illustrating the activation energy E_0 required to pass from the stable state 1 to the stable state 2.

For a system that has several stable states, e.g., chemical compounds, if one could plot the total energy of the system of constituent particles upon the multidimensional hyper plane consisting of the coordinates needed to specify the molecular configuration it would consist of valleys, representing stable states, or chemical compounds separated by passes. A reaction coordinate leads from one valley to another by way of a pass that is always normal to the contour lines. Plotting the energy against the reaction coordinate would give a graph of the form shown in Fig. 9-10. When the system reaches the point C it is said to form an activated complex and it is then in a position to move to a new equilibrium state. The energy required to raise the system from a stable configuration to the activated complex is called the activation energy and will here be denoted by E_0.

Consider the reaction

$$A + B \rightleftharpoons (AB)^{\ddagger} \rightarrow \text{products}$$

where $(AB)^{\ddagger}$ represents the activated complex. Assuming $(AB)^{\ddagger}$ is in equilibrium with the reactants, then the equilibrium constant $K^{\ddagger}$ is given as

$$K^{\ddagger} = \frac{n_{(AB)^{\ddagger}}}{n_A n_B} \tag{9-6-6}$$

where n_A, n_B, and $n_{(AB)^{\ddagger}}$ are the number per unit volume of the molecules A, B, and $(AB)^{\ddagger}$ respectively. It is further assumed that the activated complex spontaneously decomposes into its products when one of its vibrations has its classical energy $\varepsilon = k\theta$ and becomes a translation, or the rate of passage over the barrier becomes $k\theta/h$. Thus, by Eq. (9-6-6),

$$\begin{aligned}\frac{dn_A}{dt} &= n_{(AB)^{\ddagger}}\left(\frac{k\theta}{h}\right)\\ &= K^{\ddagger} n_A n_B \left(\frac{k\theta}{h}\right)\end{aligned} \tag{9-6-7}$$

Now

$$K^{\ddagger} = \frac{\tilde{z}_{(AB)^{\ddagger}}}{z_A z_B} \tag{9-6-8}$$

where $\tilde{z}_{(AB)^{\ddagger}}, z_A, z_B$ represent the partition functions per unit volume of $(AB)^{\ddagger}$, A, and B respectively and where all the energies are measured from the same reference state, i.e., the lowest energy level of $(AB)^{\ddagger}$, A, or B. Usually it is convenient to calculate partition functions relative to the ground state of the molecule and thus

$$\tilde{z}_{(AB)^{\ddagger}} = z_{(AB)^{\ddagger}} \exp\left(-\frac{E_0}{k\theta}\right) \tag{9-6-9}$$

In calculating $z_{(AB)^{\ddagger}}$ the vibration which has become a translation must be omitted. Combining Eqs. (9-6-7), (9-6-8), and (9-6-9) gives

$$\frac{dn_A}{dt} = \kappa\left(\frac{k\theta}{h}\right)\left(\frac{z_{(AB)^{\ddagger}}}{z_A z_B}\right)\exp\left(-\frac{E_0}{k\theta}\right) n_A n_B \tag{9-6-10}$$

where κ in the above formula is introduced to take account of quantum effects and the departure from equilibrium of the various initial states. Further discussion of κ can be found in the much more detailed discussion of the above theory given by Eyring, Henderson, Stover, and Eyring (1964).

9-6-4 The Reaction Rate Theory of Viscosity

In the approach to the transport of momentum in a liquid regarded as a weak solid, the stresses are taken as given, the object being to calculate the resulting motion. Each particle is surrounded by neighbors which restrict its motion so that to move into a vacant position the particle must first overcome a potential barrier. Identifying a molecule of a liquid with a given cell, this approach to the transport of momentum in a liquid is directly concerned with the diffusion of the molecule out of the cell.

Basic to Eyring's (1936) approach to the liquid state is the concept of a hole in the liquid. In the absence of a shear stress, molecules from left and right will have equal probability of filling the hole, this event occurring a given number of times a second. A further concept, basic to this approach, is the assumption that molecular motion in viscous flow is determined by the rate at which molecules pass the potential barrier of height E_0 formed by their nearest neighbors. In the presence of an applied force field F, a potential gradient is created as shown in Fig. 9-11. The work expended in displacing the molecule into the hole is $\frac{1}{2}\lambda F$ where $\frac{1}{2}\lambda$ is the distance from the center of the molecule to the peak of the potential barrier

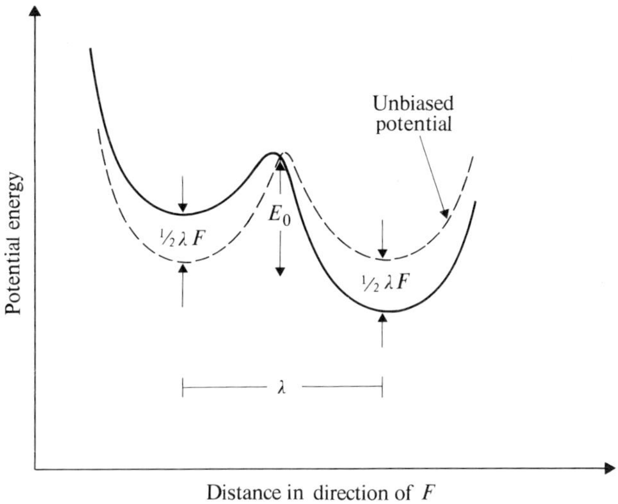

Figure 9-11 Biasing of the potential energy due to a field F; λ is the average distance between vacancies.

associated with the activation energy E_0, the distance λ being measured in the direction of the force F. It follows that the thermal energy required for a jump in the forward direction is reduced from E_0 to $E_0 - \frac{1}{2}\lambda F$, i.e., the rate of jumping is increased. Eyring proceeds to develop a theory of absolute reaction rates (see, for example, Eyring, Henderson, Stover, and Eyring, 1964), from which he obtains, for the current flow J_+ in the forward direction, the relation

$$J_+ = n\lambda\kappa\left(\frac{k\theta}{h}\right)\exp\left[-\frac{(E_0 - \frac{1}{2}\lambda F)}{k\theta}\right] \tag{9-6-11}$$

Similarly, the reverse flow J_- is given by

$$J_- = n\lambda\kappa\left(\frac{k\theta}{h}\right)\exp\left[-\frac{(E_0 + \frac{1}{2}\lambda F)}{k\theta}\right] \tag{9-6-12}$$

where it has been assumed that no density gradients exist. The total current flow is thus given by

$$J = J_+ - J_- = 2n\lambda\kappa\left(\frac{k\theta}{h}\right)\exp\left(-\frac{E_0}{k\theta}\right)\sinh\left(\frac{\lambda F}{2k\theta}\right) \tag{9-6-13}$$

If $\lambda_2\lambda_3$ represent the effective cross-sectional area of a molecule in a shear plane, then the force F acting on the molecule is given by

$$F = \lambda_2\lambda_3 T\langle xy\rangle \tag{9-6-14}$$

where $T\langle xy\rangle$ is the component of the shear stress associated with the two-dimensional laminar shear flow of the liquid. The net specific rate of forward motion per molecule, which is just the rate of shear $\Delta v\langle x\rangle/\lambda_1$, is obtained from

Eqs. (9-6-13) and (9-6-14) in the form

$$2D\langle xy\rangle = \frac{\Delta v\langle x\rangle}{\lambda_1} = 2k_0 \sinh\left(\frac{\lambda\lambda_2\lambda_3 T\langle xy\rangle}{2k\theta}\right) \tag{9-6-15}$$

where, in terms of the absolute reaction rate theory,

$$k_0 = \kappa\frac{k\theta}{h}\exp\left(-\frac{E_0}{k\theta}\right) \tag{9-6-16}$$

the quantity v_f being the fluctuation volume within which the center of gravity of the molecule carries out its thermal motions and is assumed to obey the ideal gas laws.

Equation (9-6-15) can be rearranged into the form

$$T\langle xy\rangle = \frac{2k\theta}{\lambda\lambda_2\lambda_3}\sinh^{-1}\left(\frac{D\langle xy\rangle}{k_0}\right) \tag{9-6-17}$$

which can be expressed, by way of Eq. (9-1-11), in terms of the intensity of the rate of strain $Z^{1/2}$, thus giving the nonnewtonian viscosity

$$\eta = \tfrac{1}{2}p_0\frac{\sinh^{-1}\beta_0 Z^{1/2}}{\beta_0 Z^{1/2}} \tag{9-6-18}$$

which is simply Eq. (9-1-17). The Eyring rate theory of viscosity is the only molecular-based theory of the transport of momentum in liquids which at present yields a basis for interpreting nonnewtonian viscosity in terms of molecular structure.

For the particular condition $\lambda\lambda_2\lambda_3 T\langle xy\rangle \ll 2k\theta$, Eq. (9-6-15) gives the newtonian coefficient of viscosity in the form

$$\eta = \frac{v_f^{1/3}(2\pi mk\theta)^{1/2}}{\lambda\lambda_2\lambda_3}\exp\left(\frac{E_0}{k\theta}\right) \tag{9-6-19}$$

The rate theory of viscosity has been relatively successful at the qualitative level in predicting correctly many of the properties of transport coefficients (Boon and Thomaes, 1963). Since the term in $\theta^{1/2}$ appearing in Eq. (9-6-19) varies only slowly compared with the exponential term, Eq. (9-6-19) is in general accord with the experimentally observed temperature dependence of η which can be expressed in the form

$$\frac{\partial \ln \eta}{\partial \theta} = -\frac{A}{\theta^2} \tag{9-6-20}$$

where A is a constant. The theory has also the advantage of explaining simply why the relaxation time is apparently the same for shear viscosity and bulk viscosity in associated liquids, since it is in both cases the average time between jumps of a molecule into a hole.

The reaction rate theory of viscosity, based on Eyring's hole theory of the liquid state, has been particularly successful in explaining many of the results of experiment and has also had great influence on theoretical considerations. However, certain aspects of the theory pose questions which as yet are unresolved.

The heat of activation E_0 is one particular example. In Eyring's original treatment, the heat of activation E_0 was interpreted to be the heat of activation for forming a hole plus the heat of activation for jump into the hole. In practice, E_0 was found to be ca. $\frac{1}{4}\Delta H_{\text{vap}}$ to $\frac{1}{3}\Delta H_{\text{vap}}$. A more serious criticism relates to the basic concept of flow which is difficult to interpret, since no average force acts on a molecule in simple shear flow, but only a shear stress. In simple laminar shear flow, a molecule in a given layer will be subject to a force from the layer immediately above and to an equal and opposite force from the layer immediately below; on average these forces will cancel. Hence, there is no net force acting on any of the molecules; there is, however, a resultant torque. By generalizing the argument to several species and different forces, Eyring has shown that in the linear region Onsager's reciprocal relations follow from the above theory.

For a discussion of particular systems, see, for example, Ree and Eyring (1955) and Kim, Hirai, Ree, and Eyring (1960).

GENERAL REFERENCES

Barker, J. A. (1963) "Lattice Theories of the Liquid State," Pergamon Press, Oxford.
Bird, R. B., W. E. Stewart, and E. N. Lightfoot (1960) "Transport Phenomena," Wiley, New York.
Coleman, B. D., H. Markovitz, and W. Noll (1966) "Viscometric Flows of Non-Newtonian Fluids," Springer-Verlag, Berlin.
Eyring, H., D. Henderson, B. J. Stover, and E. M. Eyring (1964) "Statistical Mechanics and Dynamics," Wiley, New York.
Ferry, J. D. (1961) "Viscoelastic Properties of Polymers," Wiley, New York.
Frenkel, J. I. (1946) "Kinetic Theory of Liquids," Oxford University Press, London.
Groot, S. R. de, and P. Mazur (1962) "Non-Equilibrium Thermodynamics," North-Holland, Amsterdam.
Landau, L. D., and E. M. Lifshitz (1958) "Statistical Physics," Pergamon Press, Oxford.
Lodge, A. S. (1964) "Elastic Liquids," Academic Press, New York.
Truesdell, C., and W. Noll (1965) The Nonlinear Field Theories of Mechanics, in S. Flügge (ed.), "Handbuch der Physik," vol. III/3, Springer-Verlag, Berlin.
Zener, C. (1948) "Elasticity and Anelasticity of Metals," University of Chicago Press, Chicago, Ill.

REFERENCES

Adams, N., and A. S. Lodge (1964) *Phil. Trans. Roy. Soc.*, vol. 256, pp. 149–184.
Andrade, E. N. da C. (1934) *Phil. Mag.*, vol. 17, pp. 497–511, 698–732.
Andrade, E. N. da C. (1952) *Proc. Roy. Soc. (London)*, vol. A215, pp. 36–43.
Arnold, J. E., and C. F. Goodeve (1940) *J. Phys. Chem.*, vol. 44, pp. 652–670.
Billington, E. W. (1960a) *Proc. Phys. Soc.*, vol. 75, pp. 40–50.
Billington, E. W. (1960b) *Proc. Phys. Soc.*, vol. 76, pp. 127–136.
Billington, E. W. (1965) *J. Sci. Instrum.*, vol. 42, pp. 569–575.
Billington, E. W. (1967) *Proc. Roy. Soc. (London)*, vol. A302, pp. 1–23.
Billington, E. W., and A. S. Huxley (1965) *Trans. Faraday Soc.*, vol. 61, pp. 2784–2793.
Billington, E. W., J. T. Jones, and D. Wyllie (1962) *J. Mech. Engng. Sci.*, vol. 4, pp. 308–314.
Bingham, E. C., and H. Green (1919) *Proc. Am. Soc. Test. Mat.*, vol. 19, pp. 640–664.
Biot, M. A. (1958) *Proc. Third U.S. Natl. Congr. Appl. Mech.*, pp. 1–18, ASME, New York.
Boon, J. P., and G. Thomaes (1963) *Physica*, vol. 29, pp. 123–128, 208–214.
Bosworth, R. C. L. (1956) "Transport Processes in Applied Chemistry," Wiley, New York.

Broadbent, J. M., A. Kaye, A. S. Lodge, and D. G. Vale (1968) *Nature*, vol. 217, pp. 55–56.
Broadbent, J. M., and A. S. Lodge (1971) *Rheol. Acta.*, vol. 10, pp. 557–573.
Buckingham, E. (1921) *Proc. Am. Soc. Test. Mat.*, vol. 21, pp. 1154–1156.
Coleman, B. D., and W. Noll (1961) *Rev. Modern Phys.*, vol. 33, pp. 239–249.
Curie, P. (1908) "Oeuvres," Société Francaise de Physique, Paris.
De Witt, T. W. (1955) *J. Appl. Phys.*, vol. 26, pp. 889–894.
Ericksen, J. L. (1960) The Behaviour of Certain Viscoelastic Materials in Lamina Shearing Motions, in J. T. Bergen (ed.), "Viscoelasticity: Phenomenological Aspects," pp. 77–91, Academic Press, New York.
Eyring, H. (1936) *J. Chem. Phys.*, vol. 4, pp. 283–291.
Fromm, H. (1948) *Z. angew. Math. Mech.*, vol. 28, pp. 43–54.
Garner, F. H., A. H. Nissan, and G. F. Wood (1950) *Phil. Trans. Roy. Soc.*, vol. A243, pp. 37–66.
Gaskins, F. H., and W. Philippoff (1959) *Trans. Soc. Rheol.*, vol. 3, pp. 181–203.
Goodeve, C. F. (1938) *Rep. Prog. Phys.*, vol. 5, pp. 20–29.
Goodeve, C. F. (1939a) *J. Sci. Instrum.*, vol. 16, pp. 19–27.
Goodeve, C. F. (1939b) *Trans. Faraday Soc.*, vol. 35, pp. 342–358.
Goodeve, C. F. (1949) *Proc. Int. Congr. Rheol. Scheveningen*, vol. IIA, pp. 5–11, North-Holland, Amsterdam.
Goodeve, C. F., and G. W. Whitfield (1938) *Trans. Faraday Soc.*, vol. 34, pp. 511–520.
Green, A. E., and J. E. Adkins (1960) "Large Elastic Deformations and Nonlinear Continuum Mechanics," Clarendon Press, Oxford.
Green, H., and R. N. Weltman (1943) *Industr. Engng. Chem.*, vol. 15, pp. 201–206.
Greensmith, H. W., and R. S. Rivlin (1953) *Phil. Trans. Roy. Soc.*, vol. A245, pp. 399–428.
Gross, B. (1953) "Mathematical Structure of the Theories of Viscoelasticity," Herman, Paris.
Higashitani, K., and A. S. Lodge (1975) *Trans. Soc. Rheol.*, vol. 19, pp. 307–335.
Higginbotham, R. S. (1950) *J. Sci. Instrum.*, vol. 27, pp. 139–141.
Jobling, A., and J. E. Roberts (1959) *J. Polym. Sci.*, vol. 36, pp. 421–431, 433–441.
Kaye, A., A. S. Lodge, and D. G. Vale (1968) *Rheol. Acta.*, vol. 7, pp. 368–379.
Kepes, M. (1956) *J. Polym. Sci.*, vol. 22, pp. 409–422.
Kim, W. K., N. Hirai, T. Ree, and H. Eyring (1960) *J. Appl. Phys.*, vol. 31, pp. 358–361.
Kotaka, T., M. Kurata, and M. Tamura (1959) *J. Appl. Phys.*, vol. 30, pp. 1705–1712.
Leaderman, H. (1958) Viscoelasticity Phenomena in Amorphous High Polymeric Systems, in F. R. Eirich (ed.), "Rheology," vol. 2, pp. 1–61, Academic Press, New York.
Lipson, J. M., and A. S. Lodge (1968) *Rheol. Acta.*, vol. 7, pp. 364–368.
Lodge, A. S. (1956) *Trans. Faraday Soc.*, vol. 52, pp. 120–130.
McDowell, C. M., and F. L. Usher (1931) *Proc. Roy. Soc. (London)*, vol. A131, pp. 409–427, 564–576.
McKennell, R. (1954) In V. G. W. Harrison (ed.), *Proc. Second Int. Congr. Rheol.*, pp. 350–359, Butterworths, London.
Markovitz, H. (1957) *Trans. Soc. Rheol.*, vol. 1, pp. 37–52.
Markovitz, H. (1960) Normal Stress Effects in Polymer Solutions, in J. T. Bergen (ed.), "Viscoelasticity: Phenomenological Aspects," pp. 133–144, Academic Press, New York.
Markovitz, H. (1965a) *Proc. Fourth Int. Congr. Rheol.*, vol. 1, pp. 189–212.
Markovitz, H. (1965b) *J. Polym. Sci.*, vol. B3, pp. 3–5.
Markovitz, H. (1965c) *Phys. Fluids*, vol. 8, p. 200.
Markovitz, H., and D. R. Brown (1964) Normal Stress Measurements on a Polyisobutylene–Cetane Solution in Parallel-Plate and Cone-Plate Instruments, in M. Reiner and D. Abir (eds.), "Second Order Effects in Elasticity, Plasticity and Fluid Dynamics," pp. 585–601, Pergamon Press, London.
Markovitz, H., L. J. Elyash, F. J. Padden, and T. W. De Witt (1955) *J. Colloid. Sci.*, vol. 10, pp. 165–173.
Markovitz, H., P. M. Yavorsky, R. C. Harper, L. J. Zapas, and T. W. De Witt (1952) *Rev. Sci. Instrum.*, vol. 23, pp. 430–437.
Mooney, M. (1931) *J. Rheol.*, vol. 2, pp. 210–222.
Mooney, M., and R. H. Ewart (1934) *Physics*, vol. 5, pp. 350–354.
Oldroyd, J. G. (1947) *Proc. Camb. Phil. Soc.*, vol. 43, pp. 100–105, 383–395, 396–405, 521–532.

Oldroyd, J. G. (1948) *Proc. Camb. Phil. Soc.*, vol. 44, pp. 200–213, 214–228.
Oldroyd, J. G. (1949) *J. Colloid Sci.*, vol. 4, pp. 333–342.
Oldroyd, J. G. (1950) *Proc. Roy. Soc.* (*London*), vol. A200, pp. 523–541.
Oldroyd, J. G. (1951) *Quart. J. Mech. Appl. Math.*, vol. 4, pp. 271–282.
Oldroyd, J. G. (1956) Non-Newtonian Flow of Liquids and Solids, in F. R. Eirich (ed.), "Rheology," vol. 1, pp. 653–682, Academic Press, New York.
Oldroyd, J. G. (1958) *Proc. Roy. Soc.* (*London*), vol. A245, pp. 278–297.
Oldroyd, J. G., D. J. Strawbridge, and B. A. Toms (1951) *Proc. Phys. Soc.*, vol. B64, pp. 44–57.
Onsager, L. (1931a) *Phys. Rev.*, vol. 37, pp. 405–426.
Onsager, L. (1931b) *Phys. Rev.*, vol. 38, pp. 2265–2279.
Ostwald, W. (1925) *Kolloid. Z.*, vol. 36, pp. 99–117.
Ostwald, W. (1929) *Kolloid. Z.*, vol. 47, pp. 176–187.
Philippoff, W. (1935) *Kolloid. Z.*, vol. 71, pp. 1–16.
Rabinowitch, B. (1929) *Z. physik. Chem.*, vol. A145, pp. 1–26.
Ree, T., and H. Eyring (1955) *J. Appl. Phys.*, vol. 26, pp. 793–800, 800–809.
Reiner, M. (1930) *J. Rheol.*, vol. 1, pp. 250–260.
Reiner, M. (1931) *J. Rheol.*, vol. 2, pp. 337–350.
Reiner, M. (1945) *Amer. J. Math.*, vol. 67, pp. 350–362.
Reiner, M. (1948) *Amer. J. Math.*, vol. 70, pp. 433–446.
Reiner, M. (1952) *Bull. Res. Council Israel*, vol. 2, pp. 65–66.
Reiner, M. (1957) *Proc. Roy. Soc.* (*London*), vol. A240, pp. 173–188.
Reiner, M. (1958) Rheology, in S. Flügge (ed.), "Handbuch der Physik. VI. Elasticity and Plasticity," pp. 434–550, Springer-Verlag, Berlin.
Reiner, M. (1960) *Phys. Fluids*, vol. 3, pp. 427–432.
Richardson, E. G., and E. Tyler (1933) *Proc. Phys. Soc.*, vol. 45, pp. 142–150.
Rivlin, R. S. (1948) *Proc. Roy. Soc.* (*London*), vol. A193, pp. 260–281.
Rivlin, R. S. (1949) *Proc. Camb. Phil. Soc.*, vol. 45, pp. 88–91.
Rivlin, R. S. (1956) *J. Rational Mech. Anal.*, vol. 5, pp. 179–188.
Rivlin, R. S., and J. L. Ericksen (1955) *J. Rational Mech. Anal.*, vol. 4, pp. 323–425.
Roberts, J. E. (1952) *U.K. Ministry of Supply Rept.* ADE 13/52.
Roberts, J. E. (1957) *Nature*, vol. 179, pp. 487–488.
Russell, R. J. (1946) "The Determination of the Basic Rheological Constants Governing the Flow of Pseudoplastic Substances," Ph.D. Thesis, University of London.
Schofield, R. K., and G. W. Scott-Blair (1930) *J. Phys. Chem.*, vol. 34, pp. 248–262.
Schofield, R. K., and G. W. Scott-Blair (1931) *J. Phys. Chem.*, vol. 35, pp. 1212–1215.
Schultz-Grünow, F. (1955) *Ver. Deut. Ing.-Z.*, vol. 97, pp. 409–416.
Schapery, R. A. (1964) *J. Appl. Phys.*, vol. 35, pp. 1451–1464.
Sisko, A. W. (1958) *Ind. Engng. Chem.*, vol. 50, pp. 1789–1792.
Taylor, G. I. (1936) *Proc. Roy. Soc.* (*London*), vol. A157, pp. 565–578.
Thom, A., and C. J. Apelt (1958) The Pressure in a Two-dimensional Static Hole at Low Reynolds Numbers, *Aeronaut. Res. Coun. Tech. Rept. R & M No.* 3090, HMSO, London.
Thornton, S. (1953) *Proc. Phys. Soc.*, vol. B66, pp. 115–119.
Thornton, S., and D. Rae (1953) *Proc. Phys. Soc.*, vol. B66, pp. 120–128.
Toms, B. A. (1949) *J. Colloid. Sci.*, vol. 4, pp. 511–521.
Toms, B. A. (1958) in F. R. Eirich (ed.), "Fundamental Techniques: Fluids, Rheology, Theory and Applications," vol. 2, pp. 475–501, Academic Press, New York.
Toms, B. A., and D. J. Strawbridge (1953) *Trans. Faraday Soc.*, vol. 49, pp. 1225–1232.
Waele, A. de (1923) *J. Oil and Colour Chem. Assoc.*, vol. 6, pp. 33–80.
Weissenberg, K. (1947) *Nature*, vol. 159, pp. 310–311.
Weissenberg, K. (1949) *Proc. Int. Congr. Rheol.*, vol. II, pp. 114–118, North-Holland, Amsterdam.
Weltman, R. N. (1943) *J. Appl. Phys.*, vol. 14, pp. 343–350.
Zarémba, S. (1937) Sur une Conception Nouvelle des Forces Interieures dans une Fluide en Mouvement, *Mem. Sci. Math. No.* 82, Gauthier-Villars, Paris.

CHAPTER

TEN

CONTINUUM PLASTICITY

10-1 PLASTICITY IN SOLIDS

Particular aspects of the deformation behavior of solids beyond the elastic range will be considered in the present chapter. Deformation of solid bodies (mainly metals) beyond the elastic range is characterized by permanent dimensional changes which result from the mechanism of slip, or from dislocations at the atomic level, this irreversible deformation being known as plastic deformation. Plastic deformations occur only at stress intensities above a certain threshold value known as the elastic limit, or in terms of an idealized theory of plasticity as the yield stress, denoted as Y for observations made in a state of uniaxial stress and as k for observations made in pure shear. In general the proportional limit, i.e., the stress at the end of the linear range, and the elastic limit do not in practice coincide. However, for the purposes of developing an idealized theory of plasticity the elastic limit and the proportional limit will be assumed to coincide.

A typical one-dimensional, i.e., uniaxial with, for example, $T_{11} = T_0$, stress-strain curve (T_0, ϵ) is shown schematically in Fig. 10-1. The yield stress Y, i.e., the point A, is seen to separate the characteristic stress-strain curve into an elastic range and a plastic range. In the initial elastic range (OA), both the loading and unloading paths coincide, consistent with the one-to-one stress-strain relationship in the elastic range. In contrast, unloading from the point B in the plastic range of deformation results in the stress-strain point (T_0, ϵ), i.e., the state point, following

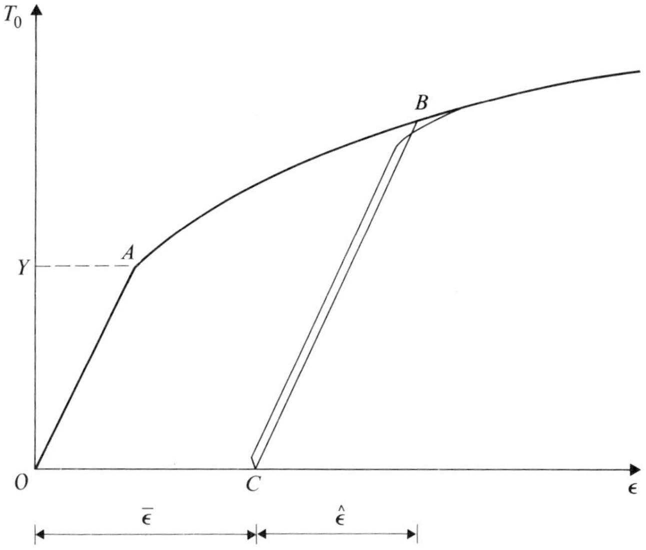

Figure 10-1 Typical uniaxial stress-strain curve.

the path (BC) which is substantially parallel with the initial elastic portion of the curve. At C, for which $T_0 = 0$, there remains a permanent plastic strain $\bar{\epsilon}$. The recoverable elastic strain from B is shown as $\hat{\epsilon}$. The elastic and plastic components of any quantity δ will be denoted by $\hat{\delta}$ and $\bar{\delta}$ respectively.

To produce a further deformation on reloading requires an increase in stress, a condition referred to as work-hardening or strain-hardening. In an unloading-reloading cycle, a reloading from C would follow a slightly different path (to BC), giving rise to a hysteresis loop, an effect which must be allowed for in cyclic loading because of the associated energy losses. The formation of a hysteresis loop during an unloading-reloading cycle implies that the stress is a history-dependent function of the strain. It is the actual change in material properties associated with the phenomenon of work-hardening, i.e., the strain-hardening arising from the cold-working involved in plastic deformation, which makes the stress depend on the history of the deformation. This is because after each unloading it is a different material which is responding to reloading. History dependence is a general characteristic feature of plastic deformation and is the principal physical difference between elastic and plastic deformation.

A specimen initially loaded in tension is observed to yield at a much reduced stress when reloaded in compression. This is known as the Bauschinger effect. For the purpose of developing an idealized theory of plasticity it will, however, be assumed that the ideal plastic body does not exhibit a Bauschinger effect. This restriction permits the yield stress to be taken to be the same in tension and compression.

Because the process of slip on a crystal plane is directional, plastic deformation must be regarded as being physically anisotropic. For this reason continuous plastic deformation is accompanied by the progressive development of anisotropy. Thus, it is to be expected that any initial isotropy which is present will usually be destroyed by continuous plastic deformation.

The classical three-dimensional theory of plastic behavior can be regarded as a generalization of the four particular idealizations of the one-dimensional stress-strain curve illustrated in Fig. 10-2. Representations of the form shown in Figs. 10-2*a* and *b* find application in studying contained plastic deformation and the representations of Fig. 10-2*c* and *d* in studying uncontained plastic flow. It is evident from Fig. 10-2 that there are two distinct features which any idealized theory of plasticity must take into account. These relate to describing the initial yield which terminates the initial elastic range of response together with a description of the post-yield behavior.

10-2 CRITERION OF YIELDING

10-2-1 Basic Assumptions

A generalization of the yield concept in one-dimensional loading to a three-dimensional state of stress, characterized by a mathematical relationship among the stress components at a point that must be satisfied if the onset of plastic behavior is to occur at the point, is referred to as a yield criterion or yield condition. Almost all rate-independent theories of continuum plasticity assume that in general the yield criterion may be expressed by the equation

$$f(\mathbf{T}) = 0 \tag{10-2-1}$$

where $f(\mathbf{T})$ is the yield function. Thus the mechanical response of the material is

1. *Elastic*:

$$\text{for } f(\mathbf{T}) < 0 \text{ or for } f(\mathbf{T}) = 0 \text{ and } \frac{\partial f}{\partial T_{ij}} dT_{ij} < 0 \tag{10-2-2}$$

2. *Plastic*:

$$\text{for } f(\mathbf{T}) = 0 \text{ and } \frac{\partial f}{\partial T_{ij}} dT_{ij} \geqslant 0 \tag{10-2-3}$$

In general, the yield function may contain several parameters characterizing the state of the material, but during plastic deformation these parameters are understood to change in such a way that f remains equal to zero while plastic deformation continues. These material parameters will also be expected to vary with temperature or other thermodynamic variables. As formulated, the general yield condition relates only to the state of stress at a point in space and, irrespective of whether the mechanical response is elastic or plastic, does not depend on the stress at neighboring points. For this reason the yield condition contains no dependence on the stress gradient.

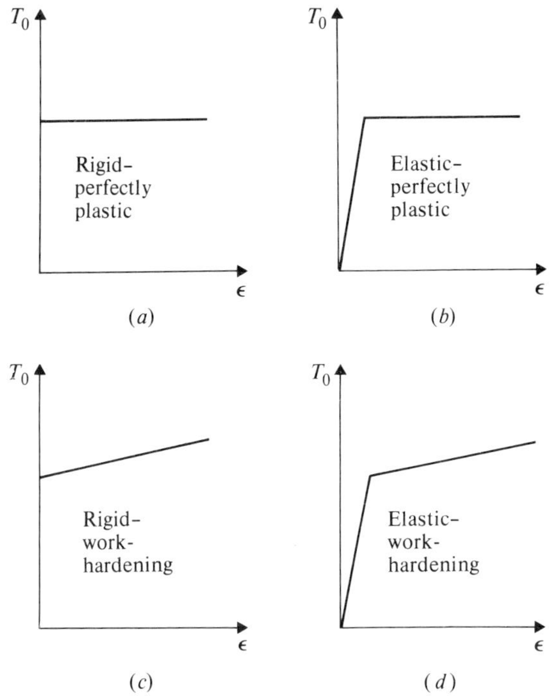

Figure 10-2 (*a*) Idealized, rigid–perfectly plastic stress-strain curve. (*b*) Idealized, elastic–perfectly plastic stress-strain curve. (*c*) Idealized, rigid–linear work-hardening stress-strain curve. (*d*) Idealized, elastic–linear work-hardening stress-strain curve.

Any yield criterion for an isotropic material can be expressed in the form

$$f(\mathrm{I_T}, \mathrm{II_T}, \mathrm{III_T}) = 0 \tag{10-2-4}$$

where the three invariants of the stress tensor **T**, that is, $\mathrm{I_T}$, $\mathrm{II_T}$, and $\mathrm{III_T}$, are expressed in terms of the principal stresses σ_a $(a = 1, 2, 3)$. The condition that an isotropic material depends only on the magnitude of the three principal applied stresses, and not on their directions, is satisfied by specifying that for an isotropic material f is to be a symmetric function of the principal stresses σ_a $(a = 1, 2, 3)$.

It is further assumed in the theory of plasticity that yield is unaffected by hydrostatic stress, a situation generally in accord with experiment for moderate stress levels. With this assumption it is possible to formulate the yield criterion for an isotropic material in terms of the invariants of the stress deviator in the form

$$f(J_2', J_3') = 0 \tag{10-2-5}$$

where the invariants

$$J_2' = \tfrac{1}{2}\operatorname{tr}\mathbf{T}'^2 \qquad J_3' = \det\mathbf{T}' \tag{10-2-6}$$

of the stress deviator $\mathbf{T}'$ are defined by the coefficients of the characteristic equations

$$-\det(\mathbf{T}'-\sigma_a'\mathbf{I}) = \sigma_a'^3+\sigma_a'\mathrm{II}_{\mathbf{T}'}-\mathrm{III}_{\mathbf{T}'} = \sigma_a'^3-\sigma_a'J_2'-J_3' = 0 \qquad (10\text{-}2\text{-}7)$$

The condition that $\mathrm{I}_{\mathbf{T}'} = J_1' = 0$ is in accord with the assumption that yield is unaffected by hydrostatic stress.

To ensure that the yield stress is the same in tension and compression, i.e., neglecting the Bauschinger effect for the ideal plastic body, it is necessary to impose the further restriction that

$$f(-\mathbf{T}) = f(\mathbf{T}) \qquad (10\text{-}2\text{-}8)$$

In applying the condition of Eq. (10-2-8) and having regard to Eq. (10-2-5) it is noted from Eq. (10-2-6) that $J_2'(-\mathbf{T}) = J_2'(\mathbf{T})$ whereas $J_3'(-\mathbf{T}) = -J_3'(\mathbf{T})$.

The additional assumptions formulated by way of Eqs. (10-2-4), (10-2-5), and (10-2-8), though limiting the generality of the yield function of Eq. (10-2-1), are all included in those theories of plasticity most widely used to describe the mechanical response of metals.

10-2-2 A Geometrical Representation of Stress

The yield criteria can be interpreted for an isotropic material in terms of a geometrical representation of the state of stress obtained by taking the principal stresses as coordinates, as shown in Fig. 10-3 (see, for example, Haigh, 1920;

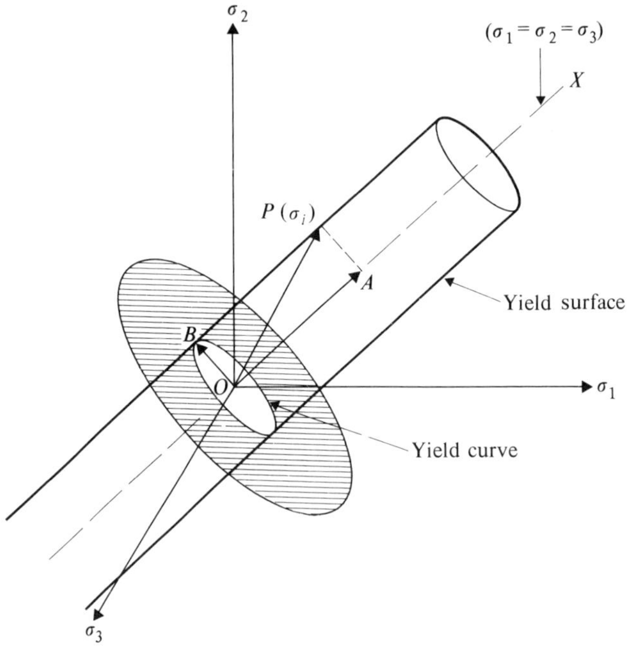

Figure 10-3 Yield surface in the principal stress space showing the yield curve in the π plane.

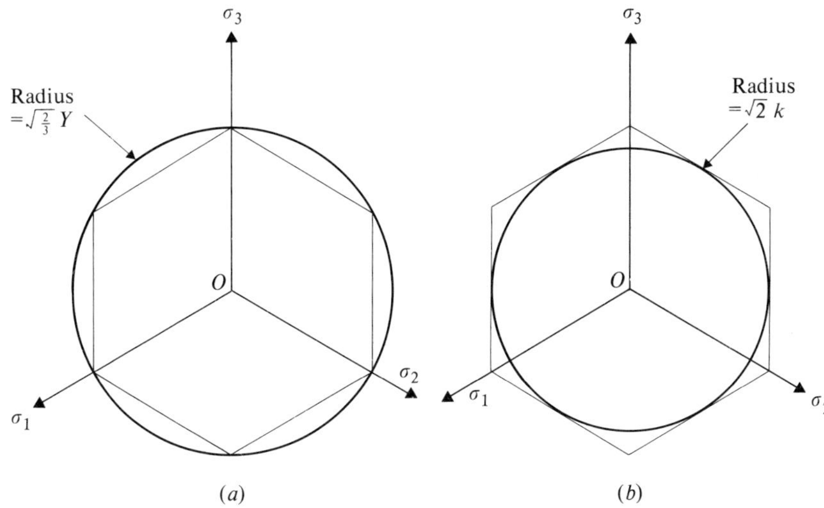

Figure 10-4 The disposition of the principal axes, and the yield curves for the Tresca and von Mises yield criteria viewed toward the origin O and along OX of Fig. 10-3.

Westergaard, 1920, 1952). Every point in this principal stress space corresponds to a state of stress, and the position vector $\boldsymbol{OP}$ of any such point $P(\sigma_1, \sigma_2, \sigma_3)$ may be resolved into a component $\boldsymbol{OA}$ along the line $\boldsymbol{OX}$, which makes equal angles with the coordinate axes, together with a component $\boldsymbol{OB}$ in a plane known as the π plane which is perpendicular to $\boldsymbol{OX}$ and passes through the origin. In stress space the yield criterion defines a surface which is generally referred to as the yield surface. The yield surfaces are parallel to the hydrostatic line OA characterized by the condition

$$\sigma_1 = \sigma_2 = \sigma_3 \qquad \text{(hydrostatic line } OA\text{)} \tag{10-2-9}$$

This follows from the assumption that yield is independent of the spherical part of the stress which requires that if $(\sigma_1, \sigma_2, \sigma_3)$ is on the yield surface, then so is $(\sigma_1 + OA, \sigma_2 + OA, \sigma_3 + OA)$ for any value of OA. This in turn implies that the yield surfaces are in general to be regarded as prisms with elements perpendicular to the deviatoric plane characterized by the condition

$$\sigma_1 + \sigma_2 + \sigma_3 = 0 \qquad \text{(deviatoric plane)} \tag{10-2-10}$$

Stress points that lie inside the yield surface represent elastic stress states, while those which lie on the yield surface represent incipient plastic stress states. The intersection of the yield surface with the π plane is generally referred to as the yield curve.

Viewed toward the origin O and along OX, the principal stress axes are seen from Fig. 10-4 to be symmetrically placed 120° apart in the deviatoric plane. The assumption of isotropy requires that f be a symmetric function of $\sigma_1, \sigma_2, \sigma_3$ such that interchanging any two of the three arguments leaves f unchanged, from which it follows that the locus in the deviatoric plane must be symmetric with respect to each of the three projected axes. The location in the π plane of the arbitrary point P can be determined by noting that each of the stress space axes makes an angle $\cos^{-1}\sqrt{2/3}$ with the π plane, and hence the projected deviatoric components are $(\sqrt{2/3}\sigma_1, \sqrt{2/3}\sigma_2, \sqrt{2/3}\sigma_3)$.

10-2-3 Formulation of the Yield Criterion

A yield function of general interest can be expressed by way of Eq. (10-2-5) in the form

$$f = J'_2 H(\omega) - \kappa L(J'_2) = 0 \tag{10-2-11}$$

where

$$\omega = \frac{27}{4}\frac{J'^2_3}{J'^3_2} \tag{10-2-12}$$

and κ is a constant characteristic of material properties.

The von Mises yield criterion

$$f = J'_2 - k^2 \tag{10-2-13}$$

is obtained from Eq. (10-2-11) by setting $H = L = 1$ and $\kappa = k^2$. For a thin-walled tubular specimen of material tested in pure torsion, where the only nonzero component of stress is specified to be $T\langle 12\rangle$, the von Mises criterion requires $J'_2 = T\langle 12\rangle^2$, from which it follows that the number k is just the yield stress in pure shear. In uniaxial tension, where the only nonzero component of stress is taken to be $T\langle 11\rangle \equiv Y$, the von Mises yield criterion of Eq. (10-2-13) gives the prediction that

$$k = \frac{1}{\sqrt{3}} Y \tag{10-2-14}$$

For the von Mises yield criterion, the yield surface $J'_2 = \frac{1}{3}Y^2$ is the right circular cylinder defined by Eq. (10-2-13), such that the locus in the deviatoric plane is symmetric with respect to each of the three projected axes. That this is so can be seen from Fig. 10-4 where the von Mises circle of radius $\sqrt{2/3}\,Y$ is shown as it appears in the π plane. The von Mises yield criterion is the simplest continuously differentiable isotropic condition of incompressible yielding.

The simplest form of a piecewise continuous-yield criterion is represented by the classical Tresca yield criterion which can be obtained from Eq. (10-2-11) by taking

$$H = (1-\omega)^{1/3} \qquad L = \left[4 - 3\left(\frac{J'_2}{k^2}\right)\right]^{2/3} \tag{10-2-15}$$

to give

$$f = J'_2(1-\omega)^{1/3} - \left[4 - 3\left(\frac{J'_2}{k^2}\right)\right]^{2/3} k^2 = 0 \tag{10-2-16}$$

which is the Reuss (1930) form of the Tresca yield condition. For pure torsion, there is only one nonzero component of stress, say $T\langle\theta z\rangle$, giving $\omega = 0$; hence, from Eq. (10-2-16) it follows that $T\langle\theta z\rangle = k$ which, as for the von Mises yield criterion, is thus the yield stress in simple shear. The formulation of the Tresca yield criterion is based on the assumption that yielding occurs when the maximum shear stress reaches the prescribed value k. Thus the Tresca yield criterion can be expressed by way of Eq. (3-6-11) in the form

$$f = (\sigma_1 - \sigma_2) - 2k = 0 \tag{10-2-17}$$

However, since the maximum shear stress yield equality is satisfied by any one of the six linear equations

$$\tfrac{1}{2}(\sigma_1 - \sigma_2) = \pm k \qquad \tfrac{1}{2}(\sigma_2 - \sigma_3) = \pm k \qquad \tfrac{1}{2}(\sigma_3 - \sigma_1) = \pm k \tag{10-2-18}$$

the Tresca yield criterion can be expressed in the more general form

$$f = [(\sigma_1 - \sigma_2)^2 - 4k^2][(\sigma_2 - \sigma_3)^2 - 4k^2][(\sigma_3 - \sigma_1)^2 - 4k^2] = 0 \tag{10-2-19}$$

which is now symmetrical with respect to the principal stresses. For uniaxial tension, $\sigma_1 = Y$ $(\sigma_2 = \sigma_3 = 0)$ and the Tresca yield criterion predicts that

$$k = \tfrac{1}{2}Y \tag{10-2-20}$$

which is to be compared with the von Mises prediction of Eq. (10-2-14). The Tresca yield surface is the regular hexagonal prism whose six plane faces are defined by Eqs. (10-2-18) with k replaced by $\frac{1}{2}Y$, thus giving the von Mises and Tresca conditions in forms which agree in uniaxial tension or compression, as can be seen from Fig. 10-4. In simple tension, i.e., Fig. 10-4*a*, the von Mises circle of radius $\sqrt{2/3}\,Y$ is seen to circumscribe the regular Tresca hexagon, whereas in simple shear, i.e., Fig. 10-4*b*, the von Mises circle is inscribed in the Tresca hexagon.

10-3 POST-YIELD BEHAVIOR

10-3-1 Formulation of Constitutive Equations

Since plastic strains are irrecoverable, plasticity has ideally to be regarded as an irreversible process. From the general discussion of constitutive theory given in Chap. 5 it is evident that a complete description of an irreversible process such as plasticity can, in principle, only be obtained by introducing response functionals. This ensures irreversibility since a history cannot be reversed. However, the theories of plasticity of present interest are formulated by way of more elementary concepts involving parameters which are effectively time dependent, i.e., irreversible parameters which are specified to be continuously increasing.

The stress-strain relations of plasticity are generally expressed in terms of strain increments, the associated theories being referred to as incremental theories.

In the formulation of incremental theories, the total natural strain increment $d\epsilon$ is expressed in the form

$$d\epsilon = \tfrac{1}{2}[\text{grad}\, d\mathbf{u} + (\text{grad}\, d\mathbf{u})^T] = d\hat{\epsilon} + d\bar{\epsilon} \tag{10-3-1}$$

where, from Eq. (6-1-28),

$$d\hat{\epsilon} = \frac{1}{2G} d\mathbf{T}' + \frac{(1-2\nu)}{3E} d(\text{tr}\,\mathbf{T})\mathbf{I}$$

is the natural elastic strain increment and $d\bar{\epsilon}$ is the natural plastic strain increment. There also follows the relation

$$d(\text{tr}\,\hat{\epsilon}) = \frac{1-2\nu}{E} d(\text{tr}\,\mathbf{T})$$

which can be obtained from Eqs. (6-1-24), (6-1-25), and (6-1-28).

Formulation of stress-strain relations by way of Eq. (10-3-1) involves the concepts of incremental loading and unloading. To consider the physical significance of incremental loading and unloading, it is necessary to define a loading function which will provide a description of the continually changing yield surface arising from plastic deformation produced by continued loading beyond the initial yield point. Work-hardening is associated with continued loading beyond the initial yield point and in simple uniaxial tension is characterized by a stress-strain curve for which the stress is a monotonically increasing function of increasing strain. Such changes in the yield surface arising from strain-hardening are accounted for by generalizing the yield function $f(\mathbf{T})$ of Eq. (10-2-1) to define subsequent yield surfaces beyond the initial one. Assuming an isothermal deformation process, a generalization can be obtained by introducing the loading function

$$f(\mathbf{T}, \bar{\epsilon}, \kappa) = 0 \tag{10-3-2}$$

which, in addition to depending on the stress, also depends upon the plastic strain $\bar{\epsilon}$ and the work-hardening characteristics represented by the parameter κ. Differentiation of Eq. (10-3-2) gives

$$\dot{f} = \frac{\partial f}{\partial T_{ij}} \dot{T}_{ij} + \frac{\partial f}{\partial \bar{\epsilon}_{ij}} \dot{\bar{\epsilon}}_{ij} + \frac{\partial f}{\partial \kappa} \dot{\kappa} \tag{10-3-3}$$

Assuming a state of stress to exist on the yield surface, unloading by an incremental amount into the elastic region does not involve the plastic strain increment $d\bar{\epsilon}_{ij}$, which can therefore be set to zero. Since work-hardening is considered to arise from plastic strains, the work-hardening parameter $d\kappa$ must also be zero. These particular conditions reduce the increment of the loading function to the form

$$\dot{f} = \frac{\partial f}{\partial T_{ij}} \dot{T}_{ij}$$

it being noted that since the change is from $f = 0$ to $f + \dot{f}\, dt < 0$, df must be less than zero for unloading. Two further loading conditions require defining. These relate to neutral loading and loading. Thus three loading conditions are to be identified and are defined as follows:

1. *Unloading*:

$$\text{for } f = 0 \qquad \frac{\partial f}{\partial T_{ij}} \dot{T}_{ij} < 0 \tag{10-3-4}$$

2. *Neutral loading*:

$$\text{for } f = 0 \qquad \frac{\partial f}{\partial T_{ij}} \dot{T}_{ij} = 0 \tag{10-3-5}$$

3. *Loading*:

$$\text{for } f = 0 \qquad \frac{\partial f}{\partial T_{ij}} \dot{T}_{ij} > 0 \tag{10-3-6}$$

The loading surface defined by Eq. (10-3-2) is the yield surface for $f = 0$, whereas $f < 0$ is a surface in the elastic region inside the yield surface. Since $f > 0$ is outside the yield surface, this particular condition has no physical significance. Regarding df as a stress increment vector, and assuming the loading surface to be a closed surface, unloading, neutral loading, or loading occurs according to whether the stress increment vector is directed inward, tangent, or outward to the yield surface respectively. It is evident that loading can only occur if the material work hardens, for which condition the yield surface simply moves outward. Ideal plasticity is characterized by a yield surface fixed in the principal stress space, for which particular condition loading cannot be sustained and plastic deformation occurs only during neutral loading.

It is evident from the yield criteria of present interest, considered in terms of uniaxial loading along one of the principal stress axes, that the yield stress corresponds to one point on the yield surface. Although this conclusion suggests that the yield surface can be considered as a generalization of the yield stress, generalization of the simple geometric definition of work- or strain-hardening of present interest does not follow directly from considerations of the simple uniaxial test. However, Drucker (1951) has proposed a generalization of the work-hardening concept which regards the stress as a monotonically increasing function of increasing strain in the simple uniaxial test. To formulate this generalization, Drucker considers the work expended by an external agency which slowly applies a set of self-equilibrating forces and then slowly removes it from a body which is initially in a state of equilibrium. The external agency is regarded as being quite distinct from the agency causing the initially existing state of stress. On completion of the cycle, the stress is returned to its initial equilibrium value, whereas the body may or may not return to its original configuration. Drucker proceeds to define a stable work-hardening material as one which satisfies both of the following conditions:

1. For all such added sets of stresses, the plastic work done by the external agency during the application of the additional stresses is positive.

2. The net total work performed by the external agency during the cycle of adding or removing stresses is either zero or positive.

The physical significance of this important concept can be examined by considering initially a program of loading referred to the idealized stress-strain diagram of the simple uniaxial tension (for example, $T_{11} = T_0$) test shown in Fig. 10-5. Loading is from an existing state of stress $T_0^* < Y$ to T_0, where Y is the stress at the yield point in Fig. 10-5. This loading is followed by incremental loading from T_0 to $T_0 + dT_0$ where the positive increment of stress dT_0 produces an increment of plastic strain $d\bar{\epsilon}_{11} = d\bar{\epsilon}$ followed in turn by unloading to T_0^*. From Fig. 10-5 it follows for this particular program of loading and unloading that

$$(T_0 - T_0^*)d\bar{\epsilon} > 0 \qquad dT_0\, d\bar{\epsilon} > 0 \tag{10-3-7}$$

The inequalities of Eq. (10-3-7) still hold for a program of loading in which the material has undergone previous stressing in excess of T_0^*, the stress T_0 now referring to states of stress at subsequent yield points such as F in Fig. 10-5. It is evident that if $T_0^* = 0$, then the first inequality of Eq. (10-3-7) simplifies to

$$T_0\, d\bar{\epsilon} > 0$$

which inequality states that the increment of plastic work expended is positive. The conclusions from this simple loading program can be generalized by considering the existing state of stress to be $\mathbf{T}^*$ on or inside a loading surface in nine-

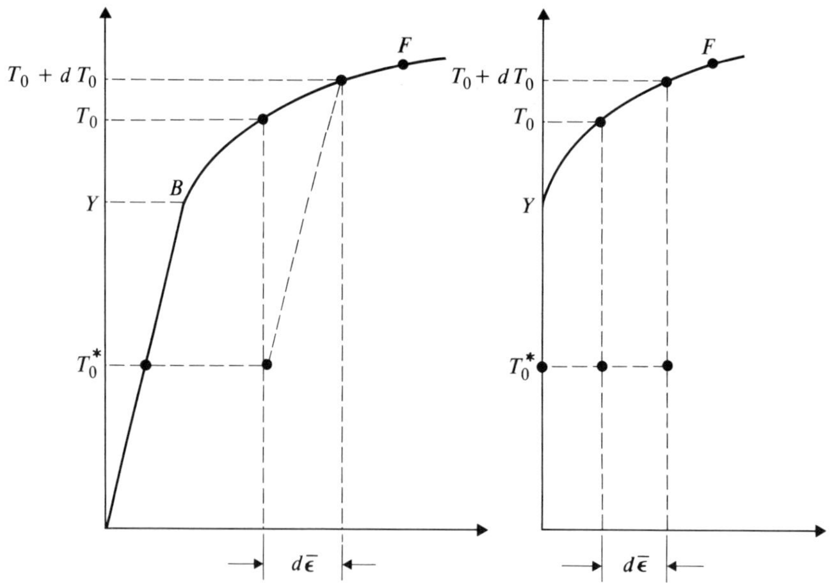

Figure 10-5 Idealized stress-strain diagram of the simple uniaxial tension test. Loading is from an existing state of stress $T_0^* < Y$ to T_0, where Y is the stress at the yield point B. This loading is followed by incremental loading from T_0 to $T_0 + dT_0$; the increment dT_0 produces an increment of plastic strain $d\bar{\epsilon}_{11} = d\bar{\epsilon}$, followed in turn by unloading to T_0^*.

dimensional stress space and that the final state is $\mathbf{T}$. Drucker's postulate can now be expressed in the form

$$\operatorname{tr}[(\mathbf{T}-\mathbf{T}^*)\dot{\bar{\epsilon}}] \geqslant 0 \qquad \operatorname{tr}(\dot{\mathbf{T}}\dot{\bar{\epsilon}}) \geqslant 0 \tag{10-3-8}$$

where the equality sign holds only during neutral loading. The inequality on the right-hand side of Eq. (10-3-8) is often referred to as the condition of uniqueness. That Drucker's definition of work-hardening implies the inequalities of Eq. (10-3-8) can be shown from the following considerations of the net work done by the external agency during the loading cycle of present interest. At time $t = 0$, the existing state of stress is $\mathbf{T}^*$. The external agency changes the stress to $\mathbf{T}^{(1)}$ on the yield surface and then to a neighboring point $\mathbf{T}^{(1)}+\delta\mathbf{T}$ outside or on the initial yield surface at time $t_1+\delta t$, followed by unloading back to $\mathbf{T}^*$ at time t^*. The net work done by the external agency during the cycle is $\Delta W_{\text{ext}} = \delta W_T - \delta W_0$ where δW_T is the total work done during the cycle and δW_0 is the work that would be done during the same deformation by the initial stress $\mathbf{T}^*$ if it were held constant. Noting that the net elastic work during the cycle is zero, it follows that

$$\begin{aligned}\delta W_T &= \int_0^{t_1} \operatorname{tr}(\mathbf{T}\dot{\hat{\epsilon}})\,dt + \int_{t_1}^{t_1+\delta t} \operatorname{tr}[\mathbf{T}(\dot{\hat{\epsilon}}+\dot{\bar{\epsilon}})]\,dt + \int_{t_1+\delta t}^{t^*} \operatorname{tr}(\mathbf{T}\dot{\hat{\epsilon}})\,dt \\ &= \int_{t_1}^{t_1+\delta t} \operatorname{tr}(\mathbf{T}\dot{\bar{\epsilon}})\,dt\end{aligned}$$

$$\delta W_0 = \int_{t_1}^{t_1+\delta t} \operatorname{tr}(\mathbf{T}^*\dot{\bar{\epsilon}})\,dt$$

and therefore

$$\Delta W_{\text{ext}} = \int_{t_1}^{t_1+\delta t} \operatorname{tr}[(\mathbf{T}-\mathbf{T}^*)\dot{\bar{\epsilon}}]\,dt \tag{10-3-9}$$

Applying the second part of Drucker's definition, which requires ΔW_{ext} to be either zero or positive for any plastic deformation during an arbitrarily short δt following a loading to an arbitrary point $\mathbf{T}$ on the yield surface, it is evident that the integrand in Eq. (10-3-9) must either be zero or positive, thus establishing the inequality on the left-hand side of Eq. (10-3-8). The first part of Drucker's definition can now be applied. For an initial $\mathbf{T}^* \equiv \mathbf{T}^{(1)}$, $d\mathbf{T} = \mathbf{T}-\mathbf{T}^*$, and the requirement that $\operatorname{tr}[(\mathbf{T}-\mathbf{T}^*)\,d\bar{\epsilon}] > 0$ then implies that $\operatorname{tr}(\dot{\mathbf{T}}\dot{\bar{\epsilon}}) > 0$, which is simply the inequality on the right-hand side of Eq. (10-3-8).

Consider ideal plasticity characterized by plastic deformation in the absence of strain-hardening. For such a system the condition of Eq. (10-3-8) has the particular form

$$\operatorname{tr}(d\mathbf{T}\,d\bar{\epsilon}) = 0 \tag{10-3-10}$$

while for an ideal plastic material it follows from Eq. (10-3-5) that any change in the stresses during plastic flow must satisfy the relation

$$f(\mathbf{T}) = 0 \qquad df = \frac{\partial f}{\partial T_{ij}}\,dT_{ij} = 0 \tag{10-3-11}$$

where $f(\mathbf{T})$ is the yield function of Eq. (10-2-1).

Following Naghdi (1960), a useful concept is to consider an increment of stress $d\mathbf{T}$ as a vector dS in the nine-dimensional stress space, and the corresponding increment of plastic strain $d\epsilon$ as the components of a vector dE in the same stress space. From Eq. (10-3-8) it then follows that

$$dS \cdot dE = |dS|\,|dE| \cos\theta \geqslant 0 \tag{10-3-12}$$

which implies that

$$-\frac{\pi}{2} \leqslant \theta \leqslant \frac{\pi}{2} \tag{10-3-13}$$

from which it is concluded that the angle between dS and dE is acute. Figure 10-6 shows a schematic representation of the plastic strain vector dE at a regular, i.e., a smooth, point B on the yield surface in stress space. The line ABC in Fig. 10-6 represents a hyperplane normal to dE. From Eqs. (10-3-12) and (10-3-13) it is evident that all interior points, i.e., the vectors dS, must lie to one side of the hyperplane ABC and opposite to the side away from which dE points. The loading vectors dS are outward vectors whose directions are bounded by the tangent plane of the yield surface, and therefore the hyperplane ABC must be tangent to the loading surface at the point B. Hence, since dE is normal to the hyperplane it must also be normal to the yield surface at the point B. Thus through every such point B on the yield surface there is a plane such that all points interior to the surface lie on the same side of the plane; i.e., the yield surface is convex.

Because the tangent plane at a regular point on a surface is unique it follows that the hyperplane ABC must be unique at a regular point of the yield surface. Hence, the direction of dE normal to the hyperplane must also be unique. From this follows the conclusion that at a smooth point of the yield surface the direction of $d\bar{\epsilon}$ is independent of the direction of $d\mathbf{T}$.

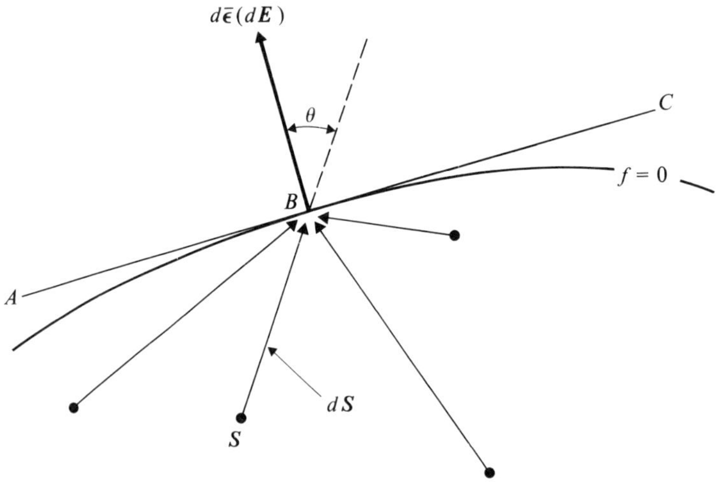

Figure 10-6 Schematic representation of the convex yield surface in stress space.

The conclusion that the direction of the plastic strain rate vector is that of the outer normal at a smooth point of the loading surface requires that

$$\dot{\epsilon} = \lambda \mathbf{G}' \qquad G'^{i}{}_{j} = \frac{\partial f}{\partial T_i^j} \tag{10-3-14}$$

where λ is a function which may depend on stress, stress rate, strain, and strain history. It is evident from the form of Eqs. (10-3-14) that the loading function f is to be regarded as a plastic potential (see, for example, Hill, 1950; Naghdi, 1960).

For a yield surface having corners, the normal vectors at these corners are not unique with regard to their directions. Thus, at a corner the inequality on the left-hand side of Eq. (10-3-8) can be satisfied by any $\dot{\epsilon}$ within or on the generalized cone formed by the limiting positions of the normals to the smooth surfaces intersecting at the corner. Koiter (1953) generalized the plastic potential theory by regarding the yield surface to be made up of several smooth surfaces, defined by several functions

$$f_1(\mathbf{T}), f_2(\mathbf{T}), \ldots, f_n(\mathbf{T}) \tag{10-3-15}$$

For elastic states all the functions are negative, while at yield at least one must equal zero. If more than one yield function is zero, the stress point is at a corner on the yield surface. Suppose $f_n, \ldots, f_m$ are all zero and all the others negative; then the plastic potential relation now takes the form

$$d\bar{\epsilon} = d\lambda_n \mathbf{G}'_n + \cdots + d\lambda_m \mathbf{G}'_m \qquad (G'^{i}{}_{j})_p = \frac{\partial f_p}{\partial T_i^j} \tag{10-3-16}$$

where $d\lambda_n, \ldots, d\lambda_m$ are arbitrary nonnegative factors of proportionality.

10-3-2 Isotropic Hardening

For the particular loading conditions in which the yield surface simply increases in size, without any change occurring in its original shape, the post-yield behavior is referred to as isotropic hardening. In the case of isotropic hardening, the yield curves in the π plane for the von Mises and Tresca yield criteria are the concentric circles and regular hexagons shown in Fig. 10-4. Basic to any description of an isotropic hardening medium are the concepts underlying the definition of the size-determining parameter and its dependence on the deformation. Two possible concepts have been proposed for this purpose which are basically different in formulation. However, when used with the von Mises yield condition, the two isotropic hardening assumptions can be shown to be equivalent.

(a) Universal plastic stress-strain curve This concept assumes the existence of a universal plastic stress-strain curve

$$\sigma'' = \tilde{H}\left[\int d\bar{\epsilon}''\right] \tag{10-3-17}$$

where the effective or generalized stress σ'' measures the size of the yield surface and $\int d\bar{\epsilon}''$ is the integral of the effective plastic strain increment $d\bar{\epsilon}''$. Consideration of the concepts underlying the formulation of general yield criteria, as, for example, Eq. (10-2-11), implies that

$$\sigma'' = \sigma''(J_2', J_3') \qquad d\bar{\epsilon}'' = d\bar{\epsilon}''(\mathrm{II}_{d\bar{\epsilon}}, \mathrm{III}_{d\bar{\epsilon}}) \tag{10-3-18}$$

where $\mathrm{II}_{d\bar{\epsilon}}$ and $\mathrm{III}_{d\bar{\epsilon}}$ are the second and third invariants respectively of the plastic strain increment tensor $d\bar{\epsilon}$. The functional form of $\tilde{H}$ is determined from a uniaxial stress-strain test of the material for which it is specified that $\sigma'' = T\langle 11\rangle$ and $d\bar{\epsilon}'' = d\bar{\epsilon}\langle 11\rangle$.

Assuming the yield criterion to be that of von Mises (1913), an effective stress σ'' consistent with Eq. (10-2-13) can be defined by the relation

$$\sigma'' = (3J_2')^{1/2} = (\tfrac{3}{2}\,\mathrm{tr}\,\mathbf{T}'^2)^{1/2} \tag{10-3-19}$$

This effective stress can be expressed in the form

$$\sigma'' = \sqrt{3}[\tfrac{1}{2}(T'\langle 11\rangle^2 + T'\langle 22\rangle^2 + T'\langle 33\rangle^2) + (T\langle 12\rangle^2 + T\langle 23\rangle^2 + T\langle 31\rangle^2)]^{1/2} \tag{10-3-20}$$

from which it is evident that in uniaxial tension or compression, that is, $T\langle 11\rangle$, the effective stress simplifies to $\sigma'' = |T\langle 11\rangle|$.

An often-used alternative to σ'' is the octahedral shear stress

$$\tau_0 = (\tfrac{2}{3}J_2')^{1/2} = \frac{\sqrt{2}}{3}\sigma'' \tag{10-3-21}$$

which, however, is not so convenient for constructing universal stress-strain curves.

The effective strain increment $d\bar{\epsilon}''$, corresponding to the use of σ'' defined by Eq. (10-3-20), is defined by the relation

$$d\bar{\epsilon}'' = (-\tfrac{4}{3}\mathrm{II}_{d\bar{\epsilon}})^{1/2} = [\tfrac{2}{3}\,\mathrm{tr}\,(d\bar{\epsilon})^2]^{1/2} \tag{10-3-22}$$

In uniaxial tension or compression $T\langle 11\rangle$, for which $d\bar{\epsilon}\langle 22\rangle = d\bar{\epsilon}\langle 33\rangle = -\tfrac{1}{2}d\bar{\epsilon}\langle 11\rangle$, the effective strain increment simplifies to $d\bar{\epsilon}'' = d\bar{\epsilon}\langle 11\rangle$.

The assumption of the von Mises yield criterion enables the multiplier $d\lambda$ to be expressed by way of Eq. (10-3-14) in the form

$$\mathrm{tr}\,(d\bar{\epsilon})^2 = (d\lambda)^2\,\mathrm{tr}\,\mathbf{T}'^2 \tag{10-3-23}$$

Equation (10-3-23) can be expressed by way of Eqs. (10-3-19) and (10-3-22) in the form

$$\tfrac{3}{2}(d\bar{\epsilon}'')^2 = (d\lambda)^2\,\tfrac{2}{3}(\sigma'')^2 \tag{10-3-24}$$

which gives

$$d\lambda = \frac{3}{2}\frac{d\bar{\epsilon}''}{\sigma''} = \frac{3}{2}\frac{d\sigma''}{\sigma''}\left(\frac{1}{\tilde{H}'}\right) \tag{10-3-25}$$

where

$$\tilde{H}' = \frac{d\sigma''}{d\bar{\epsilon}''} \tag{10-3-26}$$

is the slope of the universal plastic stress-strain curve at the current value of σ''. Hence, from Eqs. (10-3-14) and (10-3-25),

$$d\bar{\epsilon} = \frac{3}{2}\frac{d\bar{\epsilon}''}{\sigma''}\mathbf{T}' = \frac{3}{2}\frac{d\sigma''}{\sigma''\tilde{H}'}\mathbf{T}' \qquad (d\sigma'' \geqslant 0)$$

$$= \frac{3}{2\sigma''\tilde{H}'}\operatorname{tr}(\mathbf{G}'' d\mathbf{T}^T)\mathbf{T}' \qquad \left(G''^i{}_j = \frac{\partial\sigma''}{\partial T_i{}^j}\right) \tag{10-3-27}$$

from which it is seen that $d\bar{\epsilon}$ depends linearly on the stress increments $d\mathbf{T}$ during loading.

(b) Total plastic work This concept assumes that the size of the yield surface depends only on the total plastic work $\bar{W}$. Assuming the von Mises yield criterion, this particular isotropic hardening assumption can be expressed in the form

$$\sigma'' = \tilde{F}(\bar{W}) \tag{10-3-28}$$

where

$$\bar{W} = \int \operatorname{tr}(\mathbf{T}^T\, d\bar{\epsilon}) \tag{10-3-29}$$

is the total plastic work.

When the von Mises yield criterion is considered to be appropriate, the two isotropic hardening assumptions expressed by way of Eqs. (10-3-17) and (10-3-28) can be shown to be equivalent as follows.

Let $\boldsymbol{a}$ denote the vector in nine-dimensional stress space with components $\sqrt{3/2}\mathbf{T}'$ and magnitude σ''. Similarly, let $d\boldsymbol{b}$ denote the vector with components $\sqrt{2/3}\, d\bar{\epsilon}$ and magnitude $d\bar{\epsilon}''$. The increment of plastic work $d\bar{W}$ can now be expressed in the form

$$d\bar{W} = \operatorname{tr}(\mathbf{T}^T\, d\bar{\epsilon}) = \operatorname{tr}(\mathbf{T}'^T\, d\bar{\epsilon}) = \boldsymbol{a}\cdot d\boldsymbol{b} \tag{10-3-30}$$

However, the plastic potential equation (10-3-14) implies

$$d\boldsymbol{b} = \tfrac{2}{3}d\lambda\boldsymbol{u} \tag{10-3-31}$$

from which it is concluded that $d\boldsymbol{b}$ is parallel to $\boldsymbol{a}$ and hence

$$\boldsymbol{a}\cdot d\boldsymbol{b} = \sigma''\, d\bar{\epsilon}'' \tag{10-3-32}$$

From Eqs. (10-3-30) and (10-3-32),

$$d\bar{W} = \sigma''\, d\bar{\epsilon}'' \tag{10-3-33}$$

and hence $\bar{W}$, which is just the area under the universal stress-strain $(\sigma'', \int d\bar{\epsilon}'')$ curve, is a single-valued function of $\int d\bar{\epsilon}''$, thus establishing the equivalence of the two isotropic hardening assumptions of immediate interest.

The increment of plastic work $d\bar{W}$ can be expressed by way of the plastic potential equation (10-3-14) in the form

$$d\bar{W} = \operatorname{tr}(\mathbf{T}^T\, d\bar{\epsilon}) = d\lambda \operatorname{tr}(\mathbf{T}^T\mathbf{G}')$$

which can be rearranged to give the scalar multiplier

$$d\lambda = \frac{d\bar{W}}{\operatorname{tr}(\mathbf{T}^T\mathbf{G}')} \tag{10-3-34}$$

Substituting for the scalar multiplier $d\lambda$ in Eq. (10-3-14) from Eq. (10-3-34) gives the plastic potential equation in the form

$$d\bar{\epsilon} = \frac{d\bar{W}}{\operatorname{tr}(\mathbf{T}^T\mathbf{G}')}\mathbf{G}' \tag{10-3-35}$$

By redefining f the general yield condition $f(\mathbf{T}) = 0$ can be reformulated as

$$f(\mathbf{T}) - \tilde{F}(\bar{W}) = 0 \tag{10-3-36}$$

which represents a generalization of the isotropic hardening assumption of Eq. (10-3-28). From Eq. (10-3-36) can be obtained

$$\operatorname{tr}(\mathbf{G}'\,d\mathbf{T}^T) = \tilde{F}'(\bar{W})\,d\bar{W} \tag{10-3-37}$$

where $\tilde{F}'$ is the derivative of $\tilde{F}$ with respect to its argument. Equation (10-3-37) is the condition that the generalized isotropic hardening assumption expressed by Eq. (10-3-36) remains satisfied after an increment of loading. Equation (10-3-37) can be used to express Eq. (10-3-35) in the form

$$d\bar{\epsilon} = \frac{\operatorname{tr}(\mathbf{G}'\,d\mathbf{T}^T)}{\tilde{F}(\bar{W})\operatorname{tr}(\mathbf{T}^T\mathbf{G}')}\mathbf{G}' \tag{10-3-38}$$

which, compared with Eq. (10-3-27), is again seen to exhibit explicitly the linear dependence on $d\mathbf{T}$.

It is to be noted that the properties of immediate interest associated with the post-yield behavior, i.e., convexity of the loading surface $f = 0$, the normality of the plastic strain rate vector to $f = 0$ at regular regimes, and the linearity of the plastic strain rate in the stress rate, follow from Drucker's theory of strain-hardening. Drucker's theory of strain-hardening is often interpreted as a statement that the material is stable. Hence, these properties can only be regarded as being valid for the class of stable materials. The behavior of mild steel at its upper yield point is evidence for the existence of unstable materials. With regard to unstable materials, no general constitutive equations have as yet been proposed for this particular class of material.

10-4 EXPERIMENTAL APPLICATIONS

10-4-1 Some General Considerations

Consider a right circular cylindrical tube or solid rod of an incompressible material. Let (R, Θ, Z) be the cylindrical referential coordinates in the initial state of a particle that is located in the deformed configuration by the cylindrical spatial coordinates (r, θ, z). The major axis of the tube or rod is aligned with the Z axis.

The simple deformations are

$$r = \left(\frac{K+R^2}{F}\right)^{1/2} \qquad \theta = \Theta + DZ \qquad z = FZ \tag{10-4-1}$$

where D, F, and K are constants subject to the condition that F and K have values such that $(K+R^2)/F > 0$ when R is in the interval $(R_1 \geqslant R \geqslant R_2)$; R_1 and R_2 are the outer and inner radii respectively of the undeformed tube.

The plastic strain increments are

$$d\bar{\epsilon}\langle rr\rangle = -\tfrac{1}{2}d\bar{\epsilon}\langle zz\rangle - \tfrac{1}{2}d\ln\bar{\eta} \tag{10-4-2}$$

$$d\bar{\epsilon}\langle\theta\theta\rangle = -\tfrac{1}{2}d\bar{\epsilon}\langle zz\rangle + \tfrac{1}{2}d\ln\bar{\eta} \tag{10-4-3}$$

$$d\bar{\epsilon}\langle zz\rangle = \frac{d\bar{F}}{\bar{F}} \tag{10-4-4}$$

$$d\bar{\epsilon}\langle\theta z\rangle = \tfrac{1}{2}d\bar{\gamma}\langle\theta z\rangle = \frac{1}{2}\left(\frac{\bar{r}}{\bar{F}}\right)d\bar{D} \tag{10-4-5}$$

where

$$F = \hat{F}\bar{F} \qquad \hat{F} = \frac{l}{\bar{l}} \qquad \bar{F} = \frac{\bar{l}}{L} \tag{10-4-6}$$

$$\bar{\eta} = \left(\frac{\bar{r}^2}{R^2}\right)\bar{F} = 1 + \frac{\bar{K}}{R^2} \tag{10-4-7}$$

and $l, \bar{l}, L$ are the total current length, the current length corresponding to the unloaded state for a given value of the permanent twist, and the undeformed length of the specimen respectively.

For a permanently twisted tube or solid rod with its major axis aligned with the Z axis, the only nonvanishing components of stress are specified to be $T\langle rr\rangle$, $T\langle\theta\theta\rangle$, $T\langle zz\rangle$ and the shear stress $T\langle\theta z\rangle$.

10-4-2 Universal Stress-Strain Curves

The method of observing the elastic range of response of a material tested in uniaxial tension has been described in Sec. 6-5-1, the characteristic response being considered in terms of nominal stresses and conventional strains which are defined in terms of the initial, undeformed dimensions of the specimen. When there are considerable changes in the specimen dimensions, as occurs in the necking down of a specimen plastically deformed in simple tension, it is necessary to define true stresses and strains which are referred to the deformed specimen dimensions.

Consider a circular, cylindrical, solid-rod test specimen subjected to a uniform axial tension $T\langle zz\rangle$. The true stress

$$T\langle zz\rangle = \frac{F}{a} \tag{10-4-8}$$

where, in contrast to Eq. (6-5-1), a is now the actual cross-sectional area corresponding to the force F. To a good degree of approximation Bridgman

(1923, 1952) has shown that hydrostatic pressure does not affect either the initial yield or the post-yield deformation behavior of a wide range of metals, the observations of Crossland (1954) being in accord with these conclusions. For plastic deformation at constant volume

$$T\langle zz\rangle = T^{0}(1+\tilde{e}\langle zz\rangle) \qquad \text{(tension)} \tag{10-4-9}$$

where T^0 is the nominal tensile stress and

$$\tilde{e}\langle zz\rangle = F-1 \qquad \text{(tension)} \tag{10-4-10}$$

is the conventional tensile engineering strain. From Eqs. (10-4-4) and (10-4-10), the total natural strain

$$\epsilon\langle zz\rangle = \ln F = \ln(1+\tilde{e}\langle zz\rangle) \qquad \text{(tension)} \tag{10-4-11}$$

Setting $T\langle zz\rangle \equiv T_0$, the total plastic work $\bar{W}$, defined by Eq. (10-3-29), can be expressed in the form

$$\bar{W} = \int_1^F T_0\left(\frac{dF'}{F'} - \frac{dT_0}{E}\right) = \int_1^F T_0\frac{dF'}{F'} - \frac{T_0^2}{2E} \tag{10-4-12}$$

where E is Young's modulus. Hence, the isotropic hardening assumption of Eq. (10-3-28) can be expressed in the form

$$T_0 = \tilde{F}\left(\int_1^F T_0\frac{dF'}{F'} - \frac{T_0^2}{2E}\right) \tag{10-4-13}$$

the argument $\tilde{F}$ being simply the area under the $(T_0, \epsilon\langle zz\rangle - T_0/E)$ curve up to the ordinate T_0.

The difficulties referred to in Sec. 6-5-1, relating to the determination of the elastic constants from measurements in compression, also apply to observations of the plastic behavior of a material in uniaxial compression. As an example of the mechanical response of materials in compression see Altshuler and Christian (1967), who have determined the mechanical properties of pure iron tested in compression over a wide temperature range. For the uniform compression of a short cylindrical solid-rod specimen, the true stress

$$T\langle zz\rangle = T^{0}(1-\tilde{e}\langle zz\rangle) \qquad \text{(compression)} \tag{10-4-14}$$

and from Eq. (10-4-4) the total natural strain

$$\epsilon\langle zz\rangle = \ln\left(\frac{1}{F}\right) = \ln\left(\frac{1}{1-\tilde{e}\langle zz\rangle}\right) \qquad \text{(compression)} \tag{10-4-15}$$

where now $\tilde{e}\langle zz\rangle = 1-F$. As for uniaxial tension, the isotropic hardening assumption of Eq. (10-3-28) can be expressed in the form

$$T_0 = \tilde{F}\left(\int_1^F T_0\frac{dF'}{F'} - \frac{T_0^2}{2E}\right) \tag{10-4-16}$$

the argument of $\tilde{F}$ again being simply the area under the $(T_0, \epsilon\langle zz\rangle - T_0/E)$ curve up to the ordinate T_0, it being noted that for compression $\epsilon\langle zz\rangle$ is related to $e\langle zz\rangle$ by Eq. (10-4-15).

It is evident from Eqs. (10-4-13) and (10-4-16) that the isotropic hardening assumption of Eq. (10-3-28) implies that the tensile and compressive stress-strain curves coincide when expressed in terms of true stress and natural strain. Ludwik and Scheu (1925) confirmed this conclusion for annealed copper (Fig. 10-7). The comparison of the stress-strain curves, characteristic of the response of a material tested in compression and in tension, in terms of true stress and natural strain was first suggested by Ludwik (1909).

The simple torsion test described in Sec. 6-5 for observing the elastic range of response of a material can be applied directly to a material in the plastic range of deformation. Simple torsion involves subjecting the tube to a twisting moment Γ, the resulting simple deformations of immediate interest being those described by Eq. (10-4-1). It is here assumed that the only nonvanishing component of stress is the shear stress τ ($\equiv T\langle\theta z\rangle$). The mechanical response is evaluated from the (Γ, θ) curve in the form of a shear stress–shear strain $(\tau, \bar{\gamma})$ curve where the shear strain $\bar{\gamma}$ is the plastic component of the total shear strain defined by Eq. (6-5-17) and the shear stress is evaluated from Eq. (6-5-20). A typical characteristic shear stress–shear strain $(\tau, \bar{\gamma})$ curve for aluminum is shown in Fig. 10-8 (Billington, 1977a).

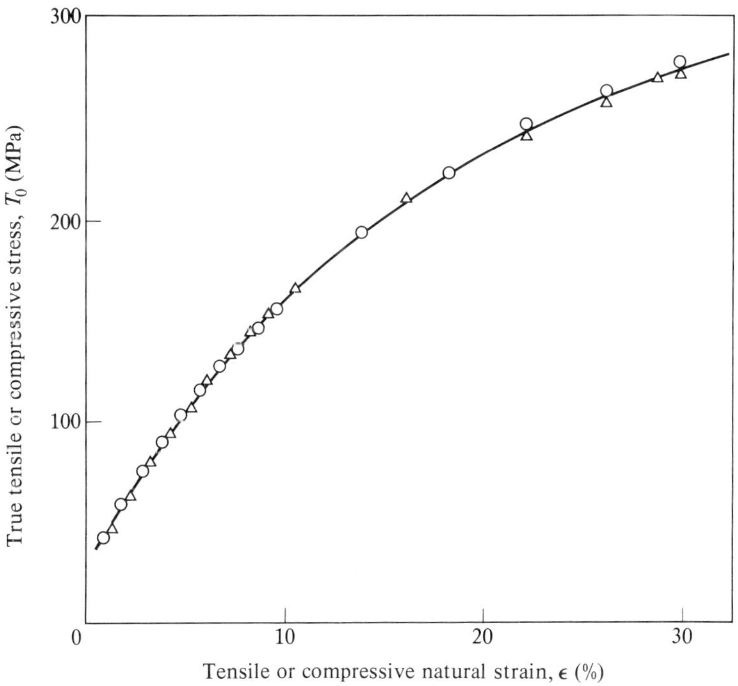

Figure 10-7 Comparison of the true stress–natural strain curves for annealed copper tested in uniaxial compression, △, and tension, ○. (*Ludwick and Scheu, 1925.*)

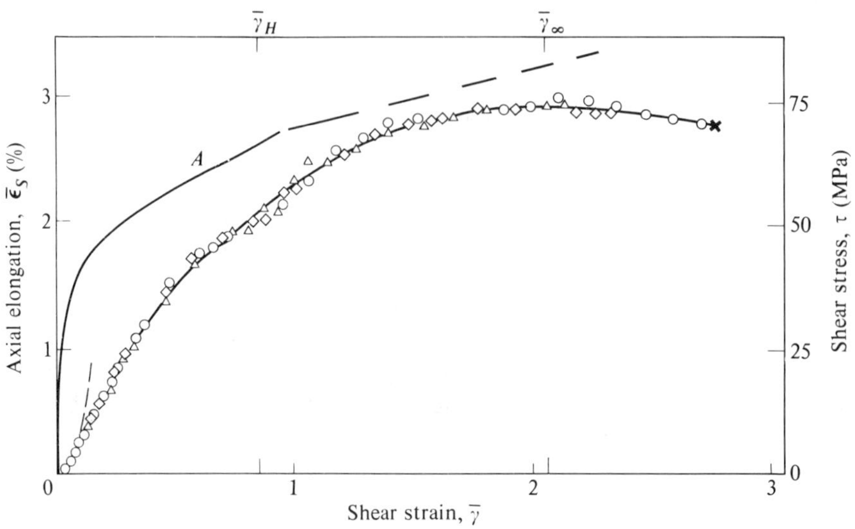

Figure 10-8 Variation of the static irrecoverable natural strain $\bar{\epsilon}_S$ with the static plastic component $\bar{\gamma}$ of the shear strain for thin-walled specimens of aluminum. Wall thickness $R_1 - R_2$: ○, 1 mm; ◇; 2 mm; △, 3 mm. Also shown is the static stress-strain $(\tau, \bar{\gamma})$ curve (curve A) for aluminum in torsion. × denotes fracture. (*Billington, 1977a.*)

Assuming for measurements in simple torsion that the only component of stress is $T\langle\theta z\rangle$ $(\equiv \tau)$, then the effective stress σ'' defined by Eq. (10-3-20) reduces to $\sigma'' = \sqrt{3}\tau$. As for uniaxial tension, the isotropic hardening assumption of Eq. (10-3-28) can be expressed in the form

$$\sqrt{3}\tau = \tilde{F}\left(\int_0^\gamma \tau\, d\gamma' - \frac{\tau^2}{2G}\right) \tag{10-4-17}$$

where G is the shear modulus of rigidity. Comparison of Eqs. (10-4-13), (10-4-16), and (10-4-17) shows that $\sqrt{3}\tau$ is the same function of $[\gamma - (\tau/G)]$ as T_0 is of $(\ln F - T_0/E)$. This conclusion is confirmed by experiment for some materials, as noted by Hill (1950), who cites as evidence the experiments of Zener and Hollomon (1946). In contrast, Taylor and Quinney (1931) show for annealed copper a comparison between measurements in tension and torsion from which it is evident that the isotropic hardening assumptions of Eqs. (10-3-17) and (10-3-28) applied by way of the von Mises yield criterion do not give a universal stress-strain curve for this particular material. This has also been observed for aluminum and copper in the fully annealed state, and iron and copper in the as-received state, the comparison being between measurements in compression, tension, and torsion (Billington, 1977b).

10-4-3 Combined Stress Measurements

The combined extension and torsion of an incompressible circular cylindrical tube can be discussed in terms of the modified Lode (1926) variables

$$\mu = \frac{3\sigma_1'}{\sigma_3' - \sigma_2'} \tag{10-4-18a}$$

$$\nu = \frac{3d\bar{\epsilon}_1}{d\bar{\epsilon}_3 - d\bar{\epsilon}_2} \tag{10-4-18b}$$

where the σ_a' $(a = 1, 2, 3)$ are the principal deviatoric stresses and the $d\bar{\epsilon}_a$ $(a = 1, 2, 3)$ are the corresponding principal incremental natural strains. For the condition

$$\sigma_3 > \sigma_1 > \sigma_2 \tag{10-4-19}$$

the characteristic equation

$$-\det(\mathbf{T}' - \sigma_a'\mathbf{I}) = \sigma_a'^3 - J_2'\sigma_a' - J_3' = 0 \tag{10-4-20}$$

can be solved in terms of the stress field of present interest to give

$$\sigma_3' = -\tfrac{1}{2}[\zeta - (\xi^2 + 4)^{1/2}]T\langle\theta z\rangle \tag{10-4-21}$$

$$\sigma_1' = \zeta T\langle\theta z\rangle \tag{10-4-22}$$

$$\sigma_2' = -\tfrac{1}{2}[\zeta + (\xi^2 + 4)^{1/2}]T\langle\theta z\rangle \tag{10-4-23}$$

where the stress ratios are

$$\xi = \frac{T\langle\theta\theta\rangle - T\langle zz\rangle}{T\langle\theta z\rangle} \qquad \zeta = \frac{2T\langle rr\rangle - T\langle\theta\theta\rangle - T\langle zz\rangle}{3T\langle\theta z\rangle} \tag{10-4-24}$$

These relations give

$$\mu = \frac{3\zeta}{(\xi^2 + 4)^{1/2}} \tag{10-4-25}$$

Similarly, ν can be expressed in the form

$$\nu = \frac{3(d\bar{\epsilon}\langle rr\rangle / d\bar{\epsilon}\langle\theta z\rangle)}{[(d\bar{\epsilon}\langle\theta\theta\rangle / d\bar{\epsilon}\langle\theta z\rangle - d\bar{\epsilon}\langle zz\rangle / d\bar{\epsilon}\langle\theta z\rangle)^2 + 4]^{1/2}} \tag{10-4-26}$$

The flow rule of Eq. (10-3-14) can be rearranged for the stress field of present interest, using Eq. (10-2-11), to give

$$\frac{d\bar{\epsilon}\langle rr\rangle}{d\bar{\epsilon}\langle\theta z\rangle} = \frac{G'\langle rr\rangle}{G'\langle\theta z\rangle} = \zeta(1 - \tfrac{4}{3}\chi) \tag{10-4-27}$$

$$\frac{d\bar{\epsilon}\langle\theta\theta\rangle}{d\bar{\epsilon}\langle\theta z\rangle} = \frac{G'\langle\theta\theta\rangle}{G'\langle\theta z\rangle} = \tfrac{1}{2}\xi - \tfrac{1}{2}\zeta(1 - \tfrac{4}{3}\chi) \tag{10-4-28}$$

$$\frac{d\bar{\epsilon}\langle zz\rangle}{d\bar{\epsilon}\langle\theta z\rangle} = \frac{G'\langle zz\rangle}{G'\langle\theta z\rangle} = -\tfrac{1}{2}\xi - \tfrac{1}{2}\zeta(1 - \tfrac{4}{3}\chi) \tag{10-4-29}$$

where

$$\chi = \left(1 - \frac{H}{\Lambda}\right)\frac{1}{\Psi} \tag{10-4-30}$$

$$\Lambda = H\left[1 - \frac{27}{4}\frac{dH}{d\omega}\frac{(1-3\Psi)(1-\Psi)\Psi}{(1-(\kappa/H)\,dL/dJ_2']H}\right] \tag{10-4-31}$$

$$\Psi = \frac{\zeta^2}{1+\frac{1}{4}\xi^2+\frac{3}{4}\zeta^2} \tag{10-4-32}$$

and ω is defined by Eq. (10-2-12).

Substituting in Eq. (10-4-26) from Eqs. (10-4-27), (10-4-28), (10-4-29), and rearranging by way of Eq. (10-4-25) gives

$$\nu = \mu(1-\tfrac{4}{3}\chi) \tag{10-4-33}$$

For the von Mises yield criterion, $\kappa = k^2$, $H = 1$, $L = 1$ in Eq. (10-2-11) and hence, from Eq. (10-4-30), $\chi = 0$. Hence, for the von Mises yield criterion, Eq. (10-4-33) reduces to

$$\nu = \mu \qquad (H = 1, L = 1, \chi = 0) \tag{10-4-34}$$

which is Lode's relation, and is characteristic of the mechanical response of an idealized isotropic incompressible elastic-plastic material.

The principal stresses can be expressed in the form

$$\sigma_3' = -\tfrac{1}{2}(\zeta+\xi\sec 2\theta)T\langle\theta z\rangle \tag{10-4-35}$$

$$\sigma_1' = \zeta T\langle\theta z\rangle \tag{10-4-36}$$

$$\sigma_2' = -\tfrac{1}{2}(\zeta-\zeta\sec 2\theta)T\langle\theta z\rangle \tag{10-4-37}$$

which, when substituted into Eq. (10-4-18a), gives

$$\mu = -3\frac{\zeta}{\xi}\cos 2\theta \tag{10-4-38}$$

Comparison of Eqs. (10-4-21) and (10-4-35) gives

$$\cot 2\theta = -\tfrac{1}{2}\xi \tag{10-4-39}$$

Similarly, ν can be expressed in the form

$$\nu = -3\frac{\zeta}{\xi}(1-\tfrac{4}{3}\chi)\cos 2\phi \tag{10-4-40}$$

For ν, the relation corresponding to Eq. (10-4-39) has the form

$$\cot 2\phi = \frac{1}{2}\left(\frac{d\bar{\epsilon}\langle zz\rangle}{d\bar{\epsilon}\langle\theta z\rangle} - \frac{d\bar{\epsilon}\langle\theta\theta\rangle}{d\bar{\epsilon}\langle\theta z\rangle}\right) \tag{10-4-41}$$

Substituting for ν in Eq. (10-4-40) from Eq. (10-4-33) and comparing the resulting relation with Eq. (10-4-38) gives

$$\cos 2\phi = \cos 2\theta \tag{10-4-42}$$

which implies the condition that the directions of the principal axes of stress and strain are the same, so that $\theta = \phi$.

Tubes of iron, copper, and nickel were subjected to combined tension and internal pressure by Lode (1926), who maintained the ratio of axial stress to circumferential stress approximately constant in each test. Lode observed deviations from the behavior predicted for the classical idealized plastic material (Reuss, 1930), which could not be completely accounted for by anisotropy. These deviations were confirmed by Taylor and Quinney (1931), who stressed tubes of aluminum, copper, and mild steel in combined tension and torsion. In these experiments the axial load was held constant, for a given test, while the torque was increased, thus giving rise to stress ratios which were not constant. The degree of anisotropy was kept sufficiently small for it to be neglected. Taylor and Quinney assumed the only nonvanishing components of stress to be the axial stress $T\langle zz\rangle$ and the shear stress $T\langle \theta z\rangle$. For these conditions, Eq. (10-4-39) reduces to

$$\tan 2\theta = 2\frac{T\langle \theta z\rangle}{T\langle zz\rangle} \tag{10-4-43}$$

which can be evaluated from the known values of $T\langle zz\rangle$ and $T\langle \theta z\rangle$. Taylor and Quinney then evaluated $\tan 2\phi$ in terms of the total components of strain, $e\langle rr\rangle$, $e\langle \theta\theta\rangle$, and $e\langle zz\rangle$, and established the condition $\theta = \phi$ within experimental error. The experimental values of ν and μ, obtained by Taylor and Quinney, are shown in Fig. 10-9. It is evident that these results are not in accord with Eq. (10-4-34). Prager (1945) has shown that these deviations can be accounted for, at least in part, by a modified von Mises yield criterion of the form of Eq. (10-2-11) (see also Handelman, Lin, and Prager, 1947). This has been confirmed by Drucker (1949) who successfully applied the same form of yield function to the combined stress

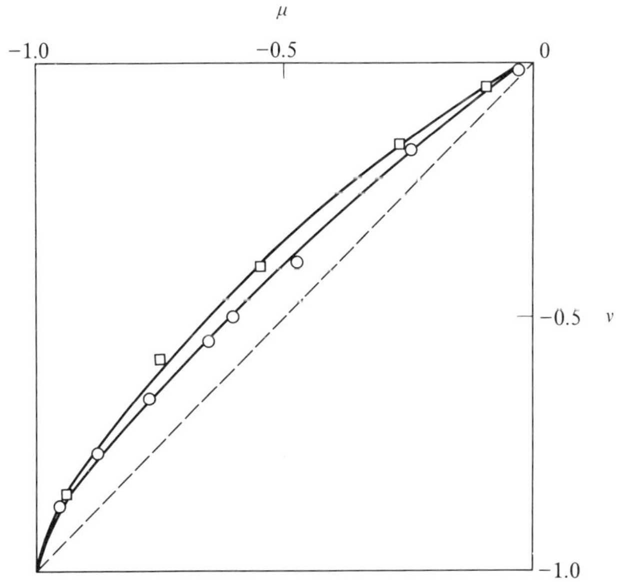

Figure 10-9 The (μ, ν) relation. Al, □; Cu, ○. (*Taylor and Quinney, 1931.*)

measurements of Osgood (1947). Morrison and Shepherd (1950) have also examined the assumptions underlying the formulation of the incremental-type constitutive equation for the idealized isotropic incompressible elastic-plastic material using thin-walled tubes loaded in combined tension and torsion. Their conclusions are in general accord with those of Taylor and Quinney. Further studies of the stress-strain relations of classical idealized plasticity have been described by Davis (1945, 1955), Peters, Dow, and Batdorf (1950), Marin, Ulrich, and Hughes (1951), Marin and Hu (1952), Phillips (1952), Drucker and Stockton (1953), Naghdi and Rowley (1954), Wiseman and Marin (1954), Naghdi, Rowley, and Beadle (1955), and Phillips and Kaechele (1956).

GENERAL REFERENCES

Bridgman, P. W. (1952) "Studies in Large Plastic Flow and Fracture," McGraw-Hill, New York.
Drucker, D. C. (1967) "Introduction to the Mechanics of Deformable Solids," McGraw-Hill, New York.
Freudenthal, A. M., and H. Geiringer (1958) The Inelastic Continuum, in S. Flügge (ed.), "Handbuch der Physik," vol. 4, pp. 229–433, Springer-Verlag, Berlin.
Fung, Y. C. (1965) "Foundations of Solid Mechanics," Prentice-Hall, Englewood Cliffs, N.J.
Hill, R. (1950) "Plasticity," Clarendon Press, Oxford.
Lubahn, J. D., and R. P. Felgar (1961) "Plasticity and Creep of Metals," Wiley, New York.
Nadai, A. (1950) "Theory of Flow and Fracture of Solids," McGraw-Hill, New York.
Prager, W. (1959) "An Introduction to Plasticity," Addison-Wesley, Reading, Mass.
Prager, W. (1961) "Introduction to Mechanics of Continua," Ginn, Boston.

REFERENCES

Altshuler, T. L., and J. W. Christian (1967) *Phil. Trans. Roy. Soc.*, vol. A261, pp. 253–287.
Billington, E. W. (1977a) *J. Phys. (D)*, vol. 10, pp. 533–552.
Billington, E. W. (1977b) *J. Phys. (D)*, vol. 10, pp. 553–569.
Boyce, W. E. (1956) *Quart. Appl. Math.*, vol. 14, pp. 277–288.
Bridgman, P. W. (1923) *Proc. Am. Acad. Art. Sci.*, vol. 58, pp. 163–242.
Budiansky, B. (1957) *J. Appl. Mech.*, vol. 24, pp. 481–482.
Crossland, B. (1954) *Inst. Mech. Eng. Proc.*, vol. 168, pp. 935–946.
Davis, E. A. (1945) *J. Appl. Mech.*, vol. 12, pp. 13–24.
Davis, E. A. (1955) *J. Appl. Mech.*, vol. 22, pp. 411–415.
Drucker, D. C. (1949) *J. Appl. Mech.*, vol. 16, pp. 349–357.
Drucker, D. C. (1951) *Proc. First U.S. Natl. Congr. Appl. Mech.*, pp. 487–491, ASME, New York.
Drucker, D. C. (1973) *Met. Trans.*, vol. 4, pp. 667–673.
Drucker, D. C., and F. D. Stockton (1953) *Proc. Soc. Expt. Stress. Anal.*, vol. 10, pp. 127–142.
Greenberg, H. J., W. S. Dorn, and A. H. Wetherell (1960) A Comparison of Flow and Deformation Theories in Plastic Torsion, in E. H. Lee and P. S. Symonds (eds.), "Plasticity," pp. 279–296, Pergamon Press, Oxford.
Haigh, B. P. (1920) *Engineering*, vol. 109, pp. 158–160.
Handelman, G. H., C. C. Lin, and W. Prager (1947) *Quart. Appl. Math.*, vol. 4, pp. 397–407.
Hencky, H. (1924) *Z. angew Math. Mech.*, vol. 4, pp. 323–334.
Hill, R. (1948) *Proc. Roy. Soc. (London)*, vol. A193, pp. 281–297.
Hill, R. (1958) *J. Mech. Phys. Solids*, vol. 6, pp. 236–249.

Hodge, P. G. (1957) *Proc. Ninth Int. Congr. Appl. Mech.*, Brussels, 1956, vol. 8, pp. 65–72.
Hodge, P. G. (1958) The Mathematical Theory of Plasticity, in J. N. Goodier and P. G. Hodge (eds.), "Elasticity and Plasticity," vol. 1, pp. 51–144, Wiley, New York.
Ishlinski, I. U. (1954) *Ukr. Mat. Zh.* (in Russian), vol. 6, pp. 314–324.
Kadashevich, I., and V. V. Novozhilov (1959) *J. Appl. Math. Mech.* (translation of *Prinkl. Mat. i. Merkh*), vol. 22, pp. 104–118.
Koiter, W. T. (1953) *Quart. Appl. Math.*, vol. 11, pp. 350–354.
Lode, W. (1926) *Z. Phys.*, vol. 36, pp. 913–939.
Ludwik, P. (1909) "Elemente der Technologischen Mechanik," Springer-Verlag, Berlin.
Ludwik, P., and A. Scheu (1925) *Stahl. u. Eisen*, vol. 45, pp. 373–381.
Marin, J., and L. W. Hu (1952) *Proc. ASTM*, vol. 52, pp. 1098–1124.
Marin, J., B. H. Ulrich, and W. P. Hughes (1951) *Tech. Note* 2425, NACA.
Miller, P. M., and L. E. Malvern (1967) *J. Appl. Mech.*, vol. 34, pp. 1005–1010.
Mises, R. von (1913) "Gottinger Nachrichten," Math-Phys. Klasse, pp. 582–592.
Morrison, J. L. M., and W. M. Shepherd (1950) *Proc. Inst. Mech. Engng.*, vol. 163, pp. 1–17.
Naghdi, P. M. (1960) Stress-Strain Relations in Plasticity and Thermoplasticity, in E. H. Lee and P. Symonds (eds.), "Plasticity," pp. 121–169, Pergamon Press, Oxford.
Naghdi, P. M., and J. C. Rowley (1954) *J. Mech. Phys. Solids*, vol. 3, pp. 63–80.
Naghdi, P. M., J. C. Rowley, and C. W. Beadle (1955) *J. Appl. Mech.*, vol. 22, pp. 416–420.
Osgood, W. R. (1947) *J. Appl. Mech.*, vol. 14, pp. 147–153.
Peters, R. W., N. F. Dow, and S. B. Batdorf (1950) *Proc. Soc. Exptl. Stress Analysis*, vol. 7, pp. 127–140.
Phillips, A. (1952) *J. Appl. Mech.*, vol. 19, pp. 496–500.
Phillips, A., and L. Kaechele (1956) *J. Appl. Mech.*, vol. 23, pp. 43–48.
Prager, W. (1945) *J. Appl. Phys.*, vol. 16, pp. 837–840.
Prager, W. (1955) *Proc. Inst. Mech. Engng.*, vol. 169, pp. 41–57.
Prager, W. (1956) *J. Appl. Mech.*, vol. 23, pp. 493–496.
Prandtl, L. (1924) *Proc. First Int. Congr. Appl. Mech.*, pp. 43–54.
Reuss, E. (1930) *Z. angew Math. Mech.*, vol. 10, pp. 266–274.
Shield, R. T., and H. Ziegler (1958) *Z. angew Math. Phys.*, vol. 9, pp. 260–276.
Taylor, G. I., and H. Quinney (1931) *Phil. Trans. Roy. Soc.*, vol. A230, pp. 323–362.
Westergaard, H. M. (1920) *J. Franklin Inst.*, vol. 189, pp. 627–640.
Westergaard, H. M. (1952) "Theory of Elasticity and Plasticity," Harvard University Press, Cambridge, Mass.
Wiseman, H. A. B., and J. Marin (1954) *Proc. ASTM*, vol. 54, pp. 1365–1382.
Zener, C., and J. H. Hollomon (1946) *J. Appl. Phys.*, vol. 17, pp. 69–82.
Ziegler, H. (1959) *Quart. Appl. Math.*, vol. 17, pp. 55–65.

CHAPTER

ELEVEN

CRYSTAL PLASTICITY

11-1 INTRODUCTION TO DISLOCATIONS

11-1-1 Strength of Single Crystals

In the case of single crystals, plastic deformation is concentrated as slip on a few parallel planes. The planes on which the deformation occurs are called slip-planes. A slip-plane is normally a crystal plane with the highest density of atoms, and the direction of slip is the direction in the slip-plane in which the atoms are most closely spaced, regardless of the orientation of the applied stress. When the applied stress is a shear stress, we shall identify it with the physical component $T\langle 12\rangle$, frequent use being made of the simpler notation τ ($\equiv T\langle 12\rangle$). The characteristic shear stress required for slip in single crystals is thus determined by the shear traction on one of the possible slip-planes reaching a critical value τ_c. This particular value, τ_c, of the stress is referred to as the critical resolved shear stress for slip.

A simple estimate of the theoretical shear strength of a perfect crystal has been given by Frenkel (1926), who considered the force required to shear two rows of atoms to be arranged as shown in Fig. 11-1. Let the spacing between the rows be a and between the atoms in a row be b. The force is zero at positions such as A and B as these are lattice sites and will be zero halfway between by symmetry. With this as a basis, Frenkel proposed the concept that a crystal undergoing plastic deformation on a crystal plane passed through equivalent configurations with

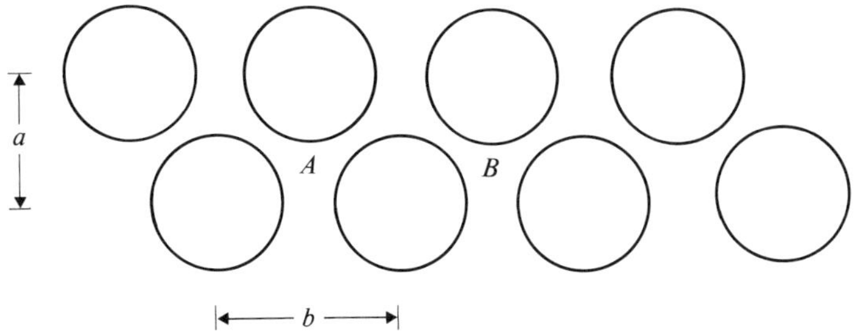

Figure 11-1 Two planes of atoms.

equal energies and with a period equal to b. Let the rows be displaced from the low-energy equilibrium position by a distance x when acted upon by a shear stress τ which is proportional to dW/dx, where W is the energy of translation per unit area of the plane. As a first approximation, Frenkel assumed the periodicity of the energy to be sinusoidal, and thus, for a homogeneously deformed crystal,

$$\tau = k \sin\left(\frac{2\pi x}{b}\right) \tag{11-1-1}$$

The constant k can be determined by specifying that the initial slope in the region of $x = 0$, where the shear strains x/a are small, corresponds to the shear modulus G of the crystal. Thus, in the limit of small shear strains x/a,

$$\tau = G\frac{x}{a} \tag{11-1-2}$$

in accordance with Hooke's law. For sufficiently small x, Eq. (11-1-1) can be expressed in the form $\tau = k2\pi x/b$, which compared with Eq. (11-1-2) gives $k = Gb/2\pi a$ and hence Eq. (11-1-1) can be expressed in the form

$$\tau = \frac{Gb}{2\pi a}\sin\left(\frac{2\pi x}{b}\right) \tag{11-1-3}$$

The maximum theoretical critical shear stress τ_m at which the system is stable is thus

$$\tau_m = \frac{Gb}{2\pi a} = k \tag{11-1-4}$$

Since b is usually nearly equal to a the maximum shear stress is about $G/2\pi$. Calculations taking account of more realistic interatomic force laws corresponding to ionic, covalent, and Van der Waals bonding for cubic lattices by Mackenzie† (1949) and Tyson (1966) only reduce this to about $G/30$ in the worst case. Although these calculations are all approximate, they serve to show that the maximum theoretical critical shear stress τ_m is many orders of magnitude greater than the observed values of the critical resolved shear stress τ_c measured in real,

† Mackenzie's work is described by Kelly (1966).

well-annealed crystals which typically cover the range $10^{-3}G$ to $10^{-4}G$. Recent experimental work indicates an even greater discrepancy between τ_m and experimental values of τ_c (see, for example, Tinder, 1964; Tinder and Washburn, 1964).

This large difference between prediction and experiment is now accounted for by the presence of particular types of defects in the crystal structure. These particular types of defect are called dislocations and have the property of reducing the force needed to cause motion in a system. Evidence that this is so is provided by crystals in the form of fibers with a small diameter, called whiskers, which are entirely free of dislocations. The strength of these whiskers approaches closely to the theoretical strength (see, for example, Kelly, 1966).

11-1-2 Dislocations

A particular type of dislocation, referred to as an edge dislocation, was proposed by Orowan (1934), Polanyi (1934), and Taylor (1934), the connection of the edge dislocation with the elastic dislocations of Volterra (1907) (see Sec. 2-8) being noted by Taylor. These separate studies provided the foundation for the modern

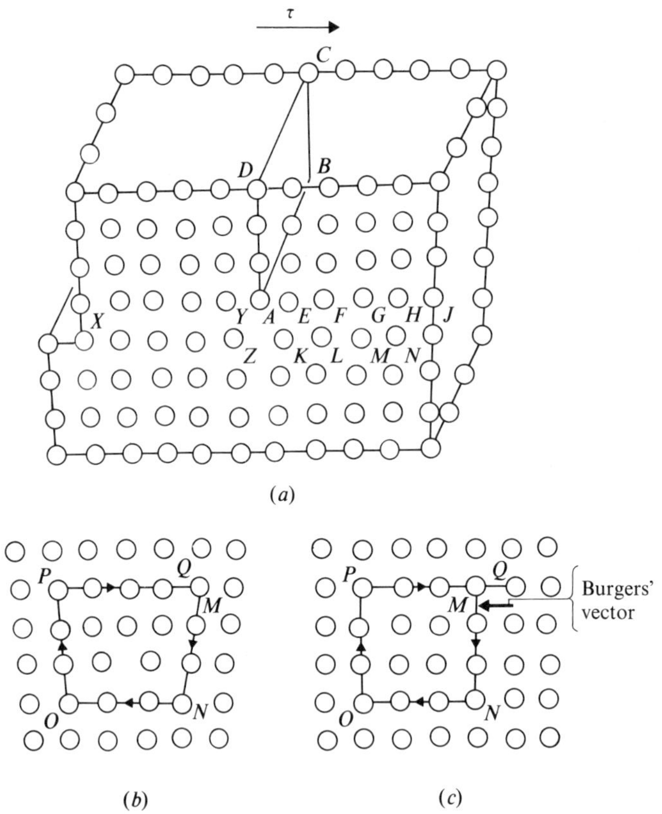

Figure 11-2 An edge dislocation.

dislocation theory of slip. A Taylor, i.e., edge, dislocation is shown in Fig. 11-2. It is called an edge dislocation because of the edge AB of the half plane $ABCD$ and, as can be seen, it matches $(n+1)$ crystal sites above the slip-plane to n crystal sites below the slip-plane. Such a dislocation is called a unit dislocation. The edge dislocation of Fig. 11-2 is frequently represented by the symbol $\perp$, the long line denoting the slip-plane and the short line denoting the extra half plane. Plastic flow is considered to occur by the microscopically inhomogeneous process of the slip of the block of atoms above a crystal plane relative to the block of atoms below it. Thus, the block of displaced atoms grows from one crystal surface until the whole of the top of the crystal has slipped. This can be seen schematically as follows. The effect of applying the shear stress shown in Fig. 11-2a is to align atom A with atom K for a very small value of τ. This alignment of A with K is followed by the successive alignment of E with L, F with M, etc., so that the very small stress τ is seen to produce slip by one atomic spacing HJ. Thus, in contrast to the assumption underlying the calculation of τ_m that slip occurs simultaneously across the whole plane XJ, the dislocation mechanism of slip occurs one row of atoms at a time which leads to a very substantial reduction in shear stress. The boundary on the crystal slip-plane (or glide-plane) between the slipped and unslipped regions is called a dislocation line. Except in the immediate vicinity of the dislocation line the crystal lattice is very little distorted. In the vicinity of the dislocation line the distortion of the glide-plane can be regarded as analogous to a ruckle in a carpet and, just as it is much easier to move a large carpet by introducing a ruckle at one side and pushing it to the other side than by trying to pull the whole carpet, so dislocations allow lamellae blocks of crystals to slip over one another at stresses much lower than it would take to move the whole plane simultaneously.

Arising from a study of the connection between Taylor's (1934) two-dimensional lattice dislocation and the elastic dislocations of Volterra (1907), Burgers (1939) proposed the description of a type of dislocation referred to as a screw dislocation. A screw dislocation can be considered as being formed by

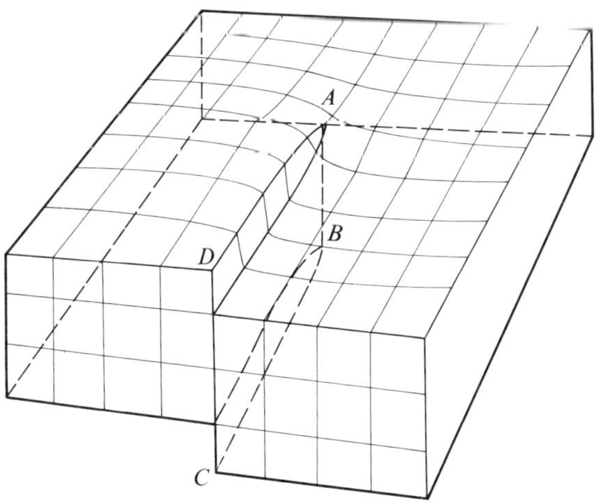

Figure 11-3 A screw dislocation.

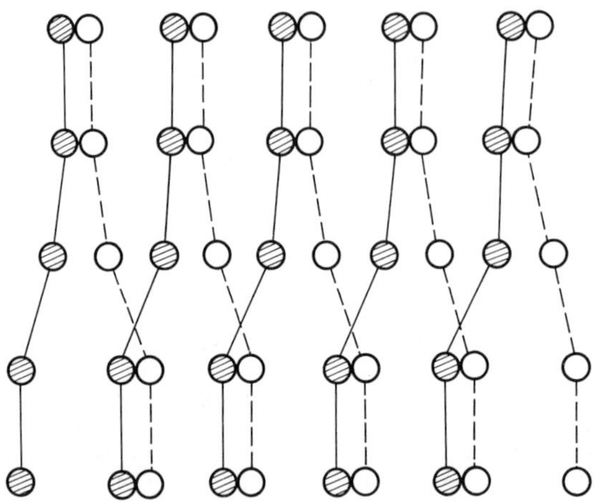

Figure 11-4 The displacement of two layers of atoms in a screw dislocation.

cutting one face *ABCD* of a crystal up to the line *AB* and then moving one cut face down past the other and rejoining the crystal one lattice spacing down, as shown in Fig. 11-3. Figure 11-4 shows in a more graphical way the spiral nature of this type of dislocation.

For a general discussion of the direct experimental evidence for the existence of dislocations in crystals, the reader is referred, for example, to Amelinckx and Dekeyser (1960), Dieter (1961), Friedel (1956), Hirsch (1959), Mitchell (1962), Read (1953), Van Bueren (1960), and Pashley (1965).

11-1-3 The Burgers Vector of a Dislocation

Consider the clockwise path *MNOPQ* shown schematically in the real crystal of Fig. 11-2*b*; the path *MNOPQ* lies entirely in good material and encloses the dislocation. The path *MNOPQ* is the algebraic sum of the lattice vectors in moving from lattice point to lattice point and is seen to form a closed loop round the dislocation. A closed circuit in the form of a lattice point to lattice point path round a dislocation, as shown in Fig. 11-2*b*, is referred to as a Burgers circuit. Shown in Fig. 11-2*c* is the same lattice point to lattice point sequence in a perfect, i.e., dislocation-free, crystal, it being noted that in a perfect crystal the circuit does not form a closed loop. Failure to close implies that the first circuit of Fig. 11-2*b* must enclose one or more dislocations. The vector required to complete the circuit is called the true Burgers vector **b**. In Fig. 11-2*c*, the closure failure, i.e., the true Burgers vector, is shown as *QM* and is seen to be at right angles to the dislocation line.

From a mapping of the closure failure back onto the real crystal, as in Fig. 11-2*b*, using the mapping of the perfect crystal there is obtained the local Burgers vector. This latter quantity has the advantage of having the same orientation as the real crystal at the point of interest. The local Burgers vector differs, in general, from the true Burgers vector, because the former is affected by the variation in the interatomic spacings around the circuit arising from the elastic strains produced by the dislocation itself. However, as the Burgers circuit is made very large and therefore lies in nearly perfect crystal, the true and local Burgers vectors approach one another.

An analogous procedure shows that for a screw dislocation the Burgers vector lies along the direction of the dislocation line.

A dislocation which is the mirror image in the slip-plane of another dislocation is said to be unlike and is accorded the opposite sign. The strength b of a dislocation having a Burgers vector $\boldsymbol{b}$ is simply the magnitude of the vector, that is, $|\boldsymbol{b}| = b$.

An equivalent definition of the Burgers vector can be given which depends on the entirely arbitrary direction of the dislocation line. However, by choosing on an arbitrary basis a unit vector ξ tangent to the dislocation line, a direction can be ascribed to the dislocation line by specifying the positive direction of ξ to be the positive direction of the dislocation line. Now describe a circuit in the crystal in a right-hand sense with respect to ξ so as to link the dislocation line once, as shown schematically in Fig. 11-5. Equivalently, the local Burgers vector $\boldsymbol{b}$ is given by the line integral, taken in a right-handed sense relative to ξ, of the elastic displacement $\boldsymbol{u}$ around the dislocation:

$$\boldsymbol{b} = \oint_c \frac{\partial \boldsymbol{u}}{\partial l}\, dl \tag{11-1-5}$$

It is evident that this approach identifies a dislocation as being characterized by a line direction ξ and a Burgers vector $\boldsymbol{b}$. If ξ is defined continuously along the line, then the Burgers vector is constant along it. This leads to the specific definitions:

$$\boldsymbol{b} \cdot \xi = \begin{cases} 0 & \text{(edge dislocation)} \\ b & \text{(right-handed screw dislocation)} \\ -b & \text{(left-handed screw dislocation)} \end{cases} \tag{11-1-6}$$

In connection with Eq. (11-1-5), there is at present no generally accepted convention for $\boldsymbol{b}$.

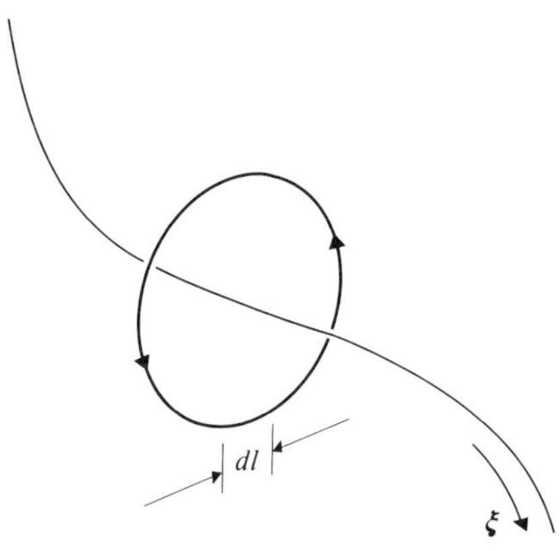

Figure 11-5 A displacement circuit.

In general, a dislocation line will lie at an angle to its Burgers vector. Those parts of the dislocation line which are parallel to the Burgers vector will be screw dislocations and those parts normal to the Burgers vector will be edge dislocations, the only condition being that if ξ is defined continuously along the line then the Burgers vector is constant along it. If these segments are sufficiently small, the dislocation line will appear as a smooth curve on a macroscopic scale. Such a curve is called a mixed dislocation. The curve of a mixed dislocation can be resolved into screw and edge components by resolving the Burgers vector into components parallel and perpendicular respectively to ξ at any point. This can be expressed in the form

$$\boldsymbol{b}_s = (\boldsymbol{b} \cdot \xi)\xi \qquad \boldsymbol{b}_e = \boldsymbol{b} - (\boldsymbol{b} \cdot \xi)\xi = \xi \times (\boldsymbol{b} \times \xi) \tag{11-1-7}$$

where $\boldsymbol{b}_s$ and $\boldsymbol{b}_e$ are the screw and edge components respectively.

It is possible to envisage a limited amount of deformation which is confined entirely within a crystal. This limited deformation would be separated from the undeformed material by a dislocation line in the form of a ring or loop. The loop must comprise a mixture of edge and screw dislocations to a varying degree. A type of dislocation loop which is entirely edge and which is referred to as a prismatic loop is considered in Sec. 11-4-3.

A Burgers circuit can be drawn large enough to enclose several dislocation lines. This is illustrated schematically in Fig. 11-6 where there is shown a Burgers circuit $ABCDA$ linking two dislocations. Since the circuit $ABCDA$ passes through the same good material as the two circuits $ABCA$ and $ACDA$ when taken together, it follows that the circuit $ABCDA$ must be equivalent to the combination of the circuits $ABCA$ and $ACDA$. Hence, the closure failure $\boldsymbol{b}_3$ for the circuit $ABCDA$ must equal the sum $\boldsymbol{b}_1 + \boldsymbol{b}_2$ of the two separate failures; that is, $\boldsymbol{b}_3 = \boldsymbol{b}_1 + \boldsymbol{b}_2$. This argument applied to a Burgers circuit enclosing several dislocation lines requires

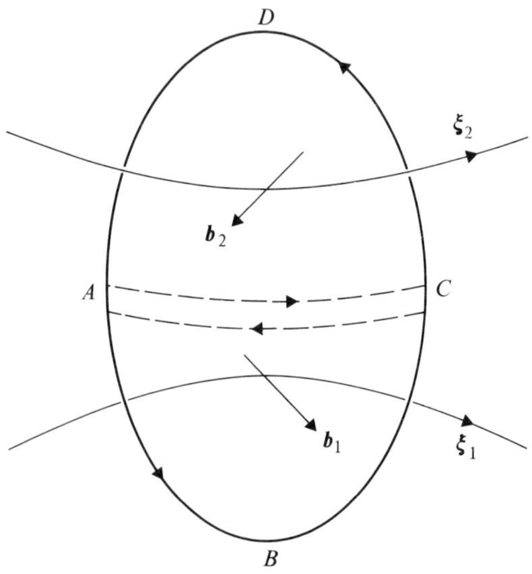

Figure 11-6 Burgers circuit enclosing two dislocations.

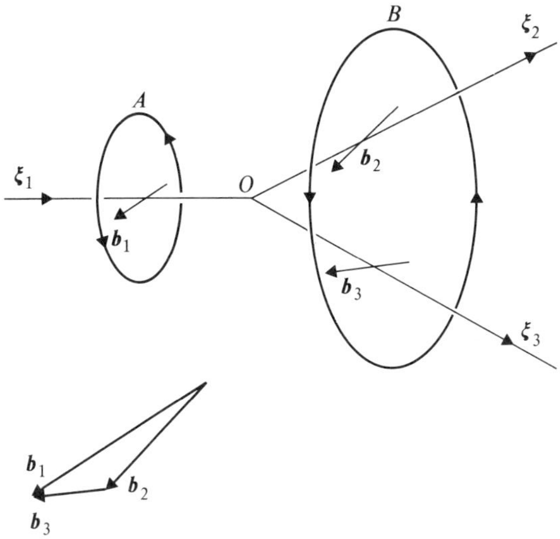

Figure 11-7 Three dislocations meeting at a node O.

that the closure failure of the resulting mapping of such a circuit shall equal the sum of the Burgers vectors of the linked dislocations. The concept of equivalent Burgers circuits can now be applied to Fig. 11-7, where there is shown three dislocations meeting at the point o, which is called a dislocation *node*. With the sense of the ξ_i $(i = 1, 2, 3)$ defined as shown in Fig. 11-7, the circuits A and B are equivalent, and therefore $\boldsymbol{b}_1 = \boldsymbol{b}_2 + \boldsymbol{b}_3$. If ξ_1 is reversed, $\boldsymbol{b}_1$ is also reversed so that $\Sigma_{i=1}^{3} \boldsymbol{b}_i = 0$. This argument can be extended to an arbitrary node, giving

$$\sum_i \boldsymbol{b}_i = \mathbf{0} \tag{11-1-8}$$

for the condition that all the ξ_i are taken as positive away from the node, the summation extending over all dislocations meeting at the node.

11-2 ELASTIC PROPERTIES OF STRAIGHT DISLOCATIONS

11-2-1 Stress Field of a Dislocation

The crystal is assumed to be an isotropic elastic continuum to which can be applied the classical linear elasticity theory of Chap. 6. Assuming sufficiently small displacements, the difference between material and spatial coordinates can be neglected. The elastic distortion around a dislocation can be described in terms of the deformation of a hollow circular cylinder of isotropic elastic material. Let (x, y, z) and (r, θ, z) be the rectangular and cylindrical spatial coordinates respectively of a particle in the deformed configuration. The principal axis of the hollow circular cylinder of inner and outer radii r_0 and r_1 respectively is aligned with the z axis.

Of particular use is the special case of plane strain discussed in Sec. 6-3-1 together with the Airy stress function for two-dimensional problems. Airy's method is based on the observation that, for zero acceleration terms, the conditions of equilibrium, i.e., Eqs. (6-3-5) and (6-3-6), appear as the divergence of a tensor. Now by analogy with hydrodynamics and in the absence of body forces, these conditions of equilibrium can be identically satisfied by introducing two functions Φ and χ in such a way that

$$T\langle xx\rangle = \frac{\partial \Phi}{\partial y} \qquad T\langle xy\rangle = -\frac{\partial \Phi}{\partial x} \tag{11-2-1}$$

$$T\langle yy\rangle = \frac{\partial \chi}{\partial x} \qquad T\langle xy\rangle = -\frac{\partial \chi}{\partial y} \tag{11-2-2}$$

which in turn can be combined by setting

$$\chi = \frac{\partial \psi}{\partial x} \qquad \Phi = \frac{\partial \psi}{\partial y} \tag{11-2-3}$$

to give

$$T\langle xx\rangle = \frac{\partial^2 \psi}{\partial y^2} \qquad T\langle rr\rangle = \frac{1}{r}\frac{\partial \psi}{\partial r} + \frac{1}{r^2}\frac{\partial^2 \psi}{\partial \theta^2} \tag{11-2-4}$$

$$T\langle yy\rangle = \frac{\partial^2 \psi}{\partial x^2} \qquad T\langle \theta\theta\rangle = \frac{\partial^2 \psi}{\partial r^2} \tag{11-2-5}$$

$$T\langle xy\rangle = -\frac{\partial^2 \psi}{\partial x \partial y} \qquad T\langle r\theta\rangle = -\frac{\partial}{\partial r}\left(\frac{1}{r}\frac{\partial \psi}{\partial \theta}\right) \tag{11-2-6}$$

where the function ψ is called the Airy stress function. The stress function ψ must generate only those stress fields that satisfy the conditions of compatibility; they will do so provided

$$\nabla^4 \psi = 0 \tag{11-2-7}$$

(a) Screw dislocation A right-handed screw dislocation of the Volterra (1907) hollow-core type is shown in Fig. 11-8 along the axis of the cylinder where ξ and $\boldsymbol{b}$ have been taken as pointing in the positive z direction. This dislocation can be produced from a perfect cylinder by shear displacements in the z direction across the xy glide-plane. For example, the dislocation can be made by cutting the cylinder on the half plane $y = 0$, $x > 0$, the top face of the cut being held fixed and the bottom face displaced by b in the z direction. It is evident from Fig. 11-8 that $u\langle x\rangle = 0$, $u\langle y\rangle = 0$, and the displacement $u\langle z\rangle$ is discontinuous at the cut surface defined by the half plane $y = 0$, $x > 0$; thus

$$\lim_{\substack{\zeta \to 0 \\ x > 0}} [u\langle z\rangle(x, -\zeta) - u\langle z\rangle(x, \zeta)] = b \qquad \text{for } \zeta > 0 \tag{11-2-8}$$

Restriction to an isotropic material is taken to imply that the displacement $u\langle z\rangle$ increases uniformly with the angle θ to give this discontinuity; thus

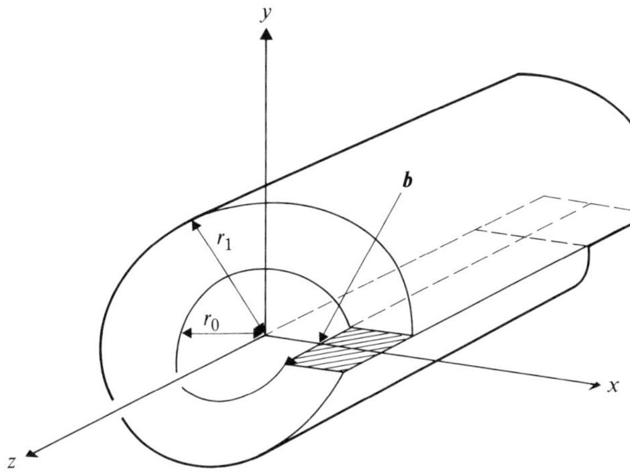

Figure 11-8 A screw dislocation.

$$u\langle z\rangle\,(r,\theta) = b\frac{\theta}{2\pi} = \frac{b}{2\pi}\tan^{-1}\left(\frac{y}{x}\right) \tag{11-2-9}$$

which in the bulk of the specimen satisfies

$$\nabla^2 u\langle z\rangle = 0 \tag{11-2-10}$$

The stresses associated with this displacement are determined from Sec. 6-3-1 in the form

$$T\langle xz\rangle = -\frac{Gb}{2\pi}\frac{y}{x^2+y^2} = T\langle zx\rangle \tag{11-2-11}$$

$$T\langle yz\rangle = \frac{Gb}{2\pi}\frac{x}{x^2+y^2} = T\langle zy\rangle \tag{11-2-12}$$

$$\begin{gathered} T\langle yx\rangle - T\langle xy\rangle = 0 \\ T\langle xx\rangle = T\langle yy\rangle = T\langle zz\rangle = 0 \end{gathered} \tag{11-2-13}$$

In cylindrical polar coordinates, the physical components of stress of a screw dislocation are

$$T\langle \theta z\rangle = \frac{Gb}{2\pi}\frac{1}{r} = T\langle z\theta\rangle \tag{11-2-14}$$

$$\begin{gathered} T\langle r\theta\rangle = T\langle \theta r\rangle = T\langle rz\rangle = T\langle zr\rangle = 0 \\ T\langle rr\rangle = T\langle \theta\theta\rangle = T\langle zz\rangle = 0 \end{gathered} \tag{11-2-15}$$

The stress field is thus seen to consist of $T\langle \theta z\rangle$ in radial planes parallel to the z direction.

In a free cylinder no stresses can act over the free surfaces. The stress field represented by Eqs. (11-2-14) and (11-2-15) fulfill the boundary condition at the free surfaces, as is evident from Eqs. (11-2-15); that is, $T\langle rr\rangle = T\langle r\theta\rangle = T\langle rz\rangle = 0$. However, on the end surfaces of a cylinder of finite length the stress field gives shear stresses $T\langle z\theta\rangle = T\langle\theta z\rangle$ which produce a torque

$$\Gamma\langle z\rangle = \int_{r_0}^{r_1}\int_0^{2\pi} rT\langle z\theta\rangle r\,dr\,d\theta = \tfrac{1}{2}Gb(r_1^2 - r_0^2) \tag{11-2-16}$$

Thus the stress field of Eqs. (11-2-14) and (11-2-15) is not the correct solution for a finite rod. Hence, for a finite rod, stresses must be superposed to cancel the stresses $T\langle\theta z\rangle$ at the ends, subject to the essential condition that these additional stresses do not introduce stresses which violate the boundary conditions on the surfaces. The quantity $-\Gamma\langle z\rangle$ can be identified with the torque associated with simple torsion (see Sec. 6-5-2); i.e.,

$$-\Gamma\langle z\rangle = \tfrac{1}{2}\pi G(r_1^4 - r_0^4)D \tag{11-2-17}$$

where D is the twist per unit length. Equations (11-2-16) and (11-2-17) give

$$D = -\frac{b}{\pi(r_0^2 + r_1^2)} \tag{11-2-18}$$

The stress $T\langle\theta z\rangle$ corresponding to the torsion displacement rzD can now be obtained in the form $T\langle\theta z\rangle = GrD$. Hence, the stress field for a cylinder of finite length can be expressed in the form

$$T\langle\theta z\rangle = \frac{Gb}{2\pi}\left(\frac{1}{r} - \frac{2r}{r_0^2 + r_1^2}\right) \tag{11-2-19}$$

it being noted, however, that the stress distribution is invalid within a distance $2r_1$ of the cylinder ends. The corresponding components of stress in rectangular

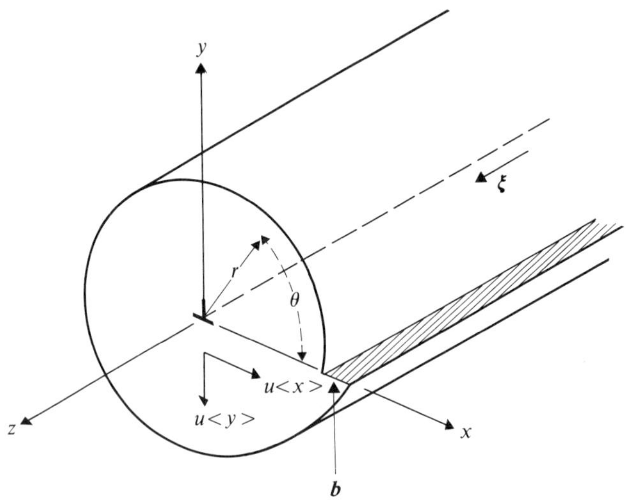

Figure 11-9 An edge dislocation showing the plane strain displacements.

cartesian coordinates are

$$T\langle zx\rangle = -\frac{Gb}{2\pi}\left(\frac{y}{x^2+y^2} - \frac{2y}{r_0^2+r_1^2}\right) \tag{11-2-20}$$

$$T\langle zy\rangle = \frac{Gb}{2\pi}\left(\frac{x}{x^2+y^2} - \frac{2x}{r_0^2+r_1^2}\right) \tag{11-2-21}$$

(b) Edge dislocation The elastic strain field associated with a Volterra (1907) edge dislocation can be produced in the hollow circular cylinder by a rigid displacement of the faces of the slit by a distance b in the x direction, as indicated schematically in Fig. 11-9. The strains in the z direction are zero and the deformation is basically plane strain defined by $u\langle z\rangle = 0$ and $\partial u\langle i\rangle/\partial z = 0$ $(i = x, y)$. The Airy stress function can be expressed in cylindrical polar coordinates in the form

$$\left(\frac{\partial^2}{\partial r^2} + \frac{1}{r}\frac{\partial}{\partial r} + \frac{1}{r^2}\frac{\partial^2}{\partial \theta^2}\right)^2 \psi = 0 \tag{11-2-22}$$

Introducing the function

$$\Psi = T\langle rr\rangle + T\langle \theta\theta\rangle = \nabla^2\psi \tag{11-2-23}$$

into Eq. (11-2-22) gives the Laplace equation

$$\nabla^2\Psi = \left(\frac{\partial^2}{\partial r^2} + \frac{1}{r}\frac{\partial}{\partial r} + \frac{1}{r^2}\frac{\partial^2}{\partial \theta^2}\right)\Psi = 0 \tag{11-2-24}$$

The Laplace equation (11-2-24) is separable, and hence the solution must be of the form

$$\Psi = \sum_n R_n(r)\Psi_n(\theta) = \nabla^2\psi \tag{11-2-25}$$

which is separable, term by term. For the condition that the solution be single valued in the plane, Eq. (11-2-25) can be expressed in the form

$$\nabla^2\psi = (\alpha_0 + \beta_0 \ln r) + \sum_{n=1}^{\infty} (\alpha_n r^n + \beta_n r^{-n}) \sin n\theta + \sum_{n=1}^{\infty} (\gamma_n r^n + \delta_n r^{-n}) \cos n\theta \tag{11-2-26}$$

which can be integrated to give

$$\begin{aligned}\psi = {} & \tfrac{1}{4}\alpha_0 r^2 + \tfrac{1}{4}\beta_0 r^2(\ln r - 1) + \tfrac{1}{8}(4\beta_1 r \ln r + \alpha_1 r^3)\sin\theta \\ & + \tfrac{1}{8}(4\delta_1 r \ln r + \gamma_1 r^3)\cos\theta + \sum_{n=2}^{\infty}\left[\frac{\alpha_n}{4(n+1)} r^{n+2} - \frac{\beta_n}{4(n-1)} r^{-n+2}\right]\sin n\theta \\ & + \sum_{n=2}^{\infty}\left[\frac{\gamma_n}{4(n+1)} r^{n+2} - \frac{\delta_n}{4(n-1)} r^{-n+2}\right]\cos n\theta + \psi_1(r,\theta)\end{aligned} \tag{11-2-27}$$

where, since ψ_1 is a harmonic function, it follows that ψ_1 also has an expansion of

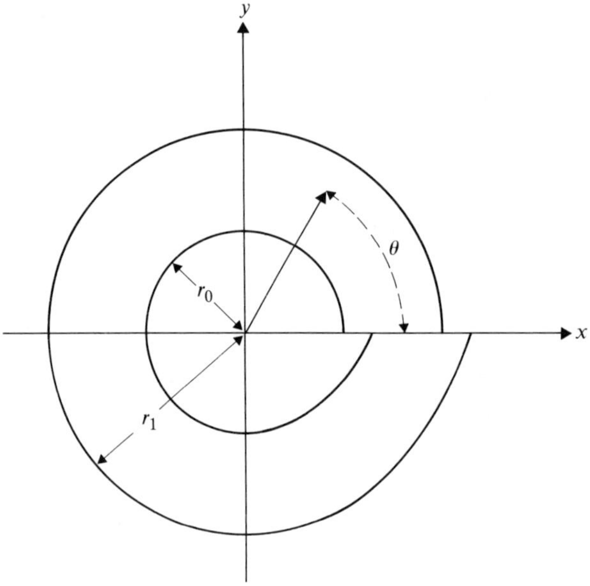

Figure 11-10 An edge dislocation formed by taking the cut on the $y = 0$ plane with $x > 0$.

the form of Eq. (11-2-26). The conditions shown schematically in Fig. 11-10 are taken to imply the inequalities:

$$T\langle\theta\theta\rangle \ll 0 \quad \text{for } \theta = \pi/2 \qquad T\langle r\theta\rangle \gg 0 \quad \text{for } \theta = 0, \pi$$
$$T\langle\theta\theta\rangle \gg 0 \quad \text{for } \theta = -\pi/2 \qquad T\langle r\theta\rangle = 0 \quad \text{for } \theta = \pm\pi/2$$

These inequalities suggest that $T\langle\theta\theta\rangle$ depends on $\sin\theta$ and $T\langle r\theta\rangle$ depends on $\cos\theta$; these conditions are satisfied by specifying ψ to be dependent on $\sin\theta$. Thus

$$\psi = \tfrac{1}{8}(4\beta_1 r \ln r + \alpha_1 r^3 + 8\alpha'_1 r + 8\beta'_1 r^{-1})\sin\theta \tag{11-2-28}$$

The last two terms in Eq. (11-2-28) arise from the term in ψ_1. Expressing the third term in Eq. (11-2-28) as $\alpha'_1 y$, it follows that since the stresses are second derivatives of ψ, this particular term cannot contribute to the stress and is therefore to be omitted.

Use of this form of the stress function allows the stresses to be determined from the second equations of (11-2-4), (11-2-5), and (11-2-6), which in turn can be used with the free surface conditions

$$T\langle rr\rangle = T\langle r\theta\rangle = 0 \qquad \text{at } r = r_0, r = r_1 \tag{11-2-29}$$

to obtain

$$\alpha_1 = -\frac{2\beta_1}{r_0^2 + r_1^2} \tag{11-2-30a}$$

$$\beta'_1 = \tfrac{1}{4}\beta_1 \frac{r_0^2 r_1^2}{r_0^2 + r_1^2} \tag{11-2-30b}$$

For the particular condition $r_1 \gg r_0$, these reduce to the approximate relations

$$\alpha_1 \simeq -2\frac{\beta_1}{r_1^2} \qquad \beta_1' \simeq \tfrac{1}{4}\beta_1 r_0^2 \tag{11-2-31}$$

These approximate values of α_1 and β_1' can be used to obtain the stress function in the form

$$\psi = \tfrac{1}{2}\beta_1\left(r\ln r - \frac{r^3}{2r_1^2} + \frac{r_0^2}{2r}\right)\sin\theta \tag{11-2-32}$$

This form of the stress function allows the stresses to be determined from the second equations of (11-2-4), (11-2-5), and (11-2-6) in the form

$$T\langle rr\rangle = \tfrac{1}{2}\beta_1\left(\frac{1}{r} - \frac{r}{r_1^2} - \frac{r_0^2}{r^3}\right)\sin\theta \tag{11-2-33}$$

$$T\langle \theta\theta\rangle = \tfrac{1}{2}\beta_1\left(\frac{1}{r} - \frac{3r}{r_1^2} + \frac{r_0^2}{r^3}\right)\sin\theta \tag{11-2-34}$$

$$T\langle r\theta\rangle = -\tfrac{1}{2}\beta_1\left(\frac{1}{r} - \frac{r}{r_1^2} - \frac{r_0^2}{r^3}\right)\cos\theta \tag{11-2-35}$$

For the special case of $r_0 \ll r \ll r_1$, the components of the stress can be expressed in the approximate form

$$T\langle rr\rangle \simeq \tfrac{1}{2}\beta_1\frac{\sin\theta}{r} \simeq T\langle\theta\theta\rangle \tag{11-2-36}$$

$$T\langle r\theta\rangle \simeq -\tfrac{1}{2}\beta_1\frac{\cos\theta}{r} \simeq T\langle\theta r\rangle \tag{11-2-37}$$

In rectangular cartesian coordinates, the components of stress are

$$T\langle xx\rangle \simeq \tfrac{1}{2}\beta_1\frac{y(3x^2+y^2)}{(x^2+y^2)^2} \tag{11-2-38}$$

$$T\langle yy\rangle \simeq -\tfrac{1}{2}\beta_1\frac{y(x^2-y^2)}{(x^2+y^2)^2} \tag{11-2-39}$$

$$T\langle zz\rangle - \nu(T\langle xx\rangle + T\langle yy\rangle) \simeq \nu\beta_1\frac{y}{x^2+y^2} \tag{11-2-40}$$

$$T\langle xy\rangle \simeq -\tfrac{1}{2}\beta_1\frac{x(x^2-y^2)}{(x^2+y^2)^2} \simeq T\langle yx\rangle \tag{11-2-41}$$

The strain components follow in the form

$$\tilde{e}\langle xx\rangle = \frac{1}{4}\frac{\beta_1}{G}\left[\frac{y(3x^2+y^2)}{(x^2+y^2)^2} - 2\nu\frac{y}{x^2+y^2}\right] \tag{11-2-42}$$

$$\tilde{e}\langle yy\rangle = -\frac{1}{4}\frac{\beta_1}{G}\left[\frac{y(x^2-y^2)}{(x^2+y^2)^2} + 2\nu\frac{y}{x^2+y^2}\right] \tag{11-2-43}$$

$$\tilde{e}\langle xy\rangle = -\frac{1}{4}\frac{\beta_1}{G}\frac{x(x^2-y^2)}{(x^2+y^2)^2} \tag{11-2-44}$$

from which the constant β_1 can be determined by way of the condition that the change in the displacement field on encircling the z axis must equal the Burgers vector. Thus,

$$b = -\int_{-\infty}^{+\infty} [\tilde{e}\langle xx\rangle(x,\zeta)-\tilde{e}\langle xx\rangle(x,-\zeta)]\,dx \qquad \text{for } \zeta\to 0, \zeta \text{ positive} \tag{11-2-45}$$

which, by way of Eq. (11-2-42), gives

$$\beta_1 = -\frac{Gb}{\pi(1-\nu)} \tag{11-2-46}$$

It is evident that the stress field associated with an edge dislocation has both dilatational and shear components.

For a cylinder of finite length, tractions on the ends of the cylinder are required to support the plane strain deformation given above. These end tractions are normal to the ends and of magnitude $\pm T\langle zz\rangle$. For plane strain, Eq. (6-3-4) expressed in cylindrical polar coordinates gives

$$\begin{aligned} T\langle zz\rangle &= \nu(T\langle rr\rangle + T\langle \theta\theta\rangle) \\ &= \nu\beta_1\left(\frac{1}{r}-\frac{2r}{r_0^2+r_1^2}\right)\sin\theta \end{aligned} \tag{11-2-47}$$

Equation (11-2-47) has been obtained using the relation for α_1 defined by Eq. (11-2-30*a*).

From Eqs. (11-2-14), (11-2-36), and (11-2-37) it is evident that the stress varies inversely as the distance from the z axis of the cylinder. This dependence of the stress on $1/r$ is such that at $r = 0$ the stress rises to infinity. Infinite stresses at the center of a dislocation are not physically acceptable, and hence it is concluded that there is a limiting radius r_0 below which these elastic solutions do not apply. This central region of radius r_0 around the dislocation is referred to as the core of the dislocation.

11-2-2 Strain Energy of a Dislocation

The existence of the core of the dislocation is taken to imply that the total potential energy U^d_{tot} of a dislocation can be expressed in the form

$$U^d_{\text{tot}} = U^d + U_{\text{core}} \tag{11-2-48}$$

where U^d is the energy stored in the elastic stress field of the dislocation and U_{core} is the energy of misfit for the material inside the core cylinder.

The total potential energy of a system can be expressed in the form

$$U = \int_{R_t} W(\tilde{\mathbf{e}})\,dv \tag{11-2-49}$$

where $W(\tilde{\mathbf{e}})$ is the strain energy function discussed in Sec. 6-1-2; see also Sec. 6-2-4, in particular Eqs. (6-2-50) and (6-2-51). Substituting in Eq. (11-2-49) for W from Eq. (6-2-51) gives

$$U = \tfrac{1}{2}\int_{R_t} \operatorname{tr}(\mathbf{T}\tilde{\mathbf{e}})\,dv \tag{11-2-50}$$

which can be expressed by way of Eq. (3-8-4) in the form

$$U = \tfrac{1}{2}\int_{\partial R_t} \boldsymbol{t}\cdot\boldsymbol{u}\,da \qquad \text{for } \boldsymbol{f} = \boldsymbol{o} \tag{11-2-51}$$

for the case of zero body force.

Consider a body $\mathscr{B}$ whose external boundary consists of two parts S_1 and S_2. The presence of a dislocation can be regarded as giving rise to a cut in the interior of the body whose two surfaces will be denoted as C_+ and C_-. The displacements and tractions will be prescribed to be as follows:

$$\begin{aligned} \boldsymbol{u} &= \boldsymbol{u}' \text{ on } S_1 & \boldsymbol{t} &= \mathbf{T}^T\boldsymbol{\nu} \text{ on } S_2 \\ \boldsymbol{u}_+ - \boldsymbol{u}_- &= \Delta\boldsymbol{u} & \boldsymbol{t}^c &= \mathbf{T}^T\boldsymbol{\nu}|_{C_+} = -\mathbf{T}^T\boldsymbol{\nu}|_{C_-} \text{ on } C_\pm \end{aligned} \tag{11-2-52}$$

where $\boldsymbol{u}$ and $\mathbf{T}$ are the displacement and stress fields respectively. In Eqs. (11-2-52), $\Delta\boldsymbol{u}$ defines the displacement discontinuity across the cut and $\boldsymbol{t}^c$ the tractions applied to the cut. The tractions $\boldsymbol{t}^c$ will be taken to be equal and opposite at corresponding points on the two faces.

Identifying the value of $\Delta\boldsymbol{u}$ with the Burgers vector of the dislocation gives a dislocation whose line is the edge of the cuts $C_\pm$. In this case the quantities $\boldsymbol{t}^c$ are then equal to the tractions necessary on the cut in order to create the dislocation. Thus for an edge dislocation, formed as in Fig. 11-10 by taking the cut on the $y = 0$ plane with $x > 0$, the tractions necessary to make the relative displacement $\Delta\boldsymbol{u} = (-b, 0, 0)$ are given by Eqs. (11-2-38) to (11-2-41) together with Eq. (11-2-46) in the form

$$\boldsymbol{t}^c = (-T\langle xy\rangle, -T\langle yy\rangle, -T\langle zy\rangle) = \left(-\frac{Gb}{2\pi(1-\nu)x}, 0, 0\right) \tag{11-2-53}$$

for an infinite isotropic medium with no applied stresses at infinity.

For the particular condition that the relative displacement of C_+ and C_- is zero, the boundary conditions on S_1 and S_2 being prescribed to be the same as in Eqs. (11-2-52), the displacements and tractions are denoted as follows:

$$\begin{aligned} \boldsymbol{u}^{(0)} &= \boldsymbol{u}' \text{ on } S_1 & \boldsymbol{t} &= (\mathbf{T}^T)^{(0)}\boldsymbol{\nu} \text{ on } S_2 \\ \boldsymbol{u}^{(0)}_+ - \boldsymbol{u}^{(0)}_- &= \boldsymbol{o} & (\mathbf{T}^T)^{(0)}\boldsymbol{\nu}|_{C_+} &= -(\mathbf{T}^T)^{(0)}\boldsymbol{\nu}|_{C_-} = \boldsymbol{t}^{(0)} \end{aligned} \tag{11-2-54}$$

where $\boldsymbol{t}^{(0)}$ are the tractions which must be applied to the faces of the cut in order to prevent their relative displacement and the displacement field $\boldsymbol{u}^{(0)}$ is the one which would be obtained by subjecting the same body, but without a cut in it, to the given conditions on S_1 and S_2.

Combination of Eqs. (11-2-52) and (11-2-54) allows the deformation corresponding to the dislocation alone to be specified in terms of the displacement field $u^d = u - u^{(0)}$ and its corresponding stress field $\mathbf{T}^d$. The displacements and tractions will be prescribed to be as follows:

$$
\begin{aligned}
u^d &= o \text{ on } S_1 \qquad (\mathbf{T}^T)^d v = o \text{ on } S_2 \\
u_+^d - u_-^d &= \Delta u \qquad t^d \equiv t^c - t^{(0)} = (\mathbf{T}^T)^d v|_{C_+} = -(\mathbf{T}^T)^d v|_{C_-}
\end{aligned} \tag{11-2-55}
$$

From Eq. (11-2-51) it is evident that the elastic energies in the states of deformation corresponding respectively to Eqs. (11-2-54) and (11-2-55) can be expressed in the form

$$U^{(0)} = \tfrac{1}{2} \int_S t^{(0)} \cdot u^{(0)} \, da \tag{11-2-56}$$

$$U^d = \tfrac{1}{2} \int_C t^d \cdot u^d \, da = \tfrac{1}{2} \int_{C_+} t^d \cdot \Delta u \, da \tag{11-2-57}$$

where S denotes $S_1 + S_2$ and C denotes $C_+ + C_-$.

Consider the energy stored per unit length in the elastic stress field of the infinite screw dislocation, for which the stress field is given by Eq. (11-2-12). The boundary ∂R_t is taken to consist of the two cylindrical surfaces plus the two faces $(y = 0, x > 0)$ of the cut. The traction on the two cylindrical surfaces vanishes and at the two faces of the cut has the values $\pm T\langle yz\rangle (x, 0)$. Hence, the energy per unit length in the z direction, as obtained from Eq. (11-2-57) for these particular conditions, can be expressed by way of Eq. (11-2-8) in the form

$$
\begin{aligned}
\frac{U^d}{L} &= \tfrac{1}{2} \lim_{\substack{\zeta \to 0 \\ x > 0}} \int_{r_0}^{r_1} T\langle yz\rangle (x, \zeta) [u\langle z\rangle (x, -\zeta) - u\langle z\rangle (x, \zeta)] \, dx \\
&= \tfrac{1}{2} \int_{r_0}^{r_1} T\langle yz\rangle b \, dx = \frac{Gb^2}{4\pi} \ln\left(\frac{r_1}{r_0}\right)
\end{aligned} \tag{11-2-58}
$$

For a cylinder of finite length,

$$
\begin{aligned}
\frac{U^d}{L} &= \tfrac{1}{2} \int_{r_0}^{r_1} T\langle yz\rangle (x, 0) b \, dx = \frac{Gb^2}{4\pi} \left[\ln\left(\frac{r_1}{r_0}\right) - \frac{r_1^2 - r_0^2}{r_1^2 + r_0^2} \right] \\
&\simeq \frac{Gb^2}{4\pi} \left[\ln\left(\frac{r_1}{r_0}\right) - 1 \right]
\end{aligned} \tag{11-2-59}
$$

Similarly, the energy per unit length of an edge dislocation can be obtained from Eq. (11-2-35) in the form

$$
\begin{aligned}
\frac{U^d}{L} &= \tfrac{1}{2} \int_{r_0}^{r_1} T\langle xy\rangle (x, 0) b \, dx = \tfrac{1}{2} b \int_{r_0}^{r_1} [T\langle r\theta\rangle]_{\theta = 0} \, dr \\
&= \frac{Gb^2}{4\pi(1 - \nu)} \left[\ln\left(\frac{r_1}{r_0}\right) - 1 \right]
\end{aligned} \tag{11-2-60}
$$

When the energy per unit length is obtained from Eq. (11-2-37), U^d has the approximate form

$$\frac{U^d}{L} \simeq \frac{Gb^2}{4\pi(1-\nu)} \ln\left(\frac{r_1}{r_0}\right) \tag{11-2-61}$$

With this approximation, it is evident from Eqs. (11-2-58) and (11-2-61) that the elastic energy of an edge dislocation exceeds that of the screw dislocation by $1/(1-\nu)$. For r_1/r_0 lying in the range $10^2 < r_1/r_0 < 10^5$, these limits corresponding to a heavily worked state and the annealed state respectively, Eq. (11-2-60) represents a 7 to 20 percent correction.

Because of the divergence of the above expressions for U^d as $r_1 \to \infty$, r_1 is in practice taken to be twice the distance of the dislocation from the nearest part of the boundary. For a crystal containing many dislocations, r_1 would be taken as half the average spacing between them. The divergence with large r_1 disappears with more complicated systems of dislocations which have no net Burgers vector, as, for example, pairs of opposite dislocations or systems of closed loops (see, for example, Hirth and Lothe, 1968; Lardner, 1974).

The divergence of the above expressions for U^d as $r_0 \to 0$, which has already been noted in relation to the stress field of a dislocation, arises from the use of linear elasticity theory which cannot deal with the gross lattice distortion near the core of the dislocation. In practice, the difficulties associated with determining the core radius r_0 raise the question of whether the Volterra hollow-core concept of a dislocation is physically valid. The total of the strain energy plus surface energy of a screw dislocation, denoted as $\hat{U}$, can be expressed in the form (Frank, 1952)

$$\frac{\hat{U}}{L} = \frac{Gb^2}{4\pi} \ln\left(\frac{r_1}{r_0}\right) + 2\pi\gamma(r_0 + r_1) \tag{11-2-62}$$

where γ is the surface energy per unit area. Assuming the dislocation to be of constant volume, $\hat{U}$ is a minimum when

$$\frac{Gb^2}{8\pi^2\gamma} = \frac{r_0}{1 - r_0/r_1} \tag{11-2-63}$$

For many metals, $\gamma \simeq Ga/10$, where a is the interatomic spacing. Hence, for $r_0 \ll r_1$, $\hat{U}$ is a minimum at the equilibrium value of $r_0 \approx 5b^2/4\pi^2 a$. For a simple dislocation, $b = a$ and $r_0 \sim a/8$ and increases to a when $b = 2\pi a/\sqrt{5}$. This small equilibrium value of r_0 is taken to imply that the hollow-core concept of a dislocation is unlikely to be physically valid in the case of a simple dislocation. However, the hollow-core concept assumes increasing physical validity for dislocations of a large Burgers vector, that is, b in excess of $2\pi a/\sqrt{5}$ for which the corresponding equilibrium core radius exceeds one lattice spacing. Equation (11-2-14) gives for the maximum strain $\hat{e}\langle\theta z\rangle$, which is attained at $r = r_0$, the value $\hat{e}\langle\theta z\rangle = b/4\pi r_0$ and hence must have $r_0 > b/4\pi\hat{e}\langle\theta z\rangle$. Using the equilibrium radius determined from Eq. (11-2-63) for $r_0 \ll r_1$ gives the condition $b \geqslant (\pi/5\hat{e}\langle\theta z\rangle)a$. This is taken to imply that the Volterra hollow-core concept of a dislocation has physical basis only in relation to composite dislocations of Burgers vector approximately $(\pi/5\hat{e}\langle\theta z\rangle)$ times the smallest possible Burgers vector. The strain $\hat{e}\langle\theta z\rangle$ is the maximum possible strain consistent with the assumption of classical, infinitesimal linear elasticity. This simple semiquantitative analysis, when applied to an edge dislocation, gives the elastic strain energy to be on the order of 8 eV for each atom plane threaded by the dislocation.

In the case of the single dislocation ($b = a$), it is concluded from this simple semiquantitative analysis that the material outside a cylinder of radius $r_0 \simeq b/4\pi\hat{e}\langle\theta z\rangle$ can be treated in the way discussed above, but that within this cylinder it is necessary to consider the displacements of, and forces between, individual atoms, the cylinder of radius r_0 no longer being a free surface. Estimates of the core energy are on the order of 0.5 to 1.5 eV for each atom plane threaded by the dislocation. Thus, the core energy is only a small fraction of the elastic energy, and for most practical purposes the total energy can be taken to be still given by the above relations for U^d/L.

Reference has been made in Sec. 11-1-3 to mixed dislocations consisting of coincident screw and edge dislocations. Because of the linearity of the theory, the stress field is to be obtained as a superposition of those of a screw dislocation of strength $b\cos\phi$ and an edge dislocation of strength $b\sin\phi$ where ϕ is the angle between the Burgers vector and the dislocation line. The energy per unit length is

$$\begin{aligned}\frac{U^d}{L} &= \tfrac{1}{2}\int_{r_0}^{r_1} [T\langle xy\rangle(x,0)b\sin\phi + T\langle xz\rangle(x,0)b\cos\phi]\,dx \\ &= \frac{Gb^2}{4\pi}\left(\frac{\sin^2\phi}{1-\nu}+\cos^2\phi\right)\left[\ln\left(\frac{r_1}{r_0}\right)-1\right] \qquad \text{for } r_0 \ll r_1 \end{aligned} \tag{11-2-64}$$

11-2-3 Forces on Dislocations

The concept of a force on a dislocation line follows from consideration of the change in energy produced by a change in the position of the dislocation line. The interaction energy U_{int} between the two states represented by Eqs. (11-2-56) and (11-2-57) can be expressed in the equivalent forms

$$U_{\text{int}} = \int_{\partial R_t} \boldsymbol{t}^d \cdot \boldsymbol{u}^{(0)}\,da = \int_{\partial R_t} \boldsymbol{t}^{(0)} \cdot \boldsymbol{u}^d\,da \tag{11-2-65}$$

where ∂R_t denotes the total boundary $S_1 + S_2 + C_+ + C_-$. In the first form of U_{int}, it follows from Eq. (11-2-55) that $\boldsymbol{t}^d$ is zero on S_2 and from Eq. (11-2-54) that, on C, $\boldsymbol{u}^{(0)}$ is continuous between the two faces, while from Eq. (11-2-55) it is seen that $\boldsymbol{t}^d$ is equal and opposite. Thus all the integrals vanish except that from S_1. For the second form of U_{int} it follows from Eq. (11-2-55) that the integral over S_1 is zero. Thus, having regard to the conditions of Eqs. (11-2-54) and (11-2-55), it is evident that Eq. (11-2-65) takes the form

$$U_{\text{int}} = \int_{S_1} \boldsymbol{t}^d \cdot \boldsymbol{u}'\,da = \int_{S_2} \boldsymbol{t} \cdot \boldsymbol{u}^d\,da + \int_{C_+} \boldsymbol{t}^{(0)} \cdot \Delta\boldsymbol{u}\,da \tag{11-2-66}$$

The energy of the state of strain described by the conditions of Eq. (11-2-52) is

$$U = U^{(0)} + U^d + U_{\text{int}}$$

Hence, from Eqs. (11-2-57) and (11-2-66),

$$\Delta U = \int_{S_2} t \cdot u^d \, da + \tfrac{1}{2} \int_{C_+} (t^{(0)} + t^c) \cdot \Delta u \, da \tag{11-2-67}$$

where $\Delta U = U - U^{(0)}$ represents the increase in the internal energy of the body, the dislocation being formed after the external loads on S_1 and S_2 have already been applied.

In addition to the energy of the body, there are external sources providing tractions on S_1 and S_2 which will also suffer an energy change upon formation of the dislocation. This change in energy is

$$\Delta U_{\text{ext}} = -\int_{S_2} t \cdot (u - u^{(o)}) \, da \tag{11-2-68}$$

since the applied traction is maintained at the constant value t on S_2, and on S_1 there is no motion of the boundary to cause any work to be done.

Thus, the total energy change upon formation of the dislocation is

$$\begin{aligned} \Delta U_{\text{tot}} = \Delta U + \Delta U_{\text{ext}} &= \tfrac{1}{2} \int_{C_+} (t^{(0)} + t^c) \cdot \Delta u \, da \\ &= U^d + \int_{C_+} t^{(0)} \cdot \Delta u \, da \end{aligned} \tag{11-2-69}$$

Consider the case when $\Delta u = b$ is constant over the cut C which is bounded by the curve Γ of the dislocation line. From Eq. (11-2-69),

$$\begin{aligned} \delta U_{\text{tot}} &= \delta U^d + \delta \left(b \cdot \int_{C_+} t^{(0)} \, da \right) \\ &= \delta U^d + b \cdot \int_{\delta C_+} t^{(0)} \, da \end{aligned} \tag{11-2-70}$$

where δC_+ is the new cut swept out as Γ moves δr. If v is the unit normal to δC_+ and $\mathbf{T}^{(0)}$ the external symmetric stress tensor, then, on δC_+, $t^{(0)} = (\mathbf{T}^T)^{(0)} v$ and

$$v \, da = -\xi \times \delta r \, dl$$

where dl is an element of arc length along Γ and ξ is a unit vector along Γ whose sense is chosen so that $\Delta u = b$ is the dislocation Burgers vector. With these definitions, Eq. (11-2-70) can be expressed in the form

$$\begin{aligned} \delta U_{\text{tot}}^- &= \delta U^d - \int_{\Gamma} (\mathbf{T}^{(0)} b) \cdot (\xi \times \delta r) \, dl \\ &= U^d - \int_{\Gamma} [(\mathbf{T}^{(0)} b) \times \xi] \cdot \delta r \, dl \end{aligned} \tag{11-2-71}$$

The concept of a general thermodynamic force per unit length F/L, in the sense of representing the change in potential of the system with displacement of the dislocation, now follows from Eq. (11-2-71). Thus,

$$\delta U_{\text{tot}} = -\int \left(\frac{\boldsymbol{F}}{L}\right) \cdot \delta \boldsymbol{r}\, dl \tag{11-2-72}$$

from which it is evident that the contribution to the force provided by the external tractions $t^{(0)}$ at any point on the dislocation line is

$$\frac{\boldsymbol{F}^{(0)}}{L} = (\mathbf{T}^{(0)}\boldsymbol{b}) \times \xi \tag{11-2-73}$$

where $\mathbf{T}^{(0)}$ is the stress tensor corresponding to these tractions evaluated at the position of the dislocation line.

In a finite medium the energy depends on the position of the dislocation relative to the external surfaces and therefore δU^d provides a contribution to the force. This leads to the concept of image forces.

An infinite medium which is nonuniform is also characterized by a δU^d which provides a contribution to the force. The crystalline solid is an important example of a nonuniform infinite medium. In the case of a crystalline solid there is a periodic variation in material constitution along the crystal axes which gives rise to a periodic variation of U^d and therefore to a periodic force. This force is called the Peierls–Nabarro force (Peierls, 1940; Nabarro, 1947) and the corresponding finite stress required to move a straight dislocation in an otherwise perfect crystal is called the Peierls–Nabarro stress τ_p. This stress is very difficult to calculate because to a first approximation it vanishes. The calculations of Sanders (1962) indicate that $\tau_p = A \exp(-\zeta)$ where ζ, the width of the dislocation line, is defined as the width in the slip-plane within which the relative displacement of the atoms above and below is greater than one-half of its maximum value. The width of the dislocation line gives an approximate measure of the extent of the core region that cannot be described by linear elasticity. The concept of image forces and the Peierls–Nabarro force are beyond the scope of the present discussion (see, for example, Nabarro, 1967; Hirthe and Lothe, 1968; Lardner, 1974).

For a straight dislocation in a uniform infinite medium, the dislocation energy U^d is independent of the dislocation position and Eq. (11-2-73) gives the total force by way of Eq. (11-2-71). Consider a screw dislocation lying along the z axis, so that $\xi = (0, 0, 1)$, $\boldsymbol{b} = (0, 0, b)$. Equation (11-2-73) gives

$$\begin{aligned}\frac{\boldsymbol{F}^{(0)}}{L} &= b(T^{(0)}\langle zx\rangle, T^{(0)}\langle zy\rangle, T^{(0)}\langle zz\rangle) \times (0,0,1) \\ &= (bT^{(0)}\langle yz\rangle, -bT^{(0)}\langle xz\rangle, 0)\end{aligned} \tag{11-2-74}$$

from which, for example, it is evident that the force tending to make the screw glide in the x direction is $bT^{(0)}\langle yz\rangle$. In the case of the edge dislocation with $\xi = (0, 0, 1)$, $\boldsymbol{b} = (b, 0, 0)$, Eq. (11-2-73) gives

$$\frac{\boldsymbol{F}^{(0)}}{L} = (bT^{(0)}\langle xy\rangle, -bT^{(0)}\langle xx\rangle, 0) \tag{11-2-75}$$

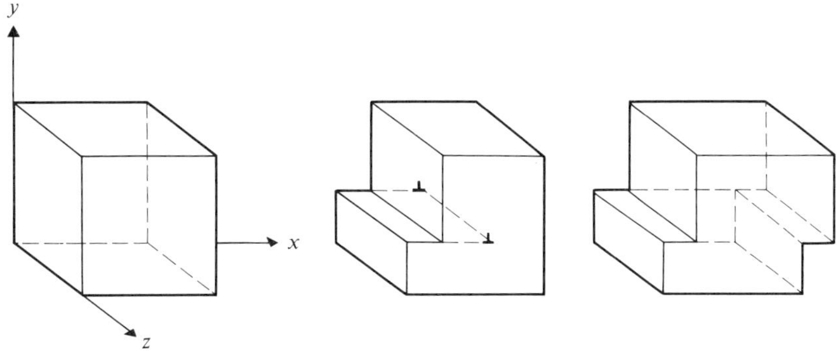

Figure 11-11 Movement of a dislocation across a crystal.

From Fig. 11-11 it is evident that when the dislocation moves across a distance λ, the top face moves a distance b in the x direction, giving a reduction in external potential energy of $(T^{(0)}\langle xy\rangle L\lambda)b$ for a distance L in the z direction. Motion of the edge dislocation in the y direction, driven by the force $-bT^{(0)}\langle xx\rangle$, is called climb, the extra half plane of atoms being squeezed out by the compressive stress $-T^{(0)}\langle xx\rangle$. Climb can usually occur only at high temperatures because there are large intrinsic forces resisting it. Thus, climb is a much more difficult motion for a dislocation than glide where the resistances to motion, such as the Peierls–Nabarro force, blocking effects of impurities and grain boundaries are comparatively small.

To examine the effect of interactions between dislocations, consider two dislocations formed by making cuts $C_{\pm}^{(1)}$ and $C_{\pm}^{(2)}$ on which certain displacement or traction conditions are imposed. The two dislocations must have

$$\begin{aligned} u_{+}^{(1)}-u_{-}^{(1)} &= \Delta u^{(1)} \text{ on } C^{(1)} & u_{+}^{(2)}-u_{-}^{(2)} &= o \text{ on } C^{(1)} \\ u_{+}^{(1)}-u_{-}^{(1)} &= o \text{ on } C^{(2)} & u_{+}^{(2)}-u_{-}^{(2)} &= \Delta u^{(2)} \text{ on } C^{(2)} \\ t_{+}^{(1)} &= -t_{-}^{(1)} & t_{+}^{(2)} &= -t_{-}^{(2)} \text{ on both } C^{(1)} \text{ and } C^{(2)} \end{aligned} \tag{11-2-76}$$

With $\partial R_t = S + C^{(1)} + C^{(2)}$, the interaction energy can be expressed in the equivalent forms

$$U_{\text{int}} = \int_{C_{+}^{(1)}} t^{(2)} \cdot \Delta u^{(1)}\, da = \int_{C_{+}^{(2)}} t^{(1)} \cdot \Delta u^{(2)}\, da \tag{11-2-77}$$

which are to be compared with the second term on the right-hand side of Eq. (11-2-69). From the formal similarity between Eq. (11-2-77) and the second term on the right-hand side of Eq. (11-2-69), it is concluded that if, for example, dislocation 2 is fixed, its effect on dislocation 1 is the same as if it were an external stress. Thus, the analysis of Secs. 11-2-2 and 11-2-3 applies, provided $u^{(0)}$ and $\mathbf{T}^{(0)}$ are everywhere replaced by $u^{(2)}$ and $\mathbf{T}^{(2)}$, i.e., by the fields corresponding to another dislocation. From this conclusion it follows that the relation of Eq. (11-2-73) for the force on a dislocation still applies, provided there is included in the stress field $\mathbf{T}^{(0)}$ the stress fields due to both external tractions and any other dislocations.

11-2-4 Forces between Dislocations

The interaction between dislocations is important because this leads to locking of dislocations within a crystal and thus to work-hardening.

(a) Parallel screw dislocations Consider two screw dislocations parallel to the z axis with Burgers vectors b_1 and b_2 and each with line direction $\xi = (0, 0, 1)$. Supposing that dislocation 2 experiences a force due to the stress field $\mathbf{T}^{(1)}$ of dislocation 1, then according to the above discussion of the interaction between dislocations, this force is given by the same relation as it would be if $\mathbf{T}^{(1)}$ were an external stress field. Therefore, from Eq. (11-2-74), it has components

$$\frac{F\langle x\rangle}{L} = b_2 T^{(1)}\langle yz\rangle \qquad \frac{F\langle y\rangle}{L} = -b_2 T^{(1)}\langle xz\rangle \tag{11-2-78}$$

Substituting for $T^{(1)}\langle yz\rangle$ and $T^{(1)}\langle xz\rangle$ in Eq. (11-2-78) from Eqs. (11-2-12) and (11-2-11) respectively gives

$$\frac{F\langle x\rangle}{L} = \frac{Gb_1b_2}{2\pi}\frac{x}{x^2+y^2} \qquad \frac{F\langle y\rangle}{L} = \frac{Gb_1b_2}{2\pi}\frac{y}{x^2+y^2} \tag{11-2-79}$$

which have a resultant force

$$\frac{F\langle r\rangle}{L} = \frac{Gb_1b_2}{2\pi}\frac{1}{r} \tag{11-2-80}$$

This resultant force $F\langle r\rangle$ is a purely radial force; it is a central repulsion if b_1 and b_2 are of the same sign and an attraction if they are of opposite signs. That interaction effects between screw dislocations are of great significance can be seen from Eqs. (11-2-74) and (11-2-80) by setting, for example, $r = 150b$, which gives the force between two straight equal screw dislocations as being equivalent to an external shear stress of the order of $G/1000$.

By comparison with Eq. (11-2-72), the interaction energy

$$\frac{U_{\text{int}}}{L} = -\int_{r_1}^{r} \frac{F\langle r\rangle}{L}\,dr = \frac{Gb_1b_2}{2\pi}\ln\left(\frac{r_1}{r}\right) \tag{11-2-81}$$

and the total energy is obtained by way of Eq. (11-2-58) in the form

$$U_{\text{tot}} = \frac{G}{4\pi}(b_1^2+b_2^2)\ln\left(\frac{r_1}{r_0}\right) + \frac{Gb_1b_2}{2\pi}\ln\left(\frac{r_1}{r}\right) \tag{11-2-82}$$

For two equal and opposite dislocations ($b_1 = -b_2$), Eq. (11-2-82) takes the form

$$U_{\text{tot}} = \frac{Gb_1^2}{2\pi}\ln\left(\frac{r}{r_0}\right) \qquad (b_1 = -b_2) \tag{11-2-83}$$

In contrast to the divergence of the expressions given in Sec. 11-2-2 for U^d as $r_1 \to \infty$, it is evident that U_{tot}, as given by Eq. (11-2-83), remains finite as $r_1 \to \infty$. This is because, with two equal and opposite dislocations, the dominant terms in the two long-range stress fields cancel one another, leading to a convergent energy integral, even in an infinite medium. This discussion of forces between straight screw dislocations is subject to the limitation $r_0 \leqslant r \leqslant r_1$ in order to minimize effects arising from the use of the stress fields of Eqs. (11-2-11) and (11-2-12) which do not exactly satisfy the free surface conditions on the two core cylinders and on the outer cylinder.

(b) Parallel edge dislocations Consider two edge dislocations where, as shown in Fig. 11-12, the x axis has been taken to be parallel to the two Burgers vectors and $\xi = (0, 0, 1)$ for each dislocation. Supposing that dislocation 2 experiences a force due to the stress field $\mathbf{T}^{(1)}$ of dislocation 1; then, according to the discussion of the interaction between dislocations, the force per unit length on dislocation 2 is given by the same relation as it would be if $\mathbf{T}^{(1)}$ were an external stress field. Therefore, from Eqs. (11-2-38), (11-2-41), and (11-2-75),

$$\frac{F\langle x\rangle}{L} = b_2 T^{(1)}\langle xy\rangle = \frac{Gb_1 b_2}{2\pi(1-\nu)} \frac{x(x^2-y^2)}{(x^2+y^2)^2} \tag{11-2-84}$$

$$\frac{F\langle y\rangle}{L} = -b_2 T^{(1)}\langle xx\rangle = \frac{Gb_1 b_2}{2\pi(1-\nu)} \frac{y(3x^2+y^2)}{(x^2+y^2)^2} \tag{11-2-85}$$

where $F\langle x\rangle$ is the component of force in the common glide direction and $F\langle y\rangle$ is the component of force perpendicular to the glide-plane. The components of this force in cylindrical polar coordinates are

$$\frac{F\langle r\rangle}{L} = \frac{F\langle x\rangle}{L}\cos\theta + \frac{F\langle y\rangle}{L}\sin\theta = \frac{Gb_1 b_2}{2\pi(1-\nu)} \frac{1}{r} \tag{11-2-86}$$

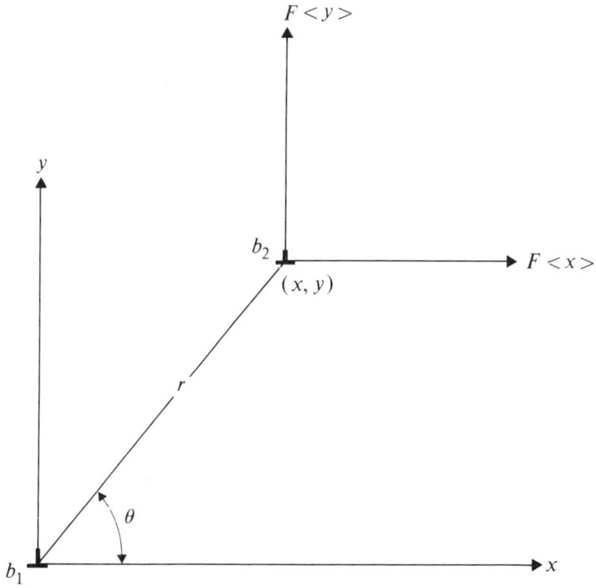

Figure 11-12 Parallel edge dislocations.

$$\frac{F\langle\theta\rangle}{L} = \frac{F\langle y\rangle}{L}\cos\theta - \frac{F\langle x\rangle}{L}\sin\theta = \frac{Gb_1b_2}{2\pi(1-\nu)}\,\frac{\sin 2\theta}{r} \tag{11-2-87}$$

where $F\langle r\rangle$ is the radial component of force between the dislocations and $F\langle\theta\rangle$ is the tangential component.

Since it is usually very difficult for an edge dislocation to execute climb, this type of dislocation tends to move by slip only in the plane contained by the dislocation line and its Burgers vector. Hence, it is only of interest to consider the force in the glide direction, which is just the $F\langle x\rangle$ of Eq. (11-2-84). In Fig. 11-13 is shown schematically the variation of the force $F\langle x\rangle$ with a distance apart of x, expressed in units of y. (Cottrell, 1953). The force $F\langle x\rangle$ is zero at $x = 0$ and $x = \pm y$. It follows that an array of edge dislocations of the same sign, if they are free to glide, is most stable when the dislocations lie vertically above one another. The situation is exactly reversed for edge dislocations of opposite signs, so that the dislocations will tend to arrange themselves in a $\pi/4$ configuration.

The interaction energy is

$$\begin{aligned}\frac{U_{\text{int}}}{L} &= -\int_{r_1}^{x}\frac{F\langle x\rangle}{L}\,dx = -\int_{r_1}^{x} b_2T^{(1)}\langle xy\rangle\,dx = \int_{r_1}^{x} b_2\frac{\partial^2\psi}{\partial x\partial y}\,dx\\ &= b_2\left[\frac{\partial\psi}{\partial y}\right]_{r_1}^{x}\end{aligned} \tag{11-2-88}$$

where ψ is the Airy stress function for the stress field of dislocation 1. From Eq. (11-2-32) for $r \geqslant r_0$,

$$\psi(x,y) = -\frac{Gb_1}{4\pi(1-\nu)}\left[y\ln(x^2+y^2) - \frac{y(x^2+y^2)}{r_1^2}\right] \tag{11-2-89}$$

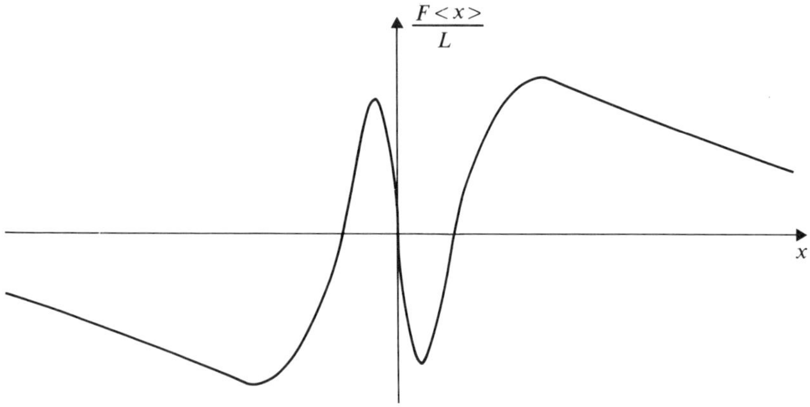

Figure 11-13 Force between parallel edge dislocations.

Thus, for $r \leqslant r_1$,

$$\frac{U_{\text{int}}}{L} = \frac{Gb_1b_2}{2\pi(1-\nu)}\left[\ln\left(\frac{r_1}{r}\right) - \sin^2\theta - \tfrac{1}{2}\right] \tag{11-2-90}$$

and the total energy is obtained by way of Eq. (11-2-60), for equal and opposite Burgers vectors, in the form (Nabarro, 1952)

$$\frac{U_{\text{tot}}}{L} = \frac{Gb^2}{2\pi(1-\nu)}\left[\ln\left(\frac{r}{r_0}\right) - \tfrac{1}{2}\cos 2\theta\right] \tag{11-2-91}$$

11-3 CONTINUOUS DISTRIBUTION OF DISLOCATIONS

11-3-1 Dislocation Density Tensor

In general, it is possible for there to be more than just one type of elementary dislocation present in a single crystal. The pth type of elementary dislocation is specified by its Burgers vector $\boldsymbol{b}^{(p)}$ and its line direction $\xi^{(p)}$. Let there be $N^{(p)}$ dislocations of the pth type crossing a unit area normal to the $\xi^{(p)}$ direction. The dislocation density tensor $\boldsymbol{\alpha}^{(p)}$ for the pth type is defined by the relation

$$\boldsymbol{\alpha}^{(p)} = N^{(p)}(\xi^{(p)} \otimes \boldsymbol{b}^{(p)}) \tag{11-3-1}$$

and the total dislocation density tensor

$$\boldsymbol{\alpha} = \sum_{(p)} \boldsymbol{\alpha}^{(p)} \tag{11-3-2}$$

The $\boldsymbol{\alpha}$ of Eq. (11-3-1) is that used by Nye (1953) and is the transpose of that used by Kroner (1958). For present purposes the choice of $\boldsymbol{\alpha}$ or its transpose is determined by the definition of the strain and rotation tensors, that is, $\tilde{\mathbf{e}}$ and $\tilde{\mathbf{R}}$ respectively of the classical, infinitesimal, linear theory of elasticity.

Consider a small planar circuit $\mathscr{C}$ of area $A(\mathscr{C})$ in the body $\mathscr{B}$. The number of dislocation lines of the pth type threading this circuit is $N^{(p)}\xi^{(p)} \cdot \nu A(\mathscr{C})$ where ν is the unit normal to $A(\mathscr{C})$. A mapping of this circuit onto a perfect reference crystal in the same way as described in Sec. 11-1-3 and Fig. 11-2 does not close, the vector required for closure being just the sum of the true Burgers vectors of all the dislocation lines threading $\mathscr{C}$. Mapping this closure failure back onto the real crystal gives the local Burgers vector $\boldsymbol{b}(\mathscr{C})$ for the circuit, which can be expressed by way of Eqs. (1-6-27) and (11-3-1) in the form

$$\begin{aligned} \boldsymbol{b}(\mathscr{C}) &= \sum_{(p)} (N^{(p)}\xi^{(p)} \cdot \nu A(\mathscr{C}))\boldsymbol{b}^{(p)} \\ &= \sum_{(p)} N^{(p)}(\boldsymbol{b}^{(p)} \otimes \xi^{(p)})\nu A(\mathscr{C}) \\ &= \boldsymbol{\alpha}^T \nu A(\mathscr{C}) \end{aligned} \tag{11-3-3}$$

Equation (11-3-3) defines the transpose of the dislocation density tensor in terms of the closure failure of Burgers circuits. This approach allows a dislocation density tensor to be defined for any elastic continuum.

In simple cubic crystals, with axes in the cube directions, the three diagonal components and six off-diagonal components represent individually the nine possible dislocation types. In general, taking one of the coordinate axes to lie along the dislocation line, screw dislocations which by definition have their Burgers vector parallel to their line direction [see Eq. (11-1-6)] must give rise to a diagonal term in the dislocation density tensor. In contrast, with the same choice of coordinates, edge dislocations give rise to off-diagonal components of the dislocation density tensor.

As discussed in Secs. 11-2-1 and 11-2-2, the presence of a dislocation in a body $\mathscr{B}$ can be regarded as giving rise to a cut in the interior of the body across which there is an appropriate displacement discontinuity. Thus, an arrangement of discrete dislocation lines can be formed by making various cuts throughout the body $\mathscr{B}$ and causing appropriate displacement discontinuities across them. The resulting deformation can be described within the limitation of the classical, infinitesimal linear theory of elasticity by the displacement field $\boldsymbol{u}(\boldsymbol{x})$, which can be defined at each point in the cut body. For these conditions, it follows from consideration of Eq. (2-7-17) that

$$d\boldsymbol{u} = (\tilde{\mathbf{e}} + \tilde{\mathbf{R}})\,d\boldsymbol{x} \tag{11-3-4}$$

With regard to the displacement field $\boldsymbol{u}$, which is well defined in the cut body, it is to be noted that $\boldsymbol{u}$ is not single valued in a body for which the cuts have all been closed. The assumption is now made that $\tilde{\mathbf{e}}$ and $\tilde{\mathbf{R}}$ do exist as single-valued fields after the cuts are closed, it being important to note, however, that although this assumption makes the right-hand side of Eq. (11-3-4) a well-defined differential form, it cannot be integrated except in the cut body. Let the circuit $\mathscr{C}$ start and finish at opposite points on one of the cuts. Substituting in Eq. (11-1-5) for $d\boldsymbol{u}$ from Eq. (11-3-4) gives, for the circuit $\mathscr{C}$,

$$\oint_{\mathscr{C}} \tilde{\mathbf{S}}\,d\boldsymbol{s} = \boldsymbol{b} \tag{11-3-5}$$

where $\tilde{\mathbf{S}} = \operatorname{grad} \boldsymbol{u}$ is the tensor of displacement gradients. Not only does Eq. (11-3-5) retain physical significance after all the cuts are closed, but there is also no need then to specify a particular starting point for $\mathscr{C}$. Now let $\mathscr{C}$ be a circuit enclosing several dislocation lines. This implies that $\mathscr{C}$ must now be equivalent to a number of circuits, one around each dislocation line. Hence, the right-hand side of Eq. (11-3-5) must be replaced by the sum of the Burgers vectors of all the dislocation lines threading $\mathscr{C}$ to give, by way of Eq. (11-3-3) and for sufficiently small $\mathscr{C}$, the relation

$$\oint_{\mathscr{C}} \tilde{\mathbf{S}}\,d\boldsymbol{s} = \boldsymbol{\alpha}^T \boldsymbol{\nu} A(\mathscr{C}) \tag{11-3-6}$$

Use of Stokes' theorem to transform the line integral into a surface integral and rearranging gives

$$\int_{\partial R_t} (\operatorname{curl}\tilde{\mathbf{S}} - \boldsymbol{\alpha}^T)\mathbf{v}\, da = 0 \tag{11-3-7}$$

The condition that Eq. (11-3-7) holds for all arbitrary circuits $\mathscr{C}$, however small, leads to the relation

$$\boldsymbol{\alpha}^T = \operatorname{curl}\tilde{\mathbf{S}} \tag{11-3-8}$$

from which it follows that

$$\operatorname{div}\boldsymbol{\alpha} = \mathbf{o} \tag{11-3-9}$$

which is a conservation law for Burgers vectors, thus recognizing the essential condition that dislocation lines cannot end within the interior of a body.

For use in experimental work, the density may be defined as either N_S, the number of dislocations crossing unit area taken at random in the solid, or as the total length of dislocation line in unit volume, denoted as N_V. Various techniques exist for the measurement of dislocation density (see, for example, Pashley, 1965).

11-3-2 Velocities of Dislocations

The velocities at which dislocations move by glide are characteristic of the applied shear stress, purity of crystal, temperature, and type of dislocation. Thus, any treatment of dislocations must have as an essential part of the theory an equation for the dislocation velocity associated with given conditions of stress and dislocation densities. Hence, the present discussion will be restricted to a description of the results of various experimental measurements of dislocation velocities.

It is to be noted, however, that, as Taylor (1956) has commented, even a fully detailed knowledge of crystal dislocation mechanics will not necessarily lead to a physically satisfactory macroscopic description of plasticity because of the mathematical complexity involved. Continuum plasticity cannot describe the situation because of inadequate variables.

Johnston and Gilman (1959) carried out stress-strain measurements on lithium fluoride crystals and by using an etch-pit technique for revealing dislocations were able to measure their velocities. In these experiments, dislocations were initially introduced into a crystal, e.g., by lightly deforming the surface. The crystal is then subjected to a constant stress pulse for a given time. The dislocation velocity is determined from the positions of the dislocations before and after the stress pulse. The dependence of dislocation velocity on stress is determined from a series of such experiments by varying both the width and height of the stress pulse. The measurements of Johnston and Gilman cover a range of velocities for both screw and edge dislocations from 10^{-8} to $5 \times 10^{2}\,\mathrm{m\,s^{-1}}$. Some typical results for lithium fluoride are shown in Fig. 11-14. From these results it was concluded that

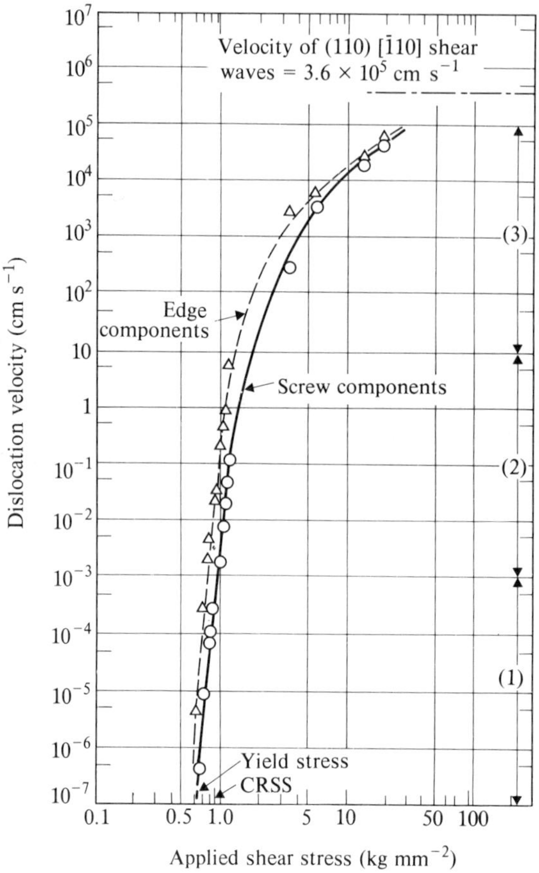

Figure 11-14 Stress dependence of the velocity of edge and screw dislocations. (*Johnston and Gilman, 1959.*)

in the range of velocities between 10^{-8} and 10^{-3} m s^{-1}, the velocity v varied with the shear stress according to the relation

$$v = v_0 \left(\frac{\tau}{\tau_0} \right)^n \tag{11-3-10}$$

which is an empirical power-law relationship in one disposable parameter, n. In Eq. (11-3-10), τ is the applied shear stress resolved in the slip-plane and from experiment it was found that $n \sim 25$ for lithium fluoride. Gilman (1960) has suggested, as an approximation to cover the whole range of velocities, the relation

$$v = v_0 \exp\left(-\frac{\tau_0}{\tau} \right) \tag{11-3-11}$$

where v_0 and τ_0 are constants, it being found by Gilman that v_0 is about half the velocity of shear-wave propagation in this material. Extrapolation of Eq. (11-3-11) gives v_0 as an upper limit to dislocation velocities. The velocity of edge and screw components were measured independently, it being concluded from these measurements that edge dislocations move an order of magnitude faster than screw dislocations over most of the range of measurement. Johnston and Gilman (1960) suggested that this is due to the onset of double cross-slip which produces a trail of loops behind a screw dislocation and has a retarding effect on its motion. Since dislocation lines are basically geometrical configurations, they are flexible and therefore localized line irregularities can occur. For example, kinks can occur in dislocation lines and are associated with local offsets lying in the glide-plane of the dislocation. A jog is a segment of dislocation line which has a component that is perpendicular to the glide-plane. The crossing of two dislocations is characterized by the formation of kinks and jogs (Heidenreich and Shockley, 1948). Weertman (1961) has pointed out that jogs on moving dislocations would have a strong retarding effect on fast-moving screw and mixed dislocations but not on pure-edge dislocations. Measurements have also been made by Gutmanas, Nadgornyi, and Stepanov (1963) on two types of sodium chloride, these results being in accord with Eq. (11-3-10) over all except very low and very high velocities, n having the values 8 and 17 for the two purities of sodium chloride.

The effect of temperature on dislocation velocity is shown in Fig. 11-15 for velocities below $10^{-4}\,\mathrm{m\,s^{-1}}$. These results were obtained by Stein and Low (1960) from studies of dislocation velocities in 3.25 percent silicon iron using the same experimental technique as that developed by Johnston and Gilman (1959). The results cover only the linear range because of the restriction to low velocities, the exponent n appearing in Eq. (11-3-10) ranging from 35 to 44, depending on the test temperature. Greenman, Vreeland, and Wood (1967) have made measurements on copper and Pope, Vreeland, and Wood (1967) on zinc. These results, using Eq. (11-3-10), require $n \sim 1$. This observation has been interpreted by these authors as evidence for the dissipative mechanisms arising from interactions between dislocations and phonons in the way discussed by Lothe (1962), who concluded from an assessment of these mechanisms that the velocity should vary linearly with the shear stress.

11-3-3 Balance Equation for Dislocations

For a perfect, unstrained crystal (i.e., no dislocations), it is possible to represent the crystalline structure by an orthonormal triad of vectors $\{\boldsymbol{a}_1, \boldsymbol{a}_2, \boldsymbol{a}_3\}$ which are constant throughout the body. Although this triad cannot in general be identified directly with the lattice vectors (with the exception of a simple cubic crystal), the triad will be some fixed linear combination of the lattice vectors. Hence, for present purposes $\{\boldsymbol{a}_k\}$ $(k = 1, 2, 3)$ will be referred to as the reference triad of lattice vectors and will be taken as the basis for a cartesian coordinate system. The changes in the triad of lattice vectors arising from distortion of the crystal and irrespective of whether it contains dislocations or not will be expressed in the form

$$\boldsymbol{e}_k(\boldsymbol{x}) = \boldsymbol{a}_k + \hat{\beta}_{kl}(\boldsymbol{x})\boldsymbol{a}_l \tag{11-3-12}$$

where $\hat{\beta}$ will be referred to as the lattice distortion. The effect of the deformation is for a small element of the material, initially located at the point $\boldsymbol{x}$, to move with the

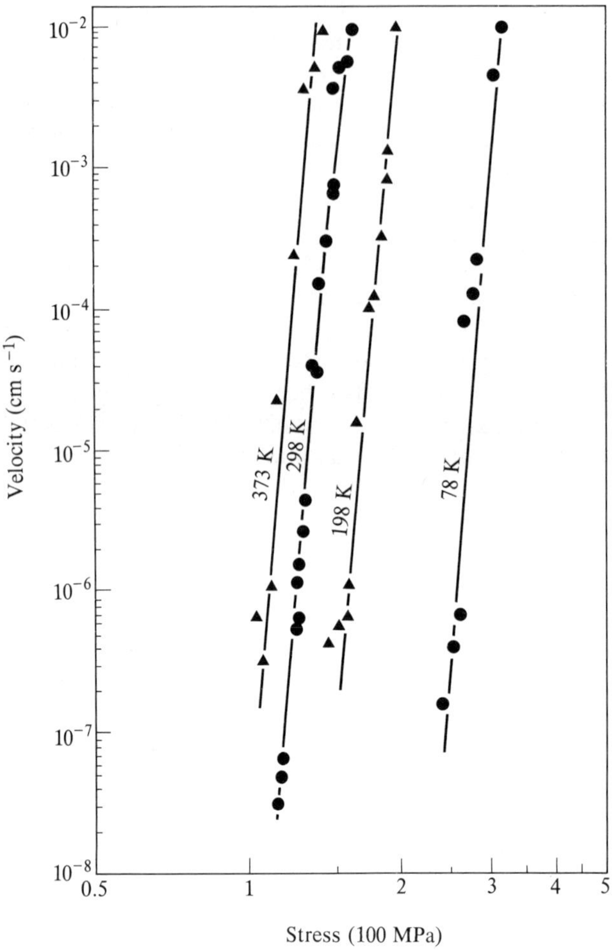

Figure 11-15 Stress dependence of the velocity of edge dislocations in 3.25 percent silicon iron at four temperatures. (*Stein and Low, 1960.*)

lattice and so become an element

$$dx_k \boldsymbol{e}_k(\boldsymbol{x}) = d\boldsymbol{x} + \hat{\beta}_{kl} dx_k \boldsymbol{a}_l \tag{11-3-13}$$

but from continuum considerations

$$dx_k \boldsymbol{e}_k(\boldsymbol{x}) = d\boldsymbol{x} + \frac{\partial u_l}{\partial x_k} dx_k \boldsymbol{a}_l \tag{11-3-14}$$

Hence, comparison of Eqs. (11-3-13) and (11-3-14) gives

$$\tilde{\mathbf{S}} = \hat{\beta}^T \tag{11-3-15}$$

Substituting in Eq. (11-3-8) for $\tilde{\mathbf{S}}$ from Eq. (11-3-15) gives

$$\alpha^T = \operatorname{curl} \hat{\beta}^T \tag{11-3-16}$$

(a) Screw dislocations The antiplane strain of Sec. 6-3-3 corresponds to the deformation field $u_1 = u_2 = 0$, $u_3 = u_3(x_1, x_2)$. For a crystal containing dislocations this particular deformation field will give rise to the lattice distortion

$$\boldsymbol{e}_1 = \boldsymbol{a}_1 + \hat{\beta}_{13}\boldsymbol{a}_3 \qquad \boldsymbol{e}_2 = \boldsymbol{a}_2 + \hat{\beta}_{23}\boldsymbol{a}_3 \qquad \boldsymbol{e}_3 = \boldsymbol{a}_3 \tag{11-3-17}$$

where $\hat{\beta}_{13} = \hat{\beta}_{13}(x_1, x_2)$ and $\hat{\beta}_{23} = \hat{\beta}_{23}(x_1, x_2)$. It is evident from Eq. (11-3-12) that the only nonzero components of strain are

$$\tilde{e}_{13} = \tfrac{1}{2}\hat{\beta}_{13} \qquad \tilde{e}_{23} = \tfrac{1}{2}\hat{\beta}_{23} \tag{11-3-18}$$

and from Eq. (11-3-16) it follows that the only nonzero component of the dislocation density tensor is

$$\alpha_{33} = \frac{\partial \hat{\beta}_{23}}{\partial x_1} - \frac{\partial \hat{\beta}_{13}}{\partial x_2} \tag{11-3-19}$$

it being noted that $\boldsymbol{\alpha}$ and $\hat{\boldsymbol{\beta}}$ are not in general symmetrical.

Let the screw dislocations move relative to the medium. In the most general case there will exist at any point a whole distribution of dislocation velocities arising, for example, from localized effects such as barriers to slip acting on only one dislocation line, its immediate neighbors being unaffected. However, the principal features of immediate interest can all be studied by way of the simplifying assumptions that the dislocations translate as straight lines with all points on any one line moving with the same velocity. The dislocation velocity is determined by the local resolved shear stress which will be the same for all positive dislocations but equal and opposite for negative dislocations. It is therefore further assumed that positive screw dislocations in the neighborhood of a given point move with the same velocity $\boldsymbol{v} = (v_1, v_2, 0)$ while the negative dislocations all have velocity $-\boldsymbol{v}$.

Let the dislocation density tensor for positive dislocations be denoted as $\boldsymbol{\alpha}^+$ and for negative dislocations by $\boldsymbol{\alpha}^-$. Expressions for the rate of change of density for each type of dislocation, that is, α_{33}^+ and α_{33}^-, can be obtained from the general balance equation (3-10-4) by setting $\rho\psi = \alpha_{33}^+$, $\rho s = (\dot{\alpha}_{33})_g$, and $\boldsymbol{j} = \boldsymbol{o}$ to give

$$\frac{\partial \alpha_{33}^+}{\partial t} = -\frac{\partial}{\partial x_1}(\alpha_{33}^+ v_1) - \frac{\partial}{\partial x_2}(\alpha_{33}^+ v_2) + (\dot{\alpha}_{33})_g \tag{11-3-20}$$

$$\frac{\partial \alpha_{33}^-}{\partial t} = \frac{\partial}{\partial x_1}(\alpha_{33}^- v_1) + \frac{\partial}{\partial x_2}(\alpha_{33}^- v_2) - (\dot{\alpha}_{33})_g \tag{11-3-21}$$

where the term in $(\dot{\alpha}_{33})_g$ represents the rate of generation per unit area of positive-negative dislocation pairs. The sum of Eqs. (11-3-20) and (11-3-21) gives

$$\frac{\partial \alpha_{33}}{\partial t} = -\frac{\partial}{\partial x_1}(v_1 \Delta\alpha_{33}) - \frac{\partial}{\partial x_2}(v_2 \Delta\alpha_{33}) \tag{11-3-22}$$

where $\Delta\alpha_{33} \equiv \alpha_{33}^{+} - \alpha_{33}^{-}$ is the total numerical density of the two types of dislocation. Equation (11-3-22) is an exact conservation law for the total Burgers vector. The difference between Eqs. (11-3-20) and (11-3-21) gives

$$\frac{\partial \Delta\alpha_{33}}{\partial t} = -\frac{\partial}{\partial x_1}(v_1\alpha_{33}) - \frac{\partial}{\partial x_2}(v_2\alpha_{33}) + 2(\dot{\alpha}_{33})_g \tag{11-3-23}$$

Equation (11-3-19) can be used to rearrange Eq. (11-3-22) into the form

$$\frac{\partial}{\partial x_1}\left(\frac{\partial \beta_{23}}{\partial t}\right) = \frac{\partial}{\partial x_2}\left(\frac{\partial \beta_{13}}{\partial t}\right) \tag{11-3-24}$$

where

$$\frac{\partial \beta_{13}}{\partial t} = \frac{\partial \hat{\beta}_{13}}{\partial t} + \frac{\partial \bar{\beta}_{13}}{\partial t} \tag{11-3-25a}$$

$$\frac{\partial \beta_{23}}{\partial t} = \frac{\partial \hat{\beta}_{23}}{\partial t} + \frac{\partial \bar{\beta}_{23}}{\partial t} \tag{11-3-25b}$$

and

$$\frac{\partial \bar{\beta}_{13}}{\partial t} = -v_2\Delta\alpha_{33} \tag{11-3-26}$$

$$\frac{\partial \bar{\beta}_{23}}{\partial t} = v_1\Delta\alpha_{33} \tag{11-3-27}$$

In Eq. (11-3-25), the $\partial\hat{\beta}_{ij}/\partial t$ represent the rate of change of the lattice distortion. It is evident from the form of Eqs. (11-3-26) and (11-3-27) that $\partial\bar{\beta}_{13}/\partial t$ and $\partial\bar{\beta}_{23}/\partial t$ are directly related to the plastic deformation. Hence Eqs. (11-3-25*a*, *b*) represent the splitting of the quantity $\partial\beta_{ij}/\partial t$ into elastic and plastic components. Let $u_3(x, y, t)$ be the overall displacement field in the body; then, having regard to Eq. (11-3-25*b*), the splitting of $\partial\beta_{ij}/\partial t$ into elastic and plastic components is taken to imply the existence of the relations

$$\frac{\partial}{\partial x_1}\left(\frac{\partial u_3}{\partial t}\right) = \frac{\partial \hat{\beta}_{13}}{\partial t} + \frac{\partial \bar{\beta}_{13}}{\partial t} \tag{11-3-28}$$

$$\frac{\partial}{\partial x_2}\left(\frac{\partial u_3}{\partial t}\right) = \frac{\partial \hat{\beta}_{23}}{\partial t} + \frac{\partial \bar{\beta}_{23}}{\partial t} \tag{11-3-29}$$

(b) Edge dislocations For a crystal containing dislocations, a state of plane strain will give rise to the lattice distortion

$$\begin{aligned} \mathbf{e}_1 &= \mathbf{a}_1 + \hat{\beta}_{11}\mathbf{a}_1 + \hat{\beta}_{12}\mathbf{a}_2 \\ \mathbf{e}_2 &= \mathbf{a}_2 + \hat{\beta}_{21}\mathbf{a}_1 + \hat{\beta}_{22}\mathbf{a}_2 \\ \mathbf{e}_3 &= \mathbf{a}_3 \end{aligned} \tag{11-3-30}$$

where the $\hat{\beta}_{ij}$ are functions of x_1 and x_2. It is evident from Eqs. (11-3-12) and (11-3-15) that the only nonzero components of strain are

$$\tilde{e}_{11} = \hat{\beta}_{11} \qquad \tilde{e}_{22} = \hat{\beta}_{22} \qquad \tilde{e}_{12} = \tfrac{1}{2}(\hat{\beta}_{12} + \hat{\beta}_{21}) \tag{11-3-31}$$

and from Eq. (11-3-16) it follows that the only nonzero components of the dislocation density tensor are

$$\alpha_{13} = \frac{\partial \hat{\beta}_{21}}{\partial x_1} - \frac{\partial \hat{\beta}_{11}}{\partial x_2} \qquad \alpha_{23} = \frac{\partial \hat{\beta}_{22}}{\partial x_1} - \frac{\partial \hat{\beta}_{12}}{\partial x_2} \tag{11-3-32}$$

For edge dislocations, the $\alpha_{13}^{\pm}$ dislocations all have the same velocity $\pm v_1$ in the x_1 direction and the $\alpha_{23}^{\pm}$ dislocations all have the same velocity $\pm v_2$ in the x_2 direction. Expressions for the rate of change of density for each type of dislocation can be obtained from the general balance equation (3-10-4) in a form similar to that of Eqs. (11-3-20) and (11-3-21). These equations give

$$\frac{\partial \alpha_{13}}{\partial t} = -\frac{\partial}{\partial x_1}(v_1 \Delta\alpha_{13}) \tag{11-3-33}$$

$$\frac{\partial \alpha_{23}}{\partial t} = -\frac{\partial}{\partial x_2}(v_2 \Delta\alpha_{23}) \tag{11-3-34}$$

and

$$\frac{\partial \Delta\alpha_{13}}{\partial t} = -\frac{\partial}{\partial x_1}(v_1 \alpha_{13}) + 2(\dot{\alpha}_{13})_g \tag{11-3-35}$$

$$\frac{\partial \Delta\alpha_{23}}{\partial t} = -\frac{\partial}{\partial x_2}(v_2 \alpha_{23}) + 2(\dot{\alpha}_{23})_g \tag{11-3-36}$$

Equations (11-3-33) to (11-3-36) are to be compared with Eqs. (11-3-22) and (11-3-23).

By the same arguments that gave Eqs. (11-3-24) to (11-3-29) there follow for edge dislocations the corresponding relations

$$\frac{\partial}{\partial x_2}\left(\frac{\partial u_1}{\partial t}\right) = \frac{\partial \hat{\beta}_{21}}{\partial t} + \frac{\partial \bar{\beta}_{21}}{\partial t} \qquad \frac{\partial}{\partial x_2}\left(\frac{\partial u_2}{\partial t}\right) = \frac{\partial \hat{\beta}_{22}}{\partial t} \tag{11-3-37}$$

$$\frac{\partial}{\partial x_1}\left(\frac{\partial u_2}{\partial t}\right) = \frac{\partial \hat{\beta}_{12}}{\partial t} + \frac{\partial \bar{\beta}_{12}}{\partial t} \qquad \frac{\partial}{\partial x_1}\left(\frac{\partial u_1}{\partial t}\right) = \frac{\partial \hat{\beta}_{11}}{\partial t} \tag{11-3-38}$$

where

$$\frac{\partial \bar{\beta}_{21}}{\partial t} = v_1 \Delta\alpha_{13} \qquad \frac{\partial \bar{\beta}_{22}}{\partial t} = 0 \tag{11-3-39}$$

$$\frac{\partial \bar{\beta}_{12}}{\partial t} = -v_2 \Delta\alpha_{23} \qquad \frac{\partial \bar{\beta}_{11}}{\partial t} = 0 \tag{11-3-40}$$

11-3-4 Equations of Motion

A further equation in the continuum theory of dislocations arises from the Cauchy equations of motion.

(a) Screw dislocations The Cauchy equation of motion for the case of screw dislocations can be expressed in the form

$$\rho \ddot{u}_3 = \frac{\partial T_{31}}{\partial x_1} + \frac{\partial T_{32}}{\partial x_2} \qquad (\mathbf{T} = \mathbf{T}^T) \tag{11-3-41}$$

which can only be used by way of a given constitutive equation relating stress to strain. For example, assuming Hooke's law, Eq. (11-3-41) can be expressed in the form

$$\rho \ddot{u}_3 = 2G\left(\frac{\partial \tilde{e}_{13}}{\partial x_1} + \frac{\partial \tilde{e}_{23}}{\partial x_2}\right) \tag{11-3-42}$$

Substituting from Eqs. (11-3-18) for $\hat{\beta}_{13}$ and $\hat{\beta}_{23}$ in Eqs. (11-3-28) and (11-3-29) respectively gives

$$\frac{\partial u_3}{\partial x_1} = 2\tilde{e}_{13} + \bar{\beta}_{13} \tag{11-3-43}$$

$$\frac{\partial u_3}{\partial x_2} = 2\tilde{e}_{23} + \bar{\beta}_{23} \tag{11-3-44}$$

which can now be used to substitute in Eq. (11-3-42) for $\tilde{e}_{13}$ and $\tilde{e}_{23}$ to give

$$\frac{1}{c_T^2}\ddot{u}_3 = \nabla^2 u_3 - \left(\frac{\partial \bar{\beta}_{13}}{\partial x_1} + \frac{\partial \bar{\beta}_{23}}{\partial x_2}\right) \tag{11-3-45}$$

where from Eq. (6-4-5*b*) it follows that $c_T = (G/\rho)^{1/2}$ is the velocity of elastic shear waves in the medium.

(b) Edge dislocations The Cauchy equation of motion for the case of edge dislocations can be expressed in the form

$$\rho \ddot{u}_1 = \frac{\partial T_{11}}{\partial x_1} + \frac{\partial T_{12}}{\partial x_2} \qquad \rho \ddot{u}_2 = \frac{\partial T_{12}}{\partial x_1} + \frac{\partial T_{22}}{\partial x_2} \tag{11-3-46}$$

The field equations are derived from Eq. (11-3-46) by substituting from some appropriate constitutive equation, as, for example, Hooke's law, and by further substitution from Eq. (11-3-31).

11-4 STRUCTURAL THEORY OF PLASTICITY: I. SINGLE CRYSTALS

11-4-1 Single Crystal Slip

Plastic deformation in single crystals is concentrated as slip on a few parallel planes which are called slip-planes. Discrete blocks of crystal between two slip-planes remain undistorted. Increasing plastic deformation is accompanied by the formation of groups of active slip-planes within the same neighborhood of the crystal. A group of near-neighbor active slip-planes is said to form a slip-band.

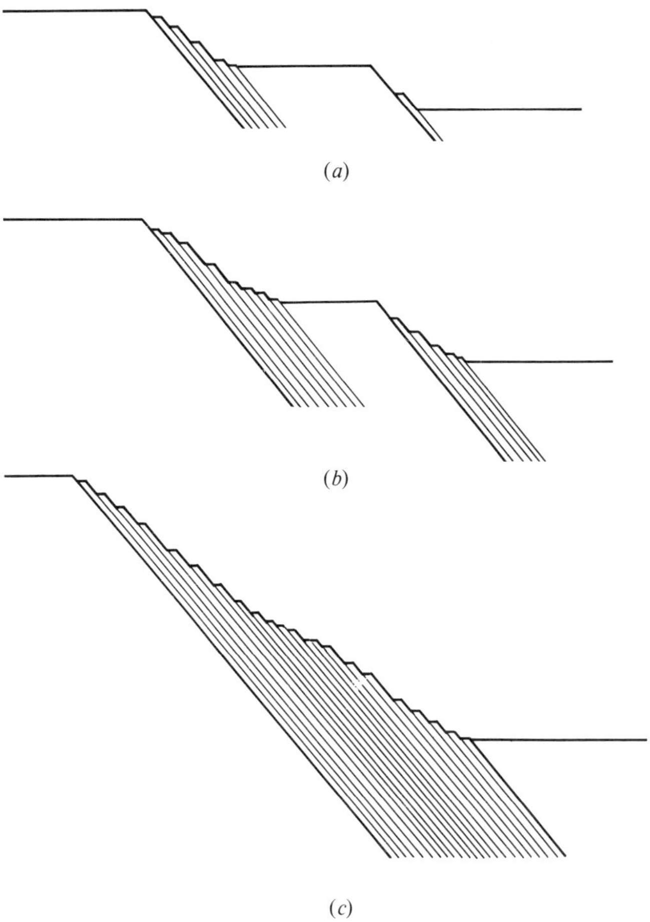

Figure 11-16 Slip-bands on single crystals. (*Heidenreich and Shockley, 1948.*)

Electron microscope studies of the surface of single crystals which have been plastically deformed provide direct evidence of this slip-band type of structure. The electron microscope studies of Heidenreich and Shockley (1947) have shown the slip-band structure to consist of a series of steps separated by thin lamellae as shown in Fig. 11-16. Brown (1949, 1950, 1952) found that the lamellae thickness remained almost constant but the average number of lamellae per slip-band increased with strain and temperature. Wilsdorf and Kuhlmann–Wilsdorf (1952) found that for aluminum deformed at room temperature, the slip-lines vary in step height from a few angstroms up to about 1200 Å. The average slip per step in the slip-bands was 300 to 400 Å with an average lamellae thickness of 400 Å. The average number of lamellae per band increased with the strain and averaged 19 at a shear strain of 32 percent. Between the slip-bands Wilsdorf and Kuhlmann–Wilsdorf observed a fine slip structure consisting of steps of generally less than

100 Å whose spacing and step heights are a function of the shear stress in the specimens. Similar results have been found for silver and copper. Noggle and Koehler (1957) have deformed single crystals of very pure aluminum in tension at 4.2, 78, and 295 K. The deformation at 4.2 K produced a uniformly distributed fine slip structure with no tendency for the lines to group into slip-bands and the associated stress-strain curve was linear from strains of about 10 to 50 percent. At 78 K up to a shear strain of 40 percent the development of slip-bands occurs slowly and the majority of the strain is associated with the fine slip structure. At higher strains a banded structure begins to appear. The strain associated with a given stress is progressively greater than at 4.2 K, giving rise to a nonlinear stress-strain curve. At room temperature a prominent slip-band structure is present and accounts for the major portion of the total strain. The strain associated with a given stress is greater than at 78 K, giving rise to an even more nonlinear stress-strain curve. At 4 K, Noggle and Koehler state that there is an approximately linear increase in average shear per step of about 20 Å kg^{-1} mm^{-2} while the lamellae thickness is constant with increasing stress at about 250 Å. At higher temperatures and increasing stress the number of slip-planes increases and they tend to cluster together into slip-bands.

11-4-2 Possible Burgers Vectors

In considering the structural theories of plasticity, the reader will be assumed to be familiar with the basic notations of crystallography (see, for example, Cullity, 1959; Taylor, 1961; Hull, 1965; Jaswon, 1965; Barrett and Massalski, 1966).

The Burgers circuit has been defined as a lattice point to lattice point movement, and hence the closure failure, which must always be between two lattice points in the perfect crystal, must be a lattice vector. That is, the Burgers vector must be a vector between some pair of corresponding lattice points and hence, conversely, any such lattice vector is a possible Burgers vector for a dislocation. A dislocation defined in this way is a *perfect* or *unit dislocation*. Thus, the Burgers vectors allowable in a given type of crystal lattice are determined by the lattice vectors and must be such that the new configuration is mechanically stable. An example of a changed, but mechanically stable, configuration is

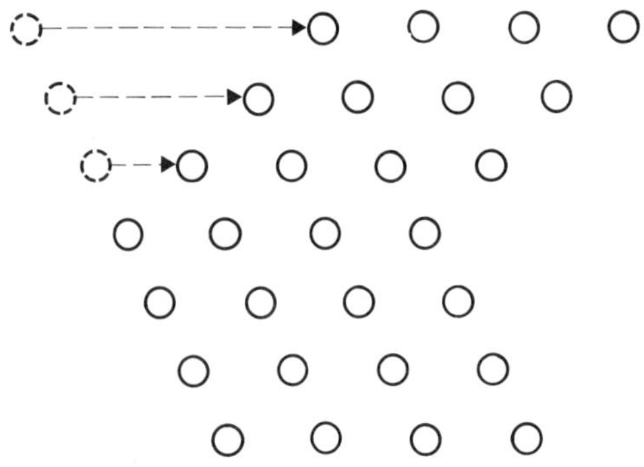

Figure 11-17 A twinned crystal in which each half has a perfect crystal structure but they are differently oriented.

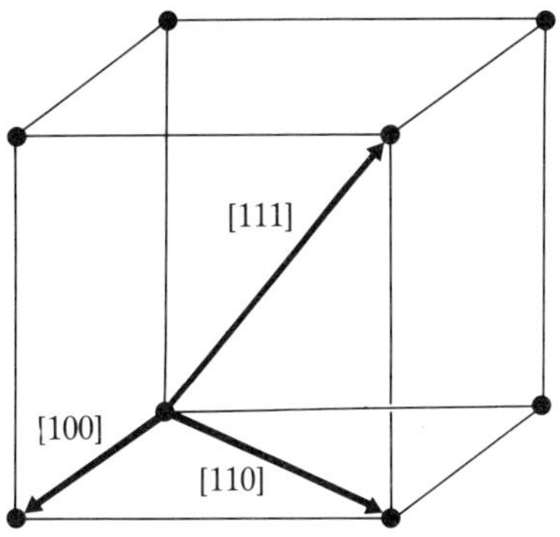

Figure 11-18 Possible Burgers vectors.

twinning, which is shown schematically in Fig. 11-17 (for a discussion of deformation by twinning, see Hall, 1954; Honeycomb, 1968). If the Burgers vector is associated with a changed configuration the dislocation is said to be *imperfect*; otherwise it is said to be perfect. Consider the three possible Burgers vectors for the simple cubic structure shown in Fig. 11-18. These vectors represent the three shortest Burgers vectors, which are [100], [110], and [111] in order of increasing length.

Dislocations may combine together or dissociate to form other dislocations. The process of combination and dissociation of dislocations is referred to as a dislocation reaction. Consider, for example, the combination of two dislocations with the Burgers vectors $\boldsymbol{b}_1$ and $\boldsymbol{b}_2$ and parallel lines to form a single dislocation of the Burgers vector $\boldsymbol{b}_3$. If such a combination occurs, then in general it is necessary that $\boldsymbol{b}_3 = \boldsymbol{b}_1 + \boldsymbol{b}_2$, a condition which can be established by taking a Burgers circuit, as in Sec. 11-1-3, which encloses both $\boldsymbol{b}_1$ and $\boldsymbol{b}_2$; the closure failure of such a circuit is unaffected by the combination of the two dislocations into $\boldsymbol{b}_3$. Conversely, it is possible for the dislocation with the Burgers vector $\boldsymbol{b}_3$ to dissociate into a pair of dislocations with the Burgers vectors $\boldsymbol{b}_1$ and $\boldsymbol{b}_2$. Taking the simple cubic structure as an example, a [110] dislocation can dissociate according to the reaction

$$[110] \rightarrow [100] + [010] \tag{11-4-1}$$

into a pair with the Burgers vectors [100] and [010].

If it is energetically favorable a dislocation reaction will proceed automatically in order to produce a reduction in energy. Frank (1949a) gives the following criterion for dissociation. The expressions given in Sec. 11-2-2 for U^d imply that in all cases a dislocation with the Burgers vector $\boldsymbol{b}$ produces a strain energy in the medium equal to $\phi \boldsymbol{b}^2$ per unit length of dislocation line. Thus the energies associated with the dislocation reaction $\boldsymbol{b}_3 \rightarrow \boldsymbol{b}_1 + \boldsymbol{b}_2$ are $\phi \boldsymbol{b}_1^2$, $\phi \boldsymbol{b}_2^2$, and $\phi \boldsymbol{b}_3^2 = \phi(\boldsymbol{b}_1 + \boldsymbol{b}_2)^2$. Frank neglected the factor of the $(1-\nu)$ difference between Eqs. (11-2-61) and (11-2-59) and on this basis stated that dissociation occurs spontaneously if

$$b_3^2 > b_1^2 + b_2^2 \tag{11-4-2}$$

If the dislocation $\boldsymbol{b}_3$ satisfies this criterion, then since the dislocation $\boldsymbol{b}_3$ must be mechanically unstable it will not contribute to the slip process.

More accurate than Frank's criterion, stability can be determined from the interaction force considered in relation to the product dislocations with the Burgers vectors $\boldsymbol{b}_1$ and $\boldsymbol{b}_2$. If the interaction force between $\boldsymbol{b}_1$ and $\boldsymbol{b}_2$ is attractive, $\boldsymbol{b}_3$ will be stable, while if the interaction force is repulsive, $\boldsymbol{b}_3$ will dissociate.

From these considerations it is concluded that any dislocation that has a Burgers vector which is a multiple of a unit dislocation Burgers vector is unstable. That is, perfect dislocations having a Burgers vector of two or more lattice spacings dissociate into unit dislocations with a Burgers vector of one lattice spacing. A unit dislocation has lowest energy when its Burgers vector is parallel to a direction of closest packing.

The question of whether a given dislocation participates in slip is decided by the ease with which the given dislocation can be made to move through the lattice. This is taken to imply that those dislocations requiring the lowest shear stress to make them move will, in general, make the largest contribution to the macroscopic plastic shear strain. This critical shear stress is a rather sensitive function of the ratio d/b, where b is the magnitude of the Burgers vector and d is the spacing between the glide-plane and the next parallel plane. The shear stress is low for small values of b and large values of d. From these considerations it is concluded that slip will arise mainly by way of dislocations with small Burgers vectors, which from Frank's energy criterion are the most stable, gliding on the planes of closest packing in the crystal since these are the planes of greatest separation.

11-4-3 Origin and Sources of Dislocations

Crystal growth is an area of study which has been shown to imply the presence of dislocations. Initial theoretical considerations of crystal growth indicated that the layer growth of perfect crystals would proceed only when supersaturations on the order of 1.5 were attained. In contrast, experimental observations indicated that crystals grow under nearly equilibrium conditions. Frank (1949b) suggested that this discrepancy between theory and experiment could be resolved by postulating that growth could proceed at low supersaturations by the propagation of ledges associated with the point of emergence of a dislocation at a surface. Following this suggestion by Frank, the direct connection between dislocations and crystal growth was observed by Griffin (1950). Thus new-grown crystals in the as-received state, i.e., before straining, contain substantial numbers of dislocations.

Plastic straining of a new-grown crystal has the effect of increasing the number of dislocations which can attain very high values. The observation that plastic deformation is concentrated as slip on a few parallel planes is taken to imply the movement of a large number of dislocations on each active slip-plane. This is an unexpected feature, since the original grown-in dislocations in the as-received state would be expected to be uniformly distributed throughout the volume of the crystal. The movement of a large number of dislocations on each active slip-plane could arise from sources of dislocations within the material. Once these sources become operative they produce many dislocations on the same glide-plane.

An important mechanism for the generation of dislocations is that due to Frank and Read (1950) which operates through stationary sources called Frank–Read sources. Dislocation lines in real materials are frequently pinned down at one or more points along their lengths. This can be due to impurity atoms in the crystal settling on the dislocation line and so inhibiting its motion. Dislocation motion can also be strongly inhibited by large precipitated particles and by nodes for which there is no force on one of the lines meeting at the junction, this immobile line restricting the motion of the other lines comprising the node. Consider a dislocation segment AB which is firmly pinned at A and B, as indicated schematically in Fig. 11-19. Since the dislocation line is firmly pinned at A and B, the effect of an applied shear stress τ is to cause the dislocation line between A and B to take up some curved configuration. The equilibrium configuration of curved dislocation lines can be estimated using the line-tension concept. This concept regards a dislocation line as being analogous to a stretched string with a certain equivalent tension T. It is evident from Sec. 11-2-2 that all the expressions for U^d/L can be written in the approximate form $\phi G b^2$, which allows the line tension T, here defined as the increase in energy per unit increase in the length of a dislocation line, to be given in the form

$$T \simeq \phi G b^2 \qquad (\phi = \text{dimensionless constant}) \qquad (11\text{-}4\text{-}3)$$

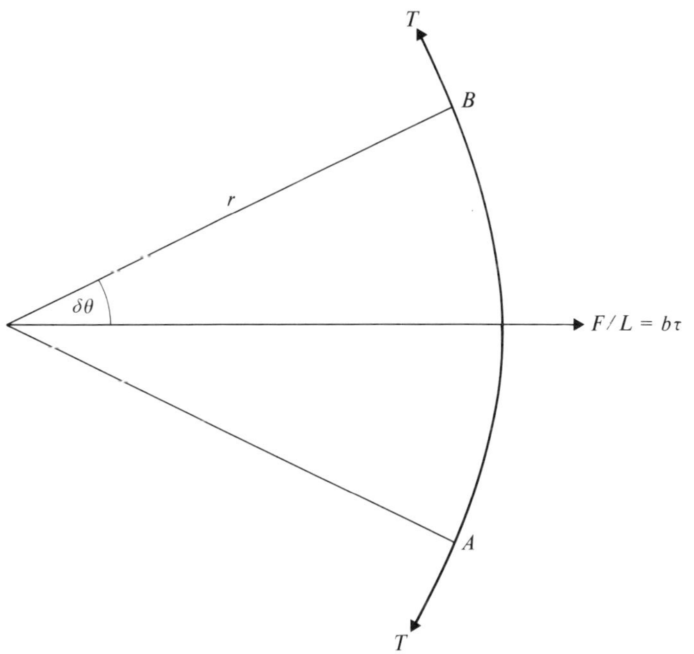

Figure 11-19 Effect of a shear stress acting on a dislocation segment shown pinned at A and B.

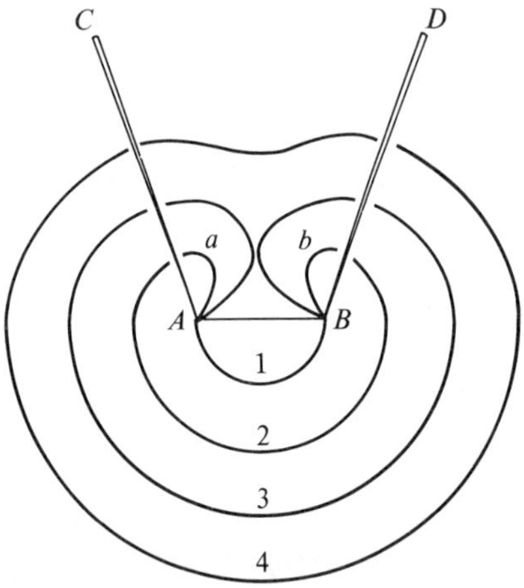

Figure 11-20 Frank–Read mechanism for dislocation generation.

The outward force acting on the dislocation can be obtained from Sec. 11-2-3 as $2\tau b r \delta\theta$ and this must balance the line tension $2T \sin \delta\theta$. The radius of curvature r of the dislocation line is thus given by

$$r = \frac{T}{\tau b} = \phi \frac{Gb}{\tau} \qquad (11\text{-}4\text{-}4)$$

Increasing τ gives rise to a series of arcs of decreasing radius, as indicated schematically in Fig. 11-20. This process continues until at some stage the minimum radius $L/2$ is attained, where L is equal to AB, this minimum radius being attained at some critical shear stress

$$\tau_c = \frac{2T}{bL} = \phi \frac{2Gb}{L} \qquad (11\text{-}4\text{-}5)$$

Further expansion of the arc leads to an increase in radius, and hence can proceed spontaneously without further increase in the shear stress. This proceeds until the points such as a and b coalesce and then break to form a closed loop, together with a new piece of dislocation line between A and B ready to repeat the process. However, the complete loop itself produces a stress field which cancels, at least in part, the original force $2\tau b r \delta\theta$ on AB, thus preventing an immediate reoperation of the source. Continued expansion of the loop reduces its effect on AB, being eventually reduced to a level at which generation can proceed. Thus, a Frank–Read source is characterized by the generation of a series of concentric loops, all on the same glide-plane, the rate of generation depending on the speed at which each loop expands. The principal feature of this mode of generation of dislocations, i.e., the series of concentric loops, have been directly observed (Dash, 1956). In practice, the bowed-out configuration of a loop is not an arc of a circle, because the local value of the line tension varies with orientation (Wit and Koehler, 1959; Ashby, 1966). Single-pinned sources involving a long dislocation segment or the ends of the double-pinned sources with large L assume spiral configurations as glide proceeds. As a first approximation, the Frank–Read glide spiral can be described in terms of an archimedian spiral. Such a spiral, when emerging from a free surface, gives rise to a slip-step several Burgers vectors in height.

From Eq. (11-1-6) it is evident that the Burgers vector b and the line direction vector ξ are parallel in the case of a screw dislocation. Hence, any plane through the line of a screw dislocation can be a possible glide-plane. Although in general slip tends to be confined to certain planes, i.e., the glide- or slip-planes, slip on other planes is still possible, and can in practice be induced by so arranging the applied stress as to make the shear stress on the given glide-plane very small in contrast to that on some transverse plane. This mechanism of slip from one plane to another is referred to as cross-slip. A suitable obstacle can inhibit slip for some central segment of a screw dislocation. However, this central segment can continue to slip by moving off the primary glide-plane onto a cross-glide-plane, provided the stress on the cross-slip-plane is high enough. Once the obstacle has been bypassed, the cross-slipping segment will tend to move into a slip-plane parallel to the original primary glide-plane by way of a second cross-slip. If the points at the extremity of the segment located at the intersection between the cross-slip-plane and the new slip-plane become fixed then the doubly cross-slipped segment can now operate exactly as a Frank–Read source in the new glide-plane. Double cross-slip as a generating mechanism was originally suggested by Koehler (1952); (see also Orowan, 1954; Amelinckx and Dekeyser, 1960). Direct evidence for double cross-slip was obtained by Gilman and Johnston (1960) from studies of dislocation multiplication in lithium fluoride. Double cross-slip is an unlikely mechanism for generating edge dislocations at ambient temperatures because climb would be involved in the process.

The double cross-slipped segment may be induced to cross-slip again, back onto its original glide-plane, by the attractive forces between itself and the outer segments, which will have continued to move on that slip-plane. With continuing movement of the dislocation there is now left behind the dislocation a pair of prismatic loops which consist entirely of edge dislocations. Nabarro (1948) suggested that prismatic loops could operate as dislocation sources, but since prismatic loops are the result of vacancy aggregate nucleation in high stacking-fault energy materials, they are likely to operate as dislocation sources only in special situations.

There is an alternative type of dislocation source which could give rise to the movement of a large number of dislocations on each active slip-plane. This behavior could arise if the motion of a given dislocation were to create others on the same slip-plane which in turn begin to move and so create other dislocations still in the same slip-plane, the process being repetitively continuous during slip, thus ensuring that once slip has been initiated on a given plane it will continue to develop extensively on the same slip-plane. An example of such a mechanism is that proposed by Frank (1948), who suggested the possibility that a moving dislocation could acquire sufficient kinetic energy to form an opposite-sign dislocation when it struck the free surface or other fast-moving dislocations. There is some evidence that cascade processes of this, and other types, become significant when a material is impact loaded, thus giving rise to very high rates of strain such as those encountered under conditions of shock loading.

11-4-4 Yielding

As already noted, new-grown crystals in the as-received state contain substantial numbers of dislocations which are not distributed entirely at random. This is because they tend to arrange themselves in a low-energy configuration. Formation of a low-energy configuration is aided by the thermal fluctuations associated with the high temperatures existing during and immediately after crystallization. These low-energy configurations consist of a three-dimensional network of dislocation segments linked at various points to form nodes having the properties discussed in Sec. 11-1-3 (Fig. 11-7). Such configurations are called Frank networks (Frank, 1949c). There are two effects arising from the presence of the Frank network which are of interest in relation to yield and plastic flow. These two effects are:

1. Some of the segments of the Frank network will act as Frank–Read sources.
2. The dislocations of a Frank network will interact with those mobile dislocations which are contributing to plastic deformation.

In this way the Frank network provides a resistance to the continuous development of slip, this resistance being observed as an increase in the yield stress of the material.

It is usually possible to observe small amounts of plastic flow, generally referred to as microplasticity at stresses which are significantly below the conventional yield stress, which is that stress at which the plasticity has spread throughout the whole of the material. Thus, upon application of a stress to a single crystal in the as-received state, any free dislocations begin to move and either glide out of the crystal or are stopped by the network. This effect is the principal source of microplasticity.

With increasing stress, Frank–Read sources begin to operate and produce a series of concentric loops. These loops expand until they are stopped by the Frank network. Further increase in stress will produce enough arrested loops from each source to allow their combined stress fields to prevent the source from operating further. Raising the stress yet further can enable the loops to break through the network, at which stage the slip propagates across the crystal and the original sources begin to operate again.

Some single crystals, particularly those containing foreign atoms, show a sharp yield drop. By this it is meant that there is an abrupt start of plastic deformation at a certain stress level, giving rise to a stress-strain curve of the form shown in Fig. 11-21. It has been proposed that this behavior arises in those single crystals in which there are initially very few mobile dislocations which can contribute to the plastic deformation. This may arise either because there are very few dislocations present, as in whiskers, or because the dislocations that are

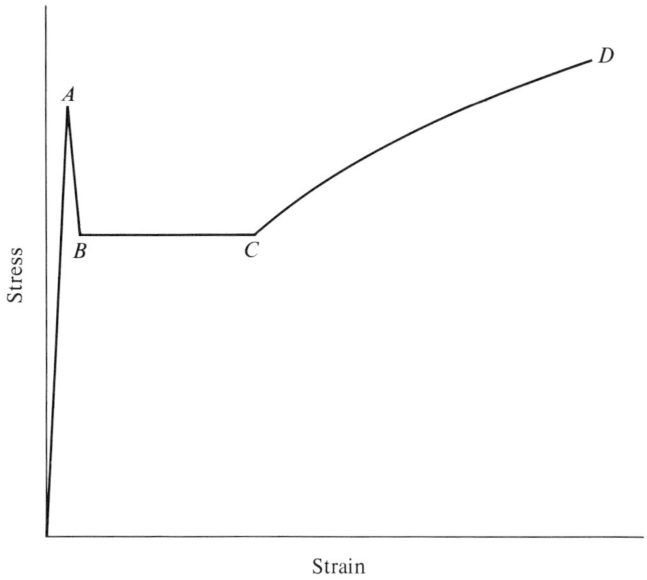

Figure 11-21 Typical stress-strain curve for mild steel showing the yield drop.

present are pinned, i.e., locked by foreign atoms that migrate to the vicinity of a dislocation line where the lattice has been slightly dilated. For example, interstitial carbon and nitrogen atoms which migrate to dislocations because the interstitial sites there can more easily accommodate them provide strong pinning in bcc metals. Thus, the sharp yield point in bcc iron is explained in terms of the locking of all mobile dislocations by carbon and nitrogen atoms. The production of a yield point is the most striking effect of dislocation locking. Johnston and Gilman (1959, 1962) and Gilman and Johnston (1960) carried out stress-strain measurements on lithium fluoride crystals and by using an etch-pit technique for revealing dislocations were able to show that two criteria were necessary for a sharp yield point. These are:

1. An increase in the number of mobile dislocations.
2. A direct relationship between the stress and velocity of dislocations.

The second of the two criteria has been discussed in Sec. 11-3-2. With regard to the first of the two criteria, it is evident from Eqs. (11-3-26), (11-3-27), (11-3-39), and (11-3-40), together with Eq. (11-3-1) and the relation $\Delta\alpha = \alpha^{+} - \alpha^{-}$, that the rate of plastic distortion, that is, $\partial\bar{\beta}/\partial t$ is directly proportional to the product $N_S v$. From this it follows that if the density increases to ϕN_S ($\phi > 1$), then to maintain a given rate of plastic distortion the velocity needed is only v/ϕ. Now assume that the lower velocity can be achieved with a lower stress; then it follows that the stress-strain curve will be characterized by a drop in stress. Calculated curves assuming different values for N_S were found by Gilman and Johnston to be in close agreement with the experimental results. The increase in the number of moving dislocations can occur either by the pinned dislocations breaking away from their atmosphere of pinning foreign atoms or by generation of new dislocations by one or more of the methods described above. The maximum stress reached before the yield drop is called the upper yield point and would be the point A in Fig. 11-21. The minimum stress in the post-yield condition is called the lower yield stress and corresponds to the point B in Fig. 11-21. Gilman and Johnston also observed for lithium fluoride that the yield drop can be suppressed if after deforming the material beyond the yield point it is unloaded and immediately reloaded. If, however, the material is aged for a few days before reloading the yield drop reappears. This observation can be explained on the basis that it takes some time for the foreign atoms to diffuse to the new sites of the dislocations and so lock them again. The various theories describing the unpinning of locked dislocations are outside the scope of this book; for a discussion of these theories the reader is referred, for example, to Hirth and Lothe (1968) or Nabarro (1967) (see also Gilman, 1969).

When the yield state is first reached at some localized point the yielding spreads, usually along the lines of maximum shear. Where these run out onto the surface of a polished specimen they distort the surface and give rise to the visible markings known as Lüders lines. The Lüders lines form a band of plastically distorted material which spreads, at approximately uniform stress, until it covers the whole surface of the specimen. Within the Lüders band there is uniform finite plastic extension, while in front there is no plastic deformation.

11-4-5 Laminar and Turbulent Plastic Flow

If the tangent modulus beyond the yield stress is nonzero the material is said to undergo work-hardening. Most metals show some degree of work-hardening which arises from the resistance to dislocation movement produced by entanglement with other dislocations and the overcoming of precipitate particles.

Once a crystal has started to deform plastically it may, if it is suitably oriented for slip to occur on one set of parallel slip-planes only, continue to deform with little or no increase in the shear stress. This type of slip has been termed laminar plastic flow and only occurs when the crystal is very pure and has an almost ideal structure. More generally, slip occurs on several intersecting families of planes either because of obstacles to dislocation motion along a certain slip-plane, such as precipitates or other locked dislocations, or because the relative orientation of the crystal planes and the shear stress encourages slip on several differently oriented planes. This type of slip is termed turbulent plastic flow and gives rise to considerable work-hardening. It would appear that although laminar plastic flow may be the basic mode of deformation it is very unstable and even small perturbations are sufficient to cause the onset of turbulent plastic flow. Because in hexagonal metals the slip is almost entirely confined to planes parallel to the basal plane these metals show a considerable region of linear plastic flow, while cubic metals which can more easily glide on intersecting slip-planes have virtually no regions of laminar plastic flow. Once turbulent plastic flow is established a region of strong work-hardening sets in, which is called stage II or linear hardening, while the laminar plastic flow mentioned above has been termed stage I. Linear hardening continues until the stress is sufficient, with the aid of thermal fluctuations, to allow dislocations to migrate from one slip-plane to the next and so avoid dislocation tangles. The lower the temperature the higher is the stress needed for the above mechanism to work, but when this cross-slip does start to operate the work-hardening decreases and stage III or parabolic hardening sets in, for which the stress-strain relation has a parabolic form $T^2 = k(\epsilon - \epsilon_0)$, although sometimes the beginning of this phase may also have a linear stress-strain relation. A detailed discussion of the above three stages of plastic flow in pure single crystals is given by Nabarro, Basinski, and Holt (1964); this paper contains an extensive bibliography of previous work.

11-5 STRUCTURAL THEORIES OF PLASTICITY: II. POLYCRYSTALS

11-5-1 Polycrystals

So far the structure and mechanisms of deformation appropriate to a single crystal have been considered. However, special conditions are required to produce single crystals and metals are more commonly produced in a polycrystalline form. The size of the small crystals, which are called "grains," is determined largely by their heat treatment. On cooling a metal from the molten or vapor state nucleation centers within the metal liquid, vapor, or solution grow when further metal atoms on their surface are deposited, the rate of deposition being determined by the degree of supersaturation.

The formation of the initial nuclei in pure metal liquids or vapors depends on the cooperative fluctuations of moving metal atoms, and solidification is greatly assisted by the presence of particulate impurities which act as additional nuclei. Growth occurs most easily at steps on the crystal surface because atoms deposited in front of such a step readily bond with two or more surface atoms. Frank (1949b) suggested that screw dislocations and certain edge dislocations can provide just such steps. The screw dislocation rotates about its point of emergence on the crystal surface as additional atoms are deposited on the step. The central portion rotates faster because fewer atoms are required to perform one revolution so the step takes up a spiral pattern which can then grow indefinitely. Such a spiral pattern has been observed by Griffin (1950), Dawson and Vand (1951), and Verma (1951), among others. By mechanisms such as those just described the crystals grow initially preferentially along certain directions giving rise to spikey or dendritic crystals. This dendritic growth is limited by meeting crystals growing from other nuclei and the final shape of the grains is determined by the competition for the remaining unprecipitated atoms. The free energy of a crystal consists of two parts, one part being proportional to the volume and the second part being proportional to the surface area. Thermodynamic equilibrium will be achieved when the free energy is a minimum; thus large crystals tend to form rather than small ones because this reduces the surface free energy while leaving the volume part unchanged. It has been observed under a microscope that at temperatures near the transformation temperature small crystals in the neighborhood of large crystals suddenly disappear. However, at temperatures well below the transition temperature this latter process proceeds at a negligible rate. The size of the grains is thus largely determined by the number of nuclei in the original melt and the rate of cooling.

For small applied stresses a polycrystalline material will deform elastically as described in Sec. 6-5. When the stress is increased sufficiently the single-crystal grains progressively plastify and ultimately rotate; however, it is usually assumed that the rotation is negligible and that all crystal grains plastically deform. Three factors dominate the possible modes of plastic deformation, namely, (1) the crystal slip-planes available, (2) the continuity of material between grains, and (3) the continuity of stress through any grain boundary. Taylor (1938), assuming that any elastic strains were negligible, considered the case of face-centered cubic metals and to satisfy condition (2) he assumed that each grain had the same homogeneous strain as the bulk material. To produce this homogeneous strain at least five differently oriented slip systems must operate out of the 12 possible slip systems for a face-centered cubic metal. To choose which slip systems would operate Taylor introduced the principle that the set would be one which produced the least work of deformation and showed that this implied that the sum of the plastic shear strains on all operative slip systems of the single crystal must be a minimum.

Bishop and Hill (1951) extended this work by showing that even if the local

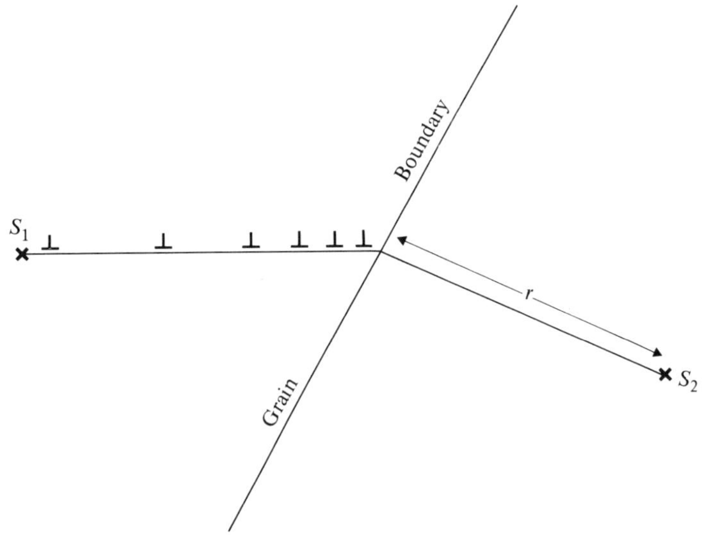

Figure 11-22 Piling of dislocations at a grain boundary.

deformations were inhomogeneous, which then allows condition (3) also to be satisfied, then the work done in plastically deforming a polycrystal is equal to that which would be done if the grains deformed equally. Thus Taylor's assumption of the same homogeneous deformation for all grains is allowable for the calculation of the work of deformation. Lin (1957) incorporated the influence of elastic strains and assumed that the total strain, elastic plus plastic, was the same in each grain. A self-consistent theory of plastic deformation of polycrystals, based on the results of Eshelby (1957) concerning elastic inclusions and an averaging technique incorporating the effects of neighboring slipped grains due to Kroner (1961), has been given by Budiansky and Wu (1962). In their work slip was allowed on all 12 slip systems of the fcc single-crystal grains and the plastic deformation of a grain was treated as if it were a sphere embedded in a homogeneous strain-hardening material whose stress-strain response was the same as that of the polycrystal. However, none of the theories described above can take account of the observation that a decrease in grain size leads to an increase in strength.

11-5-2 Upper and Lower Yield Stress in Polycrystalline Material

Because of the wide scatter in the experimental results there is no definitive trend of the upper yield stress with grain size, although there is some indication that it increases with an increase of grain size. The effect of grain size, as given by its diameter d, upon the lower yield stress has been systematically investigated by Hall (1951), Petch (1953, 1954), Conrad (1960, 1961), and Conrad and Schoeck (1960), among others, and it has been shown that the lower yield stress T_y obeys the relation

$$T_y = T_0 + k_y d^{-1/2} \tag{11-5-1}$$

Cottrell (1958) has proposed that the above relation can be accounted for in the following way. Suppose, during the initial loading, dislocations are generated at localized regions of stress concentration and move along their slip-planes until they pile up at a grain boundary, as shown in Fig. 11-22. This set of dislocations produces a stress concentration whose value at a distance r ahead of the pile-up is on the order of $(T_y - T_0)(d/r)^{1/2}$ where d is the grain diameter and T_0 represents the resistance to movement of the dislocations due to obstacles in their way. When this stress concentration is sufficient to activate a source within the next grain, at a stress of T_l say, the next grain can start yielding and the process continues. Thus

$$(T_y - T_0)\left(\frac{d}{r}\right)^{1/2} = T_l \tag{11-5-2}$$

which has the same form as Eq. (11-5-1) provided $k_y = T_l r^{1/2}$. Two principal mechanisms for yielding of the next grain can now occur: either dislocations are created at the grain boundary or dislocations pinned within the next grain may be released. Whichever mechanism gives the lower value of k_y is the one that will be preferred and this affects the temperature sensitivity because generation of dislocations at a grain boundary tends to be an athermal process whereas the unpinning of a locked dislocation is strongly temperature sensitive.

11-5-3 Tensile Instability, Adiabatic Softening, and Dispersion Hardening

The decreasing work-hardening at large strain can lead, during tensile stressing of a specimen, to instability which causes necking before the final cracking. It is the work-hardening which ensures that, at the onset of plastic strain, the specimen distorts uniformly with increasing stress; if we suppose that the specimen starts to yield at a random plane of weakness then it will extend more in the vicinity of that plane than elsewhere, and as the plastic flow is nearly incompressible the cross-sectional area will decrease and hence the tensile stress will increase. Without any work-hardening this process would lead to an instability and gross failure of the material. Consider a specimen which is extended in tension as shown in Fig. 11-23. Let the cross-sectional area be A and the stress be T and let the tensile force on the specimen be

$$F = TA \tag{11-5-3}$$

If the strain is almost entirely plastic and the deformation occurs incompressibly then the strain is given by

$$\text{Nominal strain } e = \frac{A_0}{A} - 1 \qquad \text{Natural strain } \epsilon = \ln\left(\frac{A_0}{A}\right) \tag{11-5-4}$$

where A_0 is the original cross-sectional area of the specimen. At the point of tensile instability there will be no increase in tensile force corresponding to an increase in strain

$$dF = A\,dT + T\,dA = 0 \tag{11-5-5}$$

But, by Eq. (11-5-4),

$$\frac{dA}{de} = -\frac{A}{1+e} \qquad \text{or} \qquad \frac{dA}{d\epsilon} = -A \tag{11-5-6}$$

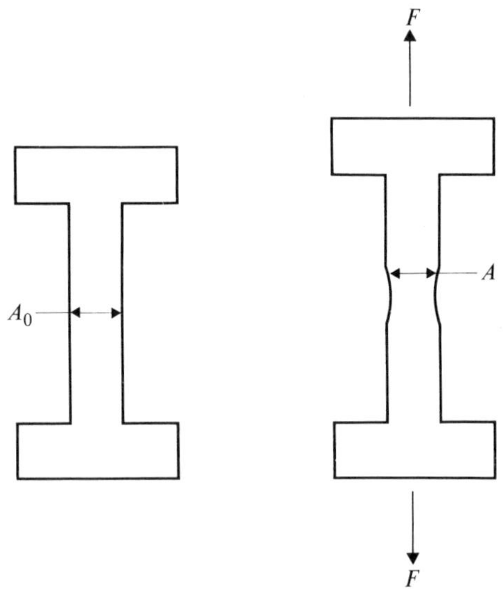

Figure 11-23 Necking of tensile specimen.

Thus

$$\frac{dT}{de} = \frac{T}{1+e} \qquad \text{or} \qquad \frac{dT}{d\epsilon} = T \tag{11-5-7}$$

The above equation for the "nominal" strain leads to Considere's (1885) construction for determining the point of instability. If the true stress is plotted against the nominal strain and a tangent is constructed from the point $e = -1$ on the strain axis to the curve, it will touch the curve at the point of instability. Uniform deformation can be maintained during tensile straining provided

$$\frac{dT}{de} > \frac{T}{1+e} \qquad \text{or} \qquad \frac{dT}{d\epsilon} > T \tag{11-5-8}$$

The rate of strain-hardening is thus critical in preventing the onset of necking. Hard steels which may sometimes be regarded as breaking in a brittle fashion in fact neck-down to about 50 percent of their original cross-sectional area before fracturing which, as Eq. (11-5-4) indicates, represents a 69 percent strain. For such steels it is the onset of the necking instability rather than any lack of ductility that ultimately leads to failure. Usually strain-hardening is primarily due to dislocation locking but this can be enhanced if, during the plastic deformation, phase transformations take place which form hard precipitates in the slip-bands and which, because of their greater strength, are more effective at blocking dislocation motion. An example of this mechanism of strain-hardening is found in steel. At high temperatures iron has a fcc structure which has a sufficiently open structure for carbon atoms to reside within the iron lattice. At low temperatures the iron transforms to a bcc structure which has no room for carbon atoms. If the iron is heated to about 900°C so that the carbon is taken into solution and is then quenched to room temperature, the carbon becomes locked within the predominantly bcc iron structure and distorts it to a body-centered tetragonal shape which is known as martensite. In pure iron-carbon alloys the martensite transformation cannot be suppressed and although they can be very strong they are also brittle. However, if nickel and chromium are added a fcc structure can be retained at room temperature which is changed to martensite during plastic deformation and therefore gives the necessary work-hardening properties to considerably delay the necking instability. A discussion of the transformations possible in various metals and alloys has been given by Christian (1965) while a general theoretical approach to transformation strains is being developed by Crocker and others, as presented, for instance, by Acton, Bevis, Crocker, and Ross (1970).

In compression the cross-sectional area tends to increase, giving rise to barreling, and so the instability discussed above does not arise. Instead, a dynamic instability called adiabatic softening occurs. The material which is being highly sheared becomes hot through the conversion of plastic work to heat. The material in the highly sheared zone therefore softens and promotes even greater strain in the sheared zone. With an increased rate of straining the less opportunity exists for heat to leak away by conduction and the more effective the above mechanism becomes.

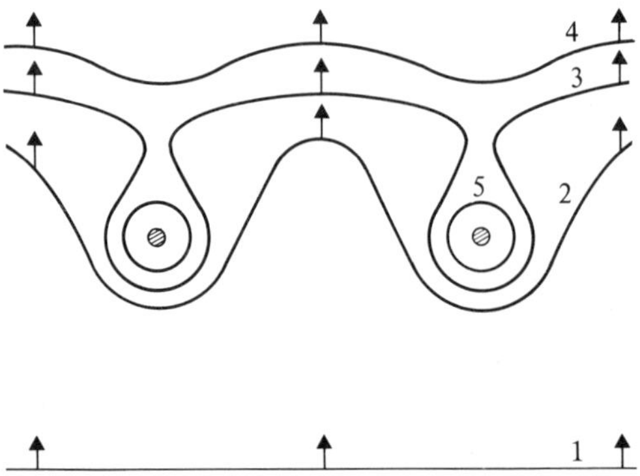

Figure 11-24 A dislocation passing between spaced obstacles. (*After Orowan, 1954.*)

In order to give ductile metals strength they are often stiffened by introducing finely dispersed particles of hard substances such as intermetallic compounds, oxides, or carbides. When a dislocation bypasses these obstacles they leave closed dislocation loops round them, as shown in Fig. 11-24; as more closed loops form round the obstacles they produce stresses within them that are much higher than those in the slip-plane as a whole. This mechanism results in the hard particles carrying all the load beyond the yield strength of the matrix and so contributes directly to the overall strength.

11-5-4 Superplasticity

In some circumstances it is possible for certain materials, when extended at slow strain rates of 10^{-1} s or less by small tensile compressive or torsional forces, to respond in a ductile manner which is much greater than the conventional behavior. Such a response is termed superplasticity and the materials concerned are called superplastics. Essentially superplastics are able to exhibit large ductile strains by avoiding the necking instability discussed earlier. There are two ways this may be achieved, corresponding to the two principal forms of superplasticity.

(a) Transformation or environmental superplasticity This is produced by temperature cycling an allotropic material above and below its transition temperature between the two different forms of crystal structure while the material is being stressed. Very large elongations on the order of 500 percent can be produced in this manner, as demonstrated, for instance, by Oelshlagel and Weiss (1966) for the ferrite-austenite transformation in plane carbon steel. It has been shown by Greenwood and Johnson (1965) using a continuum mechanics approach that the external strain rate $\dot{\epsilon}$ is linearly related to the applied stress by the equation

$$\dot{\epsilon} = 4T\left(\frac{\Delta V}{V}\right)\frac{5n/(4n+1)}{3I_0} \tag{11-5-9}$$

where $\Delta V/V$ is the volume dilation caused by the transformation, I_0 is the internal stress generated during the phase transformation when the applied stress is zero, and n is the exponent in the creep equation

$$\dot{\epsilon} = \text{constant} \times T\mathrm{II}_{\mathrm{T}}^{n-1} \exp\left(-\frac{Q}{k\theta}\right) \tag{11-5-10}$$

Q being the activation energy for creep. The linear relation (11-5-9) has been substantiated experimentally and, because the analysis involved did not contain details of the local stress and strain arrangements near the phase front, such as the dislocation behavior at the phase boundary, this suggests that such details are not very significant. The results can be shown to indicate that the flow stress of the weaker phase is of major significance and the phenomenon can most appropriately be discussed in terms of internally generated stresses.

The work-hardening capability of the material is not usually important in structural superplasticity, as is shown in the next section. Yoder and Weiss (1972), however, developed a modified form of phase transformation superplasticity in a eutectoid steel by making use of the high work-hardening capability of austenite. When the material was strained in a tensile test and simultaneously cycled through the eutectoid transformation, large elongations (490 percent) were achieved. In this instance the phase transformation acts to regenerate a relatively dislocation-free austenite on each cycle so that its work-hardening capability is preserved in an incipient necked region, thereby stopping the necking.

(b) Isothermal or structural superplasticity This form of superplasticity requires a temperature about half the absolute melting temperature θ_m of the material and thus can occur over a wide temperature range depending on the material used. Alloys of lead and tin can be superplastic at room temperature while alloys of aluminum and zinc require temperatures of 500 to 800 K and alloys of iron and nickel require temperatures greater than 1000 K. At temperatures greater than $0.5\theta_m$ the strain-hardening effect is less significant than at lower temperatures and so some other effect must be sought to explain the resistance to the onset of necking; the controlling factor in this situation appears to be strain-rate sensitivity, sometimes denoted SRS. To define SRS it is first assumed that in the temperature range of interest the constitutive equation of the material takes the form

$$T = T(\epsilon, \dot{\epsilon}) \tag{11-5-11}$$

Thus

$$dT = T_{\epsilon}\, d\epsilon + T_{\dot{\epsilon}}\, d\dot{\epsilon} \tag{11-5-12}$$

The strain-rate sensitivity index m is then defined as

$$m = T_{\dot{\epsilon}} \frac{\dot{\epsilon}}{T} \tag{11-5-13}$$

and the strain-hardening index N is given by

$$N = T_{\epsilon} \frac{\epsilon}{T} \tag{11-5-14}$$

Thus Eq. (11-5-12) can be written in the form

$$\frac{dT}{T} = N\frac{d\epsilon}{\epsilon} + m\frac{d\dot{\epsilon}}{\dot{\epsilon}} \tag{11-5-15}$$

As the tensile force F is constant throughout the specimen, Eq. (11-5-5) gives

$$\frac{dT}{T} = -\frac{dA}{A} = N\frac{d\epsilon}{\epsilon} + m\frac{d\dot{\epsilon}}{\dot{\epsilon}} \tag{11-5-16}$$

Using Eq. (11-5-4),

$$d\epsilon = -\frac{dA}{A} \tag{11-5-17}$$

and

$$\frac{d\dot{\epsilon}}{\dot{\epsilon}} = \frac{d\dot{A}}{\dot{A}} - \frac{dA}{A} \tag{11-5-18}$$

Substituting in Eq. (11-5-16) and performing the integration gives

$$\dot{A} = K\epsilon^{-N/m}A^{(m-1)/m} = KA^{(m-1)/m}\left[\ln\left(\frac{A_0}{A}\right)\right]^{-N/m} \tag{11-5-19}$$

where K is a constant. When the temperature is greater than $0.5\theta_m$, N is usually small and the multiplier of the factor $A^{(m-1)/m}$ is almost constant. As m approaches unity $\dot{A}$ becomes increasingly independent of A and large elongations are therefore possible. In a superplastic metal where m has values between 0.3 and 0.8 any local stress increase in the superplastic metal produces only a small change in strain rate and thus these regions extend at almost the same rate as the rest of the specimen; hence large elongations are possible.

Most superplastic alloys are multiphase alloys in which the phases have almost the same ductility and are present in almost equal proportions; thus many superplastic alloys are based on eutectoid systems. Superplastic alloys have a very fine grain structure (1 to 10 μm) and their deformation property appears to be associated with the presence of these small equiaxed aggregates; if two alloys of identical chemical composition are prepared in such a way that one of them has a grain size of 25 to 2500 μm it behaves in a normal way, while if the other is prepared with a grain size ~2 μm it will behave superplastically. The deformation of such superplastic alloys usually occurs at constant temperature and the microstructure remains essentially unaltered. Table 11-1, taken from Johnson (1970), gives some idea of the properties of some superplastic metals.

Several theories of superplasticity have been proposed but to date none is entirely satisfactory. They are based on three processes: vacancy creep, grain-boundary sliding, and dynamic recovery.

Vacancy creep Nabarro (1948) and Herring (1950) have proposed that under the influence of the stress any point vacancies will diffuse through the lattice and have shown that a viscous type of constitutive equation results:

$$T = \frac{b_1 d^2 k\theta\dot{\epsilon}}{D_l\Omega} \tag{11-5-20}$$

Table 11-1 Materials exhibiting fine-grain-size superplasticity

Basis metal	Alloying addition weight, %	Temperature °C	Temperature °K	Grain size, μm	SRS (maximum)
Aluminum	33 Cu	440–530	715–805	1–2	0.9
	12 Si : 4 Cu	500	775	–	0.4
Cadmium	26 Zn	20	295	1–2	0.5
Chromium	40 Ru	1280	1555	–	–
Cobalt	10 Al	1200	1475	0.4	0.3
Copper	10 Mg	700	975	–	n.d.
	10–12 Al 10 Al : 4 Fe	500	775	–	0.5
	38–50 Zn	450–650	725–925	–	n.d.
Iron	0.14 C : 1.2 Mn : 0.5 P 0.1 V	900	1175	2	0.6
	0.34 C : 0.47 Mn : 2.0 Al	900	1175	2	0.5
	0.42 C : 1.9 Mn	730	1005	1–2	0.6
	26 Cr : 6.5 Ni 30 Cr : 6.0 Ni	870–980	1145–1255	2	0.5
Lead	20 Sn	20	295	3	0.5
	5 Cd	0	275	4	0.3
Magnesium	0.2 Zr	500	775	20	0.3
	6 Zn : 0.6 Zr	270–310	545–585	0.5	0.5
	23 Ni	450	725	–	n.d.
	30 Cu	450	725	–	n.d.
	33 Al	400	675	–	n.d.
Nickel	Nil	820	1095	8	n.d.
	39 Cr : 8 Fe : 2 Ti	980	1255	2	0.5
Tin	5 B	20	295	1	0.5
	2.38 Pb	20	295	1–2	0.5
	33 Cd	20	295	1–2	0.5
Titanium	6 Al : 4 V	900–980	1175–1255	6	0.9
	5 Al : 2.5 Sn	1000	1275	18	0.7
Zinc	Nil	0–20	275–295	1–2	n.d.
	0.5 Al	20	295	1–2	n.d.
	5 Al	200–360	475–635	1–2	0.7
	22 Al	200–260	475–535	1–2	0.5
	40 Al	250	525	1–10	0.5
Zirconium	Zircaloy-4	900	1175	12	0.5

where d is the grain size, Ω is the atomic volume, D_l is the vacancy diffusion coefficient, and b_1 is a numerical constant of order 5×10^{-3} if no boundary flow is allowed and two-fifths of this value if the tangential stress is relaxed at the boundary. Coble (1963) suggested that the diffusion may occur more readily at the grain boundary than with each grain and deduced the stress-strain rate relation in this case to be

$$T = \frac{b_2 d^3 k\theta}{D_{\text{gb}} \Omega \omega} \tag{11-5-21}$$

where D_{gb} is the grain-boundary diffusion coefficient, ω is the grain-boundary width, and b_2 is constant of order 0.5. Both these mechanisms lead to a linear relationship between stress and strain rate which is not usually the case in practice, and it is implied that the diffusion elongates grains in the direction of the applied stress, an effect not observed in superplasticity where the grains remain equiaxial during flow. Several variants using more than one mechanism acting simultaneously have been introduced, such as the theory of Avery and Backofen (1965) which gives a relation of the form

$$\dot{\epsilon} = \frac{AT}{d^2} + \beta \sinh \beta T \tag{11-5-22}$$

where A and β are constants at a given temperature and the first term accounts for the Nabarro–Herring diffusional creep process while the second term accounts for climb-regulated dislocation movement. However, these theories predict a SRS of unity for small strain rates which is not observed experimentally and they do not account for the constant form of the microstructure during deformation.

Grain-boundary sliding The invariant form of the microstructure in superplastically deforming metals has suggested that most of the deformation may be due to grain-boundary sliding. The grain boundaries are then regarded as behaving like a fluid between the grains. Orowan (1947) has estimated the constitutive equation in this situation to be of the form

$$T = \frac{b_3 k\theta}{\Omega D_l} \dot{\epsilon} \tag{11-5-23}$$

Sliding must be accompanied by some accompanying deformation of the grains themselves and Gifkins (1967) has examined the deformation mechanisms which must be operative for large extensions. A strain-rate sensitivity of unity is predicted by this model, which, as mentioned before, is not usually met with in practice. As Johnson (1970) has noted, if grain-boundary sliding is a dominant mechanism it is surprising that considerable void formation has not been observed in superplastic flow. The absence of voids may be due to sintering mechanisms however.

Dynamic recovery If recrystallization of a work-hardened grain occurs through annealing, dislocations can migrate through the crystal and through its surface, so reducing the internal stress and restoring its ductility. It has therefore been suggested that a continuous process of deformation and annealing may be responsible for superplastic flow. When deforming a zinc-aluminum alloy superplastically Packer, Johnson, and Sharby (1968) found that an originally circular rod distorted into an elliptical shape but that there was no apparent change in the fine-grain structure. They infer that diffusional creep must play a secondary role, as it will not explain the change in geometry, and because of the unchanging nature of the microstructure they suggest that a recrystallization mechanism must operate. The change in cross-sectional shape would be consistent with basal slip in the zinc matrix. It has also been noted that superplastic materials, presumably because of their small grain size, show no dislocation substructure such as tangled walls of dislocations forming internal domains. Hayden, Gibson, Merrick, and Brophy (1967) show that for a nickel-iron-chromium alloy a graph of the stress versus strain rate, both scales being logarithmic, is a straight line with a slope of $\frac{1}{2}$ for strain rates from 10^{-5} to $10^{-1}\,\mathrm{s}^{-1}$. Vacancy creep and grain-boundary sliding both predict a slope of unity and the observed strain rates were much higher than these theories would predict. It has been suggested that the observed square root dependence of stress upon strain rate may arise by a recovery creep mechanism proposed by Mott (1953). Suppose that the internal stress within the grains is due to piling up of n dislocations at the grain boundary. It has been shown by Eshelby, Frank, and Nabarro (1951) that, if the grain diameter is d,

$$T = \frac{Gbn}{\pi \lambda d} \tag{11-5-24}$$

where λ is 1 for a screw dislocation and $(1-\nu)$ for an edge dislocation. If the surface moves forward a distance b then, as all the dislocations are in equilibrium, they will move forward the same distance and the work done per atom of the dislocation line in the surface is nTb^3. If W_0 is the activation energy for grain-boundary diffusion and $\tilde{\nu}$ is the frequency of atomic vibration, then the number of times per second that the boundary moves a distance b in the direction of the force is $\tilde{\nu}\exp[-(W_0 - nTb^3)/k\theta]$ and in the other direction $\tilde{\nu}\exp[-(W_0 + nTb^3)/k\theta]$. The velocity of linear growth v is thus given by

$$v = 2\tilde{\nu}b\exp\left(-\frac{W_0}{k\theta}\right)\sinh\left(\frac{nTb^3}{k\theta}\right) \tag{11-5-25}$$

and the rate of strain $\dot{\epsilon}$ by

$$\dot{\epsilon} = \text{constant} \times v = \text{constant} \times \tilde{\nu}b\exp\left(-\frac{W_0}{k\theta}\right)\sinh\left(\frac{nTb^3}{k\theta}\right) \tag{11-5-26}$$

For fine slip this can be written

$$\dot{\epsilon} = \text{constant} \times nT\exp\left(-\frac{W_0}{k\theta}\right) \tag{11-5-27}$$

and, using Eq. (11-5-24),

$$\dot{\epsilon} = \text{constant} \times T^2\exp\left(-\frac{W_0}{k\theta}\right) \tag{11-5-28}$$

As Gifkins' (1967) experiments on lead-thallium alloys lead him to the conclusion that suppression of recrystallization is a necessary prerequisite for superplasticity while Hayden's experiments suggest that this type of recovery mechanism is the only one leading to the right stress-strain rate dependence, it is clear that much theoretical and experimental work is required before a definitive theory of superplasticity will emerge. However, because of its potential practical uses, considerable effort is being devoted to this rapidly expanding field.

A recent comprehensive review of the situation at present can be found in the review article by Edington, Melton, and Cutler (1976).

11-6 DISLOCATION-BASED THEORY OF MACROSCOPIC PLASTICITY

The present discussion will be restricted to the tensile test of a single crystal in which the tensile axis is taken to make angles α, β, γ with the x, y, z axes respectively. This is shown schematically in Fig. 11-25 for slip on a single plane. The plastic strain is taken to result from the motion in the x direction of screw

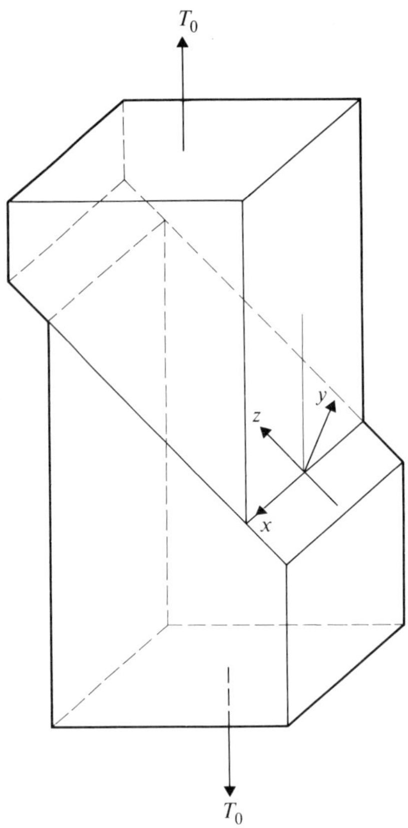

Figure 11-25 Tensile test for a single crystal.

dislocations whose lines are in the z direction. In this discussion of macroscopic plasticity, the assumption of Sec. 11-3-3 that all dislocations have the same velocity will be retained.

If T_0 is the applied tensile stress, then, from Fig. 11-25 and with the notation of Sec. 11-3-3, the shear stress driving the dislocation is

$$T_{23} = T_0 \cos\beta \cos\gamma \tag{11-6-1}$$

For dislocation motion in the x direction, it follows from Eqs. (11-3-27) and (11-3-29) that there is a nonzero component of plastic distortion $\bar{\beta}_{23}$ associated with the total deformation gradient $\partial u_3/\partial x_2$. It therefore follows that the plastic strain is

$$\bar{e}_L = \bar{\beta}_{23} \cos\beta \cos\gamma \tag{11-6-2}$$

in the direction of the tensile axis.

For a uniform deformation, $\partial\alpha_{33}/\partial t = 0$ and Eq. (11-3-22) reduces to

$$\frac{\partial}{\partial x_1}(v_1 \Delta\alpha_{33}) + \frac{\partial}{\partial x_2}(v_2 \Delta\alpha_{33}) = 0 \tag{11-6-3}$$

If equal numbers of positive and negative dislocations are present in the specimen initially, then $\alpha_{33} = 0$ during the subsequent deformation and Eq. (11-3-23) reduces to

$$\frac{\partial \Delta\alpha_{33}}{\partial t} = 2(\dot{\alpha}_{33})_g \tag{11-6-4}$$

The strain rate is still given by Eq. (11-3-27) in the form

$$\frac{\partial \bar{\beta}_{23}}{\partial t} = v_1 \Delta\alpha_{33} \tag{11-6-5}$$

Thus Eqs. (11-6-1) to (11-6-5) imply

$$(\dot{\alpha}_{33})_g = (\dot{\alpha}_{33})_g(\Delta\alpha_{33}, \alpha_{33}, v_1, T_{23}) \tag{11-6-6}$$

$$v_1 = v_1(\Delta\alpha_{33}, \alpha_{33}, T_{23}) \tag{11-6-7}$$

from which it follows that if such relations exist, then the problem reduces to solving a first-order differential equation for $\Delta\alpha_{33}$ as a function of time. It is evident from Sec. 11-3-3 that, in principle, there are sufficient independent equations to give values of the quantities α_{33}, $\Delta\alpha_{33}$, v_1, and v_2. However, in practice the two densities α_{33} and $\Delta\alpha_{33}$ can be obtained by solving Eqs. (11-3-22) and (11-3-23) only by making use of an empirical relation between velocity and stress together with an arbitrary assumption regarding the pair production rate $(\dot{\alpha}_{33})_g$.

A solution of the problem of uniform straining of a single crystal under simple tension when only a single slip system is operable has been proposed by Gillis and Gilman (1965). This solution is based on the stress-velocity relations proposed by Johnston and Gilman (1959) from their studies of lithium fluoride crystals. For a constant stress experiment, however, the dependence of v_1 on T_{23} need not be specified in order to determine the time dependence of the deformation. For this condition it is sufficient to assume that

$$v_1 = v_0 - a_0 \Delta\alpha_{33} \tag{11-6-8}$$

provided the dislocation densities are not too large. In Eq. (11-6-8), v_0 and a_0 are functions of T_{23}, the term in $\Delta\alpha_{33}$ accounting for the reduction in dislocation velocity caused by interactions with other dislocations. It is to be expected that Eq. (11-6-8) will apply to creep situations where the dislocation densities will be sufficiently small, but for higher strain rates a more general velocity relation is necessary.

In general there are a number of processes which lead to generation and decay of dislocations. These can be summarized as follows.

1. Frank–Read sources. These can be accounted for by a term $a_1 v_1$ where a_1 is proportional to the density of the sources and the rate of their operation is proportional to the velocity with which the generated dislocations move away and allow the source to operate again.
2. Generation of dislocations by moving dislocations. This process is generally associated with large rates of plastic strain and can be accounted for by a term of the form $a_2 v_1 \Delta\alpha_{33} g(v_1)$ where the factor $g(v_1)$ is the probability that this generation mechanism operates at the velocity v_1.
3. Decay arising from dislocations colliding with fixed obstacles and being made immobile. This process can be accounted for by taking the rate of such collisions to be proportional to $v_1 \Delta\alpha_{33}$. Thus it can be accounted for by a term of the form $-a_3 v_1 \Delta\alpha_{33}$.
4. Immobilization or decay of colliding pairs of dislocations. This process can also be accounted for by a term of the form $-a_4 v_1 \Delta^2\alpha_{33} f(v_1)$. This mechanism of decay has here been assumed to be proportional to the rate of pair collision, that is, $v_1 \Delta^2\alpha_{33}$, the factor $f(v_1)$ being the probability of mutual capture of a dislocation pair moving with a velocity v_1.

These two mechanisms of generation and two of decay can be included in Eq. (11-6-6) by assuming $(\dot{\alpha}_{33})_g$ to be of the form

$$(\dot{\alpha}_{33})_g = a_1 v_1 + [a_2 g(v_1) - a_3] v_1 \Delta\alpha_{33} - a_4 v_1 \Delta^2\alpha_{33} f(v_1) \tag{11-6-9}$$

Substituting from Eq. (11-6-8) for v_1 in Eq. (11-6-9) and using Taylor approximations for $g(v_1)$ and $f(v_1)$ for small dislocation densities gives

$$\frac{\partial \Delta\alpha_{33}}{\partial t} = 2(\dot{\alpha}_{33})_g = a_5 - a_6 \Delta\alpha_{33} - a_7 \Delta^2\alpha_{33} + a_8 \Delta^3\alpha_{33} \tag{11-6-10}$$

which can be integrated. In a freshly annealed specimen, the initial dislocation density will be expected to be low, and the rate of generation will be dominated by the term in $a_1 v_1$ which accounts for effects arising from Frank–Read sources. Since the velocity will remain approximately constant, $v_1 \approx v_0$, so that the dislocation density $\Delta\alpha_{33}$ and plastic strain rate will increase linearly with time. For conditions of shock loading, $\partial \Delta\alpha_{33}/\partial t = 2(\dot{\alpha}_{33})_g$ can be regarded as an increasing function of $\Delta\alpha_{33}$, and the growth of $\Delta\alpha_{33}$ becomes exponential. Whether the material is shock loaded or not, decay of dislocations in the form of immobilization will ultimately become significant, giving rise to an effective steady-state balance between the generation and decay of dislocations for which state $\Delta\alpha_{33}$ and v_1 will tend to constant, limiting values. Thus the strain rate will tend to a constant value, giving rise to an asymptotic state of steady creep. These qualitative predictions of the theory are in general accord with observation. That this is so has been shown, for example, by Li (1963) and Webster (1966) who, by suitably choosing the coefficients a_p $(p = 5, 6, 7, 8)$, were able to obtain a solution for the time dependence of the longitudinal strain $\bar{e}_L$ which agrees well with the observed creep behavior of various materials.

A similar solution to a problem of plane bending involving a nonuniform distribution of edge dislocations has been given by Lardner (1969).

GENERAL REFERENCES

Amelinckx, S. (1964) The Direct Observation of Dislocations, in "Solid State Physics," Suppl. 6, Academic Press, New York.

Barrett, C. S., and T. B. Massalski (1966) "Structure of Metals," McGraw-Hill, New York.

Cottrell, A. H. (1953) "Dislocations and Plastic Flow in Crystals," Oxford University Press, London.

Friedel, J. (1956) "Les Dislocations," Gauthier-Villars, Paris. English edition, Pergamon Press, Oxford, 1964.

Hirth, J., and J. P. Lothe (1968) "Theory of Dislocations," McGraw-Hill, New York.

Honeycomb, R. W. K. (1968) "The Plastic Deformation of Metals," Arnold, London.

Hull, D. (1965) "Introduction to Dislocations," Pergamon Press, Oxford.

Jaswon, M. A. (1965) "Introduction to Mathematical Crystallography," American Elsevier, New York.

Kelly, A. (1966) "Strong Solids," Oxford University Press, London.

Lardner, R. W. (1974) "Mathematical Theory of Dislocations and Fracture," University of Toronto Press, Toronto.

Nabarro, F. R. N. (1967) "Theory of Crystal Dislocations," Oxford University Press, London.

Nye, J. F. (1957) "Physical Properties of Crystals," Oxford University Press, London.

Read, W. T. (1953) "Dislocations in Crystals," McGraw-Hill, New York.

Schmidt, E., and W. Boas (1935) "Kristallplastizitat," Springer-Verlag, Berlin. "Plasticity of Crystals," Hughes, London, 1950, and Chapman and Hall, London, 1968.

Van Bueren, H. G. (1960) "Imperfections in Crystals," North-Holland, Amsterdam.

Weertman, J., and J. R. Weertman (1964) "Elementary Dislocation Theory," Macmillan, New York.

REFERENCES

Acton, A. F., M. Bevis, A. G. Crocker, and N. D. A. Ross (1970) *Proc. Roy. Soc. (London)*, vol. A320, pp. 101–113.

Amelinckx, S., and W. Dekeyser (1960) *Solid State Physics*, vol. 8, pp. 325–499.

Ashby, M. F. (1966) *Phil. Mag.*, vol. 14, pp. 1157–1178.

Avery, D. H., and W. A. Backofen (1965) *Trans. ASM*, vol. 58, pp. 551–562.

Bishop, J. F. W., and R. Hill (1951) *Phil. Mag.*, vol. 42, pp. 414–427, 1298–1307.

Brown, A. F. (1949) *Nature*, vol. 163, pp. 961–962.

Brown, A. F. (1950) "Metallurgical Applications of the Electron Microscope," Monograph No. 8, pp. 103–114, The Institute of Metals, London.
Brown, A. F. (1952) *Advances in Phys.*, vol. 1, pp. 427–479.
Budiansky, B., and T. T. Wu (1962) In R. M. Rosenberg (ed.), *Proc. Fourth U.S. Natl. Cong. Theor. Appl. Mech.*, pp. 1175–1185, ASME, New York.
Burgers, J. M. (1939) *Proc. K. Ned. Akad. Wet.*, vol. 42, pp. 293–325.
Christian, J. W. (1965) "The Theory of Transformations in Metals and Alloys," Pergamon Press, Oxford.
Coble, R. L. (1963) *J. Appl. Phys.*, vol. 34, pp. 1679–1682.
Conrad, H. (1960) *Phil. Mag.*, vol. 5, pp. 745–751.
Conrad, H. (1961) *J. Iron and Steel Inst. (London)*, vol. 198, pp. 364–375.
Conrad, H., and G. Schoeck (1960) *Acta Met.*, vol. 8, pp. 791–796.
Cottrell, A. H. (1958) *Trans. AIME*, vol. 212, pp. 192–203.
Cullity, B. D. (1959) "Elements of X-ray Diffraction," Addison-Wesley, Reading, Mass.
Dash, W. C. (1956) *J. Appl. Phys.*, vol. 27, pp. 1193–1195.
Dawson, I. M., and V. Vand (1951) *Nature*, vol. 167, p. 476.
Dieter, G. E. (1961) "Mechanical Metallurgy," McGraw-Hill, New York.
Edington, J. W., K. N. Melton, and C. P. Cutler (1976) *Prog. Mat. Sci.*, vol. 21, pp. 63–170.
Eshelby, J. D. (1957) *Proc. Roy. Soc. (London)*, vol. A241, pp. 376–396.
Eshelby, J. D., F. C. Frank, and F. R. N. Nabarro (1951) *Phil. Mag.*, vol. 42, pp. 351–364.
Frank, F. C. (1948) *Conf. Rept. Strength of Solids*, pp. 46–51, Physical Society, London.
Frank, F. C. (1949a) *Physica*, vol. 15, pp. 131–133.
Frank, F. C. (1949b) *Disc. Faraday Soc.*, vol. 5, pp. 48–54.
Frank, F. C. (1949c) *Proc. Phys. Soc.*, vol. A62, pp. 131–134.
Frank, F. C. (1952) *Acta Cryst.*, vol. 4, pp. 497–501.
Frank, F. C., and W. T. Read (1950) *Phys. Rev.*, vol. 79, pp. 722–723.
Frenkel, J. (1926) *Zeit. Phys.*, vol. 37, pp. 572–609.
Gifkins, R. C. (1967) *J. Inst. Metals*, vol. 95, pp. 373–377.
Gillis, P. P., and J. J. Gilman (1965) *J. Appl. Phys.*, vol. 36, pp. 3370–3380, 3380–3386.
Gilman, J. J. (1960) *Australian J. Phys.*, vol. 13, pp. 327–346.
Gilman, J. J. (1969) "Micromechanics of Flow in Solids," McGraw-Hill, New York.
Gilman, J. J., and W. G. Johnston (1960) *J. Appl. Phys.*, vol. 31, pp. 687–692.
Gilman, J. J., and W. G. Johnston (1962) *Solid State Physics*, vol. 13, pp. 147–222.
Greenman, W. F., T. Vreeland, and D. S. Wood (1967) *J. Appl. Phys.*, vol. 38, pp. 3595–3603.
Greenwood, D. W., and R. H. Johnson (1965) *Proc. Roy. Soc. (London)*, vol. A283, pp. 403–422.
Griffin, L. J. (1950) *Phil. Mag.*, vol. 41, pp. 196–199.
Gutmanas, E. Yu., E. M. Nadgornyi, and A. V. Stepanov (1963) *Soviet Phys. Solid State*, vol. 5, pp. 743–747.
Hall, E. O. (1951) *Proc. Phys. Soc.*, vol. B64, pp. 747–753.
Hall, E. O. (1954) "Twinning and Diffusionless Transformations in Metals," Butterworths, London.
Hayden, H. W., R. C. Gibson, H. F. Merrick, and J. H. Brophy (1967) *Trans. ASM*, vol. 60, pp. 3–14.
Heidenreich, R. D., and W. Shockley (1947) *J. Appl. Phys.*, vol. 18, pp. 1029–1031.
Heidenreich, R. D., and W. Shockley (1948) *Conf. Rept. Strength of Solids*, pp. 57–75, Physical Society, London.
Herring, C. (1950) *J. Appl. Phys.*, vol. 21, pp. 437–445.
Hirsch, P. B. (1959) *Metall. Rev.*, vol. 4, pp. 101–140.
Johnson, R. H. (1970) *Metall. Rev.*, vol. 4, pp. 115–134.
Johnston, W. G., and J. J. Gilman (1959) *J. Appl. Phys.*, vol. 30, pp. 129–144.
Johnston, W. G., and J. J. Gilman (1960) *J. Appl. Phys.*, vol. 31, pp. 632–643.
Koehler, J. S. (1952) *Phys. Rev.*, vol. 86, pp. 52–59.
Kroner, E. (1958) "Kontinuumstheorie der Versetzungen und Eigenspannungen," Springer-Verlag, Berlin.
Kroner, E. (1961) *Acta. Met.*, vol. 9, pp. 155–161.

Lardner, R. W. (1969) *Z. angew. Math. Phys.*, vol. 20, pp. 514–529.

Li, J. C. M. (1963) *Acta. Met.*, vol. 11, pp. 1269–1270.

Lin, T. H. (1957) *J. Mech. Phys. Solids*, vol. 5, pp. 143–149.

Lothe, J. (1962) *J. Appl. Phys.*, vol. 33, pp. 2116–2125.

Mackenzie, J. K. (1949) "A Theory of Sintering and Theoretical Yield Strength of Solids," Ph.D. Thesis, University of Bristol.

Mitchell, J. W. (1962) Direct Observations of Dislocations in Crystals by Optical and Electron Microscopy, in J. B. Newkirk and J. H. Wernick (eds.), "Direct Observation of Imperfections in Crystals," pp. 3–27, Interscience, New York.

Mott, N. F. (1953) *Phil. Mag.*, vol. 44, pp. 742–765.

Nabarro, F. R. N. (1947) *Proc. Phys. Soc.*, vol. 59, pp. 256–272.

Nabarro, F. R. N. (1948) *Conf. Rept. Strength of Solids*, pp. 75–90, Physical Society, London.

Nabarro, F. R. N. (1952) *Adv. Phys.*, vol. 1, pp. 269–394.

Nabarro, F. R. N., Z. S. Basinski, and D. B. Holt (1964) *Adv. Phys.*, vol. 13, pp. 193–323.

Noggle, T. S., and J. S. Koehler (1957) *J. Appl. Phys.*, vol. 28, pp. 53–62.

Nye, J. F. (1953) *Acta Met.*, vol. 1, pp. 153–162.

Oelshlagel, D., and W. Weiss (1966) *Trans. ASM*, vol. 58, pp. 143–154.

Orowan, E. (1934) *Z. Phys.*, vol. 89, pp. 605–659.

Orowan, E. (1947) *J. West Scot. Steel Inst.*, vol. 54, pp. 45–96.

Orowan, E. (1954) Dislocations and Mechanical Properties, in M. Cohen (ed.), "Dislocations in Metals," pp. 69–195, Institute of Metals AIME, New York.

Packer, C. M., R. H. Johnson, and O. D. Sharby (1968) *Trans. Met. Soc. AIME*, vol. 242, pp. 2485–2489.

Pashley, D. W. (1965) *Repts. Prog. Phys.*, vol. 28, pp. 291–330.

Peierls, R. E. (1940) *Proc. Phys. Soc.*, vol. 52, pp. 34–37.

Petch, N. J. (1953) *J. Iron and Steel Inst. (London)*, vol. 174, pp. 25–28.

Petch, N. J. (1954) *Prog. Met. Phys.*, vol. 5, pp. 1–52.

Polanyi, M. (1934) *Z. Phys.*, vol. 89, pp. 660–662.

Pope, D. P., T. Vreeland, and D. S. Wood (1967) *J. Appl. Phys.*, vol. 38, pp. 4011–4018.

Sanders, W. T. (1962) *Phys. Rev.*, vol. 128, pp. 1540–1549.

Stein, D. F., and J. R. Low (1960) *J. Appl. Phys.*, vol. 31, pp. 362–369.

Taylor, G. I. (1934) *Proc. Roy. Soc. (London)*, vol. A145, pp. 362–387, 388–404.

Taylor, G. I. (1938) *J. Inst. Met.*, vol. 62, pp. 307–324.

Taylor, G. I. (1956) Strains in Crystalline Aggregate, in R. Grammel (ed.), "Deformation and Flow of Solids," pp. 3–12, Springer-Verlag, Berlin.

Taylor, A. (1961) "X-ray Metallography," Wiley, New York.

Tinder, R. F. (1964) *Rev. Sci. Instrum.*, vol. 34, pp. 1049–1050.

Tinder, R. F., and J. Washburn (1964) *Acta Met.*, vol. 12, pp. 129–137.

Tyson, W. R. (1966) *Phil. Mag.*, vol. 14, pp. 925–936.

Verma, A. R. (1951) *Phil. Mag.*, vol. 42, pp. 1005–1013.

Volterra, V. (1907) *Ann. Ec. Norm. (Paris)*, vol. 24, pp. 401–517.

Webster, G. A. (1966) *Phil. Mag.*, vol. 14, pp. 775–783, 1303–1307.

Weertman, J. (1961) High Velocity Dislocations, in P. G. Shewmon and V. F. Zackay (eds.), "Response of Metals to High Velocity Deformation," pp. 205–247, Interscience, New York.

Wilsdorf, H., and D. Kuhlmann-Wilsdorf (1952) *Z. angew. Phys.*, vol. 4, pp. 361–370, 409–418, 418–424.

Wit, G. de, and J. S. Koehler (1959) *Phys. Rev.*, vol. 116, pp. 1121–1125.

Yoder, G. R., and V. Weiss (1972) *Met. Trans.*, vol. 3, pp. 675–681.

CHAPTER

TWELVE

THE YIELD CRITERION IN CONTINUUM PLASTICITY

12-1 YIELD CRITERIA FOR MATERIALS WHICH WORK HARDEN

12-1-1 General Isotropic Hardening

Consider the yield function

$$f = J'_2 H - \kappa L(J'_2) = 0 \tag{12-1-1}$$

where H is a function of stress and, in general, both H and the parameter κ may depend on the history of the deformation as discussed in Sec. 10-3-1.

Following Prager (1945), the stress dependence of H will be taken to be of the form

$$H = H(J'_2, J'_3) = H(\omega) \tag{12-1-2}$$

where

$$J'_2 = \tfrac{1}{2}\operatorname{tr}\mathbf{T}'^2 \qquad J'_3 = \det\mathbf{T}' \tag{12-1-3}$$

are the second and third invariants respectively of the deviatoric stress tensor

$$\mathbf{T}' = \mathbf{T} - \tfrac{1}{3}(\operatorname{tr}\mathbf{T})\mathbf{I} \tag{12-1-4}$$

and

$$\omega = \frac{27}{4}\frac{J'^2_3}{J'^3_2} \tag{12-1-5}$$

was used by Fromm (1933) in the form of a parameter $\bar{\mu} = \sqrt{\omega/54}$.

A concept which is fundamental to classical plasticity is the condition of convexity of the loading surface, which is taken to imply that any straight line in the deviatoric, that is, π plane, may pierce the yield surface in at most two points. Thus, the use of a given yield function is limited by the condition of convexity of the yield surface.

To obtain the yield locus in the π plane, consider a state of stress specified by the three principal stresses σ_a $(a = 1, 2, 3)$. A further system of cartesian coordinate axes x and y, which lie wholly in the π plane and whose origin is coincident with that of the principal axes of stress, are required. Let the projection of σ_2 in the π plane be coincident with the y axis. Setting

$$z = \frac{x^2}{y^2} + 1 \tag{12-1-6}$$

the substitution

$$\sigma'_1 = \frac{1}{\sqrt{2}}x - \frac{1}{\sqrt{6}}y \qquad \sigma'_2 = \frac{2}{\sqrt{6}}y \qquad \sigma'_3 = -\frac{1}{\sqrt{2}}x - \frac{1}{\sqrt{6}}y \tag{12-1-7}$$

where the σ'_a $(a = 1, 2, 3)$ are the principal stress deviator components, enables Eq. (12-1-1) to be expressed by way of Eq. (12-1-2) in the form

$$\tfrac{1}{2}y^2 zH(\omega) - \kappa L(J'_2) = 0 \tag{12-1-8}$$

where, from Eq. (12-1-5),

$$\omega = \frac{(4-3z)^2}{z^3} \tag{12-1-9}$$

From Eqs. (12-1-6), (12-1-8), and (12-1-9) there can be obtained

$$\frac{d\bar{y}}{d\bar{x}} = \pm\sqrt{z-1}\left[1 + \frac{3z(4-3z)(4-z)M(\omega)}{z^3 - 3(z-1)(4-3z)(4-z)M(\omega)}\right] \tag{12-1-10}$$

where

$$M(\omega) = -\frac{1}{1-(\kappa/H)\,dL/dJ'_2}\,\frac{1}{H}\frac{dH}{d\omega} \tag{12-1-11}$$

and $\bar{x} = x/\sqrt{2\kappa}$, $\bar{y} = y/\sqrt{2\kappa}$.

12-1-2 Stress Relations and the Yield Function

The relation on the right-hand side of Eq. (10-3-14) can be expressed by way of Eqs. (12-1-1), (12-1-2), and (12-1-5) in the form

$$\mathbf{G}' = H\left(1 - \frac{\kappa}{H}\frac{dL}{dJ'_2}\right)\left[\mathbf{T}' + \left(3\mathbf{T}' - 2\frac{J'_2}{J'_3}\mathbf{t}'\right)\omega M(\omega)\right] \tag{12-1-12}$$

where

$$\mathbf{t}' = \mathbf{T}'^2 - \tfrac{1}{3}(\mathrm{tr}\,\mathbf{T}'^2)\mathbf{I} \tag{12-1-13}$$

is the deviator of the square of the deviatoric stress tensor $\mathbf{T}'$ and M is defined by Eq. (12-1-11).

Rearrangement of Eq. (12-1-12) gives the stress relations

$$T\langle 22\rangle = T\langle 11\rangle + \frac{1}{\Lambda}(G'\langle 22\rangle - G'\langle 11\rangle + \tfrac{3}{2}\Phi G'\langle 11\rangle) - \frac{3}{4}\frac{\Phi}{\Phi-1} \times \frac{T'\langle 11\rangle\,[3\Phi(T\langle 12\rangle^2 + T\langle 31\rangle^2) - 2T\langle 12\rangle^2 - 4T\langle 31\rangle^2]}{J'_2 - 3T'\langle 11\rangle^2} \tag{12-1-14}$$

$$T\langle 33\rangle = T\langle 11\rangle + \frac{1}{\Lambda}(G'\langle 33\rangle - G'\langle 11\rangle + \tfrac{3}{2}\Phi G'\langle 11\rangle) - \frac{3}{4}\frac{\Phi}{\Phi-1} \times \frac{T'\langle 11\rangle\,[3\Phi(T\langle 12\rangle^2 + T\langle 31\rangle^2) - 4T\langle 12\rangle^2 - 2T\langle 31\rangle^2]}{J'_2 - 3T'\langle 11\rangle^2} \tag{12-1-15}$$

$$T\langle 12\rangle = \frac{1}{\Lambda}G'\langle 12\rangle - \frac{3}{2}\frac{\Phi}{\Phi-1} \times \frac{T'\langle 11\rangle\,[T\langle 31\rangle T\langle 32\rangle - T\langle 12\rangle\,(T\langle 33\rangle - T\langle 11\rangle)]}{J'_2 - 3T'\langle 11\rangle^2} = T\langle 21\rangle \tag{12-1-16}$$

$$T\langle 23\rangle = \frac{1}{\Lambda}G'\langle 23\rangle - \frac{3}{2}\frac{\Phi}{\Phi-1}\frac{T'\langle 11\rangle T\langle 12\rangle T\langle 31\rangle}{J'_2 - 3T'\langle 11\rangle^2} = T\langle 32\rangle \tag{12-1-17}$$

$$T\langle 31\rangle = \frac{1}{\Lambda}G'\langle 31\rangle - \frac{3}{2}\frac{\Phi}{\Phi-1} \times \frac{T'\langle 11\rangle\,[T\langle 12\rangle T\langle 23\rangle - T\langle 13\rangle\,(T\langle 22\rangle - T\langle 11\rangle)]}{J'_2 - 3T'\langle 11\rangle^2} = T\langle 13\rangle \tag{12-1-18}$$

where

$$\Phi = \frac{4\chi}{4\chi - 3} \tag{12-1-19}$$

$$\chi = \frac{1}{\Psi}\left(1 - \frac{H}{\Lambda}\right) \times \left[1 + 3\,\frac{(T\langle 11\rangle - T\langle 33\rangle)T\langle 12\rangle^2 + (T\langle 11\rangle - T\langle 22\rangle)T\langle 31\rangle^2 + 2T\langle 12\rangle T\langle 23\rangle T\langle 31\rangle}{(3T'\langle 11\rangle^2 - J'_2)T'\langle 11\rangle}\right]^{-1} \tag{12-1-20}$$

$$\Psi = \frac{T'\langle 11\rangle^2}{J'_2} \tag{12-1-21}$$

$$\Lambda = H\left[1 + \left(3 + 2\frac{J'_2}{J'_3}T'\langle 11\rangle\right)\omega M(\omega)\right] \tag{12-1-22}$$

where M is defined by Eq. (12-1-11). In Eqs. (12-1-14) and (12-1-15), $T\langle 11\rangle$ is to be determined from the condition of equilibrium

$$\operatorname{div}\mathbf{T} = \mathbf{0} \tag{12-1-23}$$

in the absence of body forces.

Equation (12-1-19) shows that for $\chi = \frac{3}{4}$, $\Phi \to \pm\infty$. Hence, if χ attains the value 0.75 for finite plastic deformation, then Eqs. (12-1-14) and (12-1-15) predict the possibility that both $T\langle 22\rangle$ and $T\langle 33\rangle$ independently tend to infinity as $\Phi \to \pm\infty$. This situation does not arise if $T\langle 12\rangle = T\langle 31\rangle = 0$ over the whole range of plastic deformation.

For the restricted stress system the general components of stress reduce to

$$T\langle 22\rangle = T\langle 11\rangle + \frac{1}{\Lambda}(G'\langle 22\rangle - G'\langle 11\rangle + \tfrac{3}{2}\Phi G'\langle 11\rangle) \qquad (12\text{-}1\text{-}24)$$

$$T\langle 33\rangle = T\langle 11\rangle + \frac{1}{\Lambda}(G'\langle 33\rangle - G'\langle 11\rangle + \tfrac{3}{2}\Phi G'\langle 11\rangle) \qquad (12\text{-}1\text{-}25)$$

$$T\langle 23\rangle = \frac{1}{\Lambda}G'\langle 23\rangle \qquad (12\text{-}1\text{-}26)$$

$$T\langle 12\rangle = T\langle 31\rangle = 0 \qquad (12\text{-}1\text{-}27)$$

the $T\langle 11\rangle$ component of stress again being determined from the condition of equilibrium, i.e., Eq. (12-1-23). Substituting the condition of Eq. (12-1-27) into Eq. (12-1-5) gives

$$\omega = \tfrac{27}{4}\Psi(1-\Psi)^2 \qquad (12\text{-}1\text{-}28)$$

where Ψ is defined by Eq. (12-1-21) and

$$J'_2 = \frac{T'\langle 11\rangle^2}{\Psi} \qquad J'_3 = -\frac{1-\Psi}{\Psi}T'\langle 11\rangle^3 \qquad (12\text{-}1\text{-}29)$$

Substituting in Eq. (12-1-20) for J'_2 and J'_3 and ω from Eqs. (12-1-28) and (12-1-29) gives, by way of Eq. (12-1-22), the relation

$$\chi = \frac{27(1-3\Psi)(1-\Psi)M(\omega)}{4+27(1-3\Psi)(1-\Psi)\Psi M(\omega)} \qquad (12\text{-}1\text{-}30)$$

which can be rearranged to give

$$M(\omega) = \frac{4\chi}{27(1-\chi\Psi)(1-3\Psi)(1-\Psi)} \qquad (12\text{-}1\text{-}31)$$

where M is defined by Eq. (12-1-11).

Using the relations of Eq. (12-1-27), the stress relations of Eqs. (12-1-14) to (12-1-16) can be rearranged to give

$$\frac{G'\langle 11\rangle}{G'\langle 23\rangle} = \zeta(1-\tfrac{4}{3}\chi) \qquad (12\text{-}1\text{-}32)$$

$$\frac{G'\langle 22\rangle}{G'\langle 23\rangle} = -\tfrac{1}{2}\zeta(1-\tfrac{4}{3}\chi)+\tfrac{1}{2}\xi \qquad (12\text{-}1\text{-}33)$$

$$\frac{G'\langle 33\rangle}{G'\langle 23\rangle} = -\tfrac{1}{2}\zeta(1-\tfrac{4}{3}\chi)-\tfrac{1}{2}\xi \qquad (12\text{-}1\text{-}34)$$

where the stress ratios

$$\xi = \frac{T\langle 22\rangle - T\langle 33\rangle}{T\langle 23\rangle} \qquad \zeta = \frac{2T\langle 11\rangle - T\langle 22\rangle - T\langle 33\rangle}{3T\langle 23\rangle} \tag{12-1-35}$$

12-1-3 Stress-Strain Relations

In terms of the general formulation of the yield condition, i.e., Eq. (10-3-2), substitution into Eq. (10-3-3) from Eq. (10-3-14) gives, for the scalar multiplier $\dot{\lambda}$ during loading, the relation

$$\dot{\lambda} = -\frac{\operatorname{tr}(\mathbf{G}'\dot{\mathbf{T}}^T)}{\operatorname{tr}[(\mathbf{\Phi}^T + (\partial f/\partial\kappa)\mathbf{\Psi}^T)\mathbf{G}']} \tag{12-1-36}$$

where

$$\Phi^i{}_j = \frac{\partial f}{\partial \bar{\epsilon}_i{}^j} \qquad \Psi^i{}_j = \frac{\partial \kappa}{\partial \bar{\epsilon}_i{}^j} \tag{12-1-37}$$

Equation (12-1-36) can be used to express the plastic potential equation (10-3-14) in the form

$$\dot{\bar{\epsilon}} = \tilde{G} \operatorname{tr}(\mathbf{G}'\dot{\mathbf{T}}^T)\mathbf{G}' \tag{12-1-38}$$

where the scalar multiplier

$$\tilde{G} = -\frac{1}{\operatorname{tr}[(\mathbf{\Phi}^T + (\partial f/\partial\kappa)\mathbf{\Psi}^T)\mathbf{G}']} \tag{12-1-39}$$

In an attempt to account for the deviations from the Lode relation $\mu = \nu$ [i.e., Eq. (10-4-34)], observed by Taylor and Quinney (1931) (see the present Fig. 10-9), Prager (1945) proposed a yield criterion characterized by

$$H = 1 - c\omega \qquad L = 1 \tag{12-1-40}$$

for which conditions Eq. (12-1-1) simplifies to

$$f = J'_2(1 - c\omega) - \kappa = 0 \tag{12-1-41}$$

For present purposes it will be assumed that

$$\kappa = k^2 \qquad k = k(\bar{\epsilon}'') \tag{12-1-42}$$

where $\bar{\epsilon}''$ is the effective strain, the quantity k being here interpreted as the consecutive values of the yield stress in pure shear; c is a constant for a given state of hardness, i.e., for a given value of κ.

Using the specimen geometry and notation of Sec. 10-4-1, it follows that for the simple tensile test of a solid rod ($\bar{\eta} = 1$), the only nonvanishing component of stress is $T\langle zz\rangle$. Hence it follows from Eqs. (10-3-14) and (12-1-41) that

$$d\bar{\epsilon}\langle zz\rangle = d\lambda \tfrac{2}{3}[\Lambda]_{\omega=1} T\langle zz\rangle \qquad \text{(simple tension)} \tag{12-1-43}$$

where $[\Lambda]_{\omega=1}$ is the value of Λ evaluated at $\omega = 1$ and Λ is defined by Eq. (12-1-22) subject to the conditions of Eqs. (12-1-27) and (12-1-40), it being noted that $\omega = 1$ for all plastic deformation in simple tension. For pure torsion, the only nonvanishing component of stress is $T\langle\theta z\rangle$, and the corresponding equation is

$$d\bar{\epsilon}\langle\theta z\rangle = d\lambda[\Lambda]_{\omega=0}T\langle\theta z\rangle \qquad \text{(pure torsion)} \tag{12-1-44}$$

where, again, $[\Lambda]_{\omega=0}$ is defined by Eq. (12-1-22) subject to the conditions of Eqs. (12-1-27) and (12-1-40), it being noted that $\omega = 0$ for all plastic deformation in pure torsion. For simple tension,

$$[\Lambda]_{\omega=1} = [H]_{\omega=1} = 1-c \qquad \text{(simple tension)} \tag{12-1-45}$$

and, for pure torsion,

$$[\Lambda]_{\omega=0} = [H]_{\omega=0} = 1 \qquad \text{(pure torsion)} \tag{12-1-46}$$

and Eqs. (12-1-43) and (12-1-44) reduce to

$$d\bar{\epsilon}\langle zz\rangle = d\lambda\tfrac{2}{3}(1-c)T\langle zz\rangle \tag{12-1-47}$$

$$\tfrac{1}{2}d\bar{\gamma}\langle\theta z\rangle = d\lambda T\langle\theta z\rangle \tag{12-1-48}$$

The yield condition of Eq. (12-1-41) implies that Eqs. (12-1-47) and (12-1-48) have to be rearranged into the form

$$d\bar{\epsilon}\langle zz\rangle = [\tfrac{2}{3}(1-c)d\lambda]T\langle zz\rangle \tag{12-1-49}$$

$$\frac{(1-c)^{1/2}}{\sqrt{3}}d\bar{\gamma}\langle\theta z\rangle = [\tfrac{2}{3}(1-c)d\lambda]\frac{\sqrt{3}}{(1-c)^{1/2}}T\langle\theta z\rangle \tag{12-1-50}$$

in turn implying that the effective strain increments are

$$d\bar{\epsilon}'' = \begin{cases} d\bar{\epsilon}\langle zz\rangle & \text{(simple tension)} \\ \dfrac{(1-c)^{1/2}}{\sqrt{3}}d\bar{\gamma}\langle\theta z\rangle & \text{(pure torsion)} \end{cases} \tag{12-1-51}$$

and the effective stresses are

$$\sigma'' = \begin{cases} T\langle zz\rangle & \text{(simple tension)} \\ \dfrac{\sqrt{3}}{(1-c)^{1/2}}T\langle\theta z\rangle & \text{(pure torsion)} \end{cases} \tag{12-1-52}$$

These conditions can be satisfied by defining the effective stress σ'' and the effective strain $\bar{\epsilon}''$ to be

$$\sigma'' = \left\{\frac{3}{2}\frac{[1-c(1-\Omega)]}{(1-c)^2}\operatorname{tr}\mathbf{G}'^2\right\}^{1/2} \tag{12-1-53}$$

$$\bar{\epsilon}'' = \int\frac{2}{\sqrt{3}}[1-c(1-\Omega)]^{1/2}(\dot{K}_2')^{1/2}\,dt \tag{12-1-54}$$

where

$$\Omega = \frac{27}{4}\frac{K_3'^2}{K_2'^3} \tag{12-1-55}$$

and

$$K_2' = [\int(-\text{II}_{\dot{\epsilon}})^{1/2}\,dt]^2 \qquad K_3' = [\int(\text{III}_{\dot{\epsilon}})^{1/3}\,dt]^3 \tag{12-1-56}$$

it being noted that

for *simple tension*:

$$K_2' = \tfrac{3}{4}\bar{\epsilon}\langle zz\rangle^2 \qquad K_3' = -\tfrac{1}{4}\bar{\epsilon}\langle zz\rangle^3 \qquad \Omega = 1$$

for *pure torsion*:

$$K_2' = \bar{\epsilon}\langle\theta z\rangle^2 \qquad K_3' = 0 \qquad \Omega = 0$$

In Fig. 12-1 the full-line curves are for measurements in compression and the curves represented by open symbols are for measurements in simple torsion (Billington, 1977a, 1977b). It is evident from Fig. 12-1 that universal stress-strain curves can be obtained over an initial limited range of effective strain to a useful approximation when expressed in terms of the effective stresses and effective strains defined by Eqs. (12-1-53) and (12-1-54). With increasing shear strain, the torsion curve is seen from Fig. 12-1 to fall increasingly below the compression curve. This effect may possibly arise from the progressive development of anisotropy to a greater extent in torsion than in tension or compression at corresponding effective strains, owing to transverse planes being directions of maximum shear strain throughout the twisting. Alternatively, this effect may arise from the assumption that for measurements in simple torsion the only nonzero component of stress is

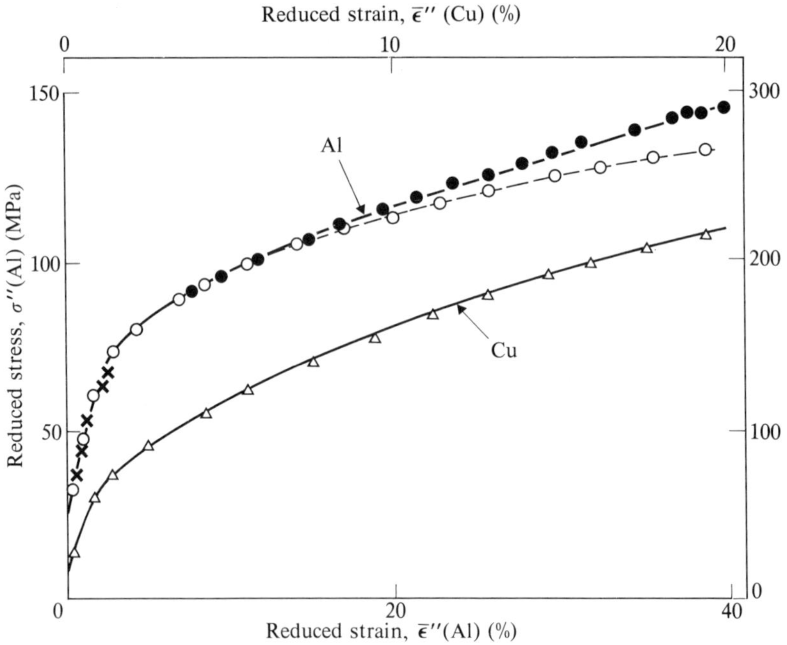

Figure 12-1 Variation of σ'' with $\bar{\epsilon}''$ for aluminum in the fully annealed state. × denotes a thin-walled tube tested in simple tension; ● a complete σ'' from measurements of Swift effect. (*Billington, 1977c.*)

the shear stress $T\langle\theta z\rangle$. It is shown in Sec. 12-2 that this assumption is not necessarily true, there being a significant contribution from the stress ratios ξ and ζ defined by Eq. (12-1-35).

Returning to the general formulation of the yield equation, i.e., Eq. (10-3-2), for a yield surface made up of several individual smooth surfaces which harden independently, Koiter's (1953) generalization of a plastic potential theory as expressed by Eqs. (10-3-15) and (10-3-16) replaces Eq. (12-1-38) with the relation

$$\dot{\epsilon} = \sum_{p=1}^{n} C_p \tilde{G}_p \operatorname{tr}(\mathbf{G}'_p \dot{\mathbf{T}}^T)\mathbf{G}'_p \tag{12-1-57}$$

where

$$\begin{aligned} C_p &= 0 \text{ for } f_p < 0 \qquad \frac{\partial f_p}{\partial T_{ij}} \dot{T}_{ij} < 0 \\ C_p &= 1 \text{ for } f_p = 0 \qquad \frac{\partial f_p}{\partial T_{ij}} \dot{T}_{ij} \geqslant 0 \end{aligned} \tag{12-1-58}$$

and the $\tilde{G}_p$ are n positive functions of stress, deformation, and deformation history, corresponding to $\tilde{G}$ in Eq. (12-1-39).

From Eqs. (12-1-38) and (12-1-57) it is evident that in general the plastic strain rate is linear in the stress rate.

12-1-4 Kinematic Hardening

Prager (1955, 1956), with the object of introducing the Bauschinger effect into the constitutive equations of the incremental theory of plasticity, proposed a model of kinematic hardening using a rigid frame in the shape of the initial yield surface to describe the loading surface in stress space. The state of loading on the π plane is represented by a small, frictionless pin initially located at the origin O of the initial yield surface. Prior to yield, the state of stress is represented by the position of the frictionless pin which is free to move within the frame. Yielding occurs when the pin first contacts the side of the frame. With continued loading the pin interacts with the frame causing it to translate. The frame is assumed to be constrained against rotation and to be smooth so that only motion normal to the surface is possible. At corners, if the motion of the pin interacts with both sides, the frame translates in the direction of the motion of the pin. Unloading occurs when the pin ceases to interact with the frame, the latter remaining stationary while the pin moves away from the frame. The model is termed kinematic since both stress and strain are represented by displacements. Thus, for a rigid work-hardening material the displacement of the center of the frame relative to O is proportional to the total strain and the state of stress is represented by the position of the pin relative to O. In this way the Bauschinger effect is represented, together with the progressive development of anisotropy due to plastic deformation.

Prager's theory of kinematic hardening requires that the general yield condition $f(\mathbf{T}) = 0$ of Eq. (10-2-1) be reformulated as

$$f(\mathbf{T} - \boldsymbol{\alpha}) = 0 \tag{12-1-59}$$

where the α are the coordinates of the center of the new yield surface. Shield and Ziegler (1958) assume linear hardening expressed in the form

$$\dot{\alpha} = c\dot{\epsilon} \tag{12-1-60}$$

where c is a constant. The scalar multiplier λ appearing in Eq. (10-3-14) can now be obtained in the same way that Eq. (12-1-36) was arrived at, the required relation being expressed in the form

$$\lambda = \frac{1}{c}\frac{\operatorname{tr}(\mathbf{G}'\dot{\mathbf{T}}^T)}{\operatorname{tr}(\mathbf{G}'\mathbf{G}'^T)} \tag{12-1-61}$$

Apparent inconsistencies have been observed (see, for example, Budiansky, 1957) in the application of kinematic hardening to plane stress problems when based on the assumption that the center of the yield surface in a subspace of less than nine dimensions moved in the direction of the outer normal to the surface in the subspace. Hodge (1957) resolved this difficulty by noting that the concept of kinematic hardening must be applied in the nine-dimensional stress space, which he called complete kinematic hardening in contrast to the direct hardening in the

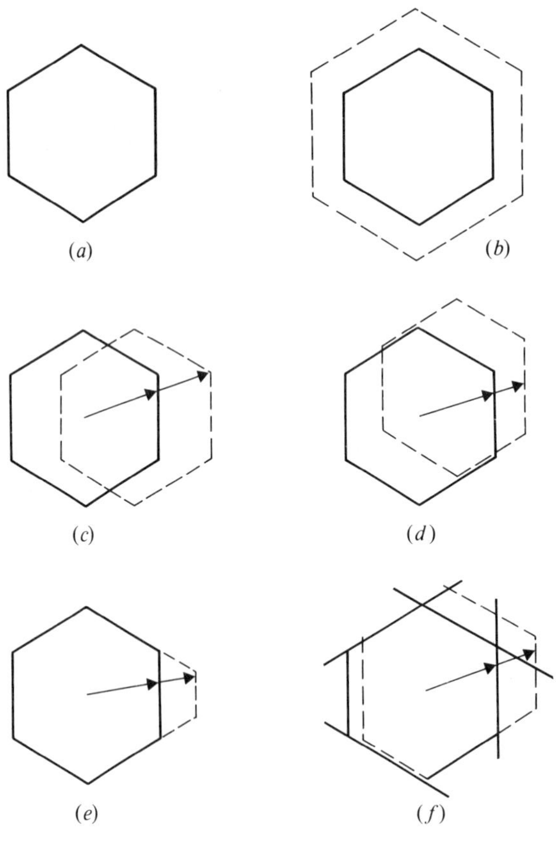

Figure 12-2 Comparison of various hardening (shown dashed) laws for Tresca's initial yield condition (shown solid): (*a*) the initial yield surface, (*b*) isotropic hardening, (*c*) kinematic hardening (Prager's rule), (*d*) kinematic hardening (Ziegler's modification), (*e*) independently acting plane loading surfaces, (*f*) interdependent plane loading surfaces. (*Naghdi, 1960.*)

subspace. In contrast to Eq. (12-1-60), the relation

$$\dot{\alpha} = \dot{\mu}(\mathbf{T} - \alpha) \qquad \text{for } \dot{\mu} > 0 \tag{12-1-62}$$

proposed by Ziegler (1959) can be applied in the subspace as well as in the nine-dimensional stress space.

A concept, similar to the concept of kinematic hardening as proposed by Prager, was introduced independently by Ishlinski (1954). The concept proposed by Ishlinski has been referred to as the ideal Bauschinger effect by Kadashevich and Novozhilov (1959). It is to be noted that Prager's approach to kinematic hardening can in principle be associated with any yield condition, in contrast to that of Ishlinski which is restricted to yield conditions involving only J'_2. Further developments and applications of the theory have been discussed by Boyce (1956), Prager (1956), Hodge (1957, 1958), and Naghdi (1960). Hodge (1957) extended the concept of kinematic hardening to include an expansion of the yield surface simultaneously with its translation. In addition to those aspects presently included in Prager's approach to kinematic hardening, a required extension to the theory is the inclusion of a change of shape, but the question of how this would depend on the deformation history is as yet unresolved.

The various hardening rules of present interest are summarized in Figs. 12-2*a* to *d* in the form of a comparison of the hardening rules for Tresca's initial yield condition (Naghdi, 1960). Two further hardening rules are shown in Fig. 12-2*e*, where the plastic deformation causes a linear segment to move, and Fig. 12-2*f*, where the plane loading surfaces change with plastic loading in some interdependent manner.

12-2 LARGE ELASTIC-PLASTIC DEFORMATION

12-2-1 Stress-Strain Relation

In the plastic range of deformation, the torsion mode is characterized by an effect which is formally similar to that observed by Poynting (1909, 1912) and which is now referred to as the Swift effect. Swift (1947) observed that a permanently twisted rod elongates in the direction of the major axis about which the rod is being twisted, this permanent elongation increasing with increasing shear strain. Similarly, a permanently twisted, thin-walled tube has been observed (Freudenthal and Ronay, 1966) to elongate in the direction of the axis about which the tube is being twisted, this permanent elongation increasing with increasing shear strain (Billington, 1977a, 1977b).

It was observed by Swift that after a reversal of the torque there is an initial, small, shortening of the specimen, but continued twisting produces, however, continued axial extension. This means that the axial extension from twisting in one direction appears to continue for large reversed shear strains. An accumulation of such lengthenings in plastic torsion has also been observed by Freudenthal and Ronay (1966) and by Ronay (1966, 1967, 1968).

A typical $(\bar{\epsilon}_s, \bar{\gamma})$ curve for a thin-walled tube of aluminum (Billington, 1978) is shown in Fig. 10-8, where $\bar{\epsilon}_s$ is the axial natural strain characteristic of the Swift effect and $\bar{\gamma}$ is the plastic component of the total shear strain defined by Eq. (6-5-17). Also shown in Fig. 10-8 is the associated $(\tau, \bar{\gamma})$ curve. It is evident from Fig. 10-8 that the three $(\bar{\epsilon}_s, \bar{\gamma})$ curves, corresponding to a wall thickness of 1, 2, and 3 mm, are indistinguishable within the limits of experimental accuracy. Just as the Poynting effect is not predicted by the theory of linear elasticity, so the Swift effect is not predicted by classical plasticity theory in the form of Eq. (10-3-14). Thus, the question arises of whether there exists an incremental-type constitutive equation for a large elastic-plastic deformation which predicts the Swift effect and which, for the appropriate approximations, reduces to the classical theory of plasticity.

The formulation of a constitutive equation for an incompressible strain-hardening medium must ideally allow for the possibility that the material exhibits the Poynting effect in the elastic range of response. From a consideration of the Poynting effect which is characterized by a component of the axial strain caused solely by geometric nonlinearity, it is to be expected that the constitutive equation for a strain-hardening medium which exhibits the Swift effect will also give a relation between the axial strain and the plastic shear strain which contains a term representing the component of axial strain caused by geometric nonlinearity alone. The physically satisfactory description of the Poynting effect, discussed in Secs. 8-5-2 and 8-7, is one of the outstanding successes of modern nonlinear continuum mechanics. Thus, assuming that the Swift effect has its origin in the Poynting effect, the question arises of whether a physically satisfactory description of the Swift effect can be obtained from consideration of elastic-plastic solids with a large deformation, assuming general isotropic hardening.

Central to any description of the mechanical response of elastic-plastic solids with a large deformation is the question of how to separately identify the elastic and plastic behavior of the material. Various attempts have been made to develop theories for elastic-plastic solids with a large deformation (see, for example, Thomas, 1955; Green, 1956; Hill, 1958; Green and Naghdi, 1965; Lee, 1969; Freund, 1970; Hahn, 1974). These theories generally assume that either the total strain can be decomposed into the sum of elastic and plastic strains or that the total deformation gradient can be decomposed into a matrix product of the local elastic and plastic deformation tensors.

Theories for elastic-plastic solids with a large deformation have been proposed by Lee (1969), Freund (1970), Hahn and Jaunzemis (1971), and Hahn (1974) by way of the matrix-product decomposition of the deformation gradients

$$\mathbf{F} = \hat{\mathbf{F}}\bar{\mathbf{F}} \tag{12-2-1}$$

where $\hat{\mathbf{F}}$ and $\bar{\mathbf{F}}$ are the local elastic and plastic deformation gradient tensors respectively. For an isotropic, incompressible strain-hardening medium, Hahn (1974) assumes the elastic behavior of the material to be characterized by a constitutive equation which admits of the Poynting (1909, 1912) effect, the plastic behavior being referred to the small strain components by way of $\mathbf{L} = \dot{\bar{\mathbf{F}}}\bar{\mathbf{F}}^{-1}$. In this way the matrix-product decomposition of Eq. (12-2-1) admits of the possibility that an isotropic incompressible strain-hardening medium may exhibit the Poynting effect, thus allowing for the frequently observed nonlinear unloading behavior of a plastically deformed body. However, referring the plastic behavior to the small strain components by way of **L** precludes the possibility that the system of constitutive equations proposed by Hahn (1974) can describe the Swift effect in the way required by the presently available observations of length changes in twisted tubes (Freudenthal and Ronay, 1966; Ronay, 1966, 1967, 1968; Billington, 1977b).

Following Hahn (1974), it has been assumed (Billington, 1977b) that the system of constitutive equations for an isotropic incompressible strain-hardening medium can be expressed in the form

$$\mathbf{T} = -P\mathbf{I} + \beta_1 \hat{\mathbf{B}} + \beta_{-1} \hat{\mathbf{B}}^{-1} \tag{12-2-2a}$$

$$\hat{\mathbf{B}} = \hat{\mathbf{F}}\hat{\mathbf{F}}^T \tag{12-2-2b}$$

$$\alpha_1 \, d\bar{\mathbf{B}}' + \alpha_{-1} \, d(\bar{\mathbf{B}}^{-1})' = d\lambda \mathbf{G}' \tag{12-2-3}$$

$$G'^i{}_j = \frac{\partial f}{\partial T_i^j} \tag{12-2-4}$$

$$\bar{\mathbf{B}}' = \bar{\mathbf{B}} - \tfrac{1}{3}\mathrm{I}\mathrm{I}_{\bar{\mathbf{B}}} \tag{12-2-5a}$$

$$\bar{\mathbf{B}} = \bar{\mathbf{F}}\bar{\mathbf{F}}^T \tag{12-2-5b}$$

$$(\bar{\mathbf{B}}^{-1})' = \bar{\mathbf{B}}^{-1} - \tfrac{1}{3}\mathrm{I}\mathrm{II}_{\bar{\mathbf{B}}} \tag{12-2-6}$$

where **T** is the stress tensor, P the pressure, $d\lambda$ is a scalar factor of proportionality, and f is the yield function. The $\hat{\mathbf{B}}$ of Eq. (12-2-2*b*) and the $\bar{\mathbf{B}}$ of Eq. (12-2-5*b*) are related to $\mathbf{B} = \mathbf{F}\mathbf{F}^T$ through the compatibility condition of Eq. (12-2-1).

There is a growing body of experimental evidence in support of Ludwik's (1909) contention that the simple tensile and simple compressive true stress–natural strain curves coincide, as, for example, the results of Ludwik and Scheu (1925), shown in Fig. 10-7. The condition that the simple tensile and simple compressive stress-strain curves are to coincide can be satisfied by taking the α_i $(i = \pm 1)$ to be of the form

$$\alpha_1 = \frac{\bar{\alpha}_1}{\bar{\alpha}_1 - \bar{\alpha}_{-1}} \frac{C_0}{\mathrm{II}_{\bar{\mathbf{B}}}} \qquad \alpha_{-1} = \frac{\bar{\alpha}_{-1}}{\bar{\alpha}_1 - \bar{\alpha}_{-1}} \frac{1}{C_0 \mathrm{I}_{\bar{\mathbf{B}}}} \tag{12-2-7}$$

where

$$C_0^2 = \frac{(\mathrm{II}_{\bar{\mathbf{B}}}^2 - 3\mathrm{I}_{\bar{\mathbf{B}}})}{(\mathrm{I}_{\bar{\mathbf{B}}}^2 - 3\mathrm{II}_{\bar{\mathbf{B}}})} = \frac{\mathrm{II}_{(\bar{\mathbf{B}}^{-1})'}}{\mathrm{II}_{\bar{\mathbf{B}}'}} \tag{12-2-8}$$

and the $\bar{\alpha}_i$ $(i = \pm 1)$ are in general infinite power series in the invariants $\mathrm{I}_{\bar{\mathbf{B}}}$ and $\mathrm{II}_{\bar{\mathbf{B}}}$ of the deformation tensor $\bar{\mathbf{B}}$, since for an incompressible solid $\mathrm{III}_{\bar{\mathbf{B}}} = 1$.

12-2-2 Resultant Longitudinal Force on Plane End of Tube

Using Eqs. (10-4-2) to (10-4-7), the following relations can be obtained from Eqs. (12-2-3), (12-2-5), and (12-2-6):

$$a_1\, d\bar{\epsilon}\langle zz\rangle + a_2\, d\ln\bar{\eta} - 16\bar{F}^2(1-2\bar{q})\bar{\epsilon}\langle\theta z\rangle\, d\bar{\epsilon}\langle\theta z\rangle = \overline{d\lambda}G'\langle rr\rangle \tag{12-2-9}$$

$$a_3\, d\bar{\epsilon}\langle zz\rangle + a_4\, d\ln\bar{\eta} + 16\bar{F}^2(2-\bar{q})\bar{\epsilon}\langle\theta z\rangle\, d\bar{\epsilon}\langle\theta z\rangle = \overline{d\lambda}G'\langle\theta\theta\rangle \tag{12-2-10}$$

$$a_5\, d\bar{\epsilon}\langle zz\rangle + a_6\, d\ln\bar{\eta} - 16\bar{F}^2(1+\bar{q})\bar{\epsilon}\langle\theta z\rangle\, d\bar{\epsilon}\langle\theta z\rangle = \overline{d\lambda}G'\langle zz\rangle \tag{12-2-11}$$

$$a_7\, d\bar{\epsilon}\langle\theta z\rangle = \overline{d\lambda}G'\langle\theta z\rangle \tag{12-2-12}$$

where $\overline{d\lambda} = d\lambda/a_0$ and

$$a_0 = \frac{1}{6}\frac{\alpha_1}{(1-\bar{q})} \qquad \bar{q} = \frac{-\alpha_{-1}/\bar{\eta}\bar{F}}{\alpha_1 - (\alpha_{-1}/\bar{\eta}\bar{F})} \tag{12-2-13}$$

$$a_1 = -\left[\frac{1}{\bar{\eta}\bar{F}} + 2\bar{\eta}^2\bar{F}^2 + \mathrm{I}_{\bar{\mathbf{B}}} + (1-2\bar{q})\left(\frac{5}{\bar{\eta}\bar{F}} - 2\bar{\eta}^2\bar{F}^2 + 6\bar{F}^2 - 3\mathrm{I}_{\bar{\mathbf{B}}}\right)\right] \tag{12-2-14}$$

$$a_2 = -\left[\frac{1}{\bar{\eta}\bar{F}} + 2\bar{\eta}^2\bar{F}^2 + \mathrm{I}_{\bar{\mathbf{B}}} - (1-2\bar{q})\left(\frac{2\bar{\eta}\bar{F}^3 + 2\bar{\eta}^3\bar{F}^3 - 1}{\bar{\eta}\bar{F}} - \mathrm{I}_{\bar{\mathbf{B}}}\right)\right] \tag{12-2-15}$$

$$a_3 = -\left[4\mathrm{I}_{\bar{\mathbf{B}}} - \frac{5}{\bar{\eta}\bar{F}} - \bar{\eta}^2\bar{F}^2 + (1-2\bar{q})\left(\bar{\eta}^2\bar{F}^2 - \frac{1}{\bar{\eta}\bar{F}}\right)\right] \tag{12-2-16}$$

$$a_4 = -\left[\frac{1}{\bar{\eta}\bar{F}} - \bar{\eta}^2\bar{F}^2 - 2\mathrm{I}_{\bar{\mathbf{B}}} + (1-2\bar{q})\left(\frac{4\bar{\eta}\bar{F}^3 + 3\bar{\eta}^3\bar{F}^3 - 1}{\bar{\eta}\bar{F}} - 2\mathrm{I}_{\bar{\mathbf{B}}}\right)\right] \tag{12-2-17}$$

$$a_5 = -a_1 - a_3 \tag{12-2-18}$$

$$a_6 = -a_4 - a_2 \tag{12-2-19}$$

$$a_7 = \left[12\bar{F}^2 + 6(1-2\bar{q})\bar{F}^2\bar{D}\frac{d\ln(\bar{\eta}\bar{F})}{d\bar{D}}\right] \tag{12-2-20}$$

Note that

$$\mathrm{I}_{\bar{\mathbf{B}}} = \bar{F}^2 + \frac{1}{\bar{\eta}\bar{F}} + \frac{\bar{\eta}}{\bar{F}} + \bar{\eta}\frac{R^2}{\bar{F}}\bar{D}^2 \tag{12-2-21}$$

$$\mathrm{II}_{\bar{\mathbf{B}}} = \frac{1}{\bar{F}^2} + \bar{\eta}\bar{F} + \frac{\bar{F}}{\bar{\eta}} + \frac{R^2}{\bar{F}^2}\bar{D}^2 \tag{12-2-22}$$

are the first and second invariants of the deformation tensor $\bar{\mathbf{B}}$.

For a permanently twisted tube with its major axis aligned with the Z axis, the nonvanishing components of stress are in general $T\langle rr\rangle$, $T\langle\theta\theta\rangle$, $T\langle zz\rangle$, and the shear stress $T\langle\theta z\rangle$.

The components of stress can be obtained in the way described in Sec. 12-1-2:

$$T\langle rr\rangle = \int_{\bar{r}_2}^{\bar{r}}\left(\frac{T\langle\theta\theta\rangle - T\langle rr\rangle}{r}\right)dr + [T\langle rr\rangle]_{r=\bar{r}_2} \tag{12-2-23}$$

$$T\langle\theta\theta\rangle = T\langle rr\rangle + \frac{1}{\Lambda}(G'\langle\theta\theta\rangle - G'\langle rr\rangle + \tfrac{3}{2}\Phi G'\langle rr\rangle) \tag{12-2-24}$$

$$T\langle zz\rangle = T\langle rr\rangle + \frac{1}{\Lambda}(G'\langle zz\rangle - G'\langle rr\rangle + \tfrac{3}{2}\Phi G'\langle rr\rangle) \tag{12-2-25}$$

$$T\langle\theta z\rangle = \frac{1}{\Lambda} G'\langle\theta z\rangle \tag{12-2-26}$$

where Φ is defined by Eq. (12-1-19) and Λ is defined by Eq. (12-1-22).

Substituting in Eq. (8-5-15) for $T\langle zz\rangle$ from Eq. (12-2-25) and rearranging by way of Eq. (12-2-26) gives

$$N_z = 2\pi \int_{\bar{r}_2}^{\bar{r}_1} rT\langle rr\rangle\, dr + 2\pi \int_{\bar{r}_2}^{\bar{r}_1} T\langle\theta z\rangle \left(\frac{G'\langle zz\rangle}{G'\langle\theta z\rangle} - \frac{G'\langle rr\rangle}{G'\langle\theta z\rangle}\right) r\, dr$$
$$+ 3\pi \int_{\bar{r}_2}^{\bar{r}_1} \Phi \frac{G'\langle rr\rangle}{G'\langle\theta z\rangle} T\langle\theta z\rangle r\, dr \tag{12-2-27}$$

With the external surface free from applied stress, the condition of equilibrium in the radial direction can be expressed in the form

$$[T\langle rr\rangle]_{r=\bar{r}_2} = -\int_{\bar{r}_2}^{\bar{r}_1} \frac{T\langle\theta\theta\rangle - T\langle rr\rangle}{r}\, dr \tag{12-2-28}$$

Substituting for the first term on the right-hand side of Eq. (8-5-17) from Eq. (12-2-28) and for the stress difference,

$$0 = -2\pi \int_{\bar{r}_2}^{\bar{r}_1} rT\langle rr\rangle\, dr + \pi \int_{\bar{r}_2}^{\bar{r}_1} T\langle\theta z\rangle \left(\frac{G'\langle\theta\theta\rangle}{G'\langle\theta z\rangle} - \frac{G'\langle rr\rangle}{G'\langle\theta z\rangle}\right)\left(\frac{\bar{r}_2^2}{r^2} - 1\right) r\, dr$$
$$+ \tfrac{3}{2}\pi \int_{\bar{r}_2}^{\bar{r}_1} T\langle\theta z\rangle \Phi \frac{G'\langle rr\rangle}{G'\langle\theta z\rangle}\left(\frac{\bar{r}_2^2}{r^2} - 1\right) r\, dr \tag{12-2-29}$$

Adding Eqs. (12-2-27) and (12-2-29) gives, on rearrangement,

$$N_z = \pi \int_{\bar{r}_2}^{\bar{r}_1} T\langle\theta z\rangle \left[\left(\frac{G'\langle\theta\theta\rangle}{G'\langle\theta z\rangle} - \frac{G'\langle rr\rangle}{G'\langle\theta z\rangle}\right)\frac{\bar{r}_2^2}{r^2} + 3\frac{G'\langle zz\rangle}{G'\langle\theta z\rangle} + \left(\frac{\bar{r}_2^2}{r^2} + 1\right)\tfrac{3}{2}\Phi \frac{G'\langle rr\rangle}{G'\langle\theta z\rangle}\right] r\, dr \tag{12-2-30}$$

12-2-3 The Swift Effect

For sufficiently thin-walled tubes, for the limiting conditions $\bar{\gamma} \to 0$, $\bar{F} \to 1$, and taking $\bar{K} - 0$, $\bar{\eta} = 1$, Eqs. (12-2-9) to (12-2-20) give

$$\frac{d\bar{\epsilon}_s}{d\bar{\gamma}} = \tfrac{1}{3}(1+\bar{q})\bar{\gamma} \qquad \text{for } \bar{\gamma} \to 0 \tag{12-2-31}$$

and for $\bar{q}$ independent of $\bar{\gamma}$,

$$\bar{\epsilon}_s = \tfrac{1}{6}(1+\bar{q})\bar{\gamma}^2 \qquad \text{for } \bar{\gamma} \to 0 \tag{12-2-32}$$

All materials for which studies of the Swift effect are available are in general accord with Eq. (12-2-32). For these materials there is a range of $\bar{\gamma}$ for which

$$\bar{\epsilon}_s = \tfrac{1}{4}\bar{\gamma}^2 \qquad \text{for } \bar{q} = \tfrac{1}{2},\ \bar{\gamma} \to 0 \tag{12-2-33}$$

as can be seen from Fig. 12-3.

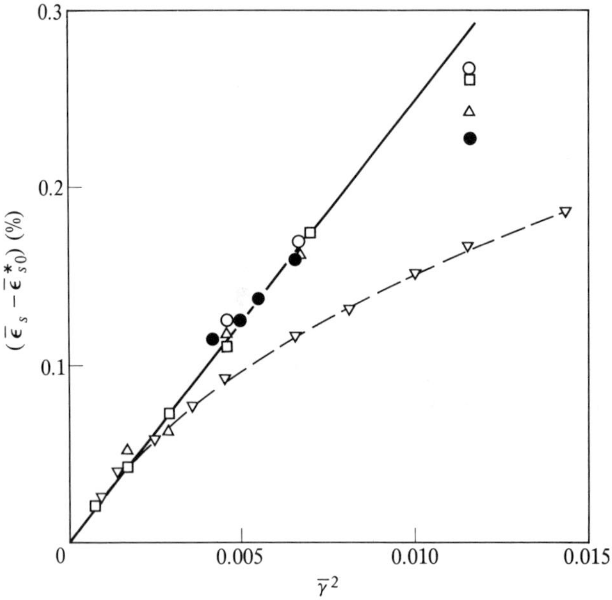

Figure 12-3 Variation of the static irrecoverable axial natural strain $\bar{\epsilon}_S$ with $\bar{\gamma}^2$, where $\bar{\gamma}$ is the static plastic component of the shear strain. ○, Al; △, Cu as-received; □, Fe as-received; ▽, Cu fully annealed; ●, Al alloy. (*Billington, 1977c.*)

Basic to the present discussion of the Swift effect is the assumption of isotropic strain-hardening. The assumption of isotropy implies the possibility that the mechanical response can be characterized by a single material coefficient $\bar{\alpha}$, which may be dependent upon the history of the deformation. Consider the α_i $(i = \pm 1)$ to be defined in terms of $\bar{\alpha}$ by the relations

$$\alpha_1 = \varphi \frac{\partial \bar{\alpha}}{\partial \mathrm{I}_{\bar{\mathbf{B}}}} \qquad \alpha_{-1} = -\varphi \frac{\partial \bar{\alpha}}{\partial \mathrm{II}_{\bar{\mathbf{B}}}} \tag{12-2-34}$$

where φ is a constant of proportionality. For the limiting conditions $\bar{F} = 1$, $\bar{K} = 0$, $\bar{\eta} = 1$, Eqs. (12-2-21) and (12-2-22) give $\mathrm{I}_{\bar{\mathbf{B}}} = \mathrm{II}_{\bar{\mathbf{B}}} = 3 + \bar{\gamma}^2$, and it thus follows from Eq. (12-2-34) that

$$\frac{\alpha_1}{\alpha_{-1}} = -1 \qquad (\mathrm{I}_{\bar{\mathbf{B}}} = \mathrm{II}_{\bar{\mathbf{B}}}) \tag{12-2-35}$$

which in turn gives, by way of the second equation of (12-2-13), the relation

$$\bar{q} = \frac{1}{1 + \bar{\eta}\bar{F}} = \tfrac{1}{2} \qquad (\bar{\eta} = \bar{F} = 1) \tag{12-2-36}$$

since for $\bar{\gamma} \to 0$, $\bar{F} \sim 1$, $\bar{\eta} \sim 1$. Thus the concept underlying the formulation of Eqs. (12-2-34) may provide a physical basis for the correlation shown in Fig. 12-3 by way of Eq. (12-2-33), which is based on $\bar{q} = \frac{1}{2}$.

With regard to the quadratic relation of Eq. (12-2-32), Freudenthal and Ronay (1966) have concluded from their studies of the total accumulative axial elongation $\Sigma\Delta\bar{\epsilon}_s(N)$ arising from N cycles of reversed torsion that $\Sigma\Delta\bar{\epsilon}_s(N)$ is proportional to $\bar{\gamma}^2$, an observation which is not reconcilable with Eq. (10-3-14) but, as noted by Freudenthal and Ronay, is in accord with the general considerations underlying the formulation of Eq. (12-2-3).

The studies described by Pukas (1976) are in general accord with the behavior predicted by Eq. (12-2-3), except for titanium, which has been shown by Pukas to contract in the axial direction with increasing shear strain.

For measurements at finite shear strains, it is necessary to assume that, in terms of the thin-walled tube approximation, the quantity Φ can be regarded as being independent of r. With this approximation, Eq. (12-2-30) can be rearranged to give

$$\Phi = \frac{N_z - \pi \displaystyle\int_{\bar{r}_2}^{\bar{r}_1} T\langle\theta z\rangle \left[\left(\frac{G'\langle\theta\theta\rangle}{G'\langle\theta z\rangle} - \frac{G'\langle rr\rangle}{G'\langle\theta z\rangle} \right) \frac{\bar{r}_2^2}{r^2} + 3\frac{G'\langle zz\rangle}{G'\langle\theta z\rangle} \right] r\,dr}{\frac{3}{2}\pi \displaystyle\int_{\bar{r}_2}^{\bar{r}_1} T\langle\theta z\rangle \frac{G'\langle rr\rangle}{G'\langle\theta z\rangle} r\,dr} \tag{12-2-37}$$

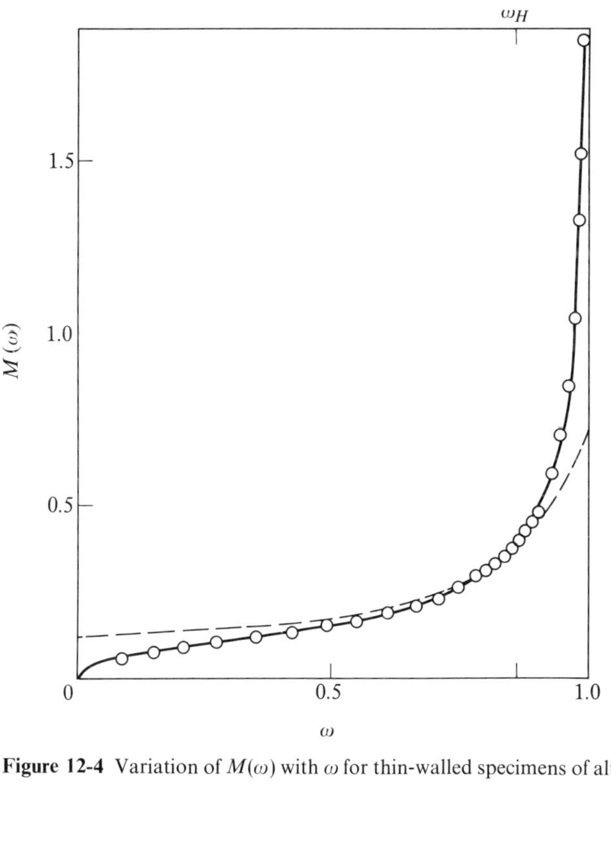

Figure 12-4 Variation of $M(\omega)$ with ω for thin-walled specimens of aluminum. (*Billington, 1979.*)

To evaluate Φ from measurements of the Swift effect it is further necessary to assume that for a sufficiently thin-walled tube, the ratio $(T\langle \theta z \rangle / r)$ and the quantity q can individually be regarded as being independent of r. With these approximations and substituting for $G'\langle rr \rangle / G'\langle \theta z \rangle$, $G'\langle \theta\theta \rangle / G'\langle \theta z \rangle$, and $G'\langle zz \rangle / G'\langle \theta z \rangle$ from Eqs. (12-2-9) to (12-2-12), values of Φ can be evaluated from the experimental $(\bar{\epsilon}_s, \bar{\gamma})$ curves for finite shear strains. These values of Φ can be inserted into Eq. (12-1-19) to give the dependence of χ on $\bar{\gamma}$. The stress ratios ξ and ζ can also be evaluated from observations of the Swift effect (Billington, 1978) and can in turn be substituted into Eq. (12-1-21) to give the variation of Ψ with $\bar{\gamma}$. Having evaluated the shear strain dependence of Ψ, the dependence of ω on $\bar{\gamma}$ can be evaluated from Eq. (12-1-28).

Once the dependence of ω on $\bar{\gamma}$ is known, the quantity $M(\omega)$ can be evaluated from Eq. (12-1-31), and hence the way in which M varies with ω is now known from measurements of the Swift effect (Billington, 1979). A typical (M, ω) curve is shown in Fig. 12-4 for a thin-walled specimen of aluminum. The broken-line curve, which for values of $\omega \geqslant \omega_H$ is indistinguishable within the limits of experimental accuracy from the full-line curve, is the (M, ω) curve corresponding to the Tresca yield criterion.

It is thus seen that, with regard to obtaining universal stress-strain curves, measurements of the Swift effect enable all the essential quantities to be evaluated. The blocked circles shown in Fig. 12-1 have been obtained in this way.

12-3 CONCLUDING REMARKS

From the above discussion of the various constitutive equations which have been suggested to represent plastic flow, it is evident that much progress is needed if the response of actual materials is to be predicted accurately for general loading histories. Included in this observation is the classical small elastic-plastic theory, since in this case the variation of the yield surface with plastic flow has generally been restricted to isotropic hardening and kinematic hardening, both of which greatly oversimplify the actual situation. The nonlinear history aspect of plastic flow requires a suitably adequate nonlinear functional relation between stress and deformation. Developments within this particular aspect of plasticity are currently being reported. However, developments in nonlinear continuum plasticity can only be regarded as definitive if they can be shown to be in accord with the basic physical theory of plastic flow as embodied in the theories of crystal dislocation mechanics as discussed in Chap. 11.

Throughout Chap. 10 and the present chapter, it has been assumed that yield is unaffected by hydrostatic stress, a situation generally in accord with experiment for a wide variety of materials when tested at moderate stress levels. However, there is an increasing body of experimental evidence showing that yield and flow stresses of certain metals and alloys are sensitive to hydrostatic pressure, but that work-hardening and the magnitude of the strength differential (SD) effect between compression and tension are relatively insensitive (see, for example, Drucker,

1973; Spitzig, Sober, and Richmond, 1975). Measurements of the volume changes occurring during plastic deformation at atmospheric pressure for these metals show that the plastic volume expansion was larger than that measured in materials which show a smaller SD effect. The SD effect can, in principle, be accounted for by way of a yield function which involves odd powers of J'_3 (Freudenthal and Gou, 1969). This is an important area of study, since these experiments provide a means of testing the normality of the plastic strain rate vector to a given yield surface.

GENERAL REFERENCES

Bridgman, P. W. (1952) "Studies in Large Plastic Flow and Fracture," McGraw-Hill, New York.

Freudenthal, A. M., and H. Geiringer (1958) The Inelastic Continuum, in S. Flügge (ed.), "Handbuch der Physik," vol. 4, pp. 229–433, Springer-Verlag, Berlin.

Hill, R. (1950) "Plasticity," Clarendon Press, Oxford.

Lubahn, J. D., and R. P. Felgar (1961) "Plasticity and Creep of Metals," Wiley, New York.

REFERENCES

Billington, E. W. (1977a) *J. Phys.* (*D*), vol. 10, pp. 519–531.

Billington, E. W. (1977b) *J. Phys.* (*D*), vol. 10, pp. 533–552.

Billington, E. W. (1977c) *J. Phys.* (*D*), vol. 10, pp. 553–569.

Billington, E. W. (1978) *J. Phys.* (*D*), vol. 11, pp. 459–471.

Billington, E. W. (1979) *J. Phys.* (*D*), vol. 12, pp. 853–874.

Boyce, W. E. (1956) *Quart. Appl. Math.*, vol. 14, pp. 277–288.

Budiansky, B. (1957) *J. Appl. Mech.*, vol. 24, pp. 481–482.

Drucker, D. C. (1973) *Met. Trans.*, vol. 4, pp. 667–673.

Freudenthal, A. M., and P. F. Gou (1969) *Acta Mechan.*, vol. 8, pp. 34–52.

Freudenthal, A. M., and M. Ronay (1966) *Proc. Roy. Soc.* (*London*), vol. A292, pp. 14–50.

Freund, L. B. (1970) *Int. J. Solids Struct.*, vol. 6, pp. 1193–1209.

Fromm, H. (1933) *Ing.-Arch.*, vol. 4, pp. 432–466.

Green, A. E. (1956) *Proc. Roy. Soc.* (*London*), vol. A234, pp. 46–59.

Green, A. E., and P. M. Naghdi (1965) *Arch. Ration. Mech. Anal.*, vol. 18, pp. 251–281.

Hahn, H. T. (1974) *Int. J. Solids Struct.*, vol. 10, pp. 111–121.

Hahn, H. T., and W. Jaunzemis (1971) *Int. J. Engng. Sci.*, vol. 11, pp. 1065–1078.

Hill, R. (1958) *J. Mech. Phys. Solids*, vol. 6, pp. 236–249.

Hodge, P. G. (1957) *Proc. Ninth Int. Congr. Appl. Mech.*, Brussels, 1956, vol. 8, pp. 65–72.

Hodge, P. G. (1958) The Mathematical Theory of Plasticity, in J. N. Goodier and P. G. Hodge (eds.), "Elasticity and Plasticity," Wiley, New York.

Ishlinski, I. U. (1954) *Ukr. Mat. Zh.* (in Russian), vol. 6, pp. 314–324.

Kadashevich, I., and V. V. Novozhilov (1959) *J. Appl. Math. Mech.* (translation of *Prinkl. Mat. i. Merkh*), vol. 22, pp. 104–118.

Koiter, W. T. (1953) *Quart. Appl. Math.*, vol. 11, pp. 350–354.

Lee, E. H. (1969) *J. Appl. Mech.*, vol. 36, pp. 1–6.

Ludwik, P. (1909) "Elemente der Technologischen Mechanik," Springer-Verlag, Berlin.

Ludwik, P., and A. Scheu (1925) *Stahl. u. Eisen*, vol. 45, pp. 373–381.

Naghdi, P. M. (1960) Stress-Strain Relations in Plasticity and Thermoplasticity, in E. H. Lee and P. Symonds (eds.), "Plasticity," pp. 121–169, Pergamon Press, Oxford.

Poynting, J. H. (1909) *Proc. Roy. Soc.* (*London*), vol. A82, pp. 546–559.

Poynting, J. H. (1912) *Proc. Roy. Soc.* (*London*), vol. A86, pp. 534–561.

Prager, W. (1945) *J. Appl. Phys.*, vol. 16, pp. 837–840.
Prager, W. (1955) *Proc. Inst. Mech. Engng.*, vol. 169, pp. 41–57.
Prager, W. (1956) *J. Appl. Mech.*, vol. 23, pp. 493–496.
Pukas, S. (1976) "The Non-Linear Mechanical Response of Real Materials in Torsion," Ph.D. Thesis, Thames Polytechnic.
Ronay, M. (1966) *J. Inst. Metall.*, vol. 94, pp. 392–394.
Ronay, M. (1967) *Int. J. Solids Struct.*, vol. 3, pp. 167–175.
Ronay, M. (1968) *Int. J. Solids Struct.*, vol. 4, pp. 509–515.
Shield, R. T., and H. Ziegler (1958) *Z. angew. Math. Phys.*, vol. 9, pp. 260–276.
Spitzig, W. A., R. J. Sober, and O. Richmond (1975) *Acta Metall.*, vol. 23, pp. 885–893.
Swift, H. W. (1947) *Engng.*, vol. 163, pp. 253–257.
Taylor, G. I., and H. Quinney (1931) *Phil. Trans. Roy. Soc.*, vol. A230, pp. 323–362.
Thomas, T. Y. (1955) *Proc. Nat. Acad. Sci. U.S.A.*, vol. 41, pp. 720–726.
Ziegler, H. (1959) *Quart. Appl. Math.*, vol. 17, pp. 55–65.

CHAPTER

THIRTEEN

IMPACT PROPERTIES OF SOLIDS

13-1 DYNAMIC PLASTICITY

13-1-1 Wave Propagation in Finite-Length Specimens

Associated with the dynamic mechanical response of a material subjected to dynamic loading below the high-energy level of shock loading is the development of inertial stresses which must be taken into account. That is, in formulating constitutive equations, wave propagation in the material must be allowed for. Theories of plastic wave propagation in a rod have been proposed by Donnell (1930), Rakhmatulin (1945), Taylor (1946), and Karman and Duwez (1950) which are based on the scalar rate-independent stress-strain relation

$$T = F(\epsilon) \tag{13-1-1}$$

where T and ϵ are here understood to be the physical components appropriate to the mode of deformation. For these theories, F is a monotonically increasing function of ϵ and $dT/d\epsilon$ is a monotonically decreasing function of ϵ. From Sec. 6-4-1, the longitudinal wave speed c_L can be obtained in the form

$$c_L = \left(\frac{1}{\rho}\frac{dF}{d\epsilon}\right)^{1/2} \tag{13-1-2}$$

where ρ is the density of the material. For elastic wave propagation in the interior of an elastic solid it has been shown that

$$c_L = \left(\frac{\lambda + 2G}{\rho}\right)^{1/2} = \left(\frac{\kappa + \frac{4}{3}G}{\rho}\right)^{1/2} \tag{13-1-3}$$

where the presence of the shear modulus G recognizes that, as longitudinal expansions and contractions are not accompanied by transverse motion, there is both a change in the shape of an element and its volume as the wave passes. Thus the resistance of the medium to shear as well as its compressibility is called into play. The longitudinal expansions and contractions of a rod will, however, necessarily result in lateral contractions, resulting in a nonuniform distribution of stress across each section of the rod. However, for cylindrical pressure bars which satisfy the condition that the wavelengths of the elastic waves are large compared with the diameter of the bars, it can be assumed to a good degree of approximation that plane transverse sections of the rod remain plane during the passage of the stress waves, and the stress acts uniformly over each section. This approximation has been shown in Sec. 6-4 to lead, in the first order, to the relation

$$c_L = \left(\frac{E}{\rho}\right)^{1/2} \tag{13-1-4}$$

The wave speed of Eq. (13-1-2) is thus seen to decrease rapidly from the one-dimensional value of Eq. (13-1-4) as the transverse motion due to the rod surface displacements increases.

Consider a right circular rod of incompressible material. Let a particle be located by the cylindrical current coordinates (r, θ, z). The principal axis of the rod is aligned with the z axis. For the case of a plane wave traveling in the direction of decreasing z, the displacement u can be obtained from Sec. 6-4-2 in the form

$$u = f(z + c_L t) \tag{13-1-5}$$

where u is the displacement of an element of the pressure bar. Differentiation of Eq. (13-1-5) with respect to z gives

$$\frac{\partial u}{\partial z} = f'(z + c_L t) \tag{13-1-6}$$

where f' denotes differentiation with respect to the argument $(z + c_L t)$. Similarly, differentiation of Eq. (13-1-5) with respect to t gives

$$\frac{\partial u}{\partial t} = c_L f'(z + c_L t) \tag{13-1-7}$$

which combined with Eq. (13-1-6) gives

$$v = \frac{\partial u}{\partial t} = c_L \frac{\partial u}{\partial z} = c_L \epsilon(t) \tag{13-1-8}$$

or, on integrating, the displacement

$$u = c_L \int_0^t \epsilon \, dt' \tag{13-1-9}$$

From Eqs. (13-1-2) and (13-1-8) it follows that

$$T = \rho \int_0^\epsilon c_L^2 \, d\epsilon' = \rho \int_0^v c_L \, dv' \tag{13-1-10}$$

where v is the particle velocity.

The function $F(\epsilon)$ of Eq. (13-1-1) can be determined from Eq. (13-1-10) by way of experiments in which the speed of propagation is measured as a function of strain or particle velocity. Use of this method has been described, for example, by Malvern (1955), Bell (1960), and Rakhmatulin and Demyanov (1966). The dynamic behavior of many materials differs from the behavior at low strain rates. For example, the high strain rate stress levels of many materials exceed the corresponding low strain rate stress levels for the same strain. However, provided the dynamic mechanical response is characterized by a single dynamic stress-strain curve, then the method can be applied. Using different forms of constitutive relation it has been shown by Ripperger and Watson (1968) that the technique does not reliably identify the type of relation required to describe the dynamic mechanical response of the material.

Consider incremental wave propagation resulting from the application of a small incremental stress to a material while it is prestrained by an amount ϵ_0. Equation (13-1-2) predicts that the incremental wave will propagate at a speed $[F'(\epsilon_0)/\rho]^{1/2}$. The results of this type of experiment (see, for example, Bell, 1951; Sternglass and Stuart, 1953; Malyshev, 1961; Bianchi, 1964) have shown that the speed of the wave front is effectively that of elastic waves given by Eq. (13-1-4). To avoid the difficulties discussed by Craggs (1957) arising from the use of one-dimensional theory in studies of incremental wave propagation in rods, incremental shear wave propagation in thin-walled cylinders has been investigated by Convery and Pugh (1968), Yew and Richardson (1969), and Campbell and Dowling (1970). The results of these experiments indicate that for various metals and alloys tested under this type of loading, for which the above simple theory is exact, the wave front is propagated at a speed which approximates very closely to that of elastic shear waves. These studies thus appear to indicate that the dynamic response is rate dependent, an aspect which must be included in the formulation of any constitutive equation describing the mechanical response of a strain-rate sensitive material.

An example of a constitutive equation which predicts strain-rate sensitivity is that of Malvern (1951a, 1951b), in which the plastic strain rate is a function of stress and strain, this particular constitutive equation implying that an incremental wave propagates at the elastic wave speed in accordance with experiment. More general relations have been proposed, an example being the constitutive equation

$$\frac{\partial \bar{\epsilon}}{\partial t} = \Phi(T, \epsilon)\frac{\partial T}{\partial t} + \Psi(T, \epsilon) \tag{13-1-11}$$

which has been discussed by Lubliner (1964) and Cristescu (1967). In Eq. (13-1-11) the function $\Phi(T, \epsilon)$ determines the instantaneous plastic flow and the function $\Psi(T, \epsilon)$ the noninstantaneous flow. Equation (13-1-11) predicts that an incremental wave propagates at a speed $c_0/(1+E\Phi)^{1/2}$ where c_0 is the elastic wave speed. Campbell and Dowling (1970), using thin-walled cylinders subjected to incremental torques, have shown that for aluminum Φ is essentially zero while for copper it increases with plastic strain but is always small compared with the elastic compliance.

It is important to note that for the wave propagation methods of observing the dynamic response of a material so far discussed, the only direct measurement of stress is at the boundary of the specimen. Hence studies based only on consideration of wave propagation are not entirely suitable for the determination of material behavior. In an attempt to reduce the limitations imposed by wave propagation phenomena in finite-length specimens, techniques have been developed which are based on the small-specimen philosophy which assumes that wave effects in these specimens can be neglected.

13-1-2 Split-Hopkinson Bar Technique

It is convenient to identify three ranges of strain rate. These are low rates of strain (10^{-4} to $10^{-1}\,s^{-1}$), medium rates of strain (10^{-1} to $10^{2}\,s^{-1}$), and high rates of strain (10^{2} to $10^{4}\,s^{-1}$). In the range of low rates of strain, the mechanical response of a material can be experimentally characterized by way of conventional screw-driven or hydraulic test machines. Standard methods of test are available for this range of strain rates, the technical difficulties being primarily those arising from the need to evaluate the mechanical response of new materials over as wide a range of environmental parameters as possible in the quasi-static range (10^{-4} to $10^{-2}\,s^{-1}$). Test methods for low and medium rates of strain are now well established, so the present discussion will therefore be restricted to high rates of strain.

The experiments involving the propagation of longitudinal waves in iron wires by J. Hopkinson (1872) and B. Hopkinson (1914) would appear to be the earliest known studies of the dynamic mechanical properties of a material. Dynamic measurements, which are essentially transient in nature, can be obtained using a modified form of the Hopkinson (1914) pressure bar apparatus described by Davies (1948) and Kolsky (1953). Test facilities of this type employ what is generally known as the split-Hopkinson bar method.

The split-Hopkinson bar technique is an example of the application of the one-dimensional, nondispersive, wave propagation theory of Sec. 6-4-1 to a study of the interaction of a stress wave with the test material. For the three simple modes of deformation, the split-Hopkinson bar technique is concerned with elastic wave propagation in circular, cylindrical rods or tubes. Consider a right circular rod, or tube, of incompressible material. Let a particle be located by the cylindrical current coordinates (r, θ, z). The principal axis of the rod is aligned with the z axis. For a deformed tube or solid rod with its principal axis aligned with the z axis, the nonvanishing components of stress are in general $T\langle rr\rangle$, $T\langle\theta\theta\rangle$, $T\langle zz\rangle$, and the shear stress $T\langle\theta z\rangle$.

(a) Compression A cylindrical specimen is sandwiched between two rods and deformed under the action of a transient stress wave induced by dynamically loading the free ends of one of the rods, as shown schematically in Fig. 13-1. The stress wave passes down the loading, or incident, pressure bar with part of the wave being reflected at the specimen and part being transmitted into the inertia, or

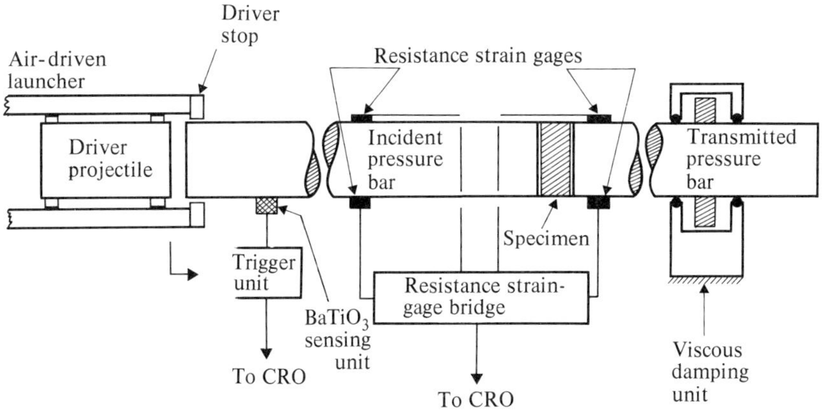

Figure 13-1 Schematic diagram of split-Hopkinson pressure bar apparatus. (*Billington and Tate, 1972.*)

transmitter, pressure bar. The use of strain gages mounted on the incident and transmitter bars to analyze the effects of the stress waves distinguishes this form of the apparatus from that described by Davies (1948), Kolsky (1949), and Davies and Hunter (1963) in which capacitance methods were used. Bell (1956, 1966) has developed an experimental method in which the strain is determined by measuring the angle of diffraction of a grating engraved onto the surface of the specimen.

For experiments of this type, to give stress-strain curves characteristic of the mechanical response of materials account must be taken of the principal limiting features of the experiment which can be summarized as follows:

1. The cylindrical pressure bars must satisfy the condition that the wavelengths of the elastic waves are large compared with the diameter of the bars.
2. The pressure bars must not be loaded beyond the elastic limit.
3. There will be frictional restraints on the loaded surfaces of the specimen.
4. Nominal stress must be distinguished from true stress.
5. There must be verification, or otherwise, of internal stress equilibrium within the specimen.
6. The effect of variable strain rate must be noted.
7. The effect of variable strain acceleration must be noted.

The essential condition that the pressure bars must not be loaded beyond the elastic limit determines the maximum stress level which can be observed for any given test material. Frictional restraints can be reduced to a minimum by applying a thin layer of lubricant to the pressure bar–specimen interfaces, yielding deformed specimens which exhibit negligible barreling.

Consider a compressive strain pulse to be propagated down the incident pressure bar. When the stress wave reaches the specimen a fraction is propagated through the specimen into the transmitter pressure bar and a fraction is reflected from the interface between the incident pressure bar and the specimen as a tensile wave in the incident pressure bar. These three pulses will be identified as the incident loading pulse $\tilde{e}_I(t)$, the transmitted pulse $\tilde{e}_T(t)$, and the reflected pulse $\tilde{e}_R(t)$ respectively, at any given instant of time t measured from the initiation of the compressive strain pulse in the incident pressure bar. From these three strain-time histories and application of the one-dimensional nondispersive, wave propagation theory, as considered in Sec. 6-4-1, it is possible to determine the displacement u, velocity v, and pressure P at both faces of the specimen as functions of the time t.

Denoting by the subscripts 1 and 2 the incident and transmitter side of the specimen respectively, the following relationships for the displacements u, velocities v, and forces F can be derived from the elastic propagation theory:

$$u_1 = c_b \int_0^t (\tilde{e}_I - \tilde{e}_R)\, dt' \qquad u_2 = c_b \int_0^t \tilde{e}_T\, dt' \tag{13-1-12}$$

$$v_1 = c_b(\tilde{e}_I - \tilde{e}_R) \qquad v_2 = c_b \tilde{e}_T \tag{13-1-13}$$

$$F_1 = E_b A_b(\tilde{e}_I + \tilde{e}_R) \qquad F_2 = E_b A_b \tilde{e}_T \tag{13-1-14}$$

where E_b is the Young modulus, c_b the wave velocity, and A_b the cross-sectional area of the pressure bars.

Taking compressive strains to be positive, and considering only specimens and pressure bars having the same cross-sectional area A_b, the average specimen stress $T\langle zz\rangle$, strain $\tilde{e}\langle zz\rangle$, and strain rate $\dot{\tilde{e}}\langle zz\rangle$ in the specimen are given by

$$T\langle zz\rangle = \frac{F_1 + F_2}{2A_b} = \tfrac{1}{2}E_b(\tilde{e}_I + \tilde{e}_R + \tilde{e}_T) \equiv T_0 \tag{13-1-15}$$

$$\dot{\tilde{e}}\langle zz\rangle = \frac{v_1 - v_2}{L} = \frac{c_b}{L}(\tilde{e}_I - \tilde{e}_R - \tilde{e}_T) \equiv \dot{\tilde{e}} \tag{13-1-16}$$

$$\tilde{e}\langle zz\rangle = \frac{u_1 - u_2}{L} = \frac{c_b}{L}\int_0^t (\tilde{e}_I - \tilde{e}_R - \tilde{e}_T)\, dt' \equiv \tilde{e} \tag{13-1-17}$$

where L is the initial, undeformed, length of the specimen.

Typical examples of the three strain histories are shown in Fig. 13-2. These have been copied from a photographic record (Billington and Tate, 1972). The Pochhammer–Chree oscillations (see Kolsky, 1953, and the present Sec. 6-4-5) following the initial rise of the incident pulse can be seen. It is evident that the instantaneous values of the three strain histories satisfy the approximation

$$\tilde{e}_I(t) + \tilde{e}_R(t) \approx \tilde{e}_T(t) \tag{13-1-18}$$

which can be used to eliminate $\tilde{e}_I$ and $\tilde{e}_R$ from Eq. (13-1-15) and $\tilde{e}_I$ and $\tilde{e}_T$ from Eqs. (13-1-16) and (13-1-17), thus giving Eqs. (13-1-15) to (13-1-17) in the simplified form

$$T_0(t) = E_b \tilde{e}_T(t) \tag{13-1-19}$$

$$\dot{\tilde{e}}(t) = -2\frac{c_b}{L}\tilde{e}_R(t) \tag{13-1-20}$$

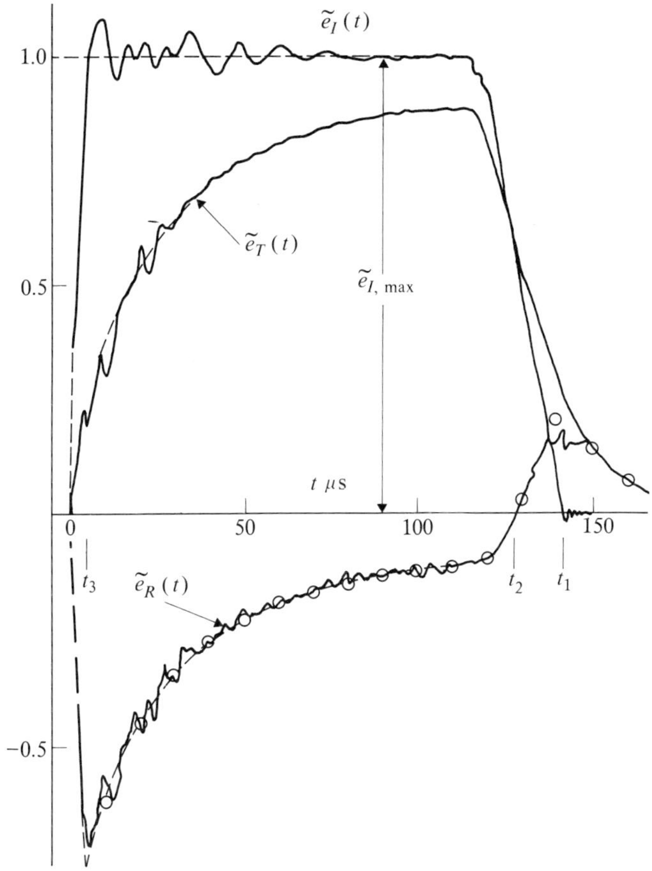

Figure 13-2 Comparison of measured pulse amplitudes confirms approximate equivalence of the forces on both faces of the specimen. ○ denotes $[\tilde{e}_I(t) - \tilde{e}_T(t)]$. (*Billington and Tate, 1972.*)

$$\tilde{e}(t) = -2\frac{c_b}{L}\int_0^t \tilde{e}_R\,dt' \tag{13-1-21}$$

The equivalence of the stress on the boundaries, which at best can only be an approximation ($F_1 \approx F_2$), implies the existence of internal stress equilibrium within the specimen, i.e., the stress gradients due to internal wave reflections are at most a second-order effect.

Measurements of true stress are obtained by using specimens having an initially smaller diameter than that of the pressure bars. Allowing for the area miss-match at the pressure bar–specimen interfaces, Eq. (13-1-19) now has the form

$$T_0(t) = E_b\tilde{e}_T(t)\left\{\frac{A_b}{A_0}[1 - \tilde{e}(t)]\right\} \tag{13-1-22}$$

where A_b and A_0 are the cross-sectional areas of the pressure bars and the undeformed specimen respectively and $T_0(t)$ is now the true stress at time t of deformation, the instantaneous value of the specimen strain being $\tilde{e}(t)$.

(b) Tension Using different types of specimen geometry, measurements in tension have been made by Harding (1965) and Lindholm and Yeakley (1968). The principles underlying the measurements of Lindholm and Yeakley in tension parallel closely those in compression. In this latter type of experiment, the transmitter pressure bar of a split-Hopkinson pressure bar apparatus is replaced by a tube having the same cross-sectional area as that of the loading bar. This modification permits the use of a cylindrical, thin-walled specimen of the type described by Lindholm and Yeakley. The thin-walled tubular specimen is loaded in simple tension by way of the loading or incident pressure bar of the modified split-Hopkinson pressure bar apparatus.

(c) Torsion One end of the loading bar is clamped in such a way that it can be twisted through a given angle. The principle of the method is to store a given amount of torsional strain energy in a fixed fraction of the length of the loading rod, measured from the end which can be twisted through the given angle. This is achieved by rigidly clamping the rod at some convenient point along its length and rotating one end through the given angle. A transient torsional strain pulse is generated by rapidly releasing the mechanical clamp. Methods of clamping and rapidly releasing the system have been described by Pope, Vreeland, and Wood (1964), Baker and Yew (1966), Nicholas and Campbell (1972), and Clyens and Campbell (1974).

In contrast to the measurements in compression and tension, the quantities to be measured are the shear stress τ ($\equiv T\langle\theta z\rangle$), by way of the applied torque Γ, and the shear strain γ ($\equiv\gamma\langle\theta z\rangle$). The quantities Γ and γ can be evaluated from a knowledge of the instantaneous values of the angular displacement θ and the angular velocity ω. The treatment of the split-Hopkinson bar for dynamic torsion measurements follows closely that for dynamic measurements in compression and tension. For dynamic torsion measurements, the three strain histories are the incident loading pulse $\gamma_I(t)$, the transmitted pulse $\gamma_T(t)$, and the reflected pulse $\gamma_R(t)$ respectively, at any given instant of the time t of deformation. From these three strain histories and application of the one-dimensional elastic wave propagation theory it is possible to determine the angular displacement θ in the form

$$\theta = C_b \int_0^t \gamma \, dt' \tag{13-1-23}$$

and the angular velocity ω by differentiation of the displacement θ with respect to t. From Eq. (6-4-5),

$$c_T = \left(\frac{G_b}{\rho_b}\right)^{1/2} \equiv C_b \tag{13-1-24}$$

the shear stress being obtained from the relation

$$\tau \equiv T\langle \theta z \rangle = G_b \gamma \langle \theta z \rangle \equiv G_b \gamma \tag{13-1-25}$$

where G_b, ρ_b are the shear modulus of rigidity and the density respectively of the torsion bar material. It is important to note that, in torsion, dispersion of the elastic waves in the bars is absent for propagation in the first mode, and the effect of higher modes is very small (Campbell and Tsao, 1972).

For the particular condition that the stresses on both faces of the specimen are approximately equivalent (that is, $\tau_1 \approx \tau_2$), the instantaneous values of the three strain histories must satisfy the approximation

$$\gamma_I(t) + \gamma_R(t) \approx \gamma_T(t) \tag{13-1-26}$$

which can be verified experimentally. Application of Eq. (13-1-26) yields for thin-walled cylindrical specimens the relations

$$\tau(t) = G_b \frac{\bar{r}}{R_b} \frac{\bar{I}_b}{\bar{I}_s} \gamma_T(t) \tag{13-1-27}$$

$$\dot{\gamma}(t) = -2 \frac{\bar{r}}{R_b} \frac{C_b}{L} \gamma_R(t) \tag{13-1-28}$$

$$\gamma(t) = -2 \frac{\bar{r}}{R_b} \frac{C_b}{L} \int_0^t \gamma_R \, dt' \tag{13-1-29}$$

where $\bar{r} = (r_1 + r_2)/2$ is the mean radius of the tubular specimen, R_b the radius of the torsion bars, $\bar{I}_b$ and $\bar{I}_s$ the polar moments of area for the torsion bars and the tubular specimens respectively, and C_b the wave speed of Eq. (13-1-24). Equations (13-1-27), (13-1-28), and (13-1-29) are to be compared with Eqs. (13-1-19), (13-1-20), and (13-1-21) respectively.

Clyens and Johnson (1977) have used a torsional split-Hopkinson bar apparatus to determine the dynamic mechanical properties of particulate materials using Eqs. (13-1-27) and (13-1-29) to evaluate the shear stress and shear strain respectively. The powder is sandwiched between the flat faces of two bars mounted in the vertical. The extent to which the powder is compacted can be controlled by applying a constant axial compressive force during the period of dynamic loading. Clyens and Johnson have also shown that the torsional split-Hopkinson bar apparatus can, in principle, be used for determining the transient response of high-viscosity liquids. For this type of measurement, the flat face of the loading bar is replaced by a shallow-angled cone to yield, in effect, a cone and plate viscometer, the analysis proceeding in the way described in Secs. 7-3-4 and 9-3-3, the shear stress and rate of shear being evaluated from Eqs. (13-1-27) and (13-1-28) respectively.

(d) Expanding ring technique In this method the specimen is made in the form of a thin-walled tubular specimen to which a symmetric radial motion is imparted so as to give a uniform state of hoop tension. Wave propagation effects in the radial direction are minimized by using sufficiently thin-walled tubular specimens. The radial motion can be imparted by explosive techniques, such as those described by

Hogatt, Orr, and Recht (1967), or by electromagnetic techniques, such as that described by Niordsen (1965). Shock-wave techniques have also been used by Reed (1975) to produce a radial motion of the specimen.

13-1-3 Dynamic Stress-Strain Curves

Having evaluated the average specimen stress by way of Eq. (13-1-22) and the average specimen strain by way of Eq. (13-1-21) for various values of the time t of deformation, the characteristic stress-strain curves for the test material can be obtained by eliminating t as the parameter. A typical example of a dynamic stress-strain $(T_0(0), \epsilon(0))_c$ curve for aluminum tested in simple compression (Billington and Tate, 1972) is shown in Fig. 13-3 where $\epsilon(0)$ $(\equiv \epsilon\langle zz \rangle)$ is the total natural strain [see Eq. (10-5-15)]. Annealing of a material prior to dynamic test has been observed to alter the mechanical response curves in the same general way for various metals. The effect of annealing is observed as a decrease in the yield stress, the post-yield stress level rising rapidly with increasing strain, the stress-strain curve approaching the stress-strain curve of the material in the as-received state. In Fig. 13-4 is shown the effect on the ambient dynamic compression stress-strain curve of varying the time for which specimens of copper were maintained at a fixed temperature of 550°C. It is evident that at high strains the stress-strain curves of the annealed specimens approach closely that of the metal in the as-received state, but do not coincide within the available range of strain. From Fig. 13-4 it can be seen that there is a particular value of the strain $\epsilon(0) = \epsilon_H(0)$ at which there occurs a marked change in the stress-strain curve for the annealed metal. Above a

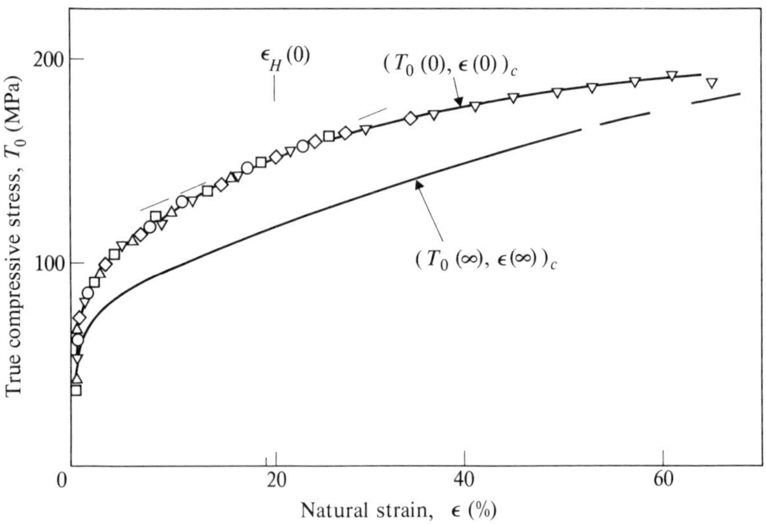

Figure 13-3 Static and dynamic stress-strain curves for aluminum showing the effect of varying the impact velocity V_I: □, 9.1 m s^{-1}; △, 13.0 m s^{-1}; ◇, 13.7 m s^{-1}; ○, 19.2 m s^{-1}; ▽, 22.8 m s^{-1}. (*Billington and Tate, 1972.*)

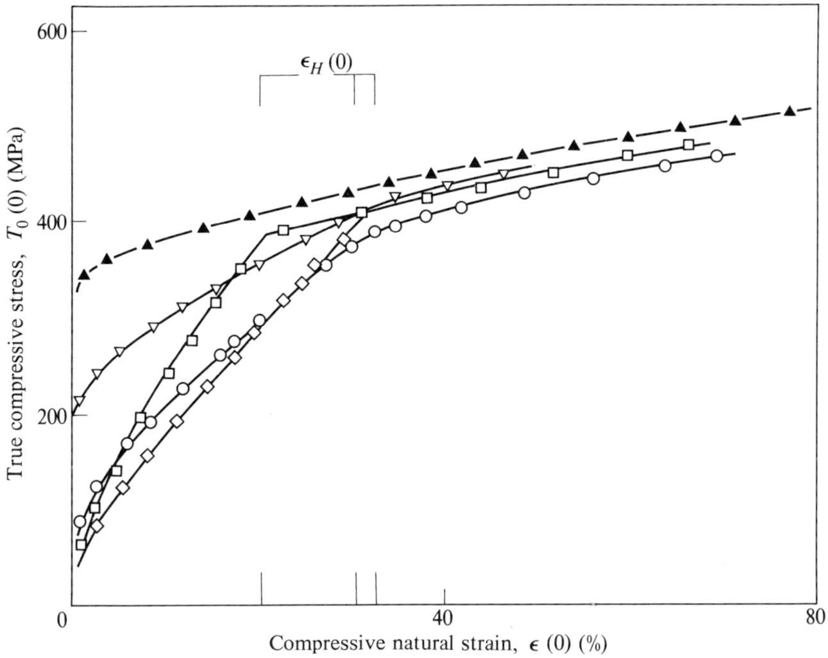

Figure 13-4 Dynamic stress-strain curves for copper tested in compression at ambient temperature showing the effect of varying the duration t_H of the heat treatment, keeping the temperature constant at 550°C: t_H, ▲, 0 h; ▽, 1 h; ○, 3.5 h; ◇, 10 h; □, 16.5 h. (*Billington, 1977.*)

particular value of the compressive engineering strain $\tilde{e}(0) = \tilde{e}_a(0)\,[\ll \tilde{e}_H(0)]$, the mechanical response of copper, and several other metals, considered in terms of the engineering strain, can be regarded as being approximately linear up to the critical value $\tilde{e}_H(0)$ of the compressive engineering strain. The way in which the limiting dynamic tangent modulus $E_{ap}(0)$ of this linear region and its intercept $T_0 = T_a$ at $\tilde{e} = 0$ vary with the time t_H for which the copper was maintained at a fixed temperature of 550°C is shown in Fig. 13-5 for copper tested in dynamic compression. It is evident from Figs. 13-4 and 13-5 that small changes in the time for which the metal is maintained at a fixed temperature have a marked effect on the mechanical response. Hence, it is essential to ensure that test specimens used to compare the mechanical response in the simple modes of deformation have all received identical treatment prior to test.

The method for obtaining the dynamic stress-strain curves for materials deformed in simple tension, in the way described by Lindholm and Yeakley (1968), parallels that for measurements in compression. Dynamic stress-strain curves for metals tested in simple tension, using thin-walled tubular specimens tested in the way described by Lindholm and Yeakley, are formally similar to the dynamic stress-strain curves for materials tested in simple compression.

For measurements in torsion, the average specimen shear stress is evaluated by way of Eq. (13-1-27) and the average specimen shear strain by way of Eq.

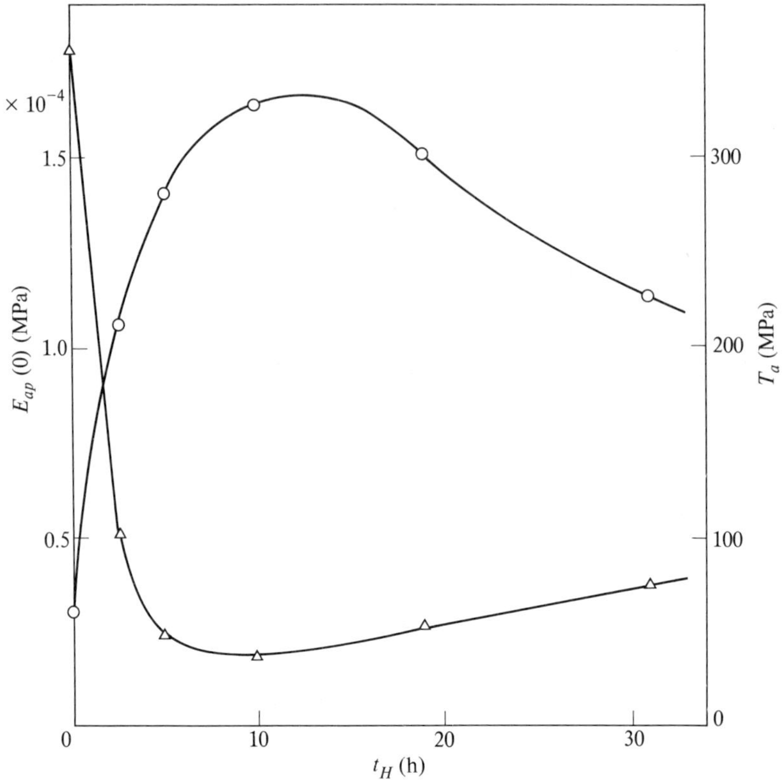

Figure 13-5 Variation of the material parameters $E_{ap}(0)$ and T_a for copper with the time of heating at a temperature of 550°C: ○, $E_{ap}(0)$; △, T_a. *(Billington, 1977.)*

(13-1-29) for various values of t. Characteristic dynamic stress-strain $(\tau(0), \gamma(0))$ curves are then obtained by eliminating t as the parameter. An example of a dynamic stress-strain $(\tau(0), \gamma(0))$ curve for aluminum tested in simple torsion (Billington, 1977) is shown in Fig. 13-6. Pukas (1976) has obtained markedly similar results for aluminum, titanium, and iron tested in simple torsion. Clyens and Johnson (1977) have used a torsional split-Hopkinson bar apparatus to determine the dynamic mechanical properties of particulate materials. A typical characteristic dynamic stress-strain $(\tau(0), \gamma(0))$ curve is shown in Fig. 13-7 for a copper powder having a uniform particle size of $-100 + 150$ Tyler mesh. Electron microscopic examination of the dynamically sheared particles revealed cold-welding. The jump-condition, shown in Fig. 13-7 as occurring at a dynamic strain of approximately 30 percent, is attributed to oxide being stripped from the individual particles and the observed occurrence of cold-welding.

It is of interest to note the marked formal similarity between the dynamic and static stress-strain curves shown in Figs. 13-3 and 13-6 and the shear stress-rate of strain hysteresis loops shown in Figs. 9-4 and 9-5 which are characteristic of the phenomenon of thixotropy.

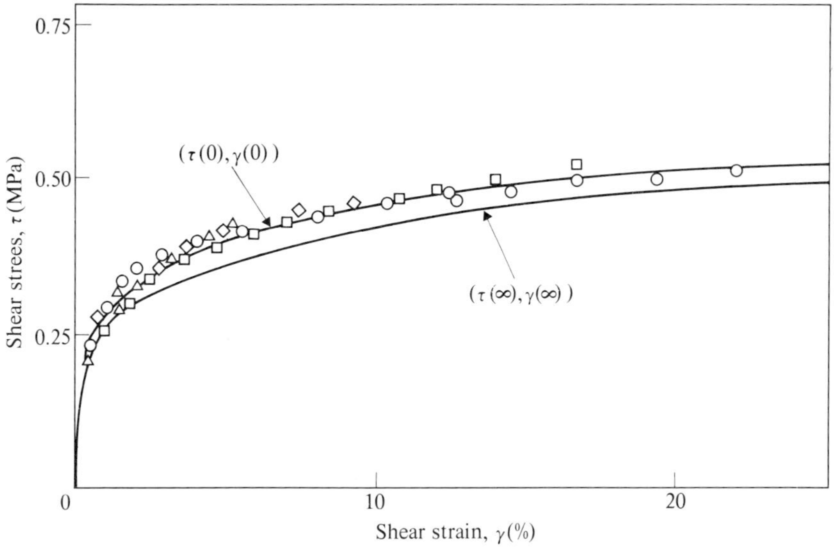

Figure 13-6 Dynamic and static stress-strain curves for aluminum in torsion. Maximum shear strain rate $\dot{\gamma}_{max}$ ($10^3\ s^{-1}$): ◇, 0.40; ○, 0.89; △, 0.95; □, 1.10. (*Billington, 1977.*)

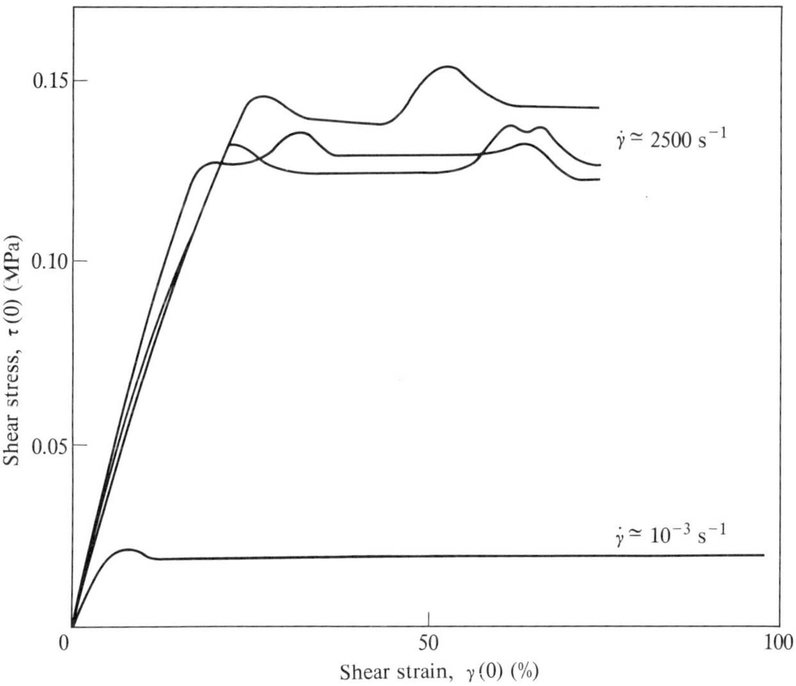

Figure 13-7 Dynamic stress-strain curve for a copper powder tested in simple torsion. (*Clyens and Johnson, 1977.*)

Also shown in Figs. 13-3, 13-6, and 13-7 are the corresponding static stress-strain $(T_0(\infty), \epsilon(\infty))_i$ $(i = C, T)$ and $(\tau(\infty), \gamma(\infty))$ curves respectively. It is evident for these metals, which are typical of many materials, that for all three simple modes of deformation there is a difference between the high strain-rate stress levels and the corresponding effectively static stress levels for the same value of the strain. It is to be noted that, although many materials exhibit this difference between the dynamic and static characteristic stress-strain curves, there are some materials for which there is no detectable difference between the dynamic and static stress-strain curves. Static measurements in compression and tension do not exhibit a marked change in the $(T_0(\infty), \epsilon(\infty))$ curves corresponding to the marked change in the dynamic $(T_0(0), \epsilon(0))$ curves at $\epsilon(0) = \epsilon_H(0)$. This observation of the static measurements applies to materials in both the as-received and annealed states. In this connection, dynamic torsion measurements do not exhibit a marked change in the characteristic torsion response curve within the limited range of available dynamic shear strain. However, the static torsion curves do exhibit a marked change in the characteristic torsion response curve at static shear strains of the order unity, which are of particular interest in relation to the dependence, or otherwise, of the Swift effect on work-hardening.

13-1-4 Strain-Rate Effects

For measurements with a split-Hopkinson bar type apparatus, Eq. (13-1-18) can be expressed in the form

$$\dot{\tilde{e}}(t) = \frac{\tilde{e}_I - \tilde{e}_T(t)}{\tau_b} \qquad (\tau_b = L/2c_b) \tag{13-1-30}$$

where in Eq. (13-1-18) the relation for $\dot{\tilde{e}}$ given in Eq. (13-1-20) has been substituted for $\tilde{e}_R(t)$. It is evident that no single one of the $(\tilde{e}_T, t)$ curves taken in isolation can provide any information with regard to material strain-rate dependence. For a given pair of Hopkinson bars, there is an upper limit to the range of stress and strain which can be obtained with this type of apparatus, this upper level being determined by the proportional limit of the bar material and the undeformed

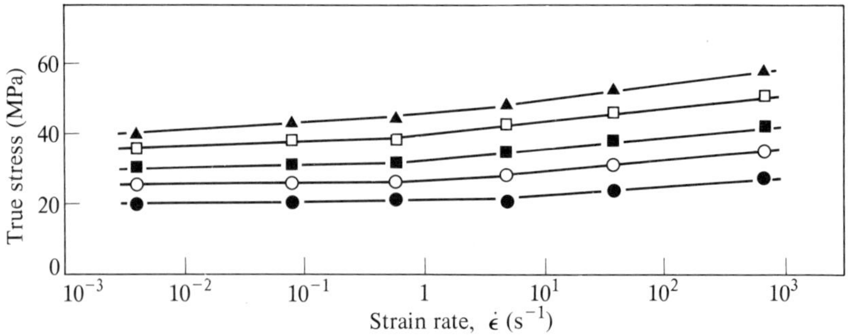

Figure 13-8 Strain-rate effects in 99.999% purity aluminum, ϵ: ●, 2%; ○, 3%; ■, 4%; □, 6%; ▲, 8%. (*Holt, Babcock, Green, and Maiden, 1967.*)

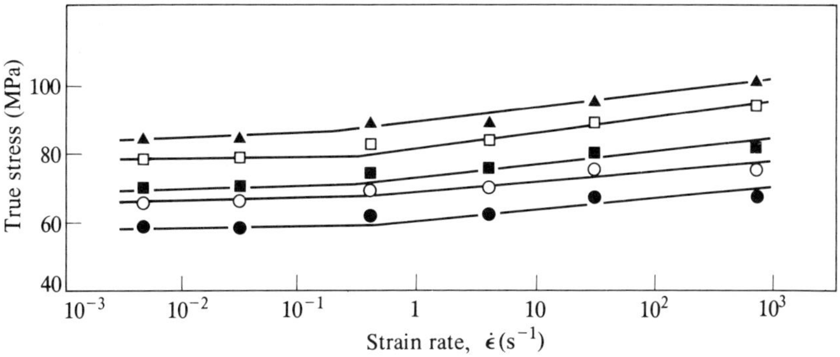

Figure 13-9 Strain-rate effects in some aluminum alloys, ϵ: ●, 2%; ○, 3%; ■, 4%; □, 6%; ▲, 8%. (*Holt, Babcock, Green, and Maiden, 1967.*)

length L of the specimen. The upper level of the strain rate, that is, $\dot{\tilde{e}}_M$, is determined by setting $\tilde{e}_T = 0$ in Eq. (13-1-30) to give $\dot{\tilde{e}}_M = \tilde{e}_{IM}/\tau_b$ and the maximum value of $\tilde{e}_T$ is determined by setting $\dot{\tilde{e}} = 0$ to yield $\tilde{e}_{TM} = \tilde{e}_{IM}$ where $\tilde{e}_{IM}$ is the maximum permitted value of the loading pulse as determined by the proportional limit. For a rectangular pulse, $\tilde{e}_I$ is constant for all t and a straight line can be constructed with the intercepts $\dot{\tilde{e}}_M$ and $\tilde{e}_{IM}$, this maximum loading line having slope $1/\tau_b$ which increases with decreasing L. From Eq. (13-1-30) it is evident that the smallest value of the strain rate $\dot{\tilde{e}}$ which can be resolved is directly proportional to the difference $(\tilde{e}_I - \tilde{e}_T)$. This discussion for measurements in compression also applies to measurements in tension and torsion.

It is evident from the stress-strain curves for the materials shown in Figs. 13-3 and 13-6 that these materials exhibit a dependence of the flow stress on strain rate for strain rates less than some critical value particular to a given material, the observed behavior being in general accord with the results reviewed by Campbell (1973). With increasing strain rate the flow stress, for a given strain, appears to approach a saturation value such that for a limited range of high strain rates the individual dynamic stress-strain curves, one for each value of the loading pulse, tend to compound together in all three modes of deformation. Although the stress-strain curves shown in Figs. 13-3 and 13-6 are typical for a wide range of materials, questions relating to the strain-rate sensitivity of the mechanical response of a material, which are of fundamental importance, cannot be regarded as resolved. This is so even for materials such as aluminum and copper in a high state of purity. For example, Holt, Babcock, Green, and Maiden (1967) find marked strain-rate dependence in 99.999 percent purity aluminum, as is evident from the results shown in Fig. 13-8, in contrast to Bell (1966) who states that aluminum and many other metals are all strain-rate independent at medium (10^{-1} to $10^2\ s^{-1}$) and high (10^2 to $10^4\ s^{-1}$) strain rates. Some aluminum alloys have been shown by Holt, Babcock, Green, and Maiden (1967) to be strain-rate dependent over a wide range of strain rates (*ca.* 10^{-3} to *ca.* $10^3\ s^{-1}$) while others are shown by these authors to be strain-rate independent at very low strain rates, becoming strain-rate dependent at high strain rates, these results being shown in Fig. 13-9. In contrast, Maiden and

Green (1966) and Lindholm and Yeakley (1968) have observed various aluminum alloys to be strain-rate independent over the range *ca.* 10^{-3} to *ca.* $10^3\ s^{-1}$. Marked dependence of the flow stress on strain rate has been observed for titanium by Maiden and Green (1966), this dependence on strain and strain rate being shown

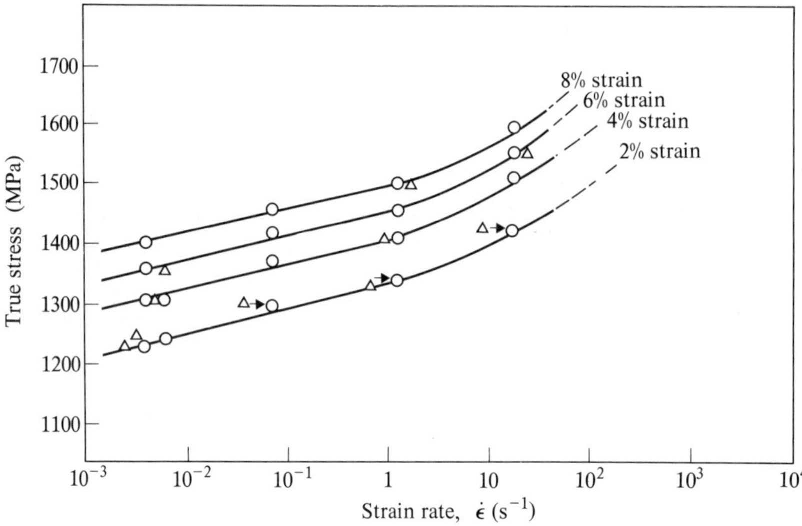

Figure 13-10 Strain-rate dependence of the flow stress of titanium: ○, average rate; △, instantaneous rate. (*Maiden and Green, 1966.*)

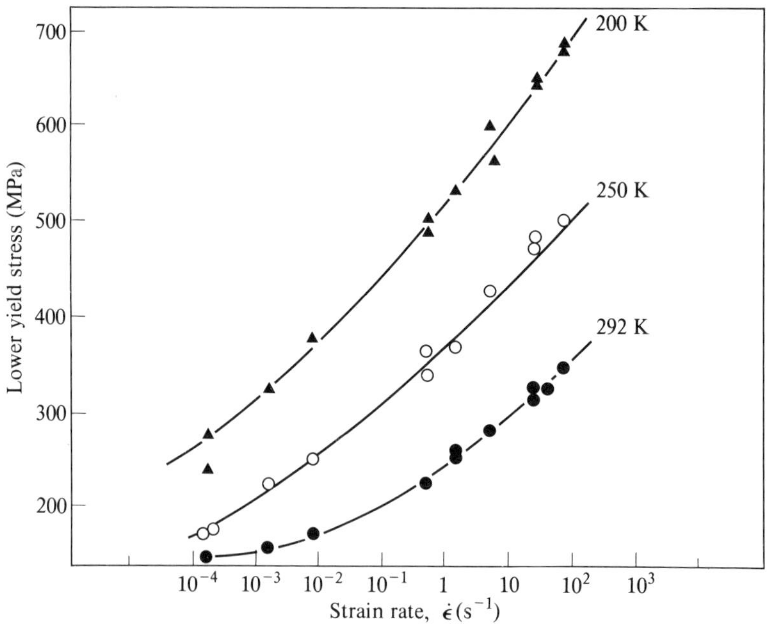

Figure 13-11 Temperature and strain-rate dependence of the lower yield stress of niobium. (*Briggs and Campbell, 1972.*)

in Fig. 13-10. The results obtained by Briggs and Campbell (1972) for the temperature and strain-rate dependence of the lower yield stress of niobium and of molybdenum are shown in Figs. 13-11 and 13-12 respectively. Shown in Fig. 13-13 are the results of Campbell and Harding (1961) showing the effect of grain size on the strain-rate dependence of the lower yield stress of high-purity iron.

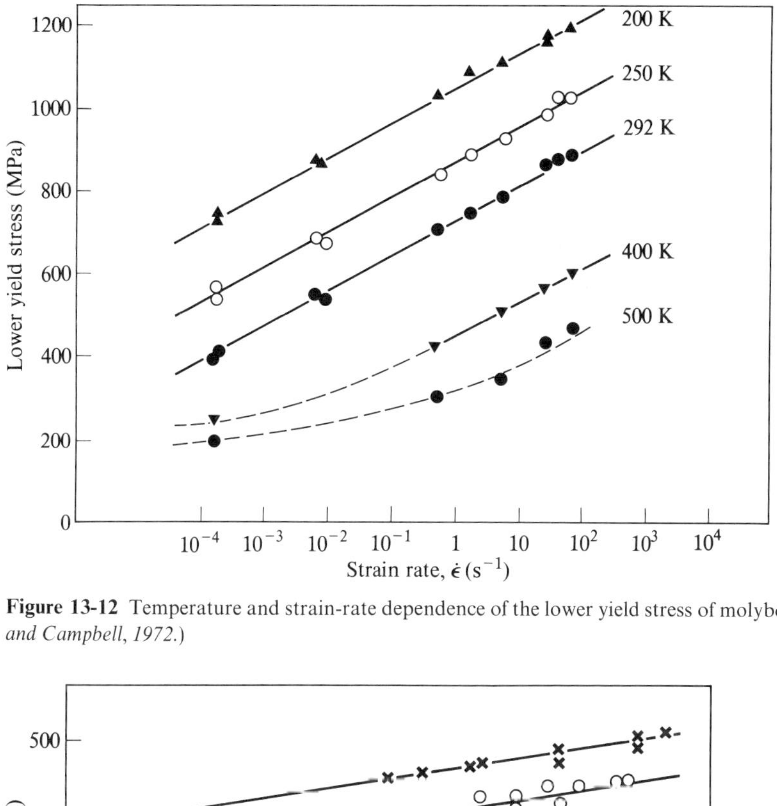

Figure 13-12 Temperature and strain-rate dependence of the lower yield stress of molybdenum. (*Briggs and Campbell, 1972.*)

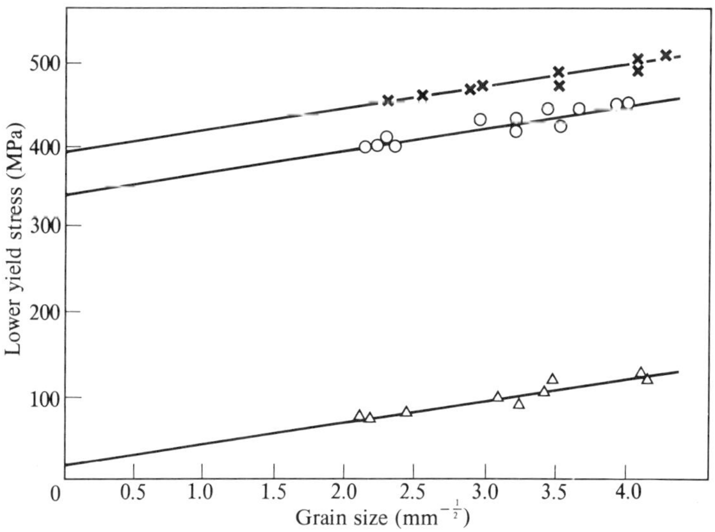

Figure 13-13 Grain-size and strain-rate dependence of the lower yield stress of high-purity iron. Mean stress rates: △, $10^{-3}\,s^{-1}$; ○, $9.6 \times 10^{-2}\,s^{-1}$; ×, $2.6 \times 10^{3}\,s^{-1}$. (*Campbell and Harding, 1961.*)

These results serve to illustrate the complexity of questions relating to strain-rate effects in solids. In particular, it does not at present appear possible to develop macroscopic constitutive equations having a wide degree of generality (see, for example, Cristescu, 1967). This aspect tends to preclude the possibility of developing macroscopic constitutive equations directly from a consideration of micromechanisms. In this connection the reader is referred to the discussion by Klahn, Mukherjee, and Dorn (1970).

13-2 SHOCK WAVES

13-2-1 One-Dimensional Normal Shock-Wave Relations

The study of shock waves in materials is a topic of great and growing interest as it is the simplest way of observing matter when subject to conditions of high pressure, temperature, and density, and is thus the most suitable tool for many high-energy density experiments in physics and chemistry.

For strong waves the shock transition between the values of the material state variables ahead and behind the shock wave usually occurs in a very short distance, of the order of several atomic spacings. On a macroscopic scale it thus can be regarded as a discontinuous change and the change of values across the shock wave can be derived on the assumption that the flow has effectively a one-dimensional character. Similarly, the time for particles to cross the shock front is small compared with most macroscopic times and the change in state variables across the shock can be estimated, assuming steady flow.

Consider, then, a plane shock wave moving at a constant velocity U_s and take axes moving with the shock wave as shown in Fig. 13-14. Let conditions upstream of the shock wave be denoted by the suffix 0 and conditions downstream be

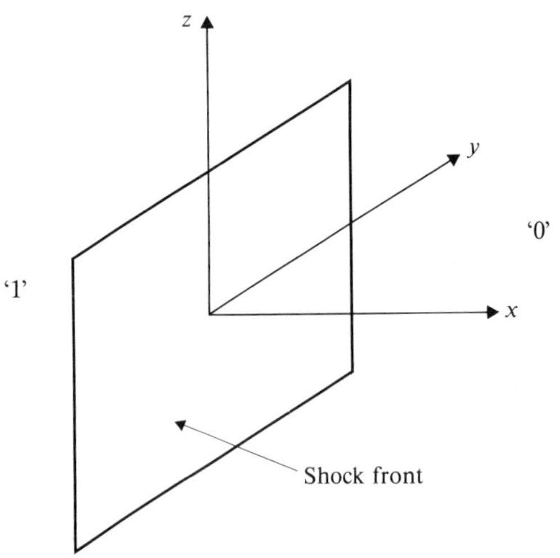

Figure 13-14 Disposition of coordinate axes and the shock front.

denoted by the suffix 1. Let the velocity components in the directions x, y, and z be u, v, and w respectively. In this frame of reference the flow is steady and it will be assumed that as $x \to \pm\infty$ the gradients of all state variables will tend to zero. Under these conditions the conservation equations of Chap. 3 take the following simple form:

Mass:

$$\frac{\partial \rho u}{\partial x} = 0 \qquad \rho u = \text{constant} = m, \text{ say} \tag{13-2-1}$$

Momentum:

$$\frac{\partial}{\partial x}(mu - T_{xx}) = 0 \qquad \rho u^2 - T_{xx} = \text{constant} \tag{13-2-2}$$

$$\frac{\partial}{\partial x}(mv - T_{xy}) = 0 \qquad \rho uv - T_{xy} = \text{constant} \quad \text{(body forces have been ignored)} \tag{13-2-3}$$

$$\frac{\partial}{\partial x}(mw - T_{xz}) = 0 \qquad \rho uw - T_{xz} = \text{constant} \tag{13-2-4}$$

Energy:

$$\frac{\partial}{\partial x}[m\varepsilon + \tfrac{1}{2}m(u^2 + v^2 + w^2) - (uT_{xx} + vT_{xy} + wT_{xz})] = \frac{\partial q_x}{\partial x} + \rho r \tag{13-2-5}$$

Therefore,

$$\varepsilon + \tfrac{1}{2}(u^2 + v^2 + w^2) - \frac{1}{\rho}\left(T_{xx} + \frac{v}{u}T_{xy} + \frac{w}{u}T_{xz}\right) - q_x - \frac{1}{m}\int \rho r\, dx = \text{constant}$$

It is convenient to refer the velocities to a stationary axis rather than one moving with the shock speed. This merely entails replacing u by $u - U_s$, and if the upstream and downstream conditions are taken sufficiently far away from the shock front that the gradients of state variables are zero then, as the thermal conductivity q_x depends on the temperature gradient, this becomes zero. The general form of the shock-wave equations thus become

$$\rho_1(u_1 - U_s) - \rho_0(u_0 - U_s) = m, \text{ say} \tag{13-2-6}$$

$$m(u_1 - u_0) = T_{xx_1} - T_{xx_0} \tag{13-2-7}$$

$$m(v_1 - v_0) = T_{xy_1} - T_{xy_0} \tag{13-2-8}$$

$$m(w_1 - w_0) = T_{xz_1} - T_{xz_0} \tag{13-2-9}$$

$$m(\varepsilon_1 - \varepsilon_0) = \tfrac{1}{2}[(u_1 - u_0)(T_{xx_1} + T_{xx_0}) + (v_1 - v_0)(T_{xy_1} + T_{xy_0}) + (w_1 - w_0)(T_{xz_1} + T_{xz_0})] + \Delta r \tag{13-2-10}$$

where Δr is the total rate of energy input per unit area of the shock front between the initial and final states. In many cases the shear stresses in the initial and final states are zero, for instance, in a viscous liquid the shear stress depends on the velocity gradient and these gradients vanish in the initial and final states. When the initial and final shear stress is zero, Eqs. (13-2-8) and (13-2-9) show that the transverse velocity components do not change on crossing the shock front and the shock equations take the simplified form

$$\rho_1(u_1 - U_s) = \rho_0(u_0 - U_s) = m \tag{13-2-11}$$

$$m(u_1 - u_0) = T_{xx_1} - T_{xx_0} \qquad v_1 = v_0 \qquad w_1 = w_0 \tag{13-2-12}$$

$$\varepsilon_1 - \varepsilon_0 = \frac{1}{2}\left(\frac{1}{\rho_1} - \frac{1}{\rho_0}\right)(T_{xx_1} + T_{xx_0}) + \frac{\Delta r}{m} \tag{13-2-13}$$

A further simplification can be made if the stress is assumed to be isotropic, in which case it is usual to express the stress in terms of the pressure and the equations take the standard form

$$\rho_1(U_s - u_1) = \rho_0(U_s - u_0) \tag{13-2-14}$$

$$(U_s - u_0)^2 = \frac{\rho_1(P_1 - P_0)}{\rho_0(\rho_1 - \rho_0)} \tag{13-2-15}$$

$$\varepsilon_1 - \varepsilon_0 = \frac{1}{2}\left(\frac{1}{\rho_0} - \frac{1}{\rho_1}\right)(P_1 + P_0) + \frac{\Delta r}{m} \tag{13-2-16}$$

where use has been made of the relation $T_{xx} = -P$. Usually the absorption of external energy such as radiation is small and occurs over times which are long compared with the time it takes a material particle to cross the shock wave; consequently the term Δr can be neglected. For this case Eq. (13-2-16) is called the "Rankine–Hugoniot" or simply the "Hugoniot" relation. It is known from thermodynamics that the state of the above system is defined by only two thermodynamic quantities, say ε and ρ, the pressure being related to these through the equation of state. In the P, ρ, ε space the equation of state defines a surface while Eq. (13-2-16) also defines a surface through the point $(P_0, \rho_0, \varepsilon_0)$. These two surfaces intersect in a line which is called the Hugoniot line and this connects the points on the equation of state surface which can be reached from the initial state through one shock wave. Experiments on the compression of solids by strong shock waves indicate that the measured P, ρ, ε states are often states of thermodynamic equilibrium (see Rice et al., 1958). The positive and negative roots of Eq. (13-2-15) correspond to forward- and backward-facing shocks.

In most shock-wave work the specific volume V $(=1/\rho)$ is used in preference to the density, and with slight rearrangement the shock-wave relations take the form

Forward-facing shocks:

$$u_1 - u_0 = [(P_1 - P_0)(V_0 - V_1)]^{1/2} \tag{13-2-17}$$

$$U_s - u_0 = V_0\left(\frac{P_1 - P_0}{V_0 - V_1}\right)^{1/2} \tag{13-2-18}$$

Rearward-facing shocks:

$$u_1 - u_0 = -[(P_1 - P_0)(V_0 - V_1)]^{1/2} \tag{13-2-19}$$

$$U_s - u_0 = -V_0\left(\frac{P_1 - P_0}{V_0 - V_1}\right)^{1/2} \tag{13-2-20}$$

For both types of wave

$$\varepsilon_1 - \varepsilon_0 = \tfrac{1}{2}(V_0 - V_1)(P_1 + P_0) \tag{13-2-21}$$

13-2-2 Weak Shock Waves

For a very weak wave propagating into a material at rest, Eqs. (13-2-17), (13-2-18), and (13-2-19) take the forms

$$U_s^2 = -V^2\frac{dP}{dV} = \frac{dP}{d\rho} = c^2 \tag{13-2-22}$$

$$du = \frac{dP}{\rho c} \tag{13-2-23}$$

$$d\varepsilon = -P\,dV \tag{13-2-24}$$

where c is the thermodynamic speed of sound and Eq. (13-2-24) is the thermodynamic identity showing that sonic processes are isentropic and thus adiabatic.

Following Bethe (1942), shock waves of finite but still small amplitude will now be considered. Let the difference between final and initial values of a variable such as ε be denoted by $\Delta\varepsilon$. Then, choosing volume and entropy S as independent thermodynamic variables and expanding $\Delta\varepsilon$ as a Taylor series

$$\Delta\varepsilon = \left(\frac{\partial\varepsilon}{\partial V}\right)_S\Delta V + \tfrac{1}{2}\left(\frac{\partial^2\varepsilon}{\partial V^2}\right)_S\Delta V^2 + \tfrac{1}{6}\left(\frac{\partial^3\varepsilon}{\partial V^3}\right)_S\Delta V^3 + \cdots + \left(\frac{\partial\varepsilon}{\partial S}\right)_V\Delta S\cdots \tag{13-2-25}$$

where all derivatives are evaluated relative to the initial conditions. Using the thermodynamic relations

$$\left(\frac{\partial\varepsilon}{\partial V}\right)_S = -P \tag{13-2-26}$$

and

$$\left(\frac{\partial\varepsilon}{\partial S}\right)_V = \theta \tag{13-2-27}$$

$$-\frac{\Delta\varepsilon}{\Delta V} = P + \tfrac{1}{2}\left(\frac{\partial P}{\partial V}\right)_S\Delta V + \tfrac{1}{6}\left(\frac{\partial^2 P}{\partial V^2}\right)_S\Delta V^2 + \cdots - \theta\frac{\Delta S}{\Delta V} \tag{13-2-28}$$

Similarly,

$$\tfrac{1}{2}(P_1 + P_0) = P + \tfrac{1}{2}\left(\frac{\partial P}{\partial V}\right)_S\Delta V + \tfrac{1}{4}\left(\frac{\partial^2 P}{\partial V^2}\right)_S\Delta V^2 + \cdots + \tfrac{1}{2}\left(\frac{\partial P}{\partial S}\right)_V\Delta S + \cdots \tag{13-2-29}$$

The last term in Eq. (13-2-29) is usually much less than the last term in Eq. (13-2-28) and comparing Eqs. (13-2-21), (13-2-28), and (13-2-29) gives

$$\Delta S = S_1 - S_0 = -\left(\frac{\partial^2 P}{\partial V^2}\right)_S \frac{\Delta V^3}{12\theta} \tag{13-2-30}$$

In general $(\partial^2 P/\partial V^2)_S$ is positive and, as the wave is thermodynamically stable only if $\Delta S > 0$, this implies $\Delta V < 0$ and the wave must be compressive. Expansive or rarefaction waves imply a negative entropy and therefore are thermodynamically unstable and break up into a train of infinitesimal waves. At phase boundaries $(\partial^2 P/\partial V^2)_S$ may not be positive or $(\partial P/\partial V)_S$ may be discontinuous and at such places rarefaction shocks might exist.

To find the speed of the shock-wave relative to the speed of propagation of sound waves in the material on either side let ΔP be expanded in the two forms

$$\Delta P = P_1 - P_0 = P(V_0 + \Delta V) - P(V_0) = \left(\frac{\partial P}{\partial V}\right)_{S_0} \Delta V + \tfrac{1}{2}\left(\frac{\partial^2 P}{\partial V^2}\right)_{S_0} \Delta V^2 + \cdots \tag{13-2-31}$$

and

$$\Delta P = P(V_1) - P(V_1 - \Delta V) = \left(\frac{\partial P}{\partial V}\right)_{S_1} \Delta V - \tfrac{1}{2}\left(\frac{\partial^2 P}{\partial V^2}\right)_{S_1} \Delta V^2 + \cdots \tag{13-2-32}$$

where the suffixes denote that the derivatives are evaluated at either conditions 0 or 1. Now Eq. (13-2-18) can be written in the form

$$U_S^2 = -V_0^2 \frac{\Delta P}{\Delta V} = -V_0^2 \left(\frac{\partial P}{\partial V}\right)_{S_0} - \tfrac{1}{2} V_0^2 \left(\frac{\partial^2 P}{\partial V^2}\right)_{S_0} \Delta V \tag{13-2-33}$$

where use has been made of Eq. (13-2-31) and u_0 has been taken to be zero. Using Eq. (13-2-22), this can be written

$$U_S^2 = c_0^2 - \tfrac{1}{2} V_0^2 \left(\frac{\partial^2 P}{\partial V^2}\right)_{S_0} \Delta V \tag{13-2-34}$$

If $(\partial^2 P/\partial V^2)_S > 0$ then ΔV is negative and thus the shock moves faster and overtakes sound waves moving ahead of it in the same direction. Multiplying Eq. (13-2-32) by $-V_1^2$ and noting $c_1^2 = -V_1^2(\partial P/\partial V)_{S_1}$,

$$-V_1^2 \left(\frac{\partial P}{\partial V}\right)_{S_1} = c_1^2 + \tfrac{1}{2} V_1^2 \left(\frac{\partial^2 P}{\partial V^2}\right)_{S_1} \Delta V \tag{13-2-35}$$

Using Eqs. (13-2-18) and (13-2-14),

$$(U_S - u_1)^2 = c_1^2 + \tfrac{1}{2} V_1^2 \left(\frac{\partial^2 P}{\partial V^2}\right)_{S_1} \Delta V \tag{13-2-36}$$

For a compressive wave in which $(\partial^2 P/\partial V^2)_{S_1} > 0$,

$$U_S < u_1 + c_1 \tag{13-2-37}$$

Thus sound waves move faster than a shock wave traveling ahead in the same direction and will eventually overtake it. By considering infinitesimal waves it is clear that a train of infinitesimal compression waves moving in the same direction will tend to move together and thus tend to make the compression pulse more steep fronted until it forms a shock wave. By reversing the sign of ΔV and thus, in general, the sign of ΔP, Eq. (13-2-36) shows that a train of rarefaction waves will gradually spread out and so, in general, no rarefaction shocks can form.

13-2-3 Shock-Wave Interactions

Once the equation of state is known, Eqs. (13-2-19) and (13-2-21) can be used to relate P_1 and u_1 to P_0 and u_0 and so construct a curve of states accessible from passing through one shock wave as shown in Fig. 13-15: $\mathscr{S}^+$ represents a forward propagating shock wave and $\mathscr{S}^-$ a rearward-facing wave. Consider now two waves A, B approaching one another as shown in the distance-time diagram of Fig. 13-16. They collide at the point α corresponding to distance x_α and time t_α and, within each region 0, 1, 2, and 3, P and u must be constant. After the collision two new shocks C and D are formed and to find their strength the P–u diagram of Fig. 13-17 can be used. The vertices of the quadrilateral correspond to pressure and velocity in the corresponding region and the lines a, b, c, d correspond to the change across the corresponding shock wave. The intersection corresponding to the forward- and rearward-facing curves through the states 2 and 1 respectively defines the intermediate state 3 and also fixes the shocks D and C by the lines d and c.

A problem of considerable interest is the reflection and transmission of an incident shock wave at the boundary of two different materials. The distance-time diagram of such an event is as shown in Fig. 13-18. To find the conditions across

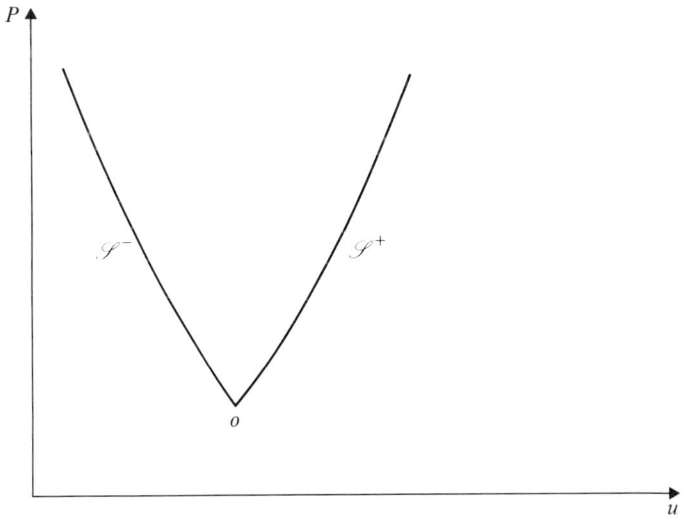

Figure 13-15 Pressure and material velocities attainable on passing through one forward S^+ or backward S^- facing shock.

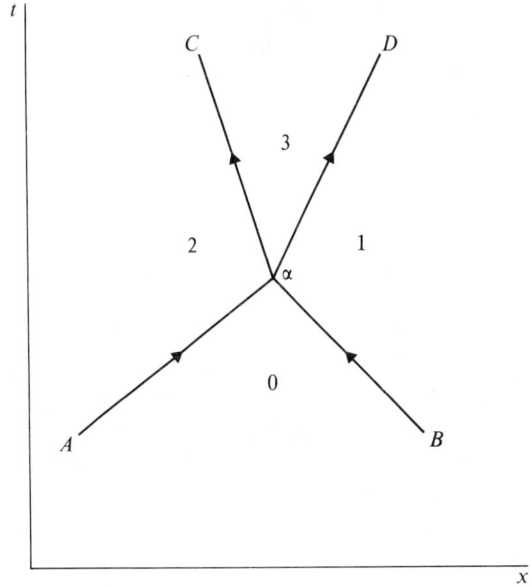

Figure 13-16 Distance-time diagram of two colliding shock waves.

the reflected and transmitted shocks refer to the P–u diagram of Fig. 13-19. The conditions of the incident shock must lie in the forward-facing shock line for material a through the point o corresponding to the initial conditions. Similarly, the conditions for the transmitted shock must lie on the forward-facing shock line for material b and, as the pressure and material velocity must be continuous across

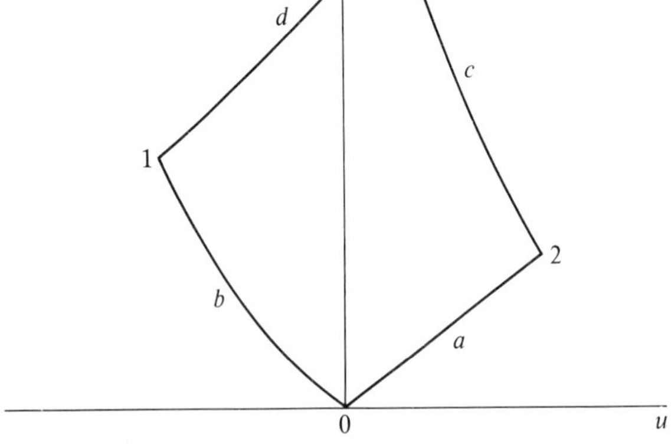

Figure 13-17 Pressure material–velocity diagram corresponding to the wave diagram for two colliding shock waves.

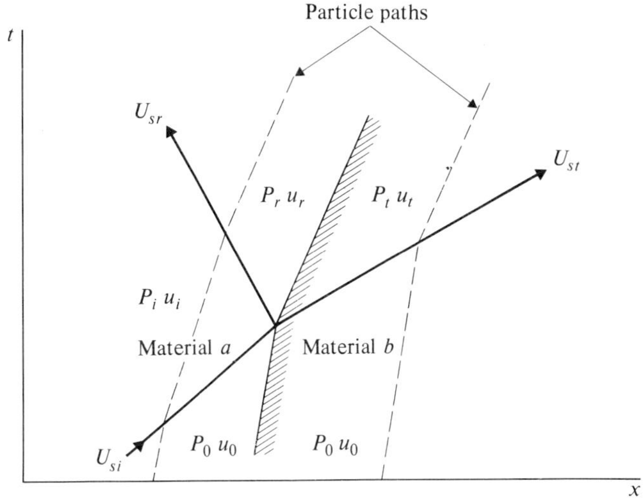

Figure 13-18 Distance-time wave diagram of the reflection of a shock wave at the boundary between two materials.

the interface, the initial point must correspond to *o*. The reflected shock must now lie on the rearward-facing shock line through *i* for material *a* and as the pressure and material velocity are again continuous across the interface the conditions are represented by the intersection of the lines *or* and *it*. As was noted in Sec. 13-2-2 a train of infinitesimal rarefaction waves tend to spread out giving a distance-time diagram as shown in Fig. 13-20; from its appearance this is generally referred to as a rarefaction fan. From Eqs. (13-2-22), (13-2-23), and (13-2-24) a unique line can be constructed in the P–u plane corresponding to forward- and rearward-facing rarefaction fans as shown in Fig. 13-21. By using combinations of shocks and rarefaction fans, all one-dimensional dynamic problems can be dealt with. For instance, a shock wave striking the free surface of a material reflects as an

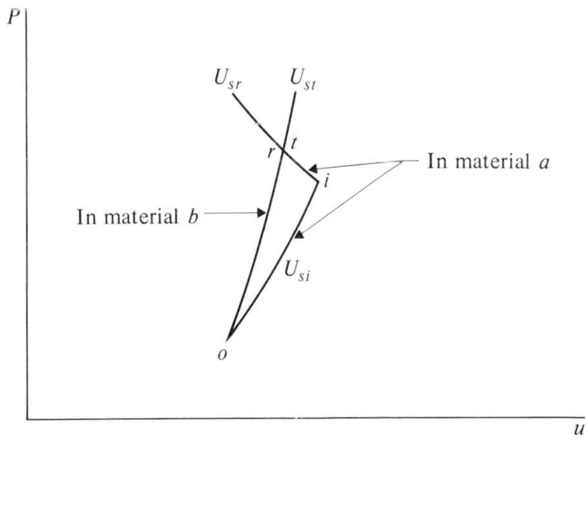

Figure 13-19 Pressure material–velocity curves for reflection of a shock wave at the boundary between two materials.

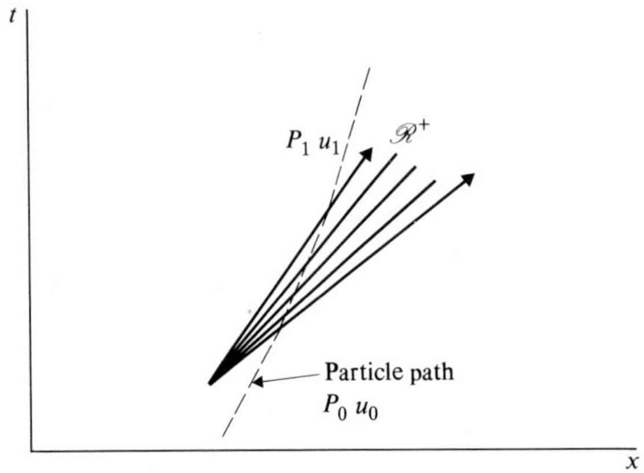

Figure 13-20 Distance-time diagram of a train of rarefaction waves forming a rarefaction fan.

expansion fan, as shown in Fig. 13-22, and has an associated P–u diagram as shown in Fig. 13-23. Usually P_0 is much less than P_1 in this case and has been set to zero.

It was shown in Eq. (13-2-30) that the entropy increase was proportional to ΔV^3 and thus, provided $(V_0 - V_1)/V_0$ is less than about 15 percent, the waves can be treated as isentropic and the shock wave and expansion fans lie on the same curves in the P–u plane and their mirror images. In these circumstances it is clear from Fig. 13-23 that the free surface velocity after a weak shock wave has reflected as an expansion fan is twice the material velocity behind the incident shock wave. For a weak wave reflecting from the interface between two materials a and b the P–u diagram has the form shown in Fig. 13-19. From Eq. (13-2-23),

$$P_i = \rho_a c_a u_i \tag{13-2-38}$$

$$P_r = \rho_b c_b u_r \tag{13-2-39}$$

$$P_r - P_i = \rho_a c_a (u_i - u_r) \tag{13-2-40}$$

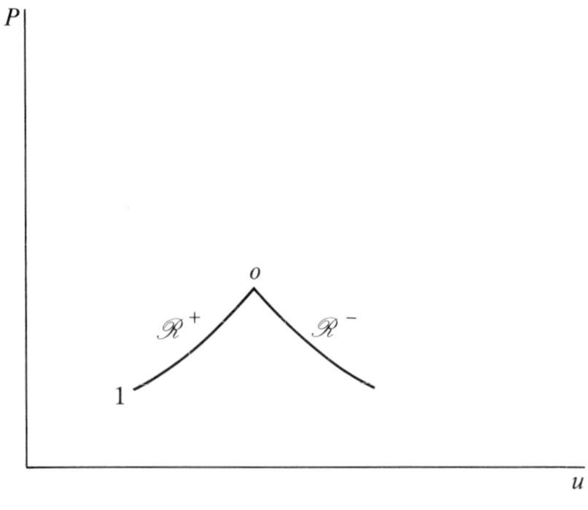

Figure 13-21 Pressure material–velocity curves representing forward $\mathscr{R}^+$ and rearward $\mathscr{R}^-$ facing rarefaction fans.

Eliminating u_i and u_r gives

$$\frac{P_r}{P_i} = \frac{2\rho_b c_b}{\rho_a c_a + \rho_b c_b} \tag{13-2-41}$$

The terms $\rho_a c_a$ and $\rho_b c_b$ are called the acoustic impedances of the materials a and b respectively.

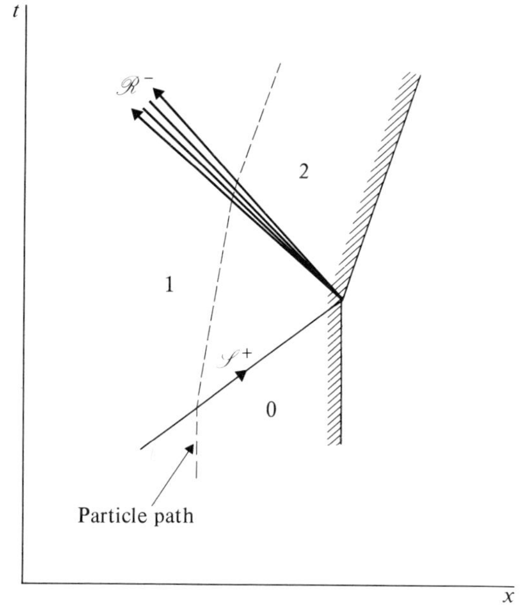

Figure 13-22 Distance-time diagram of the reflection of a shock wave from a free surface.

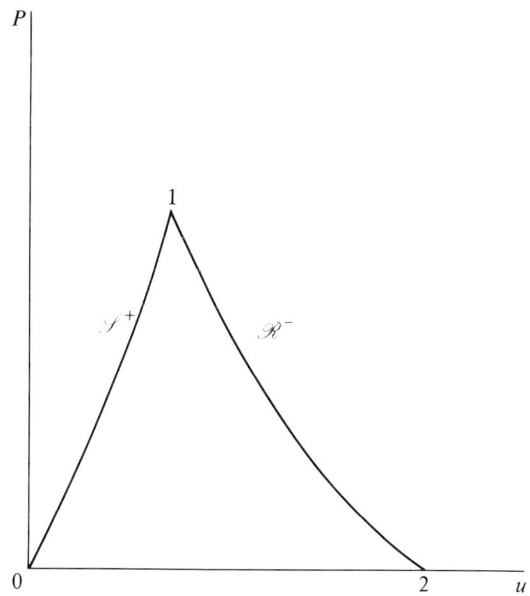

Figure 13-23 Pressure material–velocity curves for the reflection of a shock wave from a free surface.

13-2-4 Stability of Waves, Rarefaction Shocks, and Two-Wave Structures

In order that shock waves form at all it is necessary that a train of infinitesimal compressive or rarefaction waves moving in the same direction should tend to overtake one another and by running together form a steeper wave front whose gradient is limited by dissipative processes such as viscosity. It was shown in Sec. 13-2-2 that sound waves propagate isentropically and hence adiabatically. Considering first compressive waves, they will tend to run together provided

$$\left(\frac{\partial[u+c]}{\partial\rho}\right)_S > 0 \tag{13-2-42}$$

Using Eqs. (13-2-22) and (13-2-23) this can be written in the form

$$\frac{1}{\rho c}\left(\frac{\partial P}{\partial\rho}\right)_S + \frac{1}{2c}\left(\frac{\partial^2 P}{\partial\rho^2}\right)_S > 0 \tag{13-2-43}$$

or, equivalently,

$$\tfrac{1}{2}V^4\left(\frac{\partial^2 P}{\partial V^2}\right)_S > 0 \tag{13-2-44}$$

Thus

$$\left(\frac{\partial^2 P}{\partial V^2}\right)_S > 0 \tag{13-2-45}$$

Now the isentropes form a family of curves covering the P–V plane. Within those regions of the P–V plane where Eq. (13-2-45) is satisfied compressive trains of waves will tend to form shocks. In a similar manner it can be shown that in those regions of the P–V plane where $(\partial^2 P/\partial V^2)_S < 0$ trains of rarefaction waves will tend to run together to form rarefaction shocks. It is also clear from Eq. (13-2-42) that if $(\partial^2 P/\partial V^2)_S > 0$ successive rarefaction waves will travel slower than the preceding one, giving rise to a rarefaction fan, while if $(\partial^2 P/\partial V^2)_S < 0$ compressive waves will form a compression fan.

When waves run together to form shocks the stable shock shape is governed by dissipative processes at the shock front which generate entropy increases and thus, although the shock Hugoniot and adiabat through the initial (P_0, V_0) point have a second-order contact, the Hugoniot deviates from the adiabat at higher shock strengths. In order that only *one* shock is necessary to relate an initial state *o* to a final state 1 each infinitesimal wave behind the shock wave must be able to catch it up; thus

$$u_1 + c_1 > U_S \tag{13-2-46}$$

or, using Eqs. (13-2-17) and (13-2-18),

$$-V_1^2\left(\frac{\partial P}{\partial V}\right)_{S_1} > V_1^2\frac{P_1 - P_0}{V_0 - V_1} \tag{13-2-47}$$

The condition that only one shock should connect the initial and final states is thus

$$-\left(\frac{\partial P}{\partial V}\right)_{S_1} > \frac{P_1 - P_0}{V_0 - V_1} \tag{13-2-48}$$

In general ε can be expressed as a single-valued function of P and V and differentiating Eq. (13-2-21) it is found that the gradient of the Hugoniot $-(\partial P/\partial V)_H$ is

$$-\left(\frac{\partial P}{\partial V}\right)_H = \frac{\varepsilon_V + P - \frac{1}{2}(P - P_0)}{\varepsilon_p - \frac{1}{2}(V_0 - V_1)} \tag{13-2-49}$$

where

$$\varepsilon_V = \left(\frac{\partial \varepsilon}{\partial V}\right)_P = -P - \theta\left[\frac{(\partial P/\partial V)_S}{(\partial P/\partial S)_V}\right] \tag{13-2-50}$$

and

$$\varepsilon_p = \left(\frac{\partial \varepsilon}{\partial P}\right)_V = \frac{\theta}{(\partial P/\partial S)_V} = \frac{c_V}{(\partial P/\partial \theta)_V} > 0 \tag{13-2-51}$$

Similarly, from Eqs. (13-2-50) and (13-2-51),

$$-\left(\frac{\partial P}{\partial V}\right)_S = \frac{\varepsilon_V + P}{\varepsilon_p} \tag{13-2-52}$$

Using Eqs. (13-2-48), (13-2-49), and (13-2-52), it can be shown that

$$-\left(\frac{\partial P}{\partial V}\right)_{H_1} > -\left(\frac{\partial P}{\partial V}\right)_{S_1} \tag{13-2-53}$$

and thus

$$-\left(\frac{\partial P}{\partial V}\right)_{H_1} > \frac{P_1 - P_0}{V_0 - V_1} \tag{13-2-54}$$

The entropy change along the Hugoniot is given by the thermodynamic relation

$$\theta\, dS_H = d\varepsilon_H + P_H\, dV_H \tag{13-2-55}$$

$$\theta\, dS_H = \tfrac{1}{2}[(V_0 - V_H)\, dP_H + (P_H - P_0)\, dV_H] \tag{13-2-56}$$

where use has been made of Eq. (13-2-21). From Eqs. (13-2-54) and (13-2-56) it can be seen that the entropy increases along the Hugoniot for a simple shock wave and thus waves that are dynamically stable are also thermodynamically stable. Equation (13-2-53) indicates that the converse is not necessarily true.

Although the regions where $(\partial^2 P/\partial V^2)_S > 0$ and $(\partial^2 P/\partial V^2)_S < 0$ identify where trains of compressive waves run together and spread out respectively, these do not distinguish where shock waves may or may not form. To examine this aspect in more detail consider a material having a phase change whose isentropes have in consequence the form shown in Fig. 13-24 where the lines AE and BD mark the boundaries of the phase transition. Suppose the shock is generated by

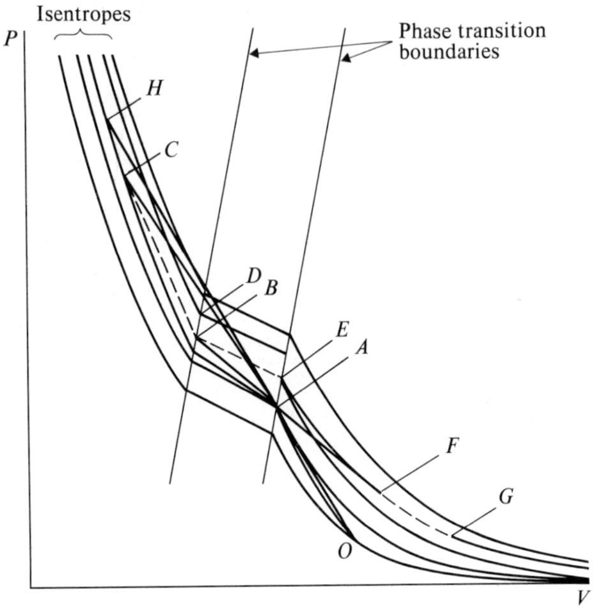

Figure 13-24 Pressure-volume diagram for the compression and expansion waves in a material exhibiting a phase change.

applying a gradually increasing pressure and then the pressure is reduced to the initial pressure through an expansion process. A distance-time diagram of the wave propagations in such a process is given in Fig. 13-25. As the compressive waves o to a run together they gradually increase the shock strength so that the state behind the shock wave follows the Hugoniot line OIA in Fig. 13-25 so forming the shock $\mathscr{S}$. Further increases of pressure are transmitted through the compression waves a to b at lower wave speeds so they form another shock whose end states lie along the Hugoniot AB. As this latter shock travels at a much slower speed than $\mathscr{S}$ a two-wave structure is formed and in general whenever the isentropes have a cusp on them the shock waves exhibit a two-wave structure. On increasing the pressure further through the compression waves b to c the final states lie on the Hugoniot B to C and form the shock $\mathscr{S}'$ whose speed of propagation is proportional to the gradient of AC in Fig. 13-25. If AH is the continuation of OA then when C lies above the line AH the shocks can coalesce into one shock again directly linking the states O and C.

If the pressure is now decreased, as $(\partial^2 P/\partial V^2)_S > 0$, the waves c to d form an expansion fan and consequently the state behind the waves decreases along the isentrope CD. A further decrease in pressure is transmitted through the even slower moving expansion fan d to e and the end states follow the same isentrope from D to E. When the pressure is reduced still further through the expansion fan e to g the first waves of the fan travel faster than the waves d to e and so they tend to overtake them. On coalescing they form a negative shock $\mathscr{S}_-$ directly relating the

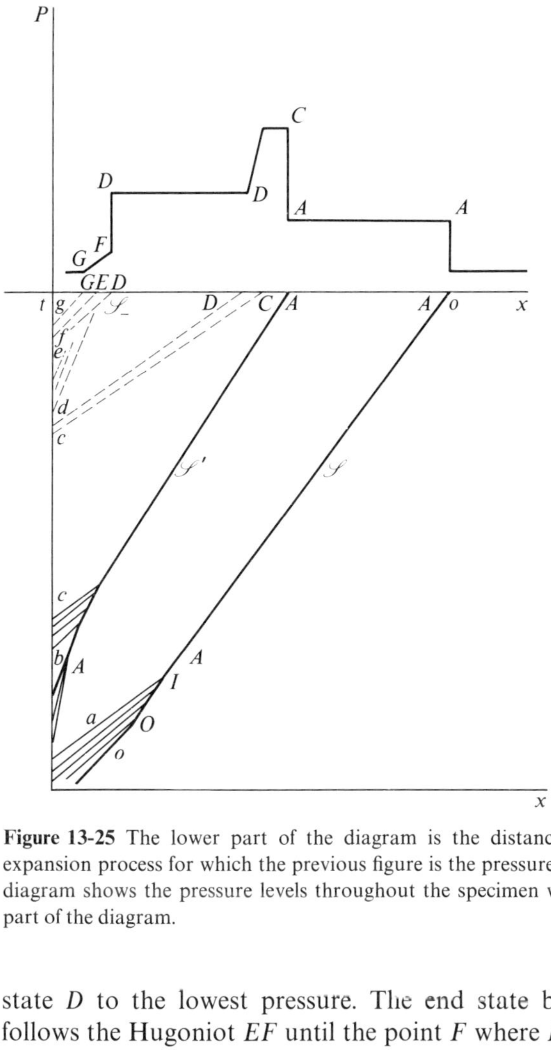

Figure 13-25 The lower part of the diagram is the distance-time diagram of a compression and expansion process for which the previous figure is the pressure-volume diagram. The upper part of the diagram shows the pressure levels throughout the specimen with the latest time shown on the lower part of the diagram.

state D to the lowest pressure. The end state behind the negative shock then follows the Hugoniot EF until the point F where DF is tangential to the isentrope through F. For any further decrease in pressure the expansion waves travel more slowly than the shock wave and so the final decrease to the initial pressure through the waves f to g occur through an expansion fan. The upper diagram of Fig. 13-25 shows the spatial distribution of pressure just after the end of the above compression and expansion process and illustrates both the two-wave shock structure and the negative shock.

It was shown in Sec. 13-2-2 that the entropy change across a weak shock wave varies as the cube of the density change and thus for compressions of less than about 15 percent the isentropes and Hugoniot are almost coincident. Strictly, however, the Hugoniot crosses the isentropes and on reducing the pressure to its initial value the density is reduced because of an increase of temperature due to the dissipative processes in the shock waves.

13-2-5 Shocks in Elastic–Perfectly Plastic Solids

The discussion will be restricted to plane waves moving into a solid at rest so that all the material displacements will be uniaxial and the only nonzero stresses are T_{xx} and $T_{yy} = T_{zz}$. As is usual the yield criterion will be taken to be the von Mises criterion (see Sec. 10-2-3), which for uniaxial strain becomes

$$|T_{xx} - T_{yy}| \leqslant Y \tag{13-2-57}$$

For stresses for which the inequality applies the material is assumed to be elastic and Hooke's law applies:

$$\mathbf{T} = \lambda(\operatorname{tr}\tilde{\mathbf{e}})\mathbf{I} + 2G\tilde{\mathbf{e}} = (\kappa - \tfrac{2}{3}G)(\operatorname{tr}\tilde{\mathbf{e}})\mathbf{I} + 2G\tilde{\mathbf{e}} \tag{13-2-58}$$

where κ is the bulk modulus and G is the modulus of rigidity. For uniaxial strain

$$\tilde{e}_{xx} = -\frac{V_0 - V}{V_0} \qquad \tilde{e}_{yy} = \tilde{e}_{zz} = 0 \tag{13-2-59}$$

When the equality holds in Eq. (13-2-57) the material is in a plastic state:

$$d\tilde{\epsilon} = d\hat{\epsilon} + d\bar{\epsilon} \tag{13-2-60}$$

there is no plastic dilation:

$$d(\operatorname{tr}\bar{\epsilon}) = 0 \tag{13-2-61}$$

the stress is supported by the elastic strain only:

$$d\mathbf{T} = \lambda\, d(\operatorname{tr}\hat{\epsilon})\mathbf{I} + 2G\, d\hat{\epsilon} \tag{13-2-62}$$

for low stresses the response is elastic, and

$$\begin{aligned} T_{xx} &= (\lambda + 2G)\tilde{e}_{xx} \\ T_{yy} &= \lambda\tilde{e}_{xx} \\ T_{zz} &= \lambda\tilde{e}_{xx} \end{aligned} \tag{13-2-63}$$

Thus

$$T_{xx} - T_{yy} = 2G\tilde{e}_{xx} = \frac{(1-2\nu)T_{xx}}{1-\nu} \tag{13-2-64}$$

The stress when the material first yields is called the Hugoniot elastic limit and will be denoted by T_{HEL}. From Eqs. (13-2-57) and (13-2-64),

$$-T_{\mathrm{HEL}} = \frac{1-\nu}{1-2\nu}Y \tag{13-2-65}$$

Above this stress the material behaves plastically and

$$-T_{xx} = P + \tfrac{2}{3}Y \tag{13-2-66}$$

where P is the hydrostatic pressure $(= -\frac{1}{3}\operatorname{tr}\mathbf{T})$ and is assumed to be a function of density and internal energy alone. In the elastic region all waves travel with the same velocity; thus the wave profile does not change with time. Beyond the

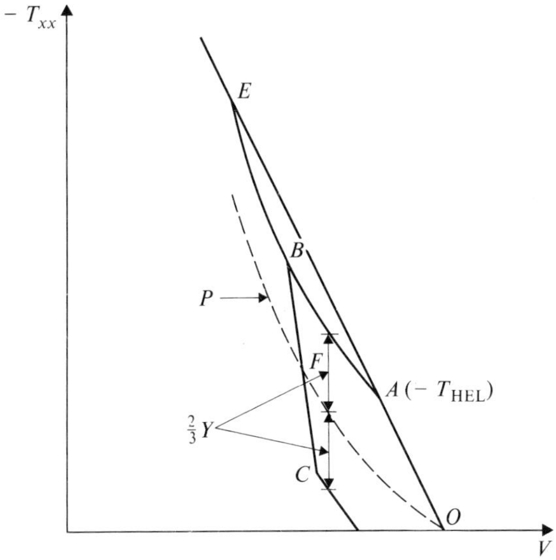

Figure 13-26 Stress-volume relation for an elastic-plastic solid.

Hugoniot elastic limit, shown as point A in Fig. 13-26, the material deforms plastically along a P–V curve parallel to the hydrostatic compression curve. The hydrostatic portion of the curve AB has an initial gradient given by

$$-\left(\frac{dP}{dV}\right)_A = \frac{\kappa_A}{V_A} \tag{13-2-67}$$

corresponding to a wave speed $[\kappa_A V_A]^{1/2}$ and as this is usually much less than the elastic wave speed a two-wave structure develops for stresses in the range A to E, where E is the intersection of the projection of OA and the curve AB. During compression, $-T_{xx}$ increases more rapidly than $-T_{yy}$ until the elastic limit is reached. In unloading, $-T_{xx}$ decreases faster than $-T_{yy}$ until yielding occurs once more. The decrease in BC thus occurs elastically and BC is very nearly parallel to OA because

$$\left(\frac{\partial T_{xx}}{\partial V}\right)_{\text{along } BC} = -\frac{\partial P_F}{\partial V_F} + \frac{4}{3}\frac{G_F}{V_F} = \frac{\kappa_F}{V_F} + \frac{4}{3}\frac{G_F}{V_F} \tag{13-2-68}$$

For the portion CD the yielding occurs again and

$$-T_{xx} = P - \frac{2Y}{3} \tag{13-2-69}$$

The unloading wave may also be expected to consist of two waves, an elastic wave and a plastic rarefaction fan, the elastic waves in both the loading and unloading regimes being called "elastic precursors." By measuring the maximum stress in the elastic precursor the Hugoniot elastic limit can be determined and the inferred dynamic yield stress is usually considerably higher than the corresponding static yield strength.

13-2-6 Equation of State

One of the most extensive applications of shock studies has been in examining the equation of state of materials and their physical properties at high-energy density. Detailed reviews of work on the equation of state have been given by Al'tshuler (1965), Skidmore (1965), and McQueen et al. (1970), and the work on physical properties has been reviewed by Keeler and Royce (1971).

It has been found that the Hugoniots of most materials examined appear to exhibit a linear relationship between shock velocity and the particle velocity behind the shock:

$$U_S = c_0 + bu_1 \tag{13-2-70}$$

In crystalline solids at pressures in excess of 5 GPa the effect of elastic stresses can usually be neglected and the stress behaves isotropically. By using the Debye theory for crystalline solids and assuming that the final state is one of thermodynamic equilibrium which is not too far removed from the initial state, Swan (1971) has derived the above linear relation. He shows that c_0 is the thermodynamic sound speed in the uncompressed material and

$$b = \tfrac{3}{4}\Gamma_0 \tag{13-2-71}$$

where the Gruneisen index

$$\Gamma = V\left(\frac{\partial P}{\partial E}\right)_V = V\left(\frac{\partial P}{\partial \theta}\right)_V c_V^{-1} \tag{13-2-72}$$

where c_V is the specific heat at constant volume, and the suffix 0 signifies conditions in the uncompressed material. The comparison of b calculated from the Gruneisen index and form-fits to the linear relationship by Rice, McQueen, and Walsh (1958) are given in Table 13-1 and show good agreement. Γ_0 can be calculated from the thermodynamic identity

$$\Gamma = \frac{3\alpha V}{c_V \kappa} \tag{13-2-73}$$

where α is the coefficient of linear expansion at constant pressure and κ is the isothermal compressibility modulus.

It was noted in Sec. 13-2-1 that the Hugoniot determines one line on the

Table 13-1

Metal	Γ_0	$b\ (=\tfrac{3}{4}\Gamma_0)$	Analytic fit
Ag	2.47	1.85	1.643
Cd	2.27	1.70	1.718
Cu	2.04	1.53	1.478
Pb	2.78	2.08	1.517
Zn	2.14	1.60	1.576

Taken from Swan (1971).

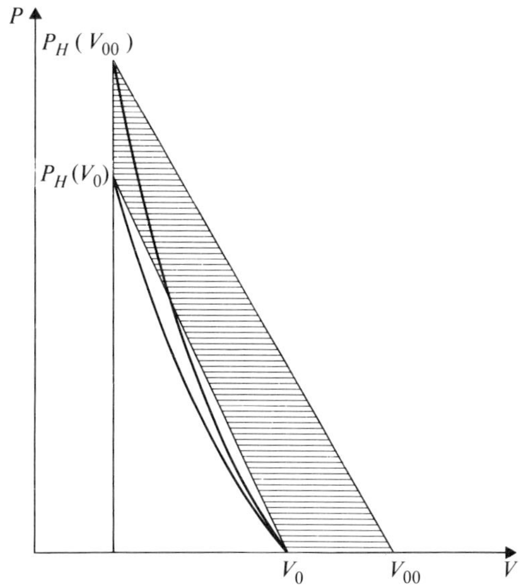

Figure 13-27 Pressure-volume curves for normal density and porous material. The shaded area indicates the extra thermal energy required in compressing the porous material.

equation of state surface in (P, V, ε) space and if this surface is to be examined in more detail some means must be found to supplement the Hugoniot data. This can be done in several ways:

1. Alter the initial state. In this respect V_0 has been altered by using porous samples as proposed by Zel'dovich (1957) and studied by Walsh and Rice (1957), Al'tshuler et al. (1958), Kormer, Funtikov, Urlin, and Kolesnikova (1962), and Krupnikov, Brazhnik, and Krupnikova (1962). It is assumed that a very small pressure is sufficient to crush the sample back to its nonporous condition, that the extra surface energy is negligible, and that any trapped gases that may have diffused into the interstices have no appreciable effect. The material behind the shock front therefore obeys the same equation of state as the nonporous material and the Hugoniot has the form shown in Fig. 13-27 where V_{00} is the initial porous volume density. It can be seen that large deformations of porous samples in which $V_{00} = mV_0$ are accompanied by a large increase in thermal energy, and as shown by Walsh and Rice (1957) the thermal expansion associated with this thermal energy can outweigh the hydrostatic compression due to the pressure increase, giving rise to the Hugoniot shown schematically in Fig. 13-28.
2. The isotropes associated with the thermodynamic state behind a shock wave may be estimated by finding the sound speed. This is done by measuring the amount of unattenuated wave front as shown in Fig. 13-29 by the area *BB* and thus the angle α. The sound speed behind the shock wave can easily be shown to be given by

$$c_1 = U_S\left[\tan^2\alpha + \frac{(U_S - u_1)^2}{U_S^2}\right]^{1/2} \tag{13-2-74}$$

An alternative method for determining c_1 is to choose the dimensions of the specimen and shock loading plates so that the rarefaction wave from the back of the loading plate overtakes the shock wave before either the free surface is reached or side rarefactions have destroyed the wave front. Probes in the specimen show the decreased shock velocity and the free surface can also be measured. Thus the value of c_1 can be deduced as shown by Al'tshuler et al. (1960). Reflected shock waves can also be used to give extra Hugoniot data.

3. A form of the equation of state can be chosen that has a few arbitrary parameters which can then be determined from the available experimental shock-wave data. The equation usually chosen as a model is the Mie–Gruneisen equation. In the early work, for which the highest pressures attained were about 25 GPa, it was assumed that the only significant contributions to the internal energy arose from the mutual potential energy of the atoms and from the lattice vibrations. Assuming the contributions are uncoupled, that the thermal energy can be adequately described by a set of simple harmonic

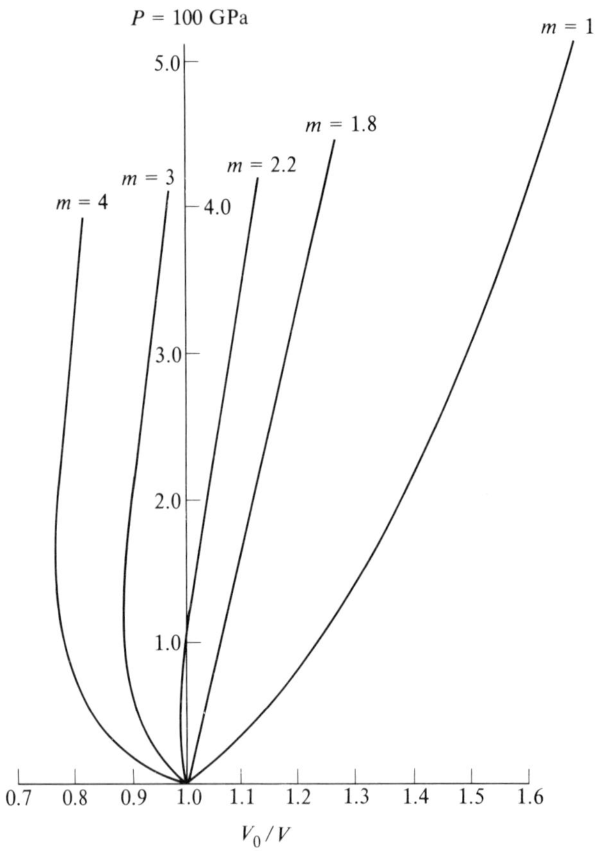

Figure 13-28 Schematic diagram of the Hugoniots for different porosities m $(= V_{00}/V_0)$. A very small pressure recompresses the material to its normal specific volume V_0.

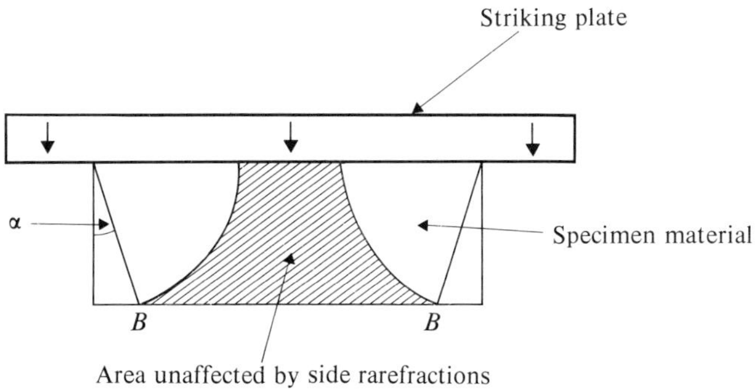

Figure 13-29 Determination of the speed of sound by determining the extent of the area unaffected by side rarefaction waves.

oscillators, and that all the lattice frequencies ν have the same logarithmic derivative with respect to volume, it can be shown that

$$\Gamma = -\frac{d \ln \nu}{d \ln V} = -\frac{d \ln \theta_D}{d \ln V} \tag{13-2-75}$$

where θ_D is the Debye temperature. The equation of state in the Mie–Gruneisen form is then

$$P = P_K + P_L \tag{13-2-76}$$

$$\varepsilon = \varepsilon_K + \varepsilon_L \tag{13-2-77}$$

where the suffix K denotes the potential energy contributions and can be evaluated at 0 K and the suffix L denotes the contribution from lattice vibrations. Clearly

$$P_K = \frac{d\varepsilon_K(V)}{dV} \tag{13-2-78}$$

and

$$P_L = \Gamma(V)\frac{\varepsilon_L}{V} \tag{13-2-79}$$

$$\varepsilon_L = \frac{3k\theta D(\theta_D/\theta)}{\mathscr{M}} \tag{13-2-80}$$

where k is Boltzmann's constant, D is the Debye function, and $\mathscr{M}$ is the mass of a vibrator, usually an atom. There are several ways of relating Γ and P_K which Al'tshuler (1965) has shown can all be represented by

$$\Gamma = -\left(\frac{2}{3} - \frac{t}{3}\right) - \tfrac{1}{2}V\frac{(d^2/dV^2)(P_K V^{2t/3})}{(d/dV)(P_K V^{2t/3})} \tag{13-2-81}$$

The value $t = 0$ corresponds to the Landau and Stanyukovich (1945)–Slater (1939) theory, the value $t = 1$, which holds for metals, corresponds to the Dugdale and McDonald (1953) theory, and the value $t = 2$, which holds for ionic crystals, corresponds to the theory of Zubarev and Vashchenko (1963).

Since the 1960s the pressures attained in shock-wave investigations have gradually been extended up to about 1 TPa and with the consequent increase in temperature it has now become necessary to extend the above theory by including terms arising from the thermal energy of the electron gas. The equation of state thus takes the form

$$P = P_K + P_L + P_e \tag{13-2-82}$$

$$\varepsilon = \varepsilon_K + \varepsilon_L + \varepsilon_e \tag{13-2-83}$$

P_L and ε_L need modifications to take account of anharmonic effects while

$$\varepsilon_e = \tfrac{1}{2}\beta\theta^2 \qquad P_e = \Gamma_e \frac{\beta\theta^2}{2V} \tag{13-2-84}$$

where $\beta = \beta_0(V/V_0)$ is the coefficient of electronic specific heat, β_0 is its experimental value at $V = V_0$, and Γ_e is the electronic analog of the Gruneisen coefficient. These equations have been used by Al'tshuler et al. (1960), Al'tshuler,

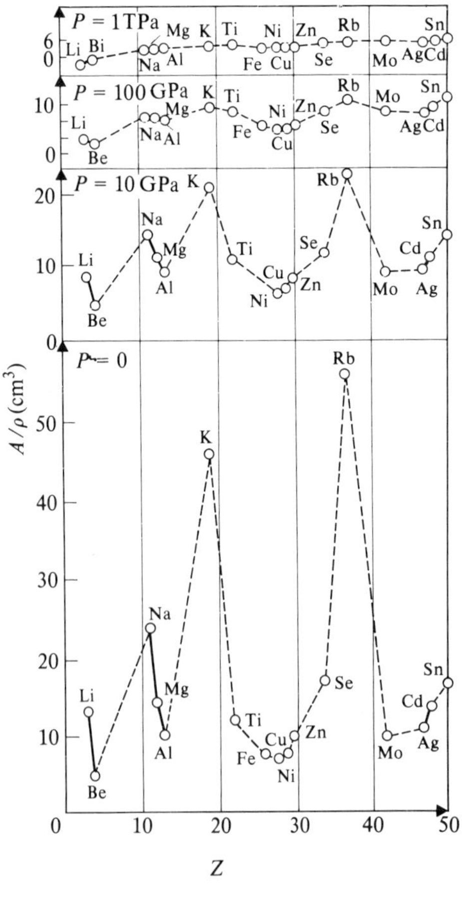

Figure 13-30 Curves of atomic volumes for normal conditions at pressures of 0, 10 GPa, 100 GPa, and 1 TPa; A = atomic weight, Z = atomic number, ρ = density. (*After Al'tshuler, 1965.*)

Bakanova, and Trunin (1962), and Bakanova, Dudoladov, and Trunin (1965) to represent their data up to 1 TPa. Figure 13-30, taken from Al'tshuler (1965), shows the gradual evening out of the periodicity in the atomic volume density of elements with small and medium values of atomic number Z as the pressure is increased. The periodicity of volume density is due to the shell structure of the atom and the change of this periodicity reflects a change in electronic structure of the material which in turn affects its electrical and magnetic properties, its ability to transmit and absorb electromagnetic and acoustic waves, and its stable chemical phases. A review of all these aspects has been given by Keeler and Royce (1971).

13-3 HIGH-SPEED IMPACT PHENOMENA

13-3-1 Identification of Different Regimes of Penetration

The problem of the penetration of a target by a projectile is one of long standing. The earliest recorded methodical tests were presented by Benjamin Robins (1742) and in the intervening centuries many empirical formulas have been put forward which, however, are only applicable for a small variation in the parameters controlling penetration. This lack of fundamental understanding was not occasioned by lack of interest but because of its intractable nature for the reasons described below. More than most branches of physics or mechanics the penetration problem involves different regimes of material response. The main velocity regimes were delineated by Hopkins and Kolsky (1960) and more recently have been classified relative to a "damage number," as introduced by Johnson (1972). The damage number is defined as the nondimensional number $\rho V^2/\bar{Y}$, which is larger because, between target and projectile materials where the relative speed at normal impact is V, the density is ρ and $\bar{Y}$ is the mean flow stress. If both the projectile and target are of steel having a $\bar{Y}$ value of 200 MPa, then the regimes are approximately as shown in Table 13 2.

(a) Elastic behavior For low impact velocities the stresses nowhere exceed the yield stress and the motion of both the target and projectile is determined in detail

Table 13-2 Impact regimes

Velocity, $m\,s^{-1}$	$\rho V^2/\bar{Y}$	Regime
0.5	10^{-5}	Elastic behavior
5	10^{-3}	Plastic behavior
1000	40	Hydrodynamic behavior
5000	10^3	Penetration at supersonic velocities
15,000	10^4	Explosive impact

by elastic wave propagation which in turn is determined by the elastic moduli only. For the normal case of a small projectile striking a large target the duration of the impact is very long compared with the time it takes the elastic waves to traverse the projectile and a quasi-static stress distribution is set up. In the target three types of elastic waves are generated, namely, a spherical dilatational wave, a spherical distortional wave, and a Rayleigh surface wave. Lamb (1904) examined the problem of waves generated in a semi-infinite medium and using his method Hunter (1957) showed that, for the normal impact of a spherical projectile on the surface of a semi-infinite solid, by far the greatest proportion (over 99 percent) of the kinetic energy of the projectile is stored as elastic strain energy in the target, where, as the projectile is brought to rest, a hertzian (Hertz, 1896) stress distribution is set up in the contact region.

(b) Plastic behavior On increasing the impact velocity a point is eventually reached when the stresses exceed the yield stress in either the projectile or target, or perhaps both. This point defines the first transitional velocity and for impacts in excess of this some plastic deformation will occur. The velocity at which noticeable deformation occurs is, however, dependent on the nose geometry as well as the material properties, as was shown by Davies (1949) who noted that whereas for a steel ball striking a steel surface the transition velocity was $0.15\,\mathrm{m\,s^{-1}}$, for a steel cylinder striking a steel surface the transitional velocity is above $0.6\,\mathrm{m\,s^{-1}}$.

For further increases in impact velocity more of the projectile kinetic energy is used in doing plastic work and the influence of the elastic deformations on the penetration is diminished. However, the region of high stresses is confined to the region around the penetrating nose of the projectile and outside a region of a few projectile diameters from the nose region the material response remains elastic. Two extreme situations have been studied, either the projectile is much stronger than the target and so remains undeformed during penetration, or the target is much stronger than the projectile and hence it remains undeformed after impact. In the latter case the gross plastic deformation can spread progressively backward through the impinging projectile and the subsequent distortions can be used to determine the plastic response of the material. We will return to these problems in more detail in later sections.

(c) Hydrodynamic regime On increasing the velocity further still the stresses will eventually become much greater than the yield stress in both the projectile and the target and both begin to behave as a fluid in the nose region of the projectile. Using this model some success has been achieved in determining the penetration over quite a wide range of impact velocities.

(d) Penetration at supersonic velocities When the impact velocity is sufficiently high for its associated penetration velocity to exceed the speed of sonic waves in the target no waves can precede the penetrating tip of the projectile which is then surrounded by an attached shock wave in the target material. This region is of particular interest for brittle materials because the maximum speed that cracks can propagate is about half the velocity of dilatational waves and consequently the hoop tensions causing radial cracks cannot propagate very far before the projectile passes and the stresses are relieved.

(e) Explosive impact At a high enough velocity the temperature generated in the penetrating tip region of a projectile is sufficient to vaporize the material there and the impact takes on much of the character of an exploding charge at the impact point. For example, the craters produced in the target tend to lose any elongated appearance they may have had and to become hemispherical even when the target is struck at an oblique angle.

In general an impact involves several regimes simultaneously in different regions of the projectile and target. Near the penetrating projectile nose the stresses can be very high and the material behaves in a gaseous or fluid way whereas further out it will behave plastically and further out still it will behave elastically.

The problem is further complicated because, particularly for the hydrodynamic regime, the penetration can be considered as passing through different phases during which different mechanisms dominate. Eichelberger and Gehring (1962) and Christman and Gehring (1966) considered the process as consisting of four phases. First is a shock-wave-dominated initial "transient phase" of short duration during which the high stress waves expanding into the target and the rod suffer strong attenuation either through reflection at a free surface or through expanding as hemispherical waves. There then follows a period of almost steady penetration which is describable in terms of a hydrodynamic theory and has been termed the "primary" phase. The "secondary" phase occurs when sufficient of the projectile has been consumed for the stresses at the penetrating tip to cause rapid deceleration, and the fourth and last phase or "recovery" phase involves any contraction of the crater due to the release of elastic stresses in the far field.

Further subdivisions arise depending on whether the projectile aspect ratios (length/diameter) is high or low and whether the target thickness is large or small compared with the projectile diameter. For projectiles of low aspect ratio which may be described as penny shaped, the initial shock phase predominates and indeed no other phase need be considered. Most of the considerable volume of work on examining the high-pressure state of materials through shock compression has been done with very low-aspect ratio cylinders because, apart from a small circumferential edge effect, the shock waves can then be regarded as plane waves and the Hugoniot shock relations described in Sec. 13-2 can be used.

If the target thickness is also small compared to the diameter of the impacting projectile or the impact velocity is very high, then a strong shock wave can be transmitted to the near surface, which upon reflection can produce sufficiently high tensile stresses to fracture the material and produce "scabbing," as described in Sec. 14-8.

For high-aspect ratio rods all four phases are often present. We will start by describing only very thick targets and return to the problem of thinner plates at the end of the chapter. It is also, in general, convenient to restrict our attention to flat-ended cylindrical rods except in the case of a hard projectile passing through a thin, soft plate.

13-3-2 Shock-Dominated Initial Impact Phase

When the projectile first strikes the target a shock wave is propagated into both the rod and target. For a flat-ended rod the initial shock waves will also be planar and using the P–u diagrams discussed in Sec. 13-2-3 it can be seen that the impact pressure P_I is given by

$$P_I = \frac{\rho_t c_t \rho_p c_p V}{\rho_t c_t + \rho_p c_p} \tag{13-3-1}$$

where V is the striking velocity of the projectile, ρ is the density, c is the sound speed, and the suffixes p and t refer to projectile and target respectively. The product ρc is termed the acoustic impedance of the material and as this is usually high for most metals, and particularly those used for protection against high-speed particles, the impact stress rapidly becomes large with increase of striking velocity.

Unless the projectile is moving so swiftly that the surface of separation between the projectile and target is moving supersonically in the target, the shock

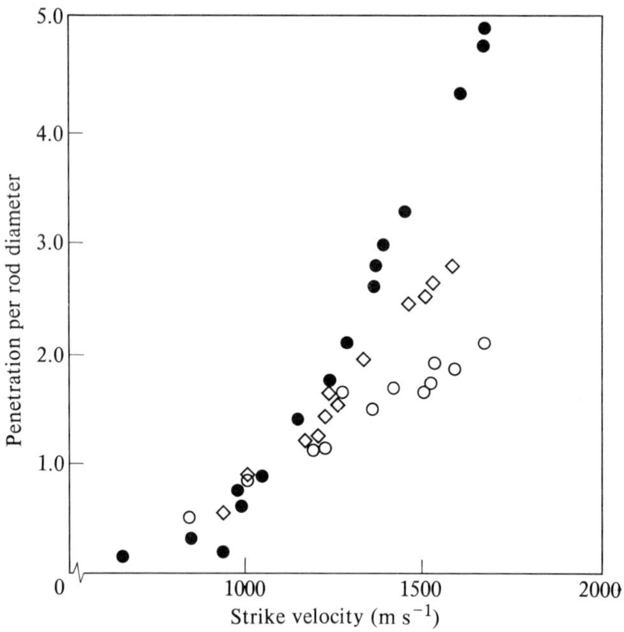

Figure 13-31 Penetration of a steel rod into a steel plate at normal obliquity. Length/diameter ratio: ○, 3; ◇, 6; ●, 12.

waves are attenuated by release waves from the free surfaces of both the projectile and target and the stress level rapidly drops to a lower level which is estimated in the next section. However, for projectiles striking with velocities in the plastic regime the importance of the initial shock plane is shown by a curious effect which is rarely mentioned in the literature. Figure 13-31 shows results obtained for the penetration of flat-ended cylindrical steel rods into a steel plate. The rods were of nine differing lengths and diameters such that their mass was either 8.5, 17, or 34 gm and their length-to-diameter ratios were 3, 6, or 12. It can be seen that at impact velocities of less than 1150 m s^{-1} the penetration is independent of the rod length. This is thought to occur because, although the initial impact stress is above the dynamic yield stress of the target, release waves from the free surface can reduce the stress to below the dynamic yield stress and thus the depth of penetration is governed by the time it takes release waves from the nearest free surface to reach the center of the rod. The velocity of the surface of separation depends only on the impact velocity and the acoustic impedances of the rod and target; immediately after impact it has the value $\rho_p c_p V/(\rho_t c_t + \rho_p c_p)$. The time the release waves take to reach the center is on the order of $d/2c_p$ and thus the penetration will depend primarily on d. As the situation is neither one dimensional nor steady any exact solution is likely to be complex, but the above explanation would account for the experimental observations.

13-3-3 Low-Velocity Mushrooming

After the initial shock phase has passed a steadier deformation phase usually ensues whose character depends upon the relative strength of the projectile and target. First we will consider the case when the target is very strong and the projectile is relatively weak. The projectile then mushrooms and the ensuing motion was first studied by Taylor (1948a). The theory has since been extended and modified by Lee and Tupper (1954), Hawkyard (1969), Wilkins and Guinan (1973), and Carley (1975).

If the impact stress exceeds the projectile dynamic yield stress Y_0 then, as shown in Sec. 13-2-5, a two-wave structure is produced, an elastic precursor wave having a velocity c_e and strength Y_0 followed by a plastic wave which usually moves at a much slower speed. The particle velocity behind the elastic wave now moves at the reduced velocity $[V-(Y_0/\rho_p c_e)]$. On reaching the rear of the rod the compression wave is reflected as a tensile wave relieving the stress back to zero and reducing the rod velocity further to $[V-(2Y_0/\rho_p c_e)]$. On reaching the plastic boundary the tension wave reflects as a compression, the situation being similar to that at impact except that the rod is shorter and its velocity is reduced to $[V-(2Y_0/\rho_p c_e)]$. As the sound speed is high, the reduction in velocity is small. The situation is also not strictly one dimensional as we have described it and thus it can in practice usually be assumed that the rear part of the rod smoothly decelerates as a rigid body. The following discussion is taken from Taylor (1948a). Let the deformation at some intermediate stage between the rod striking the plate and being brought to rest be as shown in Fig. 13-32. If H is the distance of the plastic

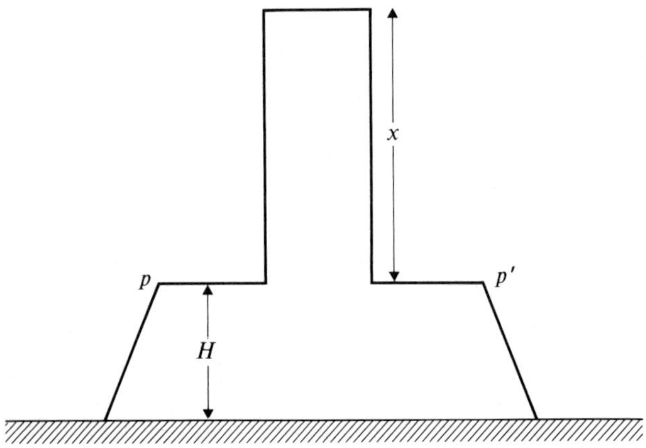

Figure 13-32 An intermediate stage in the plastic deformation of a rod striking a rigid target as it is being brought to rest.

boundary pp' from the plate, v is the velocity of the rigid rear section, x is the length that has not yet been plastically compressed, Y_0 is the projectile dynamic yield stress, and v' is the velocity of the rigid-plastic boundary from the target surface, then

$$\frac{dH}{dt} = v'$$
$$\frac{dx}{dt} = -(v'+v) \tag{13-3-2}$$
$$\frac{dv}{dt} = -\frac{Y_0}{\rho_p x}$$

The rigid-plastic boundary is imagined as a sharp discontinuity below which the plastically deformed projectile material is assumed to be at rest with respect to the target. Radial momentum is also neglected during plastic deformation so that over any cross section of the projectile the stress density and velocity are assumed to be constant. If the original area of the rod is A_0 and the area having passed through the surface pp' is A then, assuming the material to be incompressible, the continuity equation gives

$$A_0(v+v') = Av' \tag{13-3-3}$$

Choosing a frame of reference in which pp' is stationary, conservation of momentum gives

$$\rho_p A_0(v+v')^2 + A_0 Y_0 = \rho_p A v'^2 + AY \tag{13-3-4}$$

where Y is the flow stress immediately after deformation. Taylor assumed that there was no work-hardening and in this case $Y_0 = Y$. The strain $\tilde{e}$ may be defined either in the nominal form by

$$\tilde{e} = 1 - \frac{A_0}{A} \tag{13-3-5}$$

as taken by Taylor, or natural strain as considered by Carley, in which case

$$\epsilon = \ln \frac{A}{A_0} = \ln K \tag{13-3-6}$$

Using Eqs. (13-3-3) and (13-3-4) elementary manipulation gives

$$\rho_p v^2 = \frac{K-1}{K}[KY(K) - Y_0] \tag{13-3-7}$$

Let the initial value of K be K_1; then

$$\rho_p V^2 = \frac{K_1 - 1}{K_1}[K_1 Y(K_1) - Y_0] \tag{13-3-8}$$

The length of the undeformed portion of the projectile x is given by

$$\int_x^L \frac{dx'}{x'} = \frac{1}{2Y_0} \int_K^{K_1} \frac{K'}{K'-1}\left[Y(K') + (K'-1)\frac{dY(K')}{dK'} - \frac{Y_0}{K'^2}\right] dK' \tag{13-3-9}$$

where L is the undeformed length of the rod.

The position of the plastic wave front is given by

$$H = -\int_{K_1}^{K} \frac{1}{K'} \frac{dx}{dK'} dK' \tag{13-3-10}$$

and the time after impact is given by

$$t = -\int_{K_1}^{K} \frac{K'-1}{K'v} \frac{dx}{dK'} dK' \tag{13-3-11}$$

These equations are similar to Taylor's theory except that he took $Y = Y_0$. The projectile profile at any time can be obtained by numerically integrating Eqs. (13-3-9), (13-3-10), and (13-3-11). By choosing different functional forms for $Y(e)$ different profiles can be examined and those giving the best overall fit to experimental studies are chosen as characterizing the material response. Studies to date have been restricted to linear work-hardening relationships. Taylor made the further assumption that v' was constant in order to obtain an analytical expression relating dynamic yield strength to the final shape. This is, however, incompatible with the continuity equation which can be written in the form

$$v + v' = Kv' \tag{13-3-12}$$

Then using Eqs. (13-3-7) and (13-3-8) with $Y = Y_0$ shows both that

$$\rho_p v'^2 = \frac{Y_0}{K} \tag{13-3-13}$$

and that K varies from K_1 to 1 as the projectile is brought to rest. Setting $Y = Y_0$ in Eq. (13-3-9) and integrating gives

$$2\ln\left(\frac{L}{x}\right) = K_1 - K + \ln\left(\frac{K_1}{K}\right) \tag{13-3-14}$$

where

$$\rho_p V^2 = (K_1 - 1)^2 \frac{Y_0}{K_1} \tag{13-3-15}$$

and the plastically deformed length H is given by

$$\frac{H}{L} = \int_K^{K_1} \frac{K'+1}{2K'^2}\left(\frac{K'}{K_1}\right)^{1/2} \exp\left(\frac{K'-K_1}{2}\right) dK' \tag{13-3-16}$$

Hawkyard employed basically the same concept but used energy conservation instead of momentum conservation across the plastic wave. The rate of doing plastic work is given by

$$\dot{W} = A_0(v+v')Y_0 \ln K \tag{13-3-17}$$

if there is no work-hardening and the rate of loss of kinetic energy of the element and the supply of energy from the rear section due to the force $Y_0 A_0$ is $A_0(v+v')\frac{1}{2}\rho_p v^2 + Y_0 A_0 v$. Equating the rate of plastic working with the supply of energy gives

$$\dot{W} = A_0(v+v')Y_0 \ln K = A_0(v+v')\tfrac{1}{2}\rho_p v^2 + Y_0 A_0 v \tag{13-3-18}$$

and this equation replaces Eq. (13-3-4) of the previous formalism. If work-hardening does exist then $\dot{W}$ can be expressed by the relation

$$\dot{W} = A_0(v+v') \int_1^K \frac{Y(K')}{K'} dK' \tag{13-3-19}$$

The experimentally observed rod profiles are rather better predicted by the energy conservation method while the estimated dynamic yield strengths given by either method do not differ by more than 10 percent.

13-3-4 Transition to Hydrodynamic-type Flow

As the striking velocity of the projectile is further increased the stresses, even after being reduced by release waves from the neighboring surfaces, still exceed the yield stresses in both the projectile and the target. A complex plastic flow situation then arises in which the projectile is grossly distorted over most of its length while burrowing into the target in a small region near the axis, giving a profile as shown in Fig. 13-33. At even higher striking velocities the mode of projectile penetration changes character entirely and much more closely resembles a fluid jet, there being very little plastic deformation over most of the projectile length. This change in penetration mode has been examined by Weihrauch (1972) and Brooks (1974) and termed the "hydrodynamic transition velocity." This is not to be confused with the "hydrodynamic" transition velocity introduced by Bell and Suckling (1962) which

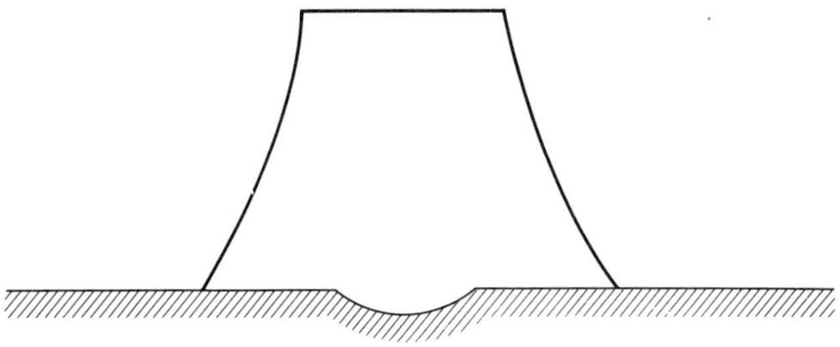

Figure 13-33 Grossly distorted projectile starting to burrow into the target.

corresponds to an impact velocity above which significant mushrooming occurs, that is, $Y_0/\rho_p c_p$.

It has been suggested by Tate (1977) that a possible reason for the change of deformation mode is that the rod erosion rate exceeds the rate of propagation of gross plastic deformation or plastic wave speed. The situation may then be analogous to the bow shock of a supersonic projectile. All the gross deformation is constrained to occur very near to the eroding tip of the rod in a region of increased entropy and temperature.

To make an estimate of the hydrodynamic transition velocity v_{HT} it may be noted that the speed of propagation of gross plastic distortion associated with high stresses is governed by the tangent modulus to the dynamic stress-strain curve for high stress levels which has been termed E_{ap} in Sec. 13-1-3. Letting the speed of propagation of gross plastic distortion be denoted by

$$\bar{w} = \left(\frac{E_{ap}}{\rho_p}\right)^{1/2} \tag{13-3-20}$$

it will be shown in the next section that the rod erosion rate w is given by

$$\tfrac{1}{2}\rho_p w^2 + Y_p = \tfrac{1}{2}\rho_t (v - w)^2 + R_t \tag{13-3-21}$$

The hydrodynamic transition velocity v_{HT} occurs when w equals $\bar{w}$ and so is given by

$$v_{\mathrm{HT}} = \left(\frac{E_{ap}}{\rho_p}\right)^{1/2} \left[1 + \left(\frac{\rho_p}{\rho_t}\right)^{1/2} \left(1 - 2\frac{R_t - Y_p}{E_{ap}}\right)^{1/2}\right] \tag{13-3-22}$$

in general. A limitation must be put on the use of Eq. (13-3-22) in the case when $R_t > Y_p$; penetration can then cease if the impact velocity drops below $[2(R_t - Y_p)/\rho_p]^{1/2} = v_c$ say. For $V < v_c$ we must write $v = w$ and if $\bar{w} < v_c$ then

$$v_{\mathrm{HT}} = \left(\frac{E_{ap}}{\rho_p}\right)^{1/2} \tag{13-3-23}$$

13-3-5 Modified Hydrodynamic Theory of Penetration

Above the hydrodynamic transition velocity the mode of penetration closely resembles that of a fluid jet; this analogy was first exploited during the period 1939

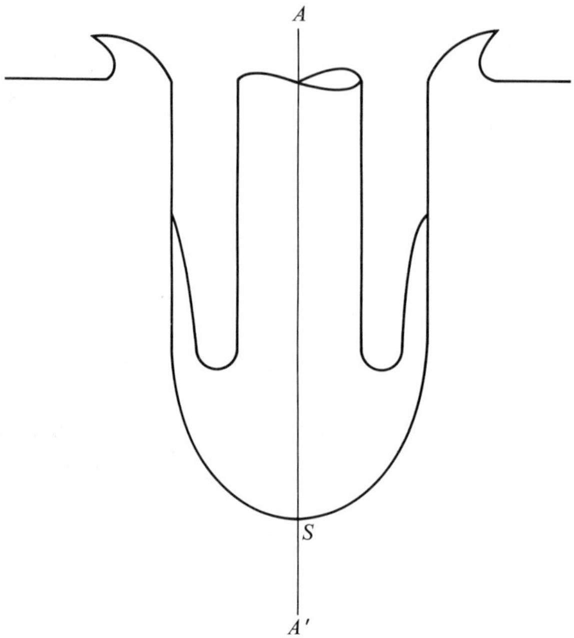

Figure 13-34 Penetration of a jet into a target.

to 1945 and published by Birkhoff, McDougall, Pugh, and Taylor (1948) and Pack and Evans (1951). In its simplest form the strength of the projectile and target are ignored and it is assumed that the flow approximates the steady penetration of an incompressible fluid jet into a semi-infinite region of another incompressible fluid, as shown in Fig. 13-34. First we take axes at rest with respect to the moving interface between the projectile and the target, S. Applying Bernoulli's equation along the axial streamline AA' for both the rod and target, ignoring atmospheric pressure compared with the stresses involved, and letting the speed of penetration be U gives

$$\tfrac{1}{2}\rho_p(V-U)^2 = \tfrac{1}{2}\rho_t U^2 \tag{13-3-24}$$

and therefore

$$U = \frac{V}{1+(\rho_t/\rho_p)^{1/2}} \tag{13-3-25}$$

If the rod has no strength then the rear cannot be decelerated and penetration will continue at a constant rate until all the rod is consumed, giving a total penetration depth L_P which is given by

$$L_P = L\left(\frac{\rho_p}{\rho_t}\right)^{1/2} \tag{13-3-26}$$

where L is the rod length.

For impact speeds greatly in excess of the hydrodynamic transition velocity the above simple formulas are good approximations to experimental observations. The penetration is only weakly dependent on the striking velocity and does vary almost linearly with rod length provided the rod aspect ratio is high. At speeds just above the hydrodynamic transition velocity, however, the situation is very different, for observation shows the penetration then to be strongly dependent on the striking velocity.

Early work by Bishop, Hill, and Mott (1945) and later by Eichelberger (1956) suggested that the penetration fell short of the depth predicted by Eq. (13-3-26) because of the strength of the target. To take account of target strength it was suggested that Bernoulli's equation should be applied only from the stagnation point to the point where it ceased to behave hydrodynamically at the dynamic yield strength of the target material. Assuming that the yield strength of the rod was either negligible or played no significant role independently of the target strength a term ϕ was introduced into Bernoulli's equation by Eichelberger (1956) and Allen and Rogers (1961):

$$\tfrac{1}{2}\rho_p(V-U)^2 = \tfrac{1}{2}\rho_t U^2 + \phi \tag{13-3-27}$$

Penetration was still calculated, however, on the assumption that all the rod was consumed during penetration, which is not in accord with experiment except for velocities very much higher than the hydrodynamic velocity. A further refinement of the theory, which allowed for deceleration of the rod, was to introduce separate strength terms for both the rod and the target, as was suggested by Alekseevski (1966) and Tate (1967, 1969), giving

$$\tfrac{1}{2}\rho_p(V-U)^2 + Y_p = \tfrac{1}{2}\rho_t U^2 + R_t \tag{13-3-28}$$

Y_p being the yield strength of the rod and R_t being the strength of the target. In the case of the target the stress required for initiating flow will be higher than for the rod because of the inertia of the surrounding material and is equivalent to the minimum pressure required to just force a hole through an infinite expanse of the material. Although the pressure R_t probably varies with depth and impact velocity the variation within the region for which the hydrodynamic phase is valid is likely to be small and it is usual to assume that R_t is constant. Empirical matching of penetration data suggests that R for a material is about $3Y$ for the same material, which is closely analogous to the enhanced pressure required to force a flat-ended die into a large block, as discussed, for instance, by Tabor (1951).

The deceleration throughout the hydrodynamic phase of penetration is usually small and to the extent that it can be neglected Bernoulli's equation applies throughout the motion. If the velocity of the rear of the rod is $v(t)$ the velocity of penetration is $u(t)$ and $x(t)$ is the length of the rigid rear section, which is assumed to extend almost to the penetrating tip of the rod. Bernoulli's equation gives

$$\tfrac{1}{2}\rho_t u^2 + R_t = \tfrac{1}{2}\rho_p(v-u)^2 + Y_p \tag{13-3-29}$$

so that

$$u = \frac{v - \mu(v^2 + A)^{1/2}}{1-\mu^2} \tag{13-3-30}$$

where

$$\mu = \left(\frac{\rho_t}{\rho_p}\right)^{1/2} \qquad A = \frac{2(R_t - Y_p)(1-\mu^2)}{\rho_t} \tag{13-3-31}$$

Now using Eq. (13-3-2) in the form

$$\frac{dv}{dt} = -\frac{Y_p}{\rho_p x} \tag{13-3-32}$$

and noting that

$$\frac{dx}{dt} = -(v-u) \tag{13-3-33}$$

gives

$$\frac{dx}{x} = \frac{\rho_p(v-u)}{Y_p}\,dv \tag{13-3-34}$$

Substituting for u from Eq. (13-3-30) then gives

$$\frac{x}{L} = \left[\frac{v+(v^2+A)^{1/2}}{V+(V^2+A)^{1/2}}\right]^{(R_t-Y_p)/\mu Y_p} \times \exp\left[\frac{\mu\rho_p}{2(1-\mu^2)Y_p}\{[v(v^2+A)^{1/2}-\mu v^2]-[V(V^2+A)^{1/2}-\mu V^2]\}\right] \tag{13-3-35}$$

where L is the initial rod length. The penetration L_P is then given by

$$L_P = \frac{\rho_p}{Y_p}\int_v^V uL\,dv \tag{13-3-36}$$

Three cases arise:

1. If $R_t = Y_p$ the penetration L_P is given by

$$\frac{L_P}{L} = \frac{1}{\mu}\{1-\exp[-B(V^2-v^2)]\} \tag{13-3-37}$$

where

$$B = \frac{\mu\rho_p}{2(1+\mu)Y_p} \tag{13-3-38}$$

and all the rod is consumed.

2. If $R_t > Y_p$ and $2(R_t - Y_p) > E_{ap}$ the rod behaves hydrodynamically until penetration ceases and this occurs when $v = [2(R_t - Y_p)/\rho_p]^{1/2}$. Once the erosion rate $w\ (=v-u)$ drops below $(E_{ap}/\rho_p)^{1/2}$ gross plastic distortion can spread from the eroding tip toward the rear.
3. If $R_t < Y_p$ then, when v drops to the value $[2(Y_p - R_t)/\rho_t]^{1/2}$, v equals u and from this point until it is brought to rest the rod behaves as a rigid body. The final length of rod left, l_s, is given by the formula

$$\frac{l_s}{L} = \left\{\frac{[A(1+\mu)/(\mu-1)]^{1/2}}{V+(V^2+A)^{1/2}}\right\}^{(R_t-Y_p)/\mu Y_p} \times \exp\left\{\frac{-\mu\rho_p}{2(1-\mu^2)Y_p}[V(V^2+A)^{1/2}-\mu V^2]\right\} \tag{13-3-39}$$

13-3-6 Development of an Asymptotic Strain Field in Hydrodynamic Flow Conditions

We have examined above the use of a hydrodynamic analogy for the case when the stresses near the penetrating tip of a rod greatly exceed the yield strengths of the rod and target materials involved. Usually the analogy is used only to estimate properties on the streamline along the rod axis and, from a knowledge of these, the penetration of the rod. Provided the rod has sufficient size and energy to pass through the target easily then its rate of penetration is almost constant and we will assume this to be the case from henceforth. Our interest will be primarily in the deformation of the target and therefore we will choose our model to give the simplest representation of this aspect of the deformation and ignore the deformation of the rod. For the assumed steady-state penetration a hydrodynamic analogy is appropriate because it is associated with incompressible behavior; to a good approximation this is true for most metals until very high stresses are attained. Of course, it must be remembered that although the inviscid hydrodynamic theory represents an incompressible solution it is not the only incompressible solution.

Once the steady flow field has been chosen the deformation history of each material element can be calculated. To do this we must use a viewpoint which is not usually adopted in the discipline of hydrodynamics but is familiar from our continuum mechanics approach. That is, we consider material elements which at a chosen initial instant occupy a given position in space $\boldsymbol{X}$. After a time interval t we suppose such a material element to have reached a current position $x(X, t)$. In particular we will be interested in the strain generated by deformation of points in the neighborhood of X as time progresses and in the deformation of the plane of the rear surface of the target plate. This deformation is given by the movement of material points, lying within this rear plane, in the assumed flow field. One practical way of examining the relevancy of the deformation field given by the approximate model proposed is by examining the deformation of etch lines in rolled plate. The rolling process usually produces a laminar structure which can be seen when it is suitably etched. In an undeformed plate a cross-section cut from the plate has the appearance of a set of parallel straight lines. A similar cross-section cut from a plate after it has been partially penetrated by a rod shows clearly a deformation of the etch lines which should correspond with those estimated from this model. Discrepancies in profile may indicate how the model could best be improved to match the real material deformation.

The steady-state hydrodynamic model which we have chosen for our representation is the simplest that gives reasonable agreement with penetration experiments and consists simply of a source of strength m in a uniform stream moving with velocity U. This situation is discussed by Milne-Thomson (1960). The situation is axisymmetric and the coordinates chosen are spherical polar coordinates as shown in Fig. 13-35. The stream function Ψ is given by

$$\Psi = -\tfrac{1}{2}Ur^2 \sin^2\theta + m\cos\theta \qquad (13\text{-}3\text{-}40)$$

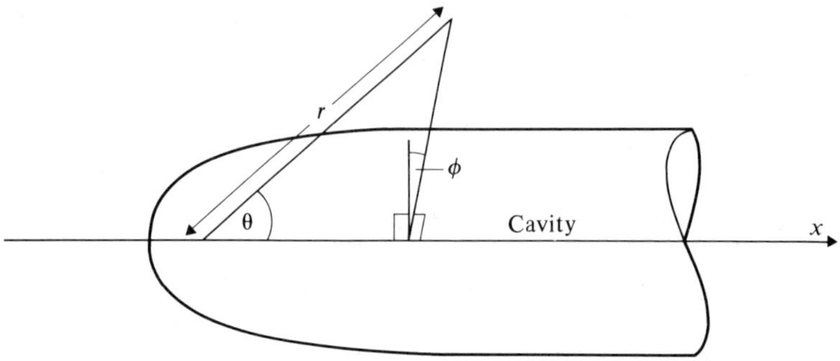

Figure 13-35 Spherical polar coordinate system in relation to the cavity. (*Tate, 1978.*)

and thus the flow velocities both radially and orthogonal to this direction are given respectively by

$$\dot{r} = -\frac{1}{r^2 \sin\theta}\frac{\partial\Psi}{\partial\theta} = U\cos\theta + \frac{m}{r^2} \tag{13-3-41}$$

and

$$r\dot{\theta} = \frac{1}{r\sin\theta}\frac{\partial\Psi}{\partial r} = -U\sin\theta \tag{13-3-42}$$

where the dot denotes a material time derivative.

From Eq. (13-3-40) it follows immediately that the dividing streamline between the source flow and the uniform stream flow, which we will take to represent the hole profile, is given by

$$r^2\sin^2\theta = 2a^2(1+\cos\theta) \tag{13-3-43}$$

where

$$a^2 = \frac{m}{U} \tag{13-3-44}$$

The hole diameter $d_{\rm H}$ is then given by

$$d_{\rm H} = 4a \tag{13-3-45}$$

From Eq. (13-3-40) it also follows that the equation of a streamline whose original distance from the axis far upstream is h is given by

$$r^2\sin^2\theta = 2a^2(1+\cos\theta)+h^2 \tag{13-3-46}$$

The problem can now be scaled by dividing all lengths by a, all velocities by U, and hence all times by a/U. The resulting equations in the scaled variables have the same form that the above equation would take if we set a, m, and U to unity. To avoid encumbering the script we will refrain from distinguishing the scaled variables but note that we may without loss of generality set a, m, and U to unity. The streamlines then take the form

$$r^2(1-\mu^2) = 2(1+\mu)+h^2 \tag{13-3-47}$$

and the rate of movement of material particles along the streamlines is given by

$$\dot{r} = \mu + \frac{1}{r^2} \tag{13-3-48}$$

and

$$r\dot{\mu} = 1 - \mu^2 \tag{13-3-49}$$

where

$$\mu = \cos\theta \tag{13-3-50}$$

Now it can easily be shown by substitution that Eq. (13-3-47) is an integral of the pair of coupled equations (13-3-48) and (13-3-49), and thus we can reduce Eqs. (13-3-48) and (13-3-49) to the integration of the following quadratures:

$$\dot{r} = \begin{cases} -\dfrac{[(r^2-1)^2 - r^2h^2]^{1/2}}{v^2} & \text{for } \mu < \mu_c \quad (13\text{-}3\text{-}51) \\ \dfrac{[(r^2-1)^2 - r^2h^2]^{1/2}}{r^2} & \text{for } \mu > \mu_c \quad (13\text{-}3\text{-}52) \end{cases}$$

where

$$\mu_c = -\frac{4}{[h+(h^2+4)^{1/2}]^2} \tag{13-3-53}$$

$$\dot{\mu} = \frac{(1-\mu^2)^{3/2}}{[2(1+\mu)+h^2]^{1/2}} \tag{13-3-54}$$

In the above relations h is to be regarded as a constant.

The above solution does have one feature which is of great practical interest and importance, namely, that in the sense described below it approaches an asymptotic solution.

Consider two material particles P' and P'' having initial radial coordinates R' and R'' which are both on the same streamline at a distance h from the axis as shown in Fig. 13-36. After time t' let the particle P' be at the position p' having a

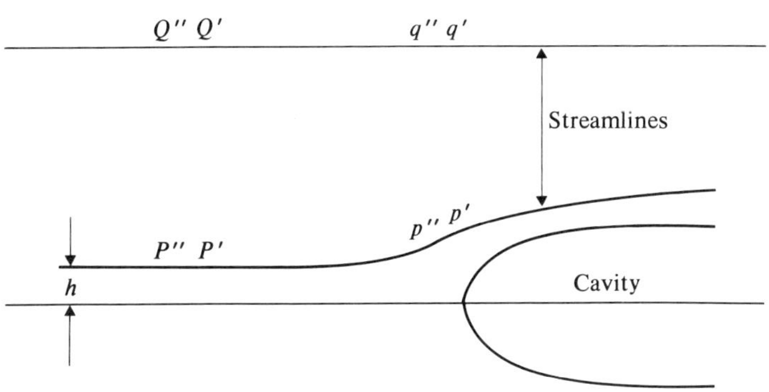

Figure 13-36 Disposition of particle positions used in proving the asymptotic property of the deformation field. (*Tate, 1978.*)

radial distance r' and let the particle P'' be at the position p'' having a radial distance r'' after a time t''. Let Q', Q'', q', and q'' be equivalent points on a streamline which is far from the axis and hence the flow velocity on this streamline is almost constant at unity. Let Eqs. (13-3-51) and (13-3-52) be integrated to symbolically give the result

$$F(r,h) = -t + \text{constant} \tag{13-3-55}$$

Then, for the particle P',

$$F(r',h) - F(R',h) = -t' \tag{13-3-56}$$

Similarly, for the particle P'',

$$F(r'',h) - F(R'',h) = -t'' \tag{13-3-57}$$

Thus it follows that

$$F(r',h) - F(r'',h) = F(R',h) - F(R'',h) - (t'-t'') \tag{13-3-58}$$

and
$$\int_{r''}^{r'} \frac{r^2\,dr}{[(r^2-1)^2 - r^2h^2]^{1/2}} = \int_{R''}^{R'} \frac{r^2\,dr}{[(r^2-1)^2 - r^2h^2]^{1/2}} - (t'-t'') \tag{13-3-59}$$

Now let us choose $R'' > R'$ and R' very much greater than h. Then to first order in h the axial distance between P'' and P' is just $(R''-R')$. Now let us choose the time interval such that, in the undisturbed flow, points such as q' and q'' coincide. In the absence of a source the same time interval would lead to the coincidence of p' and p'' and clearly this time interval is just $(R''-R')$. For these conditions Eq. (13-3-59) takes the form

$$\int_{r''}^{r'} \frac{r^2\,dr}{[(r^2-1)^2 - r^2h^2]^{1/2}} = \int_{R''}^{R'} \frac{r^2\,dr}{r^2-1} - (R'-R'') \tag{13-3-60}$$

$$= \tfrac{1}{2}\ln\left[\frac{(R'-1)}{(R'+1)}\frac{(R''+1)}{(R''-1)}\right] \tag{13-3-61}$$

As R' tends to infinity the right-hand side of Eq. (13-3-61) tends to zero and, as the integrand on the left-hand side of Eq. (13-3-61) is not zero, this implies that $(r'-r'')$ tends to zero. Thus for sufficiently large upstream initial distances the final displacement from the unperturbed motion positions in the vicinity of the hole tends to be independent of the exact initial distance. One way of regarding this is to note that etch lines will be distorted to the same extent when in the vicinity of the hole, regardless of their initial starting point. Similarly, the rear-surface distortion of a thick plate will tend to depend only on the nearness of the hole to the rear surface and not on the initial thickness of the plate. As the above results are true for an arbitrary point P', and every point in the neighborhood of P', it is clear that the strain field will also tend to an asymptotic state.

The strains experienced by a material element moving along a streamline near to the flow axis can be very large and it is therefore appropriate to choose a finite measure of strain. The finite measure which it is convenient to use is Almansi's, which is defined as

$$\mathbf{e} = \tfrac{1}{2}[\mathbf{I} - (\mathbf{F}\mathbf{F}^T)^{-1}] \tag{13-3-62}$$

where $\mathbf{F}$ is the deformation gradient. In the indicial notation this takes the form

$$e_{ij} = \frac{1}{2}\left(g_{ij} - \frac{\partial X^\alpha}{\partial x^i}\frac{\partial X^\beta}{\partial x^j} G_{\alpha\beta}\right) \tag{13-3-63}$$

and, as is well known, this strain measure tends to Cauchy's strain measure for infinitesimal strains. For the spherical polar coordinate system we have considered the following:

$$g_{rr} = 1,\ g_{\theta\theta} = r^2,\ g_{\phi\phi} = r^2 \sin^2\theta, \text{ all other } g_{ij} \text{ are zero} \tag{13-3-64}$$

$$G_{RR} = 1,\ G_{\Theta\Theta} = R^2,\ G_{\Phi\Phi} = R^2\sin^2\Theta, \text{ all other } G_{\alpha\beta} \text{ are zero} \tag{13-3-65}$$

and so

$$\frac{\partial X^\alpha}{\partial x^i}\frac{\partial X^\beta}{\partial x^j} G_{\alpha\beta} = \frac{\partial R}{\partial x^i}\frac{\partial R}{\partial x^j} + R^2\frac{\partial \Theta}{\partial x^i}\frac{\partial \Theta}{\partial x^j} + R^2\sin^2\Theta\frac{\partial \Phi}{\partial x^i}\frac{\partial \Phi}{\partial x^j} \tag{13-3-66}$$

Because the flow is axisymmetric the streamlines always lie in the same ϕ plane so that R and Θ are independent of ϕ and

$$\Phi = \phi \tag{13-3-67}$$

The last term on the right-hand side of Eq. (13-3-66) is thus always zero, except when $x^i = x^j = \phi$. Using Eqs. (13-3-63), (13-3-64), (13-3-66), and (13-3-67) it then follows that

$$e_{\phi\phi} = \tfrac{1}{2}(r^2\sin^2\theta - R^2\sin^2\Theta) \tag{13-3-68}$$

It also follows from the above that

$$e_{r\phi} = e_{\phi r} = e_{\theta\phi} = e_{\phi\theta} = 0 \tag{13-3-69}$$

To determine the remaining tensor components we note that, for a given time interval t, Eq. (13-3-55) takes the form

$$F(r, h) = F(R, h) - t \tag{13-3-70}$$

and differentiating with respect to r, remembering that h is related to r and μ (and hence θ) by Eq. (13-3-47), gives

$$\left(\frac{\partial F}{\partial r}\right)_h + \left(\frac{\partial F}{\partial h}\right)_r \left(\frac{\partial h}{\partial r}\right) = \left(\frac{\partial F}{\partial R}\right)_h \left(\frac{\partial R}{\partial r}\right) + \left(\frac{\partial F}{\partial h}\right)_R \left(\frac{\partial h}{\partial r}\right) \tag{13-3-71}$$

which can be rearranged in the form

$$\frac{\partial R}{\partial r} = \frac{f(r, h) + \{\int_R^r [\partial f(r', h)/\partial h]\, dr'\}\, \partial h/\partial r}{f(R, h)} \tag{13-3-72}$$

where

$$f(r, h) = \frac{r^2}{[(r^2-1)^2 - r^2h^2]^{1/2}} = \left(\frac{\partial F}{\partial r}\right)_h \tag{13-3-73}$$

Similarly, we can show that

$$\frac{\partial R}{\partial \mu} = \left(\int_R^r \frac{\partial f}{\partial h} dr' \right) \frac{\partial h/\partial \mu}{f(R, h)} \tag{13-3-74}$$

The equations of the streamline can be written in the form

$$r^2(1-\mu^2)-2(1+\mu) = R^2(1-M^2)-2(1+M) = h^2 \tag{13-3-75}$$

Differentiation first with respect to r and reordering the equation gives

$$\frac{\partial M}{\partial r} = \frac{R(\partial R/\partial r)(1-M^2)-r(1-\mu^2)}{1+R^2M} \tag{13-3-76}$$

and then differentiating with respect to μ gives

$$\frac{\partial M}{\partial \mu} = \frac{R(\partial R/\partial \mu)(1-M^2)+(1+r^2\mu)}{1+R^2M} \tag{13-3-77}$$

Consider now the limiting case as $R \to \infty$, $M \to -1$, and $R^2(1-M^2) \to h^2$, and let

$$\frac{\partial R}{\partial r} = R_r \qquad \frac{\partial R}{\partial \mu} = -\frac{1}{\sin\theta}\frac{\partial R}{\partial \theta} = R_\mu \tag{13-3-78}$$

Then, from Eqs. (13-3-76) and (13-3-77),

$$M_r \to \frac{r(1-\mu^2)}{R^2} \qquad M_\mu \to \frac{-(1+r^2\mu)}{R^2} \tag{13-3-79}$$

Substituting in Eq. (13-3-63) and using Eq. (13-3-66),

$$2e_{rr} \to \left\{1-R_r^2-\left[\frac{r(1-\mu^2)}{h}\right]^2\right\} \tag{13-3-80}$$

$$2e_{r\theta} \to \sin\theta\left\{R_rR_\mu-\frac{r(1-\mu^2)(1+r^2\mu)}{h^2}\right\} \tag{13-3-81}$$

$$2e_{\theta\theta} \to \sin^2\theta\left\{1-R_\mu^2+\left[\frac{(1+r^2\mu)}{h}\right]^2\right\} \tag{13-3-82}$$

The limiting forms of R_r and R_μ can then be obtained from Eqs. (13-3-72) and (13-3-73) in a straightforward manner and hence the whole asymptotic strain field is calculable and, as shown by Tate (1978), can be expressed in terms of elliptic integrals. The principal strain e_ϕ in the ϕ direction takes the particularly simple form

$$e_\phi = \frac{e_{\phi\phi}}{g_{\phi\phi}} = \frac{1}{2}\left(1-\frac{h^2}{r^2\sin^2\theta}\right) = \frac{\operatorname{cosec}^2(\theta/2)}{2r^2} \tag{13-3-83}$$

It has thus been shown that the hydrodynamic model for the deformations of material in the vicinity of a deep hole being created in a thick target by a penetrator possesses two important features:

1. The strain field scales linearly with the hole diameter.

2. The solution has an asymptotic character which takes the form that, provided the hole is deep enough, the strain field around the end of the hole is unchanging as the hole extends further into the target plate.

13-3-7 The Penetration and Perforation of Plates by Nondeforming Projectiles

If the projectile is very strong in comparison with the target plate then there can exist a region of impact velocities for which the projectile is effectively non-deforming during plate penetration. Indeed, all the earlier work on penetration was based on this concept. Even with this assumption several modes of deformation occur, depending on whether the plate is very thin ($\ll$ projectile diameter) or is on the order of the projectile diameter, on the material properties of the plate, and on the geometry of the projectile and in particular the shape of the nose.

(a) Very thin plates (i.e., plate thickness $D \ll$ rod diameter d) Taylor (1948b) examined both theoretically and experimentally the quasi-static formation of a circular hole in a thin sheet of lead. Two modes of deformation were found to exist. If the hole diameter was less than about 7 to 10 times the thickness of the plate, then a symmetrical distortion of the material on either side of the plate could occur as shown in Fig. 13-37. However, for holes of greater diameter the plate is always bent, giving rise to the asymmetric shape shown in Fig. 13-37*b* and now called "dishing."

(b) Symmetric thickening For this case the metal near the edge of the hole is thickened out to a radius r_2 beyond which the thickening is very small and outside a radius r_1 only elastic straining occurs. Taylor broadly followed an analysis of the problem by Bethe based on the following assumptions:

1. The sheet is thin and the radius of the hole b is large compared with the plate thickness, the sheet being of infinite extent.
2. Outside r_1 the sheet is in a state of elastic stress due to a radial pressure at r_1.

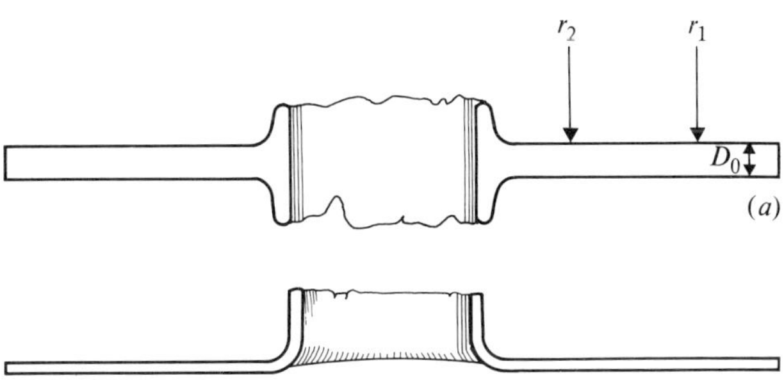

Figure 13-37 (*a*) Symmetrical material displacement around a hole in a thin plate. (*After Taylor, 1948b.*) (*b*) Dishing mode of perforation of a thin plate. (*After Taylor, 1948b.*)

3. The principal stress σ_2 perpendicular to the sheet is zero.
4. In the region $r_2 < r < r_1$,

$$\sigma_1 - \sigma_3 = Y \tag{13-3-84}$$

where Y is the yield stress.
5. The plastic deformation occurs incompressibly.

Using the equations of equilibrium and elasticity with assumptions 2 and 3 gives

$$\left.\begin{aligned} \sigma_3 &= -\frac{A(b)}{r^2} \\ \sigma_1 &= \frac{A(b)}{r^2} \\ \sigma_2 &= 0 \end{aligned}\right\} \quad \text{for } r_1 < r \tag{13-3-85}$$

where $A(b)$ is a constant for a given state of stress. Then, using the equation of equilibrium and the plasticity condition of assumption 4,

$$\left.\begin{aligned} \sigma_3 &= Y\left[\ln\left(\frac{r}{r_1}\right) - \tfrac{1}{2}\right] \\ \sigma_1 &= Y\left[\ln\left(\frac{r}{r_1}\right) + \tfrac{1}{2}\right] \\ \sigma_2 &= 0 \end{aligned}\right\} \quad \text{for } r_2 < r < r_1 \tag{13-3-86}$$

If σ_3 is continuous at $r = r_1$ then

$$A = \tfrac{1}{2} Y r_1^2 \tag{13-3-87}$$

Bethe pointed out that the maximum difference between principal stresses only occurred between σ_1 and σ_3 as in assumption 4 provided that $\sigma_1 > 0$. If $\sigma_1 < 0$ then the maximum difference is between σ_2 and σ_3. The radius r_2 at which σ_1 is zero is given from Eq. (13-3-86) by

$$r_2 = r_1 \exp\left(-\tfrac{1}{2}\right) = 0.606 r_1 \tag{13-3-88}$$

or

$$r_1 = 1.65 r_2$$

The difference between Bethe and Taylor's approach arises in the central region $b < r < r_2$, where b is the radius of the hole. Bethe assumed that in this region $\sigma_1 = 0$ and $\sigma_3 = -Y$. The equation of equilibrium of a thin sheet of variable thickness D, namely,

$$\frac{\partial}{\partial r}(D\sigma_3) + \frac{D}{r}(\sigma_3 - \sigma_1) = 0 \tag{13-3-89}$$

is satisfied provided

$$D = \frac{D_0 r_2}{r} \tag{13-3-90}$$

where D_0 is the initial thickness.

The equation of continuity gives

$$\pi r_2^2 D_0 = \int_b^{r_2} \left(\frac{D_0 r_2}{r}\right) 2\pi r\, dr$$

or

$$b = \frac{r_2}{2} \tag{13-3-91}$$

Hence

$$[D]_{r=b} = 2D_0 \tag{13-3-92}$$

Taylor, however, assumed that the strain increments were proportional to the deviatoric stress and that the expansion from a pinhole was a similarity solution in which the thickness ratio D/D_0 and the radial displacement ratio u/b were both functions of s/b only, where s was the initial radial distance of an element of the plate. He also used Tresca's stress criterion in the form

$$\begin{aligned} \sigma_1 - \sigma_3 &= Y \qquad \text{if } \sigma_1 \text{ is positive, i.e., tensile} \\ -\sigma_3 &= Y \qquad \text{if } \sigma_1 \text{ is negative, i.e., compressive} \end{aligned} \tag{13-3-93}$$

Using these conditions he found the inner stress field was very different from Bethe's form, having the form shown in Fig. 13-38. He also found a different value for r_2 of 2.12b as compared with Bethe's value of 2b. The work done in expanding the hole to a radius b, W_{HS}, is given by

$$W_{HS} = \begin{cases} 2\pi b^2 D_0 Y & \text{(Bethe)} \\ 1.33\pi b^2 D_0 Y & \text{(Taylor)} \end{cases} \tag{13-3-94}$$

(c) Dishing or unsymmetrical deformation mode Taylor also considered the unsymmetrical or dishing mode, as shown in Fig. 13-39, using the following assumptions:

1. The circumferential principal stress σ_1 is the only stress of significance, so that σ_3 and σ_2 can be taken to be zero.
2. Plastic deformation occurs under a stress $\sigma_1 = Y$.
3. The material is incompressible.

The first assumption requires that the principal strains

$$\epsilon_2 = \epsilon_3 = \epsilon \tag{13-3-95}$$

and thus

$$\frac{dz}{ds} = 1 - \epsilon = \frac{\zeta}{D_0} \tag{13-3-96}$$

From assumption 3 it follows that

$$2\pi s D_0\, ds = 2\pi b \zeta\, dz \tag{13-3-97}$$

Therefore, using Eq. (13-3-96),

$$\zeta = D_0 \left(\frac{s}{b}\right)^{1/2} \tag{13-3-98}$$

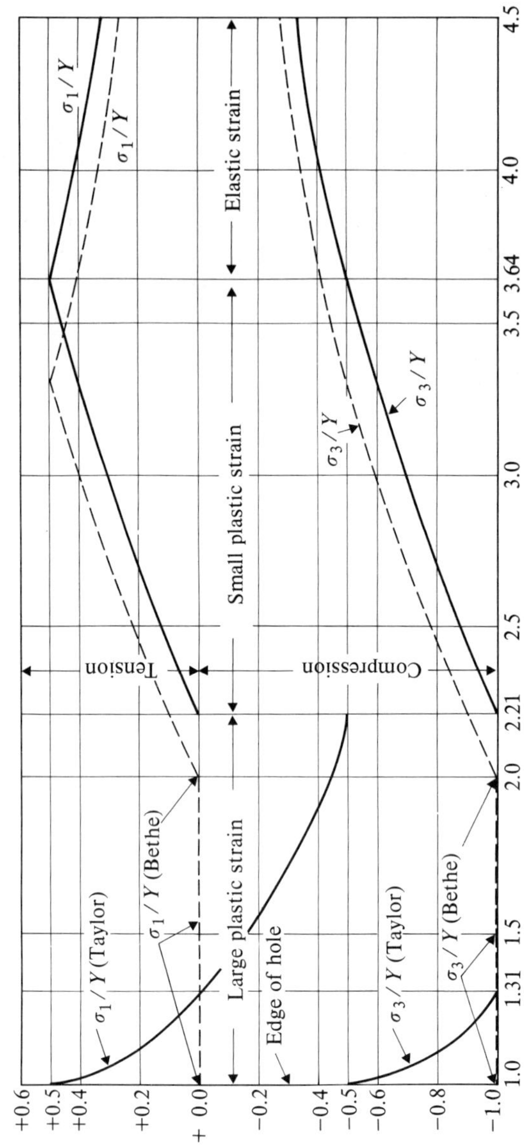

Figure 13-38 Comparison of Bethe and Taylor theories for the stress around a hole in a thin plate. (*Taylor, 1948b.*)

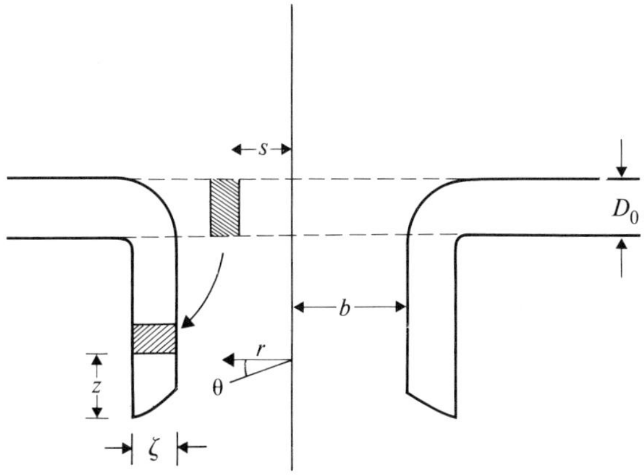

Figure 13-39 Geometrical parameters used in the theory of dishing.

Substituting into Eq. (13-3-96) and integrating over the hole radius b gives the lip height $\hat{H}$ from the equation

$$\int_0^{\hat{H}} b^{1/2}\, dz = \int_0^b s^{1/2}\, ds$$

or

$$\hat{H} = \tfrac{2}{3}b \tag{13-3-99}$$

The thickness of the lip is then given by the expression

$$\frac{\zeta}{D_0} = \left(\frac{3z}{2b}\right)^{1/3} \tag{13-3-100}$$

The work done in dishing W_D is thus given by the equation

$$W_D = \int_0^b 2\pi Y D_0 s \ln\left(\frac{b}{s}\right) ds \tag{13-3-101}$$

or

$$W_D = \tfrac{1}{2}\pi b^2 D_0 Y \tag{13-3-102}$$

Thus the dishing mode requires less energy than the symmetrical mode and as Taylor points out this is probably the reason that this mode invariably occurs if the hole is large enough. Comparison of Eqs. (13-3-94) and (13-3-102) indicate, however, that the dishing mode should always involve less energy than the symmetric mode and thus does not account for the change from one mode to the other.

The above theory was extended by Thomson (1955) to provide an energy-based theory for the penetration of thin plates by nondeforming projectiles. He added an extra term to take account of the energy used in moving the material radially. The force F_r required for radial movement is proportional to the rate of change of momentum so that

$$F_r = \frac{d}{dt}\left(M\frac{db}{dt}\right) \tag{13-3-103}$$

where M is the mass of material moved:

$$M = \pi \rho b^2 D_0 \tag{13-3-104}$$

It is here assumed that all the displaced material is at radius b. The work W_{KD} to produce an enlargement db of the hole is thus

$$\frac{dW_{KD}}{db} = F_r = \frac{d}{dt}\left(M \frac{db}{dt}\right) \tag{13-3-105}$$

If the projectile nose profile is given by

$$z = f(r) \tag{13-3-106}$$

and the projectile has a uniform velocity V, then

$$V = f'(b)\frac{db}{dt} \tag{13-3-107}$$

Substituting in Eq. (13-3-105) and integrating over the projectile radius R gives

$$W_{KD} = \pi \rho D_0 V^2 \int_0^R \left(\frac{2b}{f'^2} - \frac{b^2 f''}{f'^3}\right) db \tag{13-3-108}$$

Thus the total work done W_T in penetrating a thin plate is, as Thomson shows, predominantly just the sum of these two terms

$$W_T = W_D + W_{KD} \tag{13-3-109}$$

Zaid and Paul (1957) considered the same problem from a momentum-based point of view, letting r_{lim} be the radius of the elastic wave front as shown in Fig. 13-40. Then the absence of external forces outside this region implies conservation

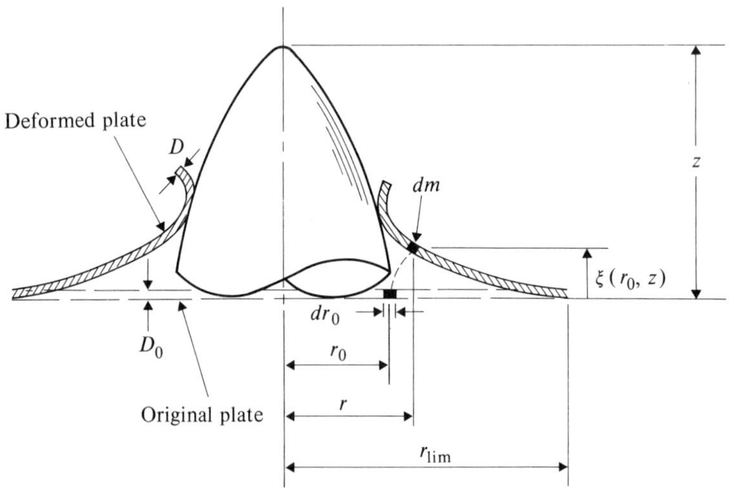

Figure 13-40 Projectile perforating a plate. (*After Zaid and Paul, 1957.*)

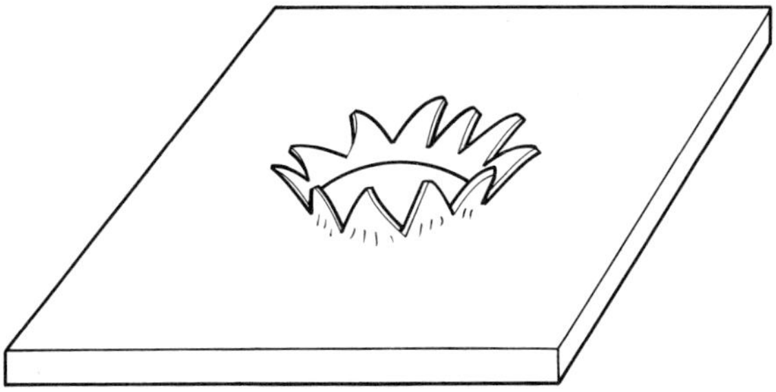

Figure 13-41 Petaling of a plate. (*After Zaid and Paul, 1957.*)

of momentum within:

$$mV = mv + p_t(z) \tag{13-3-110}$$

where m is the projectile mass, V is its striking velocity, v its velocity at any subsequent instant, z its penetration distance in the axial direction, and $p_t(z)$ is the axial component of plate momentum. The momentum in the z direction of plate elements located in the undeformed annulus at a radius r_0 is given by

$$dp_t = 2\pi\rho D_0 r_0 \dot{\xi}\, dr_0 \tag{13-3-111}$$

and the particle velocity in the z direction is

$$\dot{\xi} = \frac{\partial \xi}{\partial z}\frac{\partial z}{\partial t} = v\frac{\partial \xi}{\partial z} \tag{13-3-112}$$

The total momentum given to the plate $p_t(z)$ is thus given by

$$p_t(z) = 2\pi\rho D_0 \int_{r_{0,\min}}^{r_{\lim}} v r_0 \frac{\partial \xi}{\partial z}\, dr_0 \tag{13-3-113}$$

To progress further, Zaid and Paul assumed that the plate was either undistorted or assumed the shape of the projectile profile, behaving, as they graphically put it, like a limp rag. It was also assumed that the failure was by petaling, as shown in Fig. 13-41, and there was no radial stretching of the petals. For a conical-headed projectile having a semivertex angle α it follows from simple geometry that

$$\xi = (z \tan\alpha - r_0)\cos\alpha \tag{13-3-114}$$

and, substituting in Eqs. (13-3-110) and (13-3-113),

$$V - v = \frac{\pi\rho D_0 (z\tan\alpha)^2 v \sin\alpha}{m} \tag{13-3-115}$$

If the projectile radius is R then the total velocity drop is

$$V - V_r \approx \pi\rho D_0 R^2 V \sin\alpha / m \tag{13-3-116}$$

where V_r is the residual velocity and, for a small velocity drop, v in Eq. (13-3-115) has been replaced by V. For this model to apply the plate would have to be very thin and have a low yield point with little or no work-hardening. It is clear from

Eq. (13-3-116) that penetration is expected for all velocities, however small, which is not in accord with experimental results. It may be more applicable at higher striking velocities.

The total axial force F_A is given by

$$F_A = -mv\frac{dv}{dz} \tag{13-3-117}$$

but if $V - v$ is very small Eq. (13-3-115) can be rewritten in the form

$$v = V\left(1 - \frac{\pi\rho D_0 z^2 \tan^2\alpha \sin\alpha}{m}\right) \tag{13-3-118}$$

Substituting in Eq. (13-3-117) and neglecting the second term in the parentheses in comparison with unity gives

$$F_A \approx 2\pi\rho D_0 V^2 z \tan^2\alpha \sin\alpha \tag{13-3-119}$$

The radial velocity of the petals is $v \tan\alpha(1 - \sin\alpha)$ and the radial force F_r exerted on the petals is, to the above approximation,

$$F_r = 2\pi\rho D_0 V^2 z \tan^3\alpha(1 - \sin\alpha) \tag{13-3-120}$$

The angle $\bar{\beta}$, the resultant force on the petals, makes with the z axis is thus given by

$$\tan\bar{\beta} = \frac{F_r}{F_A} = \frac{1 - \sin\alpha}{\cos\alpha} \tag{13-3-121}$$

or

$$\bar{\beta} = \frac{\pi}{4} - \frac{\alpha}{2} \tag{13-3-122}$$

Photographs of a conical-headed projectile penetrating a thin plate did show the petals breaking away at about this angle.

In the above theories the bending of the plate has been neglected, which is clearly not in accord with experimental results except when there is a gross overmatch of the plate, i.e., the striking velocity is very much greater than that required to just perforate the plate.

Considerable theoretical work has been done on the dynamic response of thin plates, usually under one of two simplifying assumptions. Either the membrane forces governing the stretching of the plate are assumed to be dominant or the bending moments are considered to be dominant. An analysis based solely on membrane theory, such as that of Boyd (1966), leads in general to nonlinear relations requiring numerical solutions. The bending-moment approach has been explored by Hopkins and Prager (1953, 1954), Wang and Hopkins (1954), Hopkins (1957), Florence (1966), Weirzbicki (1967), and Kelly and Wilshaw (1968). Experimental data suggests that the simple bending approach overestimates the final plate deflection and a better correlation is obtained if membrane effects are included, as in the work of Jones (1968). More recently, Calder and Goldsmith (1971) have used a simplified energy approach but include work-hardening effects. Assuming thin-plate plane stress and symmetry conditions and neglecting circumferential strains, the work done in plastic straining of the plate W is given by

$$W = \frac{2\pi D_0}{(1-v+v^2)^{1/2}} \int_0^{r_p} (Y\epsilon_3 + \tfrac{1}{2}\bar{\alpha}\epsilon_3^2) r\, dr \tag{13-3-123}$$

where Y is the initial yield stress, $\bar{\alpha}$ the linear work-hardening parameter, v the Poisson ratio, D_0 the plate thickness, and r_p the plate radius. Making the very crude assumption that for large deflections the radial stretching strain can be written as

$$\epsilon_3 = \frac{1}{2}\left(\frac{\partial \xi}{\partial r}\right)^2 \qquad \text{(only true if } \partial\xi/\partial r \gg 1\text{)}$$

where ξ is the plate deflection and r is the radial coordinate, they obtain the work done in the form

$$W = \frac{\pi D_0}{(1-v+v^2)^{1/2}} \int_0^{r_p} \left[Y\left(\frac{\partial \xi}{\partial r}\right)^2 + \tfrac{1}{4}\bar{\alpha}\left(\frac{\partial \xi}{\partial r}\right)^4 \right] r\, dr \tag{13-3-124}$$

Assuming a permanent deflection profile $\xi(r) = \xi_0 f(r)$ and setting the kinetic energy loss of the projectile $(\Delta \mathrm{KE})_{\mathrm{proj}}$ equal to the work done gives

$$\left[\int_0^{r_p} \left(\frac{\partial f}{\partial r}\right)^4 r\, dr\right]\xi_0^4 + \left[\frac{4Y}{\bar{\alpha}} \int_0^{r_p} \left(\frac{\partial f}{\partial r}\right)^2 r\, dr\right]\xi_0^2 - \frac{4(1-v+v^2)^{1/2}}{\bar{\alpha}\pi D_0}(\Delta \mathrm{KE})_{\mathrm{proj}} = 0 \tag{13-3-125}$$

By taking a specific form for $f(r)$, ξ_0 can be evaluated. Calder and Goldsmith took $f = \exp(-r)$ for their particular experiments and in this way obtained good agreement with experimental results. However, as the form of f chosen depends on the units of measurement and is independent of material or geometric parameters specifying the impact it is unlikely to have very general application in its present form.

(c) Plates of about a projectile diameter thick The need for a practical penetration formula has led to the development of some purely empirical equations. Of these one of the most widely used has been that of de Marre (1886) who found that if V_b is the minimum velocity required to perforate the plate

$$\tfrac{1}{2}mV_b^2 = \bar{A}\, d^{1.5} D_0^{1.4} \tag{13-3-126}$$

where m is the projectile mass, d is its diameter, and D_0 is the plate thickness. Because the constant $\bar{A}$ then has such a peculiar set of units as to have no physical significance it is now more usually expressed in the form

$$\frac{1}{2}\frac{mV_b^2}{d^3} = \bar{A}\left(\frac{D_0}{d}\right)^{1.4} \tag{13-3-127}$$

Extensive tests quoted by Richter (1946) showed that $\bar{A}$ depended on both V_b and d and the formula was therefore unsuitable for extrapolating outside the range of values for which it had been established. Other more complicated formulas have also been tried but their applicability suffers the same restrictions as the de Marre formula.

An early form of physically based theory was proposed by Martel (1897) who assumed that the work necessary for the penetration of any type of punch into a plastically deformable material was proportional to the penetrated volume. The ballistic limit velocity V_b is then given by

$$\frac{1}{2}\frac{mV_b^2}{d^3} = C_M \frac{D_0}{d} \qquad (C_M \text{ is a constant}) \qquad (13\text{-}3\text{-}128)$$

which is not in general satisfied experimentally.

A significant advance was made by Gabeaud (1935) who related the static stress necessary to produce a radial expansion, so producing a hole, to the Brinell hardness Δ and then, by adding an inertia term dependent on the square of the velocity, determined the total resistive force acting on the projectile during penetration. Burkhardt, an account of whose work can be found in the report by Richter (1946), expanded this idea by considering the resistive force to consist of four components:

1. A compression force, which is taken as a constant multiplied by the difference in area between the front and rear cross-sectional areas of the projectile in contact with the front and rear of the plate.
2. An expansion force, which is proportional to the projectile surface area in contact with the plate.
3. A friction force.
4. An inertia force proportional to the projectile velocity squared.

Using the resulting formulas he found fair agreement with his experiments. However, the opening of the hole in the target was considered to occur only radially and effects such as back-surface distortion or plugging were ignored.

When blunt-ended projectiles or fragments are used the above type of approximation clearly breaks down. In such cases a plug of material is usually ejected and a theory taking account of this fact was suggested by Recht and Ipson (1963). They considered normal and oblique impact conditions as shown in Fig. 13-42. Considering first normal impact and assuming inelastic impact between the projectile and the plugged-out slug as if the slug were entirely free gives the final velocity V_f from the equation of conservation of momentum as

$$V_f = \frac{m}{m + M_{sn}} V \qquad (13\text{-}3\text{-}129)$$

where m is the projectile mass, M_{sn} is the slug mass at normal impact, and V is the projectile striking velocity. The energy E_{fn} lost to deformation and heating is thus

$$E_{fn} = \frac{1}{2}\frac{M_{sn}}{m + M_{sn}} mV^2 \qquad (13\text{-}3\text{-}130)$$

During perforation of a plate there will be an additional energy loss W_n due to the peripheral shear zone and thus conservation of energy requires

$$\tfrac{1}{2}mV^2 = E_{fn} + W_n + \tfrac{1}{2}(M_{sn}+m)V_r^2 \tag{13-3-131}$$

where V_r is the residual velocity of the projectile and the plug. At the ballistic limit velocity V_b, V_r is zero and, using Eq. (13-3-131),

$$W_{bn} = \frac{1}{2}\frac{m}{m+M_{sn}}mV_{bn}^2 \tag{13-3-132}$$

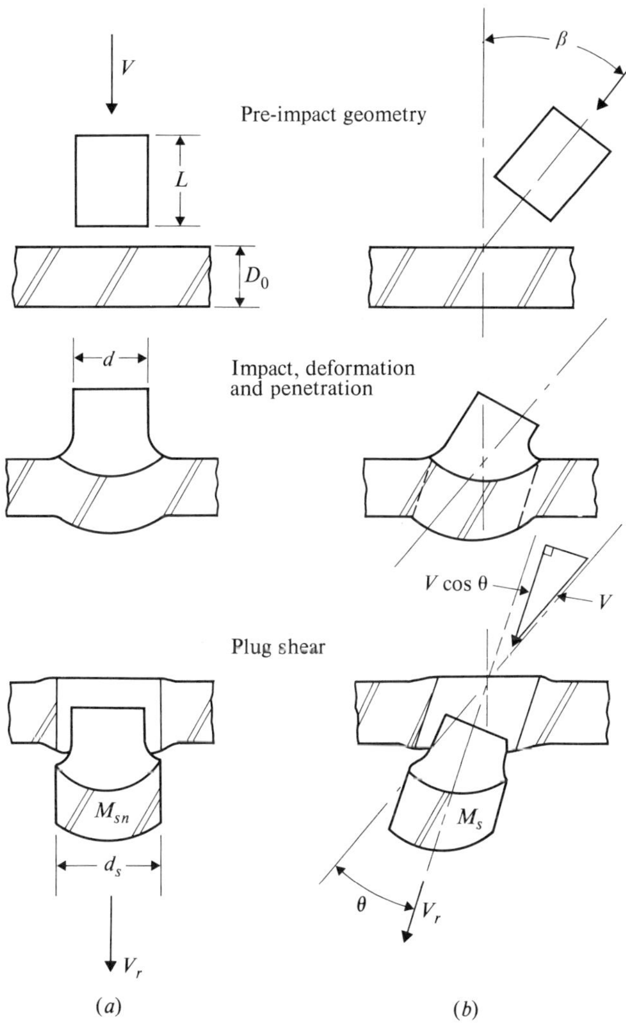

Figure 13-42 Plugging in normal and oblique impact as envisaged by Recht and Ipson. (*Recht and Ipson, 1963.*)

Assuming W_n is independent of velocity, this gives the residual velocity as

$$V_r = \frac{m}{m + M_{sn}} (V^2 - V_{bn}^2)^{1/2} \tag{13-3-133}$$

where V_{bn} is given by Recht and Ipson as

$$V_{bn} = \frac{4\mu^2 D_0^2 \Psi\eta}{Ld} \left\{ 1 + \left[\frac{L + \mu^2 D_0}{\mu^2 D_0} \left(1 + \frac{d}{4\rho_s D_0 \eta \Psi^2} \right) \right] \right\} \tag{13-3-134}$$

where μ^2 is the ratio of the plate to projectile material density, L is projectile length, D_0 is plate thickness, η is a constant having dimensions of stress, and

$$\Psi = \frac{\rho_p c_p + \rho_s c_s}{\rho_p c_p \rho_s c_s} \tag{13-3-135}$$

where

ρ_p, ρ_s = projectile and plate material densities

c_p, c_s = longitudinal acoustic wave velocities in the projectile and plate materials

Oblique perforation is taken to be approximately equivalent to normal penetration but V is replaced by $V \cos\theta$, V_{bn} by $V_b \cos\theta$, and M_{sn} by $M_{sn}/\cos\beta$, giving

$$V_r = \frac{\cos\theta (V^2 - V_b^2)^{1/2}}{1 + (M_{sn}/m \cos\beta)} \tag{13-3-136}$$

Making the assumption that the impulse producing the deflection of the projectile is proportional to the time for penetration, and this in turn is proportional to the average shear velocity, it was eventually shown that the deflection angle θ was given by

$$\sin 2\theta = \frac{\sin 2\theta_b}{(V/V_b)^2 + (V/V_b^2)(V^2 - V_b^2)^{1/2}} \tag{13-3-137}$$

where θ_b is the deflection at the ballistic limit velocity V_b. V_b and θ_b must be determined experimentally.

Recent work by Awerbuch (1970), Goldsmith and Finnegan (1971), and Awerbuch and Bodner (1974) is basically an extension of Burkhardt's methods by postulating a more detailed mechanism of penetration. The force resisting penetration is considered to have three main components:

1. An inertial component

$$F_i = \tfrac{1}{2}\rho_s k A_0 v^2 \tag{13-3-138}$$

where ρ_s is the target material density, A_0 is the projectile cross-sectional area, and k is a constant dependent on the nose geometry:

$$k = \begin{cases} 1 & \text{for a flat nose} \\ \tfrac{1}{2} & \text{for a hemispherical nose} \\ \sin^2\alpha & \text{for a conical nose whose half-apex angle is } \alpha \end{cases}$$

2. A compressive force

$$F_c = \sigma_u A_0 \tag{13-3-139}$$

where σ_u is the ultimate compressive strength of the target material.

3. A shearing force

$$F_s = \sigma_s \pi\, d_s D_s g\left(\frac{z}{D_s}\right) \tag{13-3-140}$$

where σ_s is a shear strength, D_s is the thickness of slug pushed out, z is the penetration depth, d_s is the slug diameter, and $g(z/D_s)$ is an empirical factor.

The perforation process is divided into two parts: an initial indentation or compression phase followed by a plugging phase. In the compression phase the shearing force is taken to be zero and, if penetration depth is denoted by z, the momentum equation is

$$\begin{aligned}\frac{d}{dt}(\rho_s A_0 z + m)v &= -F_i - F_c \\ &= -\tfrac{1}{2}\rho_s k A_0 v^2 - \sigma_u A_0 \qquad \text{for } 0 \leqslant z \leqslant (D_0 - D_s)\end{aligned} \tag{13-3-141}$$

where the term $\rho_s A_0 z$ takes account of the inertia of the displaced target material.

Integration of Eq. (13-3-141) gives

$$v = \left\{\left[V^2 + \frac{\sigma_u}{\rho_s(1+\frac{1}{2}k)}\right]\left(\frac{m}{\rho_s A_0 z + m}\right)^{2+k} - \left[\frac{\sigma_u}{\rho_s(1+\frac{1}{2}k)}\right]\right\}^{1/2} \tag{13-3-142}$$

In the second phase only a shearing force is assumed to be active and to operate over a surface determined by the slug thickness D_s and slug diameter d_s. In Awerbuch (1970), $g(z/D_s)$ was taken to be unity, but later work suggests that a more realistic assumption is to assume that the force is proportional to the width of the plug still in contact with the target. The quantities d_s and D_s must be determined from experiment. Awerbuch and Bodner considered further modifications to take account of rate effects on σ_s and divided the perforation process into three stages, as shown in Fig. 13-43, where τ is the shear stress and s the radial clearance within which the shear stress acts on the slug during its formation and its ejection. The resulting theory appeared to give reasonable agreement with experiment.

Inspection of the above diagrams raises some conceptual difficulties. For example, if some material is being swept ahead of the projectile then what happens to the material even further in front? The compressibility of the material at the stress levels attained in projectile impacts is usually quite small. Although blunt-ended projectiles do sometimes form small conical caps of target material ahead of the projectile these are always very small, never being much more than a projectile radius in height.

Another explanation of the terms involved may arise as follows (Tate, 1979).

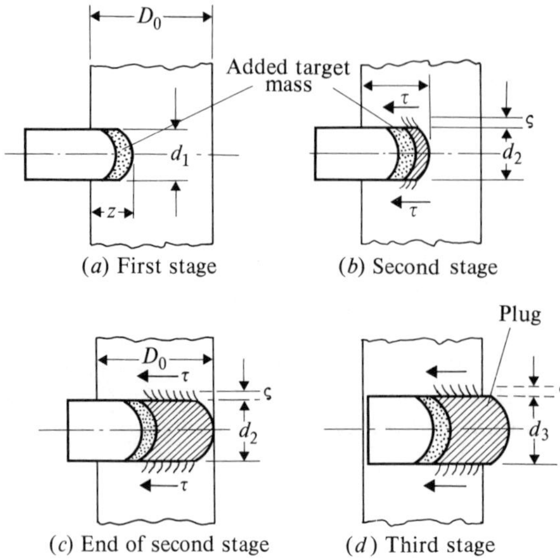

Figure 13-43 Stages in the perforation of a plate as envisaged by Awerbuch and Bodner. (*Awerbuch and Bodner, 1974.*)

Consider the deceleration of a rigid projectile as shown in Fig. 13-44 where the starred coordinates move with the projectile velocity $v(t)$ thus:

$$z^* = z - \int_0^t v(\bar{t})\, d\bar{t} \qquad y^* = y \qquad x^* = x \qquad t^* = t \tag{13-3-143}$$

The target material flow velocities $\mathbf{w}^*$ and $\mathbf{w}$ relative to the x and x^* coordinates respectively are then given by

$$w_z^* = w_z - v(t) \qquad w_y^* = w_y \qquad w_x^* = w_x \tag{13-3-144}$$

The momentum equation for the projectile can be written in the form

$$\frac{d}{dt}(mv) = -\int_\Sigma \mathbf{T}^T \mathbf{v}\, da \tag{13-3-145}$$

where $v = (0, 0, v)$ is the projectile velocity, $\mathbf{T}$ the stress tensor, $\mathbf{v}$ the outward normal to the target material, and Σ the common surface area of the projectile and target. Under the transformation of Eq. (13-3-143) let $\Sigma \rightarrow \Sigma^*$ and $\mathbf{T} \rightarrow \mathbf{T}^*$. It follows from the transformation rules of a tensor that

$$\mathbf{T} = \mathbf{T}^* \tag{13-3-146}$$

and Eq. (13-3-145) can be written as

$$m\frac{dv}{dt} = -\int_{\Sigma^*} (\mathbf{T}^T)^* \mathbf{v}^*\, da^* \tag{13-3-147}$$

under the above transformation

$$\frac{\partial}{\partial t} + \mathbf{w} \cdot \text{grad} = \frac{\partial}{\partial t'} + w^* \cdot \text{grad}^* \tag{13-3-148}$$

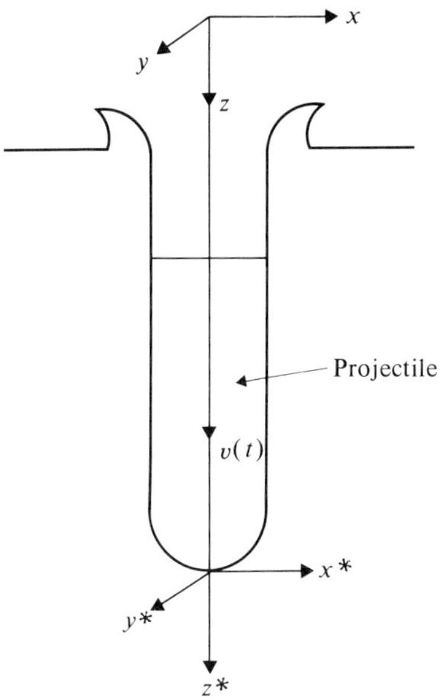

Figure 13-44 Coordinate system moving with the projectile.

and the momentum equation of motion in the target material transforms to

$$\frac{\partial \mathbf{w}^*}{\partial t^*} + (\mathbf{w}^* \cdot \text{grad}^*)\mathbf{w}^* = \frac{1}{\rho}\text{div}^*\,\mathbf{T}^* - \frac{dv(t^*)}{dt^*} \tag{13-3-149}$$

Assuming that the stress is isotropic (this will tend to be more nearly satisfied as the pressure increases) and that the target material flow is irrotational implies that

$$\mathbf{T}^* = -P^*\mathbf{I} \tag{13-3-150}$$

and a velocity potential exists such that

$$\mathbf{w}^* = -\text{grad}^*\,\phi^* \tag{13-3-151}$$

Substituting in Eq. (13 3-149) and integrating,

$$P^* = \rho\frac{\partial \phi^*}{\partial t^*} - \tfrac{1}{2}\rho w^{*2} - z^*\frac{dv}{dt^*} + f(t^*) \tag{13-3-152}$$

Now far from the crater $w^* \to -v(t')$ and if the stream surface is to coincide with the projectile surface and be independent of $v(t^*)$ then ϕ^* takes the form

$$\phi^* = v(t^*)\left[z^* + \hat{\phi}^*(z^*)\right] \tag{13-3-153}$$

where $\hat{\phi}^* \to 0$ at large distances from the crater. Substituting in Eq. (13-3-152) gives

$$P^* = \rho\hat{\phi}^*\frac{dv}{dt^*} - \tfrac{1}{2}\rho w^{*2} + f(t^*) \tag{13-3-154}$$

If a pressure σ_u is required before motion can take place, then assuming this occurs far from the crater where $\hat{\phi}^*$ is zero and $w^{*2} = v^2$ implies that f is given by

$$f = \sigma_u + \tfrac{1}{2}\rho v^2 \tag{13-3-155}$$

P^* takes the form

$$P^* = \sigma_u + \tfrac{1}{2}\rho(v^2 - w^{*2}) + \rho\hat{\phi}^* \frac{dv}{dt^*} \tag{13-3-156}$$

where $\hat{\phi}^*$ is a harmonic function satisfying

$$\text{grad}^* \, \hat{\phi}^* \cdot \mathbf{v}^* = \sin\psi \tag{13-3-157}$$

on the crater surface, ψ being the angle subtended with the z axis. Now substitute Eqs. (13-3-150) and (13-3-156) into Eq. (13-3-147). Considering the z component only and remembering that $\mathbf{v}^*$ is directed out of the target material and so $da_z^* = -v_z^* \, da^*$,

$$m\frac{dv}{dt} = -F_c - F_i - \left(\int_{\Sigma^*} \rho\hat{\phi}^* \, da^* \right) \frac{dv}{dt} \tag{13-3-158}$$

where

$$F_c = \sigma_u \int_{\Sigma^*} da_z^* = A_0 \sigma_u \tag{13-3-159}$$

and

$$F_i = \tfrac{1}{2}\rho \int_{\Sigma^*} (v^2 - w^{*2}) \, da_z^* \tag{13-3-160}$$

To proceed further we need to know $\hat{\phi}^*$, which is difficult to find for an arbitrary surface. Awerbuch and Bodner effectively assume w^* has a velocity distribution on the surface

$$w^* = v\cos\psi \tag{13-3-161}$$

and F_i then becomes identical to Eq. (13-3-138). Along a streamline at any instant

$$w^* = -\frac{d\phi^*}{dl} \tag{13-3-162}$$

where l is measured along the streamline. The projectile surface is a streamline and thus, using Eq. (13-3-161),

$$d\phi^* = -w^* \, dl = -v\cos\psi \, dl = v \, dz^* \tag{13-3-163}$$

Thus

$$\phi^* = vz^* + \text{constant} \tag{13-3-164}$$

at any instant.

Comparing Eqs. (13-3-164) and (13-3-153) shows that $\hat{\phi}^*$ is constant on the projectile surface and thus Eq. (13-3-158) can be written as

$$(m + \rho A_0 \hat{\phi}^*)\frac{dv}{dt} = -F_c - F_i \tag{13-3-165}$$

which is to be compared with Eq. (13-3-141); $\hat{\phi}^*$ must be determined by the potential flow ahead of the projectile. Comparison with the flow around a source in a uniform stream indicates that, for a blunt-ended projectile, $\hat{\phi}^*$ is on the order of half the maximum crater radius. Thus the added mass $\rho A_0 \hat{\phi}^*$ would be nearer to the mass of the target material displaced by the nose section of the projectile in contact with the target rather than with the whole displaced crater mass $\rho A_0 z$.

For a very detailed description of the mechanics of penetration together with an extensive set of references the reader is referred to the review article of Backman and Goldsmith (1978).

GENERAL REFERENCES

Campbell, J. D. (1972) "Dynamic Plasticity of Metals," Springer-Verlag, Berlin.
Cristescu, N. (1967) "Dynamic Plasticity," North-Holland, Amsterdam.
Goldsmith, W. (1960) "Impact," Edward Arnold, London.
Johnson, W. (1972) "Impact Strength of Materials," Edward Arnold, London.
Kolsky, H. (1953) "Stress Waves in Solids," Cambridge University Press, London.
Milne-Thomson, L. M. (1960) "Theoretical Hydrodynamics," 4th ed., Macmillan, London.
Slater, J. C. (1939) "Introduction to Chemical Physics," McGraw-Hill, New York.
Tabor, D. (1951) "The Hardness of Metals," Oxford University Press, London.

REFERENCES

Alekseevski, V. P. (1966) *Fizika Goreniya Vzryra*, vol. 2, pp. 99–106.
Allen, W. A., and J. W. Rogers (1961) *J. Franklin Inst.*, vol. 272, pp. 275–284.
Al'tshuler, L. V. (1965) *Soviet Phys. Uspekhi*, vol. 8, pp. 52–91.
Al'tshuler, L. V., A. A. Bakanova, and R. F. Trunin (1962) *Soviet Phys. JETP*, vol. 15, pp. 65–75.
Al'tshuler, L. V., S. B. Kormer, M. I. Brazhnik, L. A. Vladimirov, M. P. Speranskaya, and A. I. Funtikov (1960) *Soviet Phys. JETP*, vol. 11, pp. 766–775.
Al'tshuler, L. V., K. K. Krupnikov, B. N. Ledenev, V. I. Zhuckikhin, and M. I. Brazhnik (1958) *Soviet Phys. JETP*, vol. 34, pp. 606–614.
Awerbuch, J. (1970) *Israel J. Tech.*, vol. 8, pp. 375–383.
Awerbuch, J., and S. R. Bodner (1974) *Int. J. Solids Struct.*, vol. 10, pp. 671–684, 684–699.
Backman, M. E., and W. Goldsmith (1978) *Int. J. Engng. Sci.*, vol. 16, pp. 1–99.
Bakanova, A. A., I. P. Dudoladov, and R. F. Trunis (1965) *Soviet Phys. Solid State*, vol. 7, pp. 1307–1313.
Baker, W. E., and C. H. Yew (1966) *J. Appl. Mech.*, vol. 33, pp. 917–923.
Bancroft, D., E. L. Peterson, and S. Minshall (1956) *J. Appl. Phys.*, vol. 27, pp. 291–298.
Bell, J. F. (1951) *Dept. Mech. Engng., The John Hopkins University, Baltimore Tech. Rept. No.* 5.
Bell, J. F. (1956) *J. Appl. Phys.*, vol. 27, pp. 1109–1113.
Bell, J. F. (1960) *J. Appl. Phys.*, vol. 31, pp. 277–282.
Bell, J. F. (1966) *J. Mech. Phys. Solids*, vol. 14, pp. 309–327.
Bell, J. F., and J. H. Suckling (1962) In R. M. Rosenberg, *Proc. Fourth U.S. Nat. Congr. Theor. Appl. Mech.*, pp. 877–883, ASME, New York.
Bethe, H. A. (1942) The Theory of Shock Waves for an Arbitrary Equation of State, *Office of Scientific Research and Development Rept. No.* 545, Serial No. 237.

Bianchi, G. (1964) Some Experimental and Theoretical Studies on the Propagation of Longitudinal Plastic Waves in a Strain-Rate Dependent Material, in H. Kolsky and W. Prager (eds.), "Symp. IUTAM, Stress Waves in Anelastic Solids," pp. 101–117, Springer-Verlag, Berlin.
Billington, E. W. (1977) *J. Phys. D: Appl. Phys.*, vol. 10, pp. 519–531, 553–569.
Billington, E. W., and A. Tate (1972) *Proc. Roy. Soc.* (*London*), vol. A327, pp. 23–46.
Birkhoff, G., D. P. McDougall, E. M. Pugh, and G. I. Taylor (1948) *J. Appl. Phys.*, vol. 19, pp. 563–582.
Bishop, R. F., R. Hill, and N. F. Mott (1945) *Proc. Phys. Soc.*, vol. 57, pp. 147–159.
Boyd, D. E. (1966) *J. Engng. Mech. Div. Am. Soc. Civ. Engrs.*, vol. 92, pp. 1–16.
Breed, B. R., C. L. Mader, and D. Venable (1967) *J. Appl. Phys.*, vol. 38, pp. 3271–3275.
Briggs, T. L., and J. D. Campbell (1972) *Acta. Met.*, vol. 20, pp. 711–724.
Brooks, P. N. (1974) Unpublished DREV Rept.
Butcher, B. M., L. M. Barker, D. E. Munson, and C. D. Lundergam (1964) *Am. Inst. Aeronautics and Astronautics J.*, vol. 2, pp. 977–990.
Calder, C. A., and W. Goldsmith (1971) *Int. J. Solids Struct.*, vol. 7, pp. 863–881.
Campbell, J. D. (1973) *Mat. Sci. Engng.*, vol. 12, pp. 3–21.
Campbell, J. D., and A. R. Dowling (1970) *J. Mech. Phys. Solids*, vol. 18, pp. 43–63.
Campbell, J. D., and J. Harding (1961) The Effect of Grain Size, Rate of Strain and Neutron Irradiation on the Tensile Strength of α-Iron, in P. G. Shewmon and V. F. Zackay (eds.), "Response of Metals to High Velocity Deformation," pp. 51–76, Interscience, New York.
Campbell, J. D., and M. C. Tsao (1972) *Quart. J. Mech. Appl. Math.*, vol. 25, pp. 173–184.
Carley, D. J. (1975) Unpublished AWRE Rept.
Christman, D. R., and J. W. Gehring (1966) *J. Appl. Phys.*, vol. 37, pp. 1579–1587.
Clyens, S., and J. D. Campbell (1974) The Behaviour of Copper and Lead-Tin Eutectic in Torsion at High Strain Rates, in J. Harding (ed.), "Mechanical Properties at High Rate of Strain," pp. 62–71, Institute of Physics, London.
Clyens, S., and W. Johnson (1977) *Int. J. Mech. Sci.*, vol. 19, pp. 745–752.
Convery, E., and H. Ll. D. Pugh (1968) *J. Mech. Engng. Sci.*, vol. 10, pp. 153–164.
Craggs, J. W. (1957) *J. Mech. Phys. Solids*, vol. 5, pp. 115–124.
Davies, R. M. (1948) *Phil. Trans. Roy. Soc.*, vol. 240, pp. 375–457.
Davies, R. M. (1949) *Proc. Roy. Soc.* (*London*), vol. 197, pp. 416–432.
Davies, E. D. H., and S. C. Hunter (1963) *J. Mech. Phys. Solids*, vol. 14, pp. 155–179.
Donnell, L. H. (1930) *Trans. ASME*, vol. 52, pp. 153–167.
Dugdale, J. S., and D. K. C. McDonald (1953) *Phys. Rev.*, vol. 89, pp. 832–834.
Eichelberger, R. J. (1956) *J. Appl. Phys.*, vol. 27, pp. 63–68.
Eichelberger, R. J., and J. W. Gehring (1962) *Am. Rocket Soc. J.*, vol. 32, pp. 1583–1591.
Erkman, J. O. (1961) *J. Appl. Phys.*, vol. 32, pp. 939–944.
Florence, A. L. (1966) *Int. J. Solids Structs.*, vol. 2, pp. 319–335.
Gabeaud, M. L. (1935) *Mem. de l'Art franc.*, vol. 14, pp. 85–92.
Goldsmith, W., and S. A. Finnegan (1971) *Int. J. Mech. Sci.*, vol. 13, pp. 843–866.
Harding, J. (1965) *J. Mech. Engng. Sci.*, vol. 7, pp. 163–176.
Hawkyard, J. B. (1969) *Int. J. Mech. Sci.*, vol. 11, pp. 313–333.
Hogatt, C. R., W. R. Orr, and R. F. Recht (1967) *First Int. Conf. Center for High Energy Forming* (*Denver*), vol. 2, p. 7.4.1.
Holt, D. L., S. G. Babcock, S. J. Green, and C. J. Maiden (1967) *Trans. ASM*, vol. 60, pp. 152–159.
Hopkins, H. G. (1957) *Proc. Roy. Soc.* (*London*), vol. A241, pp. 153–179.
Hopkins, H. G., and H. Kolsky (1960) Mechanics of Hypervelocity Impact of Solids, in *Fourth Hypervelocity Impact Symposium*, April, 1960, vol. 1, Paper No. 12, Eglin Air Force Base.
Hopkins, H. G., and W. Prager (1953) *J. Mech. Phys. Solids*, vol. 2, pp. 1–13.
Hopkins, H. G., and W. Prager (1954) *J. Appl. Math. Phys. ZAMP*, vol. 5, pp. 317–330.
Hopkinson, B. (1914) *Phil. Trans. Roy. Soc.*, vol. A213, pp. 437–456.
Hunter, S. C. (1957) *J. Mech. Phys. Solids*, vol. 5, pp. 162–171.
Jones, N. (1968) *J. Appl. Mech.*, vol. 35, pp. 59–65.
Karman, T. von, and P. E. Duwez (1950) *J. Appl. Phys.*, vol. 21, pp. 987–994.

Keeler, R. N., and E. B. Royce (1971) In P. Caldirola and H. Knoepfel (eds.), "Physics of High Energy Density," pp. 51–150, Academic Press, New York.

Kelly, J. M., and T. R. Wilshaw (1968) *Proc. Roy. Soc.* (*London*), vol. A306, pp. 435–447.

Klahn, D., A. K. Mukherjee, and J. E. Dorn (1970) Strain Rate Effects, in *Second Int. Conf. Strength of Metals and Alloys*, Asilomar, Calif., ASM, vol. 3, pp. 951–982.

Kolsky, H. (1949) *Proc. Phys. Soc.*, vol. B62, pp. 676–700.

Kormer, S. B., A. I. Funtikov, V. D. Urlin, and A. N. Kolesnikova (1962) *Soviet Phys. JETP*, vol. 15, pp. 477–488.

Krupnikov, K. K., M. I. Brazhnik, and V. P. Krupnikova (1962) *Soviet Phys. JETP*, vol. 15, pp. 470–476.

Lamb, H. (1904) *Phil. Trans. Roy. Soc.*, vol. A203, pp. 1–42.

Landau, L. D., and K. P. Stanyukovich (1945) *Doklady Acad. Nauk SSSR*, vol. 46, pp. 399–402.

Lee, E. H., and S. J. Tupper (1954) *J. Appl. Mech.*, vol. 21, pp. 63–70.

Lethaby, J., and I. C. Skidmore (1959) *SWA Branch Note No.* 3/59, AWRE, Aldermaston.

Lindholm, U. S., and L. M. Yeakley (1968) *Expt. Mech.*, vol. 8, pp. 1–9.

Lubliner, J. (1964) *J. Mech. Phys. Solids*, vol. 12, pp. 59–65.

McQueen, R. G., S. P. Marsh, J. W. Taylor, J. N. Fritz, and W. J. Carter (1970) In R. Kinslow (ed.), "High Velocity Impact Phenomena," pp. 293–417, Academic Press, New York.

Maiden, C. J., and S. J. Green (1966) *J. Appl. Mech.*, vol. 33, pp. 496–504.

Malvern, L. E. (1951a) *Quart. Appl. Math.*, vol. 8, pp. 405–411.

Malvern, L. E. (1951b) *J. Appl. Mech.*, vol. 18, pp. 203–208.

Malvern, L. E. (1955) In N. J. Huffington (ed.), "Behavior of Materials under Dynamic Loading," pp. 81–92, ASME, New York.

Malyshew, B. M. (1961) *J. Prikl. Mech. Tech. Fiz.*, vol. 2, pp. 104–110.

Nicholas, T., and J. D. Campbell (1972) *Expt. Mech.*, vol. 12, pp. 441–447.

Niordsen, F. I. (1965) *Expt. Mech.*, vol. 5, pp. 29–32.

Pack, D. C., and W. M. Evans (1951) *Proc. Phys. Soc.*, vol. B64, pp. 298–303.

Pope, D. P., T. Vreeland, and D. S. Wood (1964) *Rev. Sci. Instr.*, vol. 35, pp. 1351–1355.

Pukas, S. R. (1976) Ph.D. Thesis, Thames Polytechnic, London.

Rakhmatulin, Kh. A. (1945) *Prik. Mat. Mech.*, vol. 9, pp. 91–100, 449–462.

Rakhmatulin, Kh. A., and A. Yu. Demyanov (1966) "Strength under High Transient Loads," Israel Program for Scientific Translations, Jerusalem.

Recht, R. F., and T. W. Ipson (1963) *J. Appl. Mech.*, vol. 30, pp. 384–390.

Reed, P. E. (1975) In J. Harding (ed.), *Proc. Conf. Mech. Prop. Mat. at High Rates of Strain*, pp. 339–343, Institute of Physics, London.

Rice, M. Y., R. G. McQueen, and J. M. Walsh (1958) In F. Seitz and D. Turnbull (eds.), "Solid State Physics," vol. 6, pp. 1–37, Academic Press, New York.

Richter, H. (1946) *DEFA and LRSTA Inst. for Ballistics Rept.* 3/46.

Rinehart, J. S., and J. Pearson (1954) "Behavior of Metals under Impulsive Loads," ASM, Cleveland, Ohio.

Ripperger, E. A., and H. Watson (1968) The Relationship between the Constitutive Equation and One-dimensional Wave Propagation, in U. S. Lindholm (ed.), "Mechanical Behavior of Materials under Dynamic Loads," pp. 294–313, Springer, New York.

Skidmore, I. C. (1965) *Appl. Mat. Res.*, vol. 4, pp. 131–147.

Sternglass, E. J., and D. A. Stuart (1953) *J. Appl. Mech.*, vol. 20, pp. 427–434.

Swan, G. W. (1971) *J. Phys. D: Appl. Phys.*, vol. 4, pp. 1077–1082.

Tate, A. (1967) *J. Mech. Phys. Solids*, vol. 15, pp. 387–399.

Tate, A. (1969) *J. Mech. Phys. Solids*, vol. 17, pp. 141–150.

Tate, A. (1977) *Int. J. Mech. Sci.*, vol. 19, pp. 121–123.

Tate, A. (1978) *Int. J. Engng. Sci.*, vol. 16, pp. 845–858.

Tate, A. (1979) *Int. J. Engng. Sci.*, vol. 17, pp. 341–344.

Taylor, G. I. (1946) *J. Inst. Civil Engng.*, vol. 26, pp. 486–519.

Taylor, G. I. (1948a) *Proc. Roy. Soc.* (*London*), vol. 194, pp. 289–299.

Taylor, G. I. (1948b) *Quart. J. Mech. Appl. Math.*, vol. 1, pp. 103–124.
Thomson, W. T. (1955) *J. Appl. Phys.*, vol. 26, pp. 80–82.
Tupper, S. J. (1961) *ARDE Rept.* (B) 12/61.
Walsh, J. M., and M. H. Rice (1957) *J. Chem. Phys.*, vol. 26, pp. 815–823.
Wang, A. J., and H. G. Hopkins (1954) *J. Mech. Phys. Solids*, vol. 3, pp. 22–37.
Weihrauch, G. (1972) *ISL Rept.* 40/72.
Weirzbicki, T. (1967) *Int. J. Solids Struct.*, vol. 3, pp. 635–647.
Wilkins, M. L., and M. W. Guinan (1973) *J. Appl. Phys.*, vol. 44, pp. 1200–1260.
Yew, C. H., and H. A. Richardson (1969) *Expt. Mech.*, vol. 9, pp. 366–373.
Zaid, M., and B. Paul (1957) *J. Franklin Inst.*, vol. 264, pp. 117–126.
Zel'dovich, Ya. B. (1957) *JETP*, vol. 5, pp. 1287–1288.
Zubarev, V. N., and V. Y. Vashchenko (1963) *Soviet Phys. Solid State*, vol. 5, pp. 653–655.

CHAPTER

FOURTEEN

FRACTURE

14-1 SOME GENERAL CONSIDERATIONS

14-1-1 The Stress Function and Complex Variables

The complex variable z is defined as

$$z = x_1 + jx_2 = r\exp(j\theta) \qquad \text{for } j = \sqrt{-1} \tag{14-1-1}$$

where $r^2 = x_1^2 + x_2^2$ and $\theta = \tan^{-1}(x_2/x_1)$. Let $f(z)$ be a single-valued function and analytic insofar as it has a unique derivative with respect to both x_1 and x_2 at every point of the Argand plane. Now express $f(z)$ in the form

$$f(z) = g(x_1, x_2) + jh(x_1, x_2) \tag{14-1-2}$$

where g and h are two new analytic functions. There now follow the relations

$$\frac{\partial f}{\partial x_1} = \frac{df}{dz} = f'(z) = \frac{\partial g}{\partial x_1} + j\frac{\partial h}{\partial x_1} \tag{14-1-3}$$

$$\frac{\partial f}{\partial x_2} = j\frac{df}{dz} = jf'(z) = \frac{\partial g}{\partial x_2} + j\frac{\partial h}{\partial x_2} \tag{14-1-4}$$

Equating the real and imaginary parts of df/dz from Eqs. (14-1-3) and (14-1-4) gives

$$\frac{\partial g}{\partial x_1} = \frac{\partial h}{\partial x_2} \qquad \frac{\partial h}{\partial x_1} = -\frac{\partial g}{\partial x_2} \tag{14-1-5}$$

These equations are the Cauchy–Riemann conditions which the components g and h of an analytic function $f(z)$ must satisfy. Differentiation of Eqs. (14-1-5) successively with respect to x_1 and x_2 and rearranging gives

$$\frac{\partial^2 g}{\partial x_1^2}+\frac{\partial^2 g}{\partial x_2^2}=0 \qquad \nabla^2 g=0 \tag{14-1-6}$$

$$\frac{\partial^2 h}{\partial x_1^2}+\frac{\partial^2 h}{\partial x_2^2}=0 \qquad \nabla^2 h=0 \tag{14-1-7}$$

Equations (14-1-6) and (14-1-7) are just two-dimensional forms of Laplace's equation, the functions g and h being referred to as conjugate harmonic functions.

Now, if Ψ is any harmonic function of x_1 and x_2, then

$$\begin{aligned}\nabla^2(x_1\Psi) &= x_1\nabla^2\Psi+2\frac{\partial\Psi}{\partial x_1}\\ &= 2\frac{\partial\Psi}{\partial x_1}\end{aligned} \tag{14-1-8}$$

and also

$$\nabla^2\left(\frac{\partial\Psi}{\partial x_1}\right)=\frac{\partial}{\partial x_1}(\nabla^2\Psi)=0 \tag{14-1-9}$$

Hence,

$$\nabla^2[\nabla^2(x_1\Psi)]=2\nabla^2\left(\frac{\partial\Psi}{\partial x_1}\right)=0 \tag{14-1-10}$$

which can be expressed in the form

$$\nabla^4(x_1\Psi)=0 \tag{14-1-11}$$

From a comparison of Eq. (14-1-11) with Eq. (11-2-7) it is evident that $x_1\Psi$ can be used as a stress function when Ψ is harmonic. It follows that similar arguments will establish that both $x_2\Psi$ and $r^2\Psi$ can also be used as stress functions, as can Ψ itself.

The above argument can be generalized in such a way that any stress function Φ can be expressed in the form

$$\Phi=\mathrm{Re}[(x_1-jx_2)\Psi(z)+\chi(z)] \tag{14-1-12}$$

where $\Psi(z)$ and $\chi(z)$ are suitably chosen analytic functions and are termed complex potentials.

Replacing j in Eqs. (14-1-1) and (14-1-2) by $-j$ gives

$$\bar{f}(\bar{z})=g-jh \tag{14-1-13}$$

which defines the conjugate function $\bar{f}(\bar{z})$. The sum of Eqs. (14-1-2) and (14-1-13) gives

$$f(z)+\bar{f}(\bar{z})=2g=2\mathrm{Re}f(z) \tag{14-1-14}$$

Equation (14-1-12) can be expressed in the form

$$\Phi = \mathrm{Re}[\bar{z}\Psi(z) + \chi(z)] \tag{14-1-15}$$

and it follows that

$$2\Phi = \bar{z}\Psi(z) + z\bar{\Psi}(\bar{z}) + \chi(z) + \bar{\chi}(\bar{z}) \tag{14-1-16}$$

From Eq. (14-1-16) there can be obtained the relation

$$\frac{\partial \Phi}{\partial x_1} + j\frac{\partial \Phi}{\partial x_2} = \Psi(z) + z\bar{\Psi}'(\bar{z}) + \bar{\chi}'(\bar{z}) \tag{14-1-17}$$

By replacing the Ψ of Eqs. (11-2-4), (11-2-5), and (11-2-6) by Φ and suitably combining there can be obtained the relations

$$T_{11} + T_{22} = 4\mathrm{Re}\,\Psi'(z) \tag{14-1-18}$$

$$T_{22} - T_{11} + 2jT_{12} = 2[\bar{z}\Psi''(z) + \chi''(z)] \tag{14-1-19}$$

Let the displacement in the x_1 direction be u_1 and in the x_2 direction be u_2; then from Secs. 6-3-1 and 6-3-2 there can be obtained the relations

$$2G(u_1 + ju_2) = \frac{3-\nu}{1+\nu}\Psi(z) - z\bar{\Psi}'(\bar{z}) - \bar{\chi}'(\bar{z}) \qquad \text{(plane stress)} \tag{14-1-20}$$

$$2G(u_1 + ju_2) = (3-4\nu)\Psi(z) - z\bar{\Psi}'(\bar{z}) - \bar{\chi}'(\bar{z}) \qquad \text{(plane strain)} \tag{14-1-21}$$

from which the individual values of u_1 and u_2 can be obtained.

14-1-2 Elliptical Coordinates

For a cartesian coordinate system, a complex number z can be expressed in the form

$$z = x_1 + jx_2 \qquad \text{for } j = \sqrt{-1} \tag{14-1-22}$$

where x_1, x_2 are cartesian coordinates. Consider the complex variable

$$p = \phi + j\mu \tag{14-1-23}$$

and let p be related to z by the relation

$$z = a\cosh p \tag{14-1-24}$$

where a is a constant. Separating and equating the real and imaginary parts of Eq. (14-1-24) gives

$$x_1 = a\cosh\phi\cos\mu \qquad x_2 = a\sinh\phi\sin\mu \tag{14-1-25}$$

which are the elliptical coordinates of the point x_1, x_2 in the plane $x_3 = 0$. Eliminating μ between the relations of Eq. (14-1-25) gives

$$\frac{x_1^2}{\cosh^2\phi} + \frac{x_2^2}{\sinh^2\phi} = a^2 \tag{14-1-26}$$

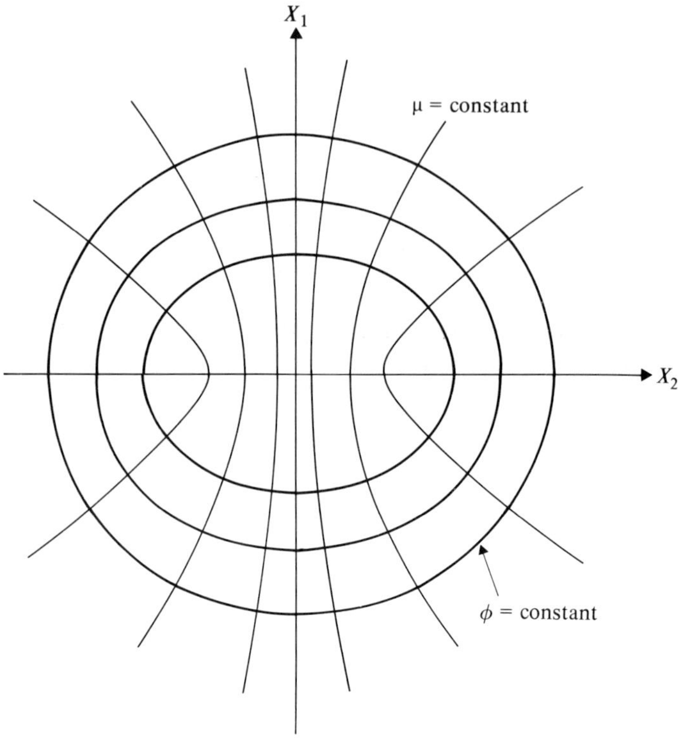

Figure 14-1 Conformal mapping of conics.

from which it follows that lines of constant ϕ with μ varying from 0 to 2π are confocal ellipses. Lines of constant μ are confocal hyperbolas. These two sets of curves intersect orthogonally, as shown schematically in Fig. 14-1.

Let the components of stress T_{ij} be referred to axes oX_1, oX_2 and the components of stress $T_{\langle\mu\phi\rangle}$ be referred to coordinate axes ϕ, μ rotated through angle θ from the axes oX_1 and oX_2. The components T_{ij} are related to the components $T_{\langle\phi\mu\rangle}$ by the transformation formulas

$$\begin{aligned} T_{\langle\phi\phi\rangle} &= T_{11}\cos^2\theta + 2T_{12}\sin\theta\cos\theta + T_{22}\sin^2\theta \\ T_{\langle\mu\mu\rangle} &= T_{11}\sin^2\theta - 2T_{12}\sin\theta\cos\theta + T_{22}\cos^2\theta \\ T_{\langle\phi\mu\rangle} &= (T_{22} - T_{11})\sin\theta\cos\theta + T_{12}(\cos^2\theta - \sin^2\theta) \end{aligned} \tag{14-1-27}$$

which can be expressed in the form

$$T_{\langle\phi\phi\rangle} + T_{\langle\mu\mu\rangle} = T_{11} + T_{22} \tag{14-1-28}$$

$$T_{\langle\mu\mu\rangle} - T_{\langle\phi\phi\rangle} + 2jT_{\langle\phi\mu\rangle} = (T_{22} - T_{11} + 2jT_{12})\exp(2j\theta) \tag{14-1-29}$$

it being noted that the sum of Eq. (14-1-28) is just the first invariant of the stress tensor in two dimensions.

The inclination θ is given by $\arg[f'(p)]$ since

$$dx_1 + j\,dx_2 = f'(p)\,d\phi \qquad \text{for } d\mu = 0$$

This may be written in the form

$$\exp(2j\theta) = \frac{f'(p)}{\bar{f}'(\bar{p})} = \frac{\sinh p}{\sinh \bar{p}} \tag{14-1-30}$$

In terms of the complex potentials Ψ and χ of Eq. (14-1-12), the stresses are given by reference to Eqs. (14-1-18) and (14-1-19) in the form

$$T_{\langle\phi\phi\rangle} + T_{\langle\mu\mu\rangle} = 4\,\text{Re}\,\Psi'(z) \tag{14-1-31}$$

$$T_{\langle\mu\mu\rangle} - T_{\langle\phi\phi\rangle} + 2jT_{\langle\phi\mu\rangle} = 2\exp(2j\theta)\left[\bar{z}\Psi''(z) + \chi''(z)\right] \tag{14-1-32}$$

14-1-3 Modes of Cracking

The states of stress frequently encountered are those corresponding to plane strain, plane stress, and antiplane strain, which have been described in Sec. 6-3. The associated modes of cracking are:

Mode I This is a crack in an infinite body under an applied uniaxial tension $T_{11} = T_0$ with $T_{22} = 0$. This mode of cracking, which is shown schematically in Fig. 14-2, occurs frequently in low-stress brittle fractures of large components.

Mode II This crack is subjected to a shear stress T_{12}. This mode is shown schematically in Fig. 14-3 and is referred to as shear because the crack is given a shearing displacement u_2 by the applied shear stress T_{12}.

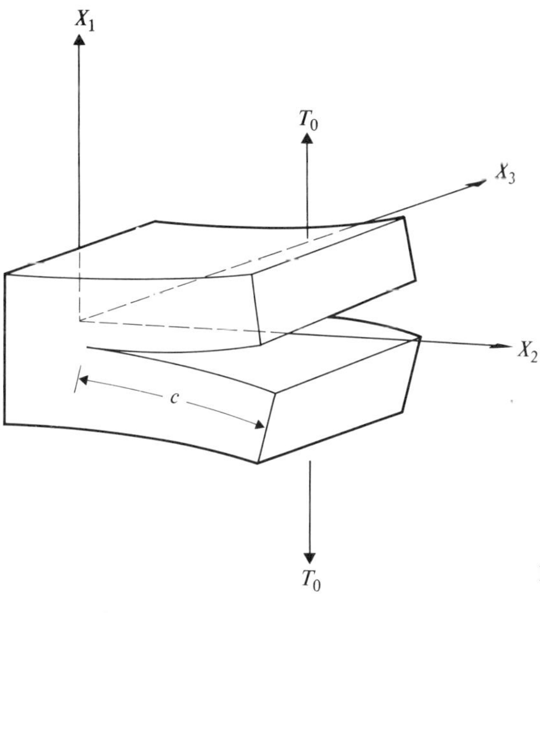

Figure 14-2 Mode I; crack under tension.

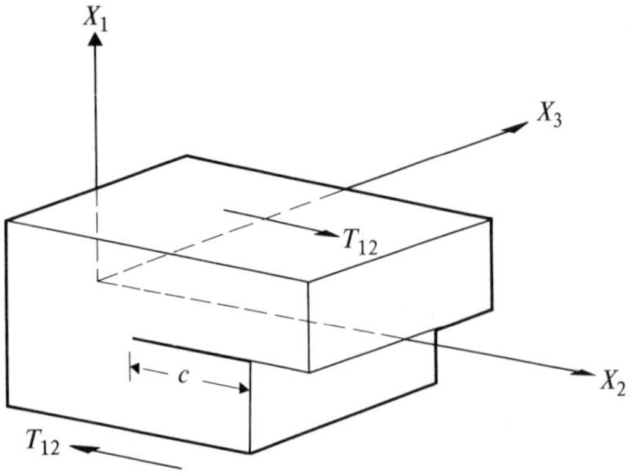

Figure 14-3 Mode II; crack subjected to a shear stress.

Mode III An alternative mode of shearing is that of a crack in antiplane strain. This mode is shown schematically in Fig. 14-4 and is referred to as shear in antiplane strain because the crack is given a shearing displacement u_3 by the applied shear stress T_{13} in the X_3 direction.

14-2 CRACKS IN EXTERNAL STRESS FIELDS

The boundary-value problem of present interest concerns a body $\mathscr{B}$ whose external boundary consists of two parts S_1 and S_2, on one of which the displacement is prescribed while on the other the traction is prescribed. The discussion of Sec. 11-2-2 can be applied directly to a body with a crack, the present discussion being concerned with its extension to two particular cases. To extend and apply Sec. 11-2-2 to fracture mechanics, consider the body $\mathscr{B}$ to contain one or

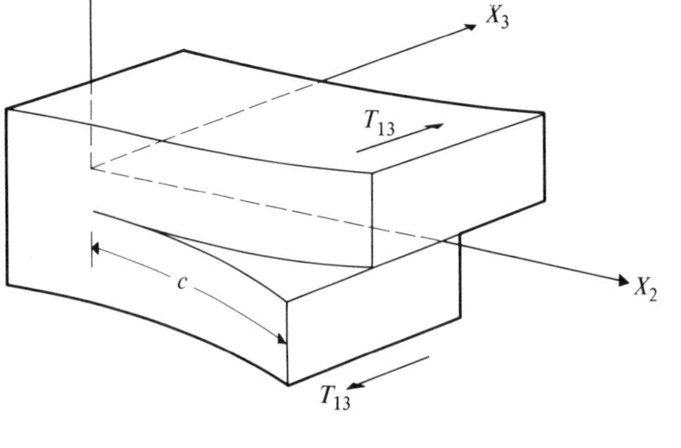

Figure 14-4 Mode III; shear in antiplane strain.

more cracks whose surfaces are $C_{\pm}$. For this situation, the traction t^c on the faces of the crack is usually prescribed, while the relative displacement $\Delta \boldsymbol{u}$ of the faces is not known a priori. There are many frequently occurring situations for which the crack surfaces would be free so that $t^c = \boldsymbol{o}$. Thus, the conditions of Eq. (11-2-52) can be used to describe a body which is externally loaded and containing a crack which may be a free surface or may have prescribed internal loads on it. For a body with a crack, the conditions of Eq. (11-2-55) correspond to the problem of an internally loaded crack, with zero displacements or traction at the external boundary. For the crack case, $t^{(0)}$ occurring in Eq. (11-2-69) is required and is found by solving the boundary-value problem of Eq. (11-2-54). The traction t^c of Eq. (11-2-69) is prescribed, it being required to find $\Delta \boldsymbol{u}$, the displacement discontinuity across the crack, by solving the boundary-value problem of Eq. (11-2-52).

For a body with a crack, there are two contributions to the ΔU_{tot} of Eq. (11-2-69):

1. *Fixed-load case.* Here the crack is taken to be unloaded ($t^c = o$) and the boundary conditions to be entirely of traction type, so that S_1 is empty. For these conditions the right-hand side of Eq. (11-2-66) is zero, and hence the resulting relation can be used to eliminate the integrals over S_2 in Eqs. (11-2-67) and (11-2-68). Comparison of the resulting expressions with Eq. (11-2-69) gives

$$\Delta U = -\Delta U_{\text{tot}} \qquad \Delta U_{\text{ext}} = 2\Delta U_{\text{tot}} \tag{14-2-1}$$

   Since ΔU_{tot} is negative, it is evident that the formation of a crack in a body under fixed loads leads to an increase in the strain energy of the material but a decrease of double the magnitude in the external energy.
2. *Fixed-grips case.* Here the boundary is taken to be either stress free or to have prescribed displacements. Thus, either S_2 is empty or, if not, $t = o$ at each point of S_2. From Eqs. (11-2-68) and (11-2-69) there can be obtained the relations

$$\Delta U = \Delta U_{\text{tot}} \qquad \Delta U_{\text{ext}} = 0 \tag{14-2-2}$$

   from which it is concluded that there is no change in the external energy, the entire change in total energy coming from a decrease in strain energy in the body upon formation of the crack, a conclusion which holds whether or not there are internal loads on the crack.

Consider the simple tensile test of a cylindrical specimen with generators parallel to the z axis. The sides are prescribed to be stress free while the ends are subject to constant normal tractions which with no crack give rise to a state of stress for which T_{zz} is a constant. The fixed-load case results if a crack is formed while maintaining the end tractions at their original constant value. For this case the elastic energy in the cylinder increases, but the cylinder ends move a little and reduce the external energy. Alternatively, the crack is formed while holding the cylinder ends fixed in their deformed positions, corresponding to the fixed-grips case, and the elastic energy in the cylinder decreases upon forming the crack.

14-3 STRESS-RAISING PROPERTIES OF HOLES

14-3-1 Elliptical Hole in a Uniformly Stressed Plate

Inglis (1913) showed that the stress in the immediate neighborhood of a hole regarded as an elliptical crack could be many times the stress in the body as a whole. The hole regarded as an elliptical crack is assumed to lie in an infinite plate with its major axis normal to a uniform tensile stress, T_0, as shown schematically in Fig. 14-5.

For an elliptical crack, set

$$a \cosh \phi = c \qquad a \sinh \phi = b \tag{14-3-1}$$

where a is a constant and c, b are the semimajor and minor axes respectively. Substituting for ϕ in Eq. (14-1-26) from Eq. (14-3-1) gives

$$\frac{x_1^2}{c^2} + \frac{x_2^2}{b^2} = 1 \tag{14-3-2}$$

from which it is evident that in the limit as $\phi \to 0$ the ellipse becomes a crack of length $2a = 2c$, and when $b = c$ the ellipse becomes a circle.

A point on the boundary of the crack goes once round the ellipse as μ varies from 0 to 2π, and hence the continuity of stress and displacement requires that they be periodic also in μ with period 2π, so that they each have the same value at 0 and at 2π.

It will now be specified that at infinity $T_{11} = 0$, $T_{22} = T_0$, $T_{12} = 0$, so that, from Eqs. (14-1-18) and (14-1-19),

$$\left.\begin{aligned} 4\,\mathrm{Re}\,\Psi'(z) &= T_0 \\ 2[\bar{z}\Psi''(z) + \chi''(z)] &= T_0 \end{aligned}\right\} \quad \text{at infinity} \tag{14-3-3}$$

At the boundary of the hole, $T_{\langle\phi\phi\rangle} = 0$, $T_{\langle\mu\phi\rangle} = 0$.

These boundary conditions were satisfied by Inglis (1913) who obtained the relations

$$4\Psi(z) = cT_0\{[1 + \exp(2\phi)] \sinh p - \exp(2\phi) \cosh p\} \tag{14-3-4}$$

$$4\chi(z) = -c^2 T_0 \left\{ [\cosh(2\phi) - \cosh \pi] p + \tfrac{1}{2} \exp(2\phi) - \cosh 2\left(p - \phi - j\frac{\pi}{2}\right) \right\} \tag{14-3-5}$$

Since $T_{\langle\phi\phi\rangle}$ is zero at the surface of the hole, $T_{\langle\mu\mu\rangle}$ can be obtained at the surface of the crack from Eq. (14-1-31) in the form

$$T_{\langle\mu\mu\rangle} = T_0 \frac{\sinh(2\phi) - 1 + \exp(2\phi)\cos(2\mu)}{\cosh(2\phi) - \cos(2\mu)} \tag{14-3-6}$$

The maximum values of $T_{\langle\mu\mu\rangle}$ correspond to those at the ends of the major axis of the crack where $\mu = 0$ or π. Hence, for these conditions $\cos 2\mu = 1$ and, since the tangent to the ellipse is parallel to X_2 at the ends of the major axis, it must

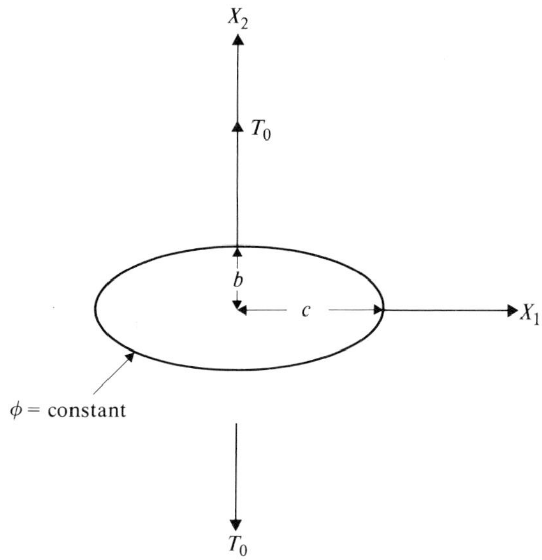

Figure 14-5 Elliptical hole in an infinite plate; crack loaded by a uniform tensile stress T_0.

have $T_{\langle\mu\mu\rangle} \equiv T_{22}$, which for these conditions can be expressed by way of Eq. (14-3-6) in the form

$$T_{\langle\mu\mu\rangle} \equiv T_{22} = T_0 \frac{2\sinh(2\phi)+\cosh(2\phi)-1}{\cosh(2\phi)-1} \qquad \text{for } \mu = 0, \pi \tag{14-3-7}$$

Equation (14-3-1) gives $a^2 = c^2 - b^2$ and hence from Eqs. (14-1-26) and (14-3-2) there follows

$$\sinh(2\phi) = \frac{2bc}{a^2} \qquad \cosh(2\phi) = \frac{b^2+c^2}{a^2} \tag{14-3-8}$$

Substituting for ϕ in Eq. (14-3-7) from Eq. (14-3-8) gives at the crack tip

$$T_{22} = T_0\left(1+2\frac{c}{b}\right) \tag{14-3-9}$$

where T_0 is the uniform tensile stress at infinity.

Taking the radius of curvature at the tip of the ellipse to be R, then

$$R^2 = \frac{(dx_1^2+dx_2^2)^3}{(dx_1\,d^2x_2+dx_2\,d^2x_1)^2} \tag{14-3-10}$$

Setting $x_2 = b\sin t$, $x_1 = c\cos t$, and substituting in Eq. (14-3-10) for x_1 and x_2 gives

$$R^2 = \frac{1}{b^2c^2}(b^2\cos^2 t + c^2\sin^2 t)^3 \tag{14-3-11}$$

At the tip of the ellipse, $x_2 = 0$, $x_1 = \pm c$, and hence $\sin t = 0$, $\cos t = 1$, which conditions give $R = b^2/c$, a condition which can now be introduced into Eq. (14-3-9) to give

$$T_{22} = T_0\left(1+2\sqrt{\frac{c}{R}}\right) \tag{14-3-12}$$

which is the relation obtained by Inglis (1913) for the stress at the tip of an elliptical crack. The greatest shear stresses are also at the ends of the major axis and make an angle of $\pi/4$ to it. This follows from Sec. 3-6 because the stress along the major axis at the point is zero and from Eq. (3-6-11) the maximum shear stress is $\frac{1}{2}T_0(1+2\sqrt{c/R})$. Inglis also established that the maximum stress was very insensitive to the shape of the crack but depended almost entirely on the length of the crack and radius of curvature at its ends, a result which is in accord with St. Venant's principle.

If the plate with the elliptical crack is sectioned on a line through the minor axis of the ellipse a notched plate results and if the notch is very deep and narrow, corresponding to an ellipse of large eccentricity, a further application of St. Venant's principle indicates that the normal stress distribution along the cut surface will have very little effect on the stress at the bottom of the notch. Thus, if the notch has a depth c at a radius of curvature R at the bottom of the notch, the maximum tensile stress there will be given by Eq. (14-3-12). Methods similar to that of Inglis (1913) have been used by Neuber (1938) to calculate stress concentrations near various internal and external notch profiles. Neuber's approach assumes that the profiles approximate closely to the elliptical or hyperbolic form.

14-3-2 Stresses near a Line Crack

Consider the stress function

$$\Gamma = \operatorname{Re}\int\left[\int\Phi(z)\,dz\right]dz + x_2 \operatorname{Im}\int\Phi(z)\,dz \tag{14-3-13}$$

where the function $\Phi(z)$ is harmonic and Γ satisfies the biharmonic equation (11-2-7). By replacing the Ψ of Eqs. (11-2-4), (11-2-5), and (11-2-6) by Γ there can be obtained the components of stress

$$T_{22} = \frac{\partial^2\Gamma}{\partial x_1^2} = \frac{\partial}{\partial z}\left(\frac{\partial\Gamma}{\partial z}\right) = \operatorname{Re}\Phi(z) + x_2 \operatorname{Im}\Phi'(z) \tag{14-3-14}$$

$$T_{11} = \frac{\partial^2\Gamma}{\partial x_2^2} = \operatorname{Re}\Phi(z) - x_2 \operatorname{Im}\Phi'(z) \tag{14-3-15}$$

$$T_{12} = -\frac{\partial^2\Gamma}{\partial x_1 \partial x_2} = -x_2 \operatorname{Re}\Phi'(z) \tag{14-3-16}$$

which satisfy both the equilibrium and the compatibility equations.

The displacements u_1 and u_2 are related to the strain by Eq. (6-2-21) which for plane strain and the limit of classical infinitesimal linear elasticity gives

$$\tilde{e}_{11} = \frac{\partial u_1}{\partial x_1} \qquad \tilde{e}_{22} = \frac{\partial u_2}{\partial x_2} \tag{14-3-17}$$

and the corresponding stress-strain relations can be expressed by way of Eqs. (6-1-30) and (6-3-3) in the form

$$2G\tilde{e}_{11} = (1-\nu)T_{11} - \nu T_{22} \tag{14-3-18}$$

$$2G\tilde{e}_{22} = (1-\nu)T_{22} - \nu T_{11} \tag{14-3-19}$$

The relations of Eq. (14-3-17) for the strain and the stress-strain relations of Eqs. (14-3-18) and (14-3-19) are satisfied if the displacements are taken to be of the form

$$2Gu_1 = (1-2\nu)\,\mathrm{Re}\int \Phi(z)\,dz - x_2\,\mathrm{Im}\,\Phi(z) \tag{14-3-20}$$

$$2Gu_2 = 2(1-\nu)\,\mathrm{Im}\int \Phi(z)\,dz - x_2\,\mathrm{Re}\,\Phi(z) \tag{14-3-21}$$

Equations (14-3-14), (14-3-15), and (14-3-16) reduce for the limiting condition $x_2 = 0$ to the form

$$T_{12} = 0 \qquad T_{11} = T_{22} \tag{14-3-22}$$

Similarly, for $x_2 = 0$, Eqs. (14-3-20) and (14-3-21) reduce to

$$u_1 = \frac{(1-2\nu)(1+\nu)}{E}\,\mathrm{Re}\int \Phi(z)\,dz \tag{14-3-23}$$

$$u_2 = \frac{2(1-\nu^2)}{E}\,\mathrm{Im}\int \Phi(z)\,dz \tag{14-3-24}$$

Consider a crack of length $2c$ in an infinite body under an applied uniform biaxial tension $T_{11} = T_{22} = T_0$ at infinity. In Westergaard's (1939) original solution to the problem of stresses in the immediate neighborhood of the tip of a sharp crack, the length of the crack, which was taken to lie along $\pm X_1$, is normal to X_2. The harmonic function $\Phi(z)$ must be formulated in such a way as to satisfy the conditions, for $x_2 = 0$,

$$T_{22} = \begin{cases} 0 & \text{for } x_1 \leqslant \pm c \\ T_0 & \text{for } x_1 \to +\infty \end{cases} \tag{14-3-25}$$

It is to be expected that there will be a stress concentrating effect at the tip, implying that $T_{22} > T_0$ in the immediate neighborhood of the crack tip. For $x_2 = 0$, Eq. (14-3-14) gives

$$T_{22} = \mathrm{Re}\,\Phi(z) \tag{14-3-26}$$

These conditions are most simply satisfied by taking $\Phi(z)$ to be of the form

$$\Phi(z) = \frac{T_0}{(1 - c^2/z^2)^{1/2}} \tag{14-3-27}$$

Substituting for $\Phi(z)$ in Eq. (14-3-26) from Eq. (14-3-27) gives, for $x_2 = 0$,

$$T_{22} = \frac{T_0}{(1 - c^2/x_1^2)^{1/2}} \tag{14-3-28}$$

which is seen to satisfy all the required conditions including the condition that, when $-c < x_1 < c$, $T_{22} = 0$.

For the displacement u_2 in the x_2 direction, Eq. (14-3-24) gives

$$u_2|_{x_2=0} = \frac{2(1-\nu^2)T_0}{E}(c^2 - x_1^2)^{1/2} \tag{14-3-29}$$

from which it is concluded that the crack is an ellipse of the form

$$\frac{u_2^2}{\text{constant}} + \frac{x_1^2}{c^2} = 1 \tag{14-3-30}$$

The stress distribution ahead of the crack tip can be obtained from Eq. (14-3-14) in the form

$$T_{22} = \frac{T_0\sqrt{\pi c}}{\sqrt{2\pi a}}\left[1 + \frac{3}{4}\left(\frac{a}{c}\right) - \frac{5}{32}\left(\frac{a}{c}\right)^2 + \cdots\right] \tag{14-3-31}$$

where $a = x_1 - c$ is the distance ahead of the crack tip.

By setting $\zeta = z - c$, where $\zeta/c \ll 1$, and $x_2 = r\sin\theta$, the angular dependency of the stresses near the crack tip can be obtained from Eqs. (14-3-14), (14-3-15), and (14-3-16) in the form

$$T_{11} = T_0\sqrt{\frac{c}{2r}}\cos\frac{\theta}{2}\left(1 - \sin\frac{\theta}{2}\sin\frac{3\theta}{2}\right) + \cdots \tag{14-3-32}$$

$$T_{22} = T_0\sqrt{\frac{c}{2r}}\cos\frac{\theta}{2}\left(1 + \sin\frac{\theta}{2}\sin\frac{3\theta}{2}\right) + \cdots \tag{14-3-33}$$

$$T_{12} = T_0\sqrt{\frac{c}{2r}}\cos\frac{\theta}{2}\sin\frac{\theta}{2}\cos\frac{3\theta}{2} + \cdots \tag{14-3-34}$$

As $\theta \to 0$, that is, $x_2 \to 0$,

$$T_{22} \simeq T_0\sqrt{\frac{c}{2r}} \tag{14-3-35}$$

The uniaxial case, for which $T_{22} = T_0$, $T_{11} = 0$ is simply obtained by superimposing a pressure in the X_1 direction equal to $-T_0$. The superimposition of this pressure leaves the crack-tip values of T_{11} and T_{22} of Eqs. (14-3-32) and (14-3-33) unaltered.

The mode I opening of a crack, as defined in Sec. 14-1-3, is obtained by interchanging the subscripts 1 and 2 in Eqs. (14-3-28), (14-3-29), and (14-3-32) to (14-3-35). Thus, the crack-tip displacement is now

$$u_1|_{x_1=0} = \frac{2(1-\nu^2)T_0}{E}(a^2 - x_2^2)^{1/2} \tag{14-3-36}$$

which represents an opening in the X_1 direction and hence the definition of mode I opening.

For a crack subjected to a shear stress $T_{21} \equiv p$, at infinity Westergaard's solution is obtained by way of the stress function

$$\Gamma = -x_2 \operatorname{Re} \int \Phi_{\mathrm{II}}(z)\, dz \tag{14-3-37}$$

which gives

$$T_{11} = 2\operatorname{Im} \Phi_{\mathrm{II}}(z) + x_2 \operatorname{Re} \Phi'_{\mathrm{II}}(z) \tag{14-3-38}$$

$$T_{22} = -x_2 \operatorname{Re} \Phi'_{\mathrm{II}}(z) \tag{14-3-39}$$

$$T_{21} = \operatorname{Re} \Phi_{\mathrm{II}}(z) - x_2 \operatorname{Im} \Phi'_{\mathrm{II}}(z) = T_{12} \tag{14-3-40}$$

Taking

$$\Phi(z) \equiv \Phi_{\mathrm{II}}(z) = \frac{p}{(1 - c^2/z^2)^{1/2}} \tag{14-3-41}$$

satisfies the essential conditions that, along the line $x_2 = 0$, T_{22} vanishes and that T_{21} vanishes everywhere within the crack. In the immediate neighborhood of the crack tip, Eq. (14-3-41) gives

$$T_{21} \simeq p \sqrt{\frac{c}{2r}} \tag{14-3-42}$$

which is to be compared with Eq. (14-3-35). The mode II opening of a crack, as defined in Sec. 14-1-3, is obtained by again interchanging the X_1 and X_2 axes, i.e., by interchanging the subscripts 1 and 2, the crack now being given a shearing displacement u_2 by an applied stress T_{12}.

Consider now Westergaard's (1939) solution for a crack of length $2c$, with its length aligned with the X_1 direction and which is sheared by a stress $T_{23} \equiv q$ applied at infinity in the X_3 direction. These conditions are just those of antiplane strain described in Sec. 6-3-3 for which classical linear elasticity gives

$$\frac{\partial u_3}{\partial x_1} = \frac{T_{13}}{G} \qquad \frac{\partial u_3}{\partial x_2} = \frac{T_{23}}{G} \tag{14-3-43}$$

since displacements only occur in the X_3 direction. Taking

$$u_3 = \frac{1}{G} \operatorname{Im} \Phi(z) \tag{14-3-44}$$

gives, by way of Eq. (14-3-43),

$$T_{13} = \operatorname{Im} \Phi'(z) \qquad T_{23} = \operatorname{Re} \Phi'(z) \tag{14-3-45}$$

The conditions that at infinity $T_{23} \equiv q$ and that everywhere within the crack $T_{23} = 0$ can be most simply satisfied by taking

$$\Phi(z) \equiv \Phi_{\mathrm{III}}(z) = q(z^2 - c^2)^{1/2} \tag{14-3-46}$$

Equations (14-3-45) and (14-3-46) give the stress in the immediate neighborhood of the crack tip in the form

$$T_{23} \simeq q \sqrt{\frac{c}{2r}} \tag{14-3-47}$$

which is to be compared with Eqs. (14-3-35) and (14-3-42). The mode III opening of a crack, as defined in Sec. (14-1-3), is obtained simply by interchanging the X_1 and X_2 axes, i.e., by interchanging the subscripts 1 and 2 so that for the mode III opening the shear stress is T_{13}.

14-4 LINEAR ELASTIC FRACTURE MECHANICS

14-4-1 Ideal Fracture Strength

There are two modes by which gross deformation can occur in a solid. It may either slip along the crystal planes in the manner discussed in Sec. 11-4-1 or the atomic planes may be torn apart or "cleaved" so leading to fracture. Orowan (1949) has given the following simple way of estimating the cleavage stress. Suppose the exact form of the interatomic forces gives rise to a stress T which is approximated by a sine curve of wavelength λ as shown in Fig. 14-6 so that $\lambda/2$ is a measure of the range of the interatomic force. If the equilibrium atomic spacing is b, then

$$T = K \sin \frac{2\pi}{\lambda}(x-b) \tag{14-4-1}$$

from which K can be found by equating the initial slope to Young's modulus E. From Fig. 14-6, the increment of strain $d\tilde{e} = dx/b$, and hence there follows from Eq. (14-4-1) the relation

$$b\left(\frac{dT}{dx}\right)_{x=b} = \frac{2\pi b K}{\lambda} = E \tag{14-4-2}$$

which can be rearranged to yield

$$K = \frac{\lambda}{2\pi}\frac{E}{b} \tag{14-4-3}$$

The work of fracture must be at least the surface energy of the two new surfaces created. Letting the surface energy be γ, an estimate of λ can be obtained by setting

$$\int_{b}^{b+(\lambda/2)} T\,dx = 2\gamma \tag{14-4-4}$$

and substituting T from Eq. (14-4-1) gives

$$\lambda = \frac{2\gamma\pi}{K} \tag{14-4-5}$$

Substituting for λ in Eq. (14-4-5) from Eq. (14-4-3) gives

$$K^2 = \frac{E}{b}\gamma \tag{14-4-6}$$

The theoretical breaking stress T_{max} is thus given by

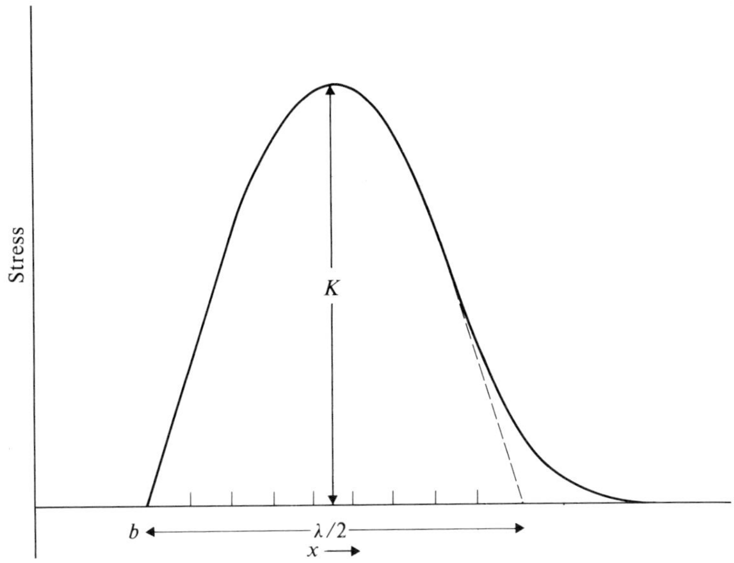

Figure 14-6 Form of interatomic force law and approximating sine curve, shown dashed.

$$T_{\text{max}} = K = \sqrt{\frac{E\gamma}{b}} \tag{14-4-7}$$

Thus strong solids will have a high elastic modulus, a large surface energy, and a small distance between atomic planes. As $\lambda/2$ will approximately equal b this would give $T_{\text{max}} \approx E/\pi$ and $\gamma \approx Eb/10$. Such high values of the breaking strength are rarely reached in practice except for very fine fibers; in general, brittle solids break at stresses of about $10^{-3}E$.

In applying Eq. (14-4-7) to the calculation of the maximum shearing stress of crystals account must be taken of the anisotropy of the material properties and E must be replaced by the appropriate elastic compliance $c_{\alpha\beta\gamma\delta}^{-1}$. It is more convenient to use the compliance than the elastic constant $c_{\alpha\beta\gamma\delta}$ because the sides of the rod which are being broken are assumed to be unconstrained. The values of γ and b will also depend on the orientation of the cleavage plane relative to the crystal planes.

More exact calculations using either the Morse function (see, for example, Gilman, 1963), for the energy of interaction U_{int} between two atoms at a separation r,

$$U_{\text{int}} = U_0\{\exp[-2a(r-r_0)] - 2\exp[-a(r-r_0)]\} \tag{14-4-8}$$

or the Lennard–Jones potential

$$U_{\text{int}} = U_0\left[\left(\frac{r_0}{r}\right)^{12} - 2\left(\frac{r_0}{r}\right)^{6}\right] \tag{14-4-9}$$

have been made by Zwicky (1923), Born and Furth (1940), and Kelly (1966). These calculations have shown that $T_{\max}$ for nonmetallic crystals lie between 5 and 16 percent of Young's modulus, depending on the type of atomic bonding and the crystal orientation. This cleavage stress can be shown to be about one-third of the heat of sublimation per unit volume and thus materials stressed to these levels will tend to show greater chemical reactivity and are more susceptible to chemical attack. A recent review of such calculations has been given by Macmillan (1972).

14-4-2 Griffith Theory of Fracture

To account for the discrepancy between the theoretical and experimentally determined values of the cleavage stress Griffith (1920) proposed that macroscopically homogeneous materials contained internal defects. These internal defects or microcracks enabled the stress to be concentrated in the way discussed in Sec. 14-3, the stress attaining in this way a sufficiently high value for the ideal fracture stress to be attained in small localized regions of the material. Consider an infinite plate in which there is a narrow elliptical crack with its major axis of length $2c$ aligned normal to the applied stress T_0. Assuming the Griffith defect to be crack-like, the radius of the tip of the ellipse will be approximated to the lattice spacing b. This assumption implies $2\sqrt{c/b} \gg 1$ and hence the Inglis solution in the form of Eq. (14-3-12) reduces to the approximate relation

$$T_{11} \approx 2T_0\sqrt{\frac{c}{b}} \qquad 2\sqrt{\frac{c}{b}} \gg 1 \tag{14-4-10}$$

Identifying the value of T_{11} given by Eq. (14-4-10) with the $T_{\max}$ of Eq. (14-4-7) gives

$$2T_0\sqrt{\frac{c}{b}} = \sqrt{\frac{E\gamma}{b}} \tag{14-4-11}$$

At fracture, $T_0 = T_F$, and hence Eq. (14-4-11) gives, on rearrangement,

$$T_F = \sqrt{\frac{E\gamma}{4c}} \tag{14-4-12}$$

Taking as typical values, $\gamma \sim 0.01Eb$, $2c \sim 5000b$, Eq. (14-4-2) gives $T_F \sim 10^{-3}E$. At fracture, the Inglis solution gives $T_{11} \sim 100T_F$, which implies crack-tip stresses well outside the range of classical infinitesimal linear elasticity.

An approach to failure criteria which is not subject to the difficulties described above, and in particular does not involve the atomic force–displacement laws, is that of Griffith (1920), who recognized that the driving force for crack extension was the difference between the energy which could be released if the crack length were increased and that needed to create new surfaces.

Consider an infinite plate in which there is a narrow elliptical crack with its major axis of length $2c$ aligned normal to the applied stress T_0. Let the crack be opened a small amount by the uniform applied stress T_0. The distortion of the lines

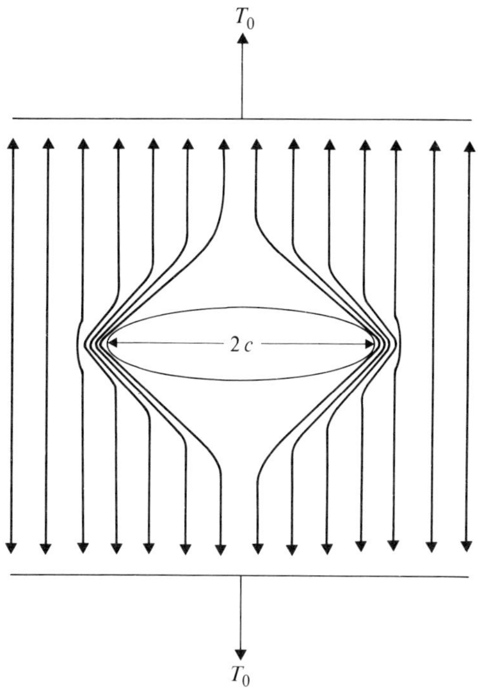

Figure 14-7 Distortion of stress field by an elliptical hole in a plate.

of force by the presence of the elliptical hole is shown schematically in Fig. 14-7. The stress-free areas above and below the crack are assumed to be approximately triangular in shape with height $k_1 c$. With this approximation, the strain energy release per unit thickness U_c is given in plane stress as [see Eq. (6-2-56)]

$$U_c = -\tfrac{1}{2} T_0 \frac{T_0}{E} k_1 c^2 \qquad \text{(plane stress)} \tag{14-4-13}$$

or, for an increase dc,

$$dU_c = -\frac{T_0^2}{E} k_1 c\, dc \qquad \text{(plane stress)} \tag{14-4-14}$$

For plane strain, it follows from Eq. (6-3-3) that the corresponding relation is

$$\frac{\partial U_c}{\partial c} = -\frac{T_0^2(1-\nu^2)}{E} k_1 c \qquad \text{(plane strain)} \tag{14-4-15}$$

From Sec. 14-2 it follows that the values of $\partial U_c/\partial c$ given by Eqs. (14-4-14) and (14-4-15) represent the decrease in potential energy when a crack extends by an infinitesimal amount dc under constant load. Griffith's more accurate calculations give $k_1 = \pi$.

The total mechanical work done is

$$W = 2U_c + U_\gamma = -\frac{T_0^2}{E}\pi c^2 + 4\gamma c \qquad \text{for } k_1 = \pi \tag{14-4-16}$$

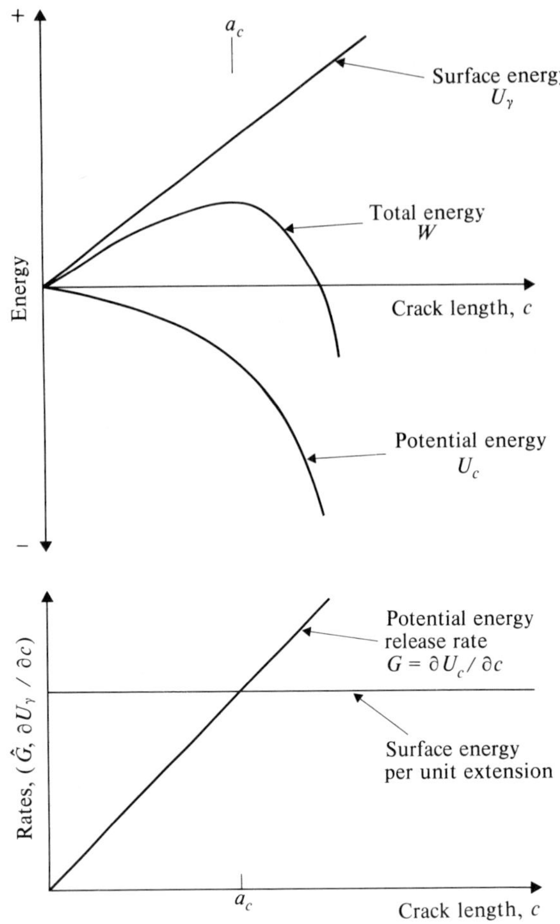

Figure 14-8 Variation of energy and rates of energy change with crack length.

where γ is the surface energy. A schematic representation of the variation of W, $2U_c$, and U_γ with crack length $2c$ is shown in Fig. 14-8. The maximum in the total energy W occurs at the critical value $c = a_c$ of crack length, and for this particular condition $\partial W/\partial c = 0$ giving, by way of Eq. (14-4-16),

$$\frac{T_0^2 \pi a_c}{E} = 2\gamma \tag{14-4-17}$$

which is Griffith's criterion of neutral stability in plane stress.

The positive value of the slope $\partial U_c/\partial c$ is generally referred to as the strain energy release rate and is given the symbol G in most of the literature. However, to avoid confusion with the shear modulus of rigidity the strain energy release rate will be given the symbol $\hat{G}$ throughout this chapter. It is evident from Eq. (14-4-13) with $k_1 = \pi$ that

$$\hat{G} = \frac{T_0^2 \pi c}{E} \tag{14-4-18}$$

which can be expressed by way of Eq. (14-4-17) in the form

$$\hat{G} = 2\gamma \qquad (c \equiv a_c) \tag{14-4-19}$$

For a given crack length $2c$, the Griffith fracture stress T_F follows from Eq. (14-4-17) in the form

$$T_F = \begin{cases} \left(\dfrac{2E\gamma}{\pi c}\right)^{1/2} & \text{(plane stress)} \qquad (14\text{-}4\text{-}20) \\ \left[\dfrac{2E\gamma}{\pi c(1-\nu^2)}\right]^{1/2} & \text{(plane strain)} \qquad (14\text{-}4\text{-}21) \end{cases}$$

The two physically significant aspects of the Griffith approach to fracture can be summarized as follows:

1. The theory is based on a thermodynamic criterion of fracture which considers the total change in energy of a cracked body as the crack extends.
2. The theory is a continuum theory in that it is not dependent on having available any detailed knowledge of the microstructural properties in the immediate neighborhood of the highly strained crack tip.

Griffith (1920) established the significance of defects in decreasing fracture strength by way of experiments on glass rods containing defects of various lengths. These experiments also confirmed that the fracture stress varied as the inverse of the square root of the crack length. The experiments of Hull, Beardmore, and Valintine (1965) on brittle refractory metals and those of Maitland and Chadwick (1969) on basal-plane cleavage of zinc containing long initial cracks are in general accord with the Griffith theory. It is now widely accepted that the Griffith theory provides a physically satisfactory description of the failure behavior of materials which effectively fracture in the classical, linear, infinitesimal range of elasticity. That is, it applies to materials where only a very small amount of plasticity occurs prior to fracture. Typical materials are glasses, where no slip-planes are present, other layer silicates, and basal-plane cleavage of zinc, diamond, tungsten, and other refractory metals.

The Griffith concept of treating fracture in terms of the change in energy remote from the immediate neighborhood of the crack tip has provided the basis for most of the later developments in the general field of fracture mechanics.

14-4-3 Cracks in Single Crystals

The Griffith theory is based on the continuum concepts of linear elasticity, taking no account of the periodic lattice structure, yet for a perfectly brittle material, i.e., one in which no plastic flow can occur, the atoms adjacent to the cleavage plane must have the form shown in Fig. 14-9 and contain atomic bonds in every stage of failure. As Sinclair and Lawn (1972) point out, an understanding of the atomic configuration at the crack tip is important in relation to such problems as the

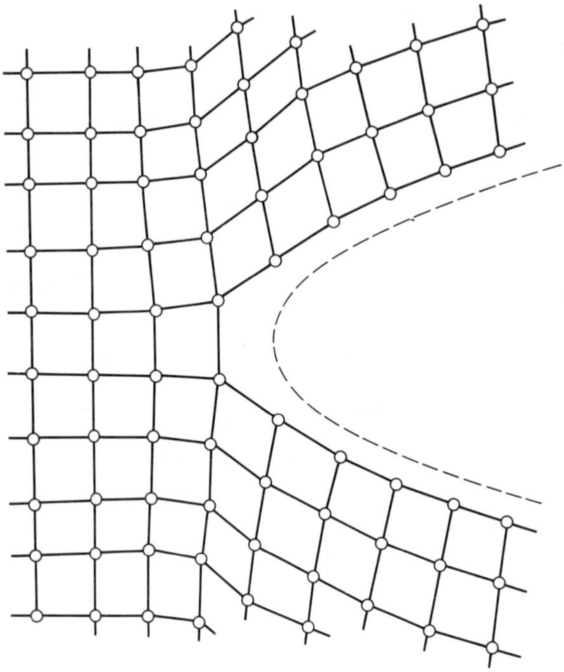

Figure 14-9 Atomic positions adjacent to the cleavage plane near the tip of a crack.

influence of a reactive environment fracture strength and the path taken by a crack as it grows in an arbitrary stress field. It has also been shown by Thomson, Hsieh, and Roma (1971) that introduction of the discrete character of the lattice into the theory of fracture leads to a stress range for which the crack is stable against extension and closure over a wide range of applied tensile loads, in contrast to the single critical stress predicted by Griffith (1920). Detailed calculations for various crystal systems and interatomic force laws have been made by Tyson (1966), Chang (1970), Gehlen and Kanninen (1970), Kanninen and Gehlen (1971, 1972), and Sinclair and Lawn (1972).

An important modification to the crack-tip geometry considered by Griffith was made by Barenblatt (1962), who suggested that for brittle materials the linear continuum theory could be employed even near to the crack tip if the atomic forces, acting across the narrow crack interface, cause the faces to close smoothly in a cusp form. When such a form is assumed the stresses and strains are everywhere finite and a necessary condition for the model to apply is that neither the crack shape nor the stress distribution should vary significantly over atomic distances. Bearing this in mind the approach used in the calculations mentioned above is to divide the material surrounding a crack into two regions. One region called the "core" or "region I" consists of the atoms near one tip of a long crack and within this region the interatomic forces are considered explicitly. The rest of the crystal comprises "region II" and is assumed to behave as a linear elastic continuum. In earlier calculations the displacement of atoms in region II was given directly from known solutions for a crack in an elastic medium. Having fixed the displacements in region II the displacement of atoms within region I, comprising some three or four atoms on either side of the crack tip, is chosen to give the minimum potential energy by a relaxation procedure usually performed by a high-speed computer. More recently, Sinclair (1972) has refined the above method by allowing the boundary between regions I and II to be flexible so that the elastic solution in region II is modified during the relaxation process in region I leading to a self-consistent picture. Sinclair finds that for a crack in silicon, taking 242 free atoms in region I, his calculations predict lattice trapping of the crack within a stress range about 10 percent of the Griffith critical stress, a spread which is much less than that originally estimated by Thomson but which is nevertheless considerable.

14-4-4 Rate of Crack Growth

In problems involving transient loading the Griffith criterion ceases to be valid because of the static nature of the theory on which it is based. If, for instance, the stress pulse is so short that the deflection process is discontinued before the necessary stress concentration occurs at the crack tip, then the crack will not extend. To examine the growth of cracks the Griffith theory must be augmented to take account of the kinetic energy K of the material.

Mott (1948) estimated the kinetic energy by the following argument. Suppose the width of the crack is increasing; then so long as dc/dt is small compared with the velocity of sound in the material, it can be assumed that, if $u(c)$ is the displacement at any point (x, y) in the static problem, the kinetic energy

$$K = \tfrac{1}{2}k\rho\dot{c}^2 \int\int \left(\frac{du}{dc}\right)^2 dx\, dy \tag{14-4-22}$$

where ρ is the density. The value of u near the surface of the crack is on the order of $(T_0 c/E)$ and thus the integral is of the form

$$K = \tfrac{1}{2}k\rho\dot{c}^2 c^2 \frac{T_0^2}{E^2} \tag{14-4-23}$$

where k is a numerical factor.

Now c can be treated as a generalized coordinate and the lagrangian L of the system, which is the difference between the kinetic and potential energies, is given by

$$L = K - (2U_c + U_\gamma) = K - W \tag{14-4-24}$$

where W is given by Eq. (14-4-16). The equation of motion of this system can be expressed by way of Eq. (4-2-9) in the form

$$\frac{d}{dt}\left(\frac{\partial L}{\partial \dot{c}}\right) - \frac{\partial L}{\partial c} = 0 \tag{14-4-25}$$

Substitute for W in Eq. (14-4-24) from Eq. (14-4-16) and then substitute the resulting equation for L in Eq. (14-4-25) to give

$$\frac{d}{dt}\left(k\rho\dot{c}\dot{c}^2\frac{T_0^2}{E^2}\right)-\left\{\left(k\rho\dot{c}^2c\frac{T_0^2}{E^2}\right)-4\gamma+\left[2(1-\nu^2)\pi c\frac{T_0^2}{E}\right]\right\}=0 \qquad (14\text{-}4\text{-}26)$$

which can be rearranged to give

$$\ddot{c}+c\dot{c}^2+\left(\frac{4\gamma E^2}{k\rho T_0^2}\right)-\left[\frac{2(1-\nu^2)\pi E}{k\rho}\right]c=0 \qquad (14\text{-}4\text{-}27)$$

For incipient cracking at $c = a_c$ then $\ddot{c} = \dot{c} = 0$, which gives the Griffith criterion of neutral stability in plane strain:

$$a_c=\frac{2\gamma E}{\pi(1-\nu^2)T_0^2} \qquad \text{(plane strain)} \qquad (14\text{-}4\text{-}28)$$

Integrating Eq. (14-4-27) and using the boundary conditions given above there is obtained the relation

$$\dot{c}^2=\left[\frac{2(1-\nu^2)\pi E}{k\rho}\right]\left(1-\frac{a_c}{c}\right)^2 \qquad (14\text{-}4\text{-}29)$$

Thus the crack, once it starts growing, rapidly accelerates and for crack lengths much greater than a_c the velocity of crack propagation is almost constant, the limiting velocity being $[2(1-\nu^2)\pi/k]^{1/2}(E/\rho)^{1/2}$. Extensions of the theory including cleavage fracture treated by simple cantilever beam theory have been given by Gilman (1959), Berry (1960), Craggs (1963), and Steverding and Lehnigk (1970, 1972). Several experimental investigations, notably by Schardin (1959), have confirmed the existence of a limiting velocity for propagation of a single crack and several theoretical attempts have been made to relate the limiting velocity to the Rayleigh surface wave speed. Experiments show that for brittle solids the limiting velocity is often about half the shear wave speed and can attain 80 percent of the Rayleigh wave speed (Field, 1971), although usually it is a lower percentage. The reason why the Rayleigh speed is not attained is not known but in part it may be due to the bifurcation or forking of the cracks at high velocities. In experiments on glass it has been shown that if notches are made in a specimen such that the cracks start propagating at different stresses T_F, then the length c_b they propagate before bifurcation is given by

$$T_F^2c_b=\text{constant} \qquad (14\text{-}4\text{-}30)$$

From Eqs. (14-4-14), (14-4-15), and (14-4-16) it would appear that bifurcation occurs at a critical velocity, which is in accord with the predictions of Yoffe (1951) who showed that when the velocity of crack propagation is low the crack proceeds in a direction normal to the applied stress axis but when the fracture velocity increases to about 0.6 of the velocity of an elastic lateral wave a stress wave is generated ahead of the moving crack and the stresses at the crack tip become almost equal over a wide arc. Yoffe suggests that at this point the crack bifurcates. The explanation is not entirely satisfactory as the predicted velocity of bifurcation is in general too high and the predicted angle of the bifurcated sections of the crack to the initial crack is too large. Johnson and Holloway (1966) have suggested that an energy criterion for bifurcation may be more consistent with the observed change in fracture surface structure.

14-4-5 Ductile Fracture: Some General Considerations

Thus far the discussion has been restricted entirely to elastic deformation of the material. However, the stress concentration at the crack tip is nearly always sufficient to induce some plastic flow even for *brittle* materials like glass, as shown by Marsh (1964). In brittle materials, however, the work expended on plastic flow is relatively small compared with the other energy terms, whereas in *ductile* fracture the plastic work can exceed the surface energy by orders of magnitude.

Orowan (1945, 1949) and Irwin (1948) independently proposed that linear elastic fracture mechanics could be applied to elastic-plastic fractures, provided the macroscopic fracture stress remained well below the general yield stress of the body. The Griffith approach was extended by Orowan (1949) who introduced a term to take account of the plastic work during deformation. Thus, in addition to the surface energy γ, Orowan showed that a plastic work γ_P per unit area of the crack surface must be supplied from the release of elastic energy. The total effective surface energy is thus $\gamma + \gamma_P$ and Griffith's criteria in the form of Eqs. (14-4-20) and (14-4-21) must be modified accordingly, the critical stress now being

$$T_F = \begin{cases} \left[\dfrac{2(\gamma+\gamma_P)E}{\pi c}\right]^{1/2} & \text{(plane stress)} \qquad (14\text{-}4\text{-}31) \\ \left[\dfrac{2(\gamma+\gamma_P)E}{\pi c(1-\nu^2)}\right]^{1/2} & \text{(plane strain)} \qquad (14\text{-}4\text{-}32) \end{cases}$$

The relations for crack growth given by Eq. (14-4-27) shows that, even if a crack is expanding rapidly because of release of elastic energy, growth can be slowed and even stopped by an increase in γ_P. Thus ductile materials can act as very efficient arrestors of crack growth. Whereas the surface energy γ is a well-defined property, both theoretically and experimentally, the plastic energy γ_P is very difficult to determine because both the plastic distortion and the appropriate constitutive equation for the plastic regime cannot as yet be satisfactorily described. However, at the point of incipient crack growth the condition $\partial W/\partial c$ still applies and, provided γ_P can be regarded as almost independent of the crack size, the dependence of the critical stress required to fracture a material on the crack dimensions will depend only on the elastic energy.

Thus, even where plastic deformation is involved, an energy approach similar to that of Griffith can still be employed, provided account is taken of the energy of plastic deformation. However, as Cottrell (1964) has noted, two distinct conditions must be satisfied for crack growth. First, the thermodynamics must be favorable in that the release of stored elastic energy must more than compensate the energy absorbed in the creation of new surfaces and in plastic deformation. Second, there must exist an atomic mechanism for crack opening. Thus a blunt notch in a brittle solid cannot grow even when the Griffith criterion is satisfied because the stress concentration at the crack tip is too small to break the atomic bonds. The thermodynamic condition is thus a necessary but not sufficient condition to ensure fracture. If, however, the parting of the atomic bonds can be achieved by other means than the application of stress alone, such as the diffusion of chemically reactive atoms along the crack surface, then the thermodynamic condition is also sufficient to ensure crack growth. When nearly all the energy for creation of new surfaces can be supplied by such chemical reactions, cracks can grow at very low values of the applied stress: such a fracturing process is called "stress corrosion."

14-4-6 The Stress Intensity Approach to Fracture

Irwin (1957a) noted that Westergaard's (1939) solution for the stresses in the neighborhood of the ends of elliptical cracks depended on one constant factor only, now called the stress intensity factor or simply fracture toughness, which Irwin denoted by K. However, to avoid confusion with the kinetic energy, the stress intensity factor will be denoted throughout this chapter by $\hat{K}_{\mathrm{I}}$, $\hat{K}_{\mathrm{II}}$, and $\hat{K}_{\mathrm{III}}$, corresponding to the modes I, II, and III respectively.

Westergaard's solutions, i.e., Eqs. (14-3-32), (14-3-33), (14-3-34), (14-3-42), and (14-3-47), can be expressed by way of the appropriate change of axes in the form:

Mode I:

$$\left.\begin{aligned} T_{11} &= \hat{K}_{\mathrm{I}}\frac{\cos(\theta/2)}{\sqrt{2\pi r}}\left(1+\sin\frac{\theta}{2}\sin\frac{3\theta}{2}\right)+\cdots \\ T_{22} &= \hat{K}_{\mathrm{I}}\frac{\cos(\theta/2)}{\sqrt{2\pi r}}\left(1-\sin\frac{\theta}{2}\sin\frac{3\theta}{2}\right)+\cdots \\ T_{21} &= \hat{K}_{\mathrm{I}}\frac{\cos(\theta/2)}{\sqrt{2\pi r}}\sin\frac{\theta}{2}\cos\frac{3\theta}{2}+\cdots \end{aligned}\right\} \tag{14-4-33}$$

Mode II:

$$T_{12} = \hat{K}_{\mathrm{II}}\frac{1}{\sqrt{2\pi r}} \tag{14-4-34}$$

Mode III:

$$T_{13} = \hat{K}_{\mathrm{III}}\frac{1}{\sqrt{2\pi r}} \tag{14-4-35}$$

where

$$\hat{K}_{\mathrm{I}} = T_0\sqrt{\pi c} \qquad \hat{K}_{\mathrm{II}} = \tau\sqrt{\pi c} \qquad \hat{K}_{\mathrm{III}} = q\sqrt{\pi c} \tag{14-4-36}$$

are the stress intensity factors. The quantity $\hat{K}'_{\mathrm{I}} = T_0\sqrt{c}$ was originally termed the stress intensity factor (Irwin, 1958; Irwin, Kies, and Smith, 1958), but the stress intensity factor $\hat{K}$ defined by Eq. (14-4-36) is now the most widely used form.

Irwin (1957a) has modified the definition of $\hat{K}_{\mathrm{I}}$ for the case of specimens of finite dimensions. In this procedure, the solution is regarded as that for an infinite plate, modified by an algebraic or trigonometric function which is chosen to make the appropriate surface forces zero. For example, in the case of a central crack of length $2c$ in a plate of width b, Irwin (1957a), using Sneddon's (1946) solution for a penny-shaped crack, obtained the relation

$$\hat{K}_{\mathrm{I}} = T_0\sqrt{\pi c}\left[\frac{2b}{\pi c}\tan\left(\frac{\pi c}{2b}\right)\right] \tag{14-4-37}$$

A list of known configurations for which $\hat{K}$ has been evaluated together with an extensive bibliography is given by Kenny and Campbell (1968) (see also Paris and Sih, 1965; Tada, Paris, and Irwin, 1973).

Irwin (1957b) evaluated the change in energy in the body as a whole by calculating the work done by the surface forces acting across the length δc when the crack is closed from length $(c+\delta c)$ to c. This energy change can be expressed in the form

$$\hat{G}\delta c = \int_0^{\delta c} T_{11}(s)u_1(s)\,ds \tag{14-4-38}$$

where the thickness is taken as unity and, from Eq. (14-3-36),

$$u_1 = 2(1-\nu^2)\frac{T_0}{E}\sqrt{2c}\,(\delta c - a)^{1/2} \qquad \text{for } a/c \ll 1 \tag{14-4-39}$$

where $a = x_2 - c$ is the distance ahead of the crack tip. The stress distribution ahead of the crack in mode I loading can be obtained from Westergaard's (1939) solution, with an appropriate change of axes, in the form

$$T_{11}|_{x_1=0} = \frac{T_0\sqrt{\pi c}}{\sqrt{2\pi a}}\left(1 - \frac{a}{4c} + \cdots\right) \tag{14-4-40}$$

which, for $a/c \ll 1$, reduces to the approximate relation

$$T_{11} \simeq \frac{T_0\sqrt{\pi c}}{\sqrt{2\pi a}} \qquad \text{for } a/c \ll 1 \tag{14-4-41}$$

Integration of Eq. (14-4-38) after substituting for u_1 from Eq. (14-4-39) and T_{11} from Eq. (14-4-41) gives

$$\hat{G}\delta c = \frac{T_0^2\pi c}{E}(1-\nu^2)\delta c \tag{14-4-42}$$

Equation (14-4-42) can be expressed, by way of Eq. (14-4-36), in the form

$$\hat{G} = \frac{\hat{K}_{\mathrm{I}}^2}{E}(1-\nu^2) \qquad \text{(plane strain)} \tag{14-4-43}$$

An identical argument gives, for plane stress,

$$\hat{G} = \frac{\hat{K}_{\mathrm{I}}^2}{E} \qquad \text{(plane stress)} \tag{14-4-44}$$

In Eqs. (14-4-43) and (14-4-44), $\hat{G}$ is just the strain energy release rate.

In terms of the stress intensity factors,

$$\hat{K}_\mathrm{I}^2 = \hat{G}_\mathrm{I} E \qquad \text{(plane stress)} \tag{14-4-45}$$

$$\hat{K}_\mathrm{I}^2 = \hat{G}_\mathrm{I} \frac{E}{1-\nu^2} \qquad \text{(plane strain)} \tag{14-4-46}$$

$$\hat{K}_\mathrm{II}^2 = \hat{G}_\mathrm{II} \frac{E}{1-\nu^2} \tag{14-4-47}$$

$$\hat{K}_\mathrm{III}^2 = \hat{G}_\mathrm{III} \frac{E}{1+\nu} \tag{14-4-48}$$

it being noted that $\hat{G}_\mathrm{I} = 2(\gamma + \gamma_P)$.

At fracture, the Griffith criterion in plane strain can be expressed by way of Eq. (14-4-15) in the form

$$\hat{G}_\mathrm{crit}\, dc = \frac{\hat{K}_\mathrm{Ic}^2}{E}(1-\nu^2)\, dc = 2\gamma\, dc \tag{14-4-49}$$

for unit thickness, that is, $G_\mathrm{crit} = 2\gamma$, which is simply Eq. (14-4-9).

To take explicit account of the plastic flow that occurs at the tips of a sharp crack, Irwin (1964) concluded that the plastic region, provided it is small, can be represented in plane stress by a cylinder of radius

$$r_p = \frac{\hat{K}_\mathrm{I}^2}{2\pi Y^2} \tag{14-4-50}$$

where Y is the yield stress in simple uniaxial tension. Irwin therefore proposed that the effects of the plastic region could be accounted for by assuming that the actual crack length be increased by r_p. Thus, the first equation of (14-4-36) could be modified by the plastic correction to give

$$\hat{K}_\mathrm{I} = T_0 \sqrt{\pi(c + r_p)} \tag{14-4-51}$$

For fractures well below yield, that is, $T_F \ll Y$, the correction factor is negligible. It is evident that, when compared against Eq. (14-4-51), the relation of the first equation of (14-4-36) underestimates the fracture toughness by ca. 4 percent when $T_F/Y = 0.4$, rising to ca. 15 percent when $T_F/Y = 0.8$.

Reviewing the concepts underlying the stress intensity factor, it is evident that Irwin's (1957a, 1957b) approach to fracture mechanics is based on the concept that linear elastic fracture mechanics can be applied to elastic-plastic fractures, provided the macroscopic fracture stress remains well below the yield stress of the material. In particular, the Griffith criterion is here equivalent to an instability resulting from a critical stress distribution ahead of the crack. It is this critical stress distribution ahead of the crack which gives rise to the stress intensity factor approach. The stress intensity factor approach is concerned primarily with the crack-tip region which, though small relative to the body as a whole, is regarded as being still large enough to ignore atomic dimensions. Hence, continuum mechanics and classical elasticity theory still apply.

Standard procedures have been developed for measuring reproducible values of plane strain fracture toughness $\hat{K}_{Ic}$. Two types of specimen of widely differing geometry are recommended for use with the standard methods of test. Both are single-edge-notched (SEN) specimens, one type being bent under three-point loading, the other type, referred to as the compact tension specimen (CTS), being subjected to point loading through pins above and below the crack faces. For a comparison of CTS, SEN bend, and SEN tension specimens with an infinite plate, see Wilson (1966). Satisfactory reproducibility has been obtained between some 20 different laboratories for the values of $\hat{K}_{Ic}$ determined on a steel and an aluminum alloy, using both SEN bend and CTS test pieces. Dimensions of the test specimens are important if satisfactory reproducibility is to be obtained (see, for example, Brown and Srawley, 1966, 1971, 1974). In practice, measurements of $\hat{K}_{Ic}$ are most likely to be restricted to comparatively high strength alloys, since it is only for these that the necessary conditions of test can be satisfied. Various aspects of testing in terms of the concept of a stress intensity factor have been considered by Morton and Ritter (1973).

14-4-7 Ductile to Brittle Transition

Once a crack is formed on a cleavage plane, the only work required for brittle fracture is that needed to break the interatomic bonds at the crack ends and this can usually be provided by the elastic energy of the specimen. As the crack does not require thermal activation the velocity of crack propagation can be very high, about half the speed of shear waves in the material. Because of the large amount of work expended in plastic deformation, ductile fracture propagates at much lower velocities and in certain cases the fracture can be stopped. Thus brittle and ductile fracture can be distinguished by their mode of propagation, even though in both cases the deformation may be of a ductile nature prior to the start of crack propagation, as noted by Orowan (1949). Many solids can exhibit both brittle and ductile fracture characteristics depending on the mode of stressing, the rate of application of the load, and the temperature. For instance, Bridgman (1947) has shown that most materials can be induced to crack in a ductile fashion if a sufficiently large hydrostatic pressure is applied to the specimen as it is pulled in tension. At high rates of strain or low temperatures most materials fracture in a brittle manner.

An explanation of the change from brittle to ductile fracture was originally proposed by Ludwik (1924), and later developed by Davidenkov (1936) and Orowan (1949), in which brittle fracture stress T_F was introduced such that for tensile stresses greater than T_F the material would cleave in a brittle manner. Plastic flow is assumed to occur independently of the brittle fracture at the shear yield stress k; thus if $T_F < k$ the fracture will be brittle, while if $T_F > k$ it will be ductile. For many materials k decreases rapidly with temperature while T_F, being independent of thermal activation, does not vary much with temperature, as shown schematically in Fig. 14-10. The Ludwik hypothesis also accounts for the effect of notches in increasing the brittle ductile transition because, as de-

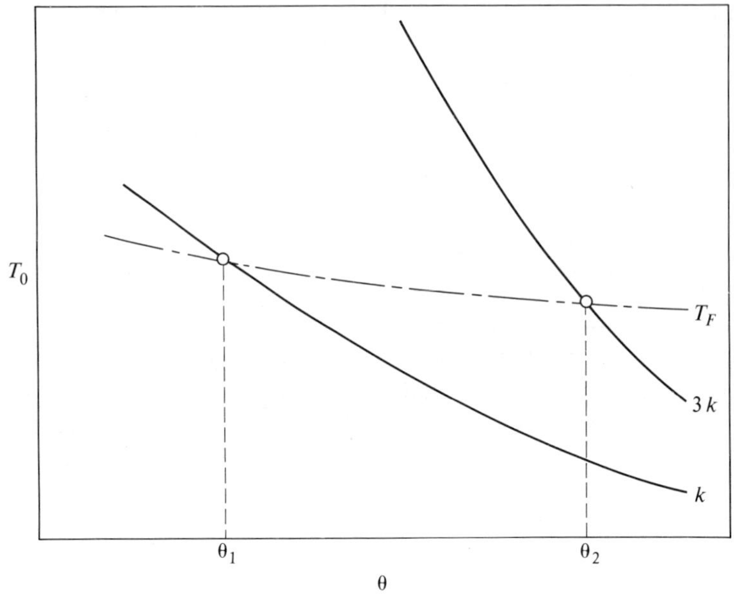

Figure 14-10 The change from brittle to ductile fracture with increase of temperature as measured by the work of fracture. (*Orowan, 1949.*)

monstrated by Orowan (1949) from the solutions of the stress distribution near an axially symmetric notch given by Hencky (1923) and a long flat crack given by Prandtl (1923) and Hill (1949), the maximum tensile stress is about $3k$ and thus, as Fig. 14-10 shows, the introduction of a notch raises the ductile-brittle transition temperature. Other contributing factors which can help to raise the yield-stress curve of notched specimens are the increasing strain rate and work-hardening. Referring to Fig. 14-10, it can be seen:

1. If $T_F < k$, i.e., at temperatures $<\theta_1$, the material is completely brittle.
2. If $k < T_F < 3k$, that is, $\theta_1 <$ temperature $< \theta_2$, the material is notch brittle.
3. If $3k < T_F$, i.e., at temperatures $>\theta_2$, the material is ductile.

The above theory also explains the relatively sharp rise in the work of fracture as measured by impacting the specimen with a pendulum hammer, as indicated in Fig. 14-11. To relate the ductile-brittle transition to more basic material parameters, in the case of single crystals Kelly, Tyson, Cottrell (1967) have proposed the criterion that if (largest tensile stress/largest shear stress) near the crack tip > (ideal cleavage stress/ideal shear stress) then fully brittle fracture is possible; otherwise a crack always exhibits some plastic flow.

Variation of the grain size d at a given temperature can also produce ductile to brittle transitions, as Petch (1953) has shown in experiments for the fracture of iron, the variation of the fracture stress T_F having the same form as for the yield stress, namely,

$$T_F = T_0 + k_f d^{-1/2} \tag{14-4-52}$$

Comparing this with the formula for the yield stress given by Eq. (11-2-1) it could be seen that there is no reason for a transition as $k_f > k_y$ implies $T_F > k$ for all d, and vice versa. However, as shown in Fig. 14-12, it is found experimentally that there is a change from brittle fracture at large grain sizes to a ductile-type fracture for small grain sizes, which would imply that $k_f > k_y$ and T_0 for fracture must be less than T_0 for yielding.

14-4-8 Biaxial and Triaxial Stresses

Griffith (1924) obtained a fracture criterion for the biaxial state of stress from the Inglis (1913) solution of the stress distribution around an elliptical hole on the

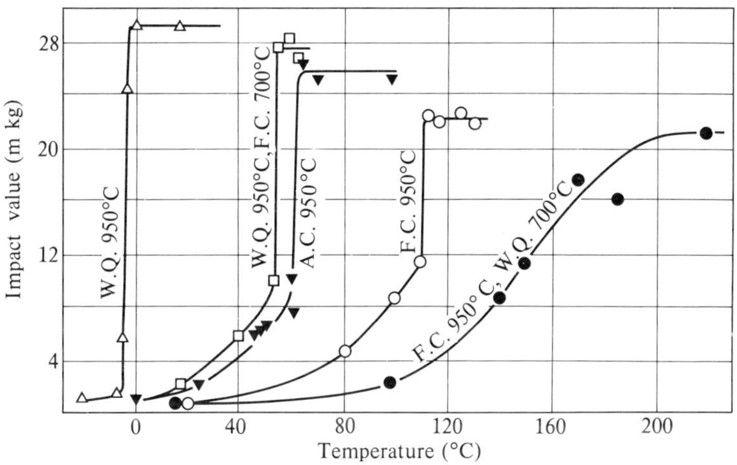

Figure 14-11 Typical ductile-brittle transitions. (*Allen et al., 1953.*)

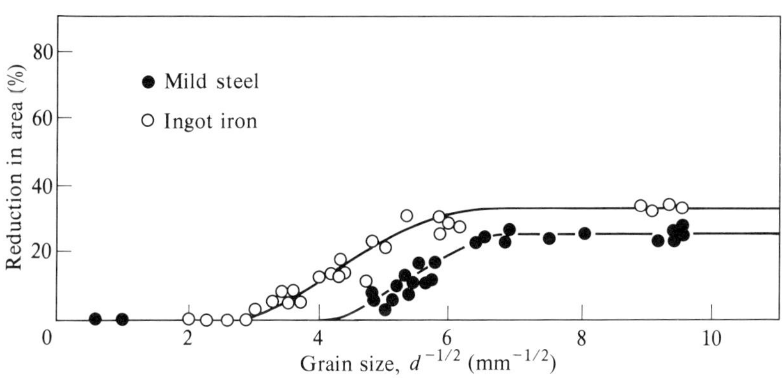

Figure 14-12 Variation of ductility as measured by area reduction before final rupture. (*Petch, 1953.*)

assumption that the material contains an isotropic distribution of similar cracks. Griffith assumes that fracture occurs when the highest local tensile stress of the largest crack in the most dangerous orientation exceeds the cohesion strength of the material. In terms of the fracture strength $\hat{K}_I$ for uniform tension Griffith's criterion takes the form:

If $3\sigma_1 + \sigma_2 > 0$, fracture occurs when $\sigma_1 = \hat{K}_I$ (14-4-53)

If $3\sigma_1 + \sigma_2 < 0$, fracture occurs when $(\sigma_1 - \sigma_2)^2 + 8\hat{K}_I(\sigma_1 - \sigma_2) = 0$ (14-4-54)

where $\sigma_1 > \sigma_2$ and the biaxial stress system is shown in Fig. 14-13. Orowan (1949) produced a graphical interpretation of these results by noting that Eq. (14-4-54) represents a parabola through the origin if σ_1 and σ_2 are taken as rectangular coordinates, while Eq. (14-4-53) represents a tangent to this parabola. If all combinations of σ_1 and σ_2 are considered the resulting fracture curve is as shown by the full line in Fig. 14-14. Fracture occurs for stress states lying within the hatched area. Equation (14-4-54) shows that the hydrostatic pressure needed to prevent fracture at the shear stress $\tau = (\sigma_1 - \sigma_2)/2$ is $\tau^2/4\hat{K}_I$. This explains why many materials that are brittle in ordinary tension and compression tests become ductile under a sufficiently high hydrostatic pressure. For uniaxial compression Eq. (14-4-54) shows the strength is $8\hat{K}_I$; thus the compressive strength in brittle fracture is eight times higher than the tensile strength, a relationship which is found to be approximately true for many rocks and concrete. Griffith's theory also predicts that the most dangerous cracks in uniaxial compression would be those at

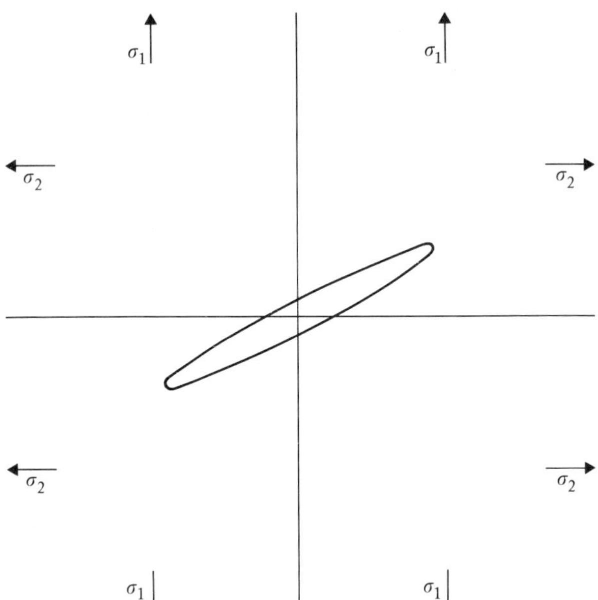

Figure 14-13 Biaxial stress system around a Griffith crack.

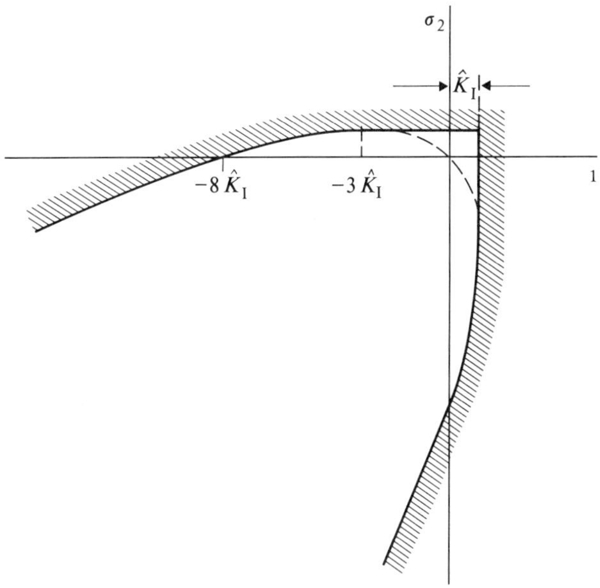

Figure 14-14 Orowan's graphical illustration of Griffith's failure criterion for biaxial stresses. (*Orowan, 1949.*)

30° to the stress direction; however, it is sometimes found that glasses and stones crack parallel to the direction of compression and this latter type of cracking occurs more frequently if the compression platens are lubricated. The reason for cracking parallel to the stress is not well understood.

The Griffith biaxial criterion can also be applied in triaxial stress states since the normal and shear stresses in the plane perpendicular to the crack edge have little significance.

14-5 MICROSTRUCTURAL FEATURE OF CRACKS IN GLASS

In an effort to obtain a better understanding of the mechanism of crack propagation considerable effort has recently been devoted to studying the morphology or characteristic features associated with a crack. One of the first features to be noticed were the "Wallner lines" (Wallner, 1939; see also Freminville, 1914) which consisted of ripples on the otherwise smooth fracture surfaces caused by the superposition of elastic waves, originally radiating from the crack tip and then reflecting from neighboring surfaces, and the applied stress. The observation of Wallner lines suggested a means of measuring crack velocity in which an ultrasonic wave of known frequency is passed through the fracturing specimen and the wavelength of the resulting ripples on the surface can be related to the speed of crack growth.

The appearance of tensile fracture surfaces of glass and other brittle polymers usually consists of three regions. In the region nearest the source of the crack the surfaces are very smooth and consequently this is known as the "mirror" region. Within the "mirror" region the crack is gaining speed and the rate of energy release is increasing. This region is then followed by a lightly roughened region which has been termed "mist" by Andrews (1959) and whose features have been examined in detail by Johnson and Holloway (1968). The roughening is thought to be due to the incipient formation of new fracture surfaces out of the plane of the main crack which, however, fails to propagate very far. As the rate of energy release becomes

Figure 14-15 Regions of fracture surfaces from a glass rod broken in simple tension. (*Johnson and Holloway, 1966.*)

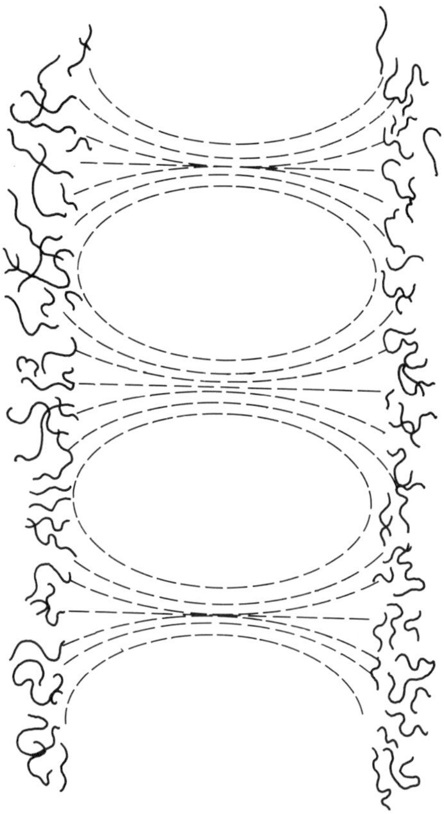

Figure 14-16 Strings connecting the bulk polymeric material in the craze region ahead of a crack. (*Doyle, Moranci, Orowan, and Stork, 1972.*)

greater new fracture surfaces can be created sufficiently far ahead of the main crack front to gain sufficient energy that they produce forking of the crack front and produce a heavily striated surface known as the hackle region. All three regions can be seen in Fig. 14-15.

Doyle, Moranci, Orowan, and Stork (1972) have examined the fracture of glassy polymers with the electron microscope and their results indicate that a rather more complex mode of fracture exists than that envisaged for the Griffith theory. Ahead of the ruptured region of the crack is a wedge-shaped region of "craze" having a porous or "expanded" texture which is caused by the separation of bundles of molecules as shown in Fig. 14-16. Within the strings connecting the two surfaces of the bulk material across the craze layer the molecules become aligned in the direction of tensile stress and separation is by viscous sliding of molecules past each other. Because of the nonalignment of the strings where they join the bulk material and because of the inability of the rupture of one string to initiate rupture in an adjacent string, the craze tends to be weakest where it joins the bulk material. Consequently, if the tensile stress rises to a sufficiently high value the craze will strip off the bulk material causing the broad striations found ahead of the smooth crack surface in polystyrene. When the craze wedge suddenly peals off, the adjacent bulk polymer either multiple-crazes and cracks arise radiating from its edge or a new craze wedge is initiated. Since the fracture mechanism includes a velocity-dependent viscous process Orowan concludes that the Griffith theory cannot be applied to glassy polymers, even as an approximation, and in support of this contention he notes that:

1. The measured plastic fracture work is abnormally high (Murray and Hull, 1971) and depends greatly on the conditions of the test.
2. The Griffith theory cannot account for advance fractures which are observed at low velocities of crack propagation when the inertia effects are negligible. Nor can they be ascribed to flaws as they do not occur close to the region of fracture but their density increases with increasing fracture velocity.
3. As the Inglis stress field predicts the highest stress at the crack edge, advance fractures in front of the crack edge are incompatible with the Griffith theory.
4. The suggestion that surface crazes can develop to observable depths before a fracture is initiated is incompatible with the Griffith theory, for as soon as a crack begins to propagate it can do so more easily and will run across the specimen before the next critical crack can propagate.
5. The craze wedge can grow to macroscopic lengths while its tip remains so sharp that its radius of curvature cannot be seen with the microscope, which means that it cannot be regarded as an extra extension of the crack.

Thus the detailed motion of a propagating crack would appear to be much more complex than envisaged in the simple Griffith theory and this would seem to preclude the identification of the energy terms with such simple properties as surface energy alone. No satisfactory basic theory incorporating the presently known fracture structure is at present available.

14-6 YIELDING FRACTURE MECHANICS

14-6-1 Crack-Opening Displacement

Plastic deformation in the immediate neighborhood of the tip of a crack allows the two faces of the crack to move apart without any increase in the length of the crack. When the plastically deformed region becomes large, classical elasticity theory can no longer be used to predict the local stresses and strains. It was independently suggested by Wells (1961) and Cottrell (1961) that the amount of crack opening prior to crack extension, regarded as a consequence of yielding at the crack tip, might be a characteristic property of a given material tested under a given set of conditions. The crack-opening displacement (COD) of present interest, although based on linear elasticity neglecting effects arising from extensive plasticity, yields a fracture criterion which has relevance to fractures after restricted amounts of plastic flow.

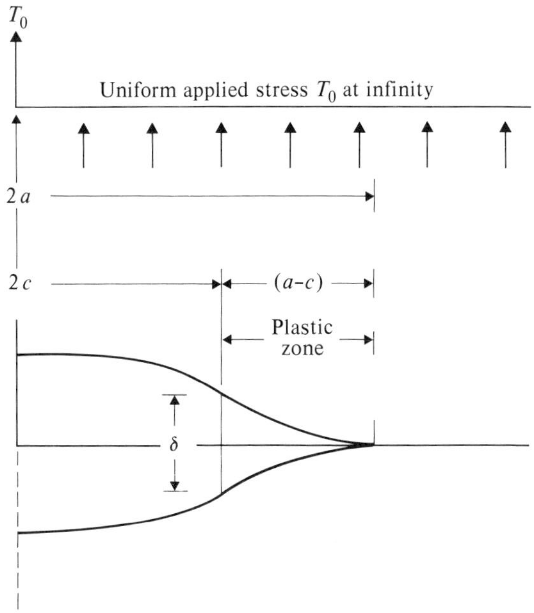

Figure 14-17 Crack-opening displacement.

Consider an infinite plate with a central crack of length $2c$ subjected to a uniform applied stress T_0 at infinity. Plastic deformation occurring at the crack tip is regarded as giving rise to an apparent increase in the crack length to some value $2a$. Associated with this concept is a restraining stress T_R, acting directly on the crack faces, which has the effect of partly restricting the opening of the crack over the distances $(a-c)$ at both ends. This is illustrated in Fig. 14-17.

Wells (1961) developed the fracture criterion by way of the observation that in practice the tip of a slot subjected to plastic deformation opened with a near square-ended contour, giving a well-defined tip opening, referred to as the crack-opening displacement. Assuming that the COD, δ, of Fig. 14-17, is a measure of crack-tip deformation, Wells proposed the criterion that fracture occurs when a critical value of δ, that is, δ_c, is attained.

Burdekin and Stone (1966) have developed the COD concept by way of an approach to yielding at the ends of a crack under tension which was proposed independently by Dugdale (1960) and Barenblatt (1962). By equating T_R to the uniaxial yield stress Y, Burdekin and Stone obtained the relation

$$\frac{c}{a} = \cos\left(\frac{\pi}{2}\frac{T_0}{Y}\right) \tag{14-6-1}$$

The COD δ at the tip of the real crack was then evaluated as

$$\delta = \frac{8}{\pi}\frac{Y}{E}c\,\text{logsec}\left(\frac{\pi}{2}\frac{T_0}{Y}\right) \tag{14-6-2}$$

where E is Young's modulus. Expanding the logsec term in Eq. (14-6-2) gives

$$\delta = \frac{\pi T_0^2 c}{EY}\left[1 + \frac{\pi^2}{24}\left(\frac{T_0}{Y}\right)^2 + \cdots\right] \tag{14-6-3}$$

which can be expressed in the form

$$\hat{G} = \frac{\pi T_0^2 c}{E}\left[1 + \frac{\pi^2}{24}\left(\frac{T_0}{Y}\right)^2 + \cdots\right] \tag{14-6-4}$$

where

$$\hat{G} = Y\delta \tag{14-6-5}$$

For linear elastic fracture mechanics it follows from Eq. (14-4-44) that, for an infinite plate with center crack $2c$, the strain energy release rate

$$\hat{G}_{\mathrm{I}} = \frac{\hat{K}_{\mathrm{I}}^2}{E} = \frac{\pi T_0^2 c}{E} \tag{14-6-6}$$

from which it is evident that Eq. (14-6-4) reduces for very low stresses to linear elastic fracture mechanics for brittle materials. Using the plastic zone correction factor for plane stress, gives the strain energy release rate in the form

$$\hat{G} = \frac{\pi T_0^2 c}{E}\left[1 + \frac{1}{2}\left(\frac{T_0}{Y}\right)^2\right] \tag{14-6-7}$$

for linear elastic fracture mechanics modified to take account of the plastic flow that occurs at the ends of a sharp crack. Equation (14-6-7) is seen to differ from Eq. (14-6-4) by only a small amount in the coefficient of the second term. Assuming, on the basis of either continuum mechanics or crystal dislocation mechanics, that $\hat{G} = T_R\delta$ relates directly to the work done to close an element of crack, it would appear that, with T_R identified with Y, the concepts underlying the formulation of the COD technique form an extension of linear elastic fracture mechanics for plane stress.

Plastic flow in plane strain tension is considerably more difficult to treat in fracture mechanics. This is because allowance must be made for the effects of constraints as the yielded zone becomes larger. Attempts have, however, been made to represent plane strain by

$$\hat{G} = \phi Y\delta \tag{14-6-8}$$

where ϕ is a numerical factor. Wells (1970) has suggested, by way of the results of experiment, the value $\phi \simeq 2.1$ for compatibility with linear elastic fracture mechanics in cases of limited ductility.

Bilby, Cottrell, and Swinden (1963) independently obtained Eq. (14-6-1) by way of crystal dislocation mechanics. The plastic zone and crack are regarded as arrays of dislocations, as indicated schematically in Fig. 14-18. If these screw-type dislocations, which are distributed along the X_2 axis, were discrete, then each dislocation would have a line vector $\xi = (0, 0, \xi_3)$ and a Burgers vector $b = (0, 0, b_3)$. The dislocation array must be in the form of an "inverse" pile-up, the

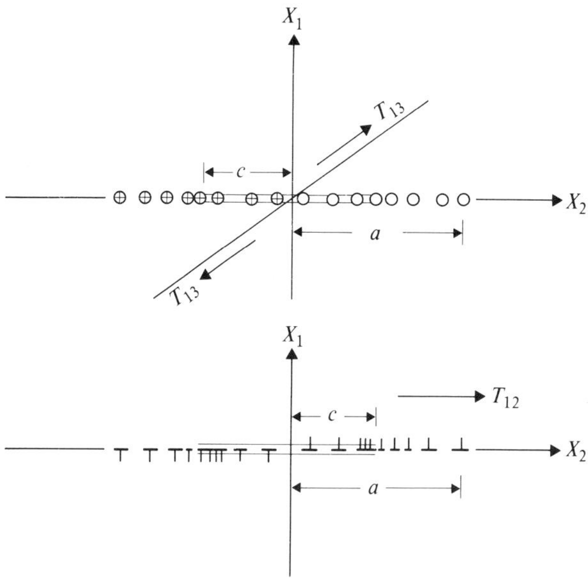

Figure 14-18 Plastic zone and crack regarded as arrays of dislocations.

dislocations being closely spaced near the crack tip and widely spaced at the extremities of the plastic zone, thus corresponding to the high and low values of the displacement gradient at the crack tip and the elastic-plastic interface respectively. Equation (14-6-1) is obtained not by regarding the dislocations as discrete but by way of the theory of continuous distributions of dislocations.

The COD approach is concerned primarily with the immediate neighborhood of the crack tip and relates directly to the micromechanisms of fracture over an area of less than 0.01 mm^2. Consistent results for the COD δ have been obtained (Smith and Knott, 1971). It is to be noted that, in general, the results depend strongly on the micromechanisms of fibrous fracture at a stress concentrator (Nichols et al., 1969). The dull fibrous appearance is characteristic of the nature of the fracture mechanism which involves the very localized, internal plastic necking of material between nonmetallic inclusions.

14-6-2 The *J*-Contour Integral Approach

The theory of fracture, discussed above, leading to the concept of fracture toughness, was developed on the basis of linear elastic material response. An extension to nonlinear elastic material behavior has been put forward by Rice (1968a, 1968b) on an energy variation basis. An elastic material is here defined as one for which a strain energy density function $W = W(e_{ij})$ exists as a single-valued function in strain space and thus we can write

$$T_{ij} = \frac{\partial W}{\partial e_{ij}} \tag{14-6-9}$$

Consider then two bodies of the same elastic material, each containing a hole or void as in Fig. 14-19*a*, *b*. Each body is identical and subject to the same system of tractions and displacements on the boundary except that the hole is larger in one body. Assume the tractions are applied on a portion of the boundary S_t and the displacements are given on a portion of the boundary S_u. The larger void may be regarded as being formed by an extension of the smaller original void by a volume increase ΔV and an associated surface increase ΔS. The body is initially loaded with tractions $t_i^{(0)}$ on S_t and displacements $u_i^{(0)}$ on S_u which produce a stress field $\mathbf{T}^{(0)}$ and a strain field $\mathbf{e}^{(0)}$. The void is then expanded but tractions $t_i^{(0)} = T_{ij}^{(0)}\nu_j$ are applied to the new surface ΔS to maintain the initial deformation state $(\mathbf{T}^{(0)}, \mathbf{e}^{(0)})$ in the body. Then, holding the tractions on S_t and the displacements on S_u fixed, the tractions on ΔS are reduced to zero to produce the new deformation state $T_{ij}^{(0)} + \Delta T_{ij}$, $e_{ij}^{(0)} + \Delta e_{ij}$. If t_i^* and u_i^* denote the loadings and displacements on ΔS during this process, the total change in potential energy may be written as

$$-\Delta U_p = \int_{\Delta V} W(e_{mn}^{(0)})\,dV - \int_{\Delta S}\left(\int_{u_i^*=u_i^{(0)},\,t_i^*=t_i^{(0)}}^{u_i^*=u_i^{(0)}+\Delta u_i,\,t_i^*=0} t_i^*\,du_i^*\right)dS \tag{14-6-10}$$

This may be interpreted as showing that the net reduction in potential energy is the strain energy of the material removed minus the work done in freeing the new surface of tractions.

Using a virtual work argument Rice has shown that the following inequalities apply:

$$0 \leqslant -\Delta U_p - \int_{\Delta V} W(e_{mn}^{(0)})\,dV \leqslant \int_{\Delta S} \Delta t_i \Delta u_i\,dS \tag{14-6-11}$$

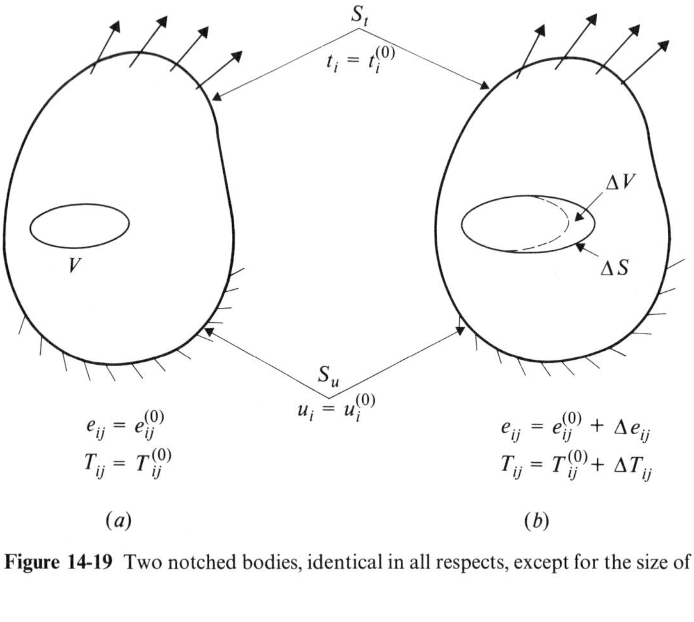

Figure 14-19 Two notched bodies, identical in all respects, except for the size of the hole.

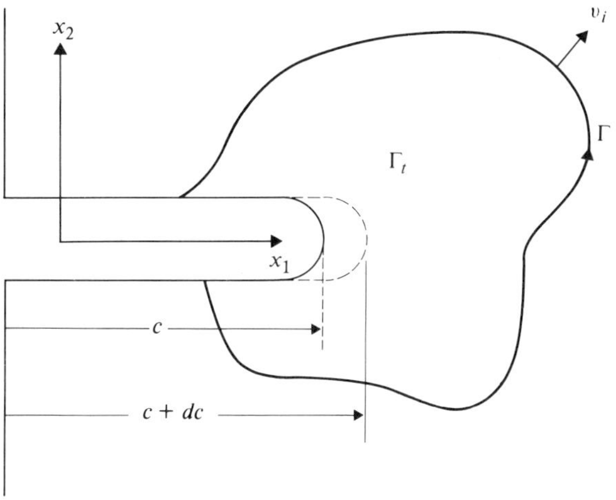

Figure 14-20 Crack configuration and path for the J-contour integral.

If the two voids differ in size by an infinitesimal amount and points on the boundary are obtainable by proceeding a distance dv normal to the boundary, then for sufficiently smoothly varying displacements and tractions Δt_i $(=\Delta T_{ij} v_j)$ and Δu_i are first-order quantities so that the upper bound of Eq. (14-6-11) is second order. Since the integral over ΔV is first order the upper bound is zero when dealing with an infinitesimal smooth change in geometry. Representing volume elements of removed material by $dS\,dv$ the potential energy decrease is given by

$$-\Delta U_p = \int_{\Delta V} W(e_{mn}^{(0)})\,dV = \int_{\Delta S} [W(e_{mn}^{(0)})\,dv]\,dS \qquad (14\text{-}6\text{-}12)$$

As a special case consider a flat-surfaced notch in a two-dimensional deformation field, as shown in Fig. 14-20. The notch has surfaces parallel to the x_1 direction and a smooth curved tip denoted by the arc Γ_t. Let us compare the potential energy of a notch of length c with that of a notch of length $c+dc$ having a geometrically identical tip. Defining U'_p as the potential energy per unit thickness in the x_3 direction and noting that ΔV can be expressed as $dc\,dx_2\,dx_3$ the rate of decrease of potential energy per unit thickness is

$$-\frac{dU'_p}{dc} = \int W(e_{mn}^{(0)})\,dx_2 \qquad (14\text{-}6\text{-}13)$$

Now consider the integral J defined as follows:

$$J = \int_\Gamma \left(W\,dx_2 - \boldsymbol{t} \cdot \frac{\partial \boldsymbol{u}}{\partial x_1}\,ds \right) \qquad (14\text{-}6\text{-}14)$$

where Γ is a curve surrounding the tip, as shown in Fig. 14-20, starting from the lower flat notch surface and ending on the upper flat notch surface, s is arc length, and Γ is traversed in an anticlockwise direction. As $t = 0$ on the notch surface we find that, for $\Gamma = \Gamma_t$, $J = -dU'_p/dc$.

To show that J is path independent consider two curves Γ_1 and Γ_2, suppose that Γ_2 encloses Γ_1, and let J_1 and J_2 be the associated values of the integral. Then $J_2 - J_1$ is the integral taken in a counterclockwise sense around the boundary of the area $S(\Gamma_2, \Gamma_1)$ enclosed by the curves and the surfaces of the notch, since both terms of the integrand vanish on the flat surfaces. Transforming to an area integral gives

$$J_2 - J_1 = \int_{S(\Gamma_1, \Gamma_2)} \left(\frac{\partial W}{\partial x_1} - T_{ij} \frac{\partial e_{ij}}{\partial x_1} \right) dx_1 \, dx_2 = 0 \tag{14-6-15}$$

since

$$\frac{\partial W}{\partial x_1} = T_{ij} \frac{\partial e_{ij}}{\partial x_1} \tag{14-6-16}$$

Thus, providing the region between Γ_2 and Γ_1 is simply connected and free of singularities, J is path independent, and, as shown above,

$$J = -\frac{dU_p'}{dc} \tag{14-6-17}$$

Because of the path independence one can choose the Γ contours which lie near the crack tip or the specimen boundaries, whichever is most convenient for purposes of calculation. The basis of the J-contour integral approach is thus conceptually equal to the Griffith (1920) term $\hat{G}$, except that $\hat{G}$ implies a linear elastic event whereas J implies a nonlinear elastic event.

The concept of a contour integral had previously been developed separately by Eshelby (1956) and Cherepanov (1967). However, its use in fracture mechanics is based upon the integration of the Rice formulation to give the stress analysis solutions of Hutchinson (1968) and Rice and Rosengren (1968). Hutchinson considered the distribution of stress occurring at the tip of a crack located in an infinite plate subjected to an applied uniaxial tensile stress T_0 at infinity for both plane stress and plane strain. The region of nonlinear response to deformation was taken to be extensive in relation to crack length but remote from the crack the stresses were specified to remain linearly elastic. In this study a total deformation theory of plasticity was used in conjunction with some suitable form of hardening stress-strain relation. For example, Hutchinson used a representation of nonlinear elastic-plastic behavior of the type suggested by Ramberg and Osgood (1943) to describe the characteristic tensile stress-strain relations of certain metals. This can be expressed in the form

$$\frac{\epsilon}{\bar{\epsilon}_Y} = A \left(\frac{T}{Y} \right)^n \tag{14-6-18}$$

where A is a constant of proportionality, T is the tensile component of stress, ϵ is the total tensile component of strain, Y is the yield stress in simple tension, and $\bar{\epsilon}_Y = Y/E$ where E is the initial slope of the stress-strain curve. Introducing the effective stress $\sigma_e = (3J_2')^{1/2}$ and corresponding effective strain ϵ_e, plastic stress and strain intensity factors were derived at the tip of the crack, the dependence on distance r from the crack tip being of the form

$$\sigma_e = a_1 \frac{1}{r^{1/(n+1)}} \sigma_e(\theta) \tag{14-6-19}$$

$$\epsilon_e = a_2 \frac{1}{r^{n/(n+1)}} \epsilon_e(\theta) \tag{14-6-20}$$

and the displacements

$$u = a_3 r^{1/(n+1)} u(\theta) \tag{14-6-21}$$

the angular distributions with respect to θ also being given by Hutchinson. Rice and Rosengren (1968) arrived separately at a similar set of relations. Hutchinson (1968) gave the above proportionality factors in terms of plastic intensity factors and showed that for small-scale yielding, for which linear elasticity dominates the regions remote from the crack tip, these plastic intensity factors could be expressed in terms of J and some tabulated constants, I, the value of which depended on n. The formulation of this approach has been discussed by McClintock (1971), the exact form depending markedly on the assumed form for the characteristic stress-strain relation. Thus, the proportionality constant for Eq. (14-6-19) follows from Eq. (14-6-18) in the form

$$a_1 = \left(\frac{JE}{Y^2 IA}\right)^{1/(n+1)} \tag{14-6-22}$$

from which it is concluded that, for any given hardening law, i.e., any value of n, the modulus E, and the limit of linearity as expressed by Y, the crack-tip stresses are a function of J. From this there follows the further conclusion that fracture, since it is controlled by conditions at the crack tip, must be describable in terms of J, in a manner similar to that by which fracture is described in linear elastic fracture mechanics by $\hat{K}$. Thus, the use of J as a crack-tip parameter in nonlinear elasticity has been established, both in terms of a thermodynamic-based criterion for fracture in the form of Eq. (14-6-17) and in terms of a characteristic relation in the form of Eq. (14-6-22).

For the plastic range of deformation, attempts to apply the J-contour integral approach encounter a particular difficulty in relation to the physical significance of J. In the plastic range of deformation, J can no longer be regarded as a true energy release rate. It has, however, been argued that the Hutchinson (1968) and Rice and Rosengren (1968) solutions can be regarded as representative, to a good approximation, of fracture in the plastic range of deformation, provided only simple monotonic loading is considered, all unloading being excluded. This argument is based on the observation that for monotonic loading in simple tension the stress-strain history of any element of material, such as elements in the immediate neighborhood of the crack tip, probably differs but little according to whether the material follows total, i.e., deformation, plasticity or whether it conforms to the incremental-type theories of plasticity. However, with the onset of crack propagation, some elements of material in the immediate neighborhood of the crack tip must be expected to unload. For the plastic range of deformation, the physical significance of J remains an open question since in either deformation or incremental plasticity the energy represented by dU'_p/dc is no longer available as potential energy for propagating the crack but has been dissipated in the plastic deformation during loading. Thus, the physically acceptable energy balance concept of Griffith (1921) no longer applies because there are now three energy terms. These are the potential energy, fracture surface energy, and the work absorbed in extensive plastic flow.

Experiments by Begley and Landes (1972) with specimens of differing geometry designed to give very different states of stress have shown that fracture initiates at an almost constant value of J, denoted by J_{Ic}. However, as Knott (1973) has pointed out, it is possible that low triaxiality induces fibrous fracture in steels whilst high triaxiality induces cleavage. In these circumstances it seems unreasonable that identical J_{Ic} values will be obtained in widely differing specimen geometries. Bucci, Paris, Landes, and Rice (1972) have investigated J-estimation procedures for several specimen geometries. For certain configurations, including the standard SEN bend and CTS specimens, Rice et al. (1973) have shown that the analysis can be reduced to a single load-displacement record by making use of J-estimation procedures (see also Hayes and Turner, 1974; Dawes, 1975).

Since the contour integral concept relates only to two dimensions, use of the associated fracture criterion is restricted to plane stress or plane strain, even in a fully plastic state. Subcritical crack growth is usually exhibited by failures occurring in plane stress, a situation not accounted for by J_{Ic}. The restriction to crack initiation arises from the use of the deformation theory of plasticity, as already noted. The uniqueness of J is attributed by Rice (1968b) to effects arising from crack-tip blunting. The major advantage of J_{Ic} over COD is the condition that J is independent of the path of the line integral. This is a major advantage because analysis at the crack tip for δ_c is a potential source of large errors.

14-7 GENERALIZED FRACTURE MECHANICS

14-7-1 Nonlinear Elastic Fracture Mechanics

The concepts underlying the failure criteria, characterized in terms of the critical values of $\hat{K}$, δ, and J, do not have a satisfactory physical basis (see, for example, Andrews, 1968; Knott, 1973; Turner, 1975). The technical importance of fracture mechanics and associated parameters is such that the deficiencies in the theories which have been discussed above must be regarded as being too restrictive. These deficiencies can be summarized as follows:

1. Linear fracture mechanics is based upon the assumptions of linear elasticity and infinitesimal strain. These assumptions are not valid in the highly strained neighborhood of the crack tip, especially when plastic zones are formed.

Evidence for the inelastic nature of the deformations is provided by the very high values of surface energy obtained by way of the Griffith approach.

2. The equations most frequently used to characterize the stress distribution around a crack are only approximate. This is so even if the material is everywhere elastic. In particular, they require zero tip radius and fail to predict the correct stress at large distances from the crack tip.
3. The energy balance approach developed by Rivlin and Thomas (1953) for rubber avoids the restriction of infinitesimal strains. However, this approach still requires the material to be reversibly elastic. Real materials, in particular metals and polymeric materials, are not reversible for large strains, i.e., for plastic deformation.

Andrews (1974) has proposed a generalized approach which removes the serious constraints imposed by the three requirements of linear elasticity, reversibility, and infinitesimal strains. This approach is concerned with the physical mechanisms of fracture and as such provides a physical basis for fracture surface energies. Certain aspects of this theory have been considered by way of experimental studies on fatigue (Andrews and Walker, 1971) in polyethylene and the adhesion of elastomers (Andrews and Kinloch, 1973a, 1973b).

Consider an infinite sheet of material containing a crack of length $2c$ and loaded at infinity by a uniform tensile stress T_0 applied at right angles to the crack axis. The material will initially be considered to be elastic. A zero tip radius will be assumed for the crack. These conditions are shown schematically in Fig. 14-21, where x_1, x_2 are cartesian coordinates in the plane of the lamina of the point P, referred to a fixed origin at the center of the crack and to the undeformed state.

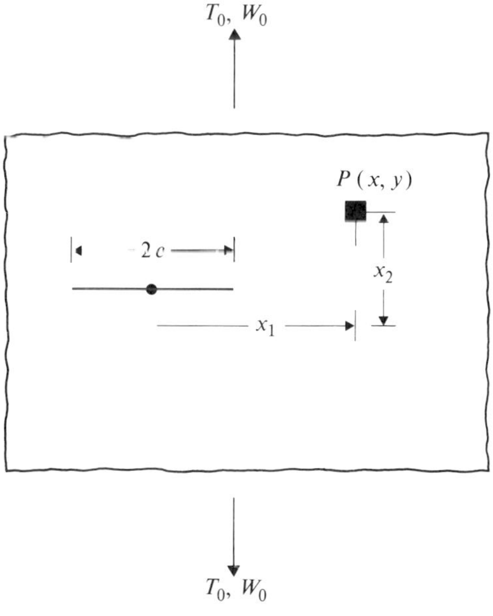

Figure 14-21 Crack of length $2c$ loaded at infinity by a uniform tensile stress T_0 applied at right angles to the crack axis.

The applied stress T_0 at infinity and the corresponding energy W_0 in the material at infinity are to be held constant during the propagation of the crack. It is important to distinguish between an elastic lamina for which W_0 will be the stored energy density and the inelastic case for which W_0 is the input energy density defined as the work done on a unit volume of material during a monotonic deformation. The same distinction must also be made when considering the local energy density W at any point in the stress field.

From dimensional considerations, Andrews (1974) assumes that

$$T_{ij}(P) = T_0 f_{ij}(x, y) \tag{14-7-1}$$

$$W(P) = W_0 f(x, y) \tag{14-7-2}$$

where

$$x = \frac{x_1}{c} \qquad y = \frac{x_2}{c} \tag{14-7-3}$$

are reduced variables and $T_{ij}(P)$ is a component of the stress tensor at the point P. The change in energy density at P due to an increment of crack growth is, in the limit,

$$\begin{aligned} \frac{dW(P)}{dc} &= W_0\left(\frac{\partial f}{\partial x}\frac{\partial x}{\partial c} + \frac{\partial f}{\partial y}\frac{\partial y}{\partial c}\right) \\ &= -\frac{W_0}{c}g(x, y, \epsilon_0) \end{aligned} \tag{14-7-4}$$

where, as for f,

$$g(x, y, \epsilon_0) = x\frac{\partial f}{\partial x} + y\frac{\partial f}{\partial y}$$

is another, as yet unspecified, function and ϵ_0 is the strain at infinity.

The total energy change in the system due to propagation of the crack is

$$\frac{dU_c}{dc} = \sum_P \frac{dW(P)}{dc}\delta v \tag{14-7-5}$$

where U_c is the total energy in the system and δv is the volume element at P, i.e.,

$$\delta v = h\delta x_1 \delta x_2 = hc^2 \delta x \delta y \tag{14-7-6}$$

where h is the undeformed lamellar thickness. Thus, from Eqs. (14-7-4), (14-7-5), and (14-7-6),

$$-\left(\frac{dU_c}{dc}\right)_{\epsilon_0} = cW_0 h \sum_P g(x, y, \epsilon_0)\delta x \delta y \tag{14-7-7}$$

Referring the energy change to unit area of crack interface and noting that $A = 4ch$, it follows that

$$-\left(\frac{dU_c}{dA}\right)_{\epsilon_0} = \tfrac{1}{4}W_0 c \sum_P g(x, y, \epsilon_0)\delta x \delta y \tag{14-7-8}$$

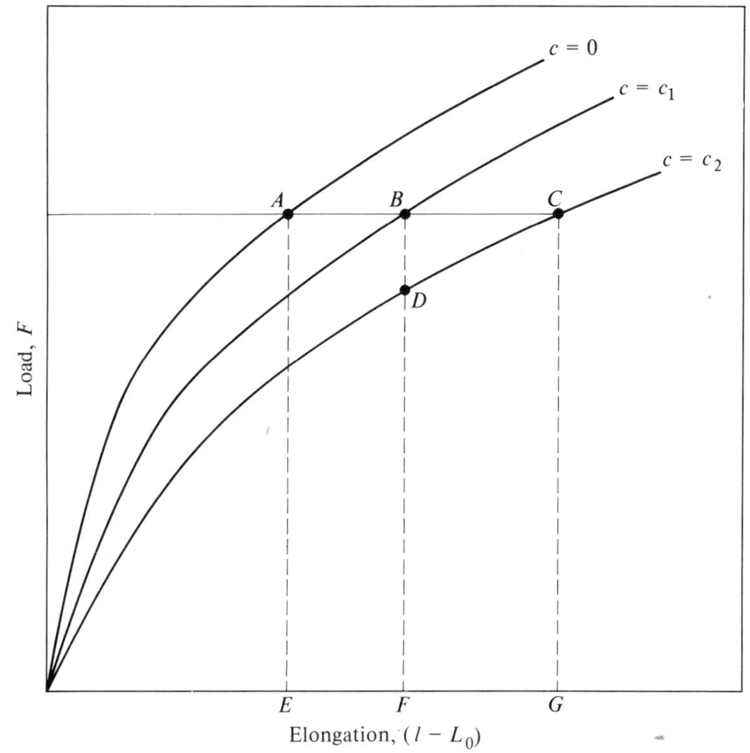

Figure 14-22 Load-deflection curves for SEN specimens containing cracks of different lengths (schematic).

A distinction must now be made between the energy change $-(dU_c/dA)_{\epsilon_0}$ associated with the unit area of crack propagation in an elastic medium at constant load, and thus at constant ϵ_0, W_0, and the energy change $-(dU_c/dA)_\delta$ associated with the unit area of crack propagation in an elastic medium at constant deflection. For this purpose consider n identical sheets of an elastic but nonlinear material, each sheet containing a center or edge crack of length $2c_i$ $(i = 0, 1, \ldots, n)$ and each crack being of a different length. The identical sheets are tested in simple tension to yield n different load-deformation curves, one for each different length of crack, it being noted that all cracks are oriented normal to the loading direction. Three typical load-deformation curves are shown in Fig. 14-22 for $i = 0, 1, 2$, it being noted that c_0 $(=0)$ corresponds to a sheet without a crack.

Propagation of the crack at constant load from a length $2c_1$ to a length $2c_2$ involves doing work on the specimen at constant ϵ_0 and W_0. This particular work is given by the area $BCGF$ in Fig. 14-22. The total elastic energy of the specimen, however, increases by the net amount ΔE such that

$$\left(\frac{\Delta U_c}{\Delta A}\right)_{\epsilon_0} = \frac{BCGF - OBC}{4h(c_2 - c_1)} \tag{14-7-9}$$

where h is the sheet thickness. Alternatively, if crack propagation occurs at constant deflection, the specimen suffers an energy loss equivalent to the area OBD. Thus,

$$\left(\frac{\Delta U_c}{\Delta A}\right)_\delta = -\frac{OBD}{4h(c_2 - c_1)} \tag{14-7-10}$$

Since, in the limit $c_2 \to c_1$, area OBD – area OBC, it follows from the geometry shown in Fig. 14-22 that

$$\left(\frac{dU_c}{dA}\right)_\delta = \frac{-p}{1-p}\left(\frac{dU_c}{dA}\right)_{\epsilon_0} \tag{14-7-11}$$

where $0 < p < 1$ and $p = 0.5$ if curves OB, OC are linear. In the general nonlinear case

$$p = p(\epsilon_0) \tag{14-7-12}$$

Since the summation appearing in Eq. (14-7-8) involves only geometrical terms and is carried out in dimensionless space where the reduced crack length is always unity, it follows that the summation is independent of stress or energy terms and also of crack length. This conclusion still applies when the boundary of integration is defined by a condition for which the only nonvanishing component of the stress tensor $\mathbf{T}(P)$ is T_0 since from Eq. (14-7-1) it is evident that this particular boundary condition is itself determined solely by x and y. Hence, Eq. (14-7-11) can be expressed by way of Eq. (14-7-8) in the form

$$-\left(\frac{dU_c}{dA}\right)_\delta = k_1(\epsilon_0)cW_0 \tag{14-7-13}$$

where

$$k_1(\epsilon_0) = \frac{-p}{1-p}\sum_P g(x, y, \epsilon_0)\delta x\delta y \tag{14-7-14}$$

is a constant. For a linear material

$$W_0 = \frac{1}{2}\frac{T_0^2}{E} \tag{14-7-15}$$

where E is Young's modulus. Hence, for a linear material, Eq. (14-7-13) simplifies by way of Eq. (14-7-10) to

$$-\left(\frac{dU_c}{dA}\right) = k_1\frac{T_0^2 c}{2E} \qquad \text{for } p = 0.5 \tag{14-7-16}$$

from which it is evident that, with k_1 set equal to π, the right-hand side of Eq. (14-7-13) reduces to Griffith's formula for a linear material. It is important to note, however, that Eq. (14-7-13) is valid for *any* elastic material regardless of linearity or strain magnitude, and as such is of far greater generality than the Griffith equation.

The fracture criterion can now be stated as follows. Crack propagation occurs when the available energy, as defined by Eq. (14-7-13), equals the energy $\mathscr{T}$

required to create unit area of new interface, i.e., when

$$\mathscr{T} = k_1(\epsilon_0)cW_{0,\,\mathrm{crit}} \tag{14-7-17}$$

it being noted that $\mathscr{T}$ is equivalent to J_{Ic}.

14-7-2 Yielding Fracture Mechanics

Consider now a material which has an unloading stress-strain curve which differs from its loading stress-strain curve. For example, in the case of metals this behavior corresponds to the onset of plastic deformation. The energy density is no longer a single-valued function of stress or strain so that, in general, it will no longer be a single-valued function of the spatial variables x, y. Thus, for elements undergoing unloading from a stress, $\mathbf{T}_P$ must now have

$$W(P) = W_0F(x, y, \mathbf{T}_P) \tag{14-7-18}$$

it being noted, however, that for monotonic loading of the material $W(P)$ is still defined by Eq. (14-7-2), in particular,

$$[F]_{\mathbf{T}_P=0} = f \tag{14-7-19}$$

must apply because as the stress tends to zero all materials ultimately behave elastically. The dependence of $W(P)$, as defined by Eq. (14-7-18), on $\mathbf{T}_P$ must be allowed because the unloading stress-strain curve is not unique but depends on the maximum stress levels achieved in the stress cycle.

These physical considerations imply the relations

$$\frac{dW(P)}{dc} = \begin{cases} -\dfrac{W_0}{c}g(x, y, \epsilon_0) & \text{(elements loading as the crack propagates)} \\ -\dfrac{W_0}{c}Q(x, y, \mathbf{T}_P) & \text{(elements unloading)} \end{cases}$$

where the function Q is to F as g is to f. Summing over the stress field gives

$$-\left(\frac{dU_c}{dA}\right)_{\epsilon_0} = cW_0\left[\sum_{PL} g(x, y, \epsilon_0)\delta x\delta y + \sum_{PU} Q(x, y, \mathbf{T}_P)\delta x\delta y\right] \tag{14-7-20}$$

where L and U stand for loading and unloading respectively.

For incremental growth of the crack, arising from monotonic loading at the point P, elastic behavior can be represented by the relation

$$\lim_{\Delta T_{ij}\to 0} \frac{d\Delta W_q}{d(\Delta T_{ij})_q} = S \qquad \text{for } q = + \text{ or } - \tag{14-7-21}$$

and inelastic behavior by the relations

$$\begin{aligned} \lim_{\Delta T_{ij}\to 0} \frac{d\Delta W_+}{d(\Delta T_{ij})_+} &= S \\ \lim_{\Delta T_{ij}\to 0} \frac{d\Delta W_-}{d(\Delta T_{ij})_-} &= \alpha S \end{aligned} \tag{14-7-22}$$

where $\Delta W, \Delta T_{ij}$ are the associated energy density increments and stress increments respectively and

$$\alpha = \alpha(\mathbf{T}_P, \theta, \dot{\epsilon}) < 1 \tag{14-7-23}$$

is the energy density recovered at the point P divided by that which would have been recovered from an elastic solid over the same negative stress increment. These physical considerations imply the condition

$$\left[\frac{dW(P)}{dc}\right]_U = \alpha\left[\frac{dW(P)}{dc}\right]_L \tag{14-7-24}$$

and hence Eq. (14-7-20) can be expressed in the form

$$-\left(\frac{dU_c}{dA}\right)_{\epsilon_0} = cW_0\left[\sum_{PL} g(x, y, \epsilon_0)\delta x\delta y + \sum_{PU} \alpha g(x, y, \epsilon_0)\delta x\delta y\right] \tag{14-7-25}$$

From Eqs. (14-7-11) and (14-7-25) it follows that

$$-\left(\frac{dU_c}{dA}\right)_\delta = cW_0\left[k_1(\epsilon_0) - \sum_{PU} \beta g(x, y, \epsilon_0)\delta x\delta y\right] \tag{14-7-26}$$

where

$$k_1(\epsilon_0) = \frac{-p}{1-p}\left[\sum_{PL} g(x, y, \epsilon_0)\delta x\delta y + \sum_{PU} g(x, y, \epsilon_0)\delta x\delta y\right] \tag{14-7-27}$$

and

$$\beta = \frac{-p}{1-p}(1-\alpha) \tag{14-7-28}$$

Since the stress $\mathbf{T}_P$ is given by Eq. (14-7-1) it follows that

$$\beta = \beta_0(T_0, \theta, \dot{\epsilon})g_1(x, y, \epsilon_0) \tag{14-7-29}$$

where g_1 is a further unspecified function. Because of the product relation of Eq. (14-7-1), Eq. (14-7-28) may, in some circumstances, impose restrictions on the form of the functions β_0 and g_1. Hence, it follows that Eq. (14-7-26) can be expressed in the form

$$\begin{aligned}-\left(\frac{dU_c}{dA}\right)_\delta &= cW_0\left[k_1(\epsilon_0) - \beta_0 \sum_{PU} g_2(x, y, \epsilon_0)\delta x\delta y\right]\\ &= k_2(T_0, \theta, \dot{\epsilon})cW_0\\ &= k_1(\epsilon_0)cW_0\frac{1}{\Phi}\end{aligned} \tag{14-7-30}$$

where

$$\Phi = \frac{k_1(\epsilon_0)}{\left[k_1(\epsilon_0) - \beta_0 \sum_{PU} g_2(x, y, \epsilon_0)\delta x\delta y\right]} \tag{14-7-31}$$

In Eq. (14-7-30), $-(dU_c/dA)_\delta$ is here the actual energy available for forming crack surfaces after deduction of all energy losses throughout the specimen. At the moment of propagation it must therefore equal the actual energy requirement per area of interface, i.e., the surface energy $\mathscr{T}_0$.

There now arises the question of distinguishing between the quantities $-(dU_c/dA)_\delta$ appearing in Eqs. (14-7-13) and (14-7-30). The first expression for $-(dU_c/dA)_\delta$, i.e., Eq. (14-7-13), is just the energy available for crack propagation in a perfectly elastic solid and hence, when the crack actually propagates, the critical value of $-(dU_c/dA)_\delta$ must, by definition, be the surface work $\mathscr{T}$ which is related to the corresponding critical value of W_0, that is, $W_{0,\text{crit}}$, by way of Eq. (14-7-17). The difference between the perfectly elastic solid and the inelastic material of present interest is that for the elastic case all this energy is available for bond rupture at the fracture plane while in the inelastic case some of this energy is dissipated in the material by inelastic deformation processes, leaving only a proportion for bond fracture. This remaining proportion is given by Eq. (14-7-30) and must attain a particular value $\mathscr{T}_0$ to break the bonds and cause propagation and hence, under conditions of crack propagation in an inelastic material, must have

$$\mathscr{T}_0 = k_2 c W_{0,\text{crit}} \tag{14-7-32}$$

Equations (14-7-17) and (14-7-32) can be combined to yield

$$\begin{aligned}\mathscr{T} &= \mathscr{T}_0 \frac{k_1}{k_2(T_0, \theta, \dot{\epsilon})} \\ &= \mathscr{T}_0 \Phi(T_0, \theta, \dot{\epsilon})\end{aligned} \tag{14-7-33}$$

where $\Phi = k_1/k_2(T_0, \theta, \dot{\epsilon})$ is defined by Eq. (14-7-31); Φ reduces to unity if the material is everywhere perfectly elastic.

The application of generalized fracture mechanics to ductile materials has been considered by Andrews and Billington (1976). In this study the deformations leading to crack propagation were established experimentally using single-edge-notched (SEN) specimens of a copper-beryllium alloy in two conditions. Condition A corresponds to specimens which had been solution-treated for one hour at 800°C and water-quenched and condition H corresponds to specimens which had received a 50 percent cold-rolling treatment. Typical mechanical properties were as shown in Table 14-1. Several load-deformation curves, typical of the behavior of wide specimens containing cracks of different lengths, are shown in Figs. 14-23 and 14-24 for conditions A and H respectively.

The left-hand side of Eq. (14-7-13) is obtained by first determining the quantity $(U_0 - U_c)$, identified from Fig. 14-22 by the area $OABF - OBF = OAB$,

Table 14-1 Mechanical properties for conditions A and H

Temper	Tensile strength, MPa	Yield strength, MPa	Strain at yield, %	VPN
A	500	235	0.9	100
H	775	725	2.9	220

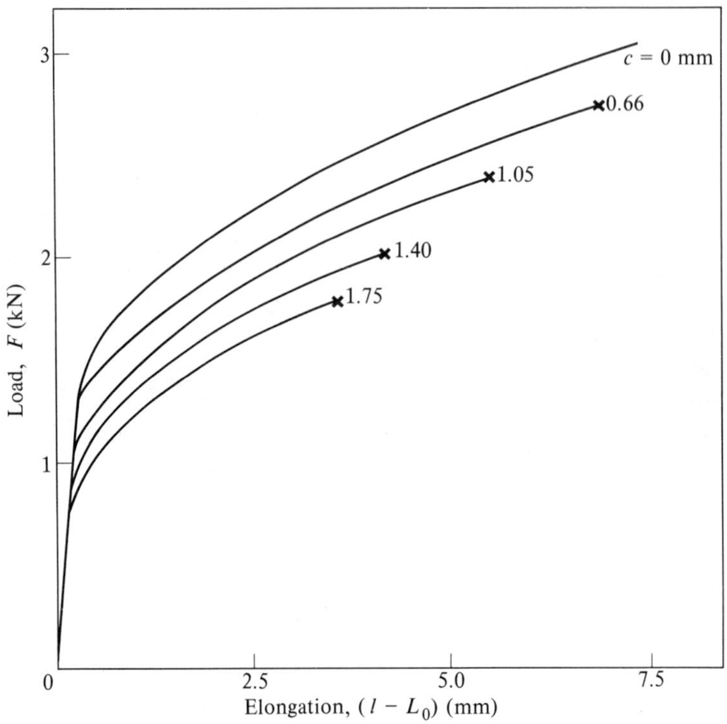

Figure 14-23 Actual load-deflection curves for narrow specimens of condition *A* containing single-edge cracks of different lengths; × denotes fracture propagation. (*Andrews and Billington, 1976.*)

for a given W_0 and various values of c. Having determined $(U_0 - U_c)$ for various crack lengths, the quantity $-(dU_c/dc)_{W_0}$ can be evaluated from the $[(U_0 - U_c), c]$ curve by graphical methods. The quantity $-(dU_c/dc)_{W_0}$ was found to vary linearly with c for constant W_0, an observation which is in accord with Eq. (14-7-13) and demonstrates its validity even where the whole specimen is subject to plastic deformation.

The slope

$$\psi = -\frac{1}{c}\left(\frac{dU_c}{dc}\right)_{W_0}$$

can be obtained directly from the linear relation between $-(dU_c/dc)_{W_0}$ and c. Having evaluated ψ, values of $k_1(W_0)$ can be calculated from the relation

$$k_1(W_0) = \frac{\psi}{2hW_0} \tag{14-7-34}$$

where h is the specimen thickness and W_0 is the input energy density for the uncracked specimen up to the strain ϵ_0.

The dependence of the function $k_1(W_0)$ on W_0 is shown in Fig. 14-25 for the

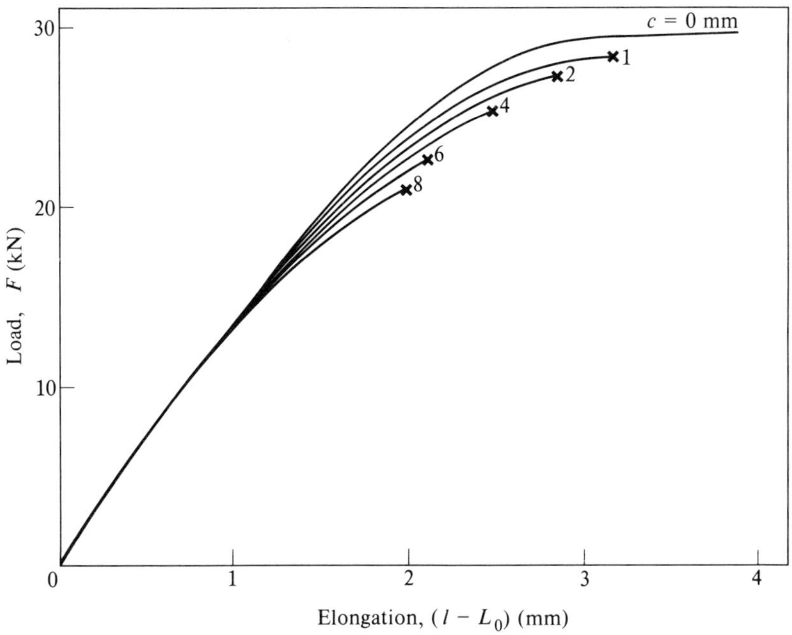

Figure 14-24 Actual load-deflection curves for wide specimens of condition H. (*Andrews and Billington, 1976.*)

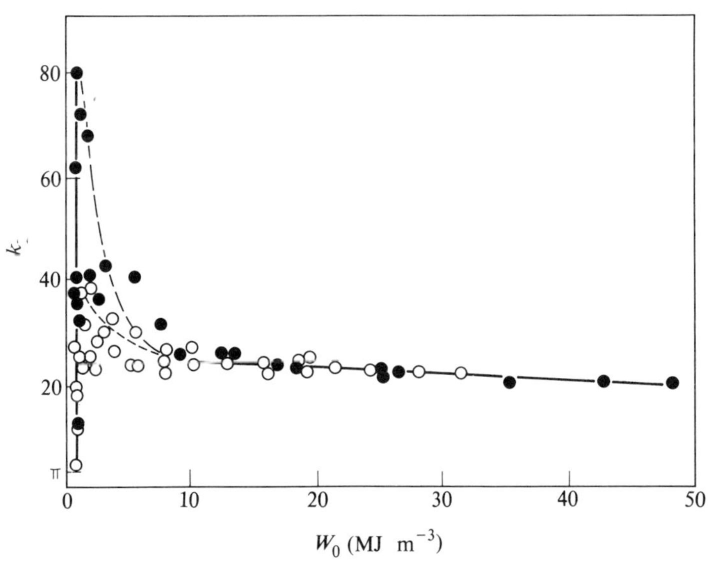

Figure 14-25 Dependence of the function k_1 on W_0 for condition A: ●, narrow specimens; ○, wide specimens. (*Andrews and Billington, 1976.*)

condition A; for the narrow specimens $h = 1.5\,\text{mm}$ and $b = 4.67\,\text{mm}$ and for the wider specimens $h = 1.0\,\text{mm}$ and $b = 42.0\,\text{mm}$, where b is the width of the specimen. It is evident that the results are in accord with the condition that the function $k_1(W_0)$ has the classical value of π for plane stress conditions as $W_0 \to 0$, i.e., for a linear elastic material and infinitesimal strains (Andrews, 1974). For increasing W_0 there is an initial, very steep rise from the value of π at $W_0 = 0$ to values in the range ca. 20 to ca. 40 for wide specimens and even higher, up to ca. 80, for the narrow specimens. The function k_1 then decreases with increasing W_0 to a much more reproducible value of around 25 and continues to decrease slowly thereafter to 20 at the highest W_0 values used.

The (k_1, W_0) curve attains a maximum at a particular value of W_0, this behavior being associated with the onset of gross yielding in the specimen. In the range $0 < W_0 < 8\,\text{MJ}\,\text{m}^{-3}$ covered by this behavior, the k_1 values are badly scattered, but are seen from Fig. 14-25 to be markedly higher for the narrow specimens. These large values of k_1 for the narrow specimens are attributed to the *hinging* mode of crack opening which characterizes the behavior of these specimens in plane bending, thus causing the crack to gape. The plastic strain field for specimens subject to the hinging mode of crack opening will be different from that in the wider specimens where no hinging can occur, a situation in accord with the observed behavior of k_1.

For the condition A, no crack growth was observed until the specimen had become fully plastic, corresponding to $W_0 > 8\,\text{MJ}\,\text{m}^{-3}$. Thus fracture occurred

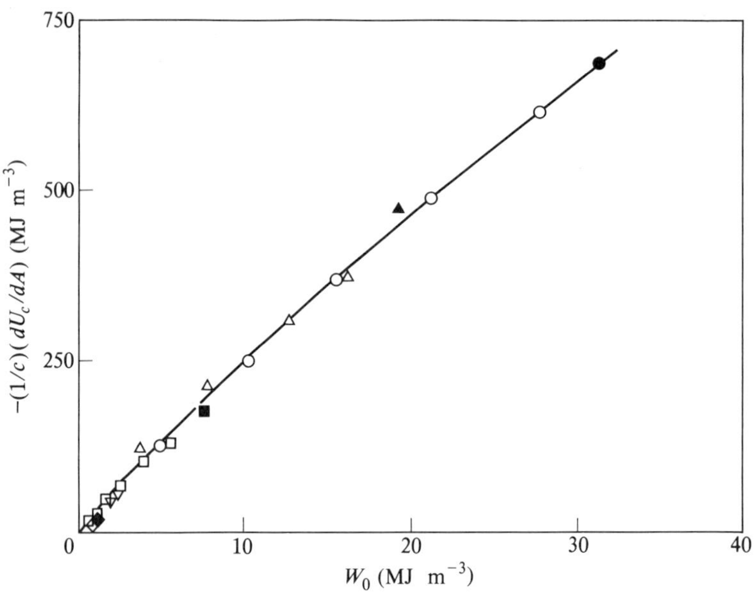

Figure 14-26 Variation of $-c^{-1}\,dU_c/dA$ with W_0 for wide specimens of condition A. c: ○, 1 mm; △, 2 mm; □, 4 mm; ▽, 6 mm; ◇, 8 mm. Blocked symbols denote fracture point. (*Andrews and Billington, 1976.*)

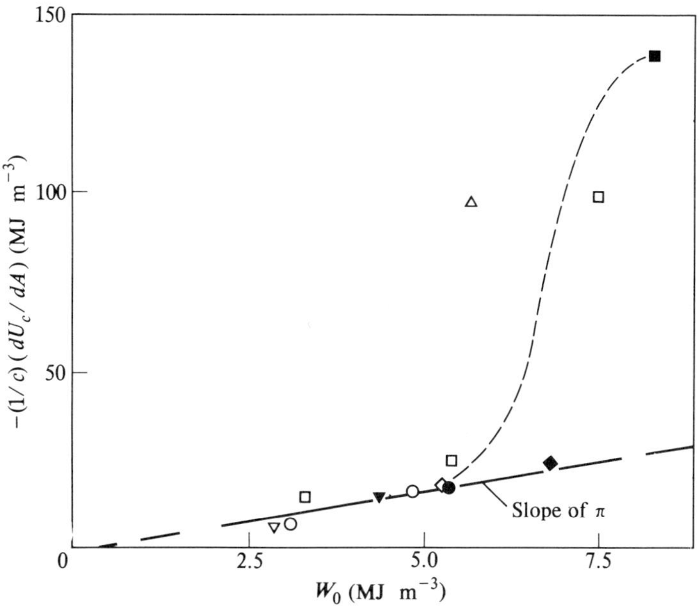

Figure 14-27 Variation of $-c^{-1}\,dU_c/dA$ with W_0 for wide specimens of condition H. c: △, 1 mm; □, 2 mm; ◇, 4 mm; ○, 6 mm; ▽, 8 mm. Blocked symbols denote fracture point. (*Andrews and Billington, 1976.*)

only where k_1 has again attained values in the range 20 to 25, the function k_1 varying only slowly in this range for all available values of $W_0 > 8\ \mathrm{MJ\,m^{-3}}$.

A consequence of the dependence of k_1 on ϵ_0 (or W_0) alone is that Eq. (14-7-13) can be expressed in the form

$$-\frac{1}{c}\left(\frac{dU_c}{dA}\right) = k_1(W_0)W_0 = \hat{f}(W_0) \tag{14-7-35}$$

implying that the left-hand side of Eq. (14-7-35) should be a unique function of W_0. It is to be noted, however, that only for linear, infinitesimal strain conditions will the left-hand side of Eq. (14-7-35) vary linearly with W_0 since only then is k_1 a constant. The variation of $-(dU_c/dA)/c$ with W_0 is shown in Fig. 14-26 for the wide specimens, the observed behavior being in accord with Eq. (14-7-35). The same behavior was observed for the narrow specimens, the only difference being a pronounced curvature for small values of W_0 due to the effect of hinging. Similar results for the cold-worked condition H are shown in Fig. 14-27. For condition H, a linear relation exists with a slope $k_1 = \pi$ up to the onset of gross yield. From this observation it follows that the material when in condition H obeys linear fracture mechanics up to the onset of gross yielding. As the specimen stress-strain curve becomes nonlinear, $\hat{f}$ increases rapidly with increasing W_0. Associated with this behavior is a rapid rise of k_1 from its initial value of π to some transitory high value, followed by a drop to a value of ~ 17 in the region of gross plasticity.

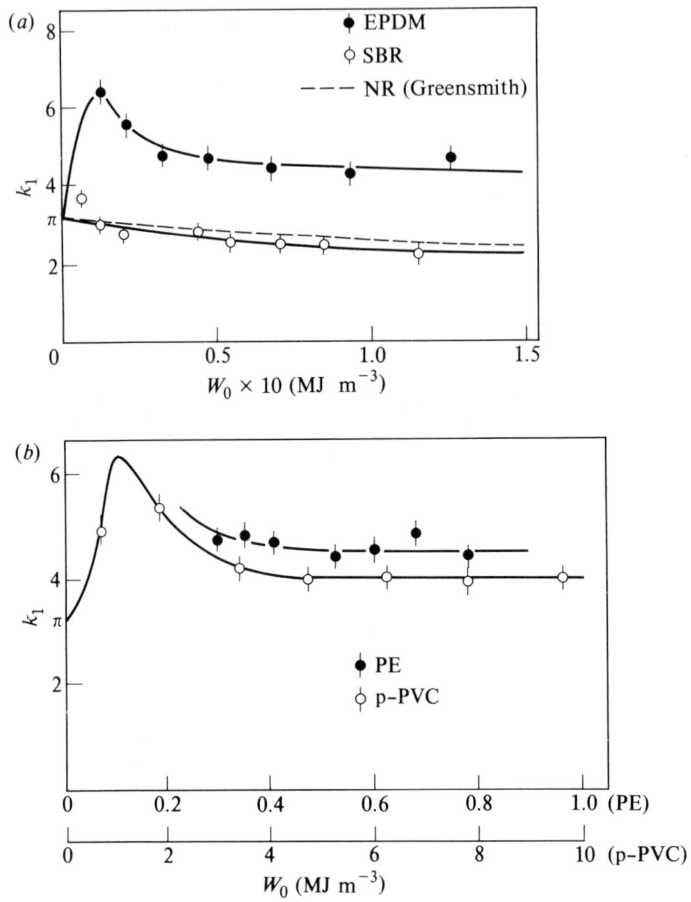

Figure 14-28 The function $k_1(W_0)$ for four materials at 23°C; (*a*) for SBR and EPDM, (*b*) for PE and p-PVC. (*Andrews and Fukahori, 1977.*)

Using the compliance method described above, values of k_1 as a function of W_0 have been evaluated by Andrews and Fukahori (1977) for several highly extensible polymeric materials. The variation of k_1 with W_0 for these materials is shown in Fig. 14-28. These results are seen to be formally similar to those obtained for the copper-beryllium alloys shown in Fig. 14-25.

14-7-3 Fracture Energies

The critical value of $-(dU_c/dA)_\delta$ is, by definition, the surface work $\mathscr{T}$ given by Eq. (14-7-17). For brittle materials $\mathscr{T}$ is sensibly constant for a given solid under given conditions of rate and temperature and provided no plane stress–plane strain transition intrudes. The same is true of elastic but highly extensible materials like cross-linked elastomers.

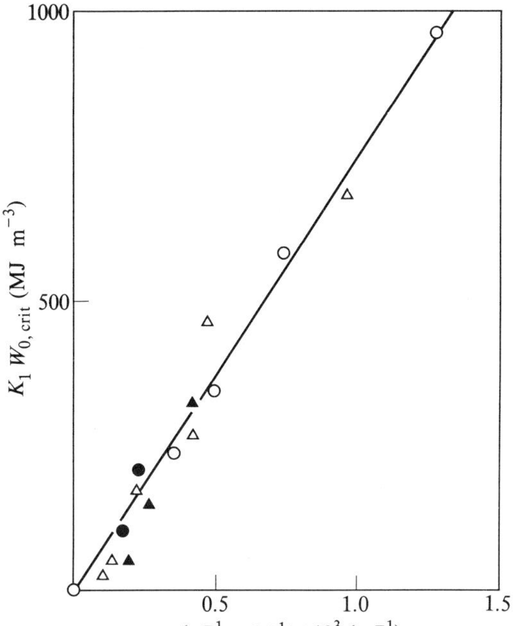

Figure 14-29 Variation of $k_1 W_{0,\text{crit}}$ with $(c^{-1} - b^{-1})$ for condition A: ○, narrow specimens, 1.0 mm tip radius; △, wide specimens, 1.0 mm tip radius; ▲, intermediate specimens, 1.25 mm tip radius; ●, intermediate specimens, 0.6 mm tip radius. (*Andrews and Billington, 1976.*)

It has not previously been possible to calculate surface work for solids which undergo substantial plastic deformation before fracture. However, Eq. (14-7-17) can be rearranged into the form

$$k_1 W_{0,\text{crit}} = \mathcal{T} c^{-1} \tag{14-7-36}$$

from which it follows that a study of the variation of the product $k_1 W_{0,\text{crit}}$ with c^{-1} should reveal whether or not a constant quantity $\mathcal{T}$ exists. Since the above equations apply specifically to specimens of infinite width, it is necessary to apply a correction for the finite width of the specimens. In the absence of an established correction comparable to the elastic corrections for finite width, c^{-1} is here simply replaced by $(c^{-1} - b^{-1})$ where b is the specimen width. This satisfies the boundary condition that $W_{0,\text{crit}} \to 0$ as $c \to b$ and that the correction itself tends to zero as $b \to \infty$.

The variation of $k_1 W_{0,\text{crit}}$ with $(c^{-1} - b^{-1})$ is shown in Figs. 14-29 and 14-30 for conditions A and H respectively. Most results are available for alloy A, it being evident from Fig. 14-29 that a linear relationship, implying constant $\mathcal{T}$, is in good accord with the results, which have been obtained from specimens for which b lies in the range 4.67 mm $\leqslant b \leqslant$ 42.0 mm and for different crack-tip radii. The value of $\mathcal{T}$ for this material, derived from the best straight line in Fig. 14-29, is 7.5×10^8 mJ m^{-2}.

Figure 14-30 reveals a more complicated situation for condition H. For cracks exceeding $c = 2$ mm in length the variation of $k_1 W_{0,\text{crit}}$ with $(c^{-1} - b^{-1})$ is linear, yielding a value of $\mathcal{T}$ of 1.2×10^8 mJ m^{-2}. However, for crack lengths of 2 and

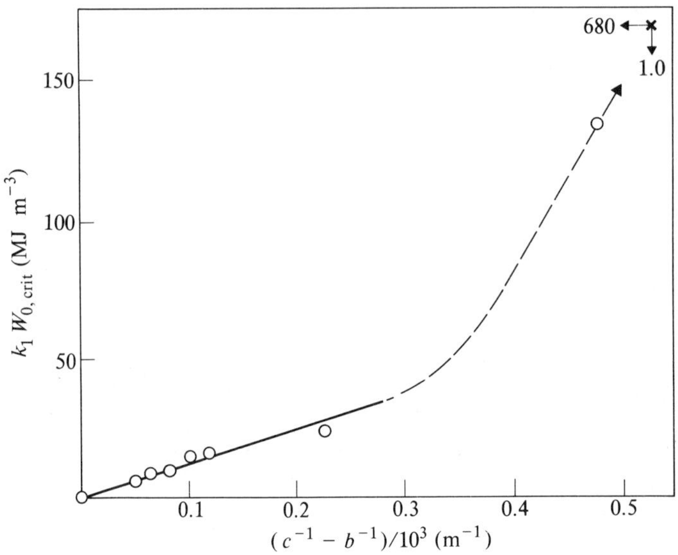

Figure 14-30 Variation of $k_1 W_{0,\text{crit}}$ with $(c^{-1} - b^{-1})$ for condition H (wide specimens). Point for 1 mm crack is off the graph but its coordinates are shown at × . (*Andrews and Billington, 1976.*)

1 mm, the $(k_1 W_{0,\text{crit}}, c^{-1} - b^{-1})$ curve is no longer linear; the apparent values of $\mathscr{T}$ increase, as is evident from Table 14-2.

Reference to Fig. 14-24 shows that specimens of condition H with crack lengths exceeding 2 mm, while displaying nonlinear behavior, do not exhibit gross yielding. Those with $c \leqslant 2$ mm show the attainment of extension at a constant load. Specimens of condition A, in contrast, all fracture after gross yield and it is significant that the value obtained for condition A (that is, 7.5×10^8 mJ m^{-2}) is close to that obtained for the shortest crack length in condition H specimens, where gross yielding is also fully established.

These observations are consistent with the suggestion that, for a given material, the surface work $\mathscr{T}$ has two regimes of constancy. The first regime is considered to apply as long as the plastic zone does not intersect the boundaries of the specimen. That is, the plastic zone is elastically contained. As elastic containment fails the specimen begins to exhibit gross yielding and the surface work $\mathscr{T}$ increases rapidly until it achieves a new consistency, identified as regime two, for the case of a fully plastic specimen as typified by condition A.

Table 14-2 Apparent values of $\mathscr{T}$ for condition H

c, mm	$\mathscr{T}$ (apparent), mJ m^{-2}
>2	1.2×10^8
2	2.7×10^8
1	6.7×10^8

Andrews and Fukahori (1977) have experimentally evaluated the loss function Φ of Eq. (14-7-31) for the four highly extensible materials, styrene-butadiene rubber (SBR), ethylene-propylene rubber (EPDM), plasticized polyvinylchloride (p-PVC), and polyethylene (PE), and at various rates of crack propagation. The reasons for selecting highly extensible polymers are that simple expressions exist for relating energy density to principal extension ratios, thus facilitating evaluation of $g(x, y)$, and that variation of Φ could be readily induced by changing the rate of crack propagation. The quantity $\mathscr{T}_0$ appearing in Eq. (14-7-33) can be calculated from existing theory and thus, in principle, a prediction could be obtained for the fracture energy. However, because the values of $\mathscr{T}_0$ are uncertain, Andrews and Fukahori chose to regard $\mathscr{T}_0$ as the unknown in Eq. (14-7-33).

The quantity $\mathscr{T}_0$ is half the minimum energy required to break the unit area of interatomic bonds across the fracture plane, which is just one definition of the surface energy of a solid. This minimum energy can, according to Lake and Thomas (1967), be calculated for polymer networks from the relation

$$\mathscr{T}_0 = \tfrac{1}{2}(\tfrac{3}{8})^{1/2}\phi l U_F \xi \tag{14-7-37}$$

where ϕ is a factor determined by the freedom of rotation about C—C bonds, l is the length of a monomer unit, U_F is the energy to rupture a C—C bond, and $\xi = Nn^{3/2}$ where N is the number of chains per unit volume and n the number of monomer units in the network chain. Equation (14-7-37) is based on the assumption that all the energy stored in the molecule between neighboring cross-links is *lost* to the elastic stress field if the molecule breaks between these cross-links. Values of $\mathscr{T}_0$ calculated from Eq. (14-7-37) are subject to large errors, and Lake and Thomas have proposed an alternative relation which gives values a

Table 14-3 Experimental values for the parameter $\mathscr{T}_0$ for various cross-linked elastomers

	$\mathscr{T}_0$, J m^{-2}	
Material	in air	*in vacuo*
Natural rubber	10–20	33†
Isomerized NR	35†	
Synthetic cis PI	35†	
Butadiene-styrene copolymer	30†	
Butadiene-acrylonitrile copolymer	< 50†	
Polychloroprene	35†	
Butyl rubber	20	20†
Polybutadiene	20–50‡	
Styrene-butadiene copolymer	45§	

† Lake and Lindley (1965).
‡ Ahagon and Gent (1975).
§ Andrews and Kinloch (1973).

Table 14-4 $\mathcal{T}_0$ (in J m^{-2})

Material	Theoretical	Experimental†	Experimental
NR	10 ~ 14		20 ~ 30‡ 30‡
SBR	16	65 ± 15	45§
EPDM	18	65 ± 15	
p-PVC	15 ± 5	100 ± 50	
PE	31 ± 6	200 ± 100	

† Andrews and Fukahori (1977).
‡ Lake and Lindley (1965).
§ Andrews and Kinloch (1973).

factor of two greater. However, it is possible to determine values of $\mathcal{T}_0$ from experiment for elastomers under certain circumstances. Examples of experimentally determined values of $\mathcal{T}_0$ are given in Table 14-3.

Corresponding values of $\mathcal{T}_0$ obtained by Lake and Thomas are a factor of two to three smaller than the *in vacuo* experimental values for the appropriate vulcanizates. Andrews and Fukahori calculated values of $\mathcal{T}_0$; these theoretical values are given in Table 14-4, which also includes slow-rate experimental results. In deriving theoretical $\mathcal{T}_0$ values for the uncross-linked systems, polyethylene and plasticized PVC, effective values of the Mooney–Rivlin constants C_1 and C_2, which are discussed in Sec. 8-8-1, were derived by Andrews and Fukahori from Mooney–Rivlin plots and averaged over their rate dependence.

Equation (14-7-33) can be rearranged by way of Eq. (14-7-30) into the form

$$k_1\left(1 - \frac{\mathcal{T}_0}{\mathcal{T}}\right) = \Pi \qquad (14\text{-}7\text{-}38)$$

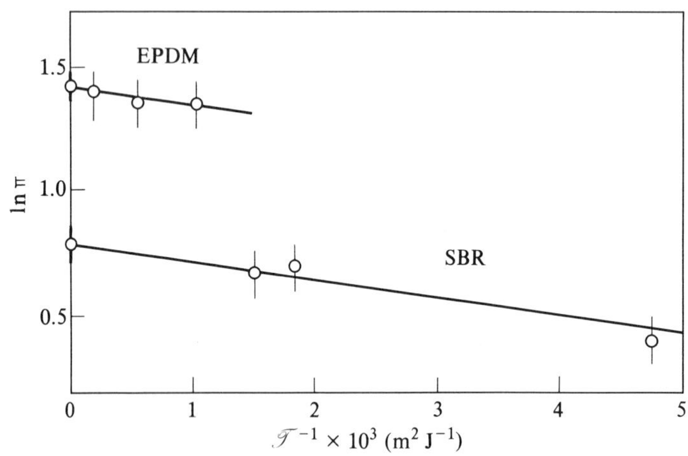

Figure 14-31 Variation of ln Π with $\mathcal{T}^{-1}$ for SBR and EPDM. (*Andrews and Fukahori, 1977.*)

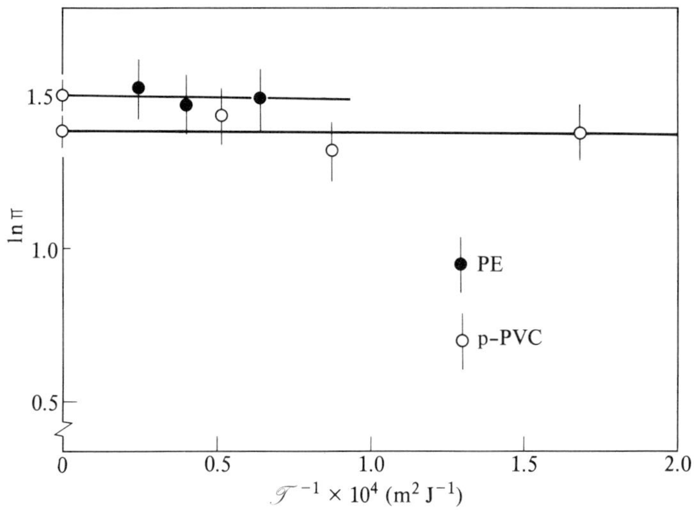

Figure 14-32 Variation of ln Π with $\mathscr{T}^{-1}$ for PE and PVC. (*Andrews and Fukahori, 1977.*)

where

$$\Pi = \beta_0 \sum_{PU} g_2(x, y, \epsilon_0)\delta x \delta y \tag{14-7-39}$$

Since $\mathscr{T}_0/\mathscr{T} \ll 1$ for most conditions, Eq. (14-7-38) can be expressed in the approximate form

$$\ln k_1 - \frac{\mathscr{T}_0}{\mathscr{T}} = \ln \Pi \tag{14-7-40}$$

Thus, ln Π should vary linearly with $\mathscr{T}^{-1}$, the slope being $-\mathscr{T}_0$ and the intercept being $\ln k_1$. Andrews and Fukahori (1977) measured $\mathscr{T}$ directly for each material at three different crack velocities. The variation of ln Π with $\mathscr{T}^{-1}$ is shown in Figs. 14-31 and 14-32 where the error bars on the points reflect the uncertainties in evaluating the summation Π. It is evident that the results of experiment are in good accord with Eq. (14-7-40), particularly for the cross-linked materials SBR and EPDM. The magnitude of the slope, that is, $\mathscr{T}_0$, is compared with theoretical and previous experimental determinations in Table 14-4. The linearity and negative slopes in Figs. 14-31 and 14-32, together with the good order-of-magnitude results for $\mathscr{T}_0$, appear at this stage to provide good confirmation of Eqs. (14-7-31) and (14-7-33).

14-8 DYNAMIC FRACTURE

When a compression wave reflects from a free surface it puts the material into tension and, if the wave is sufficiently strong, it may produce fracture parallel to the free surface so that a penny-shaped disc of material is launched from it. Such a phenomenon is known as spalling and the material ejected as the spall.

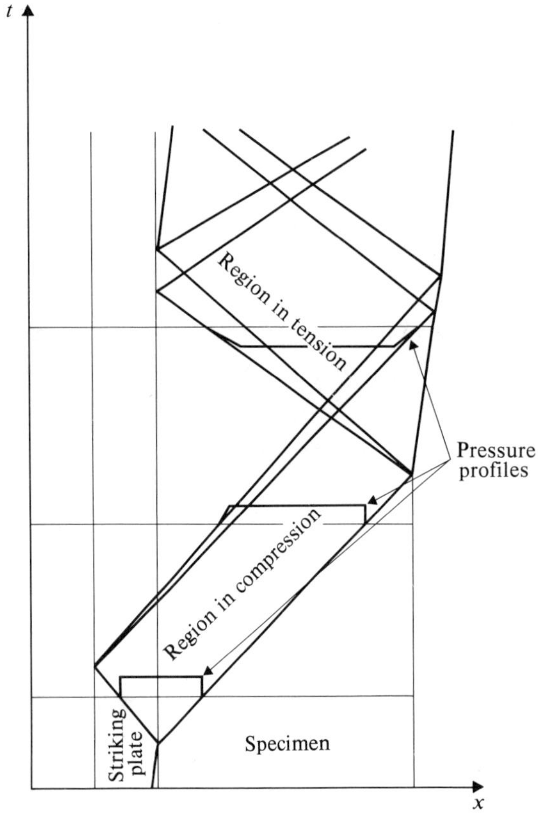

Figure 14-33 Distance-time wave diagram for the impact of a flat plate on a specimen of the same material and the pressure profile at three times during the process.

Figure 14-33 shows the wave diagram of a thin slab of material striking another thicker slab of the same material and the pressure profiles during the ensuing motion. It can be seen that the material is first put into tension where the two rarefaction waves intersect. Whether a spall will be produced or not depends on the magnitude and duration of the tensile stress. Several empirical formulas have been proposed as fracture criteria by Butcher, Barker, Munson, and Lundergam (1964), Breed, Mader, and Venable (1967), and others. None appears to be very satisfactory and a more fundamental approach is now being sought.

Iron has been observed by Bancroft, Peterson, and Minshall (1956) to exhibit a phase transition at 13 GPa and thus, as shown in Sec. 13-2-4, it has a two-wave compression shock and a rarefaction shock. The interaction of the rarefaction fan reflecting the compression shocks from a free surface and the rarefaction shock sends the iron suddenly into tension, which is much greater than the fracture stress within any grain so that the fracture surface appears very smooth. This type of fracture is called "smooth spall" and has been described by Rinehart and Pearson (1954), Lethaby and Skidmore (1959), Erkman (1961), Tupper (1961), and Skidmore (1965).

GENERAL REFERENCES

Andrews, E. H. (1968) "Fracture in Polymers," Oliver and Boyd, Edinburgh.

Cherepanov, G. P. (1979) (W. C. Cooley, A. L. Peabody, and R. de Wit, eds.), "Mechanics of Brittle Fracture," McGraw-Hill, New York.

Irwin, G. R. (1958) Fracture, in S. Flügge (ed.), "Handbuch der Physik," vol. 6, pp. 551–590, Springer-Verlag, Berlin.

Knott, J. F. (1973) "Fundamentals of Fracture Mechanics," Butterworths, London.

Tada, H., P. C. Paris, and G. R. Irwin (1973) "The Stress Analysis of Cracks Handbook," Del Research Corporation, Hellertown, Pa.

REFERENCES

Ahagon, A., and A. N. Gent (1975) *J. Polym. Sci.*, vol. 13, pp. 1903–1911.

Allen, N. P., W. P. Rees, B. E. Hopkins, and H. R. Tipler (1953) *J. Iron and Steel Inst.*, vol. 174, pp. 108–120.

Andrews, E. H. (1959) *J. Appl. Phys.*, vol. 30, pp. 740–743.

Andrews, E. H. (1974) *J. Mat. Sci.*, vol. 9, pp. 887–894.

Andrews, E. H., and E. W. Billington (1976) *J. Mat. Sci.*, vol. 11, pp. 1354–1361.

Andrews, E. H., and Y. Fukahori (1977) *J. Mat. Sci.*, vol. 12, pp. 1307–1319.

Andrews, E. H., and A. J. Kinloch (1973) *Proc. Roy. Soc.* (*London*), vol. A332, pp. 385–399, 401–414.

Andrews, E. H., and B. J. Walker (1971) *Proc. Roy. Soc.* (*London*), vol. A325, pp. 57–79.

Bancroft, D., E. L. Peterson, and S. Minshall (1956) *J. Appl. Phys.*, vol. 27, pp. 291–298.

Barenblatt, G. I. (1962) *Adv. Appl. Mech.*, vol. 7, pp. 55–129, Academic Press, New York.

Begley, J. A., and J. D. Landes (1972) *Am. Soc. Test. Mat. Spec. Tech. Pub. No.* 514, pp. 1–20.

Berry, J. P. (1960) *J. Mech. Phys. Solids*, vol. 8, pp. 194–216.

Bilby, B. A., A. H. Cottrell, and K. H. Swinden (1963) *Proc. Roy. Soc.* (*London*), vol. A272, pp. 304–314.

Born, M., and R. Furth (1940) *Proc. Camb. Phil. Soc. Math. Phys. Sci.*, vol. 36, pp. 454–465.

Breed, B. R., C. L. Mader, and D. Venable (1967) *J. Appl. Phys.*, vol. 38, pp. 3271–3275.

Bridgman, P. W. (1947) *J. Appl. Phys.*, vol. 18, pp. 246–258.

Brown, W. F., and J. E. Srawley (1966) "Plane Strain Crack Toughness Testing of High Strength Metallic Materials," STP 410, pp. 1–65, Am. Soc. Test. Mat., Philadelphia, Pa.

Brown, W. F., and J. E. Srawley (1971) "Methods for Plane Strain Fracture Toughness Testing," DD3, BSI.

Brown, W. F., and J. E. Srawley (1974) "Standard Method of Test for Plane Strain Fracture Toughness of Metallic Materials," E 399-74, Am. Soc. Test. Mat., Philadelphia, Pa.

Bucci, R. J., P. C. Paris, J. D. Landes, and J. R. Rice (1972) *Am. Soc. Test. Mat. Spec. Tech. Pub. No.* 514, p. 40.

Burdekin, F. M., and D. E. W. Stone (1966) *J. Strain Anal.*, vol. 1, pp. 145–153.

Butcher, B. M., L. M. Barker, D. E. Munson, and C. D. Lundergam (1964) *Am. Inst. Aeronautics and Astronautics J.*, vol. 2, pp. 977–990.

Chang, R. (1970) *Int. J. Fracture Mech.*, vol. 6, pp. 111–124.

Cherepanov, G. P. (1967) *J. Appl. Math. Mech.*, vol. 31, pp. 503–512.

Cottrell, A. H. (1961) *Iron and Steel Inst. Spec. Rept.* 69, pp. 281–296.

Cottrell, A. H. (1964) "The Mechanical Properties of Matter," p. 351, Wiley, New York.

Craggs, J. W. (1963) In D. C. Drucker and J. J. Gilman (eds.), "Fracture of Solids," pp. 51–63, Wiley, New York.

Crossland, B. (1954) *Inst. Mech. Engng. Proc.*, vol. 168, pp. 935–946.

Davidenkov, N. N. (1936) "Dinamicheskaya Ispytania Metallov," Moscow, Gos. Izd. Vo.

Dawes, M. G. (1975) *J. Estimation Procedures for Weldments, W.I. Res. Rept.* E/65/75, The Welding Institute.

Doyle, M. J., A. Moranci, E. Orowan, and S. T. Stork (1972) *Proc. R. Soc.* (*London*), vol. A329, pp. 137–151.

Dugdale, D. S. (1960) *J. Mech. Phys. Solids*, vol. 8, pp. 100–104.

Erkman, J. O. (1961) *J. Appl. Phys.*, vol. 32, pp. 939–944.

Eshelby, J. D. (1956) *Solid State Phys.*, vol. 3, pp. 79–144.

Field, J. E. (1971) *Contemp. Phys.*, vol. 12, pp. 1–31.

Freminville, C. de (1914) *Revue Metall. (Paris)*, vol. 11, pp. 971–1056.

Gehlen, P. C., and M. F. Kanninen (1970) An Atomic Model for Cleavage Crack Propagation in α Iron, in M. F. Kanninen et al. (eds.), "Inelastic Behavior of Solids," pp. 587–605, McGraw-Hill, New York.

Gilman, J. J. (1959) In B. L. Averbach (ed.), *Proc. Int. Conf. Atomic Mechanisms of Fracture*, Swampscott, pp. 193–224, Wiley, New York.

Gilman, J. J. (1963) Mechanical Behavior of Crystalline Solids, *National Bureau of Standards Monograph*, vol. 59, pp. 79–102.

Griffith, A. A. (1920) *Phil. Trans. Roy. Soc.*, vol. A221, pp. 163–198.

Griffith, A. A. (1924) *Proc. First Int. Congr. Appl. Mech.*, Delft, p. 55.

Hayes, D. J., and C. E. Turner (1974) *Inter. J. Fracture*, vol. 10, pp. 17–32.

Hencky, H. (1923) *Z. angew. Math. Mech.*, vol. 3, pp. 241–250.

Hill, R. (1949) *Quart. J. Mech. Appl. Math.*, vol. 2, pp. 40–52.

Hull, D., P. Beardmore, and A. P. Valentine (1965) *Phil. Mag.*, vol. 12, pp. 1021–1041.

Hutchinson, J. W. (1968) *J. Mech. Phys. Solids*, vol. 16, pp. 13–31.

Inglis, C. E. (1913) *Trans. Inst. Nav. Archit.*, vol. 55, pp. 219–230.

Irwin, G. R. (1948) Fracture Dynamics, in "Fracturing of Metals," pp. 147–166, Am. Soc. Metals, Cleveland, Ohio.

Irwin, G. R. (1957a) *J. Appl. Mech.*, vol. 24, pp. 361–364.

Irwin, G. R. (1957b) *Proc. Ninth Int. Congr. Appl. Mech.*, vol. 8, pp. 245–251, University of Brussels.

Irwin, G. R. (1964) *Appl. Mat. Res.*, vol. 3, pp. 65–81.

Irwin, G. R., J. A. Kies, and H. L. Smith (1958) *Proc. Am. Soc. Test. Mat.*, vol. 58, pp. 640–660.

Johnson, J. W., and D. G. Holloway (1966) *Phil. Mag.*, vol. 14, pp. 731–743.

Johnson, J. W., and D. G. Holloway (1968) *Phil. Mag.*, vol. 17, pp. 899–910.

Kanninen, M. F., and P. C. Gehlen (1971) *Int. J. Fracture Mech.*, vol. 7, pp. 471–474.

Kanninen, M. F., and P. C. Gehlen (1972) In P. C. Gehlen et al. (eds.), "Interatomic Potentials and Simulation of Lattice Defects," pp. 713–724, Plenum, New York.

Kelly, A. (1966) "Strong Solids," Clarendon Press, Oxford.

Kelly, A., W. R. Tyson, and A. H. Cottrell (1967) *Phil. Mag.*, vol. 15, pp. 567–586.

Kenny, P., and J. D. Campbell (1968) *Prog. Mat. Sci.*, vol. 13, pp. 135–181.

Lake, G. J., and P. B. Lindley (1965) *J. Appl. Polym. Sci.*, vol. 9, pp. 1233–1251.

Lake, G. J., and A. G. Thomas (1967) *Proc. Roy. Soc. (London)*, vol. A300, pp. 108–119.

Lethaby, J., and I. C. Skidmore (1959) *SWA Branch Note No.* 3/59, AWRE, Aldermaston.

Ludwik, P. (1924) *Z. Metallkde*, vol. 16, pp. 207–212.

McClintock, F. (1971) In H. Liebowitz (ed.), "Fracture, An Advanced Treatise," vol. 3, pp. 47–225, Academic Press, New York.

Macmillan, N. H. (1972) *J. Mat. Sci.*, vol. 7, pp. 239–254.

Maitland, A. H., and G. A. Chadwick (1969) *Phil. Mag.*, vol. 19, pp. 645–651.

Marsh, D. M. (1964) *Proc. Roy. Soc. (London)*, vol. A279, pp. 420–435.

Morton, M. E. de, and J. C. Ritter (1973) *Proc. Twenty-ninth Annual Conf. (Adelaide)*, pp. 313–342, Aust. Inst. Metals.

Mott, N. F. (1948) *Engineering*, vol. 165, pp. 16–18.

Murry, J., and D. Hull (1971) *J. Mat. Sci.*, vol. 6, pp. 1277–1285.

Neuber, H. (1938) "Kerbspannunglehre," Springer-Verlag, Berlin. Translated as "Theory of Notch Stresses," David Taylor Model Basin, Translation 74, November, 1945.

Nichols, R. W., F. M. Burdekin, A. Cowan, D. Elliot, and T. Ingham (1969) Practical Fracture Mechanics for Structural Steels, in *Proc. Conf. Culcheth*, Section F, UKAEA/Chapman and Hall.

Orowan, E. (1945) *Trans. Inst. Engng. Shipbuilders Scotland*, vol. 89, pp. 165–204.

Orowan, E. (1949) *Repts. Prog. Phys.*, vol. 12, pp. 185–232.

Paris, P. C., and G. Sih (1965) "Fracture Toughness Testing and its Applications," STP 381, pp. 30–81, Am. Soc. Test. Mat., Philadelphia, Pa.

Petch, N. J. (1953) *J. Iron and Steel Inst.*, vol. 174, pp. 25–28.

Prandtl, L. (1923) *Z. angew. Math. Mech.*, vol. 3, pp. 401–406.

Ramberg, W., and W. R. Osgood (1943) *Nat. Advis. Commit. Aeronautics*, TN 902.

Rice, J. R. (1968a) *J. Appl. Mech.*, vol. 35, pp. 379–386.

Rice, J. R. (1968b) In H. Liebowitz (ed.), "Fracture, An Advanced Treatise," vol. 2, pp. 191–311, Academic Press, New York.

Rice, J. R., and G. F. Rosengren (1968) *J. Mech. Phys. Solids*, vol. 16, pp. 1–12.

Rice, J. R., P. C. Paris, and J. G. Merkle (1973) "Progress in Flow Growth and Fracture Toughness Testing," STP 536, pp. 231–245, Am. Soc. Test. Mat., Philadelphia, Pa.

Rinehart, J. S., and J. Pearson (1954) "Behaviour of Metals under Impulsive Loads," ASM, Cleveland.

Rivlin, R. S., and A. G. Thomas (1953) *J. Polymer Sci.*, vol. 10, pp. 291–318.

Schardin, H. (1959) Velocity Effects in Fracture, in B. L. Averbach, D. K. Felbeck, G. T. Hahn, and D. A. Thomas (eds.), "Fracture," pp. 297–330, Technology Press of MIT/Wiley, New York.

Sinclair, J. E. (1972) *J. Phys. C: Solid State Phys.*, vol. 5, pp. L271–274.

Sinclair, J. E., and B. R. Lawn (1972) *Proc. Roy. Soc.* (*London*), vol. A329, pp. 83–103.

Skidmore, I. C. (1965) *Appl. Mat. Res.*, vol. 4, pp. 131–147.

Smith, R. F., and J. F. Knott (1971) *Conf. Proc. Practical Application of Fracture Mechanics to Pressure—Vessel Technology*, pp. 65–75, Inst. Mech. Engng.

Sneddon, I. N. (1946) *Proc. Roy. Soc.* (*London*), vol. A187, pp. 229–260.

Steverding, B., and S. H. Lehnigk (1970) *J. Appl. Phys.*, vol. 41, pp. 2096–2099.

Steverding, B., and S. H. Lehnigk (1972) *J. Appl. Phys.*, vol. 43, pp. 69–73.

Thomson, R., C. Hsieh, and V. Rona (1971) *J. Appl. Phys.*, vol. 42, pp. 3154–3160.

Tupper, S. J. (1961) *ARDE Rept.* (*B*), 12/61.

Turner, C. E. (1975) *J. Strain Analysis*, vol. 10, pp. 207–216.

Tyson, W. R. (1966) *Phil. Mag.*, vol. 14, pp. 925–936.

Wallner, H. (1939) *Z. Physik*, vol. 114, pp. 368–378.

Wells, A. A. (1961) *Crack Propagation Symp. Proc.*, vol. 5, pp. 210–230, Cranfield College of Aeronautics, Cranfield, England.

Wells, A. A. (1970) *J. Engng. Fracture Mech.*, vol. 1, pp. 399–410.

Westergaard, H. M. (1939) *J. Appl. Mech.*, vol. 6, pp. 49–53.

Wilson, W. K. (1966) "Plane Strain Crack Toughness Testing of High Strength Metallic Materials," STP 410, pp. 75–76, Am. Soc. Test. Mat., Philadelphia, Pa.

Yoffe, E. H. (1951) *Phil. Mag.*, vol. 42, pp. 739–750.

Zwicky, F. (1923) *Phys. Z.*, vol. 24, pp. 131–137.

AUTHOR INDEX

Numbers in bold type refer to the list of references at the end of each chapter.

SUBJECT INDEX